D1155916

On the economy of plant form and function

On the economy of plant form and function

Proceedings of the Sixth Maria Moors Cabot Symposium, "Evolutionary Constraints on Primary Productivity: Adaptive Patterns of Energy Capture in Plants," Harvard Forest, August 1983

Edited by Thomas J. Givnish
Department of Botany, University of Wisconsin

CAMBRIDGE UNIVERSITY PRESS

Cambridge
New York Port Chester
Melbourne Sydney

Published by the Press Syndicate of the University of Cambridge
The Pitt Building, Trumpington Street, Cambridge, CB2 1RP
40 West 20th Street, New York, NY 10011, USA
10 Stamford Road, Oakleigh, Melbourne 3166, Australia

First published 1986
Reprinted 1990

Printed in the United States of America

Library of Congress Cataloging-in-Publication Data

Maria Moors Cabot Symposium (6th : 1983 : Harvard Forest)
On the economy of plant form and function.
1. Primary productivity (Biology) – Congresses.
2. Botany – Morphology – Congresses. 3. Plant physiology – Congresses. 4. Plant morphogenesis – Congresses.
5. Plants – Evolution – Congresses. 6. Plants – Adaptation – Congresses. I. Givnish, Thomas J. II. Title.
QK909.5.M37 1986 581.1 85–19564

British Library Cataloguing in Publication Data

Maria Moors Cabot Symposium (6th : 1983 : Harvard Forest)
On the economy of plant form and function/ proceedings of the Sixth Maria Moors Cabot symposium, Evolutionary Constraints on Primary Productivity : Adaptive Patterns of Energy Capture in Plants, Harvard Forest, August 1983.
1. Plant physiology
I. Title. II. Givnish, Thomas J.
581.1 QK711.2

ISBN 0 521 26296 8

Contents

Acknowledgments

The Maria Moors Cabot Foundation of Harvard University, charged with the development and dissemination of knowledge related to increasing plant productivity, provided most of the financial support for the symposium that led to this volume. Additional grants, generously furnished by the E. I. Dupont de Nemours Company, Monsanto Corporation, Pfizer-Dekalb, and Cambridge University Press, helped enable the participation of several researchers from overseas. The grants from private industry provide tangible evidence of the interest that the topic of the symposium holds for plant breeders and biotechnologists, as well as for ecologists and physiologists.

I would like especially to thank Robert Robichaux, who helped review many of the chapters in this volume, and in many matters served as coeditor. Richard Olson and Mark Patterson provided useful reviews of chapters on marine topics. The staff of the Harvard Forest, particularly Ernest Gould, Barry Tomlinson, and the late Martin Zimmermann, as well as Stephen Bartz and James LaFrankie, provided cheerful and invaluable help with logistics. Richard Ziemacki and Helen Wheeler of Cambridge University Press encouraged and enthusiastically supported this project at each stage. Stephen Bartz, Victoria Elliott, Henry Horn, Sara Lewis, Otto Solbrig, and E. O. Wilson provided illuminating discussion.

Contributors

Pieter Baas
Rijksherbarium
Schelpenkade 6
P.O. Box 9514
2300 RA Leiden
The Netherlands

Martyn M. Caldwell
Department of Range Science
College of Natural Resources
Utah State University
Logan, Utah 84322

Ian R. Cowan
Department of Environmental Biology
Australian National University
P.O. Box 475
Canberra, A.C.T. 2602 Australia

James R. Ehleringer
Department of Biology
University of Utah
Salt Lake City, Utah 84112

Christopher Field
Carnegie Institute of Washington
Stanford University
Stanford, California 94305

Jack B. Fisher
Fairchild Tropical Garden
Miami, Florida 33156

Edwin L. Fiscus
United States Department of Agriculture
Agricultural Research Service
Crops Research Laboratory
Colorado State University
Fort Collins, Colorado 80523

Thomas J. Givnish
Department of Botany
University of Wisconsin
Madison, Wisconsin 53706

Guillermo Goldstein
Facultad de Ciencias
Universidad de los Andes
Mérida, Venezuela

Sherry L. Gulmon
Department of Biological Sciences
Stanford University
Stanford, California 94305

Mark E. Hay
University of North Carolina at Chapel Hill
Institute of Marine Sciences
Morehead City, North Carolina 28557

Kent E. Holsinger
Department of Botany
University of California
Berkeley, California 94720

Mimi A. R. Koehl
Department of Zoology
University of California
Berkeley, California 94720

Manfred Küppers
Lehrstuhl Pflanzenökologie
Universität Bayreuth
Postfach 30 08
Bayreuth, West Germany

David W. Lee
Department of Biological Sciences
Florida International University
Miami, Florida 33199

Rainer Matyssek
Lehrstuhl Pflanzenökologie
Universität Bayreuth
Postfach 30 08
Bayreuth, West Germany

Frederick Meinzer
Laboratory of Biomedical and Environmental Sciences
University of California
900 Veteran Avenue
Los Angeles, California 90024

Harold A. Mooney
Department of Biological Sciences
Stanford University
Stanford, California 94305

Suzanne R. Morse
Department of Botany
University of California
Berkeley, California 94720

Park S. Nobel
Department of Biology
212 Botany Building
University of California
Los Angeles, California 90024

David F. Parkhurst
School of Public and Environmental Affairs and Department of Biology
Indiana University
Bloomington, Indiana 47405

John S. Pate
Department of Botany
University of Western Australia
Nedlands, Western Australia 6009
Australia

John A. Raven
Department of Biological Sciences
University of Dundee
Dundee DD1 4HN, Scotland

James H. Richards
Department of Range Science
College of Natural Resources
Utah State University
Logan, Utah 84322

Robert H. Robichaux
Department of Botany
University of California
Berkeley, California 94720

E.-Detlef Schulze
Lehrstuhl Planzenökologie
Universität Bayreuth
Postfach 30 08
Bayreuth, West Germay

Kenneth S. Werk
Department of Biology
University of Utah
Salt Lake City, Utah 84112

Introduction

This volume represents the proceedings of a symposium held at the Harvard Forest in August 1983 under the auspices of the Maria Moors Cabot Foundation. It focuses on current efforts to quantify the impact of various plant traits on whole-plant growth and competitive ability. Natural selection should favor plants whose form and physiology tend to maximize their net rate of carbon gain in a particular competitive context, because such plants generally have the most resources with which to reproduce and compete for additional space. Thus, we must analyze how various traits contribute to whole-plant energy capture, and how the energetic costs and benefits associated with different traits vary with environmental conditions, in order to understand the adaptive significance of these traits and the distribution of species that bear them.

Already, such economic analyses have provided insights into the selective pressures on characteristics such as leaf reflectivity, effective leaf size, stomatal conductance, size of photosynthetic enzyme pools, crown form, xylem structure, nitrogen fixation, and allocation to roots versus shoots. This research has now reached an exciting stage, because it is leading to quantitative predictions of favored trends in these traits as a function of environmental parameters and fundamental physiological constraints. Such results have important implications for ecological patterns in plant form and physiology and for evolutionary constraints on photosynthesis and primary productivity; they should be of great interest to ecologists, paleobotanists, physiologists, and plant breeders. The aim of this volume is thus to summarize important recent advances in the economic analysis of plant behavior and to suggest a framework for a unified, quantitative approach to understanding photosynthetic adaptations, their integration with other aspects of plant form, and their relationships to carbon balance and ultimate constraints on plant productivity.

The twenty contributions in this volume discuss the functional significance of a wide range of characteristics in a great diversity of plants and environments, including oceanic plankton and seaweeds, desert succulents, and temperate and tropical trees, rosette shrubs, and herbs. These chapters have been grouped on the basis of the three kinds of energetic tradeoffs they involve. The first section of the book deals with the *economics of gas exchange,* and involves tradeoffs that arise from the unavoidable association of carbon gain with water loss, and from the associated costs of water and nutrient absorption needed to sustain a given rate of leaf photosynthesis. Seven chapters in this section focus on leaf adaptations that affect photosynthesis and transpiration directly, three chapters involve constraints on water and nutrient uptake in roots, and three chapters concern water transport and turgor maintenance. This section brings together the research of physiological ecologists who study leaf adaptations with that of root biologists and specialists on water status, in order to consider the integration of aboveground and belowground adaptations of the plant gas-exchange system.

The book's second section concerns the *economics of support,* which results from tradeoffs between the photosynthetic benefits and the mechanical costs of supporting and supplying different leaf and/or crown geometries. Six chapters discuss adaptation and functional constraints on support structures and their implications for light capture and competitive interactions. The final section contains a single chapter on the *economics of biotic interactions,* involving tradeoffs that result from traits that enhance a plant's potential rate of growth but increase its attractiveness to herbivores, or vice versa.

It should be understood, of course, that variations in certain traits can simultaneously affect the costs of transpiration, nutrient uptake, mechanical support, and/or herbivory. For example, a shift from a spiral leaf arrangement on erect twigs to a distichous leaf arrangement on horizontal twigs would affect both self-shading – and hence photosynthesis and transpirational costs – and mechanical efficiency (Givnish 1984). Similarly, shifts in leaf nitrogen content would affect photosynthetic capacity, the cost of nutrient absorption, and exposure to herbivory (Mooney and Gulmon 1982). Nevertheless, the three classes of tradeoffs involve conceptually distinct sets of physical and physiological processes, so that it is reasonable to discuss each separately, even if two or more of these tradeoffs are affected by variations in a given trait.

The selection of topics for this volume centered on traits related to energy capture for which quantitative cost–benefit analyses have been or are being developed. Research on some significant traits (e.g., xylem anatomy) has not yet reached this stage; in these cases, contributors were invited to discuss the functional significance and some aspects of the energetic costs associated with a given trait, both of which are prerequisite for any future economic analysis. No attempt was made to cover all traits that have been subjected to qualitative analysis only.

Two traits for which quantitative analyses are available but could not be included are leaf phenology and C_3 versus C_4 photosynthesis; excellent reviews of these subjects have appeared recently (Miller and Stoner 1979; Chabot and Hicks 1982; Pearcy and Ehleringer 1984). Analyses involving tradeoffs arising from allocation to productive versus reproductive tissue (e.g., Mirmirani and Oster 1978; King and Roughgarden 1982; Schaffer et al. 1982) were also excluded; although such models offer the hope of translating energy to fitness, they were judged likely to take us far afield from energy capture into almost every aspect of evolutionary biology. With these exceptions, the studies presented or discussed here represent virtually the entire spectrum of quantitative work now under way on economic or cost–benefit approaches to plant adaptations for energy capture in natural ecosystems.

On the use of optimality arguments

Cost–benefit models play an important role in several chapters in this volume. In essence, their aim is to provide insight into the selective basis for a given trait by considering how variation in that trait should affect plant competitive ability in a particular environment and how such effects on competitive ability should vary with environmental conditions. A comparison is then made to see which of the variations available should yield the greatest advantage in a given environment and hence be selectively or competitively favored, and to see how this best "strategy" should vary with environmental conditions. Such an optimality analysis can lead to testable predictions about ecological trends in plant form and physiology, and about the ecological distribution of species with a given constellation of traits. The validity of such an analysis clearly depends on the set of strategies being compared, on the range of environments being considered, and on the specific relation inferred between a trait and its contribution to inclusive fitness.

In recent years, this approach – which has been applied to questions involving the evolution of foraging behavior, animal biomechanics, caste structure in social insects, and growth patterns in colonial invertebrates, as well as plant form and physiology – has been criticized on a number of grounds (Lewontin 1977; Gould and Lewontin 1979; Harper 1982). Principal among these criticisms are (1) the nonfalsifiability of the underlying assumption of adaptation, (2) the post hoc nature and untestability of optimality models, (3) failure to consider nonselective factors shaping evolution, particularly the nature of available genetic variation and developmental constraints, (4) the inappropriateness of an atomistic approach, and (5) teleological and Panglossian assumptions.

Penetrating analyses by Maynard Smith (1977), Oster and Wilson (1979), Clutton-Brock and Harvey (1979), Horn (1979), and Wilson (1980) have effectively rebutted many of these criticisms. A comprehensive exposition and defense of the use of optimality models is now needed. Such a defense cannot be fully elaborated here, but because this approach underpins most of the work presented in this book, I shall summarize the main elements of such a defense.

1. *Nonfalsifiability of assumption of adaptation.* Lewontin (1977) and Gould and Lewontin (1979) suggest that the assumption of a selective basis for variation in a given trait, which underlies any optimality model, is unfalsifiable. If there is imperfect agreement between the pattern of observed variation and the predictions of a specific model, the model is often modified to incorporate other effects while continuing to assume that the trait has adaptive value. Yet, as Maynard Smith (1977) and Oster and Wilson (1979) note, the hypothesis of adaptation is not itself under test in most optimality models. Adaptation, rather than an origin based primarily on drift, founder effects, pleiotropy or epigenetic effects, is often persuasively and independently indicated by evolutionary convergence. When a functionally significant trait shows repeated ecological trends more or less independent of species ancestry or population-specific historical accidents, it is difficult to argue a cause other than selection. The main role of an optimality argument should be as a heuristic device, giving a quantitative means for testing particular hypotheses of how variation in a given trait contributes to plant competitive ability.

2. *Post hoc nature and untestability of optimality models.* Although many adaptive explanations – like many other kinds of scientific explanations – are post hoc, they usually are not ad hoc and apply with equal force to organisms other than those that led to the original hypothesis. Thus, they can be tested rigorously using the comparative approach, as advocated by Clutton-Brock and Harvey (1979), to disentangle the expected selective effects of ecological factors from other confounding factors. More precise and convincing tests can be made by using optimality arguments to produce quantitative predictions for trends in a given trait, and then experimentally varying the appropriate ecological conditions. Tests could involve comparisons of predicted versus observed behavior, or direct experimental studies on the demography, genetics, and physiology of

competing morphs. In any case, the key question in assessing the validity of a theoretical explanation for a given set of observations is not which came first, but whether means exist to test the theory beyond the original observation, how rigorous are the grounds used to include evidence to test the theory, and whether ancillary predictions generated by the theory are also confirmed (Maynard Smith 1977).

Lewontin (1977) and Gould and Lewontin (1979) argue that the general *approach* through optimality arguments leads to untestable predictions, because models can be modified repeatedly to account for discrepancies. They are certainly correct that one cannot modify a model to account for a deviation of observations from those expected from the unmodified model, and then claim that the original data support the hypothesis. However, the only goal of optimality models is to test whether a particular set of tradeoffs and contributions to fitness associated with a given trait is sufficient to account for ecological trends in that trait. Certainly, such specific hypotheses can be falsified and are of great importance.

3. Failure to consider nonselective factors. Another criticism of optimality arguments is that they overlook the roles played by pleiotropy, epigenetic effects, linkage, drift, founder effects, and developmental constraints in shaping the evolution of a given trait in a population (Lewontin 1977; Gould and Lewontin 1979; Harper 1982). Certainly, such genetic and developmental constraints play fundamental – perhaps even predominant – roles in determining where selection within a given population can lead. Geneticists, taxonomic specialists on particular groups, and ecologists studying single-species populations are likely to be sensitive to such phylogenetic constraints. Yet to ascribe heavy weight to such constraints, and to state that they will generally prevent selection from moving populations toward adaptive optima, is essentially to commit the error of orthogenesis, because it entails viewing selection as pushing a particular population toward a specific ecological role. In fact, selection in a given environment should operate on several populations simultaneously, and select for favorable traits that are favorably controlled as well. Naturalists, community ecologists, and biogeographers, whose stock-in-trade is evolutionary convergence and environmental trends in plant form and physiology, are more likely to appreciate the broad, ecologically determined selective forces that generate these trends. Conversely, they may also be more likely to underestimate the real importance of the genetic and developmental constraints that help determine which lineages evolve a particular adaptation or invade particular habitats. To understand fully the fabric of evolution, one must appreciate both its adaptive warp and phylogenetic weft.

4. Inappropriateness of atomistic approach. Gould and Lewontin (1979) criticize optimality models because many break an organism into parts that perform separate functions, and then conduct an economic analysis of each trait separately. This can entail problems because such an approach ignores genetic and developmental correlations among traits and fails to consider the integration of individual traits into the whole organism. This criticism, if softened in light of the points made in the preceding paragraph, is valid. However, comparative studies among organisms within a lineage or within a habitat can demonstrate which traits do or do not share unbreakable developmental correlations, which in turn determine the fundamental phylogenetic constraints on any optimality analysis involving those organisms. The whole thrust of this volume is to consider the integration of adaptations in various traits and their contributions to whole-plant growth and competitive ability; the cost–benefit analyses subsume the laws of correlation and range of phenotypic variation seen in flowering plants.

Stomatal conductance is a trait for which an "atomistic" analysis is well adapted, yet revelatory of the need to consider integration among adaptations. Elegant analyses by Cowan (1977), Cowan and Farquhar (1977), Schulze and Hall (1982), and others demonstrate that plants precisely adjust conductance to changes in humidity and light intensity in such a way as to maximize total daily photosynthesis for a fixed amount of transpiration. Yet their model – like all adaptive models – is a caricature of reality, designed to reveal underlying tradeoffs with unusual clarity, if not complete accuracy. It does not, for example, incorporate stomatal response time. If the environment changes much faster than stomata can open or close, then incorporating an upper limit on their response time favors a different pattern of stomatal behavior (see Chapter 5). In this regard, Mooney et al. (1983) argue that the rather flat, "non-optimal" response of conductance to increasing light intensity seen in plants found in rain-forest understories is expected, because an "optimal" pattern – unconstrained by limits on stomatal movement – would not permit harvest of most of the useful radiation there, which occurs during the passage of brief sunflecks. In addition, the original model does not yield fully quantitative predictions for optimal conductance unless the effects of root versus leaf allocation are also included (see Chapters 5 and 6). This example shows the value of an "atomistic," quantitative optimality model in uncovering additional constraints on plant behavior, whose importance can be tested, in turn, against the quantitative predictions of the model modified in the obvious ways.

5. Teleological and Panglossian assumptions. Harper (1982) argues that most optimality arguments are invalid because they are teleological and

Panglossian in nature. In fact, optimality arguments do not assume that evolution is pushing plants toward some teleological goal – such as maximal whole-plant carbon gain. Rather, they are used to generate and test hypotheses about how variations in particular traits contribute to a plant's competitive ability, and to compare the relative contributions of different variations of a given trait on components of plant fitness. The variations considered are those available in a given group, not Dr. Pangloss's best conceivable. In practice, many models assume that a plant's inclusive fitness will be maximized if its traits maximize the whole-plant rate of energy capture. Alternative hypotheses, of equal plausibility and surely of greater applicability in certain circumstances, might be that selection favors plants that consume water or mineral nutrients at "superoptimal" rates and so exclude competitors from exhaustible resources, or plants that hold a "superoptimal" density of foliage so as to smother opponents. However, it is not generally appreciated that such interesting alternatives cannot be recognized, or studied, except by comparison with the phenotypic pattern that would maximize carbon gain. Furthermore, even if selection favors either of these alternatives, a plant's net rate of energy gain places important constraints on its competitive ability. Thus, consideration of patterns that maximize energy return is likely to be of general significance.

Even a small difference in growth rate can have a dramatic effect on a plant's competitive ability. In part, this is a result of the compound-interest law: An increase in growth rate by only 1% of its magnitude can lead to a difference of 7.2% in plant mass in the time required to double the latter 10 times at the original rate of exponential growth; further increases in growth rate or length of the growth period will lead to much larger increments in final plant mass (Figure 1). Yet, these are *minimal* estimates of the effect of differences in growth rate, because they do not incorporate the impact of slower growth on the probability that a plant will be overtopped, with catastrophic effect on later growth. A significant finding to emerge from several analyses in this volume is the apparent selective importance of small differences in the rate of energy capture: Cowan (see Chapter 5) shows, for example, that the predicted and observed allocation of nitrogen between RuBP carboxylase/oxygenase and carbonic anhydrase increases photosynthesis by only 7%, and that modulation of stomatal conductance increases photosynthesis by only 2%.

Finally, it should be noted that although optimality models are often used to analyze the significance of plant behavior, in some cases the nature of competition may make them inappropriate. A game-theory approach may be required for traits in which the energetic return, or competitive ability, of a morph depends strongly on the mixture of competing morphs

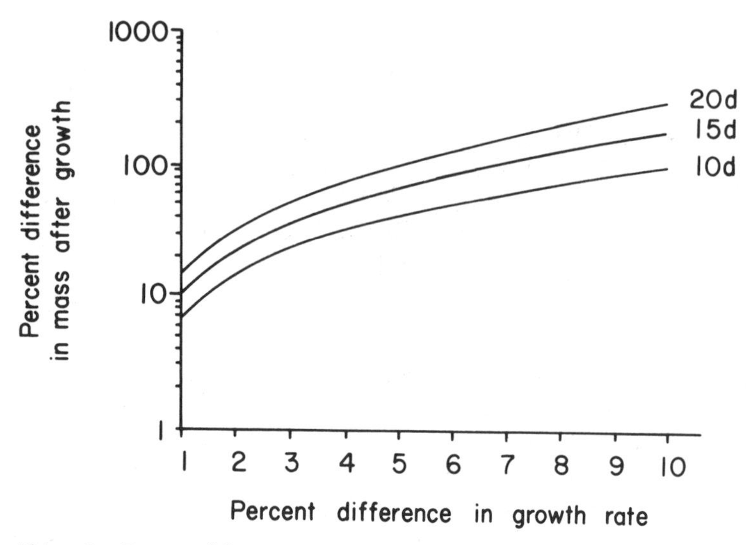

Figure 1. Percent difference in plant mass after growth, relative to control, expressed as a function of percent increments to growth rate relative to its magnitude in control. Length of growth period assumed to be the time required to achieve 10, 15, or 20 doublings (10d, 15d, and 20d, respectively) of mass in control, with exponential growth assumed.

it faces, and can in turn affect the composition of the mixture (Maynard Smith and Price 1973; Givnish 1982). Most optimality models assume that a morph with maximal carbon gain in the absence of explicit competitors will also have the highest return in their presence, but this need not always be true (see Chapters 5 and 16).

References

Chabot, B. F., and D. F. Hicks. 1982. The ecology of leaf life spans. Ann. Rev. Ecol. Syst. 13:229–259.

Clutton-Brock, T. H., and P. H. Harvey. 1979. Comparison and adaptation. Proc. R. Soc. Lond., Ser. B 205:547–565.

Cowan, I. R. 1977. Stomatal behavior and environment. Adv. Bot. Res. 4:117–228.

Cowan, I. R., and G. D. Farquhar. 1977. Stomatal function in relation to leaf metabolism and environment. Symp. Soc. Exp. Biol. 31:471–505.

Givnish, T. J. 1982. On the adaptive significance of leaf height in forest herbs. Amer. Nat. 120:353–381.

– 1984. Leaf and canopy adaptations in tropical forests. Pp. 51–84 *in* E. Medina, H. A. Mooney, and C. Vásquez-Yánes (eds.), Physiological ecology of plants of the wet tropics. Dr. Junk, the Hague.

Gould, S. J., and R. C. Lewontin. 1979. The spandrels of San Marco and the Panglossian paradigm: a critique of the adaptationist programme. Proc. R. Soc. Lond., Ser. B 205:581–598.

Harper, J. L. 1982. After description. Pp. 11–25 *in* E. I. Newman (ed.), The plant community as a working mechanism. Blackwells, Oxford.

Horn, H. S. 1979. Adaptation from the perspective optimality. Pp. 48–61 *in* O. T. Solbrig, S. Jain, G. B. Johnson, and P. H. Raven (eds.), Topics in plant population biology. Columbia University Press, New York.

King, D., and J. Roughgarden. 1982. Graded allocation between vegetative and reproductive growth for annual plants in growing seasons of random length. Theor. Pop. Biol. 22:1–16.

Lewontin, R. C. 1977. Adaptation. Sci. Am. 239:156–169.

Maynard Smith, J. 1977. Optimization theory in evolution. Ann. Rev. Ecol. Syst. 9:31–56.

Maynard Smith, J., and G. R. Price. 1973. The logic of animal conflict. Nature 246:15–18.

Miller, P. C., and W. H. Stoner. 1979. Canopy structure and environmental interactions. Pp. 428–458 *in* O. T. Solbrig, S. Jain, G. B. Johnson, and P. H. Raven (eds.), Topics in plant population biology. Columbia University Press, New York.

Mirmirani, M., and G. F. Oster. 1978. Competition, kin selection, and evolutionary stable strategies. Theor. Pop. Biol. 13:334–339.

Mooney, H. A., C. Field, C. Vásquez-Yánes, and C. Chu. 1983. Environmental controls on stomatal conductance in a shrub of the humid tropics. Proc. Natl. Acad. Sci. USA 80:1295–1297.

Mooney, H. A., and S. L. Gulmon. 1982. Constraints on leaf structure and function in reference to herbivory. BioScience 32:198–206.

Oster, G. F., and E. O. Wilson. 1979. Caste and ecology in the social insects. Princeton University Press.

Pearcy, R. W., and J. Ehleringer. 1984. Comparative ecophysiology of C_3 and C_4 plants. Plant Cell Env. 7:1–13.

Schaffer, W. M., R. S. Inouye, and T. S. Whittam. 1982. The dynamics of optimal energy allocation for an annual plant in a seasonal environment. Amer. Nat. 120:787–815.

Schulze, E.-D., and A. E. Hall. 1982. Stomatal responses, water loss and CO_2 assimilation rates of plants in contasting environments. Pp. 181–230 *in* O. L. Lange, P. S. Nobel, C. B. Osmond, and H. Ziegler (eds.), Physiological plant ecology II, Vol. 12B, Encyclopedia of plant physiology, New Series. Springer-Verlag, Berlin.

Wilson, E. O. 1980. Caste and division of labor in leaf-cutter ants (Hymenoptera: Formicidae: *Atta*). II. The ergonomic optimization of leaf cutting. Behav. Ecol. Sociobiol. 7:157–165.

Part I

Economics of gas exchange

Two fundamental energetic tradeoffs underlie many photosynthetic adaptations. The first arises from the inevitable association of carbon gain with water loss. Carbon dioxide and water vapor diffuse in and out of a leaf along the same pathway, so that increments to the conductance of this pathway increase the potential rates of both photosynthesis and transpiration. Thus, the photosynthetic benefit of any trait that increases the rate at which CO_2 can diffuse into a leaf must be weighed against the energetic costs resulting from increased water loss. These transpirational costs might include a reduction in mesophyll photosynthetic capacity caused by decreased leaf water potential, an increased allocation of energy to unproductive roots and xylem, and/or a shortened period of photosynthetic activity (Givnish and Vermeij 1976; Orians and Solbrig 1977; Givnish 1979, 1984).

A second tradeoff results from the inevitable conflict between leaf photosynthetic capacity and the energetic costs of constructing and maintaining tissue capable of high photosynthetic rates. Highly productive leaves require large inputs of nitrogen, phosphorus, and other mineral nutrients to create the enzyme and pigment pools needed to sustain high rates of CO_2 uptake, implying a tradeoff between photosynthetic benefits and the energetic costs of nutrient capture (Mooney and Gulmon 1979; Gulmon and Chu 1981).

Together, these tradeoffs create an *economics of gas exchange,* linking photosynthesis to the costs of transpiration and needed nutrients. Selection in a given environment should favor plants whose traits tend to maximize the difference between the benefits and costs of enhanced photosynthetic capacity, because such plants generally have the greatest resources with which to compete and reproduce (but see Chapters 5 and 16). Thus, adaptations in leaf form and photosynthetic physiology can be understood only in terms of their integration with other plant traits and the resulting impact on whole-plant carbon gain. The energetic benefits of leaf adaptations must be weighed against the associated costs, in the roots and xylem, of absorption and transport of water and nutrients.

Figure I.1 summarizes the effects on whole-plant carbon gain, in a constant environment, of several traits that directly affect gas exchange. Particular emphasis has been given to leaf traits known to influence photosynthesis and transpiration. Variations in these traits can influence whole-plant carbon gain by affecting photosynthesis per unit leaf biomass, through effects on (a) leaf diffusive conductance, (b) mesophyll photosynthetic capacity, (c) light interception, (d) leaf energy budget, and hence (e) leaf temperature, (f) transpiration rate, and (g) leaf water potential. Leaf traits can also influence whole-plant carbon gain by affecting the respiratory costs associated with leaf construction and maintenance.

Similarly, the fractional allocation of energy to leaves versus absorbing roots can influence whole-plant growth through its effects on (a) leaf water potential, nutrient uptake, and consequent mesophyll photosynthetic capcity, (b) the fraction of productive biomass, and (c) the balance of construction and maintenance respiration between leaves and roots. Finally, variations in several root and xylem traits undoubtedly also affect whole-plant carbon gain. For example, smaller diameters of absorbing rootlets should increase radial hydraulic conductivity per unit root mass, fostering higher levels of leaf water potential and mesophyll photosynthetic capacity for a given energetic allocation to roots. On the other hand, as rootlet diameters become smaller and smaller, at some point the diameter of root xylem elements must decrease, resulting in decreased axial hydraulic conductivity per unit root biomass. Different root diameters may be most efficient for absorbing water from different soil depths (Fowkes and Landsberg 1981), or from soils with different physical properties. Our current inability to model the balance of such effects quantitatively, because of lack of experimental evidence for the functional significance of variations in such traits as rootlet diameter, cortex thickness, width of the Casparian strip, and xylem vessel length, diameter, and wall thickness, is the reason these traits have not been portrayed explicitly in Figure I.1.

Five fundamental conclusions can be drawn from the interactions portrayed in Figure I.1 and their extention to time-varying environments:

1. There is a multiplicity of pathways by which every trait that affects gas exchange influences whole-plant carbon gain. Variation in stomatal conductance, for instance, not only influences the resistance to CO_2 diffusion but also affects transpiration and leaf heat exchange, and hence leaf temperature, leaf water potential, and mesophyll photosynthetic capacity (Figure I.1). Thus, optimal conductance in a constant environment is set

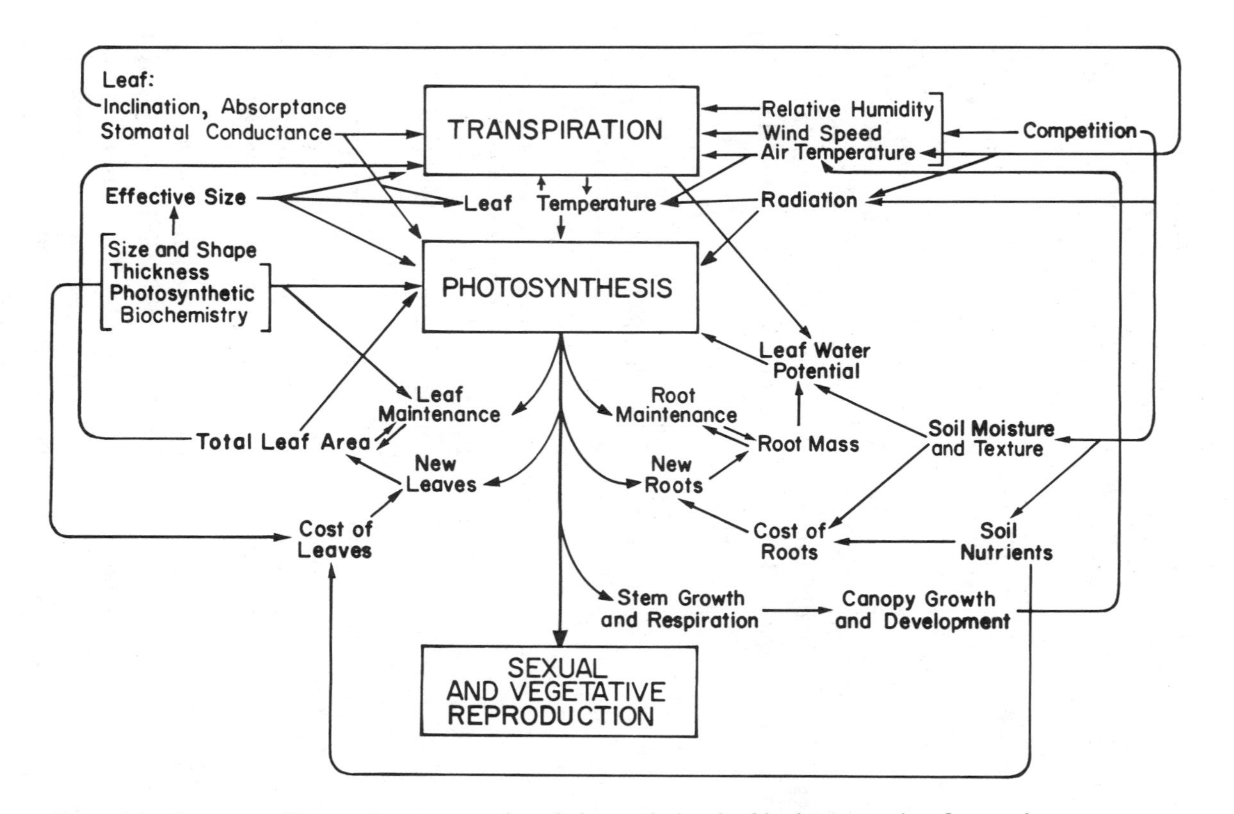

Figure I.1. Summary of interactions among selected plant traits involved in the economics of gas exchange.

by the energetic tradeoffs among its effects on gaseous diffusion, leaf water potential, leaf temperature, and mesophyll photosynthetic capacity.

2. There are strong interactions among plant traits in determining the optimal pattern of behavior in each. Let us again consider stomatal conductance. The degree to which an increase in conductance reduces leaf water potential depends on root hydraulic conductivity and the relative allocation of energy between roots and leaves. At high root allocations and/or conductivities, the optimal stomatal conductance should be higher than at lower root allocations and/or conductivities, given the smaller negative impact on photosynthesis – through its reduced effects on leaf water potential – of an increase in stomatal conductance. Similarly, the energetic impact of different root allocations depends, at least in part, on stomatal conductance. Because the relative values of stomatal conductance and root allocation determine the sizes of each other's effects on photosynthesis through leaf water potential, any analysis of optimal behavior in either trait must address the effects of both traits simultaneously.

The same conclusion should apply to other leaf traits – such as effective size, thickness, internal architecture, nitrogen content, inclination, or absorptance – whose effects on photosynthesis and whole-plant carbon gain depend intimately on stomatal conductance, root versus leaf allocation, and each other's values. Selection should operate on each trait simultaneously to maximize carbon gain. This implies a close integration among plant adaptations that affect gas exchange. Successful models for optimal plant behavior in a given trait must thus incorporate not only the multiple paths by which that trait affects carbon gain but also all other traits that have a substantial influence on that trait's effects. Carbon gain must be considered not as a function of a single parameter while others remain fixed, but as a function of several interactive parameters that are all allowed to vary.

3. The coadaptation of traits involved in gas exchange implies a "command-and-control system" of proximal physiological mechanisms by which such integration is sensed and maintained. The most likely candidate for such a system is one based on hormonal control. For example, production of abscisic acid by shoots during periods of relative water shortage induces not only stomatal closure, but also root growth and (ultimately) leaf abscission. Cytokinin production by roots, on the other hand, stimulates shoot growth and leaf expansion, and retards leaf senescence. The actions of these two opposed responses could serve to adjust stomatal conductance, leaf size, and root versus leaf allocation to co-optimal levels based on environmental conditions (Cowan and Farquhar 1977; Carmi and Van Staden 1979; Raschke 1979; Cowan et al. 1982; Farquhar

and Sharkey 1982). Hormonal control could provide a mechanism for adjustment and integration of many photosynthetic adaptations, and more research is needed on this potential link between proximal and ultimate causes of adaptation.

Direct effects of plant hormones, or source–sink relations (Gifford and Evans 1981), on photosynthesis have been explicitly excluded from Figure I.1, based on the strong relationship of photosynthetic capacity to leaf nitrogen content in different habitats (Chapter 1) and at different stages of plant development (Evans 1983).

4. Plastid sensitivity to leaf water potential is central to the economics of gas exchange. The reduced photosynthetic capacity of partially desiccated mesophyll is the ultimate cause of the costs of transpiration. Without such a reduction, plants could invest little in roots for water absorption and suffer few consequences. With it, the benefits of increased gas exchange must be balanced against either decreased mesophyll photosynthetic capacity or increased allocation to unproductive roots, with a cascading series of consequences for almost every aspect of plant form and physiology involved in energy capture.

What causes the tradeoff between leaf water potential and plastid photosynthetic capacity? Surprisingly, even though this tradeoff is a crucial constraint linking the adaptive properties of shoots and roots, we know very few details of the mechanisms underlying it (see Chapter 6). More research is urgently needed on plastid ultrastucture and physiology, particularly in poikilohydric plants adapted to varying levels or amounts of drought, to determine the basis for this relationship between photosynthetic capacity and leaf water potential and/or turgor.

5. In time-varying environments, the adjustment of different plant traits to environmental conditions must be based not only on the web of interactions portrayed in Figure I.1 but also on the cost of varying these traits at different time scales. The frequency of adaptive modulation of a trait should reflect both the temporal scale of environmental fluctuation to which it is a response and the energetic costs inherent in altering that trait. Other things being equal, more frequent adjustments are expected in traits that require less energy to modify.

Plant traits respond to environmental changes at very different rates. Stomata respond to changes in humidity, light, and temperature over time scales of seconds to hours, apparently in such a way as to maximize daily photosynthesis for a given amount of water loss (Schulze and Hall 1982, Chapters 5 and 6). Over periods of days to weeks, leaf nitrogen levels are adjusted, presumably to maximize the difference between consequent photosynthetic benefits and costs of nutrient acquisition (Mooney and

Gulmon 1979; Gulmon and Chu 1981; Mooney and Chiariello 1984). Allocation to roots versus leaves should vary over similar time periods to maximize the difference between photosynthetic benefits and transpirational costs (Givnish and Vermeij 1976, Chapter 6). Over periods of weeks to months, leaf nitrogen is retranslocated from older, shaded leaves to new leaves in such a way as to maximize net canopy carbon gain (Field 1981, 1983). Finally, over periods of growing seasons to years, allocation between reproductive tissue and leaf tissue occurs and should be adjusted so as to maximize Darwinian fitness (see Cohen 1971; Schaffer 1977; Mirmirani and Oster 1978; King and Roughgarden 1982; Schaffer et al. 1982; Chiariello and Roughgarden 1984; Mooney and Chiariello 1984).

The tempos of these changes do seem to be inversely related to their costs. Changes in stomatal aperture require small amounts of ATP to drive ion pumps in guard cells, whereas construction of roots and leaves requires considerably more energy. Allocation to reproductive tissue is tremendously costly, in terms of the direct energetic costs of producing flowers and fruits, and also the opportunity costs of growth foregone as a result of diversion of energy from productive tissue to unproductive or less productive tissue. As Cowan (Chapter 5) points out, we would expect reallocation of nitrogen between dark and light reactions to proceed at the same tempo as changes in stomatal conductance, were the costs of nitrogen reallocation comparable to those of varying stomatal aperture. The same would be true for all other leaf traits.

The modulation of plant traits at different frequencies has consequences for the integration of these traits at different temporal scales (Mooney and Chiariello 1984). For example, leaf phenology and photosynthetic characteristics should be coadapted. The relative productivity of evergreen versus deciduous leaves in a seasonal environment depends on the photosynthetic rates of each during the favorable and unfavorable seasons, the root costs associated with maintaining their activity during each season, and the amortized costs of leaf construction. Deciduous leaves should be favored when the potential net returns from leaves adapted to different seasons are very different, and when the costs of leaf construction are low – that is, on fertile sites in seasonally arid habitats, or in thermally seasonal habitats with aseasonal or summer rainfall. Restriction of leaf activity to the warm and/or moist time of year has numerous consequences for traits that can be modulated over shorter time periods, and the advantage of such a restricted phenology depends, in turn, on the energetic balance that results from such modulation over the favorable season versus the entire year.

The 13 chapters that follow address various aspects of plant form and physiology involved in the economics of gas exchange. These chapters have been divided into three groups, dealing with photosynthetic adaptations, water and nutrient acquisition, and water transport and maintenance of turgor. Each of these addresses only a few aspects of the tradeoffs implied by the preceding discussion, but together they encompass many of the interactions and constraints on which a more general, synthetic approach must be based.

Photosynthetic adaptations

Field and Mooney (Chapter 1) document a linear relationship between leaf nitrogen content and maximum photosynthetic rate in natural vegetation, in plants ranging from highly productive desert annuals to slow-growing evergreen shrubs of nutrient-poor fynbos. This chapter thus provides important background for analyzing optimal patterns of nitrogen (N) investment in leaves. The basis for the observed relationship is intriguing because a curvilinear relationship between leaf N content and photosynthesis would be expected if other factors – such as stomatal conductance – were to remain constant and limit carbon gain at high N levels (e.g., Evans 1983). Field and Mooney critically examine trends expected in other potential limits to photosynthesis as leaf N content varies. The analyses in Chapters 5 and 6 are suggestive in that they predict that under fixed environmental conditions, stomatal conductance should vary with mesophyll photosynthetic capacity in such a way as to lead to nearly linear relationships among photosynthesis, stomatal conductance, and mesophyll conductance. If it were shown that mesophyll conductance (carboxylation conductance *sensu* Farquhar and Sharkey 1982) varies linearly with leaf N content, the observed P-versus-N relationship might be partly explained.

Ehleringer and Werk (Chapter 2) explore the energetic tradeoffs associated with differences in leaf absorptance, inclination, compass orientation, and solar-tracking behavior. Their analysis emphasizes the impact of these traits on light interception, heat load, photosynthetic rate, and potential period of photosynthetic activity. They present an excellent review of research on leaf absorptance in the genus *Encelia* and provide data on the energetic cost of pubescence associated with different levels of reflectivity. Although hairs on the most reflective leaves represent only a few days of photosynthesis, the presence of an additional 50% nonproductive biomass roughly halves the rate at which a given unit of leaf mass can replicate itself – a major impact on potential growth rate.

Nobel (Chapter 3) discusses constraints on the orientation of photosynthetic surfaces in cacti and other desert succulents with Crassulacean acid metabolism (CAM) photosynthesis. This issue entails a somewhat different set of tradeoffs than those discussed in Chapter 2, because CAM plants exchange gases by night and capture light by day. This behavior reduces transpiration and may be advantageous in arid sites, but it limits photosynthesis based on the amount of malate that can be stored overnight in cell vacuoles. Nobel demonstrates that the massive, often vertical photosynthetic surfaces of cacti and other succulents often intercept so little light that their photosynthesis is on the verge of being light-limited, even in open desert habitats. He then shows that the nonrandom orientation of *Opuntia* pads varies with latitude and seasonal growing conditions in such a way as to maximize carbon gain. Nobel also discusses the effects of ribs, spines, canopy form, and shading from competitors on photosynthesis in desert succulents, and their possible relationships to species distributions. It is interesting to note that columnar *Copiapoa* cacti of the Atacama Desert incline their major axis steeply in a direction that *minimizes* light interception (Ehleringer et al. 1980) and consequent heat load and cuticular water loss. Presumably, under these most arid conditions, daytime transpirational costs become as important for cacti as for other plants.

Lee (Chapter 4) discusses three unusual adaptations for light capture found in rain-forest herbs. Such plants confront an entirely different set of tradeoffs involving light interception: unlike desert plants, they face too little light, not too much, and heat loads are transient and not as substantial as those in open habitats. Red leaf undersurfaces, lens-shaped epidermal cells, and blue iridescence all serve to increase light interception by a given amount of leaf tissue, reduce its effective light compensation point, and thus help permit growth under very low light levels. Each would yield little or no advantage in sunny habitats and might instead lead to photoinhibition or photobleaching.

Cowan (Chapter 5) analyzes three tradeoffs affecting carbon gain at the level of chloroplast, leaf, and whole plant. First, a quantitative model is presented for optimal allocation of nitrogen between two photosynthetic enzymes: carbonic anhydrase (CA) and RuBP carboxylase/oxygenase (Rubisco). CA facilitates the liquid-phase transport of CO_2 across the cytoplasm to the chloroplasts. The predicted allocation to CA versus Rubisco is close to that observed, and it increases photosynthesis by 5% over what would be expected were all nitrogen invested in Rubisco. A similar model is presented for optimal allocation between enzymes involved in light and dark reactions. These analyses of optimal adjustments of enzyme pools are ground-breaking and suggest many further studies to explain why, for example, the relative allocation to PEP carboxylase versus Rubisco in C_4 plants shifts with nitrogen availability (Sugiyama et al. 1984).

Second, the Cowan-Farquhar model for optimal stomatal conductance is reviewed. It assumes that conductance $g(t)$ should vary so as to minimize daily transpiration $\int E \cdot dt$ for a given total amount of photosynthesis $\int A \cdot dt$, implying that g should vary so that $\partial E / \partial A = (\partial E / \partial g)/(\partial A / \partial g)$ remains constant. This model explains many aspects of stomatal behavior, but it leaves unspecified the value of $\partial E / \partial A$ that would maximize whole-plant carbon gain. This issue is addressed in a final model that asks how different levels of stomatal conductance, combined with different levels of investment in root versus leaf tissue, affect the expected time until a plant exhausts its supply of soil water in areas with unpredictable rainfall. This model predicts how the optimal pattern of decline in conductance with soil water content should vary with root allocation, and vice versa, to maximize net carbon gain. It does not, however, specify a unique combination of root allocation and stomatal behavior as optimal in a particular environment.

Givnish (Chapter 6) provides a different approach to this question. A quantitative model is presented for optimal stomatal conductance and allocation to leaves versus roots in a constant environment, based on the effects of each on leaf water potential and mesophyll photosynthetic capacity. If photosynthesis is assumed to be a linear function of the CO_2 concentration in the mesophyll, then the ratio r_m / r_s of mesophyll to stomatal resistance should equal the ratio of energetic allocation to leaves versus roots.

A computer simulation is used to predict optimal plant behavior as a function of eight environmental and physiological parameters, using data from *Phaseolus vulgaris* on root hydraulic conductivity (see Chapter 9) and the sensitivity of photosynthetic parameters to leaf water potential. The predictions generally concur, qualitatively and quantitatively, with the observed trends.

The model also accounts for a linear relationship between photosynthesis and stomatal conductance as mesophyll capacity varies, and for a curvilinear relationship as transpirational costs vary. The marginal cost of transpiration is defined as the decrease in photosynthesis caused by an increment in transpiration, through its effects on leaf water potential and plastid photosynthetic capacity. This marginal cost of transpiration is shown to be identical in value with the root cost of transpiration used by Givnish and Vermeij (1976) and to the "shadow price" of water in terms of carbon, $\partial A / \partial E$, in the Cowan-Farquhar model. Furthermore, a tabulation of observed values of $\partial A / \partial E$ shows that herbs generally operate at values of this parameter close to that predicted for *Phaseolus*, which would in turn maximize whole-plant carbon gain. Methods are suggested for extending the analysis to other traits and to temporally varying environments. The similarity of actual leaf widths to those predicted suggests that leaf size may also have evolved so as to maximize whole-plant carbon gain. This chapter and the preceding one are unique in attempting to establish a direct, quantitative connection between leaf and root adaptations.

Parkhurst (Chapter 7) assesses the adaptive value of leaf internal architecture using a detailed mathematical model for the three-dimensional pattern of CO_2 diffusion within leaves. Many C_3 leaves have an upper layer of densely packed palisade cells, a lower layer of spongy mesophyll with extensive intercellular air spaces, and large substomatal cavities near the stomata at the bottom. Parkhurst demonstrates that this architecture enhances CO_2 diffusion and leaf carbon gain for a given amount of water loss, compared with hypothetical leaves composed solely of palisade cells or spongy mesophyll, or lacking substomatal cavities. These findings are based on a more general form of a photosynthetic model used to predict optimal leaf thickness and stomatal distribution, and likely to be of quite general use in analyzing various aspects of leaf internal structure.

Water and nutrient acquisition

Caldwell and Richards (Chapter 8) review general aspects of root function in relation to water and nutrient uptake. They emphasize the functional significance of different rooting patterns for water and nutrient absorption under different soil conditions and in the presence of different kinds of competitors. They present an excellent case study of competition in relation to rooting pattern in two *Agropyron* bunchgrass species growing in mixture with the shrub *Artemisia tridentata*. The authors conclude with a discussion of the potential significance of such root traits as rootlet thickness, spacing, branching behavior, and pattern of regrowth.

Fiscus (Chapter 9) presents a detailed analysis of the energetic costs of water absorption by roots. This study, the first of its kind and conducted on *Phaseolus vulgaris*, provides a framework for analyzing the impact of a given investment on root hydraulic conductivity. This, in turn, provides the basis for analyzing the effect of root investment on leaf water potential and consequent effects on photosynthesis, and thus is a vital bridge linking studies of leaf and root adaptation (see Chapter 6).

Pate (Chapter 10) provides an elegant balance-sheet analysis of the metabolic costs and benefits associated with nitrogen fixation. Nitrogen is an important limit on photosynthetic capacity (see Chapter 1), and many plants in nitrogen-poor environments form symbioses with prokaryotes that can fix atmospheric nitrogen. Why don't all plants (even in groups in which N-fixing symbioses have evolved) possess this seemingly advantageous trait? Presumably, fixation should be favored only where it yields a net energetic advantage. In this light, Pate analyzes the costs of obtaining nitrogen through symbiotic fixation, compared with those associated with carnivory or with reduction from ammonia or nitrate in the soil.

Water transport and maintenance of turgor

Baas (Chapter 11) reviews ecological trends in the anatomy of xylem elements used to transport water from roots to leaves in higher plants. Repeated patterns occur in such traits as vessel length and diameter, presence of perforation plates, and nature of fibers and rays. Although some of the qualitative tradeoffs underlying these trends are understood, quantitative explanations have remained elusive. The problem has been a lack of quantitative, experimental data on the effects of xylem diameter, length, and wall thickness on cavitation rates as a function of water potential. With such data and Poiseuille's law, optimal xylem geometry might be calculated from the tradeoff between the greater conductive capacity of broad, long, thin-walled vessels and their inherently higher exposure to cavitation. The method of Tyree et al. (1984) for ultrasonic detection of individual cavitation events may allow the crucial data to be obtained through comparative studies on species with different xylem characteristics.

Robichaux, Holsinger, and Morse (Chapter 12) present a novel series of studies on mechanisms of turgor maintenance in Hawaiian species of *Dubautia* (Compositae). Turgor is essential in most higher plants; when it is lost, growth and photosynthesis are inhibited, and irreversible mechanical damage to cells and membranes can occur. As a result, plant growth under conditions of low moisture availability should be enhanced by mechanisms that promote the maintenance of high turgor pressures as tissue water content decreases. The authors examine the nature and significance of variations in tissue osmotic and elastic properties for turgor maintenance in *Dubautia* species ranging from rain forest to semidesert. Species adapted to dry habitats have lower osmotic potentials and elastic moduli than those found in moister sites, with the most striking difference found in elastic properties. Adaptive radiation into dry sites appears related to a reduction in elastic modulus accompanying a reduction in chromosome number. Three hypotheses are evaluated concerning the mechanistic basis of variation in tissue elastic properties, including possible effects of cell size, cell wall composition, and apoplasmic water loss.

Meinzer and Goldstein (Chapter 13) present an important series of studies on the impact of three traits (stem pith volume, marcescent leaves, and leaf pubescence) on turgor maintenance and carbon balance in unbranched rosette shrubs of the genus *Espeletia,* native to Andean paramo. Plants with similar, striking growth forms occur on high mountains in tropical west Africa and Hawaii. All face the unusual challenge of enduring winter every night, summer every day. As a result, paramo plants confront unique problems of water balance.

The research reported is significant in that it addresses the adaptive significance of the capacitance (Nobel and Jordan 1983), as opposed to conductance, of the water transport system of plants in a temporarily variable environment, and helps explain several characteristic features of paramo plants. Early in the morning, the stomata of these plants open in response to light and transpiration commences, but root water absorption is slow because the soil is frozen or nearly so. During this period, *Espeletia* species draw water from a fleshy pith with interconnections to the xylem, and thereby enhance daily carbon gain. "Morning drought" becomes more severe at higher elevations where nighttime frosts are more frequent, and *Espeletia* species show the expected elevational increase in pith volume and hydraulic capacitance, possibly related to their unusual reverse elevational cline in height.

The dead, marcescent leaves that adhere to *Espeletia* stems serve to insulate the stem core and buffer it against large thermal fluctuations. Removal of dead leaves leads to freezing of xylem and pith, low morning leaf water potentials, and eventual plant death. Finally, *Espeletia* leaves have a pubescence unlike that of *Encelia* leaves (see Chapter 2), in that it consists of live hairs that allow penetration of considerable light and heat load. Thus, it enhances carbon gain by raising leaf temperature without greatly increasing leaf diffusive resistance. Pubescence also serves to insulate the sun-warmed terminal bud of an *Espeletia* when its furry leaves curl about it at night.

References

Carmi, A., and J. Van Staden. 1979. Role of roots in regulating the growth rate and cytokinin content in leaves. Plant Physiology 73:76–78.

Chiariello, N., and J. Roughgarden. 1984. Storage allocation in seasonal races of an annual plant: optimal vs. actual allocation. Ecology 65:1290–1301.

Cohen, D. 1971. Maximizing final yield when growth is limited by time or by limiting resources. Journal of Theoretical Biology 33:299–307.

Cowan, I. R., and G. D. Farquhar. 1977. Stomatal function in relation to leaf metabolism and environment. Symposia of the Society for Experimental Biology 31:471–505.

Cowan, I. R., J. A. Raven, W. Hartung, and G. D. Farquhar. 1982. A possible role for abscisic acid in coupling stomatal conductance and photosynthetic carbon metabolism in leaves. Australian Journal of Plant Physiology 9:489–498.

Ehleringer, J., H. A. Mooney, S. L. Gulmon, and P. Rundel. 1980. Orientation and its consequences for *Copiapoa* (Cactaceae) in the Atacama desert. Oecologia 46:63–67.

Evans, J. R., 1983. Nitrogen and photosynthesis in the flag leaf of wheat (*Triticum aestivum* L.). Plant Physiology 72:297–302.

Farquhar, G. D., and T. D. Sharkey. 1982. Stomatal conductance and photosynthesis. Annual Review of Plant Physiology 33:317–345.

Field, C. 1981. Leaf age effects on the carbon gain of individual leaves in relation to microsite. Pp. 41–50 *in* N. S. Margaris and H. Mooney (eds.), Components of productivity of mediterranean regions – basic and applied aspects. Dr. Junk, The Hague.

– 1983. Allocating leaf nitrogen for the maximization of carbon gain: leaf age as a control on the allocation program. Oecologia 56:341–347.

Fowkes, N. D., and J. J. Landsberg. 1981. Optimal root systems in terms of water uptake and movement. Pp. 109–125 *in* D. A. Rose and D. A. Charles-Edwards (eds.), Mathematics and plant physiology. Academic Press, New York.

Gifford, R. M., and L. T. Evans. 1981. Photosynthesis, carbon partitioning and yield. Annual Review of Plant Physiology 32:485–509.

Givnish, T. J. 1979. On the adaptive significance of leaf form. Pp. 351–380 *in* O. T. Solbrig, S. Jain, G. B. Johnson, and P. H. Raven (eds.), Topics in plant population biology. Columbia University Press, New York.

– 1984. Leaf and canopy adaptations in tropical forests. Pp. 51–84 *in* E. Medina, H. A. Mooney, and C. Vásquez-Yánes (eds.), Physiological ecology of plants of the wet tropics. Dr. Junk, The Hague.

Givnish, T. J., and G. J. Vermeij. 1976. Sizes and shapes of liane leaves. American Naturalist 100:743–778.

Gulmon, G. L., and C. C. Chu. 1981. The effects of light and nitrogen on photosynthesis, leaf characteristics, and dry matter allocation in the chaparral shrub, *Diplacus aurantiacus*. Oecologia 49:207–212.

King, D., and J. Roughgarden. 1982. Graded allocation between vegetative and reproductive growth for annual plants in growing seasons of random length. Theoretical Population Biology 22:1–16.

Mirmirani, M., and G. F. Oster. 1978. Competition, kin selection, and evolutionary stable strategies. Theoretical Population Biology 13:334–339.

Mooney, H. A., and N. R. Chiariello. 1984. The study of plant function – the plant as a balanced system. Pp. 305–323 *in* R. Dirzo and J. Sarakhán (eds.), Perspectives on plant population ecology. Sinauer, Sunderland, Mass.

Mooney, H. A., and S. L. Gulmon. 1979. Environmental and evolutionary constraints on the photosynthetic characteristics of higher plants. Pp. 316–337 *in* O. T. Solbrig, S. Jain, G. B. Johnson, and P. H. Raven (eds.), Topics in plant population biology. Columbia University Press, New York.

Nobel, P. S., and P. W. Jordan. 1983. Transpiration analysis of desert species: resistances and capacitances for a C_3, C_4, and a CAM plant. Journal of Experimental Biology 34:1379–1391.

Orians, G. H., and O. T. Solbrig. 1977. A cost-income model of leaves and roots with special reference to arid and semi-arid areas. American Naturalist 111:677–690.

Raschke, K. 1979. Movements of stomata. Pp. 383–441 *in* W. Haupt and M. E. Feinleib (eds.), Physiology of movements, vol. 7, Encyclopedia of plant physiology, new series. Springer-Verlag, New York.

Schaffer, W. M. 1977. Some observations on the evolution of reproductive rate and competitive ability in flowering plants. Theoretical Population Biology 11:90–104.

Schaffer, W. M., R. S. Inouye, and T. S. Whittam. 1982. The dynamics of optimal energy allocation for an annual plant in a seasonal environment. American Naturalist 120:787–815.

Schulze, E.-D., and A. E. Hall. 1982. Stomatal responses, water loss and CO_2 assimilation rates of plants in contrasting environments. Pp. 181–230 *in* O. L. Lange, P. S. Nobel, C. B. Osmond, and H. Ziegler (eds.), Physiological plant ecology II, vol. 12B, Encyclopedia of plant physiology, new series. Springer-Verlag, New York.

Sugiyama, T., M. Miyuno, and M. Hayashi. 1984. Partitioning of nitrogen among ribulose-1,5-biphosphate carboxylase/oxygenase, phosphoenolpyruvate carboxylase, and pyruvate orthophosphate dikinase as related to biomass productivity in maize seedlings. Plant Physiology 77:665–669.

Tyree, M. T., M. A. Dixon, E. L. Tyree, and R. Johnson. 1984. Ultrasonic acoustic emission from the sapwood of cedar and hemlock: an examination of three hypotheses regarding cavitations. Plant Physiology 75:988–992.

1 The photosynthesis–nitrogen relationship in wild plants

CHRISTOPHER FIELD AND
HAROLD A. MOONEY

Worldwide, nitrogen is one of the mineral nutrients most limiting to plant growth. Though it is the most abundant element in the atmosphere, nitrogen becomes available to plants largely through the recycling of organic matter or through the energetically expensive reduction of dinitrogen gas (see Chapter 10). Nitrogen is easily lost from ecosystems by leaching or conversion to N_2 gas by denitrifying bacteria.

At the system level, the importance of nitrogen is underscored by the sensitivity of managed and natural ecosystems to nitrogen fertilization. Agricultural grain yield is highly correlated with the level of nitrogen application. This trend extends across a wide variety of crops and farming systems and includes the record yields of several crops (Ritchie 1980). Natural ecosystems respond to nitrogen fertilization with increased productivity or changes in species composition or both (Lee et al. 1983).

Nitrogen is a limiting resource in many ecosystems, but levels of leaf nitrogen reflect the relative partitioning among multiple sinks, as well as total availability. From an evolutionary perspective, the problem of nitrogen limitation has two components. How should a plant allocate a given nitrogen pool between reproduction, leaves, roots, and stems for the maximization of fitness? And, in any ecological setting, how big should the nitrogen pool be? Answering these questions requires knowledge of the costs of nitrogen acquisition and the benefits of alternative nitrogen deployment patterns. Mooney and Gulmon (1979, 1982) have established a conceptual framework for analyzing the costs and benefits of nitrogen acquisition and deployment, but the specific shapes of the cost and benefit functions remain somewhat conjectural. The relationship between photosynthesis and leaf nitrogen is one of the most important benefit functions, because photosynthesis provides the energy and structural substrates necessary for reproduction, growth, or foraging for additional nitrogen. Here we shall explore aspects of this benefit function, examining the generality, the mechanism, and some of the implications of the relationship between the photosynthesis and the nitrogen content of leaves.

Photosynthesis–nitrogen relationships are intrinsically complex, because photosynthesis represents the integrated operation of a series of processes sensitive to environmental factors as well as leaf physiology and structure. The ecologically relevant measure of photosynthesis is time-integrated photosynthesis in the natural environment, but this index reveals little about leaf function. Maximum photosynthetic capacity under optimum conditions may not always scale simply with time-integrated CO_2 fixation, but measurements of photosynthetic capacity are important for three reasons. They indicate the maximum rates of CO_2 fixation that occur in nature and also the maximum possible benefits from a given investment in photosynthetic machinery. In addition, if constructing and maintaining photosynthetic machinery is expensive, photosynthetic capacities should be tuned to the constraints of the environment, and unusable capacity should be trimmed by natural selection. Thus, photosynthetic capacity provides a useful starting point for analyzing the relationship between photosynthesis and leaf nitrogen.

All of the biochemical and photobiological processes of photosynthesis require nitrogenous compounds. In addition to the proteins (typically 16% nitrogen) that catalyze the reactions of CO_2 fixation and the regeneration of the CO_2 acceptor, photosynthesis requires reducing equivalents (NADPH) and ATP, produced by light-driven electron-transport and proton-transport reactions. Nitrogenous compounds that provide the basis for these reactions include chlorophyll (6% nitrogen), chlorophyll proteins, electron-transport proteins, and the ATP-synthesizing enzyme. The nitrogen investment in many of these compounds is not precisely known, but the proportion of the total leaf nitrogen allocated to photosynthetic reactions is undoubtedly large. As a first approximation, the commitment of leaf nitrogen to photosynthesis in C_3 plants is given by the 75% of the leaf nitrogen that can be recovered from the chloroplasts (Stocking and Ongun 1962), minus the probably small proportion of the chloroplast nitrogen invested in the reactions of nitrite reduction and amino acid synthesis, plus the nitrogen incorporated in the peroxisome and mitochondrial enzymes that recycle the products of photorespiration, plus the nitrogen in that fraction of the nucleic acids and protein-synthesizing machinery necessary to produce and recycle the components of the photosynthetic reactions. These compounds may account for well over three-quarters of the total leaf nitrogen.

A number of studies have reported correlations between some measure of photosynthetic capacity and (a) total leaf nitrogen, (b) the nitrogen content of some protein fraction, or (c) the activities of particular enzymes, with emphasis on variation within single species or among related species.

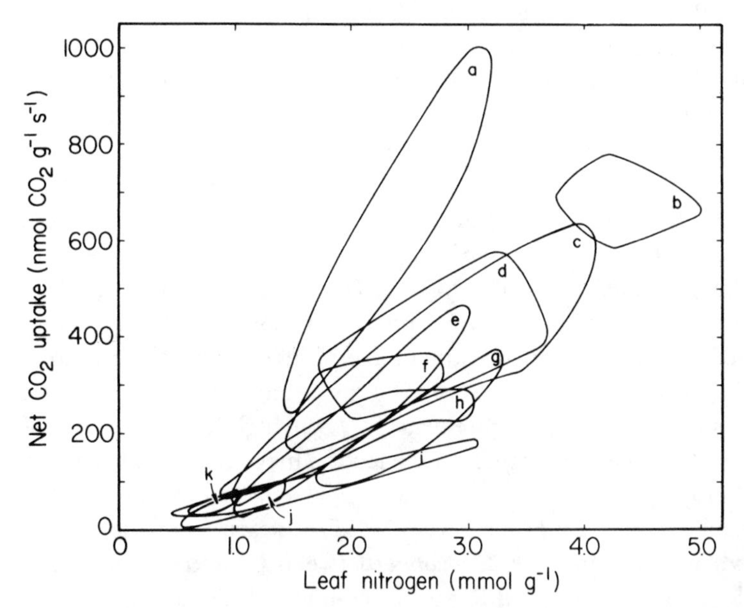

Figure 1.1. A_{max}–N relationships for a wide variety of plant species. Each outline circumscribes all of the data reported in each study. The study species and the experimental conditions are described in Table 1.1. Included are data from 33 C_3 species and 11 C_4 species. Some of the studies present data from naturally grown plants, whereas others report data from plants grown in controlled environments. The experimental variables manipulated to introduce variations in A_{max} and N include nitrogen availability, light availability, and leaf age. Those in group a, with the greatest A_{max} per unit of N, are the only C_4 data in this summary.

Nátr (1975), Yoshida and Coronel (1976), Björkman (1981), and Hesketh et al. (1983) have recently reviewed many of these studies. We approach the photosynthesis–nitrogen relationship in a broader context, examining general trends that cut across taxonomic groups, habitats, and life forms.

To provide an overview of the generality in the photosynthesis–nitrogen relationship, we have summarized a series of studies reporting photosynthetic capacity and leaf nitrogen content on the basis of leaf dry weight (Figure 1.1). In each of these studies, photosynthetic capacity (A_{max}) is the photosynthetic rate measured under saturating light intensity, optimum temperature, relatively high humidity, and the CO_2 concentration typical of normal air. Leaf nitrogen (N) is that measured on the same or matched leaves by Kjeldahl analysis. This procedure is sensitive to all forms of organic nitrogen, of which 70–80% in a typical leaf is in proteins, 10% is in nucleic acids, 5–10% is in chlorophyll and lipoproteins, and the remainder is mostly in free amino acids (Chapin and Kedrowski 1983). For the data summarized in Figure 1.1, correlation coefficients between A_{max} and N range from 0.51 to 0.97, with a median value of 0.90 (Table 1.1). The slopes of linear-regression equations expressing A_{max} as a function of N vary substantially (Table 1.1), but data from all but one of the studies cluster around a single straight line. The outlying data in Figure 1.1 are the results of Pearcy et al. (1982) and represent the only C_4 plants included in the figure. The general comparison provided here confirms the taxonomically limited conclusions of Brown (1978), Bolton and Brown (1980), Schmitt and Edwards (1981), Brown and Wilson (1983), and Wilson and Brown (1983) that for a given investment in leaf nitrogen, C_4 plants tend to realize a higher A_{max} than do C_3 plants. To the extent that environmental conditions permit leaves to operate near A_{max}, the higher A_{max} per unit of leaf nitrogen in C_4 species suggests that selection may favor these plants in nitrogen-limited habitats. This hypothesis is not supported by much of the available data (Pearcy and Ehleringer 1984), but deserves further study. Possible explanations for this difference between C_3 and C_4 plants are discussed by Brown (1978) and Raven and Glidewell (1981).

The fact that all the C_3 plants in Figure 1.1 cluster tightly around a single straight line suggests a fundamental relationship that is relatively insensitive to differences among species or growth conditions. Included in the figure are data from herbs, shrubs, and trees, from evergreens and deciduous species, from leaves ranging in age from a few weeks to more than three years, from plants growing naturally in the field, and from plants maintained under a variety of greenhouse and growth-chamber conditions.

In analyzing these and other reports of A_{max} – N relationships, we consider examples in which both A_{max} and N are presented on the basis of leaf dry weight or leaf area, but we largely ignore the substantial literature in which A_{max} is presented as an area-based quantity and N is presented on a leaf-weight basis. Because leaf-specific weight (LSW) or weight per unit of leaf area tends to change in response to variation in nutrient availability (Loveless 1961; Gulmon and Chu 1981) or variation in light intensity during growth (Björkman 1981), the results of studies reporting A_{max} on an area basis and N on a weight basis can be very confusing.

A survey: the photosynthesis – nitrogen relationship in natural vegetation

For a detailed examination of the A_{max} – N relationship, we restrict our attention to naturally growing C_3 plants. The emphasis on C_3 plants is

Table 1.1. *Summary of selected studies reporting a relationship between photosynthetic capacity at light saturation (A_{max}) and leaf organic nitrogen (N), both expressed per unit of leaf weight*

Reference	Species	Equation[a]	r	n	Notes	Label in Figure 1.1
Björkman and Holmgren (1963)	*Solidago virgaurea*	$A_{max} = -255 + 195 \cdot N$	0.93	12	Twelve clones grown in growth chambers at high light	*g*
Mooney et al. (1978)	6 *Eucalyptus* species	$A_{max} = -9.5 + 66.2 \cdot N$	0.97	18	One-year-old nursery-grown plants and 2-month-old phytotron-grown plants	*i*
Field (1981)	*Lepechinia calycina*	$A_{max} = -45 + 161 \cdot N$	0.93	15	Leaf age and shade varied: irrigated plants grown outside	
Gulmon and Chu (1981)	*Diplacus aurantiacus*	$A_{max} = -82 + 164 \cdot N$	0.96	23	Growth-chamber plants under several light and nitrogen treatments	*c*
Medina (1981)	*Nicotiana glauca* and *Eucalyptus camaldulensis*	$A_{max} = -123 + 175 \cdot N$	0.96	32	Excised branches of naturally growing plants	*e*
Mooney et al. (1981)	5 species of Death Valley annuals	$A_{max} = 171 + 80 \cdot N$	0.64	20	Leaf age series on naturally growing plants	*d*
Mooney et al. (1981)	4 species of old-field annuals	$A_{max} = 12 + 123 \cdot N$	0.78	21	Leaf age series on naturally growing plants	*f*
Pearcy et al. (1982)	11 *Euphorbia* species (C_4)	$A_{max} = -180 + 361 \cdot N$	0.90	11	Greenhouse-grown plants: each A_{max} and N represents mean of several leaves	*a*
Mooney et al. (1983)	6 fynbos species	$A_{max} = -47 + 113 \cdot N$	0.77	13	One-year-old leaves on naturally growing plants	*k*
Field and Mooney (1983)	*Lepechinia calycina*	$A_{max} = 8 + 93.8 \cdot N$	0.89	27	Leaf age series on naturally growing plants	*h*
Field et al. (1983)	5 species of California evergreens	$A_{max} = 14 + 28.6 \cdot N$	0.51	53	Leaf age series on naturally growing plants	*j*
Mooney (unpublished)	*Raphanus sativus*	n.s.	0.04	5	Growth-chamber-grown plants	*b*

[a] A_{max} is in nmol CO_2 g^{-1} s^{-1}, and N is in mmol g^{-1}.

in the interest of brevity. The focus on naturally growing plants arises from the hypothesis that an A_{max} – N relationship reflects evolutionary responses to natural habitats and may be altered by growing plants under modified nutrient, water, light, or CO_2 availability. Here we shall consider A_{max}, leaf nitrogen, LSW, and stomatal conductance from four studies. Collectively, these studies reported data from 137 leaves of 21 plant species, including trees, shrubs, and herbs from environments varying widely in nitrogen availability and productivity. Five species are winter annuals of Death Valley, California; four are summer annuals of central Illinois (all described by Mooney et al. 1981); one is a drought-deciduous shrub of the California chaparral (Field and Mooney 1983); five are evergreen shrubs and trees from coastal central California (Field et al. 1983); six are shrubs from South African mountain fynbos (Mooney et al. 1983). In this combined data set, which we refer to as the VINE (vegetation in natural environments) survey, maximum leaf durations vary 50-fold, LSW varies by a factor of 5, and A_{max} varies more than 20-fold in measurements based on a standardized protocol and instrumentation. The desert annuals have some of the highest photosynthetic capacities of any C_3 plants (Mooney et al. 1976), and the evergreen sclerophylls have some of the lowest (Mooney and Gulmon 1979). The VINE survey does not include plants that grow under deep shade, and none of the species has a symbiotic nitrogen-fixing microorganism.

In order to explore the widest range of naturally occurring nitrogen levels, we include measurements from leaves of different ages, ranging from the youngest fully expanded leaf to the oldest nonnecrotic leaf. Leaf nitrogen declines with increasing leaf age in essentially all plants and provides a completely natural source of intraspecies (and intraplant) variation. Though leaf aging is a complex process, many aspects of the gradual changes in leaf physiology following full expansion are consistent with the interpretation that the dominant phenomenon of leaf aging is nitrogen mobilization and export (Field and Mooney 1983), a process that may increase whole-plant photosynthesis when old leaves are increasingly shaded by younger leaves (Field 1983).

Over the 137 leaves in the VINE survey, A_{max} is highly correlated with leaf N. This conclusion applies when both parameters are expressed on the basis of either leaf dry weight (Figure 1.2) or leaf area (Figure 1.3). As observed in other studies (Gulmon and Chu 1981; Medina 1981; Field and Mooney 1983), the correlation coefficient is higher for weight-based measurements ($r = 0.92$) than for area-based measurements ($r = 0.53$). Subsets within the survey, however, show similar weight-based and area-based trends. For the 10 nonevergreen species (annuals and deciduous peren-

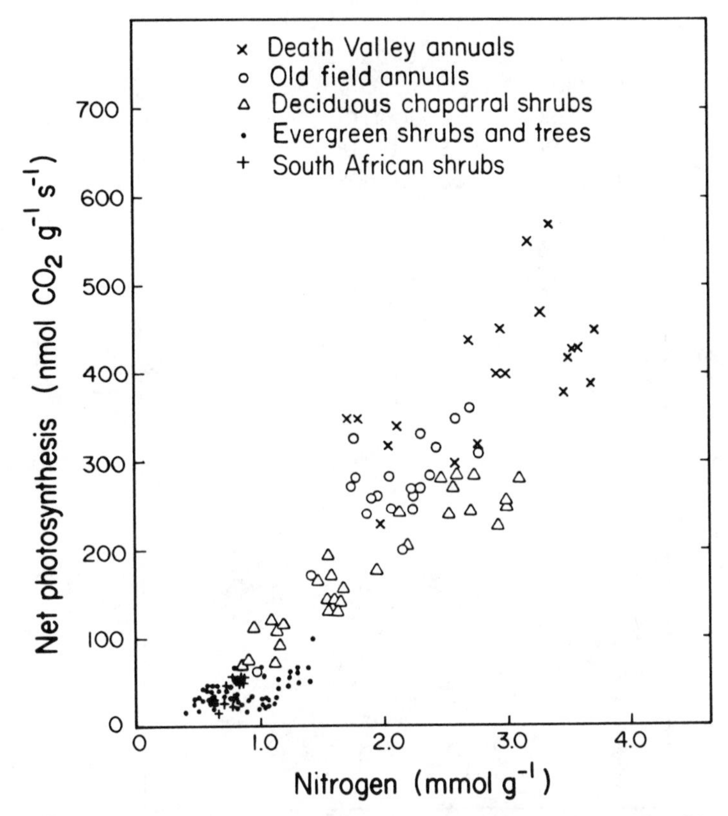

Figure 1.2. A_{max}–N relationships for the VINE survey, representing 21 species grown under natural conditions. A_{max} was measured in the field. N was determined with a micro-Kjeldahl assay on the same leaves used for the photosynthesis measurements. A_{max} and N are both expressed on a leaf-weight basis. $A_{max} = -76.1 + 149 \cdot N$, $n = 137$, $r = 0.92$, $P < 0.001$.

nials), the correlation coefficients are similar for area-based ($r = 0.86$) and weight-based ($r = 0.84$) measurements. For the 11 species of evergreen sclerophylls, the area-based ($r = 0.51$) and weight-based ($r = 0.50$) correlations are also similar. The greater variance in A_{max}–N relationships for sclerophylls than for nonsclerophylls has been reported earlier (Medina 1981; Field et al. 1983) and is an important element in the continuing discussion of the implications of sclerophylly. We shall review aspects of this discussion when we consider nitrogen-use efficiency and the causal basis of the A_{max}–N relationship.

If we focus on the weight-based A_{max}–N relationship, the correlation is striking for both its linearity and its limited scatter (Figure 1.2). The limited scatter suggests that over a wide diversity of C_3 plants, photosyn-

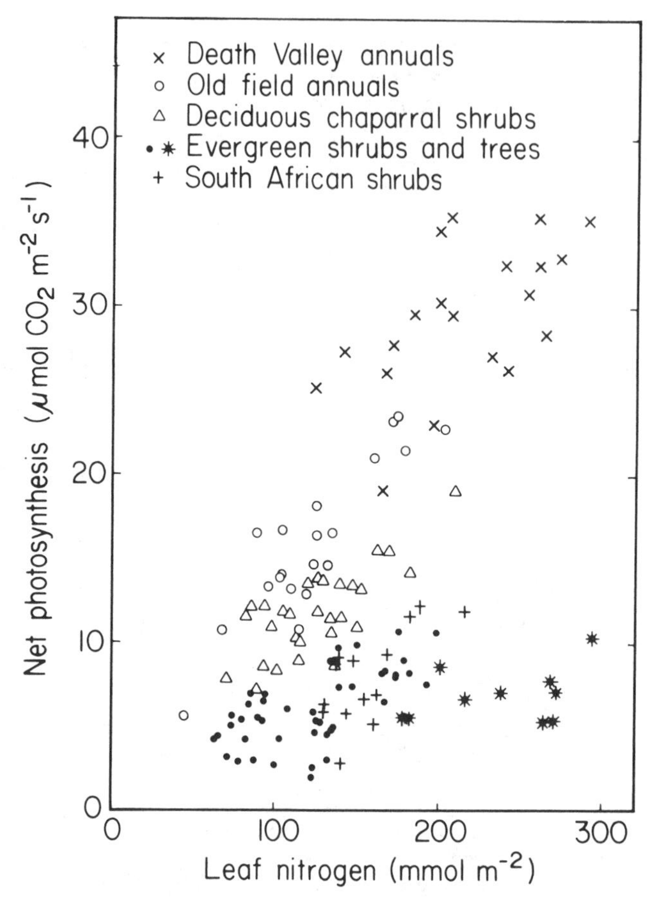

Figure 1.3. $A_{max}-N$ relationships for the VINE survey, with A_{max} and N expressed on a leaf-area basis. Data from *Prunus ilicifolia* are marked by asterisks. $A_{max} = 0.342 + 0.0814 \cdot N$, $n = 137$, $r = 0.53$, $P < 0.001$.

thetic capacity is strongly regulated by leaf nitrogen, without large effects due to habitat, growth form, or interspecies differences. The linearity, a feature not observed in some single-species analyses (Evans 1983), indicates important constraints on the relationship between nitrogen-based limitations and other limitations to A_{max}. We shall address both the linearity and the scatter in considering the causal basis of the $A_{max}-N$ relationship.

The area-based $A_{max}-N$ relationship (Figure 1.3) is somewhat confusing. The general rankings of species groups are similar for the weight-based and area-based analyses, but with one important exception. Some of the evergreen sclerophylls have high N per unit area, but low A_{max} (Figure

1.3). The most striking example of these species is *Prunus ilicifolia,* a shrub of the California chaparral. This species is a special case in that its leaves contain cyanogenic glucosides. These nitrogen-containing secondary compounds almost certainly play no direct functional role in determining A_{max} and may represent allocation of nitrogen away from compounds functionally related to A_{max} and toward defense. We shall assess the possibility that sclerophylls, in general, allocate proportionally less nitrogen to photosynthesis than do nonsclerophylls when we consider nitrogen-use efficiency. Some of the ecological factors controlling the allocation of nitrogen to defensive compounds have been considered by Bryant et al. (1983) and by Gulmon and Mooney (Chapter 20).

Which basis for expressing the A_{max} – N relationship is more significant functionally? Compelling arguments can be used to support either expression. Because light capture and CO_2 exchange with the atmosphere are intrinsically area-based phenomena, the area-based analysis provides a resource-harvesting framework for understanding the A_{max} – N relationship. On the other hand, a weight-based analysis yields more information on the economics of nitrogen and carbon allocation. Each expression gives important information, and the sources of the differences between them can contribute to the elucidation of the functional and ecological controls on photosynthetic capacity and leaf nitrogen.

Differences between the weight-based and area-based relationships result largely from variation in LSW (leaf weight/leaf area). In the VINE survey, LSW varies inversely with the weight-based measure of leaf nitrogen (Figure 1.4). The evergreen sclerophylls have the highest LSWs and also the survey's lowest photosynthetic capacities, either weight-based or area-based. The Death Valley annuals have the highest photosynthetic capacities on both measurement bases and a narrow range of LSW, somewhat above the lowest values. Converting from a weight-based (Figure 1.2) to an area-based (Figure 1.3) A_{max} – N relationship requires only multiplying each value for A_{max} and N by the LSW for that leaf. The consequence of the inverse relationship between LSW and N per unit weight is to increase small values and decrease large values of area-based A_{max} and N, relative to the weight-based parameters. Thus, the transformation from a weight-based to an area-based analysis tends to compress the total range of variation in A_{max} and N, and to increase the variability among the leaves with the highest LSWs.

The A_{max} – N relationship is not fundamentally changed by the choice of measurement basis (as long as A_{max} and N are expressed in the same units), but the choice of units does alter the prominence of various segments of the relationship. On any measurement basis, A_{max} and N are highly corre-

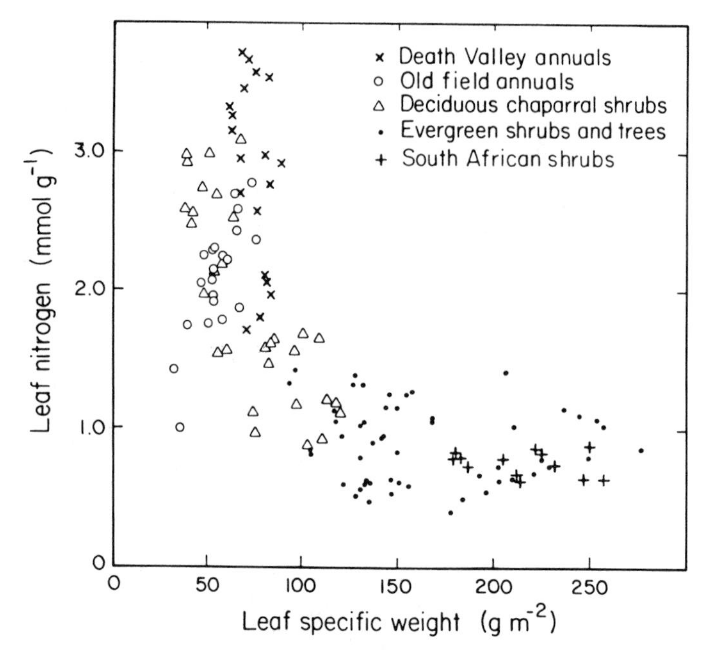

Figure 1.4. Relationship between leaf nitrogen per unit of leaf weight (N) and leaf-specific weight (LSW) for the VINE survey.

lated, and A_{max} per unit of leaf nitrogen is lower for the evergreen sclerophylls than for the other species.

Nitrogen-use efficiency

To simplify discussion of the A_{max} realized for a given level of N, it is useful to eliminate the confounding variable LSW and consider the ratio of A_{max} to N, which we term potential photosynthetic nitrogen-use efficiency (PPNUE). As long as A_{max} and N are both expressed on the same basis, PPNUE is independent of the measurement basis. We can view PPNUE as an index of potential performance under defined conditions that allows direct comparison among species. PPNUE is not an ecologically complete definition of nitrogen-use efficiency, but it is an important component of a more general, ecological definition, as provided by Rundel (1982) or Vitousek (1982). In addition to photosynthesis or growth per unit of nitrogen in tissue, these more complete definitions account for the critical roles in nitrogen-use efficiency played by leaf duration, nitrogen recovery from leaves, and whole-plant patterns of nitrogen allocation (Chapin 1980). PPNUE is currently useful for initiating the mechanistic

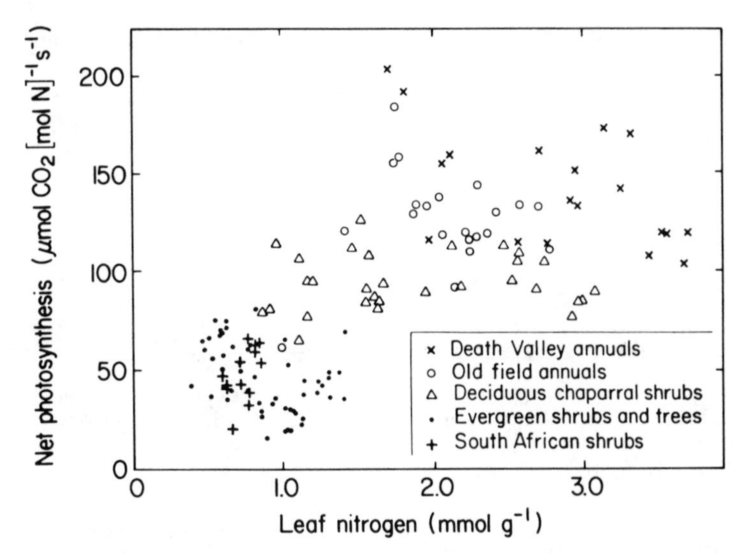

Figure 1.5. Relationship between photosynthetic capacity per unit of leaf nitrogen (PPNUE) and the weight-based measure of leaf nitrogen for the VINE survey.

interpretation of nitrogen-use efficiency. Its potential utility for approaching broader questions about nitrogen and growth will be increased as this index is modified to reflect long-term photosynthesis under natural conditions, and as photosynthetic data are combined with other components of nitrogen-use efficiency.

For the VINE survey, the analysis of PPNUE does provide several insights into the A_{max}–N relationship. PPNUE is lowest in the plants of lowest A_{max}, and increases with A_{max}. From values of less than 30 μmol CO_2 [mol N]$^{-1}$ s^{-1} for leaves of less than 1 mmol N g^{-1}, PPNUE rises to a plateau at around 125 μmol CO_2 [mol N]$^{-1}$ s^{-1} for leaves with nitrogen contents above about 2 mmol g^{-1} (Figure 1.5). Reviewing the data of Medina (1981) and others, Grubb (1984) also reached the conclusion that PPNUE is positively related to A_{max}. In the VINE survey, the relationship can be seen more precisely as a staircase of nearly constant PPNUE within major species types (annuals, deciduous shrubs, evergreen sclerophylls) and stepwise increases between them (Table 1.2). Although not in subgroups within the VINE survey, an increase in PPNUE with increasing N per unit weight is suggested by some studies on individual species (e.g., Gulmon and Chu 1981).

Several hypotheses may potentially explain the increase in PPNUE with increasing A_{max}. Observing that A_{max} and leaf longevity tend to be inversely related, Grubb (personal communication) and others have postulated that

Table 1.2. *Potential photosynthetic nitrogen-use efficiency (PPNUE) and nitrogen per unit of leaf weight (N/wt) for the species groups in the VINE survey*[a]

Species group	PPNUE (μmol CO_2[mol N]$^{-1}$ s^{-1})	N/wt (mmol g^{-1})	PPNUE vs. N/wt r	$p <$
Death Valley annuals	142 a	2.9 a	−0.58	0.01
Old-field annuals	130 a	2.2 b	0.31	n.s.
Drought-deciduous shrubs	95 b	1.9 b	0.07	n.s.
California evergreens	46 c	0.9 c	−0.42	0.01
South African shrubs	49 c	0.8 c	0.58	0.05

[a] Within a column, values followed by different letters are significantly different ($p < 0.001$). Values followed by the same letter are not significantly different ($p > 0.05$). In two of the three cases where the correlation between PPNUE and N/wt is significant, the correlation coefficient is negative.

the prerequisites for longevity constrain the options for A_{max}. From this, we can derive two hypotheses. First, Medina (1981) suggested that PPNUE is low in evergreen sclerophylls because the thick or impermeable cell walls required for longevity impose large resistances to CO_2 diffusion. Second, differences in PPNUE may result from differences in nitrogen allocation. Specifically, leaves with high A_{max} may invest a larger proportion of the leaf nitrogen in the primary carboxylating enzyme of C_3 plants, ribulose-1,5-bisphosphate carboxylase/oxygenase (Rubisco), and leaves with low A_{max} may invest a larger proportion of the leaf nitrogen in compounds required for longevity.

The data from the VINE survey do not provide a direct test of these two hypotheses for the relationship between PPNUE and N or A_{max}, but they do suggest two modifications of the second hypothesis that do not require assumptions about the prerequisites for leaf longevity. For an essentially linear weight-based A_{max} – N relationship, we can write

$$A_{max} = mN + b$$

or

$$PPNUE = A_{max}/N = m + b/N$$

If the y intercept (b) is positive, PPNUE decreases with increasing N, and if b is negative, PPNUE increases with increasing N, reaching a maximum value of m, the slope of the A_{max} – N relationship. This algebraic formulation has two implications. First, PPNUE will increase with N whenever, at least in extrapolation, some minimum nitrogen level greater than zero is required for a positive A_{max}. Because photosynthesis is only one of many nitrogen-requiring metabolic processes in leaf cells, it is intuitively reasonable that some sort of a nitrogen threshold for photosynthetic competence (A_{max} just greater than zero) does exist, and that this threshold value forces the increase in PPNUE with increasing A_{max} or N. If this hypothesis is correct, low PPNUE may result directly from low leaf nitrogen and is not necessarily mechanistically related to features required to increase leaf longevities.

In the VINE survey, the *y* intercept in the weight-based A_{max}–N relationship is negative. The regression equation suggests that a nitrogen concentration of about 0.5 mmol N g^{-1} is required for an A_{max} just equal to zero. While this value provides a first approximation of the nitrogen content at minimum photosynthetic competence, it must be considered a very rough approximation, because there is little reason to expect a constant threshold value of nitrogen independent of species or leaf type. Consistent with this caveat, extrapolated values for N at $A_{max} = 0$ vary widely among studies (data not shown).

A second implication also concerns the nitrogen invested in reactions unrelated to photosynthesis. When a high LSW dilutes a given A_{max} over a large amount of leaf mass, the nitrogen invested in reactions of intermediary metabolism, biosynthesis, and maintenance of ion gradients may not be diluted precisely in parallel. Because cell viability, independent of photosynthesis, requires nitrogen in proteins and nucleic acids, the nitrogen requirement for reactions not related to photosynthesis may scale positively with LSW. To the extent that this is the case, low PPNUE and high LSW are functionally related, because the proportion of the total leaf nitrogen invested in photosynthetic machinery decreases as LSW increases. This hypothesis not only explains the low PPNUE of evergreen sclerophylls but also is consistent with the pattern in the area-based A_{max}–N relationship (Figure 1.3), where some evergreens contain more nitrogen than would be predicted from the regression with A_{max}.

A third hypothesis is that low PPNUE results from inefficient allocation of nitrogen among photosynthetic compounds, such that some compounds are present in large excess, while the rate-limiting compounds are underrepresented for lack of nitrogen investment. Because photosynthesis is a very complex process, the problems of efficient allocation may be substantial. It is also possible that efficient nitrogen allocation under natural conditions is not efficient under the conditions employed for measurement of A_{max}. For example, shade plants invest large quantities of nitrogen in light-harvesting pigments and proteins, but make only small investments in Rubisco and other CO_2-processing enzymes (Björkman 1981).

This nitrogen allocation pattern may be efficient under low light but inefficient under saturating light. Much work remains to be done in assessing the relationship between A_{max} and the photosynthesis realized in nature.

A final hypothesis to explain the variation in PPNUE is that the kinetic properties of the photosynthesis-limiting nitrogenous compounds are not constant across species and that those plants with the highest PPNUEs have the most active or efficient rate-limiting enzymes.

To date, relatively little evidence is relevant to discriminating among the hypotheses that potentially explain the variation in PPNUE. No analysis based solely on interpretation of gas-exchange experiments can distinguish effects of cell-wall resistance to CO_2 diffusion from effects of nitrogen investment in nonphotosynthetic compounds, or from effects of variation in enzyme or allocation efficiency. However, the available information generally does not support the hypothesis that low PPNUE in sclerophylls results from unusually large internal resistances to CO_2 diffusion.

Three lines of evidence support this position. First, internal resistances should, in general, decrease as the ratio of the internal cell surface area to leaf area (S_i/S_l) increases (Nobel et al. 1975; Raven and Glidewell 1981). Though it is not always the case (Longstreth and Nobel 1980), the S_i/S_l ratio increases with LSW, if cell size and shape are constant. Thus, leaves with high LSWs should typically have low internal resistances. Second, the suggestion that extensive cell-wall development in sclerophylls imposes large limitations to CO_2 diffusion requires that these leaves contain an unusually large proportion of their biomass in cell walls. However, Merino et al. (1984) report that the biomass allocation to cell-wall components in the sclerophyllous chaparral shrub *Heteromeles arbutifolia* is 39.7% (46.7% including ash), comparable to the 38.1% (47.3% including ash) allocated to cell walls in the deciduous shrub *Lepechinia calycina,* and to the 41.4% (47.9% including ash) allocated to cell walls in the deciduous shrub *Diplacus aurantiacus,* after removal of the external resin in the latter species. All these values for leaves of perennials are lower than the 51.0% (56.0% including ash) of the total biomass allocated to cell-wall constituents in young corn plants (Penning de Vries et al. 1974). Neither of these lines of evidence addresses the permeability of the cell walls, but a third does.

Farquhar et al. (1982) argue that because diffusion and the biochemistry of carboxylation discriminate differently against the naturally occurring carbon isotope ^{13}C, then the isotopic composition of plants should reflect the relative importance of limitations to photosynthesis by diffusion and biochemical factors. In experiments where diffusional limitations are controlled, large limitations by diffusion result in high ratios of $^{13}C/^{12}C$. Small diffusional limitations lead to low ratios (Farquhar et al. 1982). For a broad survey of C_3 species, the mean $^{13}C/^{12}C$ ratio is 27.8 parts per thousand less than that in the standard limestone, with a range of 21 to 36 (Troughton et al. 1974). For the five species of California evergreens studied by Field et al. (1983), the mean ratio is 27.9 parts per thousand less than that in the standard limestone (unpublished data). The similarity between the mean values for C_3 plants and for evergreen sclerophylls indicates that the proportional limitations to photosynthesis imposed by diffusion are similar for the two groups. None of this evidence is conclusive in itself, but it all indicates that evergreen sclerophylls do not have proportionally large cell-wall resistances to CO_2 diffusion. The hypotheses favored by elimination are that PPNUE is low in the sclerophylls because proportionally less of their nitrogen is allocated to photosynthetic reactions, or that the sclerophylls have inefficient photosynthetic machinery. We shall review additional evidence concerning these hypotheses in the following section on the causal basis of the A_{max}–N relationship, but quantification of the relative importance of these hypotheses awaits further research.

Correlation or cause

Variation in leaf nitrogen can clearly explain much of the variation in photosynthetic capacity, across a wide variety of plant communities. From a purely predictive viewpoint, the correlation should prove a useful tool in the preliminary characterization of photosynthetic capacities. To understand the ecological controls on photosynthetic capacity, we need to move beyond the correlation to an assessment of cause and effect. This assessment warrants analysis at two levels. At the proximate level, establishing the functional basis of the A_{max}–N relationship will provide the foundation from which to approach a wide variety of ecological questions. How will particular plants or communities respond to changes in nitrogen availability? If nitrogen uptake is increased at the expense of additional allocation of carbon to roots, will the increased photosynthesis more than offset the costs of root maintenance and construction? At the ultimate level, important questions concern the control of nitrogen investment and the efficacy of that control. What controls nitrogen levels? How do actual nitrogen investments compare with the optimal investments for the maximization of growth, yield, or fitness? Here, we concentrate primarily on the proximate analysis of cause in the A_{max}–N relationship, providing a basis for future work at a variety of levels.

The A_{max} – N relationship suggests that N determines A_{max}, but it does not establish causation. At least three kinds of functional relationships are consistent with the correlation. Nitrogen levels may determine A_{max}; N may be controlled in response to A_{max}; or both N and A_{max} may be regulated by some other factor or factors. To summarize the evidence relevant to each of these hypotheses, we need to state them more precisely.

Hypothesis 1. Photosynthetic capacity is limited by one or more processes that operate at rates determined by their nitrogen contents. As a direct assignment of cause and effect, this hypothesis focuses attention on the physiological basis of photosynthesis. Net CO_2 fixation is potentially limited by CO_2 transport through gaseous or liquid media, by enzymatic reduction of CO_2, or by light-driven generation of NADPH and ATP. Within each process, limiting steps may operate at rates determined by nitrogen levels.

Hypothesis 1 is compatible with the possibility that A_{max} is limited by multiple nitrogenous compounds or by a combination of nitrogenous and nonnitrogenous factors. If multiple nitrogenous compounds limit A_{max}, nitrogen is efficiently distributed among the limiting factors only when all nitrogen redistributions decrease A_{max}. To the extent that inefficient nitrogen distributions decrease A_{max}, hypothesis 1 does require a relatively efficient distribution of nitrogen among limiting compounds. If a combination of nitrogenous and nonnitrogenous factors limit A_{max}, a tight A_{max} – N relationship is still possible, but it requires that the nonnitrogenous limitations be either small or a constant proportion of the total limitation to A_{max}.

Hypothesis 2. Independent of the factors that actually limit photosynthetic capacity, leaf nitrogen content is regulated to reflect photosynthetic capacity. This hypothesis emphasizes potential intraplant competition for nitrogen allocation. To the extent that the whole plant is nitrogen-limited, excess capacity in any enzyme or other nitrogenous compound represents an inefficiency, a diversion of nitrogen away from a rate-limiting step in the same leaf or away from investment in some other part of the plant where the nitrogen could contribute to growth and reproduction. With the maximum efficiency of nitrogen allocation, the level of every nonlimiting nitrogenous compound should be adjusted downward to the lowest activity still above rate limitation. At this level, the distinction between the functional constraint postulated in hypothesis 1 and an adjustment for efficient allocation fades. Hypothesis 2 does not require the assumption that some or all of the limits to A_{max} be imposed by nitrogenous compounds. If nonnitrogenous factors limit A_{max}, excess capacity in nitrogenous compounds should still be trimmed.

Hypothesis 3. The correlation between photosynthetic capacity and leaf nitrogen is not the result of a functional relationship but arises because both parameters are sensitive to or are controlled by other leaf parameters. This possibility, a null hypothesis with respect to the functional relationships suggested in hypotheses 1 and 2, should be accepted if A_{max} and N are invariant when expressed on the basis of some fundamental unit but are changed in concert when expressed on the basis of some derived unit. For example, if A_{max} and N were intrinsically area-based parameters, and if they were constants on a leaf-area basis, variation in LSW would transform these constants into a perfect correlation between the weight-based measures of A_{max} and N. Such a correlation could reflect nothing more than the differential utilization of leaves for storage, or differential allocation to defensive compounds. In the VINE survey, A_{max} and N both varied over at least a fivefold range, independent of the measurement basis. The magnitude of this variation militates against the acceptance of the null hypothesis but does not eliminate the possibility that some components of the A_{max}–N relationship result from spurious or secondary correlations.

Quantifying the importance of secondary effects resulting from potentially confounding variables like LSW, S_i/S_l, or leaf age is partially amenable to multiple-correlation analysis (e.g., Field and Mooney 1983), but the fundamental nature of the A_{max}–N relationship is too poorly known to allow much confidence in an analysis based on linear or simple curvilinear responses.

To evaluate the relative importance of each of these three hypotheses, we combine an analysis of the VINE survey with results from a variety of studies that have probed the biochemistry of photosynthesis in one or a few species. Because the data required for definitive interpretations of several aspects of the A_{max}–N relationship are not yet available, many of our comments are speculative and are intended as much to identify unanswered questions as to integrate existing information.

Deciding among the three hypotheses

The three hypotheses postulate that the A_{max}–N relationship results from substantially different mechanisms. Though the mechanisms are different, the three hypotheses are not mutually inconsistent. Deciding among them is more a problem of quantifying relative importance than of rejecting falsifiable alternatives.

Nitrogen-based limits to A_{max}

Summarizing photosynthesis as an aggregate of processes performing CO_2 transport, carboxylation, CO_2 reduction, and generation of reducing equivalents and ATP, we can search for nitrogen-based rate limitation in each of these processes. Among these, the best-studied candidate for limiting A_{max} in C_3 plants is the primary carboxylating enzyme in C_3 plants and the ultimate CO_2-fixing enzyme in all plants, ribulose-1,5-bisphosphate carboxylate/oxygenase (Rubisco) (Lorimer 1981). Circumstantial evidence that Rubisco limits A_{max} comes from its extreme abundance [it often constitutes over 40% of the leaf soluble protein (Collatz et al. 1979; Wittenbach 1979; Friedrich and Huffaker 1980) and 15–33% of the total leaf protein (Collatz et al. 1979; Seemann et al. 1981; Somerville et al. 1982)] and its low catalytic activity. Each of the eight active sites in a very large enzyme (molecular weight = 550,000) (Jensen and Bahr 1977) catalyzes only about 2 carboxylations (Badger and Collatz 1977), or 0.4 oxygenation per second (Farquhar et al. 1980). The combination of low catalytic activity and competitive inhibition by oxygen suggests that the high levels of Rubisco reflect nitrogen allocation to reduce rate limitation.

This circumstantial evidence is reinforced by a strong correlation between A_{max} and the total activity of Rubisco. Summarizing several studies of sun and shade plants grown under a variety of light regimens, Björkman (1981) found a correlation coefficient of 0.96 between A_{max} and Rubisco activity. This correlation is both general and robust. A_{max} and Rubisco activity vary in parallel in experimental treatments involving light intensity during growth (Björkman 1968), nitrogen availability (Medina 1971; Wong 1979), partial defoliation (Wareing et al. 1968; Neales et al. 1971), ploidy series (Randall et al. 1977; Molin et al. 1982), or leaf aging (Wittenbach 1979; Friedrich and Huffaker 1980).

The correlation between A_{max} and Rubisco activity is very strong, but some evidence indicates variation in the kinetic parameters of Rubisco, which could translate into variation in activity per unit of nitrogen invested in the enzyme. The CO_2 concentration required for half saturation of the enzyme varies twofold among C_3 grasses and over fourfold between C_3 and C_4 grasses (Yeoh et al. 1980). The rate of product formation per active site is about 70% greater in Rubisco from spinach than in that from soybean, and the difference in in vitro activities is reflected in measurements of whole-leaf photosynthesis (Seemann and Berry 1982). The specific activity of Rubisco during leaf aging has been reported to be relatively constant

(Wittenbach 1979; Friedrich and Huffaker 1980) or strongly decreasing (Hall et al. 1978). Though the kinetic parameters of Rubiscos from C_3 plants vary substantially, the enzyme's relative specificity for CO_2 and O_2 is nearly constant (Jordan and Ogren 1981). At this point, it appears that Rubisco activity per unit of invested nitrogen is not constant, but the variability is probably minor enough to appear as limited scatter in the VINE survey.

Some studies have indicated that Rubisco is not fully activated in vivo (Perchorowicz et al. 1980; Perchorowicz and Jensen 1983), but the evidence for incomplete activation is compelling only under conditions much different from those used for the determination of A_{max}. von Caemmerer and Farquhar (1981), Seemann et al. (1981), and Seemann and Berry (1982) all found that the activity of Rubisco is consistent with in vivo photosynthetic rates only if one assumes full activation.

While relatively strong and very general, the evidence for a functional limitation of A_{max} by Rubisco activity is largely correlative and subject to the interpretation that Rubisco varies either in response to A_{max} or in response to other factors that control A_{max}. The studies indicating that Rubisco activity is barely sufficient to account for observed photosynthetic rates begin to move beyond correlation but focus on measurements under low CO_2 concentrations, where Rubisco is most likely to be rate-limiting. One other line of evidence indicating that Rubisco activity, rather than other nitrogenous compounds, controls A_{max} comes from studies in which Rubisco was not a constant proportion of total protein and in which A_{max} varied more strongly with Rubisco than with protein (Medina 1971; Wittenbach 1983).

Although Rubisco clearly limits photosynthesis under some conditions, variation in Rubisco alone cannot account for the A_{max}–N relationship. Rubisco does constitute a substantial proportion of the total leaf protein, but the variation in nitrogen is much too large to be due to changes in one or a few enzymes. In the Death Valley annuals, Rubisco represents about 18% of the total leaf protein (Seemann et al. 1981). Eliminating all of the nitrogen in the Rubisco still leaves nitrogen contents more than an order of magnitude higher than the highest levels measured in some of the California evergreens (Figure 1.2). Levels of many nitrogenous compounds must be changing in concert.

Several examples document correlations between Rubisco and total nitrogen. In wheat, Rubisco activity scales linearly with total N (Evans 1983). In nine species of Death Valley annuals (Seemann et al. 1981), in soybean and spinach (Seemann and Berry 1982), and in aging leaves of barley (Friedrich and Huffaker 1980), soybean (Wittenbach et al. 1980), and wheat (Wittenbach 1979), Rubisco constitutes a relatively constant proportion of the total leaf protein. The proportion of total nitrogen in Rubisco appears to decrease late in leaf senescence (Wittenbach 1979; Friedrich and Huffaker 1980; Wittenbach et al. 1980) and may be sensitive to some kinds of source–sink manipulations (Wittenbach 1983).

The correlation between Rubisco and N or protein suggests three interpretations: Levels of other nitrogenous compounds may be adjusted to reflect the limitation of A_{max} by Rubisco, levels of Rubisco may be adjusted to reflect the limitation of A_{max} by other nitrogenous compounds, or A_{max} may be colimited by Rubisco and other nitrogenous compounds. Each of these possibilities is consistent with hypothesis 2: adjustment of nitrogen investment for maximum efficiency of nitrogen allocation. The third possibility represents the merger of functional limitation (hypothesis 1) with efficient investment (hypothesis 2).

Photosynthesis may be limited by Rubisco activity, but it may also be limited by CO_2 transport, CO_2 reduction capacity, or light-driven generation of reducing equivalents and ATP. Rate limitation by CO_2 transport is discussed by Cowan (Chapter 5), who demonstrates that low allocation of N to carbonic anhydrase (the enzyme that catalyzes the interconversion of CO_2 and bicarbonate) can depress A_{max} through effects on CO_2 transport. Rate limitation by either CO_2 reduction or the light reactions can be summarized under effects on regeneration of the CO_2 acceptor, ribulose-1,5-bisphosphate (RuBP) (Farquhar et al. 1980).

Evidence for limitation of A_{max} by RuBP regeneration is beginning to accumulate. The investment of nitrogen in the components of RuBP regeneration is substantial, probably greater than the investment in Rubisco (Kirk and Tilney-Bassett 1978). RuBP regeneration (Evans 1983) and electron-transport capacity (von Caemmerer and Farquhar 1981) are strongly correlated with Rubisco activity. Further, the total concentration of electron-transport components, as indicated by chlorophyll concentration, changes with A_{max} under treatments where rate limitation by Rubisco is unlikely (Terry 1983). Farquhar et al. (1980) summarize a great deal of biochemical information in a model that predicts that photosynthesis is limited by RuBP regeneration at CO_2 concentrations above some critical value. von Caemmerer and Farquhar (1981) argue that nitrogen is efficiently allocated between Rubisco and the components of RuBP regeneration when that critical CO_2 concentration is adjusted to the level at which the leaf normally operates. At this point, A_{max} is colimited by Rubisco and RuBP regeneration.

In summary, strong evidence supports the hypothesis that A_{max} is sometimes limited by Rubisco activity, but this limitation cannot account for the broad range of nitrogen variation in the VINE survey. The levels of nitrogenous compounds responsible for RuBP regeneration are certainly adjusted in concert with levels of Rubisco and may be important limitations to A_{max}.

Limitations by physical factors

In C_3 plants, photosynthesis is almost always limited by CO_2 diffusion as well as biochemical factors. Because C_3 plants are typically not CO_2-saturated in normal air (Pearcy and Ehleringer 1984), every step in the diffusion pathway from the bulk atmosphere to the sites of carboxylation in the chloroplasts represents a concentration drop that decreases photosynthesis. The concentration drop across the stomata is typically large, often about 100 μmol mol^{-1} or nearly one-third of the total CO_2 concentration (Wong et al. 1979; Sharkey et al. 1982). Farquhar and Sharkey (1982) provide a simple and elegant method for calculating the proportional limitations to photosynthesis due to effects of stomata. The concentration changes across other components of the diffusion pathway are more difficult to measure, but diffusion limitations imposed by the boundary layer, by the pathway from the substomatal cavity to the photosynthesizing cells (Parkhurst, Chapter 7 in this volume, but see Sharkey et al. 1982 for conflicting evidence), and by liquid-phase transport into the chloroplasts (Cowan, Chapter 5 in this volume; Nobel et al. 1975; Evans 1983) may all present substantial limitations to A_{max} under some circumstances.

These diffusional limitations must be balanced by at least one biochemical limitation. Photosynthetic capacity is completely limited by diffusion only when the CO_2 concentration at the sites of carboxylation drops below the CO_2 compensation point at which respiratory CO_2 evolution equals photosynthetic CO_2 uptake. Though many models have assumed these low CO_2 concentrations at the carboxylation sites, recent evidence indicates that the activity of the carboxylase is sufficient to support observed photosynthetic rates only if the CO_2 concentration at the carboxylation sites is substantially higher (von Caemmerer and Farquhar 1981; Seemann and Berry 1982; Evans 1983).

Each of the segments in the CO_2 diffusion pathway limits photosynthetic capacity, but with proportional magnitudes sensitive to other diffusional and biochemical limitations. In C_4 plants, which may be CO_2-saturated in normal air (Pearcy and Ehleringer 1984), there is no conceptual requirement for multiple limitations to photosynthetic capacity.

Stomatal limitations. Stomatal limitations to A_{max} are generally smaller than predicted by the linear analyses used for the last 25 years (Farquhar and Sharkey 1982), but they may still be 20–30% or even larger (Comstock and Ehleringer 1984). These relatively large diffusional limitations could affect the A_{max} – N relationship in three ways. If stomatal conductances were constant as biochemical limitations changed, the A_{max} – N relationship would curve downward at high levels of N, as stomatal limitation prevented large gains from increased nitrogen investment. If stomatal limitations were large but variable, they should add scatter to the A_{max} – N relationship. If stomatal conductance scales with biochemical limitations, establishing a proportionally constant stomatal limitation, the A_{max} – N relationship should have little scatter and should be linear or some other shape set by the response of the biochemistry to changes in nitrogen investment.

An accurate assessment of stomatal limitation requires an analysis of the CO_2 response of photosynthesis (Farquhar and Sharkey 1982), data not available for the VINE survey. At a slightly less sophisticated level, a proportionally constant stomatal limitation is indicated by a linear relationship between A_{max} and stomatal conductance (g). In the VINE survey, A_{max} and g are highly correlated ($r = 0.77$), even though the Illinois annuals tend to have high g values in relation to A_{max} (Figure 1.6).

The scatter in the A_{max} – g relationship suggests some variation in stomatal limitation to A_{max}. The five species of California evergreens in the VINE survey provide evidence for this. Field et al. (1983) found that the species with a higher A_{max} for a given N (a higher PPNUE) tended to be from wetter sites than species with lower PPNUE. Much of their difference in PPNUE values is attributable to higher stomatal conductance in the species of the wetter sites, and these differences generate scatter in the A_{max} – N relationship.

The mechanisms responsible for maintaining nearly constant limitations by g are not well known. Mooney and Gulmon (1979) argued that investments in photosynthetic machinery should be reduced when drought limits g, a postulate supported by evidence from Ehleringer (1983), who reported that N declines with increasing drought in the desert annual *Amaranthus palmeri*. Alternatively, the value of g may be regulated by the value of A_{max}, as determined by biochemical factors. Some of the strongest evidence for control of g by biochemical capacity comes from Wong et al. (1979), who demonstrated that stomatal conductance in C_3 and C_4 species is somehow regulated by photosynthetic capacity under a wide variety of treatments, including nutrient availability during growth, nutrient withdrawal, water stress, and chemical inhibition of photosynthesis. This type of coordination tends to eliminate scatter or curvilinearity in the A_{max} – N relationship that could result from variable stomatal limitation.

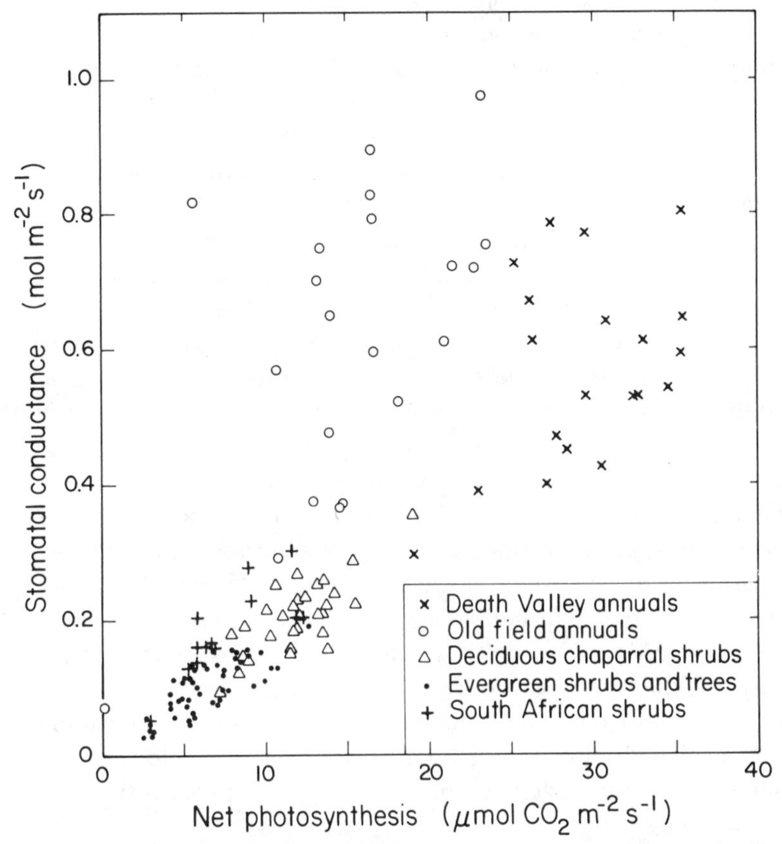

Figure 1.6. Relationship between photosynthetic capacity per unit of leaf area (A_{max}) and stomatal conductance (g) for the VINE survey. Stomatal conductances were measured simultaneously with photosynthetic capacity. $g = 0.0119 + 0.0215 \cdot A_{max}$, $n = 135$, $r = 0.77$, $P < 0.001$.

Other physical limitations. Nobel et al. (1975), Raven and Glidewell (1981), and Cowan (Chapter 5) have calculated that A_{max} typically includes a substantial limitation caused by resistance to CO_2 diffusion between the surfaces of the photosynthesizing cells and the sites of carboxylation. To the extent that this resistance is influenced by the activity of carbonic anhydrase (Cowan, Chapter 5), it acts as both a nitrogenous limitation and a physical limitation to A_{max}. Experimentally, it is very difficult to separate liquid-phase diffusion resistance from biochemical limitations. Nobel and associates demonstrated strong correlations between a measure of photosynthetic capacity and the ratio of cell surface area to leaf area (S_i/S_l) (Nobel et al. 1975; Nobel 1976, 1977, 1980), indicating that A_{max} may be limited by the internal surface area available for diffusion. However, as Björkman (1981) pointed out, the ratio of internal volume to leaf area also tends to increase with S_i/S_l, admitting the possibility that increases in A_{max} are more closely tied to increases in the amount of biochemical machinery than to increases in internal surface area. One promising approach to separating these limitations involves comparison of in vitro biochemical characteristics with in vivo photosynthetic rates. Knowing the in vitro kinetic properties of a rate-limiting enzyme, it is possible to calculate the CO_2 concentration required to support observed photosynthetic rates. From this CO_2 concentration, one can calculate the physical limitations to in vivo CO_2 diffusion.

This approach has been best developed for Rubisco. From experiments at low CO_2 concentrations, under which photosynthetic rates can be more accurately related to in vitro Rubisco kinetics, von Caemmerer and Farquhar (1981), Seemann et al. (1981), and Seemann and Berry (1982) all concluded that Rubisco activity is barely sufficient to support observed photosynthetic rates. In these studies, in vivo photosynthetic rates can be explained only by assuming no drop in CO_2 concentration across the liquid phase. Using a similar analysis, however, Evans (1983) found that photosynthesis was less than expected in leaves with the highest photosynthetic capacities, suggesting a drop in CO_2 concentration between the substomatal cavity and the sites of carboxylation. Evans's (1983) conclusion highlights the possibility that liquid-phase diffusion resistance usually imposes relatively small limitations to A_{max} but that when leaves develop exceptionally high biochemical capacities, limitations by liquid-phase resistance become important.

A general, quantitative summary of the magnitude of limitations by liquid-phase diffusion resistance is not yet possible. The very high correlation in the $A_{max}-N$ relationship is consistent with a minor limitation. If the limitation were large but constant, we would expect downward curvature at the upper end of the $A_{max}-N$ relationship, as the diffusional limitation increasingly constrained the gains of additional nitrogen investment. This kind of curvature is apparent in some single-species $A_{max}-N$ relationships (Evans 1983), suggesting that the liquid-phase diffusion resistance is less variable within species than among species. Finally, the $A_{max}-N$ relationship is consistent with large liquid-phase diffusion limitations, if the magnitude of those limitations scales linearly with the limitations imposed by biochemical factors. Unfortunately, the interface of biochemical and liquid-phase diffusional limitations to A_{max} has not been well studied. Correlations between S_i/S_l and A_{max} suggest linear scaling but have not been reinforced by independent measures of biochemical limits to A_{max}.

Integration and conclusions

In C_3 plants, A_{max} is typically limited by biochemical and diffusional factors, with the majority of the total limitation caused by biochemical factors. The biochemical limitations are imposed by nitrogen-containing compounds, and the biochemical capacity for photosynthesis scales with the nitrogen investment in these compounds. Of the nitrogenous limits, limitation by Rubisco is the best-documented case, but limitation by compounds involved in RuBP regeneration and by carbonic anhydrase is also likely. At this point, we cannot distinguish broad colimitation of A_{max} by nitrogenous factors from limitation by one or a few nitrogenous compounds and adjustment to near limitation in others. In variable natural environments, this distinction may be somewhat artificial, as changes in the environment alter the balance among limiting factors.

The mechanisms responsible for adjusting levels of nitrogenous compounds are not generally known. Schmidt and Mishkind (1983) demonstrated that the required 1 : 1 ratio between large and small subunits of Rubisco is maintained by rapid degradation of excess small subunits. They further suggested that similar proteolytic systems could establish and preserve an appropriate balance among a wide variety of photosynthetic compounds by degrading any components not integrated into a functional pathway. Little is known about the ways that environmental signals specifying the appropriate balance among photosynthetic compounds are transduced into controlling mechanisms.

The integrated balance among nitrogenous limitations to A_{max} is, in general, paralleled by a scaling between nitrogenous and diffusional limitations. Neither the mechanisms nor the ecological priorities for this scaling are well known. Is nitrogen investment curtailed when diffusional limitations become too large, or is nitrogen the primary control? This question has no single answer, and priority among multiple constraints is probably sensitive to a variety of environmental and phylogenetic factors (Mooney and Gulmon 1979). When constraints on water loss, the risk of herbivory, or limited light availability decrease the gains expected from high N, plants generally forgo those investments. In these cases, N is the major determinant of A_{max}, but it is not the primary environmental constraint. In other cases, limited N availability may act as both the major limitation to A_{max} and the major environmental constraint. An important implication of the adjustment of biochemical limitations in response to environmental constraints is that A_{max} is not an ecologically meaningless physiological parameter, but is an index of integrated natural constraints on photosynthesis.

Much remains to be learned about the mechanistic basis and the ecological implications of the $A_{max}-N$ relationship. Mechanistically, the distribution of quantitative limitations is still unknown, as is the nature of the controls on the relative levels of limiting factors. At the interface between mechanisms and implications, much work needs to be done on the sensitivity of the nature and magnitude of the limitations to changes in the environment. Ecologically, it is of great interest to quantify the relationship between A_{max} and photosynthesis under natural conditions and to use the $A_{max}-N$ relationship for cost–benefit analyses of plant structure and function.

Acknowledgments

Thanks to N. Chiariello, J. A. Berry, T. J. Givnish, and J. Comstock for useful discussions and comments on preliminary drafts of this chapter. The experiments in the VINE survey were supported by NSF grant DEB 78-02067 to H. A. M.

References

Badger, M. R., and G. J. Collatz. 1977. Studies on the kinetic mechanism of ribulose-1,5-bisphosphate carboxylase and oxygenase reactions, with particular reference to the effect of temperature on kinetic parameters. Carnegie Inst. Washington Yearbook 76:355–361.

Björkman, O. 1968. Carboxydismutase activity in shade-adapted and sun-adapted species of higher plants. Physiol. Planta. 21:1–10.

– 1981. Responses to different quantum flux densities. Pp. 57–107 *in* O. L. Lange, P. S. Nobel, C. B. Osmond, and H. Ziegler (eds.), Physiological plant ecology, I, vol. 12A, encyclopedia of plant physiology, new series. Springer-Verlag, Berlin.

Björkman, O., and P. Holmgren. 1963. Adaptability of the photosynthetic apparatus to light intensity in ecotypes from exposed and shaded habitats. Physiol. Planta. 16:889–914.

Bolton, J. K., and R. H. Brown. 1980. Photosynthesis of grass species differing in carbon dioxide fixation pathways. V. Response of *Panicum maximum, Panicum miliodes,* and tall fescue *(Festuca arundinacea)* to nitrogen nutrition. Plant Physiol. 66:97–100.

Brown, R. H. 1978. A difference in N use efficiency in C_3 and C_4 plants and its implications in adaptation and evolution. Crop Sci. 18:93–98.

Brown, R. H., and J. R. Wilson. 1983. Nitrogen response in *Panicum* species differing in CO_2 fixation pathways. II. CO_2 exchange characteristics. Crop Sci. 23:1154–1159.

Bryant, J. P., F. S. Chapin II, and D. R. Klein. 1983. Carbon/nutrient balance of boreal plants in relation to vertebrate herbivory. Oikos 40:357–368.

Chapin, F. S., III. 1980. The mineral nutrition of wild plants. Ann. Rev. Ecol. Syst. 11:233–257.

Chapin, F. S., III, and R. A. Kedrowski. 1983. Seasonal changes in nitrogen and phosphorus fractions and autumn retranslocation in evergreen and deciduous taiga trees. Ecology 64:376–391.

Collatz, G. J., M. Badger, C. Smith, and J. A. Berry. 1979. A radioimmune assay for RuP_2 carboxylase protein. Carnegie Inst. Washington Yearbook 78:171–175.

Comstock, J., and J. Ehleringer. 1984. Photosynthetic responses to slowly decreasing leaf water potentials in *Encelia frutescens.* Oecologia 61:241–248.

Ehleringer, J. 1983. Ecophysiology of *Amaranthus palmeri,* a sonoran desert summer annual. Oecologia 57:107–112.

Evans, J. R. 1983. Nitrogen and photosynthesis in the flag leaf of wheat *(Triticum aestivum* L.*).* Plant Physiol. 72:297–302.

Farquhar, G. D., M. C. Ball, S. von Caemmerer, and Z. Rosskandik. 1982. Effect of salinity and humidity on $\delta^{13}C$ value of halophytes – evidence for diffusional isotope fractionation determined by the ratio of intracellular/atmospheric partial pressure of CO_2 under different environmental conditions. Oecologia 52:121–124.

Farquhar, G. D., M. H. O'Leary, and J. A. Berry. 1982. On the relationship between carbon isotope discrimination and the intercellular carbon dioxide concentration in leaves. Aust. J. Plant Physiol. 9:121–137.

Farquhar, G. D., and T. D. Sharkey. 1982. Stomatal conductance and photosynthesis. Ann. Rev. Plant Physiol. 33:317–345.

Farquhar, G. D., S. von Caemmerer, and J. A. Berry. 1980. A biochemical model of photosynthetic CO_2 assimilation in the leaves of C_3 species. Planta 149:78–90.

Field, C. 1981. Leaf age effects on the carbon gain of individual leaves in relation to microsite. Pp. 41–50 *in* N. S. Margaris and H. A. Mooney (eds.), Components of productivity of Mediterranean-climate regions – basic and applied aspects. Dr. W. Junk, The Hague.

– 1983. Allocating leaf nitrogen for the maximization of carbon gain: Leaf age as a control on the allocation program. Oecologia 56:341–347.

Field, C., J. Merino, and H. A. Mooney. 1983. Compromises between water-use efficiency and nitrogen-use efficiency in five species of California evergreens. Oecologia 60:384–389.

Field, C., and H. A. Mooney. 1983. Leaf age and seasonal effects on light, water, and nitrogen use efficiency in a California shrub. Oecologia 56:348–355.

Friedrich, J. W., and R. C. Huffaker. 1980. Photosynthesis, leaf resistances, and ribulose-1,5-bisphosphate carboxylase degradation in senescing barley leaves. Plant Physiol. 65:1103–1107.

Gulmon, S. L., and C. C. Chu. 1981. The effects of light and nitrogen on photosynthesis, leaf characteristics, and dry matter allocation in the chaparral shrub, *Diplacus aurantiacus.* Oecologia 49:207–212.

Hall, N. P., A. J. Keys, and M. J. Merrett. 1978. Ribulose-1,5-diphosphate carboxylase protein during flag leaf senescence. J. Exp. Bot. 29:31–37.

Hesketh, J. D., E. M. Larson, A. J. Gordon, and D. B. Peters. 1983. Internal factors influencing photosynthesis and respiration. Pp. 381–411 *in* J. E. Dale and F. L. Milthorpe (eds.), The growth and functioning of leaves. Cambridge University Press.

Jensen, R. G., and J. T. Bahr. 1977. Ribulose-1,5-bisphosphate carboxylase-oxygenase. Ann. Rev. Plant Physiol. 28:379–400.

Jordan, D. B., and W. L. Ogren. 1981. Species variation in the specificity of ribulose bisphosphate carboxylase/oxygenase. Nature 291:513–515.

Kirk, J. T. O., and R. A. E. Tilney-Bassett. 1978. The plastids, ed. 2. Elsevier, Amsterdam.

Lee, J. A., R. Harmer, and R. Ignaciuk. 1983. Nitrogen as a limiting factor in plant communities. Pp. 95–112 *in* J. A. Lee, S. McNeill, and I. H. Rorison, (eds.), Nitrogen as an ecological factor. Blackwell, Oxford.

Longstreth, D. J., and P. S. Nobel. 1980. Nutrient influences on leaf photosynthesis: effects of nitrogen, phosphorus, and potassium for *Gossypium hirsutum* L. Plant Physiol. 65:541–543.

Lorimer, G. H. (1981). The carboxylation and oxygenation of ribulose 1,5-bisphosphate: the primary events in photosynthesis and photorespiration. Ann. Rev. Plant Physiol. 32:49–83.

Loveless, A. R. 1961. A nutritional interpretation of sclerophylly based on differences in the chemical composition of sclerophyllous and mesophytic leaves. Ann. Bot. 25:168–183.

Medina, E. 1971. Effect of nitrogen supply and light intensity during growth on the photosynthetic capacity and carboxydismutase activity of leaves of *Atriplex hastata* ssp. *hastata*. Carnegie Inst. Washington Yearbook 70:551–559.

– 1981. Nitrogen content, leaf structure and photosynthesis in higher plants: a report to the UNEP study group on photosynthesis and bioproductivity. IVIC, Caracas, Venezuela.

Merino, J., C. Field, and H. A. Mooney. 1984. Construction and maintenance costs of Mediterranean-climate evergreen and deciduous leaves. II. Biochemical pathway analysis. Oecologia Plantarum 4:211–229.

Molin, W. T., S. P. Meyers, G. R. Baer, and L. E. Schrader. 1982. Ploidy effects in isogenic populations of alfalfa. II. Photosynthesis, chloroplast number, ribulose-1,5-bisphosphate carboxylase, chlorophyll, and DNA in protoplasts. Plant Physiol. 70:1710–1714.

Mooney, H. A., J. Ehleringer, and J. A. Berry. 1976. High photosynthetic capacity of a winter annual in Death Valley. Science 194:322–324.

Mooney, H. A., P. J. Ferrar, and R. O. Slatyer. 1978. Photosynthetic capacity and carbon allocation patterns in diverse growth forms of *Eucalyptus*. Oecologia 36:103–111.

Mooney, H. A., C. Field, S. L. Gulmon, and F. A. Bazzaz. 1981. Photosynthetic capacity in relation to leaf position in desert versus old-field annuals. Oecologia 50:109–112.

Mooney, H. A., C. Field, S. L. Gulmon, P. Rundel, and F. J. Kruger. 1983. Photosynthetic characteristics of South African sclerophylls. Oecologia 58:398–401.

Mooney, H. A., and S. L. Gulmon. 1979. Environmental and evolutionary constraints on the photosynthetic characteristics of higher plants. Pp. 316–337 *in* O. T. Solbrig, S. Jain, G. B. Johnson, and P. H. Raven (eds.), Topics in plant population biology. Columbia University Press, New York.

– 1982. Constraints on leaf structure and function in reference to herbivory. Bioscience 32:198–206.

Nátr, L. 1975. Influence of mineral nutrition on photosynthesis and the use of assimilates. Pp. 537–556 *in* J. P. Cooper (ed.), Photosynthesis and productivity in different environments. Cambridge University Press.

Neales, T. F., K. J. Treharne, and P. F. Wareing. 1971. A relationship between net photosynthesis, diffusive resistance and carboxylating enzyme activity in bean leaves. Pp. 89–96 *in* M. S. Hatch, C. B. Osmond, and R. O. Slatyer (eds.), Photosynthesis and photorespiration. New York: Wiley-Interscience, New York.

Nobel, P. S. 1976. Photosynthetic rates of sun versus shade leaves of *Hyptus emoryi* Torr. Plant Physiol. 58:218–223.

– 1977. Internal leaf area and cellular CO_2 resistance: photosynthetic implications of variations with growth conditions and plant species. Physiol. Planta. 40:137–144.

– 1980. Leaf anatomy and water-use efficiency. Pp. 43–55 *in* N. C. Turner and P. J. Kramer (eds.), Adaptation of plants to water and high temperature stress. Wiley-Interscience, New York.

Nobel, P. S., L. J. Zaragoza, and W. K. Smith. 1975. Relation between mesophyll surface area, photosynthetic rate, and illumination during development for leaves of *Plectranthus parviflorus* Henckel. Plant Physiol. 55:1067–1070.

Pearcy, R. W., and J. Ehleringer. 1984. Comparative ecophysiology of C_3 and C_4 plants. Plant Cell Env. 7:1–13.

Pearcy, R. W., K. Osteryoung, and D. Randall. 1982. Carbon dioxide exchange characteristics of C_4 Hawaiian *Euphorbia* species native to diverse habitats. Oecologia 55:333–341.

Penning de Vries, F. W. T., A. H. M. Brunsting, and H. H. Van Laar. 1974. Products, requirements and efficiency of biosynthesis: a quantitative approach. J. Theor. Biol. 45:339–377.

Perchorowicz, J. T., and R. G. Jensen. 1983. Photosynthesis and activation of ribulose bisphosphate carboxylase in wheat seedlings. Plant Physiol. 71:955–960.

Perchorowicz, J. T., D. A. Raynes, and R. G. Jensen. 1981. Light limitation of photosynthesis and activation of ribulose bisphosphate carboxylase in wheat seedlings. Proc. Natl. Acad. Sci. USA 78:2985–2989.

Randall, D. D., C. J. Nelson, and K. H. Asay. 1977. Ribulose bisphosphate carboxylase. Altered genetic expression in tall fescue. Plant Physiol. 59:38–41.

Raven, J. A., and S. M. Glidewell. 1981. Processes limiting photosynthetic conductance. Pp. 109–136 *in* C. B. Johnson (ed.), Physiological processes limiting plant productivity. Butterworth, London.

Ritchie, J. T. 1980. Plant stress research and crop production: the challenge ahead. Pp. 21–29 *in* N. C. Turner and P. J. Kramer (eds.), Adaptation of plants to water and high temperature stress. Wiley-Interscience, New York.

Rundel, P. W. 1982. Nitrogen utilization efficiencies in Mediterranean-climate shrubs of California and Chile. Oecologia 55:409–413.

Schmidt, G. D., and M. L. Mishkind. 1983. Rapid degradation of unassembled ribulose 1,5-bisphosphate carboxylase small subunits in chloroplasts. Proc. Natl. Acad. Sci. USA 80:2632–2636.

Schmitt, M. R., and G. E. Edwards. 1981. Photosynthetic capacity and nitrogen use efficiency of maize, wheat, and rice: a comparison of C_3 and C_4 photosynthesis. J. Exp. Bot. 32:459–466.

Seemann, J. R., and J. A. Berry. 1982. Interspecific differences in the kinetic properties of RuBP carboxylase protein. Carnegie Inst. Washington Yearbook 81:78–83.

Seemann, J. R., J. M. Tepperman, and J. A. Berry. 1981. The relationship between photosynthetic performance and the levels and kinetic properties of RuBP carboxylase-oxygenase from desert winter annuals. Carnegie Inst. Washington Yearbook 80:67–72.

Sharkey, T. D., K. Imai, G. D. Farquhar, and I. R. Cowan. 1982. A direct confirmation of the standard method of estimating intercellular partial pressure of CO_2. Plant Physiol. 69:657–659.

Somerville, C. R., A. R. Portis, and W. L. Ogren. 1982. A mutant of *Arabidopsis thalina* which lacks activation of RuBP carboxylase in vivo. Plant Physiol. 70:381–387.

Stocking, C. R., and A. Ongun. 1962. The intracellular distribution of some metallic elements in leaves. Am. J. Bot. 49:284–289.

Terry, N. 1983. Limiting factors in photosynthesis. IV. Iron stress-mediated changes in light-harvesting and electron transport capacity and its effects on photosynthesis *in vivo.* Plant Physiol. 71:855–860.

Troughton, J. H., K. A. Card, and C. H. Hendy. 1974. Photosynthetic pathways and carbon isotope discrimination by plants. Carnegie Inst. Washington Yearbook. 73:768–780.

Vitousek, P. 1982. Nutrient cycling and nutrient use efficiency. Am. Nat. 119:553–572.

von Caemmerer, S., and G. G. Farquhar. 1981. Some relationships between the biochemistry of photosynthesis and the gas exchange of leaves. Planta 153:376–387.

Wareing, P. F., M. M. Khalifa, and K. J. Treharne. 1968. Rate-limiting processes in photosynthesis at saturating light intensities. Nature 220:453–457.

Wilson, J. R., and R. H. Brown. 1983. Nitrogen response of *Panicum* species differing in CO_2 fixation pathways. I. Growth analysis and carbohydrate accumulation. Crop Sci. 23:1148–1153.

Wittenbach, V. A. 1979. Ribulose bisphosphate carboxylase and proteolytic activity in wheat leaves from anthesis through senescence. Plant Physiol. 64:884–887.

– 1983. Effect of pod removal on leaf photosynthesis and soluble protein composition of field-grown soybeans. Plant Physiol. 73:121–124.

Wittenbach, V. A., R. C. Ackerson, R. T. Giaquinta, and R. R. Hebert. 1980. Changes in photosynthesis, ribulose bisphosphate carboxylase, proteolytic activity, and ultrastructure of soybean leaves during senescence. Crop Sci. 20:225–231.

Wong. S. C. 1979. Elevated atmospheric partial pressure of CO_2 and plant growth. I. Interactions of nitrogen nutrition and photosynthetic capacity in C_3 and C_4 plants. Oecologia 44:68–74.

Wong, S. C., I. R. Cowan, and G. D. Farquhar. 1979. Stomatal conductance correlates with photosynthetic capacity. Nature 282:424–426.

Yeoh, H.-H., M. R. Badger, and L. Watson. 1980. Variations in $K_m(CO_2)$ of ribulose-1,5-bisphosphate carboxylase among grasses. Plant Physiol. 66:1110–1112.

Yoshida, S., and V. Coronel. 1976. Nitrogen nutrition, leaf resistance, and leaf photosynthetic rate of the rice plant. Soil Sci. Plant Nutr. 22:207–211.

2 Modifications of solar-radiation absorption patterns and implications for carbon gain at the leaf level

JAMES R. EHLERINGER AND
KENNETH S. WERK

Introduction

At the end of the nineteenth century and during the early part of the twentieth century there was interest in determining the relationships between leaf form and function and in determining how morphology was involved in adapting plants to specific environments. Studies by Haberlandt (1884), Schimper (1903), and Warming (1909) showed that leaves of plants from arid habitats tended to possess characteristics different from those of plants from more mesic habitats. Characteristics often found in arid-zone plants included more leaf pubescence, an increase in the frequency of compound leaves, sclerophyllous anatomy, generally smaller leaves, and more steeply inclined leaves. They regarded these characters as "adaptations" for reducing water loss, but any potential effect on photosynthesis was not considered. These studies did not have an experimental basis, and although a correlation could be established between certain leaf characteristics and environmental factors, the potential functional significance has been investigated more thoroughly only in recent times.

Raschke (1956) and Gates (1962) provided a theoretical basis for the leaf energy balance, linking water loss, leaf temperature, and certain leaf characteristics such as size, spectral characteristics, and leaf orientation. The linkage between the processes of water and energy transfer and photosynthesis was made by Mooney (1972), Parkhurst and Loucks (1972), Givnish and Vermeij (1976), Cowan and Farquhar (1977), and Mooney and Gulmon (1979), in which the environmental constraints imposed on photosynthesis, transpiration, and net carbon gain were simultaneously evaluated.

In this chapter we shall focus on the consequences of variations in leaf spectral characteristics and leaf orientation for the processes of photosynthesis and transpiration and on how the importance of such morphological traits is affected by constraints imposed by the physical and biotic environ-

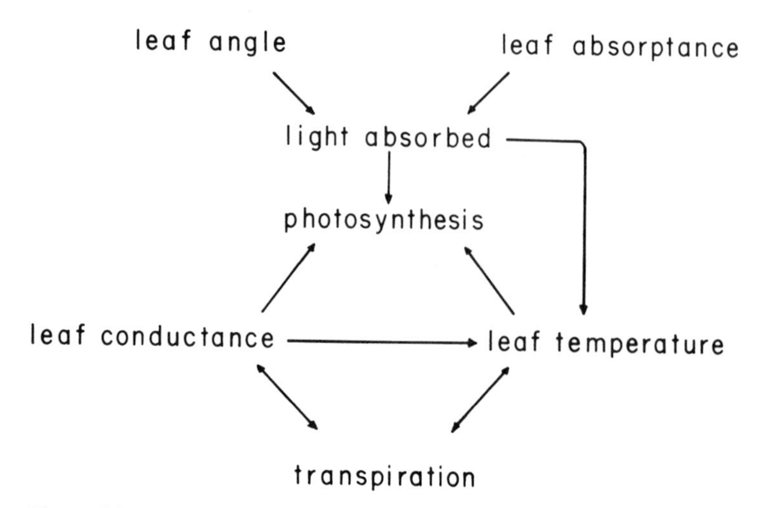

Figure 2.1. Interactions between effects of leaf angle and leaf absorptance on leaf temperature and the processes of photosynthesis and transpiration.

ment. At the leaf level, photosynthesis and transpiration are tightly coupled to each other via energy and gas exchange, which are influenced by properties of the leaf (Figure 2.1). In this chapter we shall focus on the effects of changes in leaf absorptance and leaf angle.

The leaf angle will directly influence the amount of solar radiation incident on a leaf through its effect on determining the cosine of the angle of incidence (cos i) between the normal to the leaf lamina and the direct solar beam, as outlined in equation (2.1):

$$\cos i = (\cos a_l)(\sin a_s) + (\sin a_l)(\cos a_s)[\cos(z_s - z_l)] \quad (2.1)$$

where a_l and a_s are the angles of the leaf and sun above the horizontal, and z_l and z_s are the azimuths (compass directions) of the leaf and the sun.

The leaf absorptance will determine what fraction of the incident solar radiation is absorbed by the leaf. Two different leaf absorptances to solar radiation need to be considered: the leaf absorptance to the 400–700-nm waveband for photosynthetic studies and the leaf absorptance to the 400–3,000-nm waveband for heat-balance studies. The two leaf absorptances as percentages are empirically related by equation (2.2) (Ehleringer 1981):

$$a_{400-3,000} = (0.73)(a_{400-700}) - 11.9 \quad (2.2)$$

Thus, the amount of direct-beam solar radiation absorbed by a leaf, which can influence the photosynthetic rate and/or leaf temperature, is

$$Q_k = (a_k)(\cos i)(I_p) \quad (2.3)$$

where Q_k is the total direct-beam solar radiation absorbed by the leaf (either as 400–700-nm photon flux or 400–3,000-nm thermal radiation), a_k is the absorptance to that waveband, and I_p is the intensity of the solar beam on a perpendicular to that beam.

Only about 1% of the photons in the 400–700-nm waveband are used to drive the light reactions of photosynthesis; most of the solar radiation is converted to heat, raising the leaf temperature (T_l). A change in leaf temperature will directly affect photosynthesis (A), and thus the amount of light absorbed by the leaf will have both direct and indirect effects on the rate of this process. A third parameter directly influencing the photosynthetic rate is the leaf conductance to water vapor (g), which affects the rate of carbon dioxide diffusion into the leaf, as shown in equation (2.4):

$$A = g(c_a - c_i)/1.6 \qquad (2.4)$$

where c_a and c_i are the ambient and intercellular carbon dioxide concentrations and 1.6 is the ratio of the molecular diffusion rates of water and carbon dioxide in air.

As shown in equation (2.5), the leaf conductance will also directly affect the transpiration rate E, because water vapor must diffuse through the same stomatal pore openings as does carbon dioxide:

$$E = g(\Delta w) \qquad (2.5)$$

The magnitude of E is also dependent on the water vapor concentration gradient between the leaf and the air (Δw), and thus T_l has a direct effect on E. Because E imparts a significant heat loss from the leaf, it both affects and is affected by T_l (Gates 1962).

From the preceding it should be clear that a quantitative evaluation of the "adaptive or evolutionary significance" of a change in leaf absorptance or leaf angle for plant fitness is not possible without first understanding its effects on photosynthesis, leaf temperature, and water relations.

As a reference point for our discussion of changes in leaf angle or leaf absorptance, let us use a hypothetical green leaf with a leaf angle of 0° (a horizontal leaf). If for this hypothetical green leaf we assume that air temperatures and nutrient and water availabilities are optimal for growth, the leaf photosynthetic rate will increase with increasing irradiance and may or may not become saturated by typical midday irradiance levels (see Chapter 1). Leaves with greater enzyme content and photochemical capacity should become light-saturated at proportionally higher irradiances (see Chapter 1). As irradiance levels increase, the stomata also open, resulting in increased transpiration. It is this inevitable loss of water via transpiration through the stomata in order for the leaf to take up carbon dioxide via photosynthesis that establishes a tradeoff between photosynthesis and transpiration.

Recent experimental evidence indicates that under well-watered conditions, the responses to irradiance of stomatal opening and photosynthetic rate appear coupled, such that c_i remains nearly constant, although there may be variation in c_i between species (Körner et al. 1979; Wong et al. 1979; von Caemmerer and Farquhar 1981; Farquhar and Sharkey 1982).

The question we would like to ask is what would happen to this hypothetical green leaf if g is decreased in response to less water availability (e.g., increased Δw or decreased leaf water potential)? What are the effects on A, E, and T_l? If there were no physiological changes at the leaf level other than a decreased g, then we would expect to observe three changes at the leaf level. First, A should become light-saturated at a lower irradiance level because of the decreased g. Second, as a consequence of a decreased g, T_l will be higher. This could raise Δw sufficiently that E would not be decreased in exactly the same proportion as g was decreased. Also, T_l could become sufficiently high and result in thermal damage to the photosynthetic apparatus. Third, in response to leaf exposure to supersaturating photon fluxes, photochemical damage may occur (i.e., photobleaching or photoinhibition).

What are the observed patterns in the field?

Before addressing the difficult questions of the costs and benefits of leaf angle or leaf absorptance changes, let us examine what trends are observed in the field. We analyze these trends at three levels: the community-level patterns, intrageneric patterns, and intraspecies seasonal patterns.

Community-level patterns of leaf absorptance and leaf angle

Precipitation and temperature gradients are often steep over short distances in the western United States, especially when elevated changes are involved. As a consequence, there can be large variations in habitat aridity over short geographical distances. Billings and Morris (1951) initially demonstrated that if the dominant species at two community extremes along an aridity gradient in central Nevada (saltbrush and coniferous forest) were compared, the species at the drier sites tended to have higher leaf reflectances.

In a recent survey of leaf energy-budget parameters for 192 species common to the Wasatch Front in Utah, intercommunity-level trends were analyzed for changes in leaf absorptance and leaf angle along an altitudinal

Table 2.1. *Average values of leaf absorptance to solar radiation (400–700 nm) and leaf angle for the dominant species in communities along the Wasatch Front, Utah*

Community[a]	Average leaf absorptance (%)	Average leaf angle (degrees)
Saltbush (13)	75.2	57.8
Grassland (36)	80.9	53.2
Oak-maple (17)	82.6	51.7
Juniper woodland (28)	76.4	43.8
Mountain brush (13)	84.3	23.1
Coniferous forest (28)	81.4	34.1
Alpine meadow (27)	79.5	38.7
Lower riparian (21)	82.1	46.4
Upper riparian (9)	83.3	36.1

[a] Plant communities are arranged in order of decreasing aridity. Values in parentheses are species sample sizes.

cline (Table 2.1). The lowest leaf absorptances, the greatest range of leaf absorptance values, and the steepest leaf angles were observed in the saltbush community. This plant community also occupies the most arid site along the transect. The low leaf absorptances observed in plants from this habitat resulted from increased leaf surface reflectance. Proceeding to less arid sites, the average leaf absorptance increased and reached a maximum in the mountain brush community. At higher elevations the average leaf absorptance decreased slightly because of increased leaf transmittance in the herbaceous species of the coniferous forest and alpine meadow communities. Average leaf angles showed a trend similar to that of leaf absorptance, decreasing as one proceeds from the saltbrush community up to the mountain brush community. It is interesting that there is a small but significant increase in the average leaf angle as one proceeds from the mountain brush community up through the alpine meadow community.

Although the saltbush community is the driest plant community along the Wasatch Mountains, still drier plant communities occur in the Mohave and Sonoran deserts to the south. In a survey of the common species in these deserts, Ehleringer (1981) found average leaf absorptances of the perennial vegetation to be lower than reported for the saltbush community (Figure 2.2). Thus, at the plant community level, there was a consistent trend for leaf absorptance to decrease as aridity (= decreased precipitation) increased.

Decreases in leaf absorptance can result from increased reflectance or from increased transmittance. Several surface modifications can result in increased reflectance, including waxes (Reicosky and Hanover 1978; Mulroy 1979; Ehleringer 1981), hairs (Pearman 1966; Sinclair and

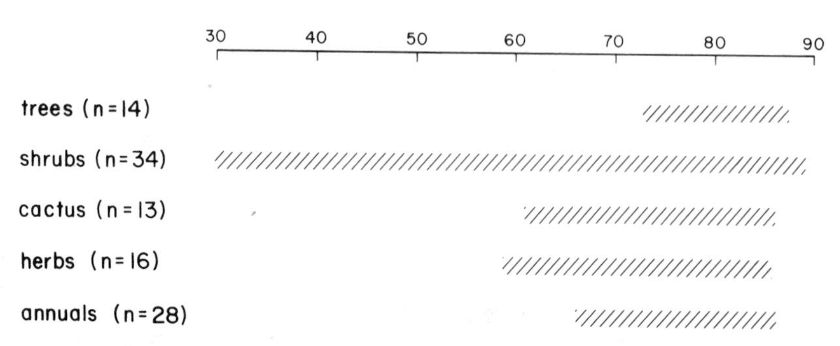

Figure 2.2. Ranges of leaf absorptance of solar radiation (400–700 nm) by leaves of common plant species in the Mohave and Sonoran deserts. (Modified from Ehleringer 1981.)

Thomas 1970; Ehleringer, Björkman, and Mooney 1976; Ehleringer 1981), and salt bladders (Billings and Morris 1951; Mooney, Ehleringer, and Björkman 1977). Although the "costs" to produce these different epidermal modifications may be different, their effects on increasing diffuse reflectance are the same.

Intrageneric patterns

Along precipitation transects, Shaver (1978) and Ehleringer (1983b) have noted that there is often a species replacement within a single genus, such that at drier sites species have leaves that are progressively more pubescent (= increased reflectance). Examples of intrageneric replacement series include *Arctostaphylos* (Ericaceae), *Encelia* (Asteraceae), *Eriogonum* (Polygonaceae), *Salvia* (Lamiaceae), and *Viguiera* (Asteraceae). Occasionally a single species will occupy habitats along a large precipitation range. In these situations there will often be ecotypes or subspecies with different leaf absorptances, such as *Acacia victoriae* (waxes or hairs) (Ehleringer, unpublished data), *Encelia canescens* (hairs) (Ehleringer 1982), and *Eucalyptus urnigera* (waxes) (Thomas and Barber 1974).

Seasonal trends in leaf absorptance

Often in species that have low leaf absorptances, the absorptance can vary seasonally in response to changes in environmental conditions. One such example of this is *Encelia farinosa*, a pubescent-leaved species, in which leaf absorptance decreases with the onset of drought (Figure 2.3).

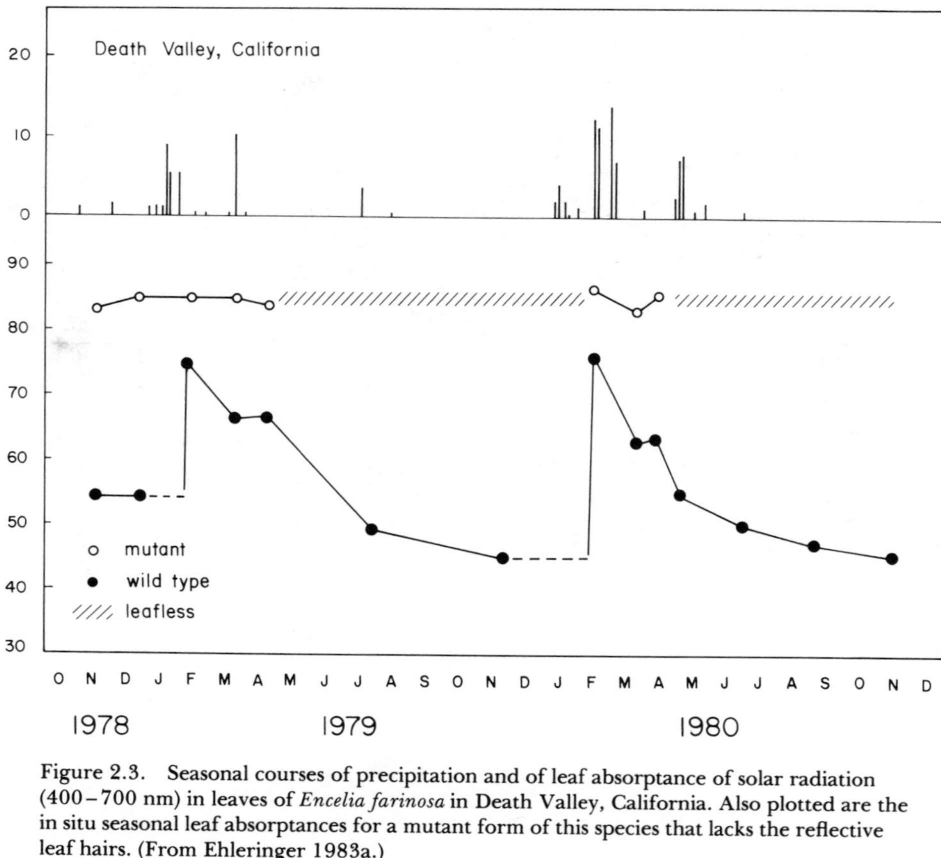

Figure 2.3. Seasonal courses of precipitation and of leaf absorptance of solar radiation (400–700 nm) in leaves of *Encelia farinosa* in Death Valley, California. Also plotted are the in situ seasonal leaf absorptances for a mutant form of this species that lacks the reflective leaf hairs. (From Ehleringer 1983a.)

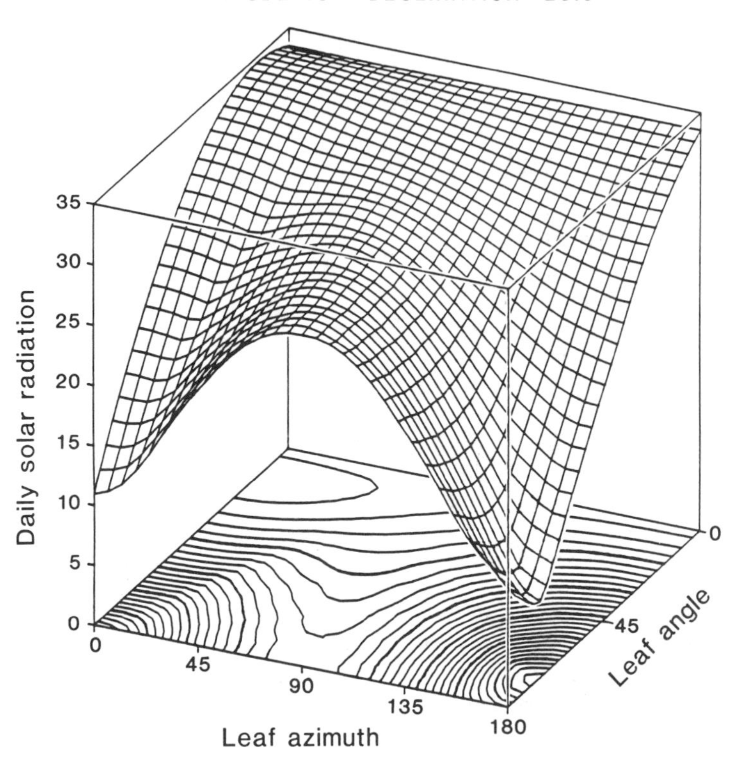

Figure 2.4. A three-dimensional plot of the total daily solar radiation incident on leaves possessing different leaf angles and leaf azimuths. Leaf azimuths are measured from south (0°). Calculations are based on a latitude of 40°, a solar declination of +23.5°, 10% diffuse solar radiation, and an atmospheric transmission coefficient of 0.8.

Green leaves (high leaf absorptance, low pubescence) are produced at the beginning of the growing season when soil water availability is high. As soil water availability decreases (leaf water potential decreases), new leaves are produced that are more pubescent and have lower absorptances (Ehleringer and Björkman 1978; Ehleringer 1982). The relationship between degree of leaf pubescence and leaf water potential is linear and without an initial threshold, so that the plant is constantly adjusting the leaf spectral characteristics in response to decreased soil water availability (as measured by decreasing leaf water potential) (Ehleringer 1982).

Variation in leaf angle and azimuth

The orientation of a leaf (angle and azimuth) affects three separate aspects of solar radiation intercepted by its lamina: (1) daily integrated

Table 2.2. *Mean leaf angles for a number of common perennial species in the Mohave and Sonoran deserts*

Species	Leaf angle (degrees)	Leaf type
Ambrosia dumosa	36.6	Drought-deciduous
Atriplex hymenelytra	67.5	Evergreen
Datura meteloides	19.9	Drought-deciduous
Encelia asperifolia	33.0	Drought-deciduous
Encelia farinosa	26.0	Drought-deciduous
Encelia frutescens	31.1	Drought-deciduous
Larrea divaricata	48.4	Evergreen
Simmondsia chinensis	84.2	Evergreen
Viguiera laciniata	31.8	Drought-deciduous
Viscainoa geniculata	67.8	Evergreen

radiation, (2) peak instantaneous irradiance, and (3) diurnal distribution of instantaneous incident irradiance. The interaction between leaf angle and azimuth is frequently ignored by investigators (because it is often assumed that the leaves have a random distribution of leaf azimuths), and most of the data on leaf orientation presented in the literature provide information on leaf inclination only.

A response surface for daily integrated solar radiation as a function of both leaf angle and leaf azimuth is presented in Figure 2.4 (for the summer solstice at 40° N latitude). In general, increasing leaf angle will result in a decrease in the amount of irradiance incident on a leaf. Thus, along the Wasatch Mountain transect, the leaves at the most arid sites will have less solar radiation incident on them. However, this holds true only if the distribution of azimuths is random (most of the species). In the warmer Mohave and Sonoran deserts, leaf angles for evergreen-leaved perennials tend to be more steeply inclined than those for deciduous-leaved perennials (Table 2.2).

Nonrandom leaf azimuths

There can be large variations in the amount of solar radiation incident on a leaf, depending on the specific leaf orientation. For instance, a vertical leaf with its lamina facing east receives as much solar radiation over the course of a day as a leaf facing northeast or northwest with an inclination of 30° from horizontal or a leaf facing southeast or southwest with an inclination of 55°. Steeper leaf angles need not necessarily reduce the solar radiation received during the summer or increase the solar radiation received during the winter if particular leaf azimuths are considered. For the Wasatch Mountain transect, the distributions of leaf azimuths

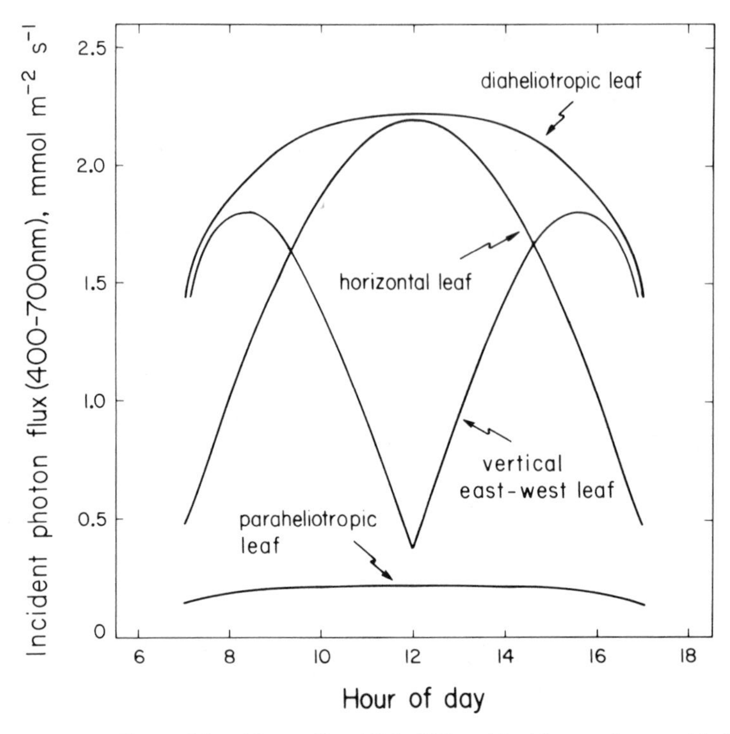

Figure 2.5. Photon flux (400–700 nm) incident on leaves with different orientations on a diurnal basis. Calculations are based on a latitude of 40° and a declination of 10°. The diffuse solar radiation was assumed to be equal to 10% of the incident perpendicular solar beam.

appear to be random, except for two notable exceptions: compass plants and solar-tracking plants, which will be discussed later.

The effect of nonrandom leaf azimuths on the solar radiation received by a plant is dependent on the specific orientation and the solar declination and latitude. East–west lamina orientations greatly reduce winter solar radiation while only slightly reducing summer radiation. The diurnal distribution on a steeply angled leaf facing east–west is heavily weighted toward the early morning and late afternoon, reducing midday irradiance (Figure 2.5). The steeper the leaf angle, the more the irradiance is shifted away from noon. The so-called compass plants (e.g., *Lactuca serriola, Silphium* species) are examples of species whose leaves are steeply inclined, with their lamina facing east–west. The frequency of species with this unusual leaf orientation is low.

North–south lamina orientations result in a simple parabolic distribution of solar irradiance over the course of a day. The amount received is reduced by steep inclinations in the summer and low inclinations during the winter. This lamina orientation can result in enhanced daily integrated irradiance if the inclination is approximately equal to the latitude minus the solar declination. Thus, low inclinations during the summer, intermediate inclinations during the spring and fall, and steep inclinations during the winter will maximize the daily integrated irradiance received by a south-facing leaf growing in the middle latitudes (see Chapter 3 for examples of this with cacti).

Leaf solar tracking

Leaves from a number of species do not remain in a fixed position diurnally, but move through the day such that the lamina remains perpendicular to the sun's direct rays (Figure 2.6). Such leaves are called diaheliotropic (= solar tracking) and receive a more or less constant irradiance throughout the day (Figure 2.5). As a consequence of this higher instantaneous solar irradiance, diaheliotropic leaves may receive as much as 35% more solar radiation over the day than a fixed leaf with a horizontal orientation (Shackel and Hall 1979; Ehleringer and Forseth 1980).

Leaf solar tracking occurs in herbaceous species and is most common in annuals. In the Wasatch Mountain transect, leaf solar tracking was effectively restricted to the grassland community. In drier sites, Ehleringer and Forseth (1980) noted that the frequency of leaf solar tracking in the annual flora of a community was inversely related to the length of the growing season and reached as high as 75% of the species in the summer annuals of the Sonoran Desert.

Costs and benefits of morphological changes for photosynthesis and transpiration

Leaf absorptance

To illustrate the costs and benefits of a leaf spectral change in plants, let us consider *Encelia*, a common genus of shrubs in the arid western United States. Members of the genus *Encelia* produce drought-deciduous leaves covered to different extents with leaf pubescence on both upper and lower surfaces. Along geographical clines of decreasing precipitation, there is a replacement of species such that leaves exhibit increases in both the density and thickness of the pubescence layer (Ehleringer 1980, 1983b). The effect of the leaf pubescence is to cause the leaf absorptance (400–700 nm) to decrease (via increased reflectance) from 85% at the

Figure 2.6. Leaves of *Oxystylis lutea* showing the diaheliotropic leaf orientation.

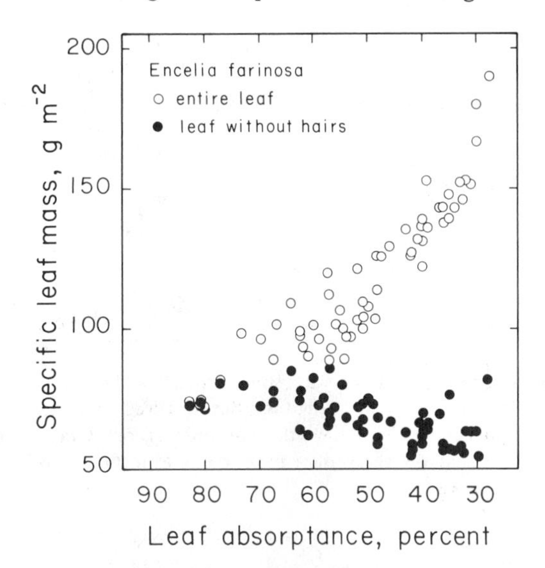

Figure 2.7. Relationship between specific leaf weight of *E. farinosa* leaves, with and without hairs, as a function of leaf absorptance of the intact leaf. (From Ehleringer and Cook 1984.)

wettest sites to as low as 29% at the driest sites (Ehleringer et al. 1976; Ehleringer and Björkman 1978).

Encelia farinosa occurs at the dry end of *Encelia*'s distribution and varies its leaf absorptance on a seasonal basis in response to changes in drought (Figure 2.3). During wet periods, leaves are green, with an absorptance of 80%. However, as leaf water potential decreases, new leaves are produced, with progressively lower absorptances (Ehleringer 1982). The relationship between midday leaf water potential and the extent of leaf pubescence development is linear, with a change in leaf absorptance of approximately 9% per MPa.

As a minimum, the caloric cost to produce an *E. farinosa* leaf with reduced leaf absorptance is the caloric cost to produce the mass of the hairs responsible for leaf absorptance changes. Figure 2.7 plots the specific leaf weight of intact leaves with pubescence and the same leaf after the pubescence has been removed as a function of leaf absorptance. These data indicate that the mass of the photosynthetic and conducting tissues remains constant at all leaf absorptances. All of the specific leaf weight differences are due to hair production. At the heavily pubescent end (low leaf absorptance), the cost to produce a leaf is high, because the hairs represent approximately 55% of the total leaf mass.

Table 2.3. *Calculated values of photosynthesis, transpiration, and leaf temperature for* Encelia farinosa *under midday summer conditions*[a]

Variable	"Green" leaf	"White" leaf
Absorptance (%)		
400–700 nm	85	40
400–3,000 nm	50	17
Leaf temperature (°C)	43.5	37.5
Transpiration (mmol m^{-2} s^{-1})	6.1	4.1
Photosynthesis (% of maximum)	36	82

[a] Energy-budget calculations assume the following values: wind speed 1 m s^{-1}, soil temperature 50°C, air temperature 40°C, 10% diffuse solar radiation, sky infrared radiation 350 W m^{-2}, air vapor density 10 g m^{-3}, leaf angle 25°, leaf width 4 cm, and leaf conductance 0.09 mol m^{-2} s^{-1}. Photosynthetic rate based on response curves from Ehleringer and Mooney (1978).

We can use the production-value approach of Penning de Vries et al. (1974) to estimate the cost to produce different leaf types in *E. farinosa*. If we assume that the pubescence consists only of cellulose and hemicellulose, we calculate a cost of 92.5 g glucose per m^2 leaf for very lightly pubescent leaves and a cost of 188 g glucose per m^2 leaf for heavily pubescent leaves. Production of pubescent leaves thus represents a significant investment on the part of the plant, and it is of interest then to understand just how the plant benefits from this additional investment in leaf structure.

One immediate benefit of the reduced leaf absorptance in pubescent *E. farinosa* leaves is a reduction in the heat load. This translates into a reduced leaf temperature and thus a lower transpiration rate (because Δw is reduced). The calculations from Table 2.3 illustrate that the pubescent leaf will have a temperature 6°C lower than the nonpubescent leaf. As a result of this lower leaf temperature, the transpiration rate will be approximately 33% less in the pubescent leaf. This saving in water loss at the single leaf could allow the plant to maintain more leaves (and thus more photosynthesis) under water-limited conditions or to maintain activity for a longer period of time into the drought period.

There are two additional benefits of the increased pubescence that directly affect photosynthesis. Ehleringer and Mooney (1978) showed (1) that the pubescence lowered leaf temperatures enough that the "upper lethal leaf temperature" was avoided and (2) that at temperatures above 30°C (thermal optimum for photosynthesis), the increase in photosynthetic rate by having a lowered leaf temperature was greater than the potential decrease in photosynthetic rate caused by increased photon reflectance. More recently, Ehleringer and Cook (unpublished data) have shown that under lower leaf water potentials, the photosynthetic rate becomes light-saturated at irradiances lower than midday levels. Thus, the pubescence is serving the additional benefit of reflecting excess photons that cannot be effectively used in photosynthesis under water-limited conditions.

We can very roughly estimate the net benefits of having the "more expensive" pubescent leaf in terms of water loss or carbon gain. Because a pubescent leaf is transpiring at a rate approximately one-third less than that of a nonpubescent leaf, it can remain photosynthetically active for a period about one-third longer. Calculating a conservative photosynthetic rate of 10 μmol m^{-2} s^{-1} over a 12-hr day will result in fixation of approximately 0.43 mol CO_2 m^{-2}. At this rate, the extra investment in leaf pubescence in a leaf can be recovered in approximately eight days (81 g m^{-2} pubescence on a low-absorptance leaf whose specific leaf weight is 150 g m^{-2} will cost 95.6 g glucose m^{-2}, divided by daily carbohydrate gain of 13.0 g glucose m^{-2}). Thus, as a rough approximation, if the pubescence allows a leaf to remain active for a period eight days longer than a nonpubescent leaf, that should result in an overall net carbon gain by the leaf.

Leaf solar tracking

The biochemical energy costs to achieve solar tracking are thought to be small, because the movements usually are accomplished by small turgor changes in the pulvinal region of the petiole. There are definite costs, though, associated with the presence of diaheliotropism. One such cost is that because only a small fraction of the incident solar radiation passes beyond the solar-tracking leaf, the extent of potential canopy development becomes restricted. In the arid western United States, diaheliotropic plants usually have very low leaf area indices (Ehleringer and Forseth 1980). A second cost associated with solar tracking is that the leaf is exposed to much higher irradiances, resulting in a greater thermal load on the leaf and thus higher leaf temperatures (Forseth and Ehleringer 1983b).

There are distinct advantages associated with diaheliotropic leaf movements, particularly in habitats with short growing seasons. A diaheliotropic leaf will receive approximately 30% more photons over the day than will a horizontally fixed leaf (Figure 2.5). In order to take advantage of these higher irradiances, leaves of solar-tracking species should not be light-saturated at midday irradiance levels. Werk et al. (1983) studied the photosynthetic rates of a large number of desert winter annuals and found that species with solar-tracking leaves tended to have higher photosyn-

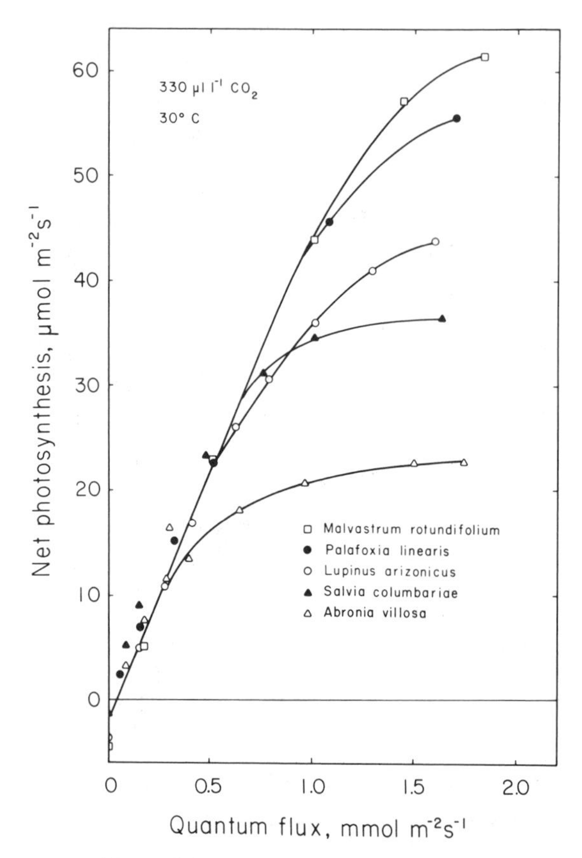

Figure 2.8. Dependence of net photosynthesis on incident photon flux (400–700 nm) in leaves of several winter desert annuals. *Lupinus arizonicus, Malvastrum rotundifolium,* and *Palafoxia linearis* exhibit diaheliotropic leaf movements. (From Werk et al. 1983.)

thetic rates than species with nontracking leaves at midday irradiances (Figure 2.8). Moreover, the solar-tracking leaves (*Lupinus, Malvastrum,* and *Palafoxia* in Figure 2.8) were not light-saturated at light levels comparable to what the leaves would receive over the course of the day in the field.

Paraheliotropic leaf movements

The higher heat load of a solar-tracking leaf places the leaf in an unfavorable water-relations position during periods of limited water availability. In this situation, a large fraction of species with diaheliotropic leaves also exhibit paraheliotropic movements (Ehleringer and Forseth 1980). That is, during periods of low water availability, the leaf will move during the day such that the leaf lamina is parallel to the sun's direct rays (Figure 2.9). As a consequence, the leaf is exposed to much lower light levels and heat loads (Figure 2.5).

One example of such a plant with both diaheliotropic and paraheliotropic leaf movements is *Lupinus arizonicus*, a common annual in the Mohave and Sonoran deserts (Forseth and Ehleringer 1980). In response to lower leaf water potentials, the cosine of the leaf's angle of incidence to the sun's direct rays decreases (Figure 2.10). Forseth and Ehleringer (1983b) have shown that at the same time as soil water becomes less available, the leaf conductance and photosynthetic rate also decrease (Figure 2.10). Even though there is also a reduction in the intercellular CO_2 concentration as the leaf water potential decreases, these results should not be interpreted as meaning that stomata are necessarily imposing a greater limitation on photosynthetic rate. Calculations of the stomatal limitation on photosynthesis (using the equation from Farquhar and Sharkey 1982) indicate that there is no increase in the stomatal limitation with decreasing leaf water potential. Rather, the parallel nature of the declines in these parameters suggests that the reduction in light incident on the leaf reduces electron-transport rates to a degree similar to the reduction in the rate of CO_2 supply by diffusion through the stomata so that the rates of these two processes remain in balance. Therefore, we can conclude that the paraheliotropic leaf movements are not advantageous in terms of increasing net productivity per se, but rather are advantageous in allowing the leaf to avoid higher leaf temperatures and exposure to high photon fluxes that cannot be used by the photosynthetic apparatus.

A major advantage of leaf solar tracking is that it allows the leaf to achieve maximal rates of photosynthesis early in the morning when Δw is lowest (Forseth and Ehleringer 1983b; Forseth et al., unpublished data). This feature will be of significant advantage in habitats with low midday humidities, because the stomata of most plants close in response to increased Δw.

Simulations comparing the performances of paraheliotropic leaves with fixed leaf orientations strongly suggest that under water-limited conditions, daily water-use efficiency (A/E) is higher in the paraheliotropic leaf (Forseth and Ehleringer 1983b). Additionally, by having a reduced heat load, the paraheliotropic leaf avoids higher leaf temperatures and thus higher transpiration rates.

Steep fixed leaf angles

When a canopy is composed of many leaves with random azimuths but nonrandom inclinations, the situation becomes complex. Individual

Figure 2.9. Leaves of *Lupinus arizonicus* showing the cupped nature of the paraheliotropic leaves under water-stressed conditions.

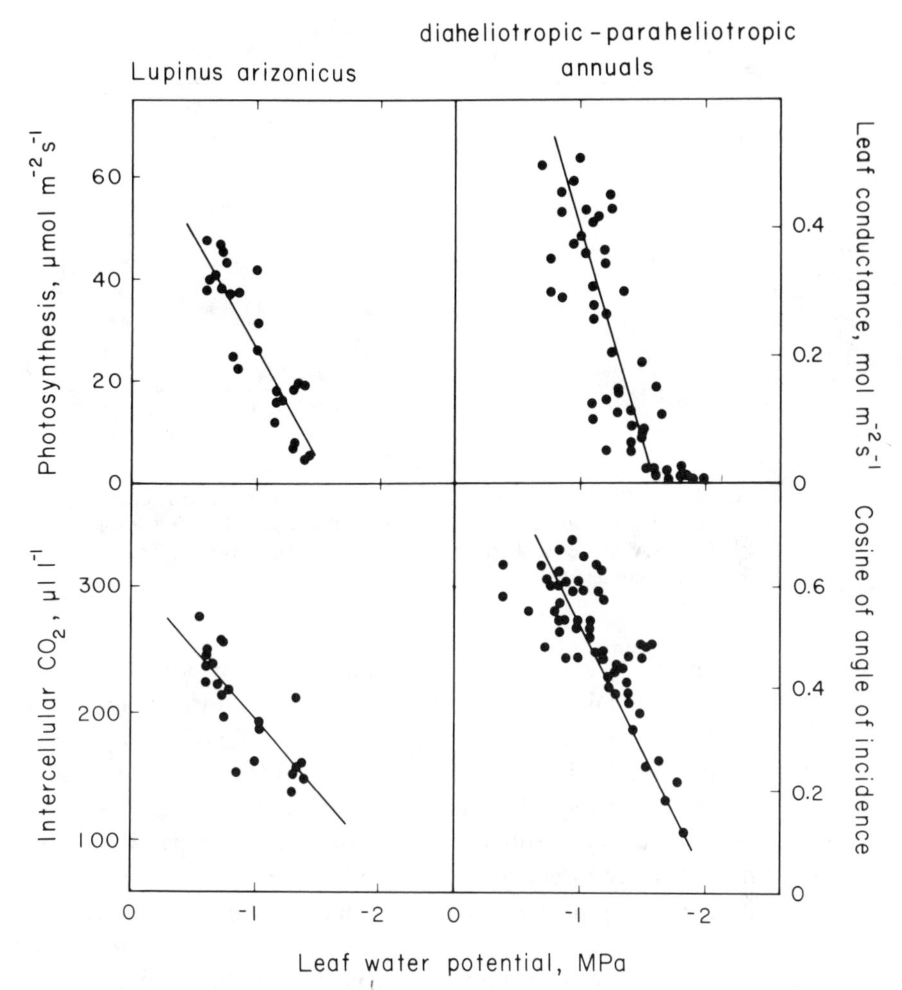

Figure 2.10. Relationships of the cosine of the angle of incidence, the photosynthetic rate at high irradiance, leaf conductance, and intercellular CO_2 concentration to leaf water potential in leaves of *Lupinus arizonicus,* a plant whose leaves exhibit both diaheliotropic and paraheliotropic movements. (Modified from Forseth and Ehleringer 1980, 1983a.)

leaves within the canopy are subject to very different patterns of solar irradiance (Figure 2.4). The simplest method used to understand these patterns is to average the irradiance received by all the leaves. This provides us with an overview of a canopy-level phenomenon, but ignores details such as the peak irradiance received by any one leaf.

On average, steep leaf inclinations reduce the light intercepted by leaves during the summer. The diurnal distribution is also changed, so that

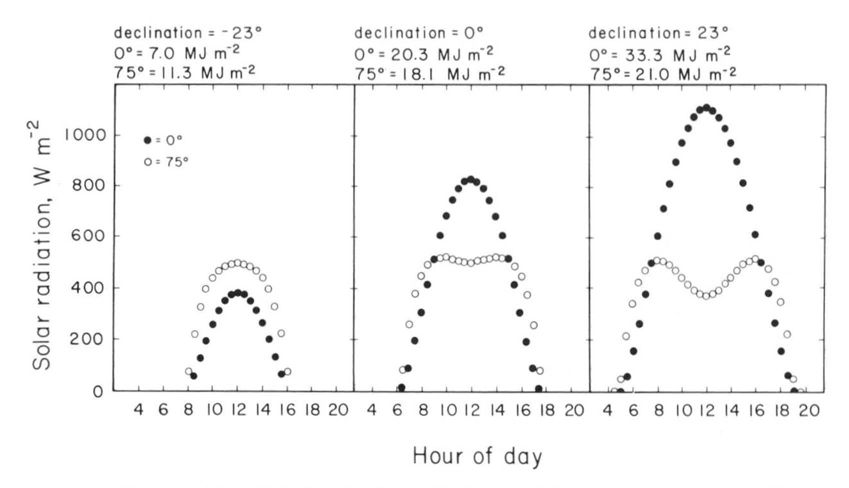

Figure 2.11. Calculated solar radiation levels incident on a horizontal leaf and on a leaf with a 70° leaf angle (and random leaf azimuth) at different solar declinations. Above each plate are the total daily solar radiations incident on the two different leaf types.

midday irradiance is reduced, whereas morning and afternoon irradiances are enhanced. As a result, the average irradiance of steeply inclined leaves is relatively flat over the course of the day (Figure 2.11). This contrasts sharply with the distribution of irradiance on leaves with low inclinations, which experience a large peak at noon (Figure 2.11). During the winter, steep inclinations actually enhance the amount of solar irradiance received. An additional consequence associated with steep leaf angles is that the peak irradiance on the average leaf remains constant throughout the year.

Atriplex hymenelytra is an evergreen-leaved desert shrub with steeply inclined leaves oriented randomly with respect to azimuth (Mooney et al. 1977). Photosynthesis is light-saturated at relatively low irradiances (Mooney et al. 1982), similar to the maximal irradiance incident on the steeply inclined leaves in Figure 2.11. *A. hymenelytra* is also characterized by having salt bladders on the leaf surface that dry out during drought. This causes the salts within the bladders to crystallize and results in decreased leaf absorptance. The implications of the steep leaf inclination for photosynthesis and transpiration can be evaluated by asking what would happen if the leaf angle were lower or if the leaf absorptance were higher. We can calculate how much photosynthesis and transpiration would be affected by these changes during the summer months, which represent the drought period for these plants. Changing both leaf absorptance and angle

Atriplex hymenelytra summer day leaf conductance, 0.02 mol m^{-2} s^{-1}

	Leaf temperature (°C)	Transpiration (mmol m^{-2} s^{-1})	Photosynthesis (μmol m^{-2} s^{-1})	WUE (μmol / mmol)
	50	1.9	4	2.1
	47	1.7	7	4.1
	43	1.4	10	7.1

Figure 2.12. Predicted relationships between leaf temperature, transpiration, photosynthesis, and water-use efficiency (ratio of photosynthesis to transpiration) as functions of different leaf absorptances and leaf angles. (Modified from Mooney et al. 1977, and based on photosynthetic data from Mooney et al. 1982.)

results in a 7°C decrease in leaf temperature from that of a horizontal green leaf (Figure 2.12). As a consequence only of this leaf temperature difference, the transpiration rate is predicted to be reduced by 34% and photosynthetic rate increased by 70%. These differences are quite large and should have a significant impact on plant performance, because these leaves are maintained throughout the prolonged drought periods. Morphological changes such as these not only may enhance plant performance in marginal habitats but also may be essential to their survival in these zones.

It is important to understand that when studying the average effect of leaf inclination on solar irradiance, we are overlooking the variation that exists within a plant. While reducing average irradiance during the summer, steeply inclined leaves also increase the range of solar irradiance incident on individual leaves. As leaf angles are increased, the difference between the highest peak irradiance received by a single leaf and the lowest is amplified (Figure 2.13). For the example presented in Figure 2.13, for a leaf inclination of 75° from horizontal, at least one leaf never receives more than 200 W m^{-2}, while another leaf receives more than 900 W m^{-2} peak irradiance. The importance of this type of variation on the overall performance of a plant is not known. It is likely that physiological and morphological differences exist between leaves with different orientations within a plant in the same way that differences exist between inner and outer leaves in dense canopies.

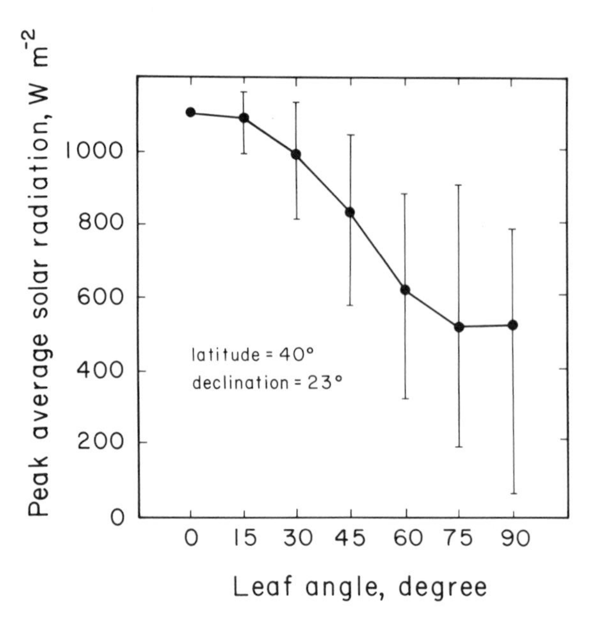

Figure 2.13. Relationship between peak average solar radiation and leaf angle. Calculations were made using a latitude of 40° and a solar declination of +23.5°. Filled circles represent average values for all possible leaf azimuths; vertical lines represent absolute ranges of values by different leaf azimuths.

Steep leaf angles and nonrandom azimuths

In contrast to the case of randomly oriented leaves, the significance of specific nonrandom lamina orientations is much simpler to understand in functional terms. The effects of a specific orientation on daily irradiance, peak irradiance, and diurnal distribution of irradiance can all be related to their effects on leaf temperature, photosynthesis, and water loss. For example, north–south lamina orientations have been studied in several species of cactus (see Chapter 3). In these examples the specific orientation has been determined to increase the solar irradiance on the growing meristems or cladodes during the winter, the active growing season. Interception of irradiance during the summer drought periods is reduced, lowering the risk of excessive heating and/or desiccation.

East–west lamina orientations have been studied in cactus (Gibbs and Patten 1970; Nobel 1980) and in herbaceous plants (Dolk 1931; Werk and Ehleringer 1984). In all cases studied thus far, the east–west orientation was associated with vertical leaves. The major effect of this is that most of the solar radiation intercepted by the lamina is received early in the morn-

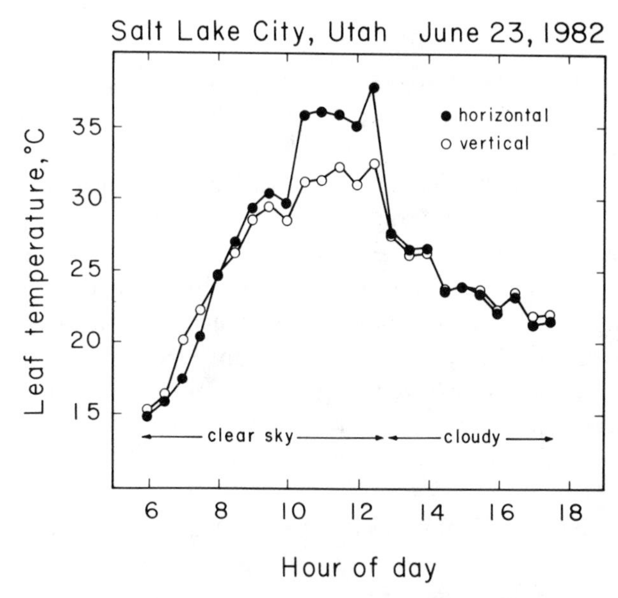

Figure 2.14. Diurnal courses of leaf temperatures on *Lactuca serriola* leaves oriented in the natural vertical position or modified so that the leaf was horizontal. (From Werk and Ehleringer 1984.)

ing and late in the afternoon. Midday irradiance is sharply reduced (Figure 2.5). Thus, the plants can be most active during the coolest time of the day when Δw is lowest. This can reduce water loss significantly, while not severely limiting photosynthesis.

Lactuca serriola is an annual that orients its leaves vertically, facing east–west (Werk and Ehleringer 1984). The effect of this leaf orientation on leaf temperature was determined by comparing the diurnal courses of leaf temperatures of adjacent leaves of a plant. One leaf was forced into a horizontal position and compared with a leaf left in its natural orientation. The patterns of leaf temperatures of these leaves closely followed the pattern of solar irradiance incident on them (Figure 2.14). The pattern of leaf-to-air vapor pressure deficits was very similar to that for leaf temperature. Leaf conductance to water vapor in *L. serriola* is reduced in response to increases in Δw (Werk, unpublished data); however, the response is not strong enough to completely offset the effect of Δw on transpiration. Thus, the horizontal leaf described in Figure 2.14 transpired approximately 10–20% more water over the course of the day than the naturally oriented leaf.

Conclusions

Leaf absorptance and leaf angle both have significant influences on leaf temperature and on the processes of photosynthesis and transpiration. In the arid western United States, leaf reflectance and leaf angle in plants increase with increased aridity. At the functional level, plants appear to use decreased leaf absorptance or increased leaf angle as a means of reducing water loss to extend their period of activity into prolonged drought periods and as a means of reducing photon flux at irradiances higher than necessary to saturate the photosynthetic process. The "cost" to produce a reflective surface may be as much as the investment "cost" to produce the glabrous leaf tissues. However, given how much longer these epidermal modifications allow the leaf to remain active into a drought period, the reflective surface is predicted to result in a positive net carbon gain to the plant.

There are two specific leaf orientations that tend to enhance photosynthetic rate on a diurnal basis. The first is solar tracking, which may enhance productivity at the leaf level because of the resulting high incident irradiances, but limits total canopy productivity by restricting the maximum canopy leaf area index. The second is fixed leaves with steep leaf angles and oriented with lamina in an east–west direction. This orientation tends to increase the incident irradiance and thus photosynthesis, early in the morning and again in the later afternoon. These are periods of the day when the transpirational demand is lowest.

References

Billings, W. D., and R. J. Morris. 1951. Reflection of visible and infrared radiation from leaves of different ecological groups. Amer. J. Bot. 38:327–331.

Cowan, I. R., and G. D. Farquhar. 1977. Stomatal function in relation to leaf metabolism and environment. Symp. Soc. Exp. Biol. 31:471–505.

Dolk, H. E. 1931. The movements of the leaves of the compass plant *Lactuca serriola*. Amer. J. Bot. 18:195–204.

Ehleringer, J. 1980. Leaf morphology and reflectance in relation to water and temperature stress. Pp. 295–308 *in* N. Turner and P. Kramer (eds.), Adaptations of plants to water and high temperature stress. Wiley-Interscience, New York.

– 1981. Leaf absorptances of Mohave and Sonoran desert plants. Oecologia 49:366–370.

– 1982. The influence of water stress and temperature on leaf pubescence development in *Encelia farinosa*. Amer. J. Bot. 69:670–675.

– 1983a. Characterization of a glabrate *Encelia farinosa* mutant: morphology, ecophysiology, and field observations. Oecologia 57:303–310.

– 1983b. Ecology and ecophysiology of leaf pubescence in North American desert plants. Pp. 113–132 *in* P. Healey and E. Rodriguez (eds.), Biology and chemistry of plant trichomes. Plenum Press, New York.

Ehleringer, J., and O. Björkman. 1978. Pubescence and leaf spectral characteristics in a desert shrub, *Encelia farinosa*. Oecologia 36:151–162.

Ehleringer, J., O. Björkman, and H. A. Mooney. 1976. Leaf pubescence: effects on absorptance and photosynthesis in a desert shrub. Science 192:376–377.

Ehleringer, J., and C. S. Cook. 1984. Photosynthesis in *Encelia farinosa* Gray in response to decreasing leaf water potential. Plant Physiol. 75:688–693.

Ehleringer, J., and I. Forseth. 1980. Solar tracking by plants. Science 210:1094–1098.

Ehleringer, J., and H. A. Mooney. 1978. Leaf hairs: effects on physiological activity and adaptive value to a desert shrub. Oecologia 37:183–200.

Ehleringer, J., H. A. Mooney. S. L. Gulmon, and P. W. Rundel. 1981. Parallel evolution of leaf pubescence in *Encelia* in coastal deserts of North and South America. Oecologia 49:38–41.

Farquhar, G. D., and T. Sharkey. 1982. Stomatal conductance and photosynthesis. Ann. Rev. Plant Physiol. 33:317–345.

Forseth, I., and J. Ehleringer. 1980. Solar tracking response to drought in a desert annual. Oecologia 44:159–163.

– 1983a. Ecophysiology of two solar tracking desert winter annuals. III. Gas exchange responses to light, CO_2, and VPD in response to long term drought. Oecologia 57:340–351.

– 1983b. Ecophysiology of two solar tracking desert winter annuals. IV. Effects of leaf orientation on calculated daily carbon gain and water use efficiency. Oecologia 58:10–18.

Gates, D. M. 1962. Energy exchange in the biosphere. Harper & Row, New York.

Gibbs, J. G., and D. T. Patten. 1970. Plant temperatures and heat flux in a Sonoran Desert ecosystem. Oecologia 5:165–184.

Givnish, T. J., and G. J. Vermeij. 1976. Sizes and shapes of Liane leaves. American Naturalist 110:743–778.

Haberlandt, G. 1884. Physiologische Pflanzenanatomie. Engelmann, Lehre.

Körner, C., J. A. Scheel, and H. Bauer. 1979. Maximum leaf diffusive conductance in vascular plants. Photosynthetica 13:45–82.

Mooney, H. A. 1972. Carbon balance of plants. Ann. Rev. Ecol. Syst. 3:315–346.

Mooney, H. A., J. A. Berry, O. Björkman, and J. Ehleringer. 1982. Comparative photosynthetic characteristics of coastal and desert species of California. Bol. Soc. Bot. Mex. 42:19–33.

Mooney, H. A., J. Ehleringer, and O. Björkman. 1977. The energy balance of leaves of the evergreen desert shrub *Atriplex hymenelytra*. Oecologia 29:301–310.

Mooney, H. A., and S. L. Gulmon. 1979. Environmental and evolutionary constraints on the photosynthetic characteristics of higher plants. Pp. 316–337 *in* O. T. Solbrig, S. Jain, G. B. Johnson, and P. H. Raven (eds.), Topics in plant population biology. Columbia University Press, New York.

Mulroy, T. W. 1979. Spectral properties of heavily glaucous and nonglaucous leaves of a succulent rosette-plant. Oecologia 38:349–357.

Nobel, P. S. 1980. Interception of photosynthetically active radiation by cacti of different morphology. Oecologia 44:160–166.

Parkhurst, D., and O. Loucks. 1972. Optimal leaf size in relation to environment. J. Ecol. 60:505–537.

Pearman, G. I. 1966. The reflectance of visible radiation from leaves of some western Australian species. Austr. J. Biol. Sci. 19:97–103.

Penning de Vries, F. W. T., A. H. M. Brunsting, and H. H. Van Laar. 1974. Products, requirements and efficiency of biosynthesis: a quantitative approach. J. Theor. Biol. 45:339–377.

Raschke, K. 1956. The physical relationships between heat-transfer coefficients, radiation exchange, temperature, and transpiration of a leaf. Planta 48:200–238.

Reicosky, D. A., and J. W. Hanover. 1978. Physiological effects of surface waxes. I. Light reflectance for glaucous and nonglaucous *Picea pungens.* Plant Physiol. 62:101–104.

Schimper, A. F. W. 1903. Plant geography upon a physiological basis. Clarendon Press, Oxford.

Shackel, K. A., and A. E. Hall. 1979. Reversible leaflet movements in relation to drought adaptation of cowpeas, *Vigna unguiculata* (L.) Walp. Aust. J. Plant Physiol. 6:265–276.

Shaver, G. S. 1978. Leaf angle and light absorptance of *Arctostaphylos* species (Ericaceae) along elevational gradients. Madroño 25:133–138.

Sinclair, R., and D. A. Thomas. 1970. Optical properties of leaves of some species in arid south Australia. Aust. J. Bot. 18:261–273.

Thomas, D. A., and H. N. Barber. 1974. Studies of leaf characteristics of a cline of *Eucalyptus urnigera* from Mount Wellington, Tasmania. II. Reflection, transmission, and absorption of radiation. Aust. J. Bot. 22:701–707.

von Caemmerer, S., and G. D. Farquhar. 1981. Some relationships between the biochemistry of photosynthesis and the gas exchange of leaves. Planta 153:376–387.

Warming, E. 1909. Oecology of plants: an introduction to the study of plant communities. Oxford University Press, London.

Werk, K. S., and J. Ehleringer. 1984. Non-random leaf orientation in *Lactuca serriola* L. Plant Cell Environ. 7:81–87.

Werk, K. S., J. Ehleringer, I. N. Forseth, and C. S. Cook. 1983. Photosynthetic characteristics of Sonoran Desert winter annuals. Oecologia 59:101–105.

Wong, S. C., I. R. Cowan, and G. D. Farquhar. 1979. Stomatal conductance correlates with photosynthetic capacity. Nature 282:424–426.

3 Form and orientation in relation to PAR interception by cacti and agaves

PARK S. NOBEL

Introduction

Cacti and agaves have many interesting features with regard to the interception of radiation and the consequences this has for productivity. For most C_3 and C_4 plants, photosynthetically active radiation (PAR) from 400 to 700 nm is generally intercepted by relatively flat translucent leaves whose orientation and arrangement are of critical importance in determining net CO_2 uptake (Monsi and Saeki 1953; Anderson 1964; Fisher and Honda 1979). On the other hand, desert succulents have opaque photosynthetic organs that generally are not flat and that exhibit Crassulacean acid metabolism (CAM), in which photosynthesis occurs during the daytime but stomatal opening and hence CO_2 uptake occur primarily during the nighttime. The three-dimensional nature of the photosynthetic organs of succulent plants (Figure 3.1) presents special challenges for modeling PAR interception, the major concern here.

The CO_2 taken up at night by CAM plants is incorporated into organic acids, leading to a gradual increase in the acidity of chlorenchyma vacuolar sap during the night. Also, because CO_2 fixed at night can come from respiration, nocturnal increases in acidity can occur in the absence of net CO_2 uptake. Beginning near dawn, organic acids are released from the vacuoles of the chlorenchyma cells and decarboxylated within the cytoplasm. The CO_2 so released, which is generally prevented from diffusing out of the plant by the daytime closure of the stomata, is then fixed into carbohydrates and other photosynthetic products by the conventional C_3 pathway of photosynthesis (Kluge and Ting 1978; Osmond 1978). Thus, for CAM plants, the absorption of PAR is temporally separated from the main period for net CO_2 uptake.

In this study, the levels of PAR incident on the stems of cacti and the leaves of agaves will be related to nocturnal CO_2 uptake and its consequent acid accumulation to help interpret the relation between plant form and productivity. CO_2 uptake and hence productivity of cacti and agaves,

Figure 3.1. Examples of four morphologically diverse succulent plants: an agave, *Agave deserti* (A), and a barrel cactus, *Ferocactus acanthodes* (B), common in the Sonoran Desert of North America; a prickly pear cactus cultivated worldwide for its fruits and cladodes (flattened stems), *Opuntia ficus-indica* (C); and a ceroid cactus from Chile that can vary considerably in height depending on shading from the surrounding vegetation, *Trichocereus chilensis* (D).

which grow naturally over large areas of North and South America and occur as cultivated plants or pests on the other continents, are found to be on the verge of being limited by the availability of PAR. The stems of cacti tend to be vertical, which leads to lower levels of PAR being available than for horizontal surfaces. This PAR limitation, which will be modeled for various latitudes and seasons, influences cactus morphology (e.g., the orientation of terminal cladodes of platyopuntias). Also, spines influence the

PAR reaching the stem surface. The basal rosettes of leaves on agaves have a different PAR interception pattern that will be briefly considered with respect to PAR availability in various canopy positions.

Materials and methods

Plant material

Morphological and physiological properties were determined for plants in their native habitats, except where indicated. *Agave deserti* Engelm. (Figure 3.1A) and *Ferocactus acanthodes* (Lemaire) Britt. & Rose (Figure 3.1B) were examined in the western Sonoran Desert at the Philip L. Boyd Deep Canyon Desert Research Center near Palm Desert, California, at 33°38′ N, 116°24′ W, 850 m. *Agave lecheguilla* Torr. was examined in the Chihuahuan Desert near Saltillo, Coahuila, Mexico, at 25°23′ N, 101°6′ W, 1,620 m. *Opuntia echios* var. *gigantea* Howell was examined at five sites on Santa Cruz Island, Galápagos Islands, Ecuador, from 0°30′ S, 90°16′ W, 30 m to 0°37′ S, 90°23′ W, 360 m. *Opuntia ficus-indica* (L.) Miller was examined at various sites in southern California and Israel (specific locations indicated in figure captions), as well as in central Chile (Figure 3.1C). *Stenocereus gummosus* (Engelm.) Gibs. & Horak was examined at 29 sites along the entire length of Baja California, Mexico, from 22°53′ N, 109°55′ W, 24 m to 31°52′ N, 116°40′ W, 60 m. *Trichocereus chilensis* (Colla) Britt. & Rose (Figure 3.10) was examined at eight sites in central Chile from 33°1′ S, 71°13′ W, 290 m to 33°4′ S, 71°5′ W, 1,540 m.

Photosynthetically active radiation

Photosynthetically active radiation (PAR) from 400 to 700 nm was measured using a Li-Cor LI-190S quantum sensor. Data are expressed either as instantaneous values (μmol m^{-2} s^{-1}) or as totals for the entire day (mol m^{-2} day^{-1}). PAR incident on horizontal and vertical surfaces was also modeled for various latitudes and seasons. The assumptions made in the model were the following: clear skies, an atmospheric transmissivity of 0.76 for an optical air mass of unity (Fitzpatrick 1979), isotropic diffuse radiation, a ground albedo of 0.2, and the proportions of PAR in solar irradiation determined by Ross (1975) for the diffuse beam and by Williams (1976) for the direct beam. A model was also used to predict the PAR on the leaves of agaves. To describe leaf area, taking into consideration differences in shape for the upper versus the lower surface, the upper surface was divided into three flat sections and the lower surface into six flat sections for all 60 leaves on the mature plant of *A. deserti* used in the simulations (Woodhouse et al. 1980).

Cladode orientation

Orientations of terminal cladodes (flattened stems) of platyopuntias were determined in the field. The angle from true north was measured within 1° for 660 vertical unshaded cladodes at each site. Data are summarized for angle classes 10° wide that are centered on the angle indicated.

Tissue acidity: CO_2 exchange

Nocturnal acid accumulation was determined by measuring tissue acidity at dawn and at dusk. Three 1.14-cm^2 samples extending from the surface to just inside the chlorenchyma (about 3 mm overall) were removed with a cork borer, immediately ground with sand in 30 ml distilled water, and then titrated to pH 6.8 using 0.010-N NaOH.

To measure CO_2 uptake by cacti, small assimilation chambers were mounted onto the stem surfaces of field plants (Nobel 1977), or attached stems were inserted into larger chambers in the laboratory (Nobel and Hartsock 1983). The total CO_2 uptake during the night was determined from approximately hourly measurements of net CO_2 exchange.

Results and discussion

Expected PAR levels: latitudinal effects

The PAR incident on a vertical surface on a clear day can be far less than that on a horizontal surface (Figure 3.2). Simulations indicate that on the summer solstice at 34° N the PAR incident on an east-facing surface is initially greater than that on a horizontal surface, but at a solar time of about 9, the PAR level becomes greater on the horizontal surface and remains so throughout the rest of the day. Indeed, only diffuse radiation is incident on the east-facing surface after solar noon. Also, north- or south-facing surfaces receive primarily diffuse radiation near the summer solstice, some direct PAR being incident on the north-facing surface (in the Northern Hemisphere) for about 3 hr at the beginning and the end of the day and on the south-facing surface for about 7 hr near the middle of the day (Figure 3.2).

The total daily PAR can be found by integrating the instantaneous values of PAR over the course of a day. For a horizontal surface on a clear day at the summer solstice at 34° N, the total daily PAR was predicted to

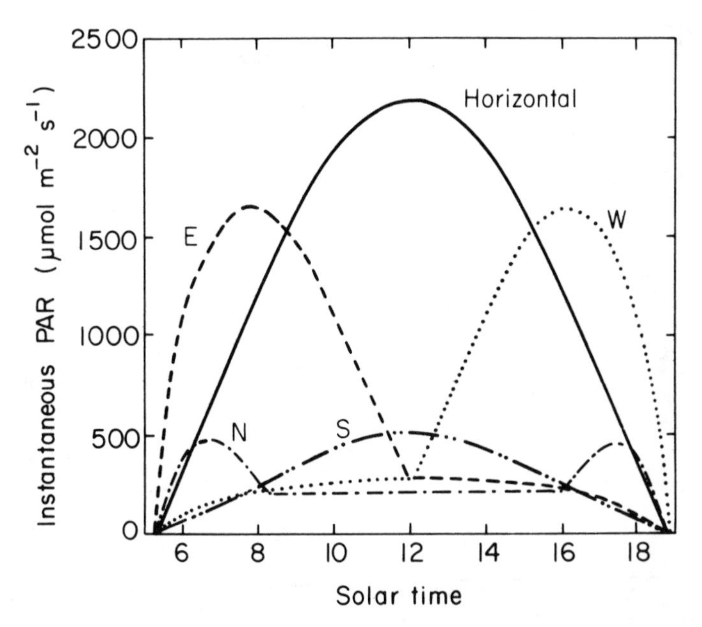

Figure 3.2. Photosynthetically active radiation predicted by the PAR model for the summer solstice at 34° N. Curves are for a horizontal surface or vertical surfaces facing in the indicated directions on a clear day.

be 66 mol m^{-2} day^{-1}, as compared with 32 mol m^{-2} day^{-1} for an east- or west-facing surface, 15 mol m^{-2} day^{-1} for a south-facing surface, and 14 mol m^{-2} day^{-1} for a north-facing surface (Figure 3.2), in close agreement with measurements made in the field (Nobel 1980; Acevedo et al. 1983). Thus, the sides of an unshaded vertical cactus stem experience a much lower PAR level than does an unshaded horizontal leaf (e.g., sides facing in the four cardinal directions receive an average of only 35% as much PAR as a horizontal surface for clear days at the summer solstice at 34° N).

In addition to predicting instantaneous PAR values at a particular latitude and season, the model was also used to predict the total daily PAR on vertical surfaces facing in the four cardinal directions at various latitudes up to 60° north of the equator at three different times of the year (Figure 3.3). At low latitudes from 0° to 20° N, east-, north-, and west-facing surfaces had similar PAR interceptions at the summer solstice, east- or west-facing surfaces had the most PAR interception on an equinox, and south-facing surfaces had the most PAR interception at the winter solstice. For mid-latitudes from 20° N to 40° N, east- or west-facing surfaces had the most PAR interception near the summer solstice, and south-facing surfaces had the most on an equinox and at the winter solstice. At relatively

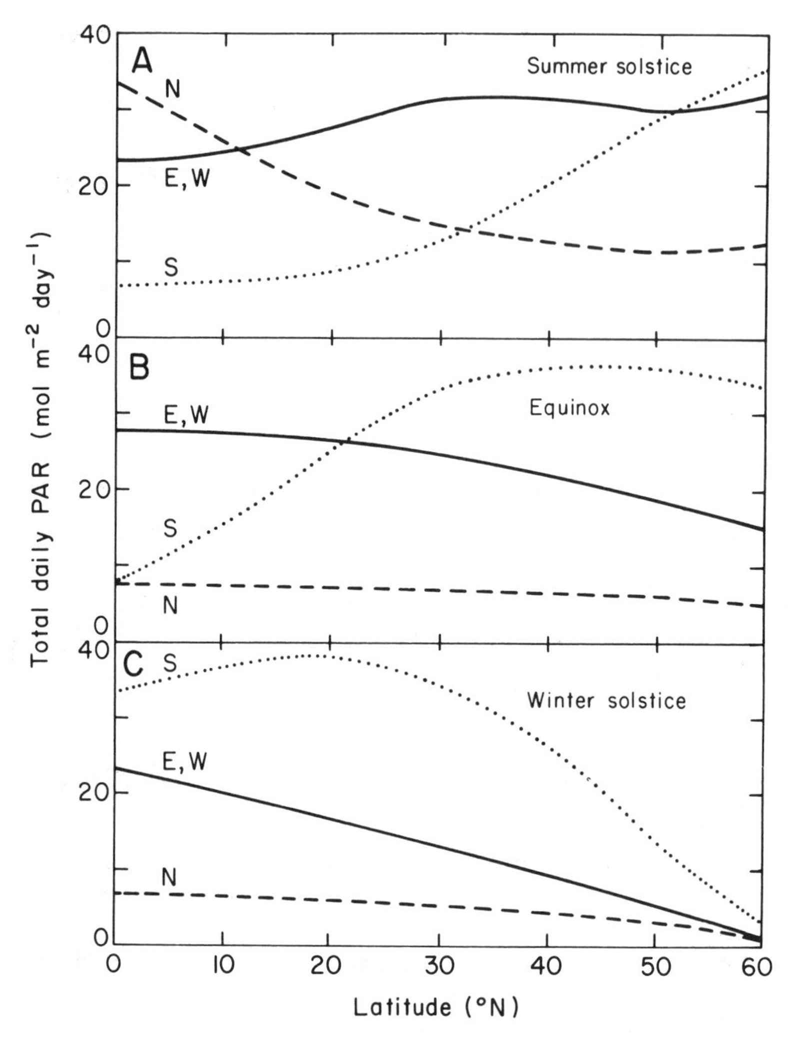

Figure 3.3. Total daily PAR predicted by the model for vertical surfaces facing in the indicated directions at various latitudes and times of the year on clear days.

high latitudes from 40° N to 60° N, east-, west-, and south-facing surfaces had similar PAR interceptions at the summer solstice, and south-facing surfaces again had the most on an equinox and at the winter solstice (Figure 3.3). Particular values at specific latitudes can be determined for an equinox or the solstices directly from Figure 3.3.

Another very important parameter for understanding PAR interception and hence productivity by the opaque photosynthetic organs of succulent plants is the average total daily PAR on unshaded vertical surfaces on clear days. For the four cardinal directions and all the latitudes considered, it was 23 mol m^{-2} day^{-1} on the summer solstice (Figure 3.3A), 20 mol m^{-2} day^{-1} on an equinox (Figure 3.3B), and 15 mol m^{-2} day^{-1} on the winter solstice (Figure 3.3C). Averaged over the year, and weighting an equinox twice as heavily as each solstice, it was 21 mol m^{-2} day^{-1} for low latitudes (0–20°), 21 mol m^{-2} day^{-1} for mid-latitudes (20–40°), and 17 mol m^{-2} day^{-1} for high latitudes (40–60°), which corresponds to 20 mol m^{-2} day^{-1} when averaged over all the latitudes considered (Figure 3.3). Comparison of these values with the PAR responses of desert succulents will show that PAR can be on the verge of limiting productivity for cacti and agaves.

PAR influences on CO_2 uptake and acidity changes

Next, the effects of total daily PAR on the metabolism of various cacti and *A. deserti* will be considered. The relationships of nocturnal CO_2 uptake to total daily PAR were very similar for *F. acanthodes* and *O. ficus-indica* (Figure 3.4A). Both exhibited PAR compensation points of 3–4 mol m^{-2} day^{-1}. Hence, surfaces receiving less than this, such as shaded surfaces or north-facing surfaces in the Northern Hemisphere over most of the year (Figure 3.3), will be a net liability for the carbon gain of the plant as a whole. PAR saturation of nocturnal CO_2 uptake occurred at about 30 mol m^{-2} day^{-1}. Indeed, PAR levels above 40 mol m^{-2} day^{-1} could lead to lower CO_2 uptake than at 30–40 mol m^{-2} day^{-1} (Nobel and Hartsock 1983), possibly reflecting photoinhibition at the higher irradiation levels. A more readily defined index than the saturation PAR, which is approached asymptotically, is the PAR level leading to 90% of maximum, which occurred for a total daily PAR of 23 mol m^{-2} day^{-1} for both species (Figure 3.4A). This is quite similar to the average PAR level on vertical surfaces on clear days in the field (Figure 3.3), suggesting that CO_2 uptake and hence productivity are on the verge of being PAR-limited.

The data for the PAR dependence of nocturnal acid accumulation, when expressed as a percentage of the maximum achieved for each species, were very similar for five species of cacti and *A. deserti* (Figure 3.4B). Dawn acidity levels were never less than dusk values, and nocturnal acid accumulation increased fairly linearly with PAR at low PAR levels. The PAR compensation point for nocturnal acid accumulation was at 0 mol m^{-2} day^{-1} (Figure 3.4B). From 0 to about 3 mol PAR m^{-2} day^{-1}, net CO_2 uptake is small in value and negative (Figure 3.4A), and yet a (positive) nocturnal accumulation of acid occurs because of CO_2 released within the plant by respiration. PAR saturation of acid accumulation was achieved

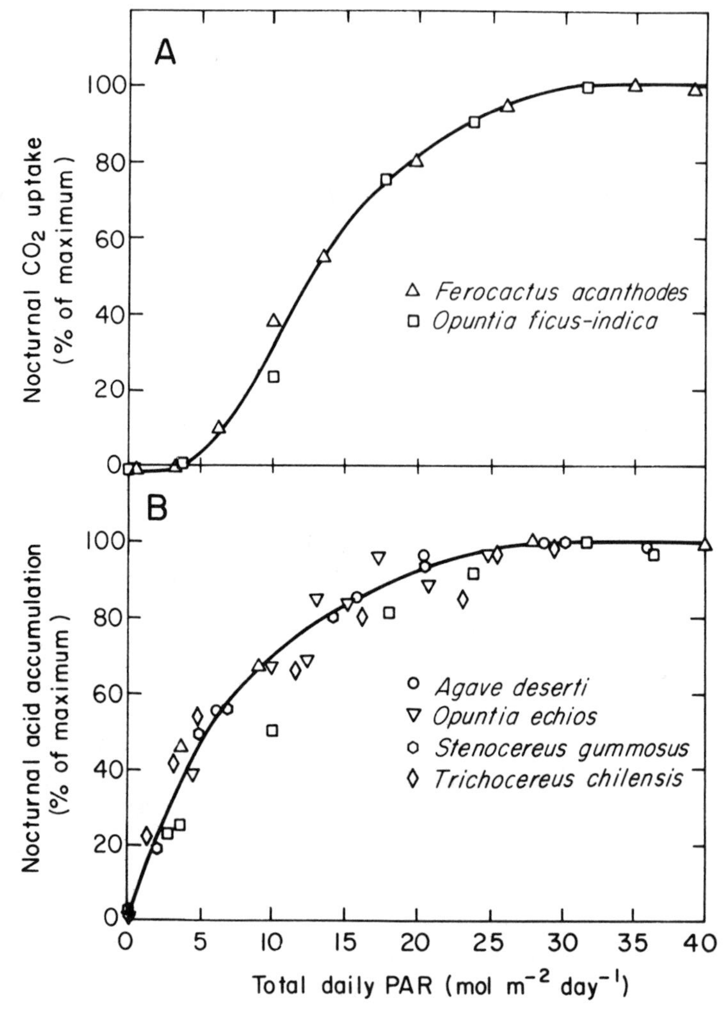

Figure 3.4. Nocturnal CO_2 uptake (A) and acid accumulation (B) at various PAR levels. Data were obtained in the field for *Ferocactus acanthodes* (Nobel 1977), *Agave deserti* (Woodhouse et al. 1980), *Opuntia echios* (Nobel 1981), *Stenocereus gummosus* (Nobel 1980), and *Trichocereus chilensis* (Nobel 1981) and in the laboratory for *Opuntia ficus-indica* (Nobel and Hartsock 1983) under wet soil conditions and near-optimal nighttime temperatures. The plants had experienced generally clear days on the day of the measurement and the week preceding (plants in the laboratory were also maintained at a given PAR for at least one week), with different PAR levels being obtained at different locations on the plants.

near 30 mol m^{-2} day^{-1}, similar to saturation of net CO_2 uptake, and 90% saturation of nocturnal acid accumulation occurred for a total daily PAR of about 19 mol m^{-2} day^{-1} (Figure 3.4B).

Nocturnal acid accumulation is considerably easier to determine in the field than is CO_2 uptake, because only the dawn and dusk acidity levels need to be measured, whereas determination of the total CO_2 uptake requires fairly constant monitoring of CO_2 uptake rates. However, total CO_2 uptake is a more relevant measure of plant productivity, because it is CO_2 that is used in the manufacture of photosynthate. Moreover, nocturnal acid accumulation may occur even when there is a net loss of CO_2, as indicated earlier. Under nonstressful conditions, and ignoring respiration, the expected ratio of H^+ accumulated to CO_2 taken up is 2.0, based on the overall net chemical reaction for converting starch or another glucan into an organic acid such as malic acid (Nobel and Hartsock 1983). However, conditions that restrict nocturnal stomatal opening, such as drought, lower daytime PAR, or even the beginning as well as the end of the dark period, cause the H^+/CO_2 ratio to increase, because some of the H^+ results from internally recycled CO_2, not CO_2 taken up from the external environment. For the CAM plant *Tillandsia usneoides*, the ratio for a 24-hr cycle changed from 2.0 after clear days to 4.3 after cloudy days (Martin et al. 1981). For *O. ficus-indica*, the H^+/CO_2 ratios after 0, 3, and 9 weeks of drought were 2.6, 6.5, and 12.3, respectively (Nobel and Hartsock 1983). Indeed, extended drought can cause stomata to remain closed, thus preventing CO_2 uptake, although daily variations in tissue acidity still occur, leading to an infinite value for the H^+/CO_2 ratio (Szarek et al. 1973; Szarek and Ting 1974).

Although total daily PAR has proved to be a useful measure of radiation to correlate with nocturnal CO_2 uptake and acid accumulation, there is some influence of instantaneous values at a given total daily PAR. For instance, a PAR of 1,200 μmol m^{-2} s^{-1} led to 30% less acid accumulation than half as much instantaneous PAR for twice as long for *O. ficus-indica*, reflecting the PAR response of C_3 photosynthesis that takes place in CAM plants during the daytime (Nobel and Hartsock 1983).

The dependence of nocturnal CO_2 uptake on PAR (Figure 3.4A) was next combined with the PAR variation with season and latitude (Figure 3.3) to predict CO_2 uptake for vertical stem surfaces (Figure 3.5). Because the flattened stems (cladodes) of platyopuntias are amenable for observing orientation effects not possible with cylindrically symmetrical stems, CO_2 uptake was calculated for flat stems facing north–south or east–west. As would be expected from a consideration of the sun's trajectory, an east–west-facing cladode will intercept more PAR and have a higher nocturnal

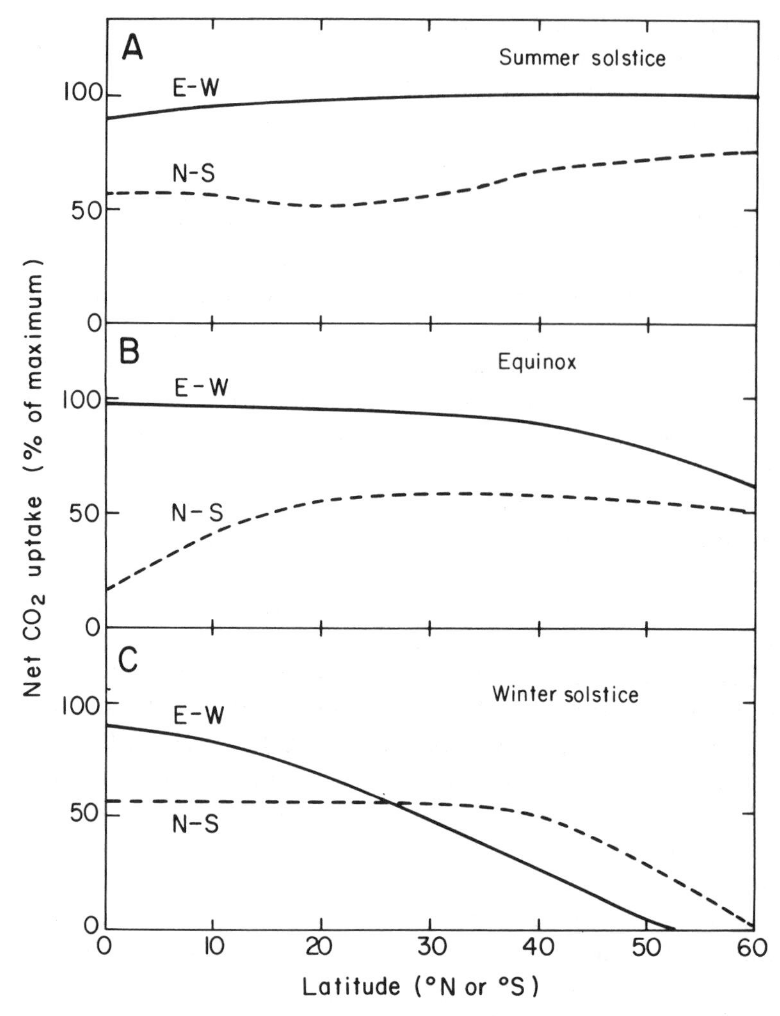

Figure 3.5. Net CO_2 uptake for vertical stem surfaces facing east–west or north–south, calculated from the PAR levels in Figure 3.3 and the CO_2 response to total daily PAR in Figure 3.4A.

CO_2 uptake than a north–south-facing cladode at the summer solstice or an equinox. Indeed, nocturnal CO_2 uptake averaged 61% greater for an east–west-facing cladode than for a north–south-facing cladode at the summer solstice (Figure 3.5A) and 79% greater at an equinox (Figure 3.5B). The situation at the winter solstice is more complicated and, morphologically speaking, more interesting. For latitudes from 0° N to 27° N, an east–west-facing cladode will again have a higher productivity. However, at higher latitudes, a north–south-facing cladode will have a higher

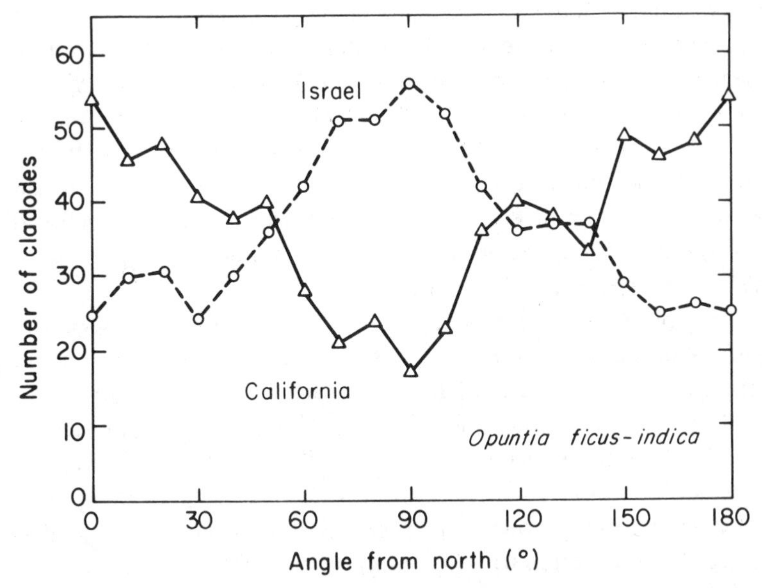

Figure 3.6. Orientation of terminal cladodes of *O. ficus-indica* at San Bernardino, California (34°7′ N, 117°30′ W, 400 m; data obtained 9 August 1980), and Elqosh, Israel (33°2′ N, 35°19′ E, 630 m; data obtained 22 January 1981).

productivity at the winter solstice (Figure 3.5C), in large measure because of the relatively high PAR level for a south-facing surface (Figure 3.3C). If development of cladodes were to occur near the winter solstice, then differences in cladode orientation as a function of latitude might be expected.

Orientation of cladodes on platyopuntias

Orientations of unshaded terminal cladodes were measured for a number of species of platyopuntias in the field. For instance, such orientations were determined for *O. chlorotica* in the Mojave Desert, where rainfall occurs predominantly in the winter, and in the Sonoran Desert, where the annual rainfall pattern can be bimodal, leading to similar winter and summer precipitations. For sites at 35° N, there was a slight tendency to face north–south in the Mojave Desert and east–west in the Sonoran Desert (Nobel 1981). Such orientations will maximize PAR interception at times of the year most favorable for growth. Such differences in orientation patterns were seen more clearly for *O. ficus-indica* (Figure 3.6). At a site in California at 34° N where year-round development of cladodes would be expected, about three times as many cladodes faced east–west as north–south. However, in Israel, at a similar latitude (33° N), the pattern was reversed, with more than twice as many cladodes facing north–south as east–west (Figure 3.6). At the Israeli site, about 80% of the annual rainfall occurs near the winter solstice (November–February) and only 1% near the summer solstice (May–August) (Katsnelson 1968/69). Moreover, these sites occur more than 27° from the equator, where productivity at the winter solstice will be greater for north–south-facing cladodes, in contrast to the favoring of east–west-facing cladodes that occurs for all latitudes (up to the 60° N tested) at the summer solstice and equinox and below 27° at the winter solstice (Figure 3.5).

The same general conclusions were arrived at in a survey of 23 species of platyopuntias on four continents (Nobel 1982a). Indeed, in all five cases in which the seasonality of rainfall favored cladode development in the winter and the site was located above 27° N, the terminal unshaded cladodes tended to face north–south, whereas in all other cases such cladodes tended to face east–west, with one interesting category of exceptions. Specifically, when topographical features or shading by surrounding vegetation affected the direction for maximum incident PAR, the orientation pattern of the cladodes was changed accordingly. For instance, at a site in central Chile, the predominant orientation for terminal cladodes on 100-year-old cultivated *O. ficus-indica* was about 20° from east–west (Nobel 1982b). Based on simulations incorporating the effects of vegetation and mountains on incident PAR, the observed predominant orientation maximized nocturnal acid accumulation at the summer solstice, an equinox, and the winter solstice for these plants. For *O. erinacea* var. *erinacea* at a site in California where mountains gave considerable blockage to the west and northwest, there was a tendency for facing northwest–southeast (Nobel 1982a). For *O. littoralis* var. *littoralis,* which occurs in southern California on steep hillsides that face the Pacific Ocean, as well as on nearby unobstructed flat sites, the predominant orientations differed by 60° for the two habitats. Taken together, the observed orientation patterns underscore the importance of available PAR on the morphology of these platyopuntias.

As a further refinement, the orientation patterns of newly developing cladodes were examined at various times of the year. For *O. basilaris* var. *basilaris* in southern California at 34° N, cladodes developing in the winter tended to face north–south, whereas those developing in spring/summer tended to face east–west (Nobel 1982a). Because slightly more cladodes occurred in the latter category, the overall tendency for mature terminal cladodes was to face east–west. Similar results were obtained with *O. ficus-indica,* also at 34° N in California, where a slight tendency for cladodes developing near the winter solstice to face north–south was overwhelmed by a strong tendency for cladodes developing near the summer solstice to face east – west (Nobel 1982b, 1982c).

Temporal changes were also seen for cladodes developing on plants of various ages for plantations of *O. ficus-indica* in central Chile. Such fields are started by placing detached mature cladodes vertically in the ground with one-third to one-half of their area below the soil surface. The cladodes developing during the first year on the planted cladodes, which are essentially all aligned with the field axis, tend also to be aligned with the field axis, but a small fraction occurs in other directions. Although cladodes developing during the second year still show a carryover of the initial planting direction, the tendency to face east – west is clearly evident. Five years after the initial planting, the orientation pattern shows a much greater tendency to face east – west than for the California *O. ficus-indica* illustrated in Figure 3.6, but evidence of the original planting direction can still be clearly perceived (Nobel 1982b). These patterns can be interpreted as a resultant of the tendency for daughter cladodes to occur in the same plane as the mother cladodes on which they develop gradually being overcome by the tendency for daughter cladodes to face in a direction that maximizes PAR interception and hence productivity.

The mechanism for producing patterns exhibiting well-defined predominant orientations apparently involves three factors: (1) phototropism, (2) PAR effects on productivity, and (3) morphological alignment of succeeding cladodes. When horizontal PAR was directed onto vertical cladodes of *O. ficus-indica* whose surface normals were at an angle of 45° to the incident beams, the surface normals of the newly developing cladodes were rotated an average of 16° in a direction that increased PAR interception (Nobel 1982b). Cladodes facing in a direction leading to more PAR interception are expected to have higher productivity (Figure 3.5), including the production of more new cladodes. Indeed, east – west-facing cladodes of *O. amyclaea* in central Mexico received more PAR and had greater dry-matter accumulation in the fall than did north – south-facing cladodes (Rodriquez et al. 1976). Finally, for many species of platyopuntias, the newly developing daughter cladodes tend to face in the same direction as the underlying mother cladode. Thus, any tendency to face in a favorable direction will be perpetuated. Indeed, orientation tendencies are more conspicuous in species with many cladodes occurring in sequence along a plant axis than for those species with just a few cladodes (Nobel 1982a). The reinforcing of predominant orientations could cause the eventual pattern to be more pronounced than the angular variation in PAR interception or nocturnal CO_2 uptake, as was found to be the case. Such preferred orientations are advantageous for productivity, because

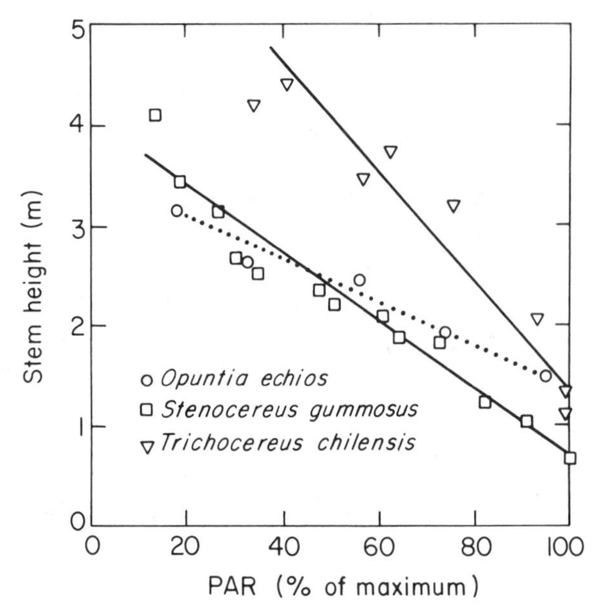

Figure 3.7. Correlation between average PAR incident on (vertical) stem surfaces and height of cacti. Data represent the total PAR received from the four cardinal directions (at 1 m above the ground for *S. gummosus* and at 2 m for the other two species) divided by the analogous PAR in an exposed region (where nothing blocked the sun's rays). Data are for 30 plants per site for *O. echios* over a 16-km transect (Nobel 1981), 15 plants for *S. gummosus* over a 1,300-km transect (Nobel 1980), and 32 plants per site for *T. chilensis* over a 12-km transect (Nobel 1981).

PAR is on the verge of being limiting, as mentioned earlier, even on clear days for unshaded terminal cladodes.

Other correlations between cactus morphology and PAR

Although ceroid cacti do not have flattened stems such that angular orientation can be identified, the heights of such cacti nevertheless constitute another morphological parameter that can be correlated with available PAR, reflecting the influence of neighboring plants (Figure 3.7). For example, *Stenocereus gummosus* occurs in open coastal scrub in northern Baja California, where its 0.7-m-tall stems often curve back to the ground and reroot, forming bramble-like thickets. It occurs essentially continuously all the way to the tip of Baja California over 1,300 km away. In southern Baja California, *S. gummosus* occurs in subtropical forests as plants with distinct trunks and fairly vertical branches, the overall stem height often exceeding 5 m. The stem height varies in concert with the height of the surrounding vegetation, which is short at the northern sites and much taller at the southern sites, leading to the observed changes in available PAR (Nobel 1980). Genetic differences that would be expected to occur over such a long transect are less likely for the other two species whose stem heights have been correlated with PAR. Specifically, mean stem heights varied twofold for *O. echios* over a 16-km transect in the Galápagos Islands and over threefold for *Trichocereus chilensis* (Figure 3.1D) over a 12-km transect in central Chile (Figure 3.7).

The observed growth patterns for all three species can be interpreted as "etiolation" responses, where increases in stem heights at low PAR allow the plants to grow into canopy positions with higher PAR. Although the appropriate field or laboratory experiments have not been performed to check this proposed etiolation response, all three of these species show the PAR dependence of nocturnal acid accumulation characteristic of desert succulents (Figure 3.4B), that is, they all reach 90% saturation of nocturnal acid accumulation near 19 mol m^{-2} day^{-1} and so are on the verge of being PAR-limited with respect to carbon balance at ambient PAR levels when no shading occurs. Indeed, *O. echios* occurred where ambient PAR was reduced to 18% of the unshaded maximum, but not in a nearby site where it was only 8% of the maximum. The latter would represent a total daily PAR of about 2 mol m^{-2} day^{-1} on clear days, which is below the value estimated for light compensation for nocturnal CO_2 uptake by cacti (Figure 3.4A).

Spines can also intercept incoming PAR and hence influence the productivity of cacti. Indeed, the ambient PAR required for 90% saturation of nocturnal acid accumulation by *O. bigelovii* was 23 mol m^{-2} day^{-1} for stems that were 32% shaded by spines and 16 mol m^{-2} day^{-1} when the spines were removed (Nobel 1983a). For *Ferocactus acanthodes* (Figure 3.1B), an ambient PAR of about 50 mol m^{-2} day^{-1} was required when the stem was 74% shaded by spines and 21 mol m^{-2} day^{-1} when the spines were removed. This suggests that the absence of spines will increase the net CO_2 uptake of cacti. For *O. bigelovii* at a total daily PAR of 9 mol m^{-2} day^{-1}, periodically removing the spines led to a 60% greater increase in stem volume over a 2.5-year period (Nobel 1983a). Thus, although spines may protect the stem from herbivory, they reduce the PAR incident on the photosynthetic surfaces and thereby reduce productivity.

Another morphological aspect influencing PAR distribution over the surfaces of some cacti is the presence of ribbing. This ribbing, which in barrel cacti populations often displays a rib-number distribution dominated by the Fibonacci sequence (Robberecht and Nobel 1983), can allow the bellows-like swelling of a cactus stem following periods of rainfall.

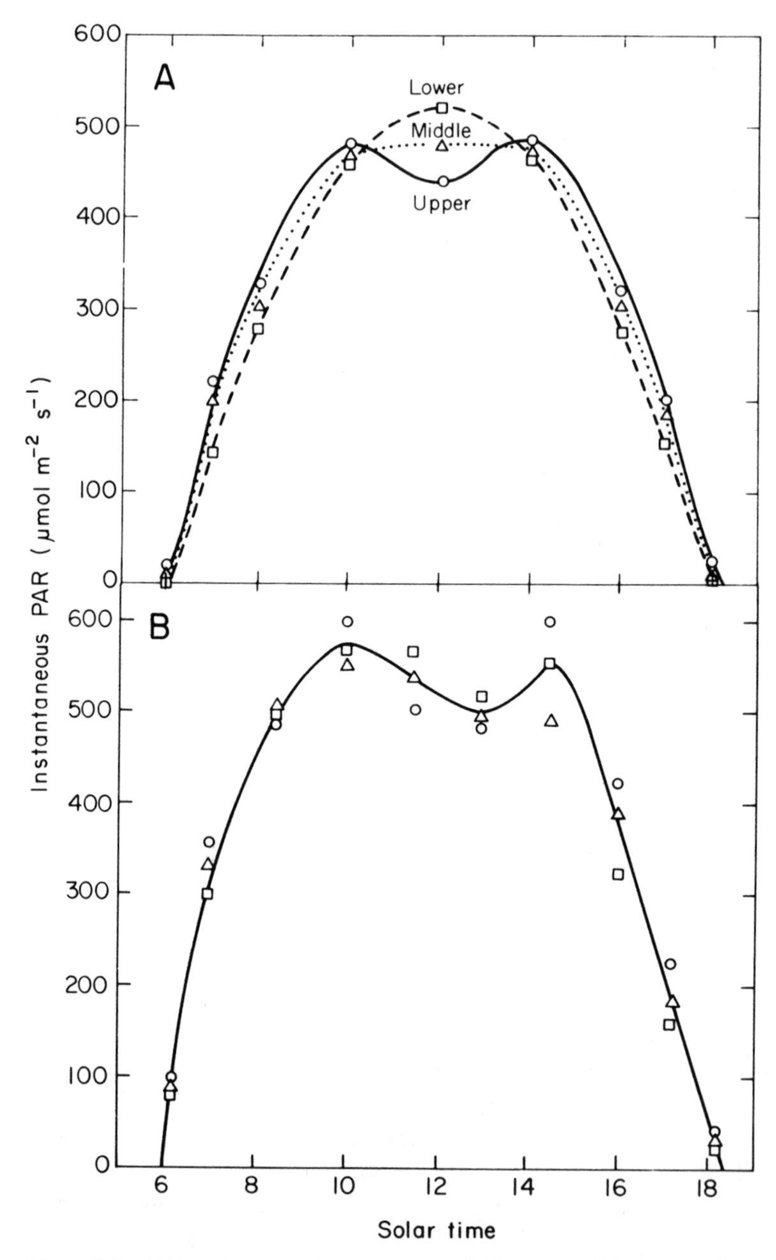

Figure 3.8. PAR values at various canopy levels for agaves: (A) averages for the entire adaxial and abaxial surfaces on a clear day determined by simulation for an *Agave deserti* with 60 leaves for an equinox at 34° N, where "Upper" refers to the centermost, youngest one-third of the leaves and "Lower" refers to the most prostrate, oldest one-third of the leaves; (B) averages of measurements one-sixth, one-half, and five-sixths of the distance from leaf base to leaf

Although increasing the stem surface area by increasing the number or depth of ribs causes the average PAR per unit area to decrease relative to that for a smooth cylindrical surface of the same diameter, the area near PAR saturation can be decreased, allowing more effective utilization of incident PAR (Geller and Nobel 1984). Also, the fraction of PAR absorbed is increased because of reflections from neighboring ribs. Consequently, increasing the rib number from 5 to 34 was predicted to increase net CO_2 uptake per unit height by an average of 24% for unshaded stems on an equinox at 34° N at rib depths typical for *F. acanthodes* and *Carnegiea gigantea.* However, such an increase would decrease net CO_2 uptake if the ribs were very deep (Geller and Nobel 1984).

PAR distribution in agave canopies

The computer model developed for vertical surfaces of cacti was modified to predict the PAR incident on surfaces of any orientation. It was then used to consider the PAR available to an agave with a basal rosette of leaves (Figure 3.1A). The geometry of an agave is actually fairly difficult to describe analytically, because the leaves unfold sequentially from a central spike and assume progressively more prostrate orientations with age. The plant chosen was a mature one with 60 leaves, and the relatively complicated pattern of shadows cast by one leaf onto another was determined for hourly positions of the sun.

The model predicted that the values for PAR averaged over both leaf surfaces would be quite similar for leaves radiating toward the upper, middle, and lower parts of the canopy (Figure 3.8A) (Nobel 1984). In the early morning and late afternoon, the more erect youngest leaves near the central spike received the most PAR, whereas near noon the upper surfaces of the oldest leaves in the lower part of the rosette received the most PAR. When totaled over a whole day at an equinox, the PAR on leaves of an unshaded *A. deserti* at 34° N ranged only from 14.7 mol m^{-2} day^{-1} in the lower part of the canopy to 15.2 mol m^{-2} day^{-1} in the middle to 15.5 mol m^{-2} day^{-1} in the upper (central) part of the canopy (Figure 3.8A). Field measurements on *A. lecheguilla* at 25° N near an equinox also indicated that a fairly similar average PAR occurred on leaves radiating toward the three canopy positions considered (Figure 3.8B), again indicat-

Caption to Figure 3.8 (*cont.*)
tip along the adaxial and abaxial surfaces of leaves facing in each of the four cardinal directions in the upper (circles), middle (triangles), and lower (squares) parts of the canopy for an *Agave lecheguilla* with 38 leaves, as determined on 22 March near 25° N.

ing the advantages of a rosette design for distributing PAR fairly uniformly over agave leaves of different ages. Such changes in leaf orientation from near vertical at the top of the canopy to near horizontal at the bottom also lead to a relatively even distribution of PAR for sugar beets (*Beta vulgaris*) (Hodáňová 1979) and certain other species (e.g., Nobel 1983b). For *A. deserti,* the relatively even distribution of PAR over the leaves tends to avoid PAR saturation of nocturnal acid accumulation (Figure 3.4B) and thus should lead to greater productivity than when some leaves have very high total daily PAR levels and others very low.

Once the model had been developed, it was used to simulate other conditions. For instance, simulated tilting of the plant by 55° such that the vertical axis pointed to the sun at solar noon on a winter day increased the PAR incident on the upper surfaces of leaves and lowered it slightly for the lower surfaces, but did not affect the total nocturnal increase in acidity by the whole plant (Woodhouse et al. 1980). Simulated removal of alternate leaves increased the PAR per unit leaf area for the remaining leaves, but it reduced the nocturnal acid accumulation of the whole plant by 31%. Although tilting the ground toward the south while maintaining the plant axis perpendicular to the ground surface had little effect on intercepted PAR, and tilting to the east or west had only moderate effects, tilting the ground toward the north had major effects on the availability of PAR. At 34° N, the average PAR incident on the leaves of *A. deserti* was reduced 50% for a slope of 68° in late summer (22 August) and for a slope of 38° in winter (19 January). Field observations of the frequencies of *A. deserti* on slopes facing in various directions at elevations at which temperature was not limiting showed that the number of plants per square meter was 26% less on north-facing slopes than on south-facing slopes in the slope-angle class of 31° to 45° and 62% less in the slope-angle class of 46° to 60° (Woodhouse et al. 1980). This decrease apparently resulted from the lower PAR for the north-facing slopes, similar to the PAR limitation for *Carnegiea gigantea* on north-facing slopes (Niering et al. 1963).

Summary and conclusions

The often massive leaves or stems of CAM plants present interesting three-dimensional photosynthetic organs, although the PAR intercepted is generally much less than for unshaded horizontal surfaces. For instance, the PAR incident on the vertical surfaces of unshaded cacti or on both sides of the leaves of unshaded agaves for various latitudes and seasons averages only about 20 mol m^{-2} day^{-1} on clear days. This is similar to the total daily PAR leading to 90% saturation of nocturnal CO_2 uptake or acid accumulation for these CAM plants. Hence, the productivity of these desert succulents will be lowered when the incident PAR is reduced by weather conditions, topographical features, or shading by other vegetation. Also, spines decrease the PAR reaching the stems of cacti and thereby lower the net CO_2 uptake, and ribs also influence PAR distribution over the stem surface.

Consideration of the PAR limitation helps to interpret the orientation of terminal cladodes of platyopuntias, which tend to face in the direction maximizing PAR interception by their surfaces. The favored direction to face is generally east–west, except at sites more than 27° from the equator, where cladodes facing north–south will intercept more PAR near the winter solstice. The variations in heights of certain cacti, in concert with the height variation and hence the shading by the surrounding vegetation, can also be interpreted as a morphological response to PAR limitations. The variation in orientation of the leaves of agaves, from almost erect near the central spike to almost horizontal near the ground, helps ensure a relatively even distribution of PAR throughout the canopy, which in turn tends to lead to a higher net productivity. Thus, the form and orientation of the opaque photosynthetic organs of agaves and cacti can be interpreted with respect to the capture of PAR, which is on the verge of limiting the productivity of such species, even in the high-radiation environments of deserts.

Acknowledgments

The author gratefully acknowledges useful comments on this manuscript by Gary N. Geller and financial support by National Science Foundation grant DEB 81-00829 and Department of Energy contract DE-AM03-76-SF00012.

References

Acevedo, E., I. Badilla, and P. S. Nobel. 1983. Water relations, diurnal acidity changes, and productivity of a cultivated cactus, *Opuntia ficus-indica*. Plant Physiol. 72:775–780.

Anderson, M. C. 1964. Studies of the woodland light climate. I. The photographic computation of light conditions. J. Ecol. 52:27–41.

Fisher, J. B., and H. Honda. 1979. Branch geometry and effective leaf area: a study of *Terminalia*-branching pattern. 1. Theoretical trees. Am. J. Bot. 66:633–644.

Fitzpatrick, E. A. 1979. Radiation. Pp. 347–371 *in* D. W. Goodall and R. A. Perry (eds.), Arid-land ecosystems: structure, functioning, and management, vol. 1. Cambridge University Press.

Geller, G. N., and P. S. Nobel. 1984. Cactus ribs: influence on PAR interception and CO_2 uptake. Photosynthetica 18:482–494.

Hodáňová, D. 1979. Sugar beet canopy photosynthesis as limited by leaf age and irradiance. Estimation by models. Photosynthetica 13:376–385.

Katsnelson, J. 1968/69. Rainfall in Israel as a basic factor in the water budget of the country. Tel Aviv University Press.

Kluge, M., and I. P. Ting. 1978. Crassulacean acid metabolism: analysis of ecological adaptation. Ecological studies series, vol. 30. Springer-Verlag, Berlin.

Martin, C. E., N. L. Christensen, and B. R. Strain. 1981. Seasonal patterns of growth, tissue acid fluctuations, and $^{14}CO_2$ uptake in the Crassulacean acid metabolism epiphyte *Tillandsia usneoides* L. (Spanish moss). Oecologia 49:322–328.

Monsi, M., and T. Saeki. 1953. Über den Lichtfaktor in den Pflanzengesellschaften und seine Bedeutung für die Stoffproduktion. Jpn. J. Bot. 14:22–52.

Niering, W. A., R. H. Whittaker, and C. H. Lowe. 1963. The saguaro: a population in relation to environment. Science 142:15–23.

Nobel, P. S. 1977. Water relations and photosynthesis of a barrel cactus, *Ferocactus acanthodes,* in the Colorado Desert. Oecologia 27:117–133.

– 1980. Interception of photosynthetically active radiation by cacti of different morphology. Oecologia 45:160–166.

– 1981. Influences of photosynthetically active radiation on cladode orientation, stem tilting, and height of cacti. Ecology 62:982–990.

– 1982a. Orientation of terminal cladodes of platyopuntias. Bot. Gaz. 143:219–224.

– 1982b. Orientation, PAR interception, and nocturnal acidity increases for terminal cladodes of a widely cultivated cactus, *Opuntia ficus-indica.* Am. J. Bot. 69:1462–1469.

– 1982c. Interaction between morphology, PAR interception, and nocturnal acid accumulation in cacti. Pp. 260–277 *in* I. P. Ting and M. Gibbs (eds.), Crassulacean acid metabolism. American Society of Plant Physiologists, Baltimore.

– 1983a. Spine influences on PAR interception, stem temperature, and nocturnal acid accumulation by cacti. Plant Cell Env. 6:153–159.

– 1983b. Biophysical plant physiology and ecology. W. H. Freeman, San Francisco.

– 1984. PAR and temperature influences on CO_2 uptake by desert CAM plants. Pp. 193–200 *in* C. Sybesma (ed.), Proceedings of the 6th International Congress on Photosynthesis. Martinus Nijhoff/Dr. W. Junk, Brussels.

Nobel, P. S., and T. L. Hartsock. 1983. Relationships between photosynthetically active radiation, nocturnal acid accumulation, and CO_2 uptake for a Crassulacean acid metabolism plant, *Opuntia ficus-indica.* Plant Physiol. 71:71–75.

Osmond, C. B. 1978. Crassulacean acid metabolism: a curiosity in context. Ann. Rev. Plant Physiol. 29:379–414.

Robberecht, R., and P. S. Nobel. 1983. A Fibonacci sequence in rib number for a barrel cactus. Ann. Bot. 51:153–155.

Rodriquez, S. B., F. B. Perez, and D. D. Montenegro. 1976. Efficiencia fotosintética del nopal (*Opuntia* spp.) en relación con la orientación de sus cladodios. Agrociencia 24:67–77.

Ross, J. 1975. Radiative transfer in plant communities. Pp. 13–55 *in* J. L. Monteith (ed.), Vegetation and the atmosphere. vol. 1. Principles. Academic Press, London.

Szarek, S. R., H. B. Johnson, and I. P. Ting. 1973. Drought adaptation in *Opuntia basilaris:* significance of recycling carbon through Crassulacean acid metabolism. Plant Physiol. 52:539–541.

Szarek, S. R., and I. P. Ting. 1974. Seasonal patterns of acid metabolism and gas exchange in *Opuntia basilaris.* Plant Physiol. 54:76–81.

Williams, J. G. 1976. Small variation in the photosynthetically active fraction of solar radiation on clear days. Arch. Meteorol. Geophys. Bioclimatol., Series B 24:209–217.

Woodhouse, R. M., J. G. Williams, and P. S. Nobel. 1980. Leaf orientation, radiation interception, and nocturnal acidity increases by the CAM plant *Agave deserti* (Agavaceae). Am. J. Bot. 67:1179–1185.

4 Unusual strategies of light absorption in rain-forest herbs

DAVID W. LEE

Introduction

By temperate standards, the leaves of many plants growing in the profound shade of humid tropical forests appear bizarre. Several species feature leaves with (1) splotches or stripes of color other than green, (2) red or purple undersurfaces, (3) a velvety or satiny surface sheen, and even, in a few cases, (4) a vivid blue iridescence. These and other characteristics of tropical understory herbs have been described at length by Richards (1952) and Burtt (1978) and more recently by Givnish (1984). However, scientific interest in the leaf structure of rain-forest herbs stretches well back into the nineteenth century.

Empirical trends in such foliar structures were first established by European botanists working at botanical gardens in the colonial tropics, especially at Peradeniya in what is now Sri Lanka and at Bogor on Java in Indonesia. These early observations became part of a body of research known as physiological plant anatomy, well summarized and popularized by Haberlandt (1912). The aim of this field of research was to understand structure in the context of what was known about plant physiological function. Despite this early research, which was responsible for identifying many of the patterns and environmental correlates in the leaf structure of tropical plants of interest today, work on such leaf traits lost its popularity early in this century. The decline in interest was partly due to the primitive experimental methods available at the time, which prevented rigorous tests of the ever-multiplying hypotheses on the adaptive value of different traits, partly because of a lack of statistical design and approach, and partly because of a temperate bias in the choice of research subjects by twentieth-century botanists.

I first became interested in rain-forest understory plants while working as an experimental taxonomist at the University of Malaya. Interest in plants of the rain-forest understory changed the direction of my career and is a source of fascination a decade later. In this chapter I shall review and analyze the various hypotheses proposed to explain the selective advantage of the last three distinctive understory leaf features mentioned earlier: reddish leaf undersurfaces, velvety surface texture, and blue iridescence. The fourth trait, leaf variegation, will not be discussed in detail here. Each of these four traits is strongly associated with extreme shade in humid tropical forests, and to a lesser degree in temperate forests. I shall examine the structural features of plants with these three characteristics that confer on advantage in harvesting light in poorly lit environments and discuss the disadvantages of such adaptations in more brightly lit habitats. Although variegation is also strongly associated with shade, it actually reduces a leaf's capacity to absorb light through the reduction of chloroplast density, decreased absorption from the silvery reflectance caused by intercellular spaces above chloroplast-containing layers, or through optical masking by anthocyanin pigments. Thus, selection pressures for variegation appear not directly related to those for the other three traits. I shall first review background material on the nature of understory light climates, and then the corresponding physiological and structural adaptations in understory plants.

Plant adapatations and understory light climates

The existence of leaves with red undersurfaces, velvety surfaces, and blue iridescence in very distantly related taxonomic groups, strongly associated with a very specific environment, argues strongly for their selective advantage. Lack of radiant energy is the most obvious selection pressure leading to their present distribution. However, the light climates within any forest, let alone those in the tropics, have been poorly characterized, partly for theoretical reasons (Anderson 1970; Idso and de Wit 1970; Reifsnyder et al. 1979; Hutchison et al. 1980) and for lack of field studies. Data on light climates in humid tropical forests are beginning to accumulate (Bazzaz and Pickett 1980; Chazdon and Fetcher 1984a). The spectral characteristics of individual leaves and the heterogeneity of the forest canopy determine the quantity, diurnal distribution, and spectral quality of light reaching the forest floor. Most terrestrial plant leaves absorb strongly in the visible wavelengths and absorb little at wavelengths above 700 nm (Gates et al. 1965; Woolley 1971; Gausman and Allen 1973). An absorption spectrum for the leaves of a tropical tree of Central America, *Bursera simaruba* (L.) Sarg., is shown in Figure 4.1. Note the strong absorptance in the visible range (somewhat less around 550 nm) and the abrupt decrease above 700 nm. Absorptance is calculated from

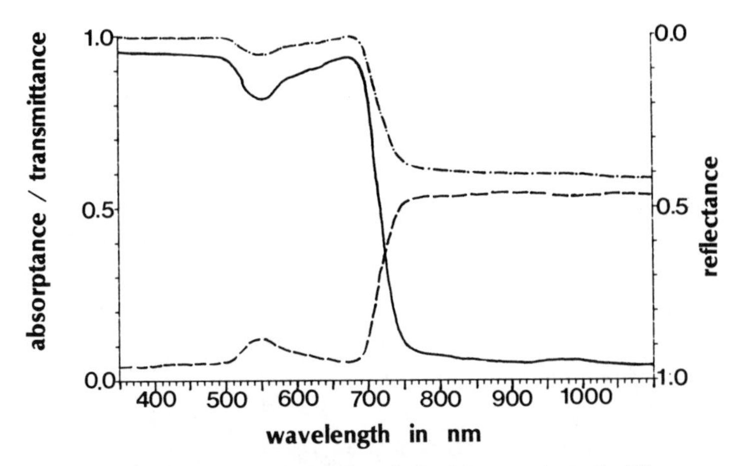

Figure 4.1. Leaf optical analysis of *Bursera simaruba*. The curves represent means for five leaves measured in an integrating sphere attached to a Li-Cor 1800 spectroradiometer: absorptance (solid line), diffuse transmittance (dash line), and diffuse reflectance (dot-dash line).

direct measurements of diffuse reflectance and diffuse transmittance in an integrating sphere (1 − reflectance − transmittance = absorptance). *B. simaruba* leaves absorb 90% of the photosynthetic photon flux density (PPFD, or quanta 400–700 nm) of full sunlight. Leaf optical characteristics are determined by the following factors. All epidermal surfaces reflect some light because of differences in refractive indices of air and the epidermal wall, and surfaces with wax deposits or pubescence reflect more (Ehleringer 1981). Chloroplast pigments absorb strongly in the visible wavelengths, and flavonoid pigments may absorb at visible wavelengths and always absorb in the ultraviolet range. The abrupt decrease in absorption in the far-red range is due both to pigment composition and to wavelength-dependent light scattering in the leaf mesophyll (Gausman 1974, 1977). Above 1,100 nm, leaves again increase their absorption, and also emittance, of heat.

The filtering effect of foliage determines the spectral quality of the light reaching the forest floor (Björkman and Ludlow 1972; Stoutjestijk 1972; Tasker and Smith 1977; Lee 1985). For a typical spectrum in the forest understory at La Selva, in Costa Rica (Figure 4.2) (Lee 1985), we see that the spectral distribution of light is complementary to the optical properties of leaves. In this shade environment, the PPFD (5.4 μmol m^{-2} s^{-1}) is only 0.4% of the PPFD reaching the canopy above (1,280 μmol m^{-2} s^{-1}). The spectral quality of the shade light is also revealed by the quantum ratio of 660 to 730 nm (R : FR, in the sense of Smith 1982). Phytochrome, which

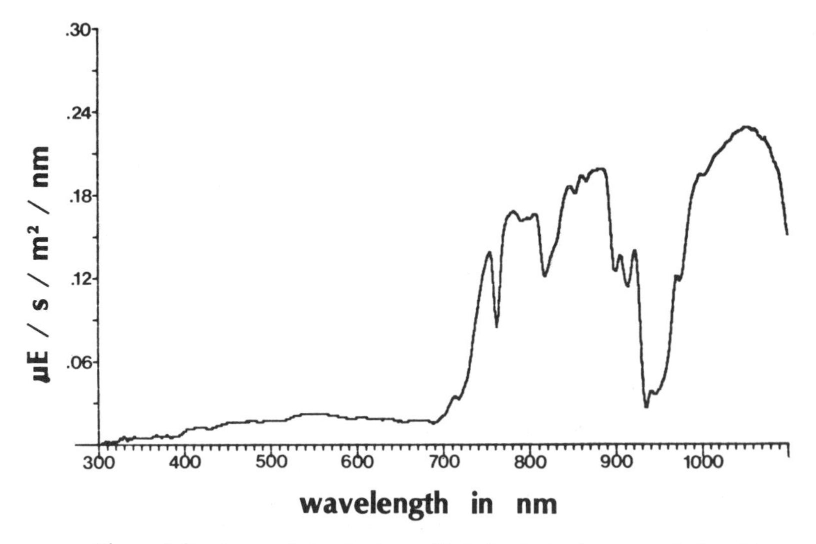

Figure 4.2. Spectral distribution of light in the understory of a humid tropical-forest site at La Selva, Costa Rica, with the sun near its zenith. PPFD is 5.4 μmol m^{-2} s^{-1}, and R : FR = 0.31. Percentage of PPFD compared with sunlight was 0.4%.

will be discussed later for its role in the "perception" of shade light by plants, absorbs maximally at 660 and 730 nm. For the shade, the R : FR is 0.31, compared with 1.27 for the sunlight above. It would certainly be incorrect to say that this is the light climate of a particular patch of forest. It varies as different flecks of light of various sizes (with different PPFDs and R : FRs) move across the floor during the day. For this particular site, the greatest proportion of total daily quanta available for photosynthesis comes from light flecks (Chazdon and Fetcher 1984b). However, the leaf structures discussed in this chapter are primarily characteristic of plants growing in the shadiest sites. These plants grow where the canopy is most closed, on the trunks of trees, or in gullies. For instance, the mean percentage of full sunlight for the site of an iridescent blue-green fern, *Trichomanes elegans* Rich., at La Selva, Costa Rica, was 0.25%, with an R : FR of 0.153 (Lee 1985). Judging from these data, light flecks are less important for extreme-shade-adapted rain-forest plants. Their light environments typically are very diffuse in origin and are spectrally altered to remove almost all of the radiation in the range of 400–700 nm.

These light environments should function as a strong selective pressure for plants growing in them in the following ways. Extreme-shade plants have been generally characterized as minimizing the capital costs of construction per energy-harvesting unit and minimizing energy-transducing catalysts per unit of light-harvesting pigment (Boardman 1977; Björkman 1981; Richardson et al. 1983; Givnish 1984). Thus, extreme-shade-adapted species tend to produce leaves that are lighter and thinner than sun-adapted ones, and the proportion of plant biomass produced as leaves may be greater. The ratio of soluble protein (much of it as ribulose bisphosphate carboxylase, or Rubisco) per amount of chlorophyll is lower in shade-adapted species, and lower chlorophyll a/b ratios provide more effective absorbances across the visible spectrum. The ultrastructural correlate of the latter is fewer and larger chloroplasts with strikingly thick grana stacks. Despite the structural differences between sun and shade plants and the greater proportion of photosystem II in shade plants, there seem to be no profound biochemical differences between them, and their quantum efficiencies are not different (Björkman 1981). Although the preponderant number of studies on shade plants have been carried out in temperate regions, there have been several detailed analyses of the structure and performance of plants in the rain-forest understory (Björkman et al. 1972; Goodchild et al. 1972; Robichaux and Pearcy 1980; Pearcy and Calkin 1983).

One badly neglected area of investigation concerns the photosynthetic spectral responses of extreme-shade plants (Mooney et al. 1980). The spectral distribution of radiation available in rain-forest understories (cf. Figure 4.2) indicates an abrupt increase in the quantum flux density near the conventionally accepted limits of photosynthesis, at 700 nm. Given the low densities of visible wavelengths in the rain-forest understory at all times of the year and the rapid flux increases above 700 nm, Björkman (1973) speculated that extreme-shade plants might possess adaptations to allow absorptance and use of these quanta. To date, confirmation of Björkman's hypothesis is lacking for terrestrial plants. However, there is a well-documented exception to the conventional wavelength limits of photosynthesis in an alga living within a large coral (Halldal 1968). A species of *Ostreobium* (Siphonales, Chlorophyta) lives within the coral and beneath incrustations of dinoflagellates that absorb visible wavelengths of light (as foliage does); the upper limit of its action spectrum is 750 nm.

In his landmark studies on the action spectra of crop plants, McCree (1972) concluded that the upper wavelength limits of photosynthesis were partially determined by the optical characteristics of leaves; in the spectrum of relative quantum efficiencies of all taxa, photosynthesis was evident beyond 700 nm (Figure 4.3). McCree (1981) predicted that leaves in very shady areas within a crown might use a greater percentage of quanta above 700 nm, perhaps by as much as 7%. To obtain a rough estimate of how important absorption of such long-wave radiation might be in a shady

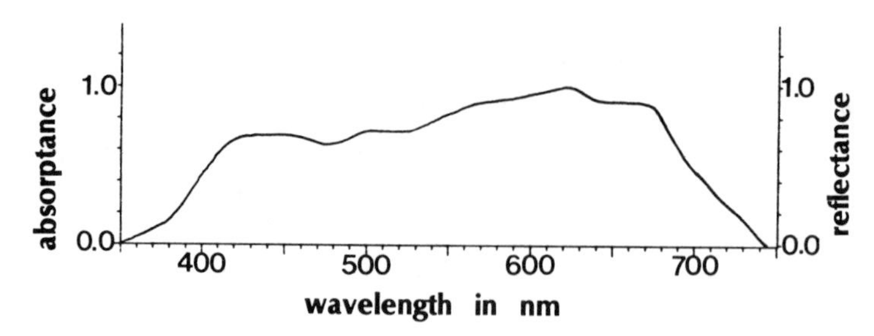

Figure 4.3. Spectrum of the relative quantum efficiencies of 10 crop plants grown under field conditions (McCree 1972). All values are in reference to the maximum, at 620 nm. The line beyond 725 nm is extrapolated to zero.

environment, consider *Bursera simaruba* (Figure 4.1). If we assume that *Bursera* has roughly the same spectrum of relative quantum efficiency as the crop plants surveyed by McCree (1972), the importance of quanta at 700–750 nm relative to those at 400–700 nm can be calculated by forming the wavelength-specific product of leaf absorptance (Figure 4.1), understory radiation (Figure 4.2), and relative quantum yield (Figure 4.3). For *B. simaruba,* this quantum integration predicts a relative contribution of 4.0% by the 700–750-nm spectrum, twice the percentage under full sunlight. Whether or not extreme-shade plants use a greater percentage of light at 700–750 nm needs to be investigated. However, a comparison of the leaf optical properties of sun and shade plants from tropical forests (Lee and Graham 1986) did not reveal significant differences between the samples, although some individual species of extreme-shade plants absorbed more quanta between 400 and 700 nm. The mean absorption for both samples was approximately 90%, and the percentage at 700–750 nm (using the calculations described earlier) was 4.5% for both. The leaves of these species absorb more PPFD than any in the large sample of desert plants examined by Ehleringer (1981).

Thus, the leaf structures discussed in the following sections may increase photosynthetic efficiencies in extreme shade in the following ways. More efficient absorptance of quanta between 400 and 700 nm could be achieved by increasing the absorptive path length of receptor pigments by more efficiently backscattering light that passes through the chloroplast layer (red leaf undersides). More efficient use of low flux densities might be achieved by focusing light onto chloroplasts at light levels well above light compensation points (epidermal lenses). Physical structures may also increase light transmittance to chloroplasts through interference (blue iridescence). Both interference and backscattering properties might also specifically improve absorption at wavelengths above 700 nm. Still an-

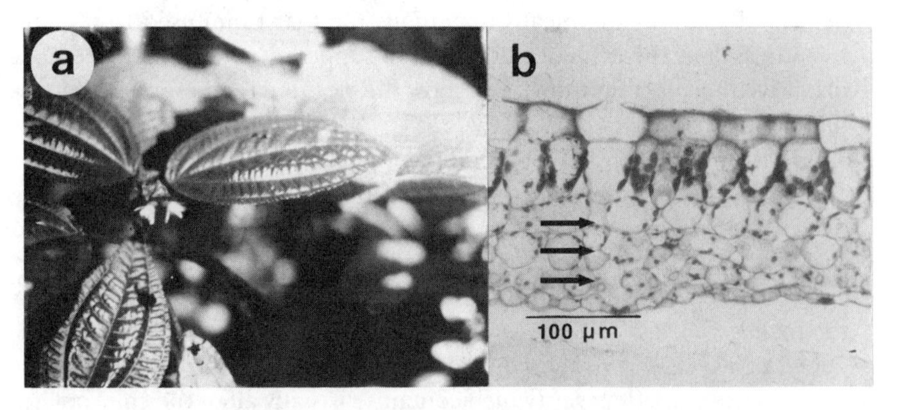

Figure 4.4. *Triolena hirsuta:* (a) plant habit; (b) red-undersurfaced leaf, transverse section, layer containing anthocyanins indicated by arrows.

other implication of the far-red-enriched light of extreme shade is its potential effect, mediated by phytochrome, on the development of structures that enhance absorption. I shall discuss this topic in the next to last section.

Red undersurfaces

Leaves with anthocyanic coloration on their undersurfaces are very common among humid tropical-forest understory plants. Their occurrence was noted by Richards (1952), and they were also studied by the physiological anatomists around the turn of the century (Kerner and Oliver 1895; Haberlandt 1912). Attempts to understand the selective advantage of this feature have been clouded by the presence of undersurface coloration in other plants, such as sun-exposed floating aquatics (Sculthorpe 1967). However, anatomical localization in these sun-adapted groups is quite different from that seen in understory species. In aquatic plants, pigmentation occurs in the lower epidermal cells, whereas in the understory species the pigment is normally in a layer of mesophyll cells immediately beneath the palisade parenchyma (Figure 4.4). There is certainly no reason that these two anatomically distinct features might not have different functions in their different environmental contexts.

The classic explanation for this coloration is that it absorbs radiation and helps to elevate the temperatures of the pigmented leaves. Such an elevated temperature might be expected to increase metabolic rates, transpiration, and nutrient uptake by these plants. Kerner and Oliver (1895) described some of the means (ingenious for that time) used to test the hypothesis and the evidence obtained to support it. Smith (1909) used a primitive thermocouple device to show the elevated temperatures of juvenile pigmented leaves.

A critical analysis of the problem requires plants that are identical, except for the undersurface pigmentation. Lee et al. (1979) studied four taxa, three polymorphic for undersurface pigmentation. Using a fine thermocouple thermometer and measuring the plants in situ, they concluded there were no significant differences in leaf temperatures between the forms.

Lee et al. (1979) proposed that such a dense anthocyanic layer just beneath the palisade layer (which contains virtually all of the chloroplasts in the species studied) could function as a reflective layer, backscattering photosynthetically usable light into the chloroplast layer and increasing the efficiency of light absorptance at wavelengths more critical for photosynthesis in a shady environment. Anthocyanins typically are transparent at wavelengths above 600 nm, with a strong absorbance peak at 525 nm and others below 320 nm. Thus, the predicted chloroplast absorptance of red and far-red wavelengths for photosynthesis would occur at the expense of greater absorbance of wavelengths below 550 nm by the anthocyanins (because these wavelengths might otherwise be absorbed by the chloroplasts). Depending on their in vivo condition, anthocyanins could be incorporated into an efficiently reflective layer at wavelengths where they are transparent (Patton 1973).

The initial test of this hypothesis was to analyze the specular reflectance of the leaf morphs at a fixed angle of 45°, both lower and upper surfaces. As predicted, the red undersurfaces reflected more light above 650 nm as compared with green undersurfaces, and the reflectance of the red-leaf upper surfaces was less. Although less reflectance should contribute to greater absorptance by these leaves, measurements of diffuse transmittance are necessary to estimate differences in total absorptance between different leaf forms. Thus, the research needs to be repeated and extended to other taxa. The following example (Lee and Graham 1986) suggests an approach and the potential significance of such research.

Triolena hirsuta Triana is a small understory herb growing in Central American rain forests. At La Selva, Costa Rica, it grows as two distinct genotypes, one with a green appearance and the other a dark reddish brown due to a layer of anthocyanins in the mesophyll beneath the chloroplast-containing palisade cells (Figure 4.4). Preliminary results (Trapp et al. 1982) suggest that the two morphs have preferences for different light levels, but these results need to be confirmed by more detailed research. According to the hypothesis, the red form should absorb more photosyn-

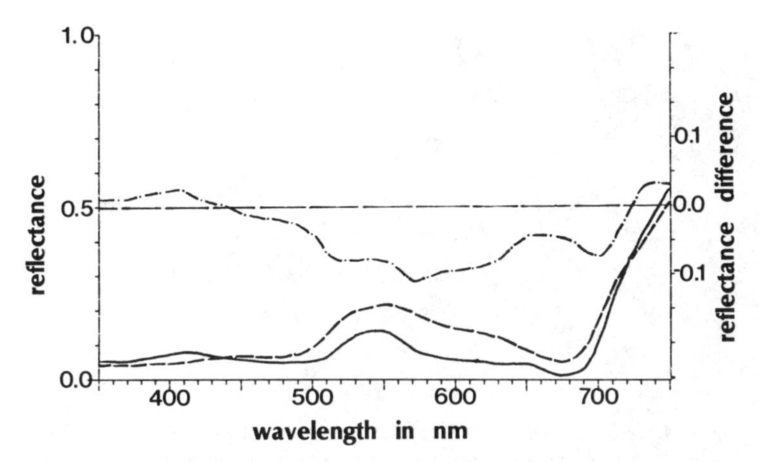

Figure 4.5. Optical comparisons of red- and green-undersurfaced leaves of *Triolena hirsuta*, each curve being the mean for five leaves. Solid line indicates absorptance of red-undersurfaced leaves; dash line indicates absorptance of green-undersurfaced leaves; dot-dash line denotes the difference in absorptance between them.

thetically active quanta than the green form. The red form's leaf undersurface reflected more light above 600 nm, just as the taxa analyzed by Lee et al. (1979). Moreover, the optical analysis (Lee and Graham 1986) (Figure 4.5) showed that the red form absorbed 90% of the PPFD, compared with 82% for the green form. In comparisons of effective absorptance of quanta 700–750/400–700 nm, the red morph used 4.4%, compared with 4.0% for the green form. When the amount used at 700–750 nm was compared with that at 400–700 nm for the green form, the percentage rose to 4.8%. Although absorbance of light by anthocyanins in the red form, unavailable for photosynthesis, exaggerates this difference, greater absorptance at wavelengths where anthocyanins do not absorb (87% versus 82% at 620 nm) supports the hypothesis. It is important that such comparisons be made for morphs identical except for the anthocyanin pigments. *Psychotria suerrensis* J.D. Sm. (Rubiaceae) is a small understory shrub of the neotropics, relatively common at La Selva. At maturity it is a 1–2-m shrub growing in forest gaps, but it is most conspicuous in its juvenile form – with orange-nerved, satiny, purple-undersurfaced leaves growing in deep shade (Figure 4.6). The juvenile leaves absorbed more light only at around 550 nm, more likely the result of absorbance of anthocyanins. Both forms absorbed more PPFD than *Bursera simaruba*, but neither absorbed a greater percentage of quanta at 700–750 nm (Lee and Graham 1986). The problem in comparing the two forms of *P. suerrensis* is that they are different in several respects, including the amount of

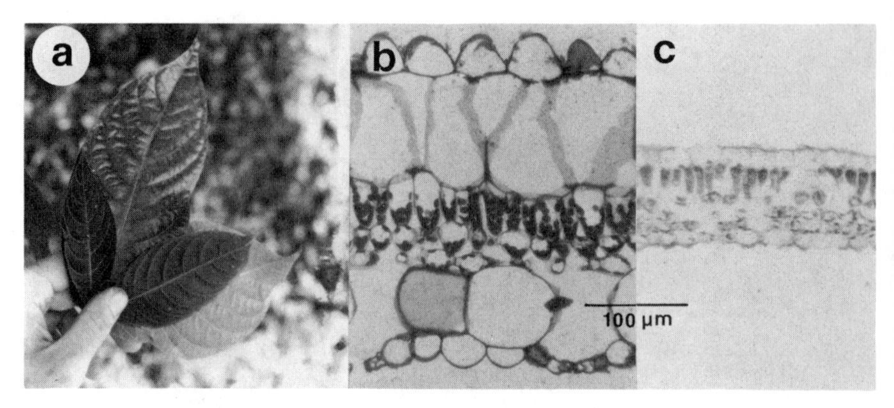

Figure 4.6. (a) Leaves of juvenile and adult plants of *Psychotria suerrensis.* The smaller leaves (length 7 cm) resting directly against the thumb are the juvenile form. (b) Transverse section of juvenile leaf. (c) Transverse section of leaf of mature plant.

dry weight per surface area (2.2 mg cm^{-2} for juvenile and 5.1 mg cm^{-2} for adult leaves), variegation along nerves, and epidermal surface characteristics (Figure 4.6). More optical comparisons and temperature measurements among shade herbs polymorphic for undersurface color would be helpful in testing the hypothesis, but other problems remain. For instance, what is the ultrastructural basis of anthocyanin accumulation? Most probably the anthocyanin pigments occur in the recently discovered "anthocyanoplast" organelles released into the central vacuole (Small and Pecket 1982). What are the optical properties of these structures in vivo? Production of anthocyanin layers in these leaves must occur at considerable metabolic costs. What are these costs? How do the optical properties of these leaves affect photosynthetic capacities, particularly spectral responses?

Epidermal lenses

Another hypothesis raised by the physiological anatomists is that the raised epidermal surfaces characteristic of many understory herbs (Figure 4.7) refract light and thus increase leaf absorptance for greater photosynthetic efficiency. Raised epidermal cells are manifest visually as a satiny or velvety sheen. Such plants are very common in extreme shade, as exemplified by the juvenile leaves of climbing aroids. This hypothesis was supported by the classic experiment of cutting sections parallel to the leaf surface, placing them under a microscope, and demonstrating the focusing of diffuse light into bright spots (Haberlandt 1912) (Figure 4.8). De-

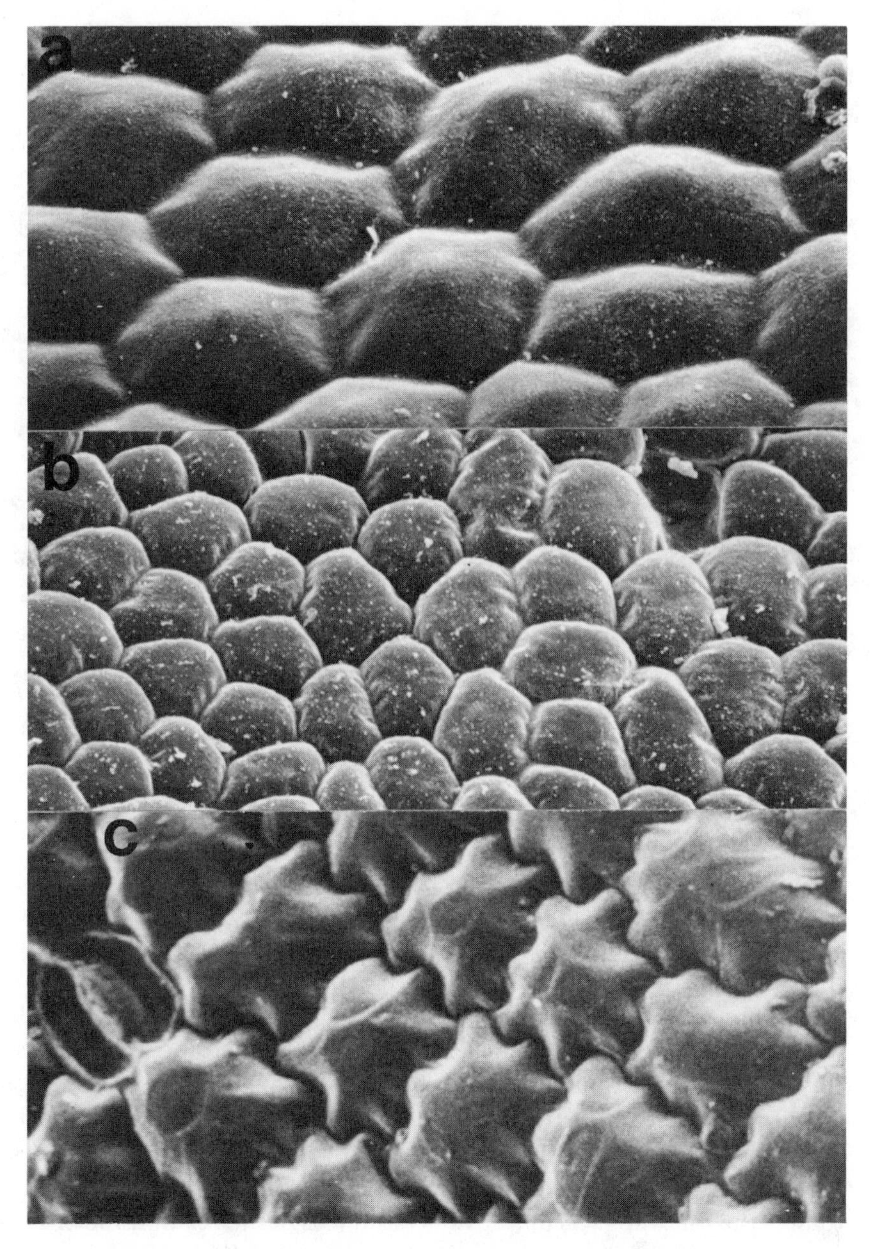

Figure 4.7. Scanning electron-micrographs of raised epidermal surfaces of extreme-shade plants. (a) *Scindapsus pictus*, ×300. (b) *Anthurium warocqueanum*, ×300. (c) *Selaginella willdenowii*, ×800.

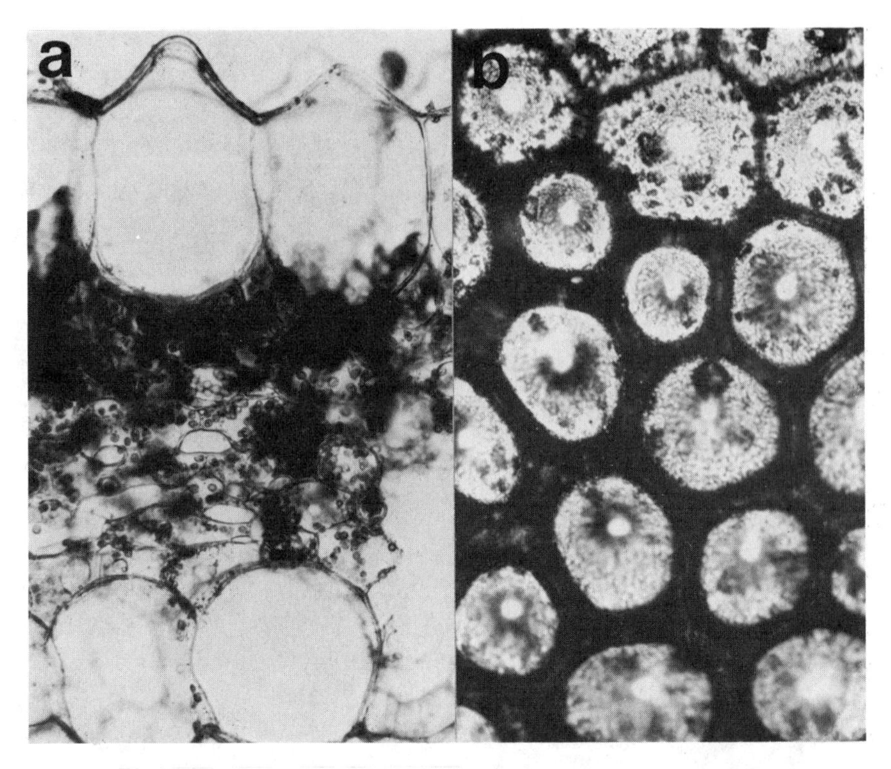

Figure 4.8. Light-microscope photographs of *Scindapsus pictus*. (a) Transverse leaf section showing the somewhat cone-shaped outer epidermal walls, ×300. (b) Section parallel to the leaf surface showing the refraction of diffuse light into bright spots, ×300.

spite this early research, the subject has been badly neglected; the studies have consisted mainly of taxonomic surveys of epidermal characteristics (Atwood and Williams 1979) and analyses of their relationship to chloroplast movements (Haupt 1982). Haberlandt (1912) suggested that the convexly curved upper leaf surfaces could direct light toward chloroplasts in very shady environments, but more than such a simple explanation is required. Despite the refraction of light, a fixed number of photosynthetic quanta arrive at the leaf surface, and the focusing of light within the leaf does not change this fact. A more rigorous testing of the hypothesis requires the following: (1) objective field surveys of the relationship of such plants to the light climates in which they row, (2) a physical analysis of the optics of raised epidermal cells, and (3) an anatomical analysis of chloroplast distribution in relationship to light refraction.

In a study of epidermal cells of eight understory species, Bone et al. (1984) described a computer model predicting the distribution of radiation within a leaf. In this research, cell outlines were traced from photo-

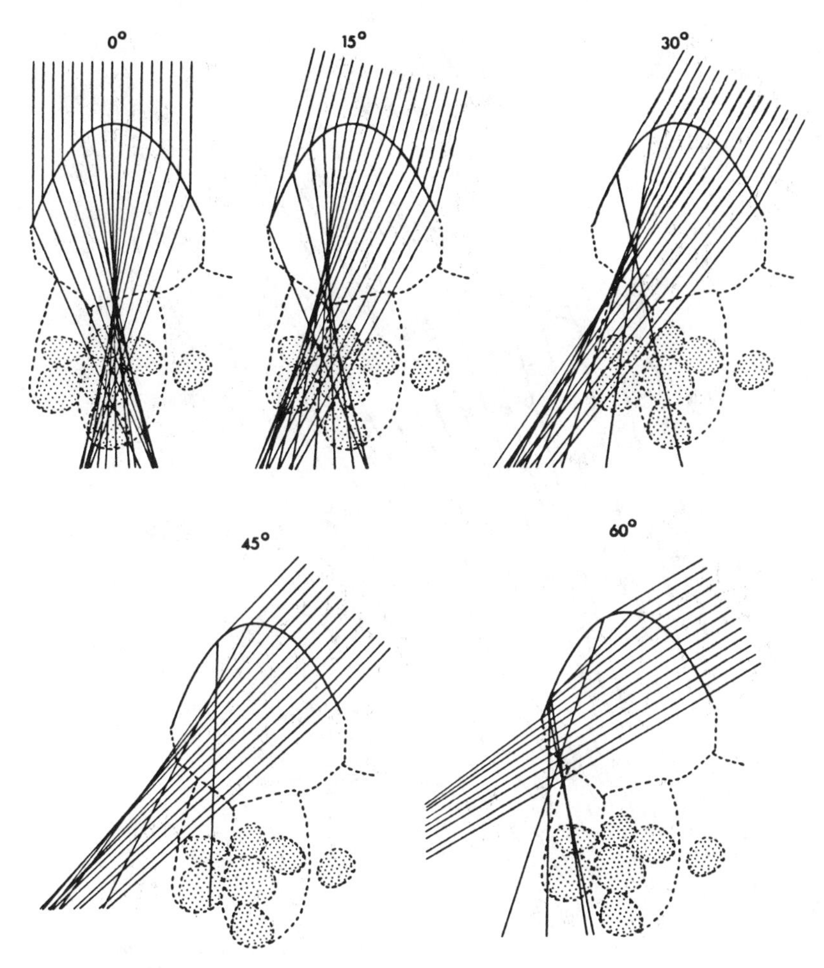

Figure 4.9. Ray-tracing diagrams of an epidermal cell (actually the mean of five measurements) of *Anthurium warocqueanum*, with light at different incidence angles.

graphs of fresh hand sections, and series of $x-y$ coordinates were obtained. A computer program was then used to fit these coordinates to a higher-order polynomial equation, using Snell's law to generate a set of refracted rays corresponding to a number of equally spaced incident rays. The concentration of ray tracings, and subsequently the PPFD inside of the leaf, could be predicted at any level in two dimensions. The solution is simple for a point source of light (such as the sun) hitting the leaf at an angle perpendicular to the surface, and concentration ratios at the upper edges of layers of principal chloroplast concentrations were determined. For *Anthurium warocqueanum* Moore (Figure 4.9), a ratio of 10.11 was calcu-

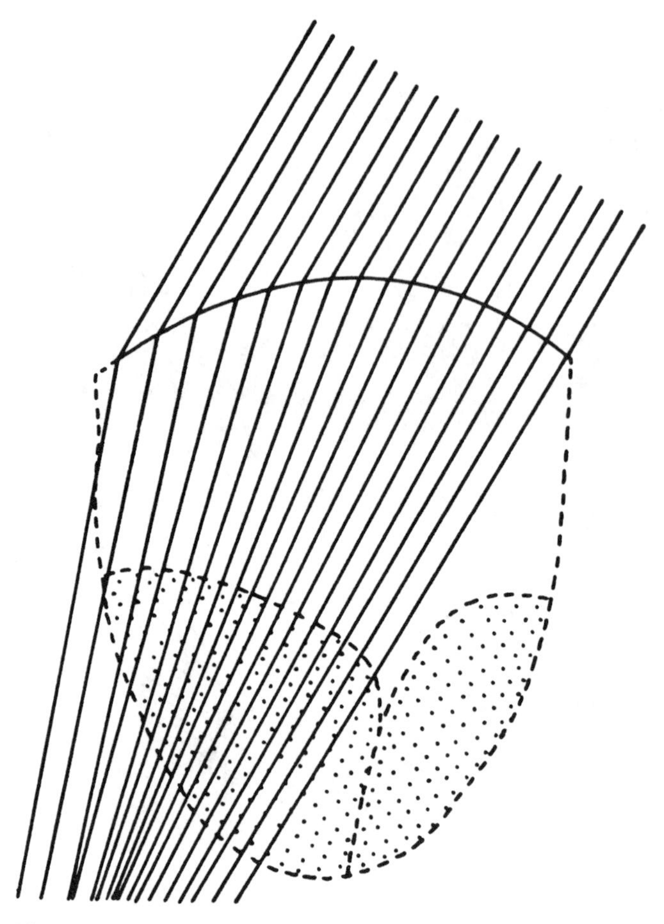

Figure 4.10. Ray-tracing diagram of an epidermal cell of *Selaginella uncinata*, with light arriving at an incidence angle of 30° to the vertical. The cell shape is the mean of five measurements.

lated, with ratios of 1.93 for *Selaginella uncinata* Spr. (Figure 4.10) and 3.49 for *Scindapsus pictus* Hassk. (Figure 4.8). However, only light flecks would provide such a point source of radiation in forest understory, and intense light-focusing effects for an area within the cell could be damaging. Assuming a typical PPFD of 200 μmol m^{-2} s^{-1} (Chazdon and Fetcher 1984b; Lee 1985) would mean a focused intensity of 2,022 μmol m^{-2} s^{-1} for *Anthurium*, a value close to that for full sunlight, which could be potentially damaging to such a shade-adapted plant. Although the focal points of the lenses (and thus the maximum light-concentrating effects) are deeper than the levels of chloroplasts, the prediction of the model is that plants with convexly curved epidermal cells would be at a selective disadvantage in understory environments subject to frequent sun flecks.

Predictions are more difficult to make for a natural-forest light climate because of its variation with different times of the day and varying conditions of the sky. Different incidence angles of light affect the concentration ratios and the level within the leaf at which maximum focusing occurs. Bone et al. (1985) used the model of Norman and Jarvis (1974), employing data for a forest site in west Malaysia (Yoda 1974). In this model, diffuse light from different incidence angles approximating that of natural forest was used to determine the refractive properties of the same epidermal cells. For *Anthurium warocqueanum,* the concentration ratio for the chloroplast level was 1.96, and the focal point was close to the level of chloroplasts. For *Selaginella* the concentration ratio was 1.45, and for *Scindapsus* 1.38. For *Anthurium* this means that the actual light level for some of the chloroplasts will be 10.6 μmol m^{-2} s^{-1}, given the light environment of Figure 4.2. This concentrating effect is significant for extreme-shade-adapted plants in at least two ways. Some species may not distribute chloroplasts uniformly in the palisade, but in discrete packets near the cone of focus, as is the case for *Selaginella* (Figures 4.10, 4.11A, and 4.11B). For such species there is an obvious advantage in distributing the photosynthetic pigments in a more efficient pattern. Jagels (1970) measured light compensation points for this plant at 1.5 – 10 μmol m^{-2} s^{-1}, depending on respiration rates for individual plants. Using the light environment of Figure 4.2, focusing would elevate the intensity of PPFD by a factor of 7.8 at the chloroplast level and could be the difference between net carbon gain and loss for the plant. However, other species (such as *Anthurium* and *Scindapsus,* Figure 4.8) produce a continuous layer of chloroplasts in the palisade cells beneath the epidermis, meaning that some chloroplasts are more intensely illuminated at certain times of the day. The advantage to these plants may be that providing fewer chloroplasts at any one time with higher flux densities increases photosynthetic efficiency by reducing the rates of leakage effects ("slippage reactions") at low light levels, caused by photosynthetic membrane properties or by unstable light-reaction (photosystem I) intermediates (Raven and Beardall 1982; Richardson et al. 1983).

For plants with the epidermal features described here, the chloroplasts typically form a discrete and dense layer at an even distance beneath the leaf surface, a distribution perhaps anatomically analogous to the monolayered tree canopy structures predicted by Horn (1971). In some species this layer is at the bottom of the epidermis, an example being *Selaginella uncinata* (Figures 4.10 and 4.11). There are two chloroplasts per cell, with

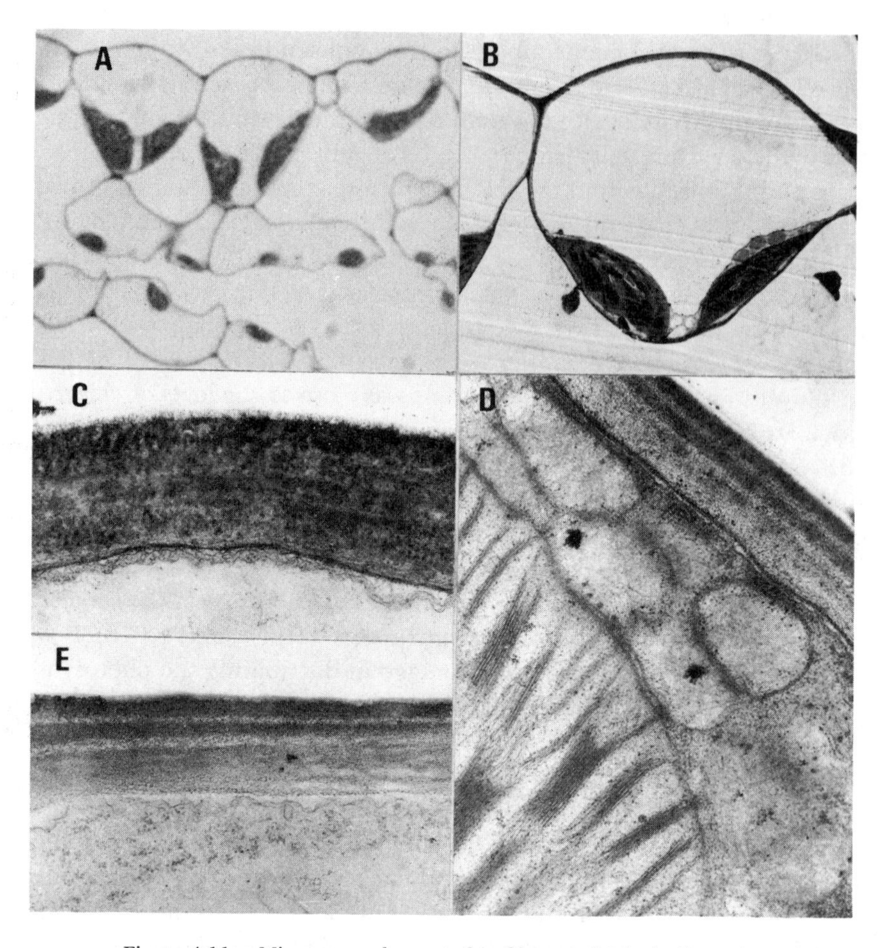

Figure 4.11. Microscope photographs of leaves of *Selaginella uncinata.* (A) Light-microscope photograph of entire leaf transverse section, ×300. (B) Electron-microscope (EM) photograph of epidermal cell with two chloroplasts, ×840. (C) EM photograph of the epidermal wall of a green leaf, ×19,000. (D) EM photograph of a chloroplast, ×6,600. (E) EM photograph of the epidermal wall of a blue leaf with two lamellae approximately 90 nm thick near the upper surface, ×55,000.

ultrastructure typical of extreme-shade plants (Figure 4.11D). It is rare for plants to develop functioning chloroplasts in epidermal cells, but not uncommon in extreme-shade plants. Epidermal chloroplasts are most frequently observed in pteridophytes but also may be found in flowering plants, such as the understory palm *Geonoma cuneata* H. Wendl ex Spruce (R. Chazdon, personal communication).

Another implication for epidermal cell surfaces is that curvatures that efficiently refract light onto specially oriented chloroplasts may not efficiently absorb light striking the leaf at angles oblique to the cell surface, and the opposite. An ideal cell surface shape for capturing light at oblique angles would be a cone, although its focusing characteristics would not be perfect. Leaves displayed vertically, as on tree trunks, might be expected to feature surfaces minimizing reflectance at oblique angles. These leaves face a light environment with approximately one-half the PPFD of a leaf facing vertically (Lee 1985), with even less likelihood of obtaining radiant energy from light flecks. *Scindapsus pictus* leaves are displayed vertically, and although its epidermal cells are not very efficient at focusing light, it does reflect significantly less light at oblique angles than leaves with flat epidermal surfaces (Bone et al. 1984).

Of course, much more work is needed to correlate these epidermal surface characteristics with chloroplast function within the leaf and with the environments in which these plants live.

Blue iridescence

Plants with blue iridescent leaves grow in only the shadiest environments of humid tropical forests around the equator (Richards 1952). Perhaps the most iridescent taxa are found in forests of Southeast Asia (Lee 1977). At La Selva, Costa Rica, there are six iridescent blue plant species, the most common being the fern *Danaea nodosa* (L.) J. Sm. Iridescent blue species are relatively most common among spikemosses (Selaginellaceae) and ferns. Angiosperm families in which iridescence is represented include the Begoniaceae, Melastomataceae, and Cyperaceae (Lee 1977). By "iridescence" I refer to color produced by a physical optical effect – similar to that produced by oil films and certain butterfly scales – not by pigmentation. In none of six blue iridescent species studied has a blue pigment been extractable (Lee 1977). The intense metallic quality of the color is itself suggestive of its physical basis (Fox 1976). In all species studied so far, the physical basis for blue iridescence is thin-film interference. Diffraction can be eliminated as the cause, because the color is relatively constant at different angles of incidence. Tyndall scattering can also be eliminated as a mechanism, because backscattered light is not polarized, and a dark background is not necessary for color production (Fox 1976; Lee 1977). Thin-film or quarter-wavelength interference (Figure 4.12) requires the presence of a layer with a different refractive index;

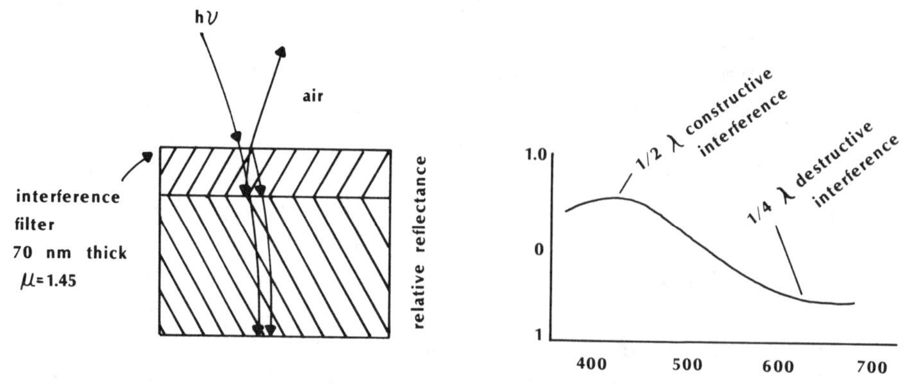

Figure 4.12. Diagram of a model interference filter.

its thickness is one-half the wavelength of light being reflected, or constructively interfered. Light at wavelengths approaching four times the filter thickness, destructively interfered, will pass through the layer. Familiar examples of thin-film interference include the sheen on an oily mud puddle, antireflection coatings in compound lenses, and peacock feathers.

In some plants, especially *Selaginella*, the physical basis for coloration must be near the surface of the leaf, because the color disappears when wetted, reappearing when dry again. In other plants, such as *Danaea nodosa*, the blue color does not disappear on wetting, and the physical basis may be in the epidermal wall away from the surface (unpublished results).

Localization of the structural basis of iridescence, as well as determination of its probable selective advantage, is aided by the physical laws of interference. If a refractive index (μ) of 1.45 is assumed for the outer epidermal cell wall – and thus the interference filter, based on measurements by Woolley (1975) – we can use measurements of the angular and wavelength dependences of the reflectance of blue leaves to predict the thickness of the filter. Following the calculations of Hébant and Lee (1984) for *Selaginella willdenowii* (Desv.) Bak. and *S. uncinata*, we obtain the following:

$$\text{filter thickness} = \lambda/4 \times \mu \times \cos \delta$$

where δ is the refracted angle of light in the filter, and λ is the peak wavelength for constructive interference.

S. willdenowii: $405/4 \times 1.45 \times 0.873 = 80$ nm

S. uncinata: $410/4 \times 1.45 \times 1.00 = 71$ nm

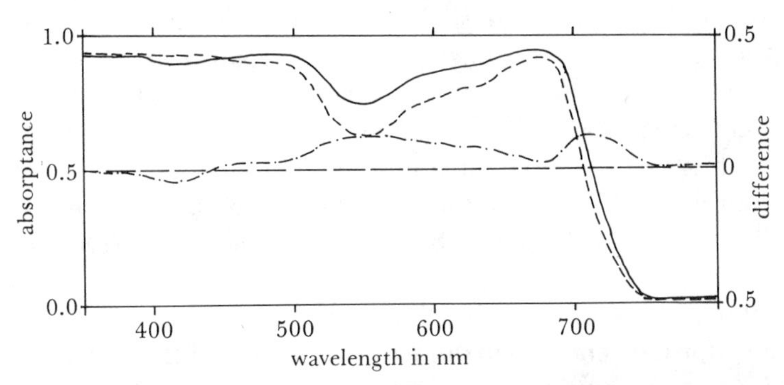

Figure 4.13. Optical comparisons of blue and green leaves of *Selaginella uncinata* grown at the same PPFD but at different R : FR, 3.0 for green-leaved and 0.35 for blue-leaved plants. Reflectance was measured following Hébant and Lee (1984) and Lee and Graham (1986). Diffuse transmittance, which allowed the estimation of the absorptance plotted here, was measured with a Li-Cor 1800 spectroradiometer attached to the photographic tube of a Zeiss microscope with a 40× neoflaur objective in place, each curve being the mean of five measurements. Solid line denotes blue leaf absorptance; dash line denotes green leaf absorptance; dot-dash line is the difference in absorptance between the two.

In both species the ultrastructural basis for the filter is provided by two lamellae of the predicted thickness in the outer portion of the upper epidermal cell wall (Figure 4.11E). The lamellae are absent from older and more sun-exposed green branches (Figure 4.11C), as well as from the outer wall of the lower epidermis.

What might be the selective advantage of leaf iridescence in an extremely shady environment, especially of a characteristic that should decrease the absorptance of some visible wavelengths? Remember that an interference filter, while constructively reflecting blue light, should also destructively absorb light at higher wavelengths. Such a filter at the leaf surface could thus function in a manner analogous to the antireflection coating of a camera lens, increasing absorptance of wavelengths more abundant in the understory (as just beyond 700 nm) or more efficiently used (550–700 nm) (Lee and Lowry 1975). Differences in reflectance between green and blue leaves (Lee and Lowry 1975; Hébant and Lee 1984) should contribute to differences in absorptance, but transmittance must also be measured in these tiny leaves to allow a direct estimate of absorptance (Figure 4.13). Not all of the differences in absorptance are due to interference, because growth conditions (discussed in the following section) could alter other leaf properties. However, the effect of interference can be discerned from the slight decrease in absorptance of blue leaves at 400–420 nm and the much greater absorptance at wavelengths above 500 nm. In a comparison of absorptance of shade PPFD (Figure 4.2), the green leaves absorbed 81%, and blue leaves absorbed 87%. For blue leaves, the percentage potential utilization of quanta at 700–750/400–700 nm was 4.3%, versus 3.0% for green leaves. Thus, the blue leaves of *S. uncinata* absorb light usable in photosynthesis more efficiently than green leaves.

This research needs to be extended to other iridescent plants. In most species studied, iridescence occurs in structures beneath the leaf surface (Lee 1977) and thus cannot reduce surface reflectance. In *Begonia pavonina* Ridl. (native to the Malayan Peninsula) and *Trichomanes elegans* Rich (native to neotropical forest understory), the ultrastructural basis for iridescence appears to be the membrane spacing of plastids (unpublished results). Does chloroplast iridescence enhance optical properties, leading to more efficient use of light in photosynthesis? Why is iridescence not encountered in other shady terrestrial environments, such as in temperate forests? Perhaps the generally higher light levels or seasonality of foliage cover select against it. I have noticed that a few temperate understory plants do have a blue sheen, especially the juvenile leaves of the Virginia creeper, *Parthenocissus quinquefolia* Planch, growing in shade. Blue iridescence is a very striking feature that certainly deserves more investigation.

Developmental control

The extreme-shade features of reddish leaf undersurfaces, velvety epidermis, and blue iridescence are associated with understory herbs in three different ways. In some species the characteristics are uniform for all individuals and populations (such as velvety leaf surfaces in *A. warocqueanum* and *S. uncinata*). Other species are polymorphic for a characteristic, chiefly pigmented leaf undersurfaces, but an individual keeps these characteristics for its life span. In addition to *Triolena hirsuta,* discussed earlier, *Ischnosiphon pruinosus* (Reg.) O. G. Peterson, an understory herb of the Marantaceae growing on Barro Colorado Island, in Panama, consists of red- and green-undersurfaced individuals, and the red-undersurfaced plants grow in the shadiest habitats (R. Morris, personal communication). Still other plants produce shade-adapted characteristics in relation to the different conditions in which the plants grow. Many of these produce juvenile foliage with features of shade adaptation, replacing them with more normal leaves later on. The juvenile leaves of *Psychotria suerrensis* are strikingly different from its mature leaves, and the velvet-surfaced leaves of many aroids (exemplified by *Scindapsus pictus*) are different from the large and frequently lobed mature leaves. In the latter case the development may be under genetic control, and/or it may be a direct response to the environment. Little is known about the role of light in the developmental control of these plants. Differences in PPFD may be one factor, and other responses may be more wavelength-dependent. Given the large spectral differences of light in different degrees of shade, a good candidate for mediating this developmental response is phytochrome. For instance, the R : FR ratios in the light and shade environments used earlier in this discussion are 1.28 and 0.30, respectively. A pronounced shift in this ratio has been shown to initiate a number of developmental responses in plants (Morgan and Smith 1981; Smith 1982; Holmes 1983). These include germination (Vásquez-Yánes 1980; Frankland 1981), rapid stem elongation (Morgan and Smith 1978), branching patterns (Deregibus et al. 1983), and leaf structure (Child et al. 1981). Thus, in those cases in which leaf structures are developmental responses to low light levels, phytochrome is the prime candidate for photoreception in this control. Phytochrome may also be involved in rapid responses in forest shade, as in the skototropic responses of aroid seedlings toward the dark (Strong and Ray 1975) and in chloroplast orientation in leaf cells (Haupt 1982).

An example of the process by which extreme-shade structures can develop in response to light quality is the appearance of blue iridescence in *Selaginella* species (Hébant and Lee 1984). In both *S. willdenowii* (native to forests of Southeast Asia) and *S. uncinata* (native to forests of south China), iridescent blue leaves appear in plants grown under foliage shade. Plants grown in greenhouse shade produce leaves that are mostly green. Blue-leaved branches eventually turn green with age. *S. uncinata* is a compact plant, very amenable to experimental analysis. It was grown in two chambers at low light levels (12 μmol m^{-2} s^{-1} for a 12-hr photoperiod, or 0.5 mol/day) in which the R : FR ratio was either 3.0 or 0.35, depending on filters and light sources. As predicted, plants grown under the extreme foliage shade ratio of 0.35 developed intensely blue leaves, and those under the ratio of 3.0 were green. Furthermore, two strikingly different forms grew under these different treatments (Figure 4.14). Those under the ratio of 3.0 grew erect, branching more equally, with greater distances between adjacent leaves. Those under the ratio of 0.35 grew totally prostrate, with a tendency for certain branches to extend very rapidly. Leaf intermodes were much shorter, and leaves were produced in a marked plagiotropic fashion. The growth pattern of the latter plants may also enhance their growth and survival in shade, and the former should be

Figure 4.14. Photograph of branches of *Selaginella uncinata* grown at the same PPFD, but with different spectral qualities. Plant on left was grown in R : FR = 0.35, simulating deep shade, and that on the right was grown in R : FR = 3.0, simulating full sunlight.

more suited to more exposed environments. Only the 3.0-grown plants produced spores.

Similar studies may reveal the importance of the spectral quality of light in controlling the development of many extreme-shade adaptations in humid tropical-forest understory plants.

Summary

The main thesis of this chapter is that plants growing in the profound shade of humid tropical forests are adapted to more efficiently absorb diffuse and spectrally altered radiation because of its passage through layers of foliage. Each of the leaf structures discussed here may provide an advantage in extreme shade, but probably puts the same plants at a disadvantage in sunlight. For epidermal lens effects, the light levels inside the leaves may damage the photosynthetic apparatus. For undersurface pigmentation, the plant may divert unnecessary metabolic energy into the production of molecules that are of no advantage in more open sun. In iridescence, there may be a developmental cost.

The principal approach I am using in this research is to predict optical properties of leaves growing in extreme shade, and then test these predictions with measurements of a sample of extreme-shade plants compared with typically sun-adapted plants. The research needs to be extended to more plants, but its ultimate shortcoming is ignorance of the detailed photosynthetic characteristics of extreme-shade plants. A shift in spectral yield of just 10 nm toward the far-red range might mean a significant increase (perhaps 10–15%) in the energy available for photosynthesis. Photosynthetic research on these plants should be performed with the following caveats in mind. It may be inaccurate to measure photosynthetic characteristics of these plants using the standard artificial light sources, making the standard assumption of the exclusive use of quanta between 400 and 700 nm. Such measurements may underestimate the productivity of these plants in their natural environments. In addition, light environments at the level of the chloroplast may be quite different from those measured at the leaf surface, and the source of illumination (whether diffuse or narrowly defined) may also be important. Finally, measurements of action spectra are critically important, but they must be done with care. Most studies have been done with filters that allow a half-peak bandwidth of 25 nm, and even studies with monochromators have had little more precision (McCree 1972; Inada 1978). To carefully document the upper end of the action spectrum for these plants it will be necessary to measure at more frequent spectral intervals with more narrowly restricted wavelengths.

It may be expecting too much to ask that our "big physiological weapons" be brought to bear on such obscure little plants when we know so little about the trees above them (which constitute the real primary productivity of tropical forests), or about numerous important tropical crops. However, the dividend may be an understanding of the limits of terrestrial plants in responding to, and growing in, the shadiest of environments. These limitations may be structural ones, such as the intercellular basis for the increase in leaf reflectance above 700 nm (Gausman 1974, 1977), or they may be biochemical, such as limits to the quantum efficiency of photosynthesis (Björkman 1981). Other chapters in this volume discuss the tradeoffs between leaf structures efficiently absorbing energy for photosynthesis versus those adding to a heat load affecting the water economy of particular plants. The species discussed here grow slowly in a wet and very dark environment. The ultimate constraints on their growth and reproduction are structural limitations on the efficiency with which the plants can produce and display a surface for photosynthesis. Velvety leaf surfaces, red undersurfaces, and blue iridescence appear to slightly enhance absorption by leaves and thus have been selected in many taxonomically unrelated groups ecologically restricted to extreme shade in tropical rain forests.

Acknowledgments

I would like to thank the Florida International University and Whitehall Foundations for support of the research described in this chapter.

References

Allan, W. A., H. W. Gausman, and A. J. Richardson. 1973. Willstatter-Stoll theory of leaf reflectance evaluated by ray tracing. Appl. Optics 12:2448–2453.

Anderson, M. C. 1970. Interpreting the fraction of solar radiation available in forest. Agric. Meteorol. 7:19–28.

Atwood, J. T., and N. H. Williams. 1979. Surface features of the adaxial epidermis in the conduplicate-leaved Cypripedioideae (Orchidaceae). Bot. J. Linn. Soc. 78:141–156.

Bazzaz, F. A., and S. T. A. Pickett. 1980. Physiological ecology of tropical succession: a comparative review. Ann. Rev. Ecol. Syst. 11:287–310.

Björkman, O. 1973. Comparative studies of photosynthesis in higher plants. Photophysiology 7:1–63.

– 1981. Responses to different quantum flux densities. Pp. 57–107 *in* O. L. Lange, P. S. Nobel, C. B. Osmond, and H. Ziegler (eds.), Physiological plant ecology, I, vol. 12A, Encyclopedia of plant physiology, new series. Springer-Verlag, Berlin.

Björkman, O., and M. M. Ludlow. 1972. Characterization of the light climate on the floor of a Queensland rainforest. Carnegie Inst. Yearbook 71:85–94.

Björkman, O., M. M. Ludlow, and P. A. Morrow. 1972. Photosynthetic performance of two rainforest species in their native habitat and analysis of their gas exchange. Carnegie Inst. Yearbook 71:94–102.

Boardman, N. K. 1977. Comparative photosynthesis of sun and shade plants. Ann. Rev. Plant Physiol. 28:355–377.

Bone, R. A., D. W. Lee, and J. M. Norman. 1984. Leaf epidermal cells function as lenses in tropical rainforest shade plants. Appl. Optics 24:1408–1412.

Burtt, L. 1978. Notes on rain-forest herbs. Gard. Bull. Singapore 29:37–49.

Chazdon, R. L., and N. Fetcher. 1984a. Light environments of tropical forests. Pp. 27–36 *in* E. Medina, H. A. Mooney, and C. Vásquez-Yánes (eds.), Physiological ecology of plants of the wet tropics. Dr. Junk, The Hague.

– 1984b. Photosynthetic light environments in a lowland tropical rainforest in Costa Rica. J. Ecol. 72:553–564.

Child, R., D. C. Morgan, and H. Smith. 1981. Morphogenesis in simulated shadelight quality. Pp. 409–420 *in* H. Smith (ed.), Plants and the daylight spectrum. Academic, London.

Deregibus, V. A., R. A. Sanchez, and J. J. Casal. 1983. Effects of light quality on tiller production in *Lolium* spp. Plant Physiol. 72:900–902.
Ehleringer, J. 1981. Leaf absorptances of Mohave and Sonoran desert plants. Oecologia 49:366–370.
Fox, D. L. 1976. Animal biochromes and structural colours. University of California Press, Berkeley.
Frankland, B. 1981. Germination in shade. Pp. 187–203 *in* H. Smith (ed.), Plants and the daylight spectrum. Academic, London.
Gates, D. M., H. J. Keegan, J. C. Schleter, and V. R. Weidner. 1965. Spectral properties of plants. Appl. Optics 4:11–20.
Gausman, H. 1974. Leaf reflectance of near infra-red. Photogrammetric Engineering 1974:183–191.
– 1977. Reflectance of leaf components. Remote Sens. Environment 6:1–9.
Gausman, H., and W. A. Allen. 1973. Optical parameters of leaves of 30 plant species. Plant Physiol. 52:57–62.
Givnish, T. J. 1984. Leaf and canopy adaptations in tropical forests. Pp. 51–84 *in* E. Medina, H. A. Mooney, and C. Vásquez-Yánes (eds.), Physiological ecology of plants of the wet tropics. Dr. Junk, The Hague.
Goodchild, D. J., O. Björkman, and N. A. Pyliotis. 1972. Chloroplast ultrastructure, leaf anatomy, and content of chlorophyll and soluble protein in rainforest species. Carnegie Inst. Yearbook 71:102–197.
Haberlandt, G. 1912. Physiological plant anatomy. Macmillan, London.
Halldal, P. 1968. Photosynthetic capacities and photosynthetic action spectra of endozoic algae of the massive coral *Favia*. Biol. Bull. 134:411–424.
Haupt, W. 1982. Light effects on chloroplast movements. Ann. Rev. Plant Physiol. 33:205–233.
Hébant, C., and D. W. Lee. 1984. Ultrastructural basis and developmental control of blue iridescence in *Selaginella* leaves. Amer. J. Bot. 71:216–219.
Holmes, M. G. 1983. Perception of shade. Phil. Trans. R. Soc. London, Series B 303:503–521.
Horn, H. S. 1971. The adaptive geometry of trees. Princeton University Press.
Hutchison, B. A., D. R. Matt, and R. T. McMillen. 1980. Effects of sky brightness distribution upon penetration of diffuse radiation through canopy gaps in a deciduous forest. Agric. Meteorol. 22:137–147.
Idso, S. B., and C. T. de Wit. 1970. Light relations in plant canopies. Appl. Optics 9:177–184.
Inada, K. 1978. Spectral dependence of photosynthesis in crop plants. Acta Hort. 87:177–185.
Jagels, R. 1970. Photosynthetic apparatus in *Selaginella*. I. Morphology and photosynthesis under different light and temperature regimes. Can. J. Bot. 48:1843–1852.
Kerner, A., and F. W. Oliver. 1895. The natural history of plants, vol. 1. Holt & Co., New York.
Lee, D. W. 1977. On iridescent plants. Gard. Bull. Singapore 30:21–29.
– 1985. The spectral distribution of radiation in two neotropical rain forests. Biotrop. (submitted).
Lee, D. W., and R. Graham. 1986. Leaf optical properties of rainforest sun and extreme shade plants. Amer. J. Bot. (in press).

Lee, D. W., and J. B. Lowry. 1975. Physical basis and ecological significance of iridescence in blue plants. Nature 254:50–51.

Lee, D. W., J. B. Lowry, and B. C. Stone. 1979. Abaxial anthocyanin layer in leaves of tropical rainforest plants: enhancer of light capture in deep shade. Biotropica 11:70–77.

McCree, K. J. 1972. The action spectrum, absorptance and quantum yield of photosynthesis in crop plants. Agric. Meteorol. 9:191–216.

– 1981. Photosynthetically active radiation. Pp. 41–55 *in* O. L. Lange, P. S. Nobel, C. B. Osmond, and H. Ziegler (eds.), Physiological plant ecology, I, vol. 12A, Encyclopedia of plant physiology, new series. Springer-Verlag, Berlin.

Mooney, H. A., O. Björkman, A. E. Hall, E. Medina, and P. B. Tomlinson. 1980. The study of the physiological ecology of tropical plants – current status and needs. BioScience 30:22–26.

Morgan, D. C., and H. Smith. 1978. Simulated sunflecks have large, rapid effects on plant stem extension. Nature 273:534–536.

– 1981. Non-photosynthetic responses to light quality. Pp. 109–134 *in* O. L. Lange, P. S. Nobel, C. B. Osmond, and H. Ziegler (eds.), Physiological plant ecology, I, vol. 12A, Encyclopedia of plant physiology, new series. Springer-Verlag, Berlin.

Norman, J. M., and P. G. Jarvis. 1974. Photosynthesis in Sitka spruce [*Picea sitchensis* (Bong.) Carr.]. V. Radiation penetration theory and a test case. J. Appl. Ecol. 12:839–878.

Patton, T. C. (ed.) 1973. Pigment handbook, vols. 1–3. Wiley, New York.

Pearcy, R. W., and H. Calkin. 1983. Carbon dioxide exchange of C_3 and C_4 tree species in the understory of a Hawaiian forest. Oecologia 58:26–32.

Raven, J. A., and J. Beardall. 1982. The lower limit of photon fluence rate for phototropic growth: the significance of "slippage" reactions. Plant Cell Environ. 5:117–124.

Reifsnyder, W. E., G. M. Furnival, and J. C. Horowitz. 1970. Spatial and temporal distribution of solar radiation beneath forest canopies. Agric. Meteorol. 9:21–37.

Richards, P. W. 1952. The tropical rainforest. Cambridge University Press.

Richardson, K., J. Beardall, and J. A. Raven. 1983. Adaptation of unicellular algae to irradiance: an analysis of strategies. New Phytol. 93:157–191.

Robichaux, R. H., and R. W. Pearcy. 1980. Photosynthetic responses of C_3 and C_4 species from cool shaded habitats in Hawaii. Oecologia 47:106–109.

Sculthorpe, C. D. 1967. The biology of aquatic vascular plants. Edward Arnold, London.

Small, C. J., and R. C. Pecket. 1982. The ultrastructure of anthocyanoplasts in red cabbage. Planta 154:97–99.

Smith, A. M. 1909. On the internal temperatures of leaves in tropical insolation, with special reference to the effect of their color on temperature: also observations on the periodicity of the appearance of young coloured leaves of trees growing in Peradeniya Gardens. Ann. Roy. Bot. Gard. Peradeniya 4:229–281.

Smith, H. 1982. Light quality, photoreception and plant strategy. Ann. Rev. Plant Physiol. 33:481–518.

Stoutjestijk, P. 1972. A note on the spectral transmission of light by tropical rainforest. Acta Bot. Neerl. 21:346–350.

Strong, D. R., and T. S. Ray. 1975. Host tree location behavior of a tropical vine (*Monstera gigantea*) by skototropism. Science 190:804–806.

Tasker, R., and H. Smith. 1977. The function of phytochrome in the natural environment. V. Seasonal changes in radiant energy quality. Photochem. Photobiol. 26:487–491.

Trapp, J., L. Gillespie, D. Gorchov, and K. Grove. 1982. Color variation in forest floor herb leaves. Organization of Tropical Studies 82:1–5.

Vásquez-Yánes, C. 1980. Light quality and seed germination in *Cecropia obtusifolia* and *Piper auritum* from a tropical rain forest in Mexico. Phyton 38:33–35.

Woolley, J. T., 1971. Reflectance and transmittance of light by leaves. Plant Physiol. 47:656–662.

– 1975. Refractive index of soybean leaf cell walls. Plant Physiol. 55:172–174.

Yoda, K. 1974. Three-dimensional distribution of light intensity in a tropical rain forest of West Malaysia. Jpn. J. Ecol. 24:247–254.

5 Economics of carbon fixation in higher plants

IAN R. COWAN

Annual income twenty pounds, expenditure twenty pounds and sixpence: the leaf is withered and the blossom is blighted.

– Charles Dickens, *David Copperfield*

Introduction

Three problems of economics are touched on. The first concerns a relationship between the amounts of two enzymes involved in carbon fixation in chloroplasts. The cost of the synthesis of the enzymes is taken to be the nitrogen incorporated, and the problem is to find the partitioning of nitrogen leading to the maximum rate of carbon fixation per unit of nitrogen. The matter is carried no further than that. There is presumably a complementary relationship between the nitrogen taken up by a plant and the cost of obtaining it in terms of carbon used in root growth or lost in respiration (see Chapter 9), but it is doubtful that these can be defined precisely enough for an application of optimization theory. The second problem concerns the manner in which stomatal movements control the assimilation of CO_2. The cost of assimilation is defined in terms of transpiration of water. The objective is to determine how a plant should arrange matters so as to take up CO_2 with the least cost. As with the first problem, there is a complementary question; in this instance it has to do with the cost, in terms of carbon, of acquiring water. The third example goes some way toward dealing with that. It is assumed that there is a penalty to the plant, in terms of carbon, if it uses soil water too rapidly and becomes desiccated. It then becomes possible to determine not only how the plant should manage its water use with time to minimize the risk of desiccation but also how it should allocate growth between tissues that fix carbon and tissues that gather water.

Use of nitrogen

Investment in two chloroplast enzymes

The enzymes are ribulose-bisphosphate carboxylase/oxygenase (Rubisco), which catalyzes the primary carboxylation reaction in the chloroplasts of C_3 plants, and carbonic anhydrase, which catalyzes the hydration and dehydration of CO_2. The question to be addressed is whether or not the relative costs of synthesizing the two enzymes reflect the relative contributions that the enzymes make to the process of carbon fixation.

The specific cost of enzyme synthesis is approximately the same for different enzymes, whether the cost be defined in terms of energy or in terms of the amounts of carbon or nitrogen incorporated. We shall choose nitrogen. Let N_c and N_a be the amounts of nitrogen (moles) incorporated in the form of carboxylase and anhydrase. Then $N = N_c + N_a$ is the total cost of synthesis of the two enzymes. It will be convenient, for the purpose of comparing prediction with observation, to express the amount of nitrogen relative to the amount of chlorophyll in the leaf. The data of Atkins et al. (1972a, 1972b) for 13 dicotyledonous species show that the mass of carbonic anhydrase per unit mass of chlorophyll varies from 0.03 to 0.58, the mean being 0.23 and the standard deviation 0.15. The nitrogen content of protein being 11.4 mmol g^{-1} and the molecular weight of chlorophyll being 897, then $N_c = 2.3$ mol per mole chlorophyll, with a standard deviation of 1.5. Data from Björkman (1968) for five "sun" species suggest (assuming that 50% of soluble protein is ribulose-bisphosphate carboxylase) that the average mass of carboxylase per unit mass of chlorophyll is about 6; see also Lyttleton and T'so (1958). The standard deviation of carboxylase activity per unit mass of chlorophyll was about 40% of the mean. In terms of nitrogen, then, $N_a = 61$ mol per mole of chlorophyll, with a standard deviation of 24. These are the data against which a criterion of optimal nitrogen allocation is to be examined, in an abbreviated version of a detailed treatment to be published elsewhere.

Functions of the two enzymes

The two enzymes have complementary functions. The role of one is to fix CO_2. That of the other is to help deliver CO_2 to the sites of fixation, by a process known as facilitated transfer (Enns 1967; Raven and Glidewell 1981). The process can be explained with aid of Figure 5.1. Chloroplasts are lens-like in shape and usually lie close to the cell wall, with very little, if any, cytoplasm in between. For the purpose of analysis, the cell wall and

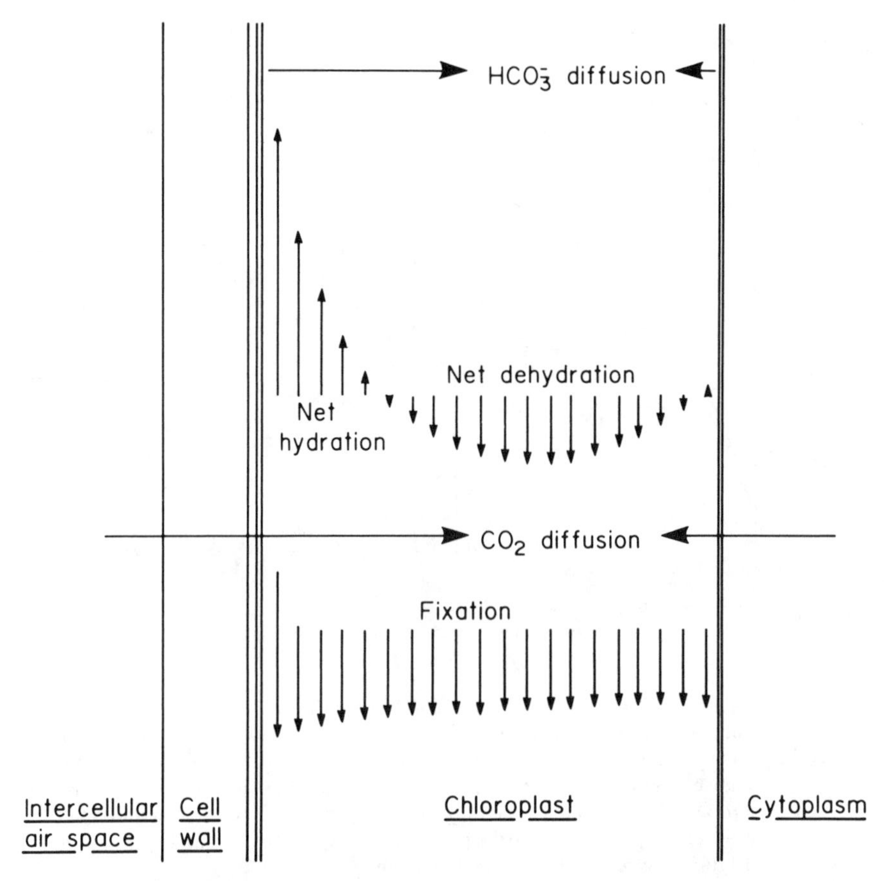

Figure 5.1. Representation of fluxes of CO_2 and HCO_3^- in the cell-wall and chloroplast system.

chloroplast are taken as plane slabs, separated by the plasma membrane and chloroplast envelope. CO_2 from the intercellular air space in the leaf diffuses across the cell wall and membranes. Also, CO_2 released from the mitochondria enters the chloroplast through the envelope remote from the cell wall. Within the chloroplast, CO_2 is fixed through the agency of carboxylase. The supply of CO_2 to the sites of fixation is maintained partly by the diffusion of CO_2 as such, but also by the diffusion of HCO_3^-. There is a positive net rate of hydration of CO_2 wherever the concentration of CO_2 exceeds the average concentration within the chloroplast, and a positive net rate of dehydration wherever the concentration is less than the average. As the product of hydration, HCO_3^-, also diffuses down its gradient of concentration, the whole process, hydration-diffusion-dehydration, forms a second means of transferring CO_2 across the chloroplast. However, the

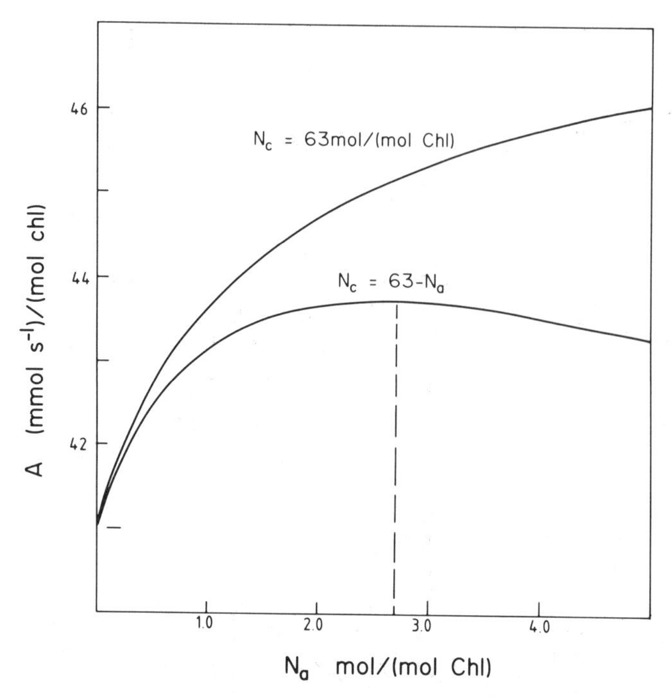

Figure 5.2. Net rate of CO_2 fixation in a chloroplast, A, as a function of the amount of nitrogen incorporated in carbonic anhydrase, N_a. The uppermost curve assumes that the amount of nitrogen incorporated in ribulose-bisphosphate carboxylase, N_c, is constant; the second curve assumes that $N_c + N_a$ is constant. The "optimal" magnitude of N_a corresponds to the maximum in the latter. The intercellular partial pressure of CO_2, p_i, is taken as 230 μbar, and the thickness of the chloroplast as 1.7 μm.

time constants of uncatalyzed hydration and dehydration are about 20 s and 1,000 s, respectively, compared with a characteristic time of about 1 ms for transfer of CO_2 or HCO_3^- across the chloroplast. Therefore, it is only in the presence of carbonic anhydrase that this second pathway is effective. This is what Enns (1967) called facilitation of transfer. Appendix I shows how the equations for the reactions catalyzed by carboxylase and anhydrase and the equations for diffusion of CO_2 and HCO_3^- can be solved to provide an expression for the net rate A of CO_2 fixation per unit of chlorophyll in terms of the nitrogen invested in the two enzymes; that is,

$$A = \text{function}(N_c, N_a, l, p_i, \ldots) \qquad (5.1)$$

in which l is the thickness of the chloroplast, p_i is the partial pressure of CO_2 in the intercellular air space, and the unspecified parameters consist of the kinetic properties of the enzymes and the coefficients of diffusion and transfer, the assumed magnitudes of which are given in Appendix I.

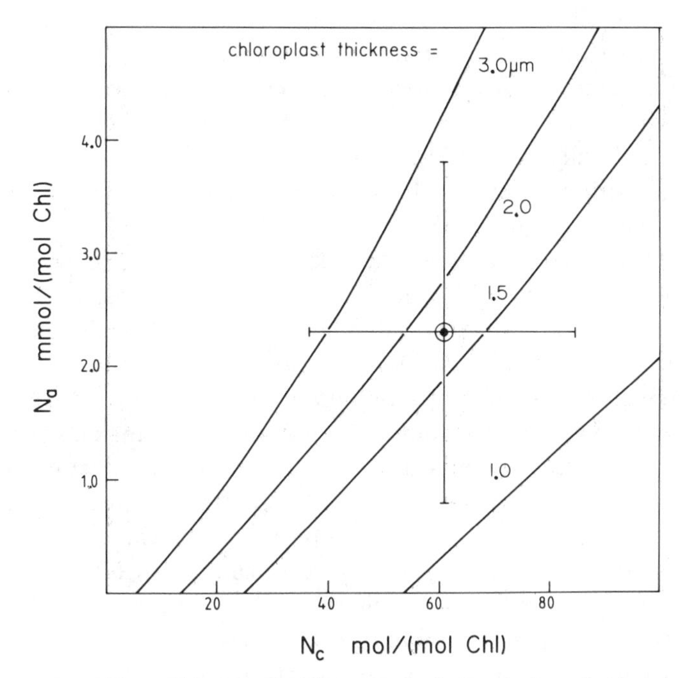

Figure 5.3. Optimal investment of nitrogen in carbonic anhydrase, N_a, as a function of nitrogen in ribulose-bisphosphate carboxylase, N_c, and chloroplast thickness, l. The intercellular partial pressure of CO_2, p_i, is taken as 230 μbar. Also indicated are the means and standard deviations of N_a and N_c derived from enzyme assays in a number of C_3 dicotyledonous plants.

Optimal partition of nitrogen

Economy demands that the photosynthetic rate A be the maximum possible corresponding to the total cost. Formally, this requires that carboxylase and anhydrase be present in proportions such that

$$\left(\frac{\partial A}{\partial N_c}\right)_N - \left(\frac{\partial A}{\partial N_a}\right)_N = 0 \tag{5.2}$$

The full curve in Figure 5.2 demonstrates how an optimal combination of N_c and N_a is located. It is computed with a constant total amount of nitrogen per unit of chlorophyll, $N = 63$. Because the molecular weight of Rubisco is three times that of CA, this implies that one molecule of the carboxylase is sacrificed for each three additional molecules of the anhydrase. The maximum, which fulfills the criterion of optimization expressed in equation (5.2), corresponds to $N_a = 2.7, N_c = 60.3$. Also seen in Figure 5.2 is the variation of A with N_a for a constant amount of carboxylase per unit of chlorophyll, equivalent to $N_c = 63$.

Figure 5.3 shows optimal combinations of N_c and N_a with several magnitudes of chloroplast thickness, l. Estimates of the means and standard deviations of N_c and N_a in dicotyledonous C_3 plants are superimposed. The combination of the mean values conforms to the criterion of optimality in chloroplasts having a thickness $l = 1.7$ μm, a thickness that is certainly quite close to that of real chloroplasts. This is very strong evidence for the truth of the optimization hypothesis. If the functions of the carboxylase or anhydrase had been misunderstood, or if the relative costs of the two enzymes had been incorrectly assessed, it would be unlikely in the extreme that an agreement between prediction and observation as close as that illustrated in Figure 5.3 would nevertheless have emerged. The closeness of the agreement is not associated with a discriminating choice of values in the analysis; I have shown elsewhere (unpublished data) that the prediction is rather insensitive to variations in diffusion and transfer coefficients and the kinetic characteristics of the enzymes over the ranges of uncertainty in the magnitudes of these quantities. Figure 5.4 shows that it is little affected by the magnitude of the intercellular partial pressure. We can conclude that the relative amounts of the two enzymes are indeed optimized, at least approximately, according to the hypothesis. Precisely how finely tuned the optimization is cannot be determined, for the data set is clearly inadequate for that. Measurements of the concentrations of carboxylase and anhydrase and of the chloroplast dimensions carried out on the same set of species will doubtless answer the question in the future.

This is a well-posed optimization problem. The primary biological constraints are the catalytic properties of the two enzymes. These are unlikely to have evolved quickly; there is evidence that they do not differ very much between unicellular aquatic algae and higher plants. It is more likely that the relative amounts of the enzymes synthesized can change rapidly in response to selection pressure. The main environmental circumstance that might have been thought to influence the optimum partitioning, and that has varied in a short period of evolutionary time, is the atmospheric concentration of CO_2 (Pearman 1984). We have seen that the optimum is not at all sensitive to CO_2 concentration.

Figure 5.2 suggests that the rate of CO_2 fixation with optimal partitioning of nitrogen is about 5% greater than it would be if the same amount of nitrogen were invested in carboxylase only. It is 7.5% greater than it would be if carbonic anhydrase were not present and the amount of carboxylase were not increased. Physiologists might consider that small; indeed, Jacobson et al. (1975) estimated the enhancement to be 7% and implied that the role of anhydrase in facilitating CO_2 transfer is therefore of negligible importance. However, evolutionary biologists will have no such prejudices. No additional justification need be sought for the presence of an

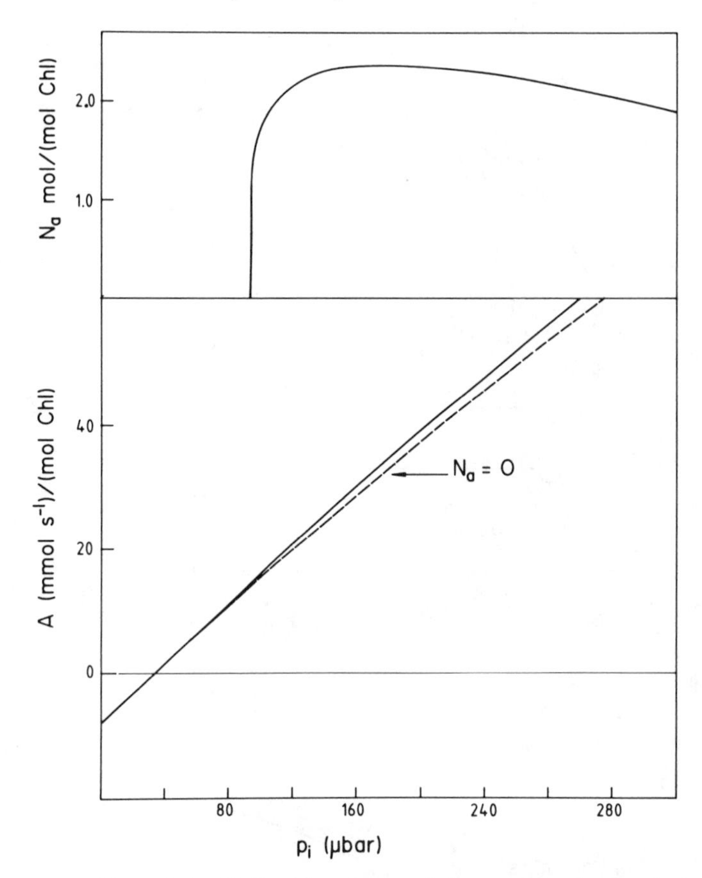

Figure 5.4. Optimal investment of nitrogen in carbonic anhydrase, N_a, and the corresponding rate of CO_2 fixation, A, as functions of intercellular partial pressure of CO_2, p_i. The amount of nitrogen incorporated in ribulose-bisphosphate carboxylase and anhydrase is taken as 63 mol N per mole chlorophyll, and the chloroplast thickness as 1.7 μm. Also shown is the rate of assimilation in the absence of anhydrase ($N_a = 0$).

enzyme that composes no more than 2% of leaf protein and yet enhances the primary function of green leaves by 7.5%.

Extension to other enzymes: interaction with water use

Doubtless, such analyses will be applied to other enzymes and components of the biochemical machinery of photosynthetic carbon fixation before long. Of particular interest, in that it has important implications in plant ecology, is the partitioning of resources between the system that sustains the dark reactions of fixation and the system that sustains the harvesting of light and photoelectron transport.

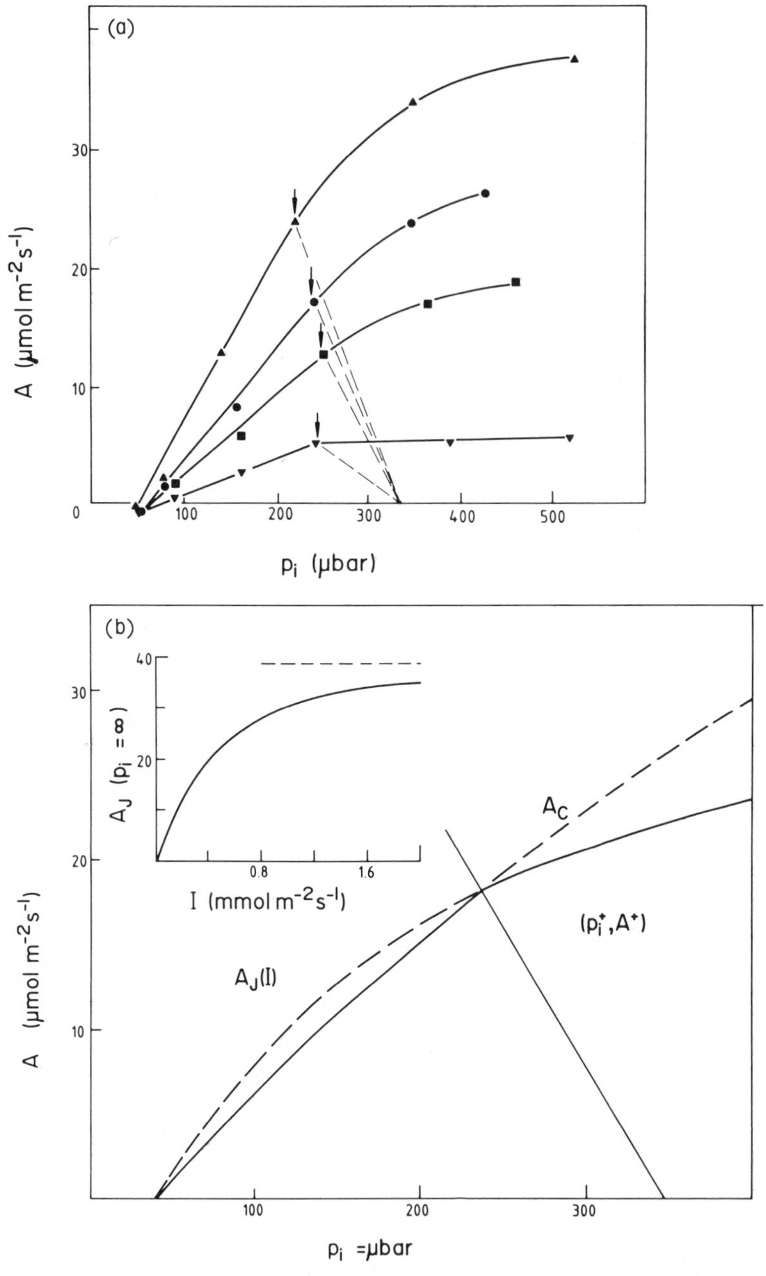

Figure 5.5a. Rate of assimilation of CO_2, A, as a function of intercellular partial pressure of CO_2, p_i, in leaves of *Phaseolus vulgaris* plants grown with 24 (triangles), 12 (circles), 4 (squares), and 0.6 (inverted triangles) mM nitrate in solution supplied to the roots (von Caemmerer and Farquhar 1981). Measurements were made with 1.5 mmol m^{-2} s^{-1} leaf irradiance and 28 °C leaf temperature. The arrows indicate data with 330 μbar external partial pressure of CO_2.

Figure 5.5a, due to von Caemmerer and Farquhar (1981), shows how rates of assimilation, A, in leaves of plants of *Phaseolus vulgaris*, supplied with different concentrations of nitrate in solution, varied with internal partial pressure of CO_2, p_i. Variation of p_i was brought about by varying the ambient partial pressure of CO_2. The point on each curve that corresponds to normal ambient partial pressure of CO_2, p_a, is indicated. The corresponding magnitude of the conductance, g, to CO_2 transfer across the external boundary layer and epidermis of the leaf is the absolute slope, multiplied by atmospheric pressure, of the straight line drawn through that point and intersecting the abscissa at $p_i = p_a$. Following the analysis of Farquhar et al. (1980a) and von Caemmerer and Farquhar (1981), each characteristic in Figure 5.5a can be represented, as illustrated in Figure 5.5b, by

$$A = \min\{A_C, A_J\} \qquad (5.3)$$

A_C being the potential net rate of assimilation at any given magnitude of p_i with unlimited rate of ribulose-bisphosphate regeneration, and A_J the potential net rate of assimilation at any given p_i and irradiance I with unlimited activity of ribulose-bisphosphate carboxylase/oxygenase. Evans (1983a, 1983b) has shown that the activities of the 12 enzymes involved in the Calvin cycle are matched, in that none of them limits carbon flow to a greater extent than the others. That being so, we can assume $A_C = A_C(p_i, N_C)$, where N_C is the amount of nitrogen per unit leaf area incorporated in Calvin-cycle enzymes (carboxylase accounting for about 90%). Similar remarks apply to components of the light-harvesting, electron-transport, and ribulose-bisphosphate regeneration system, and therefore $A_J = A_J(I, p_i, N_J)$, where N_J is the amount of nitrogen per unit leaf area incorporated in proteins in the chloroplast thylakoid membranes. It will be useful to define the intersection of the curves A_C and A_J as (p_i^+, A^+). Both p_i^+ and A^+ are functions of I, N_C, and N_J.

Let us consider the distribution of nitrogen in relation to conductance to CO_2 transfer across boundary layer and epidermis at normal ambient

Caption to Figure 5.5 (*cont.*)
Figure 5.5b. Rate of assimilation of CO_2, A, as a function of intercellular partial pressure of CO_2, p_i, derived from the model of Farquhar et al. (1980a); see also von Caemmerer and Farquhar (1981). A_C is the rate of assimilation at any given p_i assuming an unlimited rate of ribulose-bisphosphate regeneration and $N_C = 25$ mmol N per square meter of leaf area invested in Calvin-cycle enzymes; A_J is the rate of assimilation at any given p_i with irradiance $I = 1.6$ mmol m^{-2} s^{-1} assuming unlimited activity of ribulose-bisphosphate carboxylase and $N_J = 14$ mmol N per square meter of leaf area invested in proteins in the chloroplast thylakoids. The inset shows A_J at saturating p_i as a function of I.

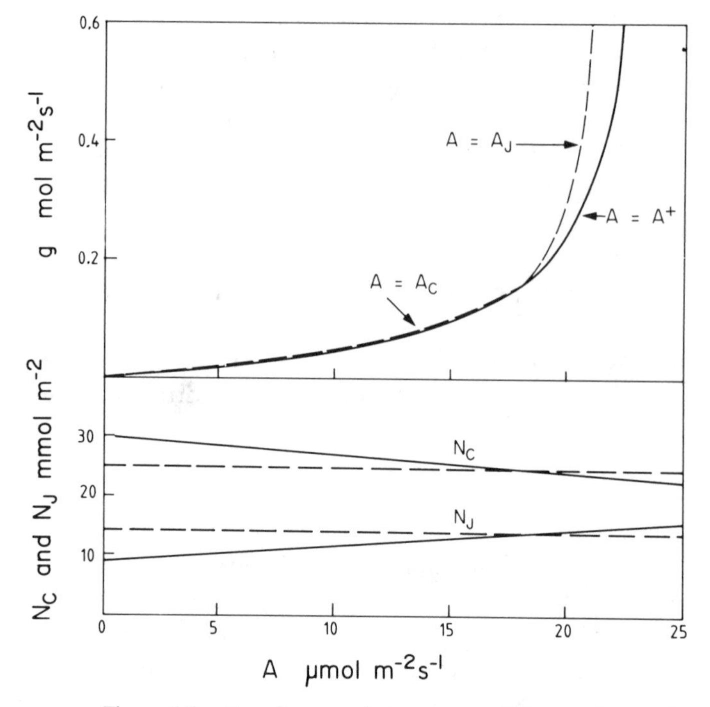

Figure 5.6. Gas-phase conductance, g, to CO_2 transfer as a function of rate of assimilation, A, with irradiance 1.6 mmol $m^{-2}\,s^{-1}$. The curve $A = A^+$ assumes that the distribution of nitrogen $N = 39$ mmol $m^{-2}\,s^{-1}$ corresponding to each magnitude of A is optimal (the variation of N_C and N_J is shown). The broken curve assumes $N_C = 25$ mmol m^{-2}, $N_J = 14$ mmol m^{-2}.

partial pressure of CO_2. Clearly, the distribution of nitrogen is optimal if $A = A^+$. Any redistribution of nitrogen would cause a decrease in A unless conductance were also changed. The data in Figure 5.5a appear to be approximately consistent with $A = A^+$; that is to say, the nitrogen distributions are approximately optimal, given that the conductances are what they are. However, the converse (that the conductances are optimal given that the distributions are what they are) does not follow, for the rate of assimilation would be enhanced if the conductances were greater. The limitation on nitrogen-use efficiency must be sought in terms of an increase in rate of water loss consequent on an increase in conductance. Figure 5.6, based on calculations described in Appendix II, shows the conductances required to sustain various rates of assimilation, it being assumed that the nitrogen allocation remains optimal with respect to conductance (i.e., that $A = A^+$). The corresponding variations in N_J and N_C are also plotted. In small, well-ventilated leaves, in which leaf temperature differs little from that of the ambient atmosphere, the rate of transpiration is almost proportional to conductance. Evidently, an increase in A^+ is then accompanied by a disproportionately large increase in the rate of transpiration. In broad leaves, the influence of the external boundary layer on leaf thermal relations affects both assimilation and transpiration and must be taken into account. However, provided the ambient temperature is not very far above the optimum for photosynthesis, a detailed analysis (not reproduced here) sustains the conclusion that any relative increase in $A = A^+$ associated with a hypothetical redistribution of nitrogen would entail a larger relative increase in the rate of transpiration. It follows that optimal nitrogen distribution is determined by a compromise between physiological requirements to take up CO_2 and avoid excessive water loss.

Of course, the foregoing argument was simplified by ignoring temporal variations in external conditions that influence the loss and gain of water and CO_2 in plants. The rate of transpiration at a given conductance varies with leaf environment, particularly with the ambient vapor-pressure deficit. Also, the relative importances of water loss and carbon gain vary with the amount of soil water available to a plant. The stomatal aperture responds to these factors. Assuming that the functions A_C and A_J are constant, then the relation between A and g will be that of the broken curve in Figure 5.6. However, variations in the environment directly affect the rate of assimilation also. In particular, the function A_J depends on irradiance. We may ask the question: Given that the distribution of nitrogen happens to be optimal with respect to conductance at one particular irradiance, does it remain so at other irradiances? Figure 5.7a shows A^+ and the conductance required to maintain $A = A^+$ as functions of I. But the actual rate of assimilation and the actual conductance are not found to vary with I in this way. They tend to vary as shown in Figure 5.7b, intercellular partial pressure of CO_2 remaining almost constant, except when I is small.

These observations show that the compromise between carbon gain and water loss in terrestrial plants could be enhanced if protein degradation and synthesis were so rapid and cheap that enzyme activities associated with A_C and A_J might vary as quickly as the stomatal aperture. Over periods of a few days or more, some reallocation of protein nitrogen might indeed occur. Over shorter periods, N_C and N_J are quasi-steady, and therefore the optimal allocation is itself a compromise determined by the variations in external factors that affect the carbon and water economy of the plant.

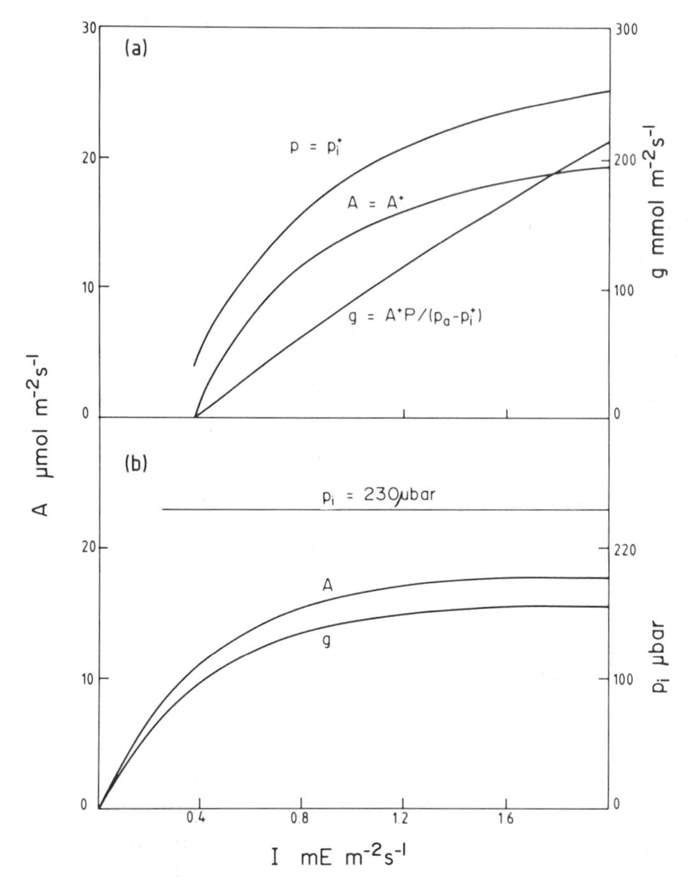

Figure 5.7. Rate of assimilation, A, intercellular partial pressure of CO_2, p_i, and the corresponding conductance to CO_2 transfer, g, as functions of irradiance, I. Both diagrams relate to the characteristics shown in Figure 5.5 and assume that the ambient partial pressure of CO_2, p_a, is 345 μbar. In (a), p and A correspond to the intersection (p_i^+, A^+). In (b), p_i is taken as 230 μbar.

Balance of carbon fixation and water loss

A paradigm of stomatal function

This topic will be touched only lightly here, having been dealt with at length elsewhere (Cowan 1977; Cowan and Farquhar 1977; Cowan 1982). It has to do with optimal stomatal regulation of the rates of assimilation and transpiration per unit leaf area in a plant during a finite interval of time (a single photoperiod is appropriate) in which there are variations in the radiation and atmospheric conditions to which the foliage is exposed,

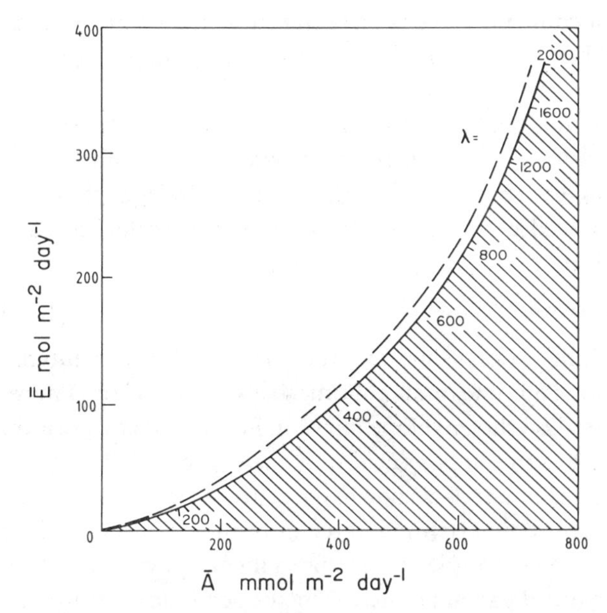

Figure 5.8. Hypothetical combinations of the average rate of assimilation, $\bar{A}$, and transpiration, $\bar{E}$, per unit leaf area per unit time in a plant during a day. Each point on the broken curve corresponds to a particular, constant stomatal conductance. Each point on the full curve corresponds to a particular variation of stomatal conductance that is optimal, in the sense that no other variation could have led to a smaller $\bar{E}$ at the same $\bar{A}$, or a larger $\bar{A}$ at the same $\bar{E}$ (Cowan 1982).

but changes in the amount of soil water available to the plant are small. The paradigm of optimal regulation can be explained by reference to Figure 5.8. The rate of CO_2 assimilation, $\bar{A}$, and the rate of transpiration, $\bar{E}$, averaged over time and over the whole surface of the foliage for a particular plant during a particular period are represented by a single point on this graph. It is the result of the physiological characteristics of the foliage and the physical conditions that the foliage encounters. Prominent among the physiological characteristics involved are those relating to stomatal behavior – temporal within-leaf variation in stomatal aperture as well as contemporaneous between-leaf variation contribute to the relationship between $\bar{A}$ and $\bar{E}$. Presumably, if the temporal variation in aperture had been different for a given leaf, the rates of assimilation and transpiration would also, in general, have been different. However, not all combinations of $\bar{A}$ and $\bar{E}$ are attainable. There is a region, represented by the shaded area in Figure 5.8, outside the range of stomatal control. Each point on the curve that bounds this region represents a unique variation of stomatal aperture that could not have been bettered – in the sense that no other variation could have led to a smaller $\bar{E}$ at the same $\bar{A}$, or a larger $\bar{A}$ at the same $\bar{E}$.

Provided we can assume that the physiological properties of a leaf are independent of the previous time course of stomatal aperture and of the pattern of stomatal aperture in the remainder of the foliage, then the paradigm will be achieved if the stomatal aperture in each leaf at each instant of time is such that

$$A - E/\lambda = \text{maximum} \tag{5.4}$$

where λ is a constant. The proof is simple: If the foregoing linear combination is everywhere and at all times a maximum, so also is $\bar{A} - \bar{E}/\lambda$. Therefore, $\bar{A}$ and $\bar{E}$ are located on the envelope curve in Figure 5.8 at a position determined by λ, which is the slope, $d\bar{E}/d\bar{A}$, of the curve.

The parameter λ is the benefit of carbon gain relative to the cost of water loss, with benefit and cost being defined in terms of their effects on plant fitness. The reason for employing λ, rather than a term equivalent to $1/\lambda$, say, is historical; Cowan and Farquhar (1977) introduced the topic in such a way that the use of λ was the more appropriate. Earlier, Givnish and Vermeij (1976) treated leaf size in relation to environment as an optimization problem by maximizing the function $A - bE$, with b being defined as "the metabolic cost of arranging for the supply of unit flow rate of water." This would be an inappropriate interpretation of the cost of E in the present context, for it can be assumed that the investment of metabolic energy in the growth of vascular tissue has already been made and is constant during the period of time being considered. A suitable interpretation, to be discussed later, relates E to the likelihood that removal of soil water will eventually cause the plant to become desiccated and metabolic function to cease altogether. Because the likelihood increases with decreases in soil water content, we can anticipate that λ will decrease with decreases in soil water content.

Estimates (Cowan 1982) suggest that the amounts of water potentially to be saved or carbon potentially to be gained as a result of appropriate variations in stomatal aperture are relatively small (Figure 5.8). However, the nature of stomatal responses to variations in environment indicates that the amounts have been significant in terms of an influence on selective adaptation.

Examples of optimal stomatal control

Cowan (1977) and Cowan and Farquhar (1977) have shown, using models of photosynthetic carbon metabolism and hypothetical variations in leaf physical environment, how the criterion of optimality is approximately reproduced in a plant in which stomatal opening is an appropriate balance between a positive response to irradiance and a negative response to external vapor-pressure deficit, and in which intrinsic differences in opening from leaf to leaf are correlated with intrinsic differences in internal capacity for CO_2 fixation. In essence, these are indeed the responses and differences that are characteristic of what is known about stomatal behavior in natural environments. Stomatal opening in response to an increase in irradiance is ubiquitous, except in plants having Crassulacean acid metabolism and in aquatic plants. Closure in response to an increase in vapor-pressure deficit, perhaps associated with peristomatal transpiration, has now been observed in many species (Schulze and Hall 1982) following a seminal experiment by Lange et al. (1971). Wong et al. (1979, 1985) and von Caemmerer (1981) have shown that conductance is positively correlated with leaf mesophyll capacity for photosynthesis, irrespective of whether differences in capacity are due to differences in previous exposure to light, differences in nutrition, or differences in age. However, the quantitative question of how closely the paradigm of optimality is reproduced in plants in their natural environments is exceedingly difficult to answer. Not only must A, E, and g be measured, but also sufficient information must be obtained to calculate how A and E would be affected by hypothetical changes in g. The physical environment of the leaves must be accurately characterized, and ancillary experiments must be done to determine the responses of the rate of assimilation to changes in intercellular partial pressure of CO_2 and leaf temperature. It is hardly surprising that investigations have, with one exception (Field et al. 1982), been confined to plants in the laboratory and have, with one exception (Hall and Schulze 1980), involved single-factor perturbations of the leaf environment. The following example is an extension of the analysis by Hall and Schulze of responses of stomata in *Vigna unguiculata* to variations in ambient humidity and temperature.

Figure 5.9a shows stomatal conductances to vapor transfer, g_s, at various ambient humidities and leaf temperatures. This is the data set against which Hall and Schulze tested the optimality hypothesis. Figure 5.9b shows rates of assimilation at various intercellular partial pressures of CO_2 and leaf temperatures. These are the data used to estimate the effects of hypothetical changes in conductance on the rate of assimilation. One of the observations of stomatal conductance (at 30°C leaf temperature and 15×10^{-3} mol mol^{-1} ambient humdity) is treated in Figure 5.10. Using Hall and Schulze's estimate of the conductance of the boundary layer to vapor transfer, it has been estimated how the rate of transpiration and leaf temperature would have varied had conductance been caused to vary without change in the external environment. Using the data in Figure

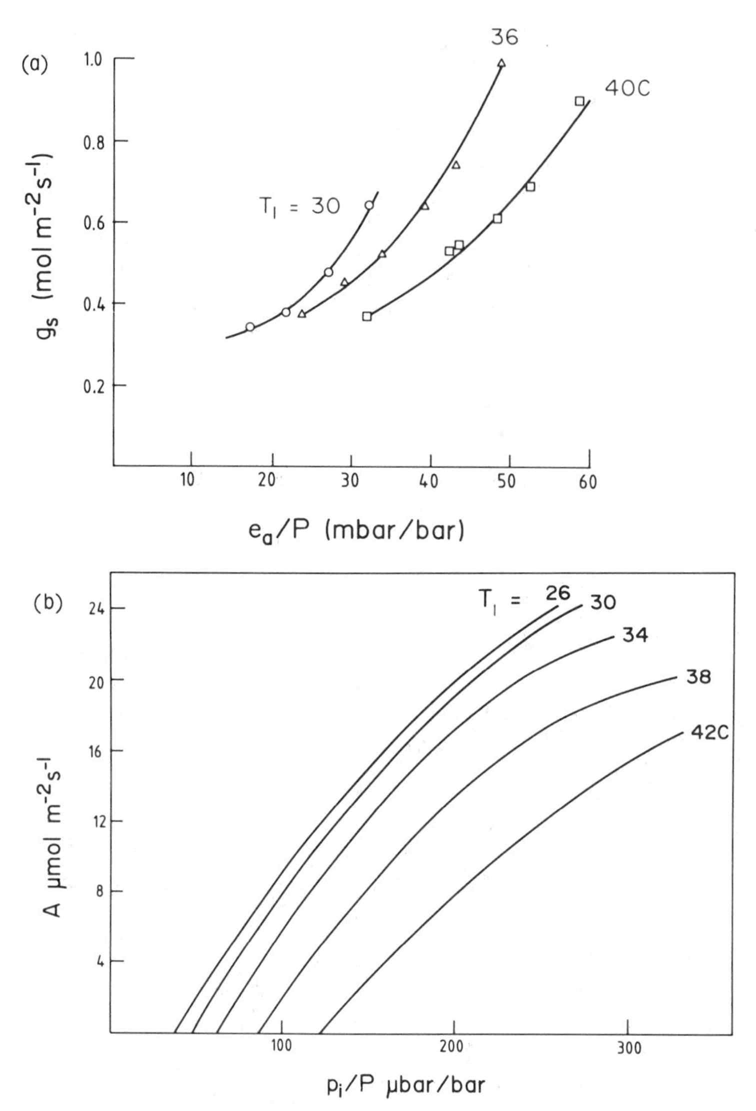

Figure 5.9(a). Stomatal conductances to vapor transfer, g_s, in a leaflet of *Vigna unguiculata* at various ambient humidities and leaflet temperatures; e_a is ambient vapor pressure, and P is total atmospheric pressure. Ambient CO_2 concentration was 300 ± 5 ppm. (b) Rates of assimilation in the same leaflet at various intercellular concentrations of CO_2 and leaflet temperatures. Irradiance was 0.9 mol m^{-2} s^{-1} for all data, and boundary-layer conductance was constant. (Data from Hall and Schulze 1980.)

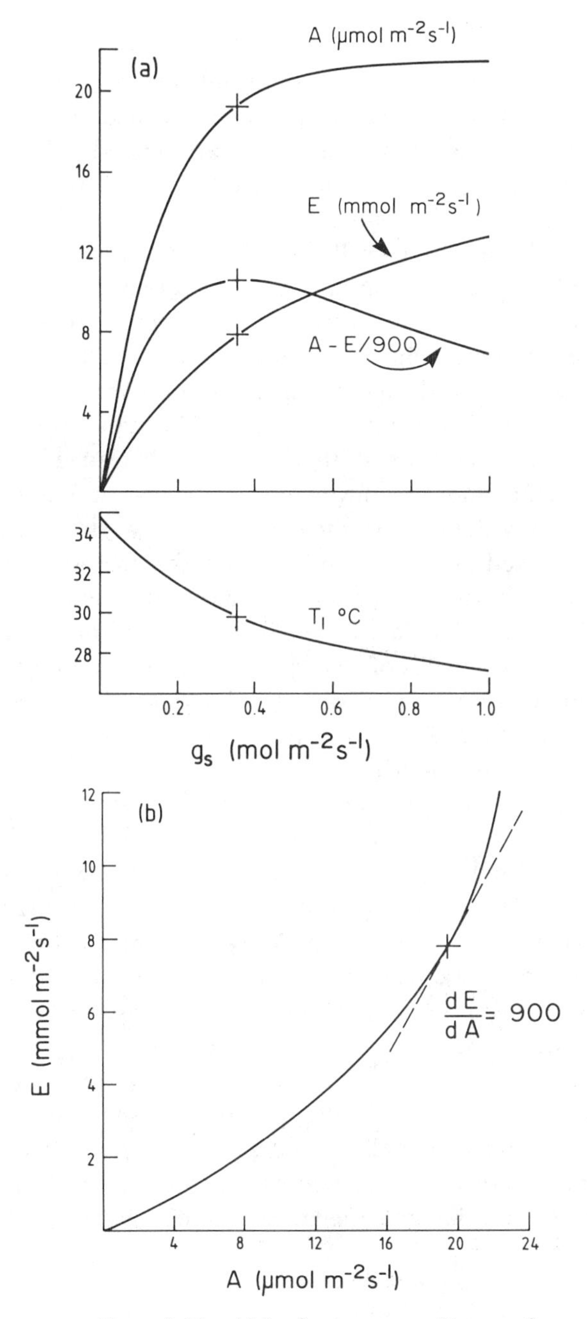

Figure 5.10. (a) Leaf temperature, T_l, rate of transpiration, E, rate of assimilation, A, and the combination $A - E/900$ as functions of assumed magnitude of conductance to vapor transfer, g_s. (b) E as a function of A. The functions were calculated using the data in Figure 5.9b and taking the environmental conditions to be those corresponding to the observation $g_s = 0.35$ mmol m^{-2} s^{-1} at $T_l = 30$°C and $e_a \div P = 15$ mbar/bar in Figure 5.9a.

5.9b, the corresponding variation in the rate of assimilation, taking into account the influence of variation in leaf temperature, has been found. The relations used in making these estimates are set out in Appendix III. It is important to note that if stomatal conductance had in reality made an excursion from its mean, then the environmental control system employed by Hall and Schulze would have automatically varied ambient air temperature so as to maintain leaf temperature constant. We shall presume that the stomata were unaware of this unusual characteristic of their environment.

Also shown in Figure 5.10a is the linear combination $A - E/900$. The value 900 has been chosen because the maximum of the curve then corresponds exactly with the observed stomatal conductance. The implication is that if the stomatal behavior was optimal, then $\lambda = 900$. Of course, the magnitude can be determined more readily by plotting E against A, as in Figure 5.10b, and measuring the slope of the curve, $\partial E/\partial A$, at the point corresponding to the observed values of A and E. When this procedure is carried out for all of the observations of stomatal conductance shown in Figure 5.9a, the magnitudes of $\partial E/\partial A$ are found to be rather uniform, the average being close to 900. Estimates of $\partial E/\partial A$ have also been found to vary very little while ambient humidity is varied in *Nicotiana glauca* and *Corylus avellana* (Farquhar et al. 1980b), *Lepechinia calycina* (Field et al. 1982), and *Pseudotsuga menziesii* (Meinzer 1982).

However, that the variation in $\partial E/\partial A$ is relatively small is not sufficient to prove that the stomatal responses conform to the paradigm of optimality. The variation that occurred in stomatal conductance should be compared with that which would have occurred if the stomata had conformed exactly with the criterion stated in equation (5.4). The comparison is shown in Figure 5.11. Evidently the variation of conductance is quite close to being optimal. The three points on the left illustrate exceptions. They are associated with conditions of ambient temperature and ambient humidity such that estimates of $A - E/900$ are negative when g_s is finite. Therefore, the maxima are $A - E/900 = 0$ at $g_s = 0$, corresponding to complete closure of the stomata. I shall discuss these exceptions again shortly, and only remark here that if the stomata had closed and the environment of the leaf had not changed, then the leaf temperature would have increased to about 48 °C.

Some constraints on control

Although investigations of the type described have been instructive, they have not added much to the qualitative assessment that stomatal responses *tend* to minimize $\overline{E}$ for a given $\overline{A}$ and maximize $\overline{A}$ for a given $\overline{E}$. Ambient humidity generally varies little during the course of a day out-of-doors. It is principally the variations in ambient temperature and irra-

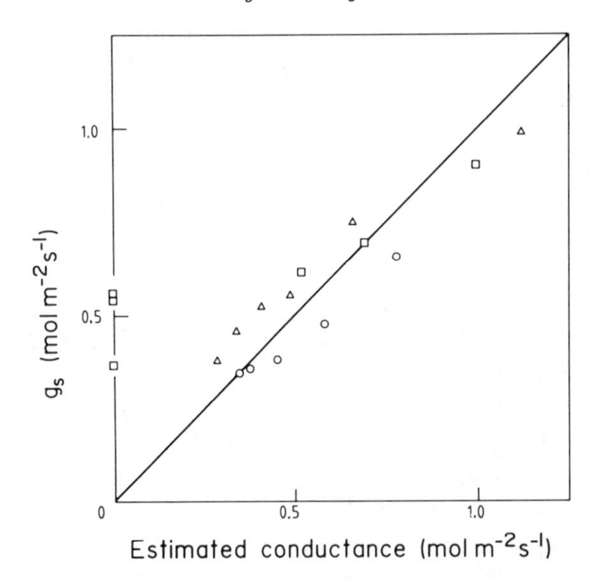

Figure 5.11. Observed magnitude of stomatal conductance to vapor transfer, g_s, plotted against the magnitude of conductance estimated to maximize the combination $A - E/900$. The observations are those shown in Figure 5.9a. The estimates were made by the means illustrated in Figure 5.10.

diance (there being a strong correlation between the two) that determine the variation in the potential rate of transpiration from foliage. No specific examination of the hypothesis of optimization with varying irradiance has yet been reported. It would require that measurements of the kind shown in Figure 5.9b be repeated at a number of irradiances and would therefore be a laborious task.

There are reasons to doubt that the paradigm as represented by equation (5.4) will be found to represent stomatal behavior very accurately. It is perhaps implausible that the stomatal complex, operating in a rapidly changing and partly unpredictable environment, has the capability of exploring the relationship between A and E and determining a maximum with respect to conflicting requirements for carbon gain and water conservation in the way we have done in the previous example. Indeed, we have assumed that it does not have this capability. But it is also unlikely that an array of open-loop and closed-loop responses to changes in environment could lead to exactly the same result. It follows that the paradigm of optimality can be no more than an approximation to the truth.

We note also that stomatal movements are much slower than some of the changes that may occur in a natural environment, particularly changes in irradiance associated with broken cloud, for example. This, too, would seem to place a constraint on the extent to which control of gas exchange is "optimal." But, alternatively, the constraint itself might have an economic basis. Perhaps the metabolic cost of stomatal movement is sufficient to make frequent movements uneconomical. If this is so, then it should be taken into account in deriving the implications of the paradigm of optimality. In effect, the net rate of carbon assimilation will be a function not only of environment and leaf conductance but also of the rate of change of leaf conductance. Equation (5.4) will define optimal control only in a slowly varying environment.

In general, the derivation of equation (5.4) presumes that the physiological properties of a leaf influencing A and E are independent of the previous time course of conductance in that leaf and of conductance in the remainder of the foliage. Comins and Farquhar (1982) considered optimization of nocturnal gas exchange in plants having Crassulacean acid metabolism. The presumption is not valid in this instance, because assimilated CO_2 is stored in the form of malic acid in the mesophyll tissue, and the capacity to assimilate CO_2 declines with the amount of acid already accumulated. A treatment similar to that used by Comins and Farquhar would be appropriate for diurnal gas exchange in other plant species if accumulation of photosynthate in leaf tissue were to inhibit the capacity for further photosynthesis. A second example of the same kind of theoretical complication – but one having a different physiological basis – relates to the influence of conductance on leaf temperature. If temperature is sufficiently high to cause permanent or transient damage to the photosynthetic metabolism of a leaf, then the presumption underlying equation (5.4) is again invalid, because conductance must affect the subsequent ability of the leaf to fix carbon. It may be that the "failure" of the stomata to close in *V. unguiculata* (*vide supra*) in circumstances in which closure would have caused leaf temperature to become very high was not inconsistent with the paradigm of optimization.

In this discussion of the difficulties in applying optimization theory to stomatal function I have not questioned the appropriateness of the paradigm, only that of the criterion in equation (5.4). The reason for assuming that a stomatal control system maximizing $\bar{A}$ for a given $\bar{E}$ and minimizing $\bar{E}$ for a given $\bar{A}$ confers a selective advantage to plants will emerge in the next section.

Avoidance of drought

Relation between carbon fixation and likelihood of desiccation

An attempt will be made to indicate how the function in Figure 5.8 involving daily rates of assimilation and transpiration might be transformed into a function of quantities more nearly related to the competitive fitness of plants. The treatment is surrounded by restrictions. It is appropriate only to a mature, perennial, evergreen plant in which the relative growth rate is small, with carbon assimilation being used primarily to make good the losses associated with maintenance respiration, senescence of leaves and roots, and herbivory and for the production of reproductive organs and propagules. It is assumed that the climate is constant from day to day except in respect to rainfall. Plant metabolism is unaffected by the potential of water available to the roots unless the total amount of water in the root zone is depleted to a particular critical level, in which case the plant dies.

Figure 5.12 shows a sequence of processes influenced by rainfall. It is most easily understood beginning with the curve representing the daily rate of assimilation, $\overline{A}$, as a function of soil water deficit, ϕ, in the bottom right-hand corner, and then following the relationships counterclockwise. The soil water deficit is zero when the soil has been thoroughly wetted by rain and is taken as unity when all of the available water, q per unit plant leaf area, say (here taken as 500 mol m^{-2}), has been removed from the root zone of the plant. The available water and the soil water deficit are old concepts, not much used today, but they are necessary simplifications in the present context. It is assumed that the decline in $\overline{A}$ with an increase in ϕ is due only to stomatal closure (later this assumption will be relaxed). Therefore, variation in $\overline{A}$ is associated with variation in the rate of transpiration, $\overline{E}$, the link being an optimal relationship of the kind discussed in the previous section.

In addition to the loss of water from the plant root zone due to transpiration, there is loss of water that is not under direct control by the plant. Water is lost by evaporation close to the soil surface and by drainage to depths beyond the root zone. It can be withdrawn by encroaching roots of neighboring plants. And the plant itself will sustain an uncontrolled loss of water through the cuticle of its aerial parts. Let the total rate of uncontrolled loss of water per day per unit area of leaf be E_0. It is taken to be a function of ϕ. It is opportune here to draw attention to a weakness in the analysis to be presented. Clearly, E_0 (and $\overline{E}$) will depend not only on ϕ but also on the distribution of water in the root zone. For example, E_0, insofar as it incorporates evaporation from the soil, will be large if it happens that because of recent light rain the available water is concentrated near the soil surface. I have not yet been able to take this complication into account. Thus, the function $E_0(\phi)$ expresses only the fact that E_0 will, *on the average,* be larger when ϕ is small than when it is large.

Given the total rate of withdrawal of water from the plant root zone, $\overline{E} + E_0$, as a function of ϕ, and given a record of rainfall, then the time

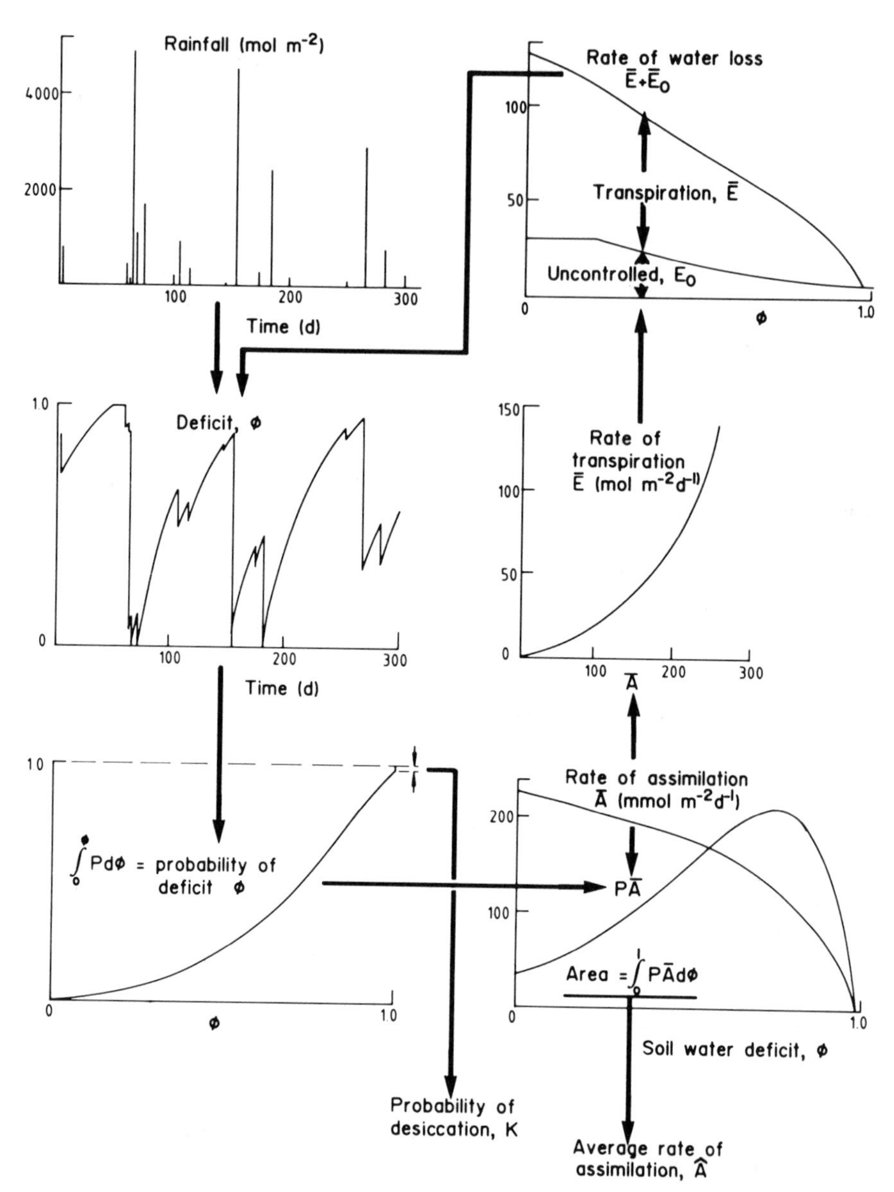

Figure 5.12. Interrelationships of processes influencing the probability of desiccation, K, and long-term average rate of assimilation, $\hat{A}$, in a plant.

course of ϕ follows. The particular sequence of rainfall in Figure 5.12 was generated using random numbers and two assumptions. The probability of rain occurring in unit time was taken to be constant, irrespective of the previous rainfall history, and the probability of a rainfall being greater than a certain amount was taken as decreasing exponentially with the amount. Therefore, the statistical distribution of rainfall is characterized by two parameters only: the average interval between successive falls, τ (taken here as 20 days), and the average size of a rainfall, ρ (taken here as 1,000 mol m^{-2}), expressed as the amount of water, per unit leaf area of the plant, falling on the soil catchment area, determined by the lateral spread of the plant roots. It follows that ρ/τ is the rate of water loss per unit leaf that can be continuously sustained if the water-holding capacity of the root zone, q, is infinitely large. By generating a sufficiently extensive rainfall record, and the corresponding variation of deficit, the probability distribution of the deficit shown in Figure 5.12 can be found. But the same result can be obtained analytically, as shown in Appendix IV. $P(\phi)\, d\phi$ is the instantaneous probability that the deficit is in the range ϕ to $\phi + d\phi$. The probability, K, that $\phi = 1$ is the complement of the integral of P over the range 0 to 1. It is a rough measure of the likelihood that a plant will become desiccated and then die within any given period of time. It will be assumed that the probability K of soil water becoming exhausted is so small that the plant is likely to survive for a period that is very long compared with the average interval between successive rainfalls, τ.

The final link in Figure 5.12 connects the probability distribution $P(\phi)$ with the variation $\overline{A}(\phi)$ first considered. The average rate of assimilation during the lifetime of the plant is approximately

$$\int_0^1 \overline{A}P\, d\phi = \hat{A} \tag{5.5}$$

which is the shaded area in the figure. Thus, we have estimates of both the average assimilation rate and the probability of death due to desiccation, K.

It will be apparent that the greater the average rate of assimilation, $\hat{A}$, then the greater the corresponding average rate of evaporation, and the greater the risk, K, that the plant will become desiccated. But the risk does not depend on the average rate of assimilation alone. It depends also on the shape of the function $\overline{A}(\phi)$, that is to say, the way in which the plant "chooses" to vary its rate of assimilation per day with an increase in soil water deficit. We may ask the question what form $\overline{A}(\phi)$ should take so that for any particular magnitude of K, the average rate of assimilation, $\hat{A}$, is the maximum possible. The $\overline{A}(\phi)$ variation in Figure 5.12 is in fact one that subscribes to this criterion.

Optimal responses of stomata to variations in soil water content

The problem is equivalent to finding $\overline{A}(\phi)$ such that

$$\hat{A} - \nu K = \text{maximum} \tag{5.6}$$

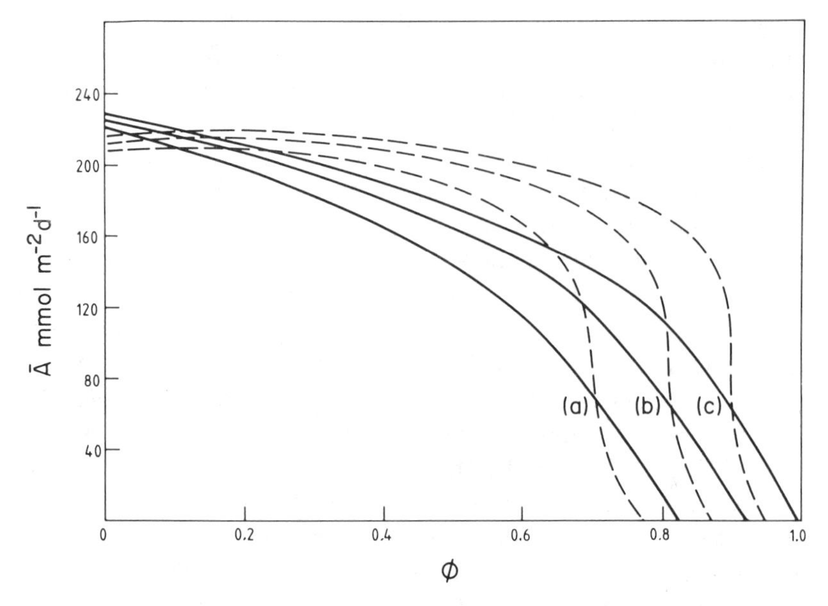

Figure 5.13. Optimal variations in daily rate of assimilation, A, with soil water deficit, ϕ. Each curve corresponds to a particular combination of probability of desiccation, K, and average rate of assimilation, $\hat{A}$. The full curves relate to plants that do not compete for water with their neighbors; the broken curves relate to plants that are identical except that their root systems occupy twice the land area uniquely available to each plant. The magnitudes of K are 2×10^{-3}, 2×10^{-2}, and 2×10^{-1} for curves (a), (b), and (c), respectively. The magnitudes of $\hat{A}$ are different for plants that do and do not compete, as shown in Figure 5.14.

where ν is an undetermined multiplier. Appendix V shows this requires that

$$\frac{d\chi}{d\phi} + \frac{q}{\rho}[\chi - \chi(0)] + \frac{q}{\tau} \cdot \frac{1}{\lambda} = 0 \tag{5.7}$$

where $\chi = \overline{A} - (\overline{E} + E_0)/\lambda$, with, as before, $\lambda = d\overline{E}/d\overline{A}$. Given q, ρ, and τ and the functions $\overline{E} = \overline{E}(\overline{A})$ and $E_0 = E_0(\phi)$, this equation can be solved numerically for any assumed boundary condition, $\chi = \chi(0)$ at $\phi = 0$, to provide the functions $\overline{A}(\phi)$ and $\overline{E}(\phi)$. The corresponding magnitudes of $\hat{A}$ and K can then be obtained.

For the purposes of example, $\overline{E}(\overline{A})$ and $E_0(\phi)$ are taken to be the functions illustrated in Figure 5.12, with $q = 5{,}000$ mol m^{-2}, $\rho = 1{,}000$ mol m^{-2}, and $\tau = 20$ days. Figure 5.13 shows optimal variations in $\overline{A}$ and $\overline{E}$. Each variation corresponds to particular magnitudes of $\hat{A}$ and K, indicated in Figure 5.14.

Why do $\overline{A}$ and $\overline{E}$ vary with ϕ as they do? The principal reason is that rainfall is unpredictable. The shape of the $\overline{E} = \overline{E}(\overline{A})$ function in Figure

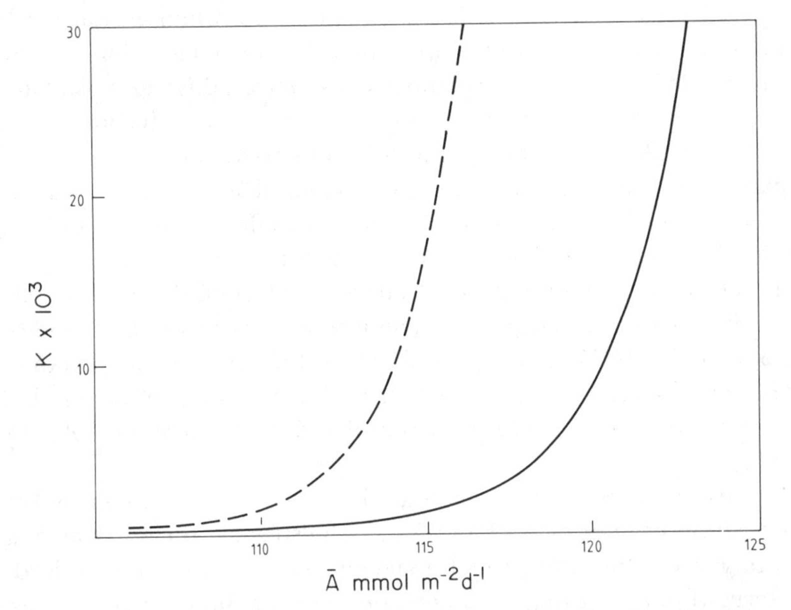

Figure 5.14. Combinations of the probability of desiccation, K, and the average rate of assimilation, $\hat{A}$. Each point on the curves corresponds to a unique variation of the daily rate of assimilation, $\overline{A}$, with soil water deficit, ϕ, that is optimal in that no other variation would lead to a smaller K at the same $\hat{A}$, or a larger $\hat{A}$ at the same K. The full curve relates to plants that do not compete for water; the broken curve relates to plants that are similar except that their root systems occupy twice the land area uniquely available to each plant.

5.12 has the following implication: If a plant were constrained to use a given amount of water in a given time, then the corresponding amount of carbon fixed would be maximized if the water were used at a constant rate. But in reality the period that a plant has to eke out the water available to it is not determinate. The optimum rate of transpiration depends on the amount of water in the soil and on a statistically based expectation of rainfall. Because the expectation of rainfall is constant, independent of the time at which rain last occurred, the optimal magnitude of $\overline{E}$ declines with an increase in ϕ.

There is another reason for the decreases in $\overline{A}$ and $\overline{E}$ with ϕ. The uncontrolled rate of water loss $E_0(\phi)$ tends to be greatest when the soil is very wet (i.e., ϕ is small). Therefore, the plant acquires a greater proportion of the water available to it if it also uses water most rapidly when ϕ is small. A note on what will happen if a part of E_0 is associated with absorption by the root systems of neighboring plants that adopt a similar strategy of water use is of interest here. The repercussions can be deduced in the following way.

Suppose Figures 5.13 and 5.14 refer to a plant in a community in which the root systems do not quite overlap, so that E_0 is not in fact influenced by the behavior of the neighboring plants. Let us contrast this with a plant in a community that is similar in every respect except that each root system occupies a lateral area n times as great. Adjustments to three parameters in our analysis are required: q and ρ are to be replaced by nq and $n\rho$, and E_0 is to be replaced by $(n - 1)\overline{E} + nE_0$. However, the effective amounts of soil surface and soil water-holding capacity available to a plant are unchanged, and therefore the substitutions should have no effect on the form of $P(\phi)$ and K as functions of $q, \rho, \overline{E}$, and E_0. This is readily confirmed by consideration of equation (5.IV.2) in Appendix IV. It follows that equation (5.6) need be changed only by the same explicit substitutions. The result is that q/τ is replaced by nq/τ, and $\overline{E} + E_0$ is replaced, in the definition of χ, by $n(E + E_0)$.

Solutions of the modified equation with $n = 2$ are illustrated by the broken curves in Figures 5.13 and 5.14. The struggle for water among competing plants (the same struggle that entices roots to intermingle) leads to patterns of water use that are quite different from those in plants whose root zones do not overlap each other. The average rate of assimilation, $\hat{A}$ corresponding to any given probability of desiccation, K, has been reduced. The community would seem to be the poorer, quite apart from the apparent inefficiency associated with the allocation of carbon for growth and maintenance of larger, but no more productive, root systems. Nevertheless, the relationship between $\hat{A}$ and K would not be bettered in any one individual if that individual were to adopt a different pattern of water use.

The analysis goes some little way to explaining why closure of stomata and declines in rates of assimilation and transpiration with decreases in soil water content may differ in different species – not mechanistically in terms of differences in intrinsic plant metabolic characteristics but in terms of differences in the statistical properties of the rainfall regimen, the leaf environment, and the nature of the competition for water to which species are adapted. The complications that have been left out of contention are too numerous to list. I shall mention only one of them here. Closure of stomata is not the only reason that the rate of CO_2 fixation decreases with a decrease in soil water content. The capability of the mesophyll tissue to fix carbon becomes inhibited. For this reason, the quantity $d\overline{E}/d\overline{A}$ is sometimes found to increase with a decrease in soil water content (Farquhar et al. 1980b). To the extent that the metabolic characteristics of carbon fixation can be expressed as a function of soil water content, this complication can readily be encompassed in the foregoing analysis. The unique relationship between $\overline{E}$ and $\overline{A}$ previously assumed is replaced by the function $\overline{E} = \overline{E}(\overline{A}, \phi)$. Equation (5.7) is unchanged, except that λ must be redefined as $(\partial \overline{E}/\partial \overline{A})_\phi$.

However, to introduce an empirical description of the influence of water stress on carbon metabolism in this way would detract from whatever merit the analysis has. We do not understand the fundamental constraints, if any, underlying the inhibition of carbon metabolism at negative potentials of water that can hardly be expected directly to influence rates of chemical reactions. Nor is there any explanation apparent in terms of economic theory, such as a reallocation of protein nitrogen with a decline in leaf conductance in the way suggested by Figure 5.6. Whereas generally it is the capability for CO_2 fixation at a high intercellular partial pressure of CO_2 that is most inhibited, there is no evidence that carboxylase activity is correspondingly enhanced; indeed, carboxylase activity sometimes declines (e.g., Mooney et al. 1977).

Optimal allocation of carbon between root and shoot

Thus far, attention has been concentrated on the cost to the plant, in terms of water use and its repercussions, of acquiring carbon. The discussion has been rather like a previous treatment of the subject (Cowan 1982). However, the previous treatment contained the assumption that every rainfall saturates the plant's root zone, a restriction that made it impossible to deal with the final topic of this chapter: the cost, in terms of carbon, of acquiring water. The available water-holding capacity, per unit area of plant leaf, of the plant root zone, q, is dependent on the size of the root system. The capacity will be greater, and the risk of becoming dehydrated will be less, in plants that have larger root–shoot ratios. Let the amount of carbon in the roots per unit area of leaf be R, and let the available water-holding capacity of the root zone relative to the amount of carbon in the roots be γ. Then

$$q = \gamma R \tag{5.8}$$

Needless to say, γ will vary with the plant's growth form, the stage of development, and the soil type. In a numerical example that follows, γ is taken as 5,000 mol mol^{-1}, corresponding approximately to 100 g dry weight of root per 1 m^3 of soil with a volumetric available-water capacity per unit volume of 0.3. It will be assumed that variation in R reflects variation in depth, rather than the lateral extent of rooting, so that no adjustment should be made to ρ, the average amount of a rainfall relative to the leaf area of the plant.

The advantage of large R in conferring security from drought is offset by a disadvantage in allocating carbon to the maintenance of nonphoto-

Table 5.1. *Hypothetical relationships between the water-holding capacity of the root zone and measures of relative investment in roots versus shoots*

Root carbon/leaf area, R (mol m^{-2})	Root carbon/shoot carbon, R/S	Plant carbon/leaf area, $R+S$	Soil water capacity/leaf area, $q \times 10^{-3}$
0.8	0.16	5.8	4
1.0	0.20	6.0	5
1.2	0.24	6.2	6
1.4	0.28	6.4	7
1.6	0.32	6.6	8
1.8	0.36	6.8	9
2.0	0.40	7.0	10

synthesizing tissue. Let the amount of carbon in the shoot (stems and leaves) per unit leaf area be S. It follows that the average rate of carbon fixation per unit mass of plant carbon is

$$G = \frac{\hat{A}}{R + S} \tag{5.9}$$

Of course, G is a first-order expression only of the ability of a plant to maintain itself, grow, and propagate. Respiration and attrition due to herbivory and senescence are certain to vary between leaves, stems, and roots. This complication is ignored. The quantity S will here be taken as 5 mol m^{-2}, perhaps appropriate for a mature, thick-leaved shrub. Table 5.1 shows the water-holding capacity of the plant root zone and the amount of plant carbon per unit leaf area for several magnitudes of root carbon per unit leaf area.

The effects of R on the interrelationship between G and K are illustrated in Figure 5.15. Each curve has been constructed in the way described in the previous section, K and $\hat{A}$ being found for the appropriate magnitude of q in Table 5.1, and $\hat{A}$ being transformed to G according to equation (5.9). The curves intersect, and there is an envelope curve representing the optimal relationship between G and K, the linking parameter being R. Each point on this curve is optimal in the following sense: Given the particular magnitude of R, then the relationship could not be bettered by adopting a different variation of rate of assimilation with soil water deficit. And given the variation of $\overline{A}$ with ϕ, it could not be bettered by adopting a different magnitude of R. Thus, a link has been established between the control of carbon fixation in relation to water loss and the allocation of carbon for growth.

This treatment of carbon allocation could also be developed to encom-

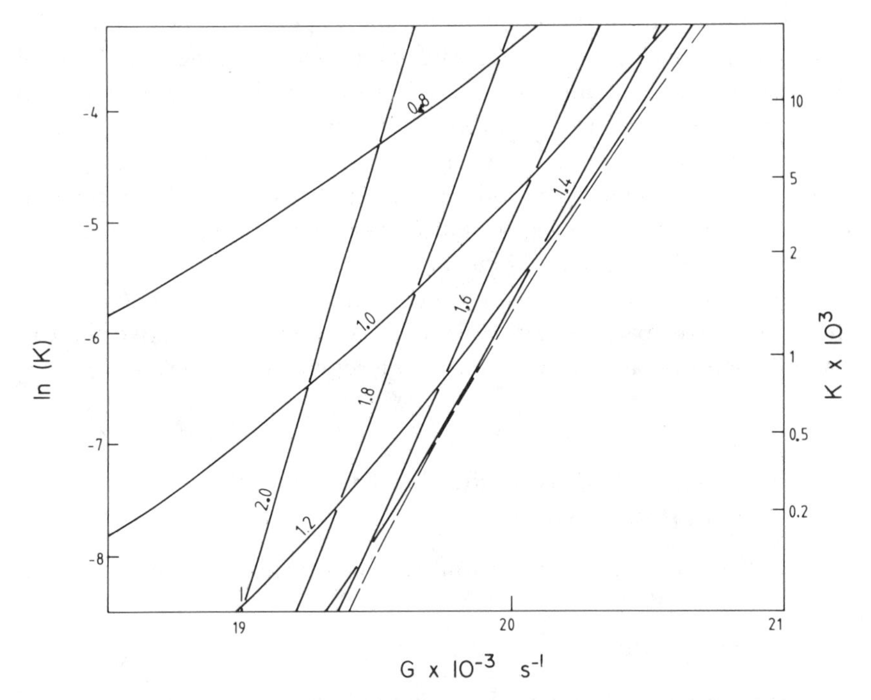

Figure 5.15. Combinations of the probability of desiccation, K, and the average relative rate of carbon fixation, G, in plants having different amounts of root carbon per unit leaf area, R (mol m^{-2}). The curve $R = 1.0$ corresponds to the full curve in Figure 5.14. Each point on the broken envelope curve corresponds to a unique alliance of an optimal variation in $\bar{A}$ with ϕ, and a magnitude of R, such that no other would lead to a smaller K at the same G, or a larger G at the same K.

pass the lateral extent of rooting. Both q and ρ would be allowed to increase with R, and the variation of E_0, including the problems associated with plant competition already referred to, would need to be taken into account. However, it must be admitted that the result would remain a much simplified description of environmental pressures and adaptive responses underlying growth and survival in water-limited environments. Without a testable prediction in immediate sight, it would perhaps be inappropriate to carry the formal analysis further.

Conclusions

As we have progressed from chloroplast to leaf to plant in attempting to describe and analyze the compromises that are made in the acquisition and use of carbon, the need for and the difficulty of making a compromise of our own become increasingly apparent. The compromise is that of maintaining touch with reality and yet keeping the complexity of the description to a minimum. Doubtless there will be improvements in what has been done here; many topics mentioned only by way of qualification will perhaps become incorporated into the conceptual framework without undue complication, and the tenuous thread that links the sections in this chapter will be strengthened. Beyond that, there is a challenge to take account of seasonality of climate, to consider other drought-avoidance mechanisms, such as deciduousness and the annual growth habit, and then to relate the balance of growth and survival in the vegetative phase to that of dispersion and germination in the reproductive phase.

Appendix I: Carbon dioxide fixation with facilitated transfer

The rate of fixation per unit volume of chloroplast is

$$f(p) = \frac{k_c X_c p}{p + K_c'} \simeq f(p_i) - k_f(p_i - p) \tag{5.I.1}$$

where p is the partial pressure of CO_2 in solution, X_c is the molar density of carboxylase in the chloroplast, k_c is the turnover rate of carboxylase, K_c' is the Michaelis constant for carboxylase expressed in terms of partial pressure (including the effect of competitive inhibition by oxygen at normal atmospheric partial pressure), p_i is the partial pressure of CO_2 in the intercellular air space, and k_f is defined as df/dp at $p = p_i$. The magnitudes of k_c and K_c' are taken, at 25°C, as 29.3 s^{-1} (Evans 1983b) and 670 μbar (Badger and Andrews 1974; Evans and Seemann 1984; Brooks and Farquhar 1985), respectively. The linear expression in p is a satisfactory approximation provided that p is not very much less than p_i.

The inhibition of carboxylase by oxygen is associated with oxygenation of ribulose-bisphosphate and subsequent release of photorespiratory CO_2 in the mitochondria. The rate of respiration expressed per unit volume of chloroplast is

$$r = f(p)\Gamma^*/p \simeq f(p_i)\Gamma^*/p_i$$

where Γ^* is the partial pressure of CO_2 at which the rate of CO_2 fixation equals the rate of photorespiration and is 41 μbar at 25°C (Brooks and Farquhar 1985). The net rate of hydration of bicarbonate per unit volume of chloroplast is, say,

$$h = \frac{k_a X_a}{K_a}\left(p - \frac{b}{s_b}\right) = k_h\left(p - \frac{b}{s_b}\right) \tag{5.I.2}$$

in which b is the molar concentration of bicarbonate in solution, s_b is the concentration of bicarbonate in equilibrium with 1 bar partial pressure of CO_2, X_a is the molar density of anhydrase in the chloroplast, k_a is the turnover rate of anhydrase, and K_a is the Michaelis constant for anhydrase expressed in terms of partial pressure of CO_2. The magnitude of s_b is taken as 1.5 mol m^{-3} bar^{-1}, assuming that the equilibrium ratio of the HCO_3^- and CO_2 densities is 45 at pH 8 (Edsall and Wyman 1958) and the solubility of CO_2 in solution, s_c, is 33.0 mol m^{-3} bar^{-1} (Linke 1958). The values of k_a and K_a are taken as 3.6×10^5 s^{-1} and 60 mbar at pH 8 and 25°C (Pocker and Ng 1973).

The convergence of the diffusive flux of CO_2 across the chloroplast is equal to the sum of the rate of fixation and the net rate of hydration; that is,

$$s_c D_c \frac{d^2p}{dz^2} = f + h \tag{5.I.3}$$

in which D_c is the effective coefficient of diffusion of CO_2 in the chloroplast, assumed (allowing for constraints due to the presence of macromolecules and membranes in the chloroplast) to be half that of CO_2 in water, 2.0×10^{-9} (Duda and Vrentas 1968). The corresponding equation for bicarbonate is

$$D_b \frac{d^2b}{dz^2} = h \tag{5.I.4}$$

where D_b is the effective coefficient for diffusion of HCO_3^-, assumed to be $0.56D_c$ (Kigoshi and Hashitani 1963).

Equations (5.II.1)–(5.II.4) can be solved for g, h, p, and b. The boundary conditions are $db/dz = 0$ at $z = 0$ and $z = l$ (because the chloroplast is believed to be impermeable to HCO_3^-) (Werdan and Heldt 1972), $F = G(p_i - p) = -S_c D_c(dp/dz)$ at $z = 0$ (G being the permeability of the cell-wall, plasma-membrane, and chloroplast envelope), and $S_c D_c(dp/dz) = lf(p_i)\Gamma^*/p_i$ at $z = l$. The permeability G is taken as 77 mmol m^{-2} s^{-1} bar^{-1} (Evans 1983a, 1983b). The solution is tedious to derive, and the resulting expressions are cumbersome. Only the expression for the flux density of CO_2 across the cell wall is required here. It is

$$F = \frac{lf(p_i)[1 - (\rho_1 + \rho_2)\Gamma^*/p_i]}{lk_f/G + \rho_1 \cosh(\alpha_1^{1/2}l) + \rho_2 \cosh(\alpha_2^{1/2}l)} \tag{5.I.5}$$

where

$$\rho_1 = \frac{(k_f - s_c D_c \alpha_2)}{s_c D_c(\alpha_1 - \alpha_2)} \frac{\alpha_1^{1/2}l}{\sinh(\alpha_1^{1/2}l)},$$

$$\rho_2 = \frac{(s_c D_c \alpha_1 - k_f)}{s_c D_c(\alpha_1 - \alpha_2)} \frac{\alpha_2^{1/2}l}{\sinh(\alpha_2^{1/2}l)}$$

and α_1, α_2 are the roots of

$$\alpha^2 - \alpha\left(\frac{k_f + k_h}{s_c D_c} + \frac{k_h}{s_b D_b}\right) + \frac{k_f k_h}{s_c s_b D_c D_b} = 0$$

With v being the volume of chloroplast per mole of chlorophyll, taken here as 0.027 $m^3\ mol^{-1}$ (Nobel 1968), then the net rate of CO_2 fixation per mole of chlorophyll is $A = Fv/l$. With M_c (= 550 kd) and M_a (= 180 kd) being the molecular masses of carboxylase and anhydrase, then the amounts of nitrogen per mole of chlorophyll invested in carboxylase and anhydrase are $N_c = 11.4vX_cM_c \times 10^{-3}$ and $N_a = 11.4vX_aM_a \times 10^{-3}$. In this way the solutions of equation (5.II.5) are expressed in terms of the variables used in the main text.

Because

$$F = \frac{lk_cX_c(\bar{p} - \Gamma^*)}{\bar{p} + K_c'} \qquad (5.I.6)$$

p, the mean partial pressure of CO_2 in the chloroplast, is readily found once F has been obtained. For the point in Figure 5.3, $N_R = 61$, $N_c = 2.3$, $l = 1.7$ μm, thought to be representative of chloroplast functioning, $F = 2.9$ $\mu mol\ m^{-2}\ s^{-1}$, and $p_i - p = 58$ μbar. It follows that the corresponding conductance $F/(p_i - p) = G_1$, say, is 50 $mmol\ m^{-2}\ s^{-1}\ bar^{-1}$, to be compared with 77 $mmol\ m^{-2}\ s^{-1}\ bar^{-1}$ for the conductance of cell wall and membranes alone. The whole-leaf "liquid-phase conductance" with the dimensions commonly used for transfer in a gas phase is $g_l = aPG_l$, where a is the area of chloroplast adjacent to the cell wall per unit area of leaf, and P is total atmospheric pressure. Taking $a = 10$, $P = 1$ bar, then $g_l = 0.50$ $mol\ m^{-2}\ s^{-1}$. This magnitude is used in calculations described in Appendix II.

Appendix II: Rate of assimilation as a function N_J and N_C

From von Caemmerer and Farquhar (1981), equation (5.3) corresponds to

$$A = (\bar{p} - \Gamma^*) \cdot \min\left(\frac{V_{cmax}}{p + K_c'}, \frac{J}{4.5p + 10.5\Gamma^*}\right) - R_d \qquad (5.II.1)$$

where V_{cmax} is the rate of carboxylation at a high partial pressure of CO_2 and an unlimited supply of ribulose-bisphosphate, J is the potential rate of electron transport, and R_d is the rate of mitochondrial respiration, all of these qualities being expressed per unit area of leaf. The definitions and assumed values of K'_c and Γ are given in Appendix I. From Farquhar and Wong (1984), J is the smaller root of

$$0.67J^2 - (J_{max} + J_I)J + J_I J = 0 \quad (5.II.2)$$

in which J_{max} is the maximum capacity of the electron-transport system, and J_I is the magnitude of J at low irradiance, being taken, following Farquhar and Wong (1984), as $0.32I$, where I is the irradiance of the leaf.

In equation (5.II.1), p is the average partial pressure within the chloroplast. To express A in terms of intercellular partial pressure of CO_2, p_i, the equation is solved simultaneously with

$$A = g_l(p_i - \bar{p})/p \quad (5.II.3)$$

in which g_l is the liquid-phase conductance and has been estimated in Appendix I as 0.50 mol m^{-2} s^{-1}.

An activity, V_{cmax}, of 1 mol s^{-1} corresponds to an investment of nitrogen in carboxylase of $11.4(M_c/k_c) \times 10^{-3} = 217$ mol. Because approximately 0.86 of the nitrogen in Calvin-cycle enzymes is accounted for by carboxylase (Evans 1983a), it follows that $N_c/V_{cmax} = 250$ s. Evans (1983a) determined the maximum rate of electron transport in wheat as 540 mol s^{-1} per mole chlorophyll and estimated that the amount of nitrogen in proteins associated with the two photosystems, the light-harvesting complex, and the electron-transport chain is 44.4 mol per mole chlorophyll. Therefore, $N_J/J_{max} = 80$ s. Typically, $V_{cmax} = 100$ μmol m^{-2} s^{-1}, and $J_{max} = 175$ μmol m^{-2} s^{-1} in *P. vulgaris* liberally supplied with nitrogen (von Caemmerer and Farquhar 1981), corresponding to $N_C + N_J = N = 39$ mmol m^{-2}. This is the figure assumed in the calculations underlying Figures 5.6 and 5.7. Of course, N is only a part, perhaps about 40% in herbaceous plants (Evans 1983a, 1983b), of total leaf nitrogen content.

Using equations (5.II.1)–(5.II.3), together with the proteinaceous nitrogen contents corresponding to V_{cmax} and J_{max}, it is possible to determine the functions $A_C = A_C(p_i, N_C)$ and $A_J = A_J(p_i, I, N_J)$. The condition $A_C = A_J$ yields the solution for the intersection, $p_i^+ = p_i^+(I, N_C, J_J)$ and $A^+ = A^+(I, N_C, N_J)$.

Appendix III: Variation of *E* and *A* with leaf conductance

The rates of transpiration corresponding to the observations in Figure 5.7a are

$$E = \frac{[e'(T_l) - e_a]/P}{1/g_s + 1/g_b} \quad (5.III.1)$$

where $e'(T_l)$ is the saturation vapor pressure at temperature T_l, and g_b is the boundary-layer conductance to vapor transfer (2.43 mol m^{-2} s^{-1} according to Hall and Schulze 1980). Rates of assimilation and intercellular partial pressures of CO_2 are found by numerical solution of

$$A = g[p_a - p_i]/P = A(p_i, T_l) \tag{5.III.2}$$

in which the function of p_i and T_l represents the data in Figure 5.9b, and g, the conductance to CO_2 transfer across boundary layer and epidermis, is found from

$$\frac{1}{g} = \frac{1.6}{g_s} + \frac{1.37}{g_b} \tag{5.III.3}$$

the numerical coefficients allowing for the effects of the differing coefficients of diffusion of water vapor and CO_2 in air. Strictly, a term should be included in equation (5.III.2) to account for the influence of vapor transfer on CO_2 diffusion (e.g., von Caemmerer and Farquhar 1981). But it was not included in deriving Figure 5.9b from the raw data and therefore has not been used in this analysis either. To some extent, the effects of these omissions cancel each other, and I estimate that the net result would not be of major significance in the present context.

It is easier to determine the effects on E and A of a hypothetical change in g_s implicitly, by assuming a change in E and finding the corresponding change in g_s. First, the change in leaf temperature ΔT_l is obtained from

$$\Delta E = [0.89 C_p g_b \Delta T_l + 2\sigma\Delta(T_l^4)]/L \tag{5.III.4}$$

in which $C_p = 29.2$ J mol^{-1} K^{-1} is the molar heat capacity of air, $\sigma = 5.9 \times 10^{-8}$ W m^{-2} K^{-4} is the Stefan-Boltzmann constant, and $L = 44$ kJ mol^{-1} is the molar heat of vaporization of water. Using the changed values of E and T_l, the corresponding values of g_s, A, and p_i are then found from equations (5.III.1) and (5.III.2).

Appendix IV: Probability distribution of soil water deficit

The subsequent remarks derive from the assumptions stated in the main text. In any short interval of time dt, the probability of rain not occurring is $(1 - dt/\tau)$. If rain does not occur, then the deficit increases by the amount $[(E + E_0)/q]\,dt$. The probability of rain occurring is dt/τ, and the probability of the amount exceeding r is $\exp(r/\rho)$. Rainfall r will decrease the deficit by r/q, unless the initial deficit is less than r/q, in which case the deficit becomes zero.

With these facts in mind, we can make use of the following statement. Over any period of time, the probability that the soil water deficit was initially less than any given magnitude, ϕ, say, and was greater than ϕ at the end of the period is equal to the probability that the deficit was initially greater than ϕ and was less than ϕ at the end of the period. For a vanishingly small period of time, this implies that

$$P(\phi)\cdot(\bar{E}+E_0)/q=\frac{1}{\tau}\left\{K\exp[-q(1-\phi)/\rho]+\int_0^1 P(\phi')\exp[-q(\phi'-\phi)/\rho]\,d\phi'\right\} \quad (5.\text{IV}.1)$$

in which $K=1-\int_0^1 P(\phi)\,d\phi$ is the probability that the deficit is unity. By differentiation and resubstitution, it is found that

$$\frac{d[P\cdot(\bar{E}-E_0)]}{d\phi}=P\cdot(\bar{E}+E_0)\cdot\left[\frac{q}{\rho}-\frac{q}{\tau}\cdot\frac{1}{\bar{E}+E_0}\right] \quad (5.\text{IV}.2)$$

from which it can be shown that

$$P=\frac{\frac{q}{\tau}\cdot\frac{1}{\bar{E}+E_0}\exp\left[\frac{q}{\rho}\phi-\frac{q}{\tau}\int_0^{\phi}\frac{d\phi'}{\bar{E}+E_0}\right]}{1+\frac{q}{\rho}\int_0^1\exp\left[\frac{q}{\rho}\phi-\frac{q}{\tau}\int_0^{\phi}\frac{d\phi'}{\bar{E}+E_0}\right]d\phi} \quad (5.\text{IV}.3)$$

Appendix V: Maximization of probable average rate of assimilation, $\hat{A}$

Let us suppose that $\bar{A}$ declines to zero at $\phi=\phi_0$, so that

$$\hat{A}=\int_0^{\phi_0}\bar{A}P\,d\phi \quad (5.\text{V}.1)$$

It can be shown that for a given $E_0(\phi)$, K is minimized by maximizing the integral of $P(\phi)$ over the range $0\leqq\phi\leqq\phi_0$. Therefore, equation (5.6) is equivalent to

$$\int_0^{\phi_0}(\bar{A}+\nu)P\,d\phi=\text{maximum} \quad (5.\text{V}.2)$$

If $\bar{A}(\phi)$ is to meet this requirement, then

$$\int_0^{\phi_0}\{(\bar{A}+\nu)\delta P+P\delta\bar{E}/\lambda\}\,d\phi\leqq 0 \quad (5.\text{V}.3)$$

where δP and $\delta\bar{E}$ are small deviations from the corresponding functions P and $\bar{E}$. From equation (5.IV.2),

$$\delta P = -\frac{P}{\bar{E} + E_0} \cdot \delta\bar{E} - \frac{q}{\tau} \cdot P \cdot \int_{\phi}^{\phi_0} \frac{\delta\bar{E}}{(\bar{E} + E_0)^2} \cdot d\phi \qquad (5.V.4)$$

Substitution in equation (5.V.3) and integration by parts leads to

$$\int_0^{\phi_0} \left\{ -\frac{\bar{A} + \nu}{E_0 + E} - \frac{q}{\tau} \cdot \frac{1}{(\bar{E} + E_0)^2} \int_0^{\phi} (\bar{A} + \nu)P \, d\phi + \frac{P}{\lambda} \right\} \delta\bar{E} \, d\phi \qquad (5.V.5)$$

Because $\delta\bar{E}$ is arbitrary, this is satisfied only if the expression in the outer braces is identical with zero, that is, if

$$\int_0^{\phi} (\bar{A} + \nu)P \, d\phi + \frac{\tau}{q} P \cdot (\bar{E} + E_0) \cdot \left[\bar{A} + \nu - \frac{\bar{E} + E_0}{\lambda} \right] = 0 \qquad (5.V.6)$$

This requires that $\nu = -\bar{A} + (\bar{E} + E_0)/\lambda$ at $\phi = 0$. Further, by differentiating equation (5.V.6) and making use of equation (5.IV.2) to eliminate P, equation (5.V.7) follows. For numerical solution, it is conveniently rewritten as

$$\frac{d(\ln\lambda)}{d\phi} = -\frac{\frac{q}{\rho}[\lambda(\bar{A} + \nu) - (\bar{E} + E_0)] + \frac{q}{\tau} - \frac{dE_0}{d\phi}}{\bar{E} + E_0} \qquad (5.V.7)$$

References

Atkins, C. A., B. D. Patterson, and D. Graham. 1972a. Plant carbonic anhydrases. I. Distribution of types among species. Plant Physiol. 50:214–217.

– 1972b. Plant carbonic anhydrases. II. Preparation and some properties of monocotyledon and dicotyledon enzyme types. Plant Physiol. 50:218–223.

Badger, M. R., and J. J. Andrews. 1974. Effects of CO_2, O_2 and temperature on a high affinity form of ribulose diphosphate carboxylase-oxygenase from spinach. Biochem. Biophys. Res. Commun. 60:204–210.

Björkman, O. 1968. Carboxydismutase activity in shade-adapted and sun-adapted species of higher plants. Physiol. Plant. 21:1–10.

Brooks, A., and G. D. Farquhar. 1985. Effect of temperature on the CO_2/O_2 specificity of ribulose-1,5-bisphosphate carboxylase/oxygenase and the rate of respiration in the light: estimates from gas-exchange measurements on spinach. Planta 165:397–406.

Comins, H. N., and G. D. Farquhar. 1982. Stomatal regulation and water economy in Crassulacean acid metabolism plants: an optimisation model. J. Theor. Biol. 98:263–284.

Cowan, I. R. 1977. Stomatal behaviour and environment. Adv. Bot. Res. 4:1176–1227.

– 1982. Regulation of water use in relation to carbon gain in higher plants. Pp. 589–613 *in* O. L. Lange, P. S. Nobel, C. B. Osmond, and H. Ziegler (eds.), Physiological plant ecology, I, vol. 12A, Encyclopedia of plant physiology. Springer-Verlag, Berlin.
Cowan, I. R., and G. D. Farquhar. 1977. Stomatal function in relation to leaf metabolism and environment. Symp. Soc. Exp. Biol. 31:471–505.
Duda, J. L., and J. S. Vrentas. 1968. Laminar liquid jet diffusion studies. Amer. Inst. Chem. Eng. J. 14:286.
Edsall, J. T., and J. Wyman. 1958. Carbon dioxide and carbonic acid. Pp. 550–590 *in* Biophysical chemistry, vol. I. Academic Press, New York.
Enns, T. 1967. Facilitation by carbonic anhydrase of carbon dioxide transport. Science 155:44–47.
Evans, J. R. 1983a. Photosynthesis and nitrogen partitioning in leaves of *Triticum aestivum* and related species. PhD thesis, Australian National University, Canberra.
– 1983b. Nitrogen and photosynthesis in the flag leaf of wheat (*Triticum aestivum* L.). Plant Physiol. 72:297–302.
Evans, J. R., and J. R. Seemann. 1984. Differences between wheat genotypes in specific activity of ribulose-1,5-bisphosphate carboxylase and the relationship to photosynthesis. Plant Physiol. 74:759–765.
Farquhar, G. D., S. von Caemmerer, and J. A. Berry. 1980a. A biochemical model of photosynthetic CO_2 assimilation in leaves of C_3-species. Planta 149:78–90.
Farquhar, G. D., E.-D. Schulze, and M. Küppers. 1980b. Responses to humidity by stomata of *Nicotiana glauca* L. and *Corylus avellana* L. are consistent with the optimisation of carbon dioxide uptake with respect to water loss. Aust. J. Plant Physiol. 7:315–327.
Farquhar, G. D., and S. C. Wong. 1984. An empirical model of stomatal conductance. Aust. J. Plant Physiol. 11:191–210.
Field, C., J. A. Berry, and H. A. Mooney. 1982. A portable system for measuring carbon-dioxide and water vapour exchange of leaves. Plant Cell Environ. 5:179–186.
Givnish, T. J., and G. J. Vermeij. 1976. Size and shapes of liane leaves. Amer. Nat. 975:743–778.
Hall, A. E., and E.-D. Schulze. 1980. Stomatal responses to environment and a possible interrelationship between stomatal effects on transpiration and CO_2 assimilation. Plant Cell Environ. 3:467–474.
Jacobson, B. S., F. Fong, and R. L. Heath. 1975. Carbonic anhydrase of spinach. Plant Physiol. 55:474–486.
Kigoshi, K., and T. Hashitani. 1963. The self-diffusion coefficients of carbon dioxide, hydrogen carbonate ions and carbonate ions in aqueous solutions. Bull. Chem. Soc. Japan 36:1372.
Lange, O. L., R. Lösch, E.-D. Schulze, and L. Kappen. 1971. Responses of stomata to changes in humidity. Planta 100:76–86.
Linke, W. F. 1958. Solubilities of inorganic and metal organic compounds, 4th ed., vol. I. Van Nostrand, New York.
Lyttleton, J. W., and P. O. P. T'so. 1958. The localisation of fraction 1 protein of green leaves in chloroplasts. Arch. Biochem. Biophys. 73:120–126.

Meinzer, F. C. 1982. The effect of vapour pressure on stomatal control of gas exchange in douglas fir (*Pseudotsuga menziesii*) saplings. Oecologia 54:236–242.

Mooney, H. A., O. Björkman, and G. T. Collatz. 1977. Photosynthetic acclimation to temperature and water stress in the desert shrub *Larrea divaricata.* Carnegie Inst. Washington Yearbook 76:328–335.

Nobel, P. S. 1968. Light-induced chloroplast shrinkage *in vivo* detectable after rapid isolation of chloroplasts from *Pisum sativum.* Plant Physiol. 43:781–787.

Pearman, G. I. 1984. Pre-industrial carbon dioxide levels: a recent assessment. Search 15:42–45.

Pocker, Y., and J. S. Ng. 1973. Plant carbonic anhydrase. Properties and carbon dioxide hydration kinetics. Biochem. 12:5127–5134.

Raven, J. A., and S. M. Glidewell. 1981. Processes limiting photosynthetic conductance. Pp. 109–136 *in* C. B. Johnson (ed.), Physiological processes limiting plant productivity. Butterworth, London.

Schulze, E.-D., and A. E. Hall. 1982. Stomatal responses, water loss and CO_2 assimilation rates of plants in contrasting environments. Pp. 181–230 *in* O. L. Lange, P. S. Nobel, C. B. Osmond, and H. Ziegler (eds.), Physiological plant ecology, II, vol. 12B, Encyclopedia of plant physiology. Springer-Verlag, Berlin.

von Caemmerer, S. 1981. On the relationship between chloroplast biochemistry and gas exchange of leaves. PhD thesis, Australian National University, Canberra.

von Caemmerer, S., and G. D. Farquhar. 1981. Some relations between the biochemistry of photosynthesis and the gas exchange of leaves. Planta 153:376–387.

Werdan, K., and H. W. Heldt. 1972. Accumulation of bicarbonate in intact chloroplasts following a pH gradient. Biochim. Biophys. Acta 283:430–441.

Wong, S. C., I. R. Cowan, and G. D. Farquhar. 1979. Stomatal conductance correlates with photosynthetic capacity. Nature (Lond.) 282:424–426.

– 1985. Leaf conductance in relation to CO_2 assimilation. I. Influence of nitrogen nutrition, phosphorus nutrition, photon flux density, and ambient partial pressure of CO_2 during ontogeny. Plant Physiol. 78:821–825.

6 Optimal stomatal conductance, allocation of energy between leaves and roots, and the marginal cost of transpiration

THOMAS J. GIVNISH

Stomata are the principal conduits through which CO_2 diffuses into a leaf and water vapor diffuses out. This shared diffusive pathway causes an inevitable association of water loss with carbon gain. Increases in the diffusive conductance of the stomata increase the potential rate at which CO_2 can diffuse to the plastids, but also increase the rate of transpirational water loss. Thus, variations in stomatal conductance entail an energetic tradeoff between potential increments to photosynthesis and the increased costs associated with the corresponding increments to transpiration. These costs include decreased photosynthetic capacity resulting from lower leaf water potential, greater allocation to unproductive root tissue, and/or shorter periods of photosynthetic activity (Givnish and Vermeij 1976; Givnish 1976, 1979, 1984).

Cowan (1977) and Cowan and Farquhar (1977) have presented an econometric model for optimal stomatal behavior that indirectly addresses this energetic tradeoff. They suggest that natural selection should favor plants whose stomatal conductance $g(t)$ varies temporally in such a way as to minimize total daily transpiration $\int E(g, t)\, dt$ for a fixed amount of total photosynthesis $\int A(g, t)\, dt$. The optimal pattern of stomatal conductance satisfying this criterion is given by the solution of

$$\frac{\partial E(g, t)}{\partial g(t)} - \lambda \cdot \frac{\partial A(g, t)}{\partial g(t)} = 0 \qquad (6.1)$$

or

$$\frac{\partial E}{\partial A} = \frac{\partial E(g, t)/\partial g(t)}{\partial A(g, t)/\partial g(t)} = \lambda \qquad (6.2)$$

where λ is an unspecified, constant Lagrange multiplier. Such behavior would minimize the costs associated with transpiration – provided they increase monotonically with total water loss – for a fixed amount of assimilation, and would thus maximize a plant's net rate of return *for that fixed amount of assimilation.*

This model has generated renewed interest in the significance of stomatal behavior. Several recent studies have attempted to determine if the observed patterns of stomatal responses to changes in various environmental factors are "optimal" in the sense of satisfying equation (6.2). Research by Farquhar et al. (1980), Hall and Schulze (1980a), Field et al. (1982), Meinzer (1982), Mooney et al. (1983b), and Ball and Farquhar (1984a, 1984b) has shown that stomatal responses to short-term changes in relative humidity or air temperature maintain $\partial E/\partial A$ relatively constant in several species, supporting the Cowan–Farquhar model. Wong et al. (1979), Schulze and Hall (1982), and others have also shown that A and g vary roughly in proportion to each other as mesophyll photosynthetic capacity varies in response to nutrient availability, leaf age, or light intensity; simulations by Farquhar (1979) have shown that such behavior is consistent with maintaining $\partial E/\partial A$ constant.

A fundamental difficulty with the Cowan–Farquhar model and attempts to test it is that the model does not produce a fully quantitative prediction for stomatal resistance: λ in equation (6.2) is undefined, and variations in its value generate an infinite, one-parameter family of optimal conductances $g(t, \lambda)$. The key question is what value of λ – and hence what fixed amount of total photosynthesis – would maximize the net difference between assimilation and the costs of transpiration in a given environment.

To attempt to answer this question, I present here an explicit analysis of the costs of transpiration, and use it to derive a quantitative model for optimal stomatal conductance and energetic allocation between leaves and roots for plants growing under constant environmental conditions. My aims are (1) to predict, for the first time quantitatively, how stomatal conductance and allocation to leaves versus roots should respond to changes in environmental and physiological parameters in order to maximize the whole-plant rate of carbon gain, (2) to account for the observed linear relationship between A and g as mesophyll photosynthetic capacity varies, and for their curvilinear relationship as humidity or water availability varies, (3) to calculate the marginal cost of transpiration, in terms of reduced photosynthesis, of variations in conductance near optimal conductance and leaf-versus-root allocation, (4) to demonstrate that this marginal cost is equal to the root cost of transpiration, as defined by Givnish and Vermeij (1976), and to $1/\lambda = \partial A/\partial E$, as defined by Cowan and Farquhar (1977), (5) to estimate the energetically optimal value of $1/\lambda$, based on physiological data, (6) to compare this estimate with observed values of

$\partial A / \partial E$ to determine if plants are actually operating near energetic optimality, and (7) to reconcile the Cowan – Farquhar approach with that used here and to propose how the model might be modified in order to apply in time-varying environments and to other traits.

Throughout, I assume that natural selection in a given environment favors plants whose form and physiology tend to maximize their net rate of carbon gain, because such plants should generally have the greatest resources with which to compete and reproduce (Horn 1971; Givnish 1979, 1982; Cowan 1982). Where mathematically convenient, trends in stomatal conductance are analyzed in terms of its inverse, stomatal resistance (cf. Cowan 1977).

Qualitative effects of stomatal resistance and leaf allocation

To simplify matters, let us consider a plant with narrow leaves all exposed to similar, temporally constant conditions in well-stirred air. In this case, leaf boundary-layer resistance will be quite small, and leaf temperature will roughly equal air temperature regardless of stomatal resistance (Gates 1965; Gates and Papian 1971). This assumption allows us to avoid the complications introduced by thermal effects; relaxing it would permit analysis of the impact of effective leaf size as well (Givnish and Vermeij 1976; Givnish 1979, 1984; Chiariello 1984).

Under the conditions stated, Figure 6.1 summarizes the qualitative effects of stomatal resistance and energetic allocation to leaves versus roots on whole-plant carbon gain, as other environmental and physiological parameters remain fixed. Increases in stomatal resistance (or, alternatively, decreases in stomatal conductance) affect whole-plant carbon gain in two conflicting ways. First, higher stomatal resistance tends to reduce photosynthesis per unit leaf mass directly by decreasing the diffusive conductance to the plastids (Farquhar and Sharkey 1982). Second, higher resistance tends to increase photosynthesis indirectly by reducing transpiration and thus increasing leaf water potential (ψ_l) and mesophyll photosynthetic capacity. Several studies have shown that low ψ_l can reduce photosynthesis independently of its effects on stomatal resistance, as measured by reduced photosynthesis of plastids subjected to osmotic shock in vitro (Boyer and Bowen 1970; Hsiao 1973; Mohanty and Boyer 1976; Berkowitz et al. 1983) or by reduced photosynthesis of intact leaves as a function of CO_2 concentration in the mesophyll [Boyer 1971 (*Helianthus*); Jones 1973 (*Gossypium*); O'Toole et al. 1977 (*Phaseolus*); Mooney et al.

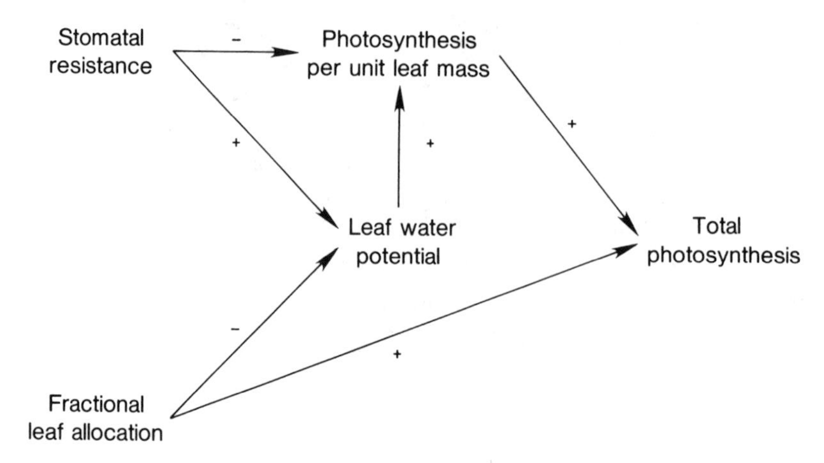

Figure 6.1. Qualitative effects of stomatal resistance and fractional allocation of energy to leaves versus roots on whole-plant carbon gain. + or − indicates sign of change in factor at arrowhead caused by increment of factor at base.

1977 (*Larrea*); Björkman et al. 1980 (*Nerium*); Ball and Farquhar 1984a (*Aegiceras* and *Avicennia*); Björkman and Powles 1984 (*Oxalis*); Ehleringer and Cook 1984 (*Encelia*)]. Recent experiments by Ball and Farquhar (1984a, 1984b), Sharkey (1984), and von Caemmerer and Farquhar (1984) demonstrate that the changes in ψ_l induced by changes in transpiration rate under short- or long-term exposure to different relative humidities are sufficient to affect mesophyll photosynthetic capacity appreciably, thus confirming the postulated trend toward higher capacity at higher ψ_l shown in Figure 6.1.

Increased allocation to leaves versus roots also affects whole-plant carbon gain in two conflicting ways. First, increases in fractional leaf allocation tend to increase whole-plant carbon gain by increasing the proportion of productive biomass. Second, such increases in fractional leaf allocation tend to decrease whole-plant carbon gain by decreasing leaf water potential (as a result of increased evaporative surface and decreased absorptive surface) and hence by decreasing photosynthesis per unit leaf mass actually constructed.

Because stomatal resistance and fractional leaf allocation interact to determine ψ_l (Figure 6.1), the relative sizes of their positive and negative impacts on whole-plant carbon gain depend on each other's value. In plants with high stomatal resistance, a given increment in leaf allocation should reduce ψ_l, and hence photosynthesis per unit leaf mass, less than in plants with lower stomatal resistance. Similarly, in plants with a large leaf allocation, a given increment in stomatal resistance should increase ψ_l, and hence photosynthesis per unit leaf mass, more than in plants which a smaller allocation to leaves. As a result, any quantitative analysis of optimal stomatal resistance or energetic allocation between leaves and roots must address the effects of both traits simultaneously.

Quantitative model for optimal stomatal resistance and fractional leaf allocation

The preceding discussion can be made quantitative by presenting equations that model the dependence of photosynthesis and leaf water potential on stomatal resistance and fractional leaf allocation. Photosynthesis per unit leaf area (g cm^{-2} s^{-1}) can be modeled as

$$A = \frac{c_a - c_i}{r_s} \tag{6.3}$$

where c_a (g cm^{-3}) is the ambient atmospheric concentration of CO_2, c_i is the corresponding concentration in the intercellular air spaces of the mesophyll, and r_s (s cm^{-1}) is the total leaf resistance to diffusion of CO_2 through the stomata and cuticle (Farquhar and Sharkey 1982). Transpiration can be modeled as

$$E = 1.6 \cdot \frac{w_a - w_i}{r_s} = \frac{\delta}{r_s} \tag{6.4}$$

where w_i (g cm^{-3}) is the saturation concentration of water vapor inside the leaf at leaf (=air) temperature, w_a is the corresponding ambient concentration, and the factor 1.6 corrects for the ratio of the diffusivities of water vapor and CO_2. The values of A, c_a, E, w_i, and r_s in equations (6.3) and (6.4) can be measured directly using standard IRGA techniques to detect CO_2 and H_2O fluxes, so that the intercellular CO_2 concentration can be inferred to be $c_i = c_a - A \cdot r_s$ (Farquhar and Raschke 1978; Wong et al. 1979; Sharkey et al. 1982).

At low c_i, photosynthesis is limited by RuBP carboxylase activity and should increase roughly linearly with internal CO_2 concentration,

$$A = \frac{c_i - \Gamma}{r_m} \tag{6.5}$$

where Γ (g cm^{-3}) and r_m (s cm^{-1}) can be interpreted as the traditional CO_2 compensation point and apparent mesophyll resistance, and in practice measure the intercept and the reciprocal of the slope of the $A(c_i)$ curve at low c_i (Farquhar et al. 1980; von Caemmerer and Farquhar 1981; Badger et al. 1984). At higher values of c_i, photosynthesis tends to be limited by

RuBP regeneration and, *inter alia,* by light capture and electron-transport capacity, so that the $A(c_i)$ curve plateaus (von Caemmerer and Farquhar 1981). It might thus be modeled more generally as

$$A = \frac{A_m \cdot (c_i - \Gamma)}{c_i - \Gamma + A_m r_m} \tag{6.6}$$

where A_m (g cm^{-2} s^{-1}) is the peak, regeneration-limited photosynthetic rate.

Under fixed environmental conditions, the parameters A_m, r_m, and Γ in equations (6.5) and (6.6) are functions of leaf water potential. Short-term exposure to low ψ_l tends to reduce electron-transport capacity and A_m, and thus the peak photosynthetic rate and the internal CO_2 level at which $A(c_i)$ begins to plateau; long-term exposure frequently also decreases RuBP carboxylase activity while increasing r_m and Γ, thus reducing the initial slope and elevation of the $A(c_i)$ curve (Mooney et al. 1977; O'Toole et al. 1977; Farquhar and von Caemmerer 1982; Ball and Farquhar 1984a; Björkman and Powles 1984; Ehleringer and Cook 1984).

In practice, many plants appear to operate at internal CO_2 levels near the breakpoint in the $A(c_i)$ curve, so that $A(c_i)$ can be closely approximated near actual c_i by equation (6.5) (Farquhar and Sharkey 1982). Combining equations (6.3) and (6.5), we thus obtain the approximation

$$A = \frac{c_a - \Gamma}{r_s + r_m} \tag{6.7}$$

where r_m and Γ are functions of the long-term ψ_l imposed by the constant environmental conditions assumed, stomatal resistance, and allocation of energy between leaves and roots. Although we shall use equation (6.7) to calculate optimal stomatal resistance and leaf allocation, calculations could also be based on the more complicated photosynthetic model implied by equation (6.6), and the implications that variations in A_m would have in the context of such a model are discussed briefly at the end of this section.

Finally, we need to determine the leaf water potential ψ_l (MPa) as a function of stomatal resistance r_s and the fraction f of energy allocated to leaves versus roots. If we assume a total biomass N in leaves and roots, then fN is total leaf biomass, and $(1 - f)N$ is total root biomass. Under constant environmental conditions, the equilibrium rate of water loss through transpiration must just balance the rate of water absorption by the roots:

$$\frac{EfN}{\rho} = k \cdot (1 - f)N \cdot (\psi_s - \psi_l) \tag{6.8}$$

where ρ is leaf mass density (g cm^{-2}), k is effective root hydraulic conductivity (g H_2O g^{-1} root MPa^{-1} s^{-1}), and ψ_s is soil water potential (MPa). The volume flux of water through the root system is known to vary nonlinearly with the potential gradient across the roots (Fiscus and Kramer 1975; Fiscus 1975, 1977, 1981; Aston and Lawlor 1979) (see Chapter 9 in this volume), so that k is potentially a function of $\Delta\psi$ and hence ψ_l. The major resistance to liquid-phase water transport in the plant lies in the root endodermis (Clarkson and Sanderson 1974; Taylor and Klepper 1975; Greacen et al. 1976; Powell 1978; Passioura 1980, 1982; Oosterhuis 1983; Sanderson 1983) (cf. Landsberg and Fowkes 1978). We have thus assumed that xylem resistance to water transport is relatively small, so that the water potential inside the root xylem can be approximated by ψ_l, and the potential differential $\Delta\psi$ across the roots is roughly $\psi_s - \psi_l$.

Equation (6.8) can be solved to obtain leaf water potential as a function of stomatal resistance, leaf allocation, and effective root hydraulic conductivity:

$$\psi_l = \psi_s - \frac{f\delta}{\rho k r_s(1-f)} \tag{6.9}$$

It should be appreciated that k is not necessarily equal to L_p, the asymptotic root hydraulic conductivity at high $\Delta\psi$ and high flow rates (Fiscus 1975) (see Chapter 9 in this volume), even when L_p is expressed in terms of root biomass. First, even in an idealized nutrient solution, volume flux $J_v = L_p(\Delta P - \sigma\Delta\pi)$, where ΔP is the pressure differential across the roots, $\Delta\pi$ is the difference in osmotic potential between the root xylem and nutrient solution, and σ is the reflection coefficient (Fiscus 1975). Thus, even in a nutrient solution, $k \simeq L_p(1 - \sigma\Delta\pi/\Delta P)$, where $\Delta\pi$ will, in general, be a function of $\Delta P \simeq \Delta\psi$ and the resulting flow rate. Second, in dry soils, the soil resistance to water flow to the roots can be substantial, as a result of decreased soil hydraulic conductivity or decreased root area exposed to saturated soil pores (Newman 1974; Sánchez-Díaz and Mooney 1979; Passioura 1982; Schulze and Hall 1982) (see Chapter 8 in this volume). Hall and Schulze (1980b) have shown that although transpiration and leaf water potential are roughly linearly related over the short term, as implied by equation (6.9), moderate- and long-term reductions in soil water content increase the slope of this relationship, as should occur if the effective hydraulic conductivity of the roots decreases.

Optimality conditions

Equations (6.4), (6.7), and (6.9) determine transpiration, photosynthesis, and leaf water potential as functions of stomatal resistance r_s and leaf allocation f. We want to know which values of r_s and f will maximize whole-plant carbon gain for a plant with total biomass N in leaves and roots. If we assume that the rate of maintenance respiration per unit mass of leaves is comparable to that for roots (or treat each as negligible), then this condition is equivalent to maximizing fNA/ρ, the product of total leaf biomass and the photosynthetic rate per unit leaf mass. Fixing N and leaf mass density ρ, and setting the partial derivatives of fA with respect to r_s and f equal to zero, we obtain from equation (6.7) the optimality conditions

$$\frac{\partial(fA)}{\partial r_s} = \left.\frac{\partial(fA)}{\partial r_s}\right|_{\psi_l} + \left.\frac{\partial(fA)}{\partial \psi_l}\right|_{r_s} \cdot \frac{\partial \psi_l}{\partial r_s} = 0 \tag{6.10a}$$

$$= -\frac{f}{r_s + r_m} \cdot \left[A + \left(A \cdot \frac{\partial r_m}{\partial \psi_l} + \frac{\partial \Gamma}{\partial \psi_l}\right) \cdot \frac{\partial \psi_l}{\partial r_s}\right] = 0 \tag{6.10b}$$

and

$$\frac{\partial(fA)}{\partial f} = \left.\frac{\partial(fA)}{\partial f}\right|_{\psi_l} + \left.\frac{\partial(fA)}{\partial \psi_l}\right|_{r_s} \cdot \frac{\partial \psi_l}{\partial f} = 0 \tag{6.11a}$$

$$= A - \frac{f}{r_s + r_m} \cdot \left[\left(A \cdot \frac{\partial r_m}{\partial \psi_l} + \frac{\partial \Gamma}{\partial \psi_l}\right) \cdot \frac{\partial \psi_l}{\partial f}\right] = 0 \tag{6.11b}$$

Setting equations (6.10b) and (6.11b) equal to each other, we obtain

$$-\frac{f}{r_s + r_m} \cdot \frac{\partial \psi_l}{\partial f} = \frac{\partial \psi_l}{\partial r_s} \tag{6.12}$$

Equation (6.12) can be solved by substituting the calculated values of $\partial \psi_l / \partial r_s$ and $\partial \psi_l / \partial f$. From equation (6.9) we derive

$$\frac{\partial \psi_l}{\partial r_s} = \frac{\delta}{\rho k r_s^2} \cdot \frac{f}{1-f} + \frac{\delta}{\rho k r_s} \cdot \frac{f}{1-f} \cdot \frac{\partial k}{\partial \psi_l} \cdot \frac{\partial \psi_l}{\partial r_s} \tag{6.13}$$

so that

$$\frac{\partial \psi_l}{\partial r_s} = \frac{\delta}{\alpha \rho k r_s^2} \cdot \frac{f}{1-f} \tag{6.14}$$

where

$$\alpha = 1 - \frac{\delta}{\rho k r_s} \cdot \frac{f}{1-f} \cdot \frac{\partial k}{\partial \psi_l} \tag{6.15}$$

Similarly, we can obtain

$$\frac{\partial \psi_l}{\partial f} = -\frac{\delta}{\alpha \rho k r_s} \cdot \frac{1}{(1-f)^2} \tag{6.16}$$

Thus, combining equations (6.14) and (6.16), we derive

$$\frac{\partial \psi_l}{\partial f} = -\frac{r_s}{f(1-f)} \cdot \frac{\partial \psi_l}{\partial r_s} \quad (6.17)$$

Substituting equation (6.17) into equation (6.12), we finally obtain the optimality condition

$$\frac{r_s}{r_s + r_m} = 1 - f \quad (6.18)$$

implying

$$\frac{r_m}{r_s + r_m} = f \quad (6.19)$$

Dividing equation (6.18) by equation (6.19), we obtain the summary relationship

$$\frac{r_m}{r_s} = \frac{f}{1-f} \quad (6.20)$$

This remarkable equation holds regardless of the detailed dependence of the photosynthetic parameters r_m and Γ on leaf water potential, and is independent of variation in effective root hydraulic conductivity at different flow rates. It crystallizes the direct relationship between leaf adaptation and allocation above and below ground. It states that the ratio of mesophyll resistance (important in determining carbon uptake) to stomatal resistance (important in determining water loss) should equal the ratio of allocation of energy to CO_2-absorbing leaves versus water-absorbing roots.

Quantitative prediction of r_s *and* f *in* Phaseolus vulgaris

In order to determine how stomatal resistance and leaf-versus-root allocation should vary with changes in environmental and physiological parameters, equation (6.20) must be combined with equation (6.10b) or (6.11b) to obtain solutions for r_s and f. In general, it is difficult to obtain an explicit solution to this system of equations, because the physiological effects of leaf water potential on apparent mesophyll resistance, CO_2 compensation point, and effective root hydraulic conductivity (i.e., $\partial r_m/\partial \psi_l$, $\partial \Gamma/\partial \psi_l$, and $\partial k/\partial \psi_l$) can vary from species to species and are potentially complex in nature (e.g., Ludlow and Björkman 1984).

However, sufficient physiological data exist for the common bean (*Phaseolus vulgaris*) to permit us to compute numerical solutions for these equations and thus obtain optimal r_s and f. O'Toole et al. (1977) measured r_m and Γ as functions of leaf water potential in *Phaseolus* plants exposed to moderate- and long-term water stress. Plants were grown in a greenhouse with a 21.0/15.5°C day/night temperature regimen; daytime relative humidity, although uncontrolled, was roughly 35%. After 24 days of growth, water was withheld to induce lower ψ_l through drying of the soil medium. Photosynthesis and transpiration were measured at saturating light intensity and 25°C, with an incoming air stream at 30% relative humidity and 340 ppm CO_2. Γ was measured directly using a closed chamber, under identical light and temperature conditions, in which respiration and photosynthesis were allowed to equilibrate before CO_2 concentration was measured. Equation (6.7), in which r_s was replaced with the sum of the observed leaf and boundary-layer resistances, was then used to calculate r_m. Based on values of Γ taken from the graph given by O'Toole et al. (1977), the CO_2 compensation point shows an exponential relationship to leaf water potential:

$$\Gamma = \Gamma^0 \exp(-a_1 \psi_l) \tag{6.21}$$

where $\Gamma^0 = 1.04 \times 10^{-7}$ g CO_2 cm^{-3} and $a_1 = 1.02$ MPa^{-1} ($r = 0.87$, $P < 0.01$ for 8 d.f.). Similarly, on the basis of the tabulated values of r_m given by O'Toole et al. (1977) and the mean ψ_l for the range of ψ_l values corresponding to each r_m value, apparent mesophyll resistance also shows an exponential relationship to leaf water potential:

$$r_m = r_m^0 \exp(-a_2 \psi_l) \tag{6.22}$$

where $r_m^0 = 5.98$ s cm^{-1} and $a_2 = 0.72$ MPa^{-1} ($r^2 = 0.99$, $p < 0.01$ for 1 d.f.).

Fiscus (Chapter 9) provides data on hydroponically grown *P. vulgaris* that make it possible to estimate the effective root hydraulic conductivity k. First, data on the total, asymptotic hydraulic conductivity L_R (g H_2O MPa^{-1} s^{-1}) at high flow rates are presented for whole root systems of plants grown to various sizes. Then, based on growth analysis and respiration measurements, Fiscus calculates the energetic equivalent E_T (g dry biomass) required to construct and maintain these root systems. Although E_T/L_R – the biomass "cost" per unit conductance – fluctuates during ontogeny in apparent response to shifts in the diameter distribution of fine roots, a mean value of 53.1 g dry mass g^{-1} H_2O MPa s emerges. If we assume that volume flow rates are high and that soil/solution resistances to liquid-phase transport are low, then k should approximately equal L_R/E_T, yielding an estimated value of $k = 1.88 \times 10^{-2}$ g H_2O g^{-1} dry mass MPa^{-1} s^{-1}.

Table 6.1. Standard values and ranges of eight environmental and physiological parameters used in calculations of optimal r_s *and* f

Parameter[a]	Standard value	Range
Relative humidity	0.50	0.00–0.95
k (g H_2O g^{-1} root MPa^{-1} s^{-1})	1.88×10^{-2}	0.4×10^{-2} to 8.0×10^{-2}
Γ^0 (g CO_2 cm^{-3})	1.04×10^{-7}	0.0×10^{-7} to 4.0×10^{-7}
r_m^0 (s cm^{-1})	5.98	1–10
a_1 (MPa^{-1})	1.02	0.2–4.0
a_2 (MPa^{-1})	0.72	0.2–4.0
ψ_s (MPa)	0	0–−1
c_a (g CO_2 cm^{-3})	6.48×10^{-7}	1.3×10^{-7} to 2.6×10^{-6}

[a] Leaf density ρ held constant in all simulations at observed $\bar{\rho} = 1.83 \times 10^{-3}$ g cm^{-2}.

The plants studied by Fiscus allocate an average of 22.2% of total dry biomass to roots, 22.6% to stems, and 55.2% to leaves, implying an allocation to leaves versus roots of roughly $f = 0.713$. Stomatal resistance is of the order 4 s cm^{-1} for these specimens, which were grown hydroponically at a ψ_s of nearly zero, in a greenhouse with uncontrolled relative humidity of very roughly 50% (Fiscus, personal communication).

These values for r_s and f can be compared with those predicted by the model if we assume the values given in the preceding paragraphs for relative humidity, k, Γ^0, r_m^0, a_1, a_2, ψ_s, and ambient CO_2 (Table 6.1). The sensitivity of the predictions to each of these eight parameters can then be studied by varying the value of each while the others are held at their "standard" values, and plotting the resulting trends in optimal r_s and f. A computer program using a standard hill-climbing technique was written in BASIC to solve for r_s and f. At the beginning of each iteration, the current estimates for optimal r_s and f were used to obtain ψ_l, Γ, and r_m from equations (6.4), (6.9), (6.21), and (6.22). The corresponding value of fP was then calculated. Next, r_s and f were independently incremented and decremented by the current step sizes for each, and the corresponding values of fP were computed. If one of the four new (r_s, f) combinations produced a greater value of fP than the current pair, it replaced that estimate, and the process recycled. If no new combination yielded a greater value of fP, the step sizes for r_s and f were halved and the process recycled until the estimates for optimal r_s and f were stable to the fourth decimal place. All simulations assumed an air temperature of 25°C and saturating light intensity; all but one assumed an ambient CO_2 concentration c_a of 6.48×10^{-7} g cm^{-3}, corresponding to 330 ppm. The 20-fold

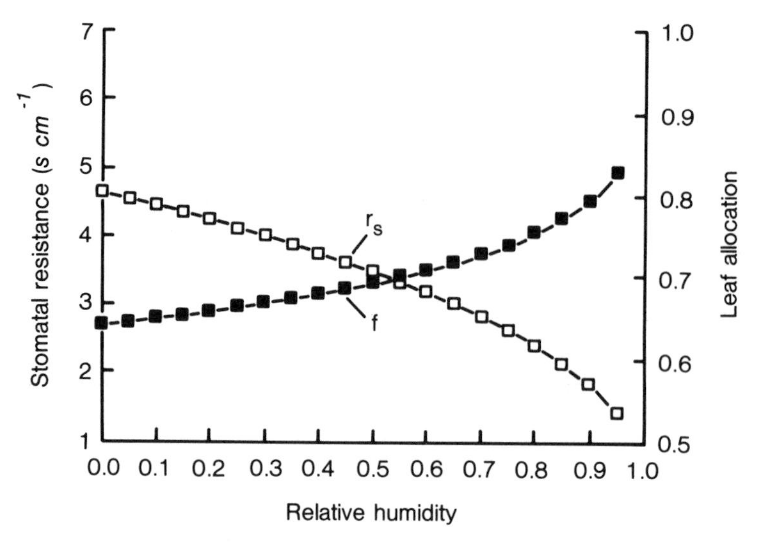

Figure 6.2. Optimal stomatal resistance and fractional leaf allocation as functions of relative humidity, with other parameters held constant at standard values.

ranges of values over which the eight environmental and physiological parameters were varied are listed in Table 6.1.

The principal trends in optimal stomatal resistance and leaf allocation are as follows:

Relative humidity. As humidity increases from 0 to 95%, optimal r_s decreases from 4.64 s cm^{-1} to 1.41 s cm^{-1}, while optimal leaf allocation f increases from 0.640 to 0.830 (Figure 6.2). These trends parallel those generally seen in moving from less humid to more humid habitats, at least in qualitative terms (Cowan 1977, 1978; Farquhar 1978; Körner et al. 1979; Schulze 1982; Schulze and Hall 1982). The expected values of $r_s = 3.48$ s cm^{-1} and $f = 0.693$ at 50% relative humidity approach quite closely the observed values of $r_s = 4$ s cm^{-1} and $f = 0.713$ under similar, though uncontrolled, levels of humidity.

Effective root hydraulic conductivity. As k increases from 0.004 to 0.08 g H_2O g^{-1} dry mass MPa^{-1} s^{-1}, optimal stomatal resistance drops dramatically from 6.69 s cm^{-1} to 1.96 s cm^{-1}, while optimal leaf allocation rises sharply from 0.568 to 0.786 (Figure 6.3), paralleling the qualitative trends expected as water becomes less costly to obtain. Such decreases in cost, due to increases in effective hydraulic conductivity, should occur not only in moist soil (as described earlier) but also in warm and/or well-

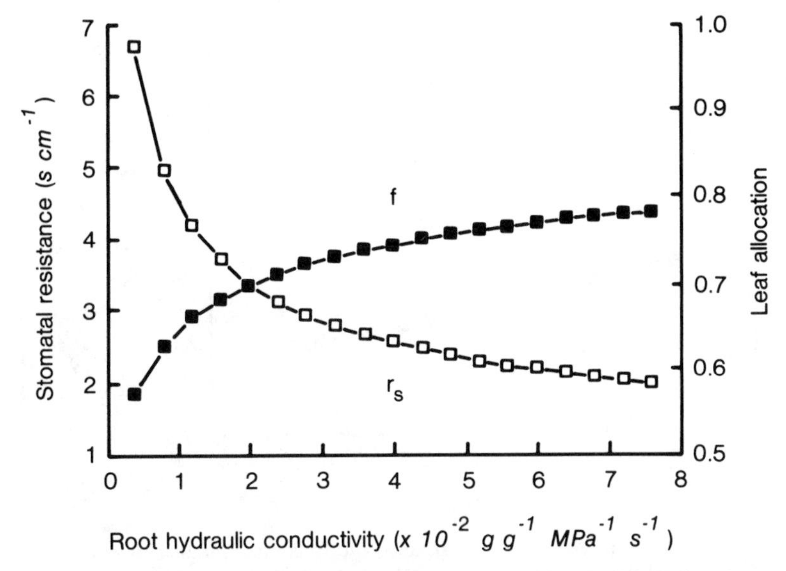

Figure 6.3. Optimal stomatal resistance and fractional leaf allocation as functions of effective root hydraulic conductivity (g H_2O g^{-1} dry root mass MPa^{-1} s^{-1}), with other parameters held constant at standard values.

drained soils, in plants with highly conductive xylem, and in xylem-tapping hemiparasites like mistletoes that obtain water and nutrients at the expense of their host's root system. Markhart et al. (1979) have shown that the asymptotic conductivity L_p of roots drops with decreasing temperature, independently of the consequent rise in water viscosity. This drop in conductivity appears to result from changes in root metabolism or membrane permeability and can be modified by acclimation or adaptation to low temperature. Similar drops in root conductivity occur at low O_2 or high CO_2 levels in the soil (Newman 1976), so that, paradoxically, water may be more costly to obtain in waterlogged, anaerobic soils than in well-drained, well aerated soils with lower moisture content.

To the extent that a substantial resistance to liquid-phase transport resides in the xylem, species with broader, more conductive vessels should achieve greater flow rates at lower cell-wall cost (and hence higher effective hydraulic conductivity overall) than species with narrow vessels or tracheids, provided the broader vessels are not more subject to cavitation under prevailing conditions (Zimmerman 1983) (see Chapters 11 and 14). Finally, in mistletoes and other photosynthetic hemiparasites that develop connections with their host's xylem, the effective hydraulic conductivity of the haustoria should be very high, given their connection via xylem to the comparatively vast root system of the host. This would favor higher sto-

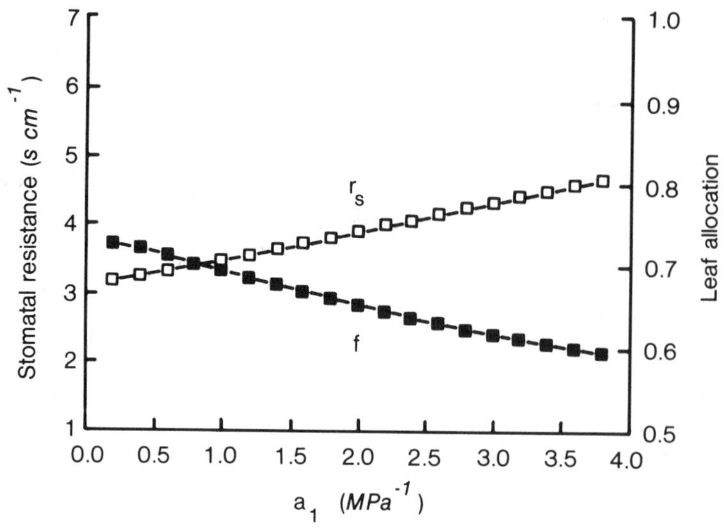

Figure 6.4. Optimal stomatal resistance and fractional leaf allocation as functions of the sensitivity a_1 of the CO_2 compensation point to leaf water potential, with other parameters held constant at standard values.

matal conductance in hemiparasites than in their hosts, other things being equal; indeed, mistletoes have stomatal conductances and transpiration rates that are often several times those of their hosts (Glatzel 1983; Schulze et al. 1984). Schulze et al. (1984) have proposed an alternative, though not mutually exclusive, explanation for this trend: Xylem hemiparasites have high stomatal conductances and transpiration rates in order to sequester enough N from the host's xylem stream for photosynthetic enzymes, whereas the host can maintain sufficient leaf N at lower transpiration rates by recycling N through the phloem. This hypothesis is intriguing and requires further research. However, their suggestion that mistletoes have liquid-phase conductances equal to but not greater than those of their hosts is unlikely to be correct. This claim is based on similar slopes relating transpiration rate to ψ_l in hosts and mistletoes. However, the expected relationship between E and ψ_l under steady-state conditions [equation (6.9)] is complicated in mistletoes by the fact that their host is transpiring simultaneously over a much larger surface, greatly reducing xylem water potential. Measurements of whole-plant hydraulic conductance (and, more important, conductance per unit allocation to haustoria) in mistletoes will require independent, experimental manipulation of the hemiparasite's and host's transpiration rates, perhaps through exposure to different water vapor-pressure deficits. Such experiments should show that

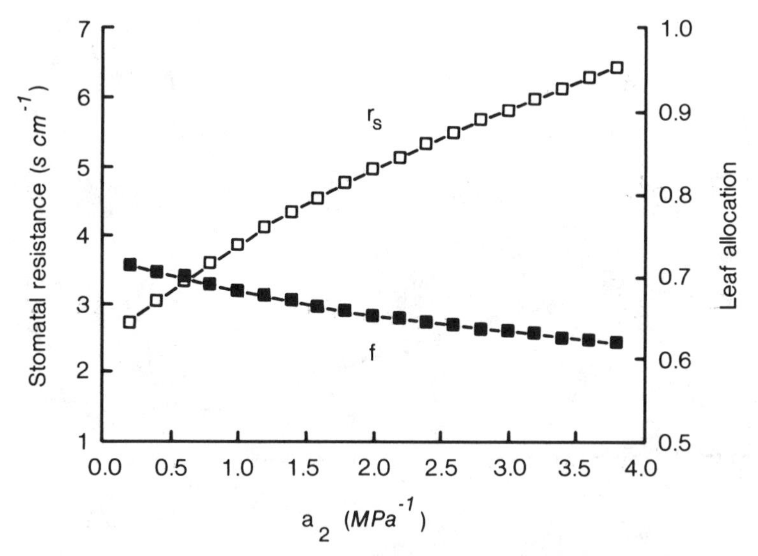

Figure 6.5. Optimal stomatal resistance and fractional leaf allocation as functions of the sensitivity a_2 of mesophyll resistance to leaf water potential, with other parameters held constant at standard values.

most of the small resistance to liquid-phase transport in mistletoes resides in the haustorial connection rather than the vast host root system, so that mistletoe r_s should be relatively independent of host root/leaf allocation in a given environment.

Sensitivity of CO_2 compensation point to ψ_l. As a_1 increases from 0.2 to 4.0 MPa^{-1}, optimal stomatal resistance increases slightly from 3.19 to 4.74 s cm^{-1}, while optimal leaf allocation decreases from 0.726 to 0.589 (Figure 6.4). This parallels the trend expected toward higher r_s and more allocation to water-gathering tissue in plants whose photosynthesis is more sensitive to decreases in leaf water potential (Schulze and Hall 1982).

Sensitivity of mesophyll resistance to ψ_l. As a_2 increases from 0.2 to 4.0 MPa^{-1}, optimal r_s rises steeply from 2.68 to 6.57 s cm^{-1}, whereas optimal f declines from 0.712 to 0.616 (Figure 6.5), paralleling the pattern just observed for variation in the sensitivity of Γ to ψ_l. The greater effect of r_m sensitivity is expected, because variations in Γ near $\Gamma = 50$ ppm have little effect on photosynthesis when intercellular CO_2 levels are near 240 ppm [equation (6.7)] (Wong et al. 1979), whereas variations in r_m affect $A(c_i)$ multiplicatively.

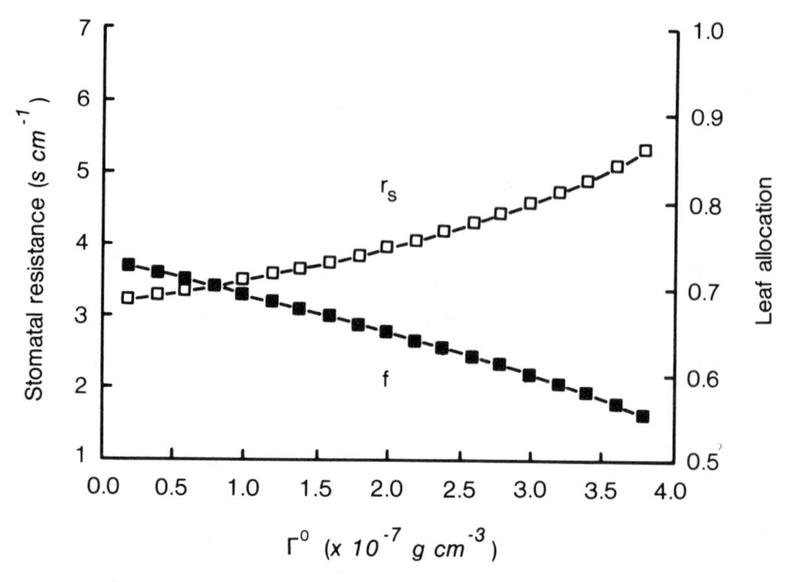

Figure 6.6. Optimal stomatal resistance and fractional leaf allocation as functions of the basal CO_2 compensation point Γ^0, with other parameters held constant at standard values.

Basal CO_2 compensation point. As Γ^0 varies from 0 to 4.0×10^{-7} g CO_2 cm^{-3} (0 to 210 ppm at sea level), optimal r_s increases from 3.21 to 5.58 s cm^{-1}, and optimal f decreases from 0.724 to 0.539 (Figure 6.6). A shift in Γ^0 from 0 to 50 ppm corresponds to the shift from C_4 to C_3 photosynthetic metabolism (Farquhar et al. 1980; Pearcy and Ehleringer 1984). Thus, under identical conditions, C_4 plants should have slightly lower stomatal resistance (3.21 vs. 3.48 s cm^{-1} at 50% relative humidity) than C_3 plants with similar mesophyll resistance. However, the expected difference is so small that it would be masked entirely in natural situations by the tendency of C_4 plants to occupy hotter, drier sites favoring high r_s values, given the greater ability of C_4 plants to avoid photorespiration and thus achieve a higher ratio of assimilation to transpiration under conditions of limited conductance and low c_i (cf. Osmond et al. 1982; Pearcy and Ehleringer 1984).

Basal mesophyll resistance. As r_m^0 increases from 0.5 to 10 s cm^{-1}, optimal r_s increases dramatically from 1.24 to 7.19 s cm^{-1}, while optimal leaf allocation f also increases sharply from 0.547 to 0.772 (Figure 6.7). The tendency for stomatal and mesophyll resistances (or conductances) to vary in parallel fashion matches the observed linear relationship

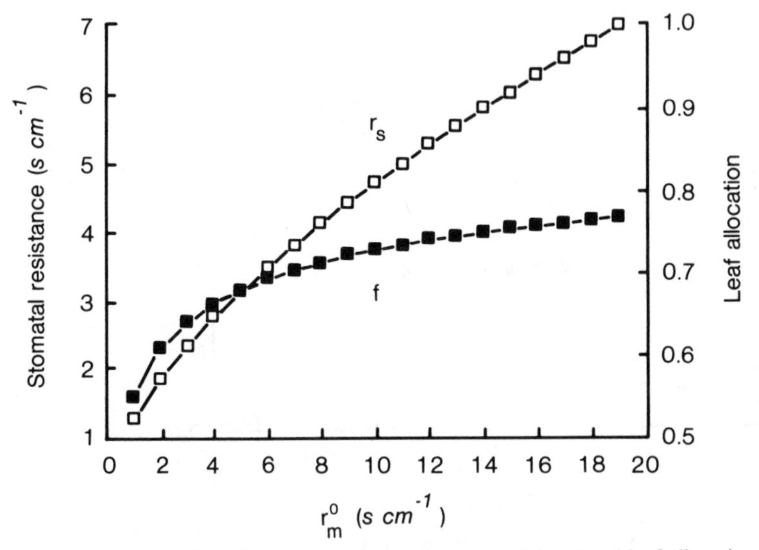

Figure 6.7. Optimal stomatal resistance and fractional leaf allocation as functions of basal mesophyll resistance r_m^0, with other parameters held constant at standard values.

between *A* and *g* (Wong et al. 1979; Schulze and Hall 1982), which implies that c_i remains roughly constant while stomatal and mesophyll conductances vary in concert (Farquhar 1978). However, the prediction that leaf allocation should also increase with mesophyll resistance is probably an error that results from not including the costs of nutrient capture in the model. Mesophyll resistance is not a wholly external variable, determined independent of environment or investment in leaves versus roots. Leaf nitrogen content – and hence carboxylation resistance, which accounts for most of total mesophyll resistance (Farquhar and Sharkey 1982) – varies considerably between species of different habitats, with the lowest leaf N levels in evergreen sclerophyll shrubs from South African fynbos and Venezuelan bana on extremely sterile soils (see Chapter 1). Plants should allocate relatively more energy to nutrient capture in sterile habitats, given the greater marginal benefit of doing so when photosynthesis is potentially limited by leaf nutrient levels (Mooney and Gulmon 1979). Thus, along a gradient of increasing site sterility, leaf N content and mesophyll resistance might decrease while allocation to roots increases, implying an inverse rather than direct relationship between leaf allocation and r_m^0.

Greater allocation to nutrient capture in sterile habitats could entail higher maintenance respiration associated with active solute transport in

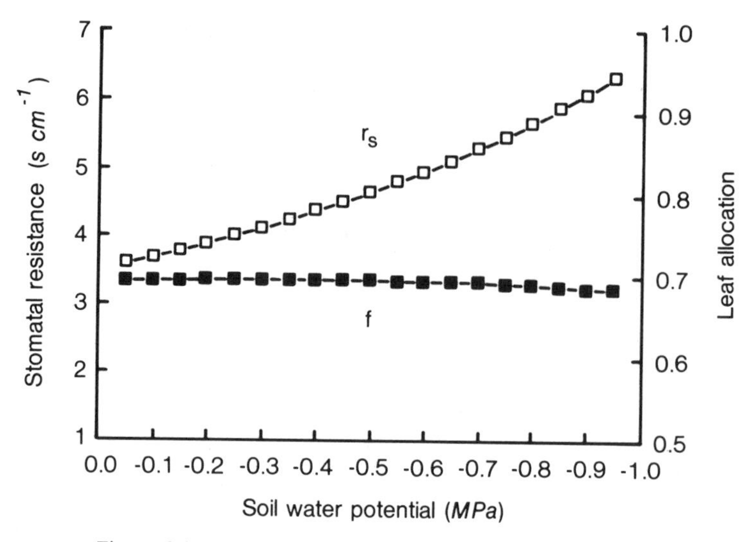

Figure 6.8. Optimal stomatal resistance and fractional leaf allocation as functions of soil water potential, with other parameters held constant at standard values.

water-absorbing roots, or more root construction and growth respiration; the latter would be expected especially for ions of low solubility or mobility (Caldwell 1979) (see Chapter 8). Although the increase in energy allocation to roots might be substantial in some quite sterile habitats, it is almost surely overestimated by the root/shoot ratio employed by Gulmon and Chu (1981), because that measure depends strongly on nonabsorptive, nonphotosynthetic stem tissue whose growth should increase sharply with habitat fertility and productivity. Givnish (1984) has suggested how the energy costs associated with water and nutrient capture might be partly separated. The close approach between observed values of r_s and f and those predicted by the foregoing model based solely on roots acting as water-absorbing tissue suggests that the separable costs of nutrient capture may be quite small, at least for plants grown hydroponically at high nutrient levels.

Soil water potential. As ψ_s decreases from 0 to -1 MPa, optimal stomatal resistance increases from 3.57 to 6.58 s cm^{-1}, while leaf allocation rises almost imperceptibly from 0.693 to 0.695 at -0.3 MPa, before falling equally slowly to 0.681 at -1.0 MPa (Figure 6.8). The trend in r_s parallels that expected in qualitative terms, but the independence of leaf allocation and ψ_s counters the usual trend toward increased root allocation in drier sites (see references cited earlier). The problem is that in this

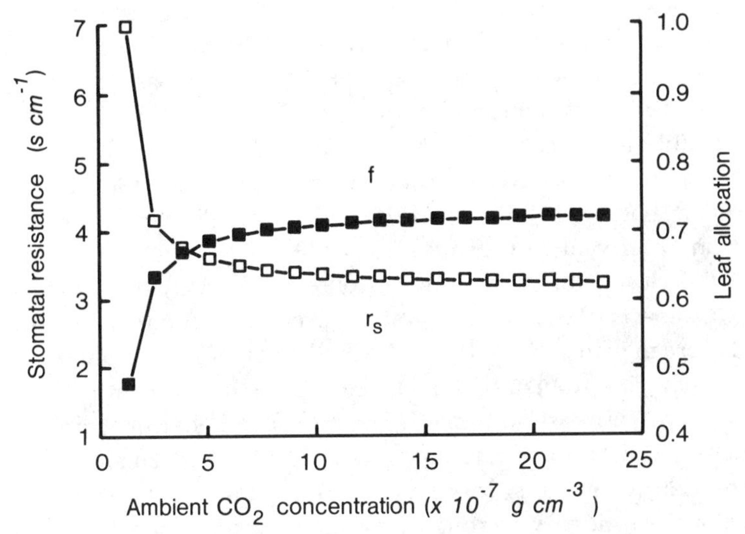

Figure 6.9. Optimal stomatal resistance and fractional leaf allocation as functions of ambient CO_2 concentration c_a, with other parameters held constant at standard values.

simplified implementation, the effect of ψ_s on root effective hydraulic conductivity k was not included, and $\partial k/\partial \psi_l$ was assumed to equal zero (Table 6.1). As a result, the primary effect of changes in ψ_s is on mesophyll photosynthetic capacity, as conditioned by ψ_l. This is a problem involving implementation, not the basic theory. Given that lower ψ_s favors higher r_s while leaving f essentially unchanged, independent of consequent drops in k in soil-grown plants, and given that lower k by itself favors higher r_s and lower f (Figure 6.3), a full analysis would indicate that decreases in ψ_s would sharply raise r_s while lowering f, as expected.

Ambient CO_2 concentration. As c_a increases from 1.30×10^{-7} to 2.59×10^{-6} g CO_2 cm^{-3} (66 to 1,320 ppm at sea level), optimal r_s drops sharply from 6.98 s cm^{-1} to 3.48 s cm^{-1} at the current ambient CO_2 level near 4.68×10^{-7} g cm^{-3}, and then decreases almost imperceptibly to 3.22 s cm^{-1} at four times the current atmospheric CO_2 level (Figure 6.9). Similarly, optimal f increases sharply from 0.474 to 0.693 at the current ambient and then plateaus, reaching only 0.722 at four times the current ambient. This would suggest that the current increase in atmospheric CO_2 should have little effect on stomatal resistance and energy allocation between leaves and roots. However, our calculations are based on current photosynthetic parameters in *Phaseolus*, notably r_m^0 and Γ^0. At higher c_a, selection may favor a higher allocation to RuBP carboxylase and electron-transport capacity, given the greater marginal benefits of these in a CO_2-rich environment, and may thus favor a lower r_m^0. Similarly, higher c_a may reduce Γ toward zero by reducing the competitive oxygenase activity of RuBP carboxylase, thus reducing photorespiration (Osmond et al. 1982; Pearcy and Ehleringer 1984). Thus, increments to current CO_2 levels might favor substantially lower stomatal resistance and higher leaf allocation, provided that increased CO_2 levels result, through selection or biochemistry, in lower values of r_m^0 and Γ^0.

The partial closure of stomata in response to *short-term* exposure to elevated CO_2 levels (Raschke 1979) may represent an (understandable) adaptive failure in plants, not a flaw in the model. As mesophyll photosynthetic capacity varies, stomatal conductance tends to increase proportionately with photosynthesis, presumably because such behavior maintains $\partial E/\partial A$ roughly constant and minimizes water loss for a given amount of assimilation (Farquhar 1979; Wong et al. 1979). This behavior has the property of maintaining c_i very roughly constant *provided that external CO_2 concentration remains constant.* Because c_a should vary little for many plants, sensing c_i and varying r_s to maintain it relatively constant could provide an inexpensive alternate to sensing the multiple parameters needed to control r_s so that $\partial E/\partial A$ remains constant (Cowan et al. 1982; Farquhar and Wong 1984). Such a sensory apparatus, operating in conjunction with other systems that detect epidermal transpiration and allow feedforward control of stomatal aperture in response to humidity (Farquhar 1978; Bunce 1984), could help adjust r_s to internal photosynthetic capacity. When c_i is too high, r_s is too low and transpiration too high relative to the current photosynthetic capacity of the plastids; when c_i is too low, r_s is too high, and net photosynthesis can be increased sharply with little additional water loss. However, unnatural increases in external CO_2 could increase c_i and trigger stomatal closure by "deceiving" the plant about the relation between r_s and plastid capacity. It will be interesting to see if understory herbs lack such inappropriate stomatal behavior, because they regularly experience large fluctuations in ambient CO_2 levels, and hence should not maintain c_i constant in order to achieve constant $\partial E/\partial A$.

Summary of effects. If we exclude c_a, then variations in mesophyll resistance, effective root hydraulic conductivity, and relative humidity have the greatest proportional effects of optimal r_s and f, whereas variations in CO_2 compensation point and its sensitivity to ψ_l have lesser effects (Table 6.2). When the effects of soil water potential on effective hydraulic conductivity are ignored, variations in ψ_s have the least effect on optimal r_s and f. The transpirational cost of decrements to r_s, in terms of reduced ψ_l

Table 6.2. *Proportional changes in optimal* r_s *and* f *induced by 20-fold variation in seven envrionmental and physiological parameters in computer model (see Table 6.1)*

Parameters (x_i)	Relative increase in optimal r_s $\{[r_s(x_i^{max}) - r_s(x_i^{min})]/\bar{r}_s\}$	Relative increase in optimal f $\{[f(x_i^{max}) - f(x_i^{min})]/\bar{f}\}$
Relative humidity	-1.07	0.86
k	-1.09	0.32
r_m^0	1.41	0.34
Γ^0	0.54	-0.29
a_1	0.39	-0.21
a_2	0.84	-0.14
ψ_s	0.59	-0.02

and consequent decreases in plastid photosynthetic capacity, increases as relative humidity and k decrease, or as the sensitivity of r_m and Γ to ψ_l increases. Therefore, optimal r_s tends to increase, while optimal f and hence the ratio r_m/r_s tend to decrease, as those transpirational costs become higher (Figures 6.2–6.8).

Interaction of the effects of variations in different physiological and environmental parameters are as expected. For example, additional simulations (not shown) indicate that at higher effective hydraulic conductivities, shifts in relative humidity have less impact on optimal r_s and f. Indeed, Bunce (1981) observed that species with higher ratios of root surface to leaf surface increase r_s less in response to drops in humidity than do species with lower ratios.

Finally, we may ask how the predictions of the model would be affected if the linear model for $A(c_i) = (c_i - \Gamma)/r_m$ were replaced with a Michaelis–Menten model [equation (6.6)], and only the parameter A_m in that model were sensitive to changes in ψ_l. In this case, equations (6.3), (6.10a), and (6.11a) could be combined to form the analogue to equation (6.12) for arbitrary $A(c_i)$:

$$\left.\frac{\partial c_i}{\partial r_s}\right|_{\psi_l} = \frac{f}{1-f} \cdot \left.\frac{\partial c_i}{\partial \psi_l}\right|_{r_s} \cdot \frac{\partial \psi_l}{\partial r_s} \tag{6.23}$$

Then, using equation (6.6) as a model for $A(c_i)$ and assuming that plants will operate at a c_i near the breakpoint $A_m r_m$ in this curve, we can derive the following equation for optimal r_s if only A_m varies with ψ_l (Farquhar and Sharkey 1982):

$$r_s^4(r_s - 1) = \frac{\delta}{\rho k A_m^2} \cdot \frac{\partial A_m}{\partial \psi_l} \cdot (c_a - A_m r_m) \tag{6.24}$$

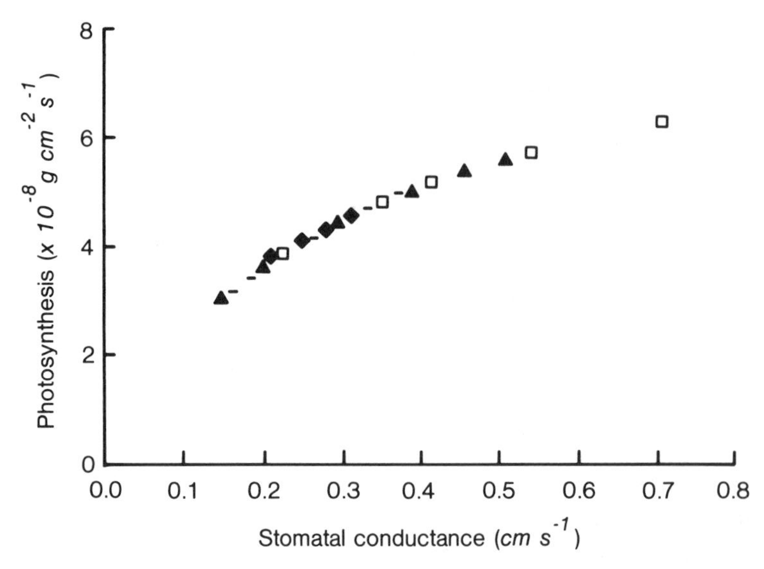

Figure 6.10. Curvilinear relationship between photosynthesis and optimal stomatal conductance, based on variations in relative humidity (open squares), effective root hydraulic conductivity (triangles), a_1 (diamonds), and a_2 (bars).

Hence, optimal r_s should increase with decreasing humidity and effective hydraulic conductivity, with decreasing RuBP regeneration capacity and peak photosynthetic rate, and with increasing sensitivity of that regeneration capacity and peak rate to changes in ψ_l. Thus, a model based on sensitivity of A_m (but not r_m or Γ) to changes in ψ_l produces predictions that are qualitatively similar to the earlier analysis based on shifts in r_m and Γ. Short-term water stress appears to affect electron-transport capacity, RuBP regeneration, and hence A_m, not r_m or Γ (Farquhar and von Caemmerer 1982; Ball and Farquhar 1984a, 1984b; von Caemmerer and Farquhar 1984). Thus, it remains unclear whether the long-term effects of ψ_l on r_m and Γ (O'Toole et al. 1977; Ehleringer and Cook 1984), which form the basis for most of the analysis in this chapter, are mechanistic constraints or the result of an economic readjustment in various enzyme pools in response to mechanistically imposed changes in A_m.

Predicted relationships between *A* and *g*

Variations in the four model parameters (relative humidity, k, a_1, a_2) that affect the transpirational cost of increments to r_s all lead to the same curvilinear relationship between stomatal conductance $g = 1/r_s$ and calculated photosynthetic rate A (Figure 6.10). As transpirational costs

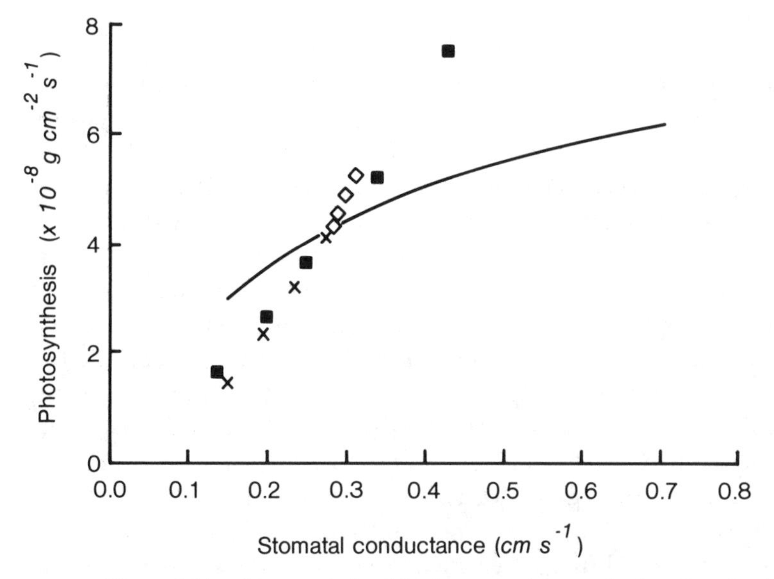

Figure 6.11. Linear relationships between photosynthesis and optimal stomatal conductance as mesophyll photosynthetic capacity varies, based on variations in r_m^0 (filled squares), ψ_s (crosses), and Γ^0 (open diamonds) (0 – 50 ppm only). Curve indicates relationship between A and g in Figure 6.10.

decrease, g increases faster than A, leading to a quasi-Michaelis – Menten relationship that appears to pass through the origin. By contrast, as basal mesophyll resistance varies, g varies almost linearly with A along a much steeper curve passing near the origin (Figure 6.11). Variations in ψ_s, which mainly affect mesophyll photosynthetic capacity in the simulations presented earlier, also lead to a nearly linear relationship between g and A along almost the same curve (Figure 6.11). Finally, variations in Γ^0 from 0 to 50 ppm lead to steep increases in A with little increase in g (Figure 6.11).

These trends parallel those actually seen. Several studies have shown that as the water vapor-pressure deficit between leaves and the surrounding air decreases, g increases more rapidly than A as $\partial E/\partial A$ remains constant or increases very slightly, so that g is a saturating function of A whose extension appears to pass through the origin (Farquhar et al. 1980; Hall and Schulze 1980a; Meinzer 1982; Schulze and Hall 1982). Changes in mesophyll photosynthetic capacity lead to a linear relationship between A and g, with slopes of roughly 1.96×10^{-7} g CO_2 cm^{-3} for C_3 species and 4.32×10^{-7} g CO_2 cm^{-3} for C_4 species under well-watered conditions (Wong et al. 1979). The predicted slope induced by variations in r_m^0 for $\Gamma^0 \simeq 50$ ppm (Figure 6.11) is 2.24×10^{-7} g cm^{-3} ($r^2 = 0.995$, $p < 0.001$ for 18 d.f.), quite close to that actually seen in C_3 species at comparable humidities. However, the predicted slope when $\Gamma^0 = 0$ is 2.64×10^{-7} g

cm^{-3}, which, although tending in the proper direction, is considerably less than that usually seen in C_4 species. This discrepancy suggests that in addition to using $\Gamma^0 = 0$, a nonlinear model for $A(c_i)$ [e.g., equation (6.6)] must be used to account quantitatively for the A-versus-g relationship in C_4 plants. $A(c_i)$ plateaus sharply at relatively low mesophyll CO_2 levels near 100 ppm (Osmond et al. 1982; Pearcy and Ehleringer 1984), so that variation in $A_m(\psi_l)$ and the resultant breakpoint in $A(c_i)$ may play an important role in determining g in C_4 species. However, it should be noted that the relatively small change in g, compared with that in A, that results when Γ^0 shifts from 50 ppm (C_3) to 0 ppm (C_4) corresponds with the trends seen in most comparisons of C_3 and C_4 species (Pearcy and Ehleringer 1984).

The qualitative relationships between A and g as transpirational costs or mesophyll photosynthetic capacity can also be understood in terms of an earlier model by Givnish and Vermeij (1976). That simplified model assumed that as r_s varied evolutionarily, ψ_l would be held constant, so that the energetic allocation to roots per unit area would scale linearly with transpiration rate. Thus, the central question was which value of r_s would maximize the difference between photosynthesis and root costs, $A - bE$. If we adopt the notation used in this chapter, the optimality condition $\partial(A - bE)/\partial r_s = 0$ implies

$$r_s = r_m \cdot \sqrt{\frac{b\delta}{c_a - \Gamma}} \Big/ \left(1 - \sqrt{\frac{b\delta}{c_a - \Gamma}}\right) = r_m\gamma/(1-\gamma) \qquad (6.25)$$

so that

$$g = g_m(1-\gamma)/\gamma \qquad (6.26)$$

and

$$A = \frac{c_a - \Gamma}{r_s + r_m} = g_m \cdot (c_a - \Gamma) \cdot (1-\gamma) \qquad (6.27)$$

From equations (6.26) and (6.27) it is obvious that as mesophyll photosynthetic capacity (g_m) varies while other parameters remain fixed, A and g will vary in proportion to each other. Similarly, as the relative transpirational cost of an increment in conductance (captured in the term $\gamma = \sqrt{b\delta/(c_a - \Gamma)}$ varies, then A varies as a Michaelis–Menten function of $g(\gamma)$:

$$A = \frac{\xi g(\gamma)}{g(\gamma) + \kappa} \qquad (6.28)$$

where $\xi = g_m(c_a - \Gamma)$ is the maximum photosynthetic rate achieved as $\gamma \rightarrow 0$ and $g(\gamma) \rightarrow \infty$, and $\kappa = g_m$ [equations (6.26) and (6.27)].

Marginal cost of transpiration

Although the term "transpirational costs" has been used somewhat loosely up to this point, such costs can be quantified in terms of the general model. The *marginal cost of transpiration* Ω associated with a slight change in stomatal resistance (or conductance), in terms of reduced photosynthesis, is perhaps the most useful concept. This marginal cost can be defined as the rate of decrease in photosynthesis caused by the effect of small changes in r_s on transpiration and hence ψ_l, divided by the rate at which small changes in r_s increase transpiration:

$$\Omega = \left.\frac{\partial A}{\partial \psi_l}\right|_{r_s} \cdot \frac{\partial \psi_l}{\partial r_s} \Big/ \frac{\partial E}{\partial r_s} \tag{6.29a}$$

$$= \left.\frac{\partial A}{\partial \psi_l}\right|_{r_s} \cdot \frac{\partial \psi_l}{\partial E} \cdot \frac{\partial E}{\partial r_s} \Big/ \frac{\partial E}{\partial r_s} \tag{6.29b}$$

$$= \left.\frac{\partial A}{\partial \psi_l}\right|_{r_s} \cdot \frac{\partial \psi_l}{\partial E} \tag{6.29c}$$

This last form of the expression for Ω [equation (6.29c)] illustrates most clearly why it is termed the marginal cost of transpiration: It is decrement to photosynthesis caused by a change in leaf water potential, times the change in leaf water potential caused by a change in transpiration rate.

The value of Ω can be calculated using equations (6.4), (6.7), (6.9), and (6.29a) to obtain

$$\Omega = -\frac{1}{r_s + r_m} \cdot \left(A\,\frac{\partial r_m}{\partial \psi_l} + \frac{\partial \Gamma}{\partial \psi_l}\right) \cdot \frac{\delta f}{\rho k(1-f)r_s^2} \Big/ \left(-\frac{\delta}{r_s^2}\right) \tag{6.30a}$$

$$= \frac{f}{1-f} \cdot \frac{1}{r_s + r_m} \cdot \frac{1}{\rho k} \cdot \left(A\,\frac{\partial r_m}{\partial \psi_l} + \frac{\partial \Gamma}{\partial \psi_l}\right) \tag{6.30b}$$

Equation (6.30b) can be put in final form using the relationship $f/(1-f) = r_m/r_s$, which holds for optimal r_s and f, to derive

$$\Omega = \frac{f}{r_s} \cdot \frac{1}{\rho k} \cdot \left(A\,\frac{\partial r_m}{\partial \psi_l} + \frac{\partial \Gamma}{\partial \psi_l}\right) \tag{6.31}$$

Calculated value

We can compute the marginal cost of transpiration using equation (6.31) and physiological data for *P. vulgaris* under the "standard" conditions assumed in the simulations for opimal r_s and f (Table 6.1). Under these conditions, Ω is roughly -2.2×10^{-3} g CO_2 g^{-1} H_2O. That is, at 25°C, 50% relative humidity, 340 ppm CO_2, and standard values for the physiological parameters, the decrease in photosynthesis caused by a small

Table 6.3 *Calculated values of the marginal cost of transpiration* (Ω) *at optimal* r_s *and* f *corresponding to extreme values of parameters in simulation (see Table 6.1)*

Parameters (x_i)	$\Omega(x_i^{min})$ (10^{-3} g CO_2 g^{-1} H_2O)	$\Omega(x_i^{max})$ (10^{-3} g CO_2 g^{-1} H_2O)	Relative increase in marginal cost ($\Delta\Omega/\overline{\Omega}$)
Relative humidity	−1.51	−7.51	0.665
k	−4.10	−1.16	−0.559
r_m^0	−4.44	−1.31	−0.544
Γ^0	−2.34	−1.56	−0.200
a_1	−2.07	−2.41	0.076
a_2	−1.45	−4.12	0.479
ψ_s	−2.22	−1.21	−0.294

increase in transpiration is roughly 2.2×10^{-3} times that increase in transpiration rate. As expected from equation (6.31), the marginal cost of transpiration increases with increasing sensitivity of the photosynthetic parameters r_m and Γ to ψ_l, and decreases with increasing effective hydraulic conductivity k (Table 6.3). Increases in r_m^0, however, lead to a decrease in the absolute value of Ω, although its magnitude relative to photosynthesis (Figure 6.11) greatly increases (Table 6.3). Increases in ψ_s, which mainly affect r_m rather than k in the simulations presented earlier, also tend to decrease the absolute value of Ω. Surprisingly, perhaps, the marginal cost of transpiration is greater at higher humidities (Table 6.3). This reflects the fact that at high humidities, plants are operating at low r_s and high f, so that the change in ψ_l caused by an infinitesimal change in transpiration rate [$\partial\psi_l/\partial E = f/\rho k r_s$, equations (6.29c) and (6.31)] is comparatively large. High values of $\partial\psi_l/\partial E$, in turn, tend to favor high values of Ω [equation (6.31)].

Equivalence with b *and* $\partial A/\partial E$

Another method of calculating the marginal cost of transpiration might be to compute how the proportion of photosynthesis diverted to roots varies with transpiration rate. The total diversion to roots, $(1 - f) \cdot A$, should equal bE, where b is the cost of transpiration (g CO_2 g^{-1} H_2O) associated with a unit rate of transpiration in the model of Givnish and Vermeij (1976). Thus, the marginal cost of transpiration according to these definitions should be

$$b = \frac{(1 - f) \cdot A}{E} \tag{6.32}$$

The value of b under energetically optimal r_s and allocation to roots versus leaves can be calculated using equations (6.4), (6.11b), (6.16), and (6.18) to substitute values for A and E in equation (6.32):

$$b = \frac{f}{r_s} \cdot \frac{1}{\rho k} \cdot \left(A \frac{\partial r_m}{\partial \psi_l} + \frac{\partial \Gamma}{\partial \psi_l} \right) \tag{6.33}$$

Clearly, by comparison of equations (6.31) and (6.33), $b = \Omega$. *Thus, the marginal cost of transpiration, calculated in terms of the reduction in leaf photosynthesis caused by a decrease in* ψ_l*, is equal to the marginal cost of transpiration, calculated in terms of the diversion of energy to roots.* As a result, the concept of transpirational cost developed by Givnish and Vermeij (1976), even though couched in terms of a highly schematic model in which r_s and f cannot vary independently and have independent effects on A through ψ_l, corresponds quantitatively to the marginal cost of transpiration Ω when r_s and f are allowed to vary independently.

Even more surprisingly, the marginal cost of transpiration Ω is equal in magnitude to $1/\lambda = \partial A/\partial E$ in the model of Cowan (1977) and Cowan and Farquhar (1977). For plants in a constant environment, the condition for optimal stomatal resistance [equations (6.10a) and (6.10b)] implies that

$$\left.\frac{\partial A}{\partial r_s}\right|_{\psi_l} = -\left.\frac{\partial A}{\partial \psi_l}\right|_{r_s} \cdot \frac{\partial \psi_l}{\partial r_s} \tag{6.34}$$

The same equation would hold if the resistance r_s were everywhere replaced by the corresponding conductance g:

$$\left.\frac{\partial A}{\partial g}\right|_{\psi_l} = \left.\frac{\partial A}{\partial \psi_l}\right|_{g} \cdot \frac{\partial \psi_l}{\partial g} \tag{6.35}$$

The partial rate of change in photosynthesis with stomatal conductance *at constant* ψ_l [the left-hand side of equation (6.35)] is just the "$\partial A/\partial g$" employed in the Cowan–Farquhar model. Both in the original theoretical development and in several subsequent tests, the term "$\partial A/\partial g$" has referred solely to the influence that stomatal conductance has on photosynthesis through its direct effects on c_i via diffusion. Every empirical study of $\lambda = (\partial E/\partial g)/(\partial A/\partial g)$ has evaluated "$\partial A/\partial g$" by varying internal CO_2 concentration through a manipulation of external CO_2 concentration at a fixed air temperature and humidity, using the resulting data to calculate $A(c_i)$, and then calculating the hypothetical effects of changes in conductance on c_i and A through effects on diffusion (Farquhar et al. 1980; Hall and Schulze 1980a; Field et al. 1982; Meinzer 1982; Mooney and Chu 1983; Mooney et al. 1983b; Williams 1983; Ball and Farquhar 1984a, 1984b). The conditions of these experiments ensure that the negative impact of increased g on photosynthesis, through consequent reduction in ψ_l, will not be detected. This is because at high c_i, induced by high c_a and supposed to correspond to the effects on $A(c_i)$ of high values of g, actual stomatal conductance is at the same value or a *lower* value than it achieves at lower c_i induced by lower c_a (Raschke 1979; Wong et al. 1979). As a result, transpiration rate will remain constant or decrease, and ψ_l will remain constant or increase, as c_i increases with increasing c_a under otherwise constant conditions. Thus, the values of $A(c_i)$ at high c_i, which are supposed to correspond to the effects of high conductance (see references cited earlier), do not reflect the diminution in plastid photosynthetic capacity that would result from lowered ψ_l at actual high conductances. As noted earlier, recent work by Sharkey (1984) and Ball and Farquhar (1984a) clearly indicates that increased transpiration reduces mesophyll photosynthetic capacity. Furthermore, this reduction in capacity is not an artifact of increased H_2O net diffusion outward interfering with CO_2 diffusion into the leaf, because both studies used the diffusion equations incorporating such ternary effects proposed by Jarman (1974) and Leuning (1983). These effects, although not included in equations (6.3) and (6.4) of the current model, are usually quite small and could easily be incorporated in later, more baroque analyses.

If we examine equation (6.35), however, we can see that *the measured values of "$\partial A/\partial g$" at roughly constant ψ_l should just equal the negative of the marginal decrease in photosynthesis caused by changes in conductance through its effects on ψ_l*. Thus, if we divide both sides of equation (6.35) by $\partial E/\partial g$ or, equivalently, divide both sides of equation (6.34) by $\partial E/\partial r_s$, we obtain the summary relationship

$$\frac{1}{\lambda} = \frac{\partial A/\partial g}{\partial E/\partial g} = -\frac{f}{r_s} \cdot \frac{1}{\rho k} \cdot \left(A\,\frac{\partial r_m}{\partial \psi_l} + \frac{\partial \Gamma}{\partial \psi_l} \right) \tag{6.36}$$

Clearly, by comparison of equations (6.31) and (6.36), $1/\lambda = -\Omega$. *Thus, in a constant environment, $1/\lambda$ should equal, except for sign, the marginal cost of transpiration, calculated in terms of the reduction in A caused by decreases in ψ_l.*

This result makes eminent sense. The Cowan–Farquhar model [equation (6.1)] assumes that $\int E(g, t)\,dt$ is minimized for a fixed value of $\int A(g, t)\,dt$ over the same (short) time period. The value of the Lagrangian multiplier λ that allows both criteria to be met measures the *effective* marginal cost $\partial E/\partial A$ [equation (6.2)] of an increase in photosynthesis in terms of increased transpiration. That is, plants that maintain $\partial E/\partial A = \lambda$ behave *as if* they are minimizing the costs associated with transpiration, *given* that λ measures the transpirational equivalent of an additional unit of photosynthesis. In other words, plants that maintain $\partial E/\partial A = \lambda$ behave *as if* $1/\lambda$ is the photosynthetic equivalent cost of an additional unit of transpiration (Intriligator 1971; Cowan 1978).

Observed versus expected values of $\partial A/\partial E$

The key question is thus whether or not the observed values of $1/\lambda = \partial A/\partial E$ correspond to those that would actually maximize whole-plant growth. That is, does the *apparent* marginal cost of transpiration $\partial A/\partial E$ equal the *actual* marginal cost of transpiration Ω, computed for plants with optimal stomatal conductance and leaf allocation and determined by the sensitivity of photosynthetic parameters to ψ_l, effective root hydraulic conductivity, and various environmental parameters?

Observed values of $\partial A/\partial E$ are listed by species and life form in Table 6.4. In several cases, specific mean values were not given in the source text and had to be taken from graphs. Williams (1983) found that *Rhamnus californica* maintained $\partial E/\partial A$ constant under natural conditions only from mid-morning to mid-afternoon; at other times, $\partial E/\partial A$ fluctuated wildly, but conductance varied in a manner very similar to that which would have held $\partial E/\partial A$ constant. This unique study suggests that plants are unable to regulate g by sensing $\partial E/\partial A$ directly. Williams (1983) suggested that the instability in $\partial E/\partial A$ observed near dawn and dusk is probably a result of the high humidity and low light intensity at those times, which would make $\partial E/\partial A$ an unstable ratio of two small numbers, $\partial E/\partial g$ and $\partial A/\partial g$. Mooney et al. (1983a) and Mooney and Chu (1983) failed to observe maintenance of constant $\partial E/\partial A$ in tropical *Piper* species and *Diplacus aurantiacus,* and their data are not included in Table 6.4.

The observed values of $\partial A/\partial E$ are strongly correlated with plant life form (Table 6.4). With the exception of the coastal bluff species *Ambrosia chamissonis,* which has several woody relatives in nearby arid and semiarid areas (Munz 1975), herbaceous species have relatively low values of $\partial A/\partial E$, ranging from 2.1×10^{-3} to 2.7×10^{-3} g CO_2 g^{-1} H_2O. Shrubs and sub-shrubs have somewhat higher values of 4.1×10^{-3} to 5.4×10^{-3} g CO_2 g^{-1} H_2O. The coniferous tree *Pseudotsuga menziesii* has a still higher value of $\partial A/\partial E$, roughly 6.7×10^{-3} g g^{-1}. Two Australian mangrove species have by far the highest value of $\partial A/\partial E$: 9.8×10^{-3} g g^{-1}.

These trends accord with those expected from the general model. First, the observed range of values for $\partial A/\partial E$ in herbs ($2.1-2.7 \times 10^{-3}$) corresponds very closely to the marginal cost of transpiration expected in the herb *Phaseolus vulgaris* at 50% relative humidity (2.2×10^{-3}). This suggests that such plants are indeed operating close to a value of $\partial A/\partial E$ that

Table 6.4. *Apparent marginal cost of transpiration* ($\partial A/\partial E$) *as a function of plant growth form*

Growth form	Species	$\partial A/\partial E$ (g CO_2 g^{-1} H_2O)	Source
Herbs	*Nicotiana glauca*	2.1×10^{-3}	Farquhar et al. (1980)
	Fragaria chiloensis[a]	2.6×10^{-3}	Mooney et al. (1983b)
	Vigna unguiculata	2.7×10^{-3}	Hall and Schulze (1980a)
	Ambrosia chamissonis[a,b]	6.5×10^{-3}	Mooney et al. (1983b)
Shrubs	*Corylus avellana*	4.1×10^{-3}	Farquhar et al. (1980)
	Lepechinia calycina	5.0×10^{-3}	Field et al. (1982)
	Eriogonum latifolium[a]	5.2×10^{-3}	Mooney et al. (1983b)
	Rhamnus californica[c]	5.4×10^{-3}	Williams (1983)
Coniferous tree	*Pseudotsuga menziesii*	6.7×10^{-3}	Meinzer (1982)
Mangroves	*Aegiceras corniculatum*	9.8×10^{-3}	Ball and Farquhar (1984a)
	Avicennia marina	9.8×10^{-3}	Ball and Farquhar (1984a)

[a] Interpolated from graph.
[b] Several congeners woody.
[c] Midday values only.

will maximize whole-plant carbon gain. This is the first instance in which such a statement can be made about observed values of $\partial A/\partial E$; however, a completely rigorous test of this view must await detailed measurements of k, a_1, a_2, and r_m^0 for the relevant species and growth conditions.

Second, the trend toward higher values of $\partial A/\partial E$ in woody plants corresponds with the expected trend (Table 6.3) in the marginal cost of transpiration as effective hydraulic conductivity decreases. The few comparative data available (e.g., Fiscus and Markhart 1979; Sands et al. 1982) suggest that herbs, as expected, have higher asymptotic radial and axial conductivities per unit root area and volume than do woody plants with more suberized root surfaces and narrower xylem elements. The high values of $\partial A/\partial E$ in *Pseudotsuga,* which lacks vessels and has narrow, inefficiently conducting tracheids, are perhaps relevant in the latter regard. The high values of $\partial A/\partial E$ observed in the mangroves *Aegiceras* and *Avicennia* probably reflect the extraordinary metabolic costs of blocking salt uptake and/or excreting salt in a highly saline environment.

Extension to time-varying environments

Extension of the model to a temporally varying environment depends on the nature of temporal variation and the time scale of plant response. Let us initially assume an idealized pattern of diurnal environmental fluctuation, with predictable shifts in aboveground conditions through a repeated daily cycle, while belowground conditions – notably moisture availability – show little day-to-day variation. These conditions are essentially those assumed by Cowan (1977) and Cowan and Farquhar (1977) in their analysis of optimal stomatal behavior in a time-varying environment. In this case, if leaf water potential equilibrates over very short time scales relative to those of environmental fluctuations and consequent shifts in stomatal conductance, then our approach can be applied directly. The optimality criterion would be to maximize $\int fA(g, t)\, dt$ over a daily cycle with respect to variation in g and f; we assume that $g(t)$ can fluctuate substantially within a day, whereas f cannot and is constant. Thus, based on the calculus of variations, the necessary conditions for maximizing $\int fA(g, t)\, dt$ are

$$\left.\frac{\partial A}{\partial g}\right|_{\psi_l} = -\left.\frac{\partial A}{\partial \psi_l}\right|_{g} \cdot \frac{\partial \psi_l}{\partial g} \tag{6.37}$$

$$\int A\, dt = -f \cdot \int \left.\frac{\partial A}{\partial \psi_l}\right|_{g} \cdot \frac{\partial \psi_l}{\partial g} \cdot dt \tag{6.38}$$

Note that $\partial A/\partial \psi_l = \partial A/\partial A_m \cdot \partial A_m/\partial \psi_l$ in equation (6.37) because we are assuming that the fluctuations in environmental conditions (and hence ψ_l) are short-term rather than long-term, and so should not affect the values of r_m and Γ.

Equation (6.37) states that at optimal conductance g, the total partial derivative of photosynthesis with respect to g must equal zero. How can this condition possibly be consistent with observed stomatal behavior? All previous tests of the Cowan–Farquhar model have shown that $\partial A/\partial E > 0$ and hence $\partial A/\partial g > 0$ for all g. However, as noted previously, these measurements of $\partial A/\partial g$ are actually measurements of $(\partial A/\partial g)|_{\psi_l}$ and do not include detrimental effects of conductance on photosynthesis through decreased ψ_l. Even if $(\partial A/\partial g)|_{\psi_l} > 0$ for every constant value of ψ_l, however, the observed pattern of stomatal behavior might lead to the total partial derivative of A with respect to g equaling zero. This is because even though A increases with c_i for every given ψ_l, the peak value A_m that A can achieve *decreases* with increasing g and c_i, because ψ_l decreases with increasing g. As a result, under fixed environmental conditions, A should achieve a maximum short-term value at an intermediate value of g, where the photosynthetic benefits of an increase in diffusive conductance just balance the photosynthetic costs of a consequent decrease in ψ_l [see equation (6.37)]. Sharkey (1984) indeed found such a peaked relationship of A to g in *Scrophularia desertorum* as it closed its stomata in response to a step decrease in relative humidity. More research is clearly needed to determine how the stomatal conductance of real plants compares with that predicted by equation (6.37). Such data would be useful not only in testing the current model but also in evaluating the importance of various factors limiting photosynthesis. Farquhar and Sharkey (1982) have suggested that plants tend to be limited equally by carboxylation activity and RuBP regeneration, because they tend to operate near the breakpoint in the $A(c_i)$ curve. This interpretation may be radically changed when it is recognized that such $A(c_i)$ curves are usually determined under roughly constant ψ_l and hence do not incorporate the effects of g on A through ψ_l. The fact that plants operate near the breakpoint in the $A(c_i)$ curve may thus reflect the fact that the right-hand, ψ_l-dependent side of equation (6.37) just begins to achieve the magnitude of the left-hand, g-dependent side as it falls at higher g. This conclusion will not necessarily be incompatible with the view that carboxylation and RuBP regeneration colimit photosynthesis, but that view must be tested in the context of the functional dependence of ψ_l on stomatal conductance.

If plant hydraulic capacitance is high, and ψ_l equilibrates over longer time scales than those of environmental fluctuations and consequent changes in stomatal conductance, then it is very difficult to incorporate the effects of transpiration on photosynthesis. This is because, as Cowan (1977) noted, these effects will depend on the integral of transpiration over some period in the past and will explicitly include some measure of the capacitance of the hydraulic system. When the response time for ψ_l is very long indeed, then the Cowan–Farquhar criterion of minimizing $\int E(g, t)\,dt$ for a given value of $\int A(g, t)\,dt$ is probably very reasonable, because it minimizes the long-term average reduction in ψ_l and plastid photosynthetic capacity.

Finally, if the aboveground and belowground environments fluctuate irregularly over multiday periods, then any analysis of optimal stomatal conductance and allocation to roots versus leaves must take into account possible adjustments in root allocation through time. Furthermore, in this case an increased allocation of energy to roots brings with it not only higher ψ_l at any one time but also an increase in the length of the period of potential photosynthetic activity following a rainfall. Cowan (1982) (see Chapter 5) has focused on the latter effect (which should be most important in xeric areas, to the exclusion of the former) in deriving his model for optimal conductance and allocation between roots and leaves. In this chapter, the former effect of g and ψ_l, which should be most important in mesic areas with a dependable water supply, has been emphasized almost to the exclusion of the latter. Real plants undoubtedly operate somewhere between these theoretical extremes, so that a complete model must take into account the effects of g and root allocation on both ψ_l and the length of the photosynthetic period. Furthermore, such a model must be phrased in terms of plant growth, so that the fractional allocation $f(t)$ of current photosynthate to leaves versus roots can vary temporally.

Extension to other traits affecting gas exchange

The approach taken in this chapter to optimal stomatal conductance and leaf-versus-root allocation in a constant environment could easily be extended to include other traits that affect leaf gas exchange. For example, Ehleringer and Mooney (1978) presented a quantitative model for optimal leaf absorptance in *Encelia farinosa* in terms of the effects of varying amounts of pubescence on leaf temperature, mesophyll light interception, diffusion, and thus photosynthesis. However, the predicted level of absorptance depends entirely on stomatal conductance, which is considered an externally determined, empirically fixed parameter for the purposes of their analysis. Incorporating analogues of equations (6.10b) and (6.11b) into the Ehleringer – Mooney model might permit a fully quantitative prediction of absorptance and stomatal conductance.

A second interesting trait for which our approach could be extended is effective leaf size (Givnish and Vermeij 1976; Givnish 1979, 1984). To do this, leaf boundary resistance r_a must be included in the denominators of equations (6.3) and (6.4), and the effects of varying r_s and r_a on leaf temperature T at nonzero r_a must be taken into account. Then, in the set of optimality conditions determined by equations (6.10a) and (6.11a), equation (6.10a) would be replaced by the analogous equations

$$0 = \frac{\partial(fA)}{\partial r_a} = f \cdot \left(\left.\frac{\partial A}{\partial r_a}\right|_{\psi_l, T} + \left.\frac{\partial A}{\partial \psi_l}\right|_{r_a, T} \cdot \left.\frac{\partial A}{\partial T}\right|_{r_a, \psi_l} \cdot \frac{\partial T}{\partial r_a} \right) \tag{6.39}$$

and

$$0 = \frac{\partial(fA)}{\partial r_s} = f \cdot \left(\left.\frac{\partial A}{\partial r_s}\right|_{\psi_l, T} + \left.\frac{\partial A}{\partial \psi_l}\right|_{r_s, T} \cdot \frac{\partial \psi_l}{\partial r_s} + \left.\frac{\partial A}{\partial T}\right|_{r_s, \psi_l} \cdot \frac{\partial T}{\partial r_s} \right) \tag{6.40}$$

For the present, we might ask if this system of equations is likely to favor leaf sizes within the quantitative range seen under particular conditions,

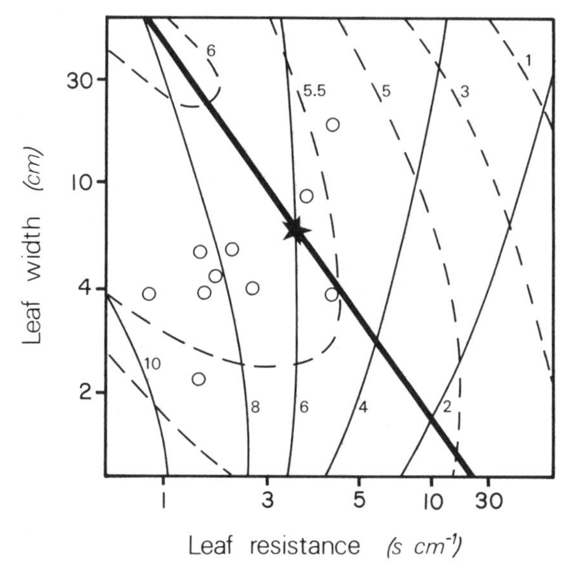

Figure 6.12. Predicted photosynthesis (broken lines; $\times 10^{-8}$ g CO_2 cm^{-2} s^{-1}) and transpiration (solid lines; $\times 10^{-6}$ g H_2O cm^{-2} s^{-1}) as functions of leaf size and stomatal resistance for typical summer conditions at the Michigan Biological Station. Circles indicate actual characteristics of each major tree species in the region (after Taylor 1975). The heavy line represents the locus of optimal leaf size and stomatal resistance as functions of root cost of transpiration b; the star indicates the leaf size and resistance that maximize $A - bE$ when $b = 5.0 \times 10^{-3}$ g CO_2 g^{-1} H_2O (after Givnish 1984).

by adopting the artifice of Givnish and Vermeij (1976) and assuming that root allocation per unit leaf area increases linearly with transpiration, so that net allocation of new photosynthate to leaves is $A - bE$. If we do this, we know that in the analogous case when $r_a = 0$, b is equal to the marginal cost of transpiration when f and ψ_l are allowed to vary independently of r_s in the more sophisticated model. Let us assume that b is also equal to the marginal cost of transpiration when $r_a \neq 0$. In this case, if we substitute the observed value of roughly 5.0×10^{-3} g CO_2 g^{-1} H_2O for $\partial A/\partial E$ in woody plants (Table 6.4) for b and use Taylor's (1975) empirical model for photosynthesis and transpiration as a function of effective leaf width and r_s for "standard" summer conditions at the Michigan Biological Station, the predicted leaf width and r_s are quite near the centroid of actual leaf widths and r_s values for woody plants native there (Figure 6.12) (Givnish 1984). This suggests strongly that leaf size, as well as stomatal resistance, is quantitatively adjusted so as to maximize whole-plant carbon gain.

Finally, the most intriguing, but conceptually troublesome, trait to which the model should be extended is leaf nitrogen concentration (Mooney and Gulmon 1979; Givnish 1984) (see Chapter 1 in this volume). On the one hand, Mooney and Gulmon (1979) showed that plants should have higher levels of leaf N in moister habitats, because the photosynthetic benefits associated with increased leaf N rise more steeply and plateau less rapidly when stomatal conductance is high. Furthermore, because the root costs associated with obtaining N should increase monotonically with leaf N level, and should be higher on N-poor soils, optimal leaf N levels should be lower on more sterile soils (Mooney and Gulmon 1979). On the other hand, in this chapter it is argued that as leaf N content increases, and r_m consequently decreases, stomatal conductance should increase. Thus, stomatal conductance should be high on fertile sites and lower on more sterile sites. Furthermore, because the root costs associated with obtaining water should be higher on drier soils, plants should have lower conductances on drier sites.

Hence, both arguments predict that stomatal conductance and leaf N level should increase toward moister and/or more fertile sites, but on very different grounds. On the one hand, an economic model for whole-plant carbon exchange is phrased in terms of the costs of obtaining N, to the exclusion of the aboveground and belowground costs associated with replacing transpirational water loss; on the other hand, the analysis is based on the latter costs to the exclusion of the costs of obtaining N and other limiting nutrients. The key question that must be answered in order to incorporate the Mooney–Gulmon effects in our approach is this: How do the metabolic costs associated with nutrient uptake through the roots vary with the total uptake rate of a particular nutrient, and with total root mass? An answer to this question requires an understanding not only of the direct metabolic costs associated with active transport of ions across concentration gradients of various magnitudes but also of the characteristic distances different ions diffuse in soils with various physical and chemical properties (see Chapter 8).

Only an approach that incorporates the costs of obtaining both water and nutrients can resolve the nettlesome question whether observed differences in gas-exchange properties are set more by the economics of nutrient capture or by the economics of water absorption. Clearly, the costs and benefits associated with both nutrient capture and water absorption should be important in every plant in determining optimal stomatal conductance, leaf N level, and allocation between leaves and roots. This is because, at least in a constant environment, the total partial derivatives $\partial(fA)/\partial g$, $\partial(fA)/\partial f$, and $\partial(fA)/\partial N$ should all equal zero when whole-plant energy gain is maximized. However, it is undoubtedly true that across some landscapes, differences between species in the cost of nutrient capture may be largely responsible for the observed differences in stomatal conductance, leaf N level, and allocation to leaves versus roots, whereas across other landscapes differences in the costs of water absorption largely determine species differences in these traits. The challenge is to develop an approach that can recognize both of these possibilities, and distinguish which applies in a particular interspecies comparison.

To illustrate the difficulties involved, consider the excellent comparative study of gas-exchange properties in mistletoes and their hosts by Ehleringer et al. (1985). That study attempted to test the hypothesis of Schulze et al. (1984) that xylem-tapping mistletoes have higher conductances than their hosts, primarily to capture more N from the host xylem stream, and thereby increase their leaf N content and photosynthetic capacity. This hypothesis is essentially that the economics of N capture is paramount in determining stomatal conductance and consequent values of water-use efficiency (A/E) in mistletoes. Ehleringer et al. (1985) initially showed that water-use efficiencies under uniform conditions in hosts and mistletoes had the same strong relationship to the δ^{13}C-isotope discrimination ratio in leaf tissue that developed under those conditions. They then demonstrated, based on δ^{13}C values, that mistletoes on three continents always have higher A/E values than their hosts, and thus a higher level of stomatal conductance at a given level of plastid photosynthetic capacity. However, the difference in δ^{13}C values, and hence A/E, between mistletoes and their hosts declines for N-fixing hosts. Ehleringer et al. (1985) interpreted this as support for their hypothesis: When N is abundant in the xylem stream, mistletoes need not transpire much more rapidly than their hosts to sequester sufficient N in their leaves. Thus, the economics of nitrogen capture dominates in determining differences in stomatal conductance between mistletoes on N-fixing hosts and nonfixing hosts.

However, close inspection of their data suggests an alternative interpretation. In fact, as Ehleringer et al. (1985) noted, N fixation in the host has no significant effect on the δ^{13}C and inferred A/E values of its mistletoes; the primary effect on the difference between host and parasite is through increased values of δ^{13}C and A/E in the host. Nitrogen-fixing hosts with higher N levels in their leaves have higher stomatal conductances and lower A/E ratios, as expected from the model presented in this chapter. The relatively constant value of δ^{13}C in mistletoes on fixing and nonfixing hosts on a given continent suggests that their high conductance may be set largely by the economics of water, not nutrients. Mistletoes may provide an excellent system for testing predictions of optimal stomatal resistance, leaf N content, and allocation between leaves and water-gathering haustoria. To the extent that there is no active transport of nutrients from host to parasite across the haustoria, the problem of separating the costs of water absorption from those of nutrient capture largely disappears. A predictive model in this case could be based solely on straightforward evaluations of $A(g, N, \psi_l)$, $\psi_l(f, g)$, and leaf nitrogen content N as a function of f, g, and mean leaf lifetime.

Summary

A quantitative model is presented for optimal stomatal conductance (or, alternatively, resistance) and energetic allocation between leaves and roots for plants growing in a constant environment. The analysis assumes that natural selection favors plants whose behavior maximizes whole-plant carbon gain, and is based on the effects of stomatal conductance and root allocation on leaf water potential, and on the effects of conductance and leaf water potential on photosynthesis per unit leaf mass. In general, if a linear relationship is assumed between photosynthesis and the CO_2 concentration in the intercellular air spaces of the mesophyll, then the ratio r_m/r_s of mesophyll to stomatal resistance should equal the ratio $f/(1-f)$ of energetic allocation to leaves versus roots.

A computer simulation is used to predict optimal stomatal resistance and fractional allocation to leaves as a function of eight environmental and physiological parameters. The predicted trends generally accord with those observed in qualitative terms; in addition, the observed values of r_s and leaf allocation in hydroponically grown *Phaseolus vulgaris* under one set of conditions closely match those predicted from available data on photosynthetic parameters and root hydraulic conductivity. The model also successfully accounts for the observed linear relationship between photosynthesis and stomatal conductance as mesophyll photosynthetic capacity varies, as well as the observed curvilinear relationship between photosynthesis and conductance as the transpirational costs associated with an increment in conductance increase.

The marginal cost of transpiration is defined in terms of the model as the rate at which photosynthesis is reduced by increased stomatal conductance via its effects on leaf water potential, divided by the rate at which stomatal conductance decreases leaf water potential. This marginal cost is also equal to the rate of decrease in photosynthesis with transpiration, due to the effects of transpiration rate on leaf water potential and plastid photosynthetic capacity. Substitution of observed values of the physiological and environmental parameters for *P. vulgaris* under standard conditions yields a marginal cost of 2.3×10^{-3} g CO_2 g^{-1} H_2O; as expected, decreases in root effective hydraulic conductivity and increases in the sensitivity of photosynthetic parameters to leaf water potential raise this marginal cost. Analysis shows that this measure of the marginal cost of transpiration is also equal to the root cost of transpiration *b*, defined in an earlier and less sophisticated model by Givnish and Vermeij (1976), and to $1/\lambda = \partial A/\partial E$ in the model of Cowan and Farquhar (1977) under constant environmental conditions. Tabulation of observed values of $\partial A/\partial E$ shows that herbs generally operate at values of λ close to that predicted for *P. vulgaris*, which would, in turn, maximize whole-plant carbon gain. Values of $\partial A/\partial E$ are higher in woody plants, especially conifers and mangroves, as expected in plants with more suberized root surfaces, narrower xylem vessels, and/or higher metabolic costs associated with exclusion or excretion of salt. Methods are suggested for extending the analysis to a temporally varying environment. The resulting criterion that the total partial derivative of photosynthesis with respect to conductance vanishes is compatible with the frequently reported results that $\partial A/\partial E > 0$ and $\partial A/\partial g > 0$, because tests of the Cowan–Farquhar model have been designed in such a way as to exclude almost all negative impacts of conductance on photosynthesis through leaf water potential. The implications of this for assessing the relative importance of stomatal and nonstomatal limits on photosynthesis are briefly discussed. Finally, extensions of the analysis to other leaf traits that affect gas exchange are outlined, with particular reference to leaf absorptance, effective leaf size, and leaf N content. The similarity between actual leaf widths and those predicted for typical summer conditions at the Michigan Biological Station, based on a simplified model using the observed value of $\partial A/\partial E$ in woody plants as the effective root cost of transpiration, suggests that leaf size may also be quantitatively adjusted so as to maximize whole-plant carbon gain.

Acknowledgments

I wish to thank my good friend and colleague Edwin Fiscus for his encouragement and help at several stages in the development of this chapter. Ian Cowan and Jack Morgan also shared in fruitful discussion and suggested a number of important references. Stan Wulfschleger helped produce the artwork and arranged several logistical details.

References

Aston, M. J., and D. W. Lawlor. 1979. The relationship between transpiration, root water uptake and leaf water potential. J. Exp. Bot. 30:169–181.

Badger, M. R., T. D. Sharkey, and S. von Caemmerer. 1984. The relationship between steady-state gas exchange of bean leaves and the levels of carbon-reduction cycle intermediates. Planta 160:305–313.
Ball, M. C., and G. D. Farquhar. 1984a. Photosynthetic and stomatal responses of two mangrove species, *Aegiceras corniculatum* and *Avicennia marina*, to long term salinity and humidity conditions. Plant Physiol. 74:1–6.
– 1984b. Photosynthetic and stomatal responses of the grey mangrove, *Avicennia marina*, to transient salinity conditions. Plant Physiol. 74:7–11.
Berkowitz, G. A., C. Chen, and M. Gibbs. 1983. Stromal acidification mediates *in vivo* water stress inhibition of nonstomatal-controlled photosynthesis. Plant Physiol. 72:1123–1126.
Björkman, O., J. S. Downton, and H. A. Mooney. 1980. Response and adaptation to water stress in *Nerium oleander*. Carnegie Inst. Washington Yearbook 79:150–157.
Björkman, O., and S. B. Powles. 1984. Inhibition of photosynthetic reactions under water stress: interaction with light level. Planta 161:490–504.
Boyer, J. S. 1971. Nonstomatal inhibition of photosynthesis in sunflower at low leaf water potentials and high light intensities. Plant Physiol. 48:532–536.
Boyer, J. S., and B. L. Bowen. 1970. Inhibition of oxygen evolution in chloroplasts isolated from leaves with low water potentials. Plant Physiol. 45:612–615.
Bunce, J. A. 1981. Comparative responses of leaf conductance to humidity in single attached leaves. J. Exp. Bot. 32:629–634.
– 1984. Effect of boundary layer conductance on the response of stomata to humidity. Plant Cell Environ. 8:55–57.
Caldwell, M. M. 1979. Root structure: the considerable cost of belowground function. Pp. 408–427 *in* O. T. Solbrig, S. Jain, G. B. Johnson, and P. H. Raven (eds.), Topics in plant population biology. Columbia University Press, New York.
Chiariello, N. 1984. Leaf energy balance in the wet lowland tropics. Pp. 85–98 *in* E. Medina, H. A. Mooney, and C. Vásquez-Yánes (eds.), Physiological ecology of plants of the wet tropics. Springer-Verlag, New York.
Clarkson, D. T., and J. Sanderson. 1974. The endodermis and its development in barley roots as related to radial migration of ions and water. Pp. 87–100 *in* J. Kolek (ed.), Structure and function of primary root tissues. Slovak Academy of Science, Bratislava.
Cowan, I. R. 1977. Stomatal behavior and environment. Adv. Bot. Res. 4:117–228.
– 1978. Water use in higher plants. Pp. 71–107 *in* A. K. McIntyre (ed.), Water: planets, plants, and people. Australian Academy of Science, Canberra.
– 1982. Water use and optimization of carbon assimilation. Pp. 589–613 *in* O. L. Lange, P. S. Nobel, C. B. Osmond, and H. Ziegler (eds.), Physiological plant ecology, II, vol. 12B, Encyclopedia of plant physiology, new series. Springer-Verlag, Berlin.
Cowan, I. R., and G. D. Farquhar. 1977. Stomatal function in relation to leaf metabolism and environment. Symp. Soc. Exp. Biol. 31:471–505.
Cowan, I. R., J. A. Raven, W. Hartung, and G. D. Farquhar. 1982. A possible role for abscisic acid in coupling stomatal conductance and photosynthetic carbon metabolism in leaves. Austral. J. Plant Physiol. 9:489–498.

Ehleringer, J. R., and C. S. Cook. 1984. Photosynthesis in *Encelia farinosa* Gray in response to decreasing leaf water potential. Plant Physiol. 75:688–693.

Ehleringer, J., and H. A. Mooney. 1978. Leaf hairs: effects on physiological activity and adaptive value to a desert shrub. Oecologia 37:183–200.

Ehleringer, J. R., E.-D. Schulze, H. Ziegler, O. L. Lange, G. D. Farquhar, and I. R. Cowan. 1985. Xylem-tapping mistletoes: water or nutrient parasites? Science 227:1479–1481.

Farquhar, G. D. 1978. Feedforward responses of stomata to humidity. Austral. J. Plant Physiol. 5:787–800.

– 1979. Carbon assimilation in relation to transpiration and fluxes of ammonia. Pp. 321–328 *in* R. Marcelle, H. Clijsters, and M. van Poucke (eds.), Photosynthesis and plant development. Dr. Junk, The Hague.

Farquhar, G. D., and K. Raschke. 1978. On the resistance to transpiration of the sites of evaporation within the leaf. Plant Physiol. 61:1000–1005.

Farquhar, G. D., E.-D. Schulze, and M. Küppers. 1980. Responses to humidity by stomata of *Nicotiana glauca* L. and *Corylus avellana* L. are consistent with the optimization of carbon dioxide uptake with respect to water loss. Austral. J. Plant Physiol. 7:315–327.

Farquhar, G. D., and T. D. Sharkey. 1982. Stomatal conductance and photosynthesis. Ann. Rev. Plant Physiol. 33:317–345.

Farquhar, G. D., and S. von Caemmerer. 1982. Electron transport limitations on the CO_2 assimilation rates of leaves: a model and some observations in *Phaseolus vulgaris* L. Pp. 163–175 *in* G. Akoyunoglov (ed.), Proceedings of the 5th International Congress on Photosynthesis. Balaban, Philadelphia.

Farquhar, G. D., and S. C. Wong. 1984. An empirical model of stomatal conductance. Austral. J. Plant Physiol. 11:191–210.

Field, C., J. A. Berry, and H. A. Mooney. 1982. A portable system for measuring carbon dioxide and water vapor exchange of leaves. Plant Cell Environ. 5:179–186.

Fiscus, E. L. 1975. The interaction between osmotic- and pressure-induced water flow in plant roots. Plant Physiol. 55:917–922.

– 1977. Determination of hydraulic and osmotic properties of soybean root systems. Plant Physiol. 59:1013–1020.

– 1981. Effects of abscisic acid on the hydraulic conductance of and the total ion transport through *Phaseolus* root systems. Plant Physiol. 68:169–174.

Fiscus, E. L., and P. J. Kramer. 1975. General model for osmotic pressure induced flow in plant roots. Proc. Natl. Acad. Sci. USA 72:3114–3118.

Fiscus, E. L., and A. H. Markhart, III. 1979. Relationships between root system water transport properties and plant size in *Phaseolus*. Plant Physiol. 64:770–773.

Gates, D. M. 1965. Energy, plants and ecology. Ecology 46:1–13.

Gates, D. M., and L. E. Papian. 1971. An atlas of leaf energy budgets. Academic Press, New York.

Givnish, T. J. 1976. Leaf form in relation to environment. PhD thesis, Princeton University, Princeton, N.J.

– 1979. On the adaptive significance of leaf form. Pp. 375–407 *in* O. T. Solbrig, S. Jain, G. B. Johnson, and P. H. Raven (eds.), Topics in plant population biology. Columbia University Press, New York.

– 1982. On the adaptive significance of leaf height in forest herbs. Amer. Nat. 120:353–381.

– 1984. Leaf and canopy adaptations in tropical forests. Pp. 51–84 *in* E. Medina, H. A. Mooney, and C. Vásquez-Yánes (eds.), Physiological ecology of plants of the wet tropics. Dr. Junk, The Hague.

Givnish, T. J., and G. J. Vermeij. 1976. Sizes and shapes of liane leaves. Amer. Nat. 975:743–778.

Glatzel, G. 1983. Mineral nutrition and water relations of hemiparasitic mistletoes: a question of partitioning. Experiments with *Loranthus europaeus* on *Quercus petraea* and *Quercus robur*. Oecologia 56:193–201.

Greacen, E. L., L. Ponsana, and K. P. Barley. 1976. Resistance to water flow in the roots of cereals. Pp. 86–100 *in* O. L. Lange, L. Kappen, and E.-D. Schulze (eds.), Water and plant life. Springer-Verlag, Berlin.

Gulmon, S. L., and C. C. Chu. 1981. The effect of light and nitrogen on photosynthesis, leaf characteristics, and dry matter allocation in the chaparral shrub, *Diplacus aurantiacus*. Oecologia 49:207–212.

Hall, A. E., and E.-D. Schulze. 1980a. Stomatal response to environment and a possible interrelation between stomatal effects on transpiration and CO_2 assimilation. Plant Cell Environ. 3:467–474.

– 1980b. Drought effects on transpiration and leaf water status of cowpea in controlled environments. Austral. J. Plant Physiol. 7:141–147.

Horn, H. S. 1971. The adaptive geometry of trees. Princeton University Press, Princeton, N.J.

Hsiao,T. C. 1973. Plant responses to water stress. Ann. Rev. Plant Physiol. 24:519–570.

Intriligator, M. D. 1971. Mathematical optimization and economic theory. Prentice-Hall, Englewood Cliffs, N.J.

Jarman, P. D. 1974. The diffusion of carbon dioxide and water vapour through stomata. J. Exp. Bot. 25:927–936.

Jones, H. G. 1973. Moderate-term water stresses and associated changes in some photosynthetic parameters in cotton. New Phytol. 72:1095–1105.

Körner, C., J. A. Scheel, and H. Baur. 1979. Maximum leaf diffusive conductance in vascular plants. Photosynthetica 13:45–82.

Landsberg, J. J., and N. D. Fowkes. 1978. Water movement through plant roots. Ann. Bot. 42:493–508.

Leuning, R. 1983. Transport of gases into leaves. Plant Cell Environ. 6:181–194.

Ludlow, M. M., and O. Björkman. 1984. Paraheliotropic movement in *Siratro* as a protective mechanism against drought-induced damage to primary photosynthetic reactions: damage by excessive light and heat. Planta 161:505–518.

Markhart, A. H., III, E. L. Fiscus, A. W. Naylor, and P. J. Kramer. 1979. Effect of temperature on water and ion transport in soybean and broccoli root systems. Plant Physiol. 64:83–87.

Meinzer, F. C. 1982. The effect of vapour pressure on stomatal control of gas exchange in Douglas fir, *Pseudotsuga menziesii*. Oecologia 54:236–242.

Mohanty, P., and J. S. Boyer. 1976. Chloroplast response to low leaf water potentials. IV. Quantum yield is reduced. Plant Physiol. 57:704–709.

Mooney, H. A., O. Björkman, and G. J. Collatz. 1977. Photosynthetic acclimation to temperature and water stress in the desert shrub *Larrea divaricata.* Carnegie Inst. Washington Yearbook 76:328–335.
Mooney, H. A., and C. Chu. 1983. Stomatal responses to humidity of coastal and interior populations of a California shrub. Oecologia 57:148–150.
Mooney, H. A., C. Field, C. Vásquez-Yánes, and C. Chu. 1983a. Environmental controls on stomatal conductance in a shrub of the humid tropics. Proc. Natl. Acad. Sci. USA 80:1295–1297.
Mooney, H. A., C. Field, W. E. Williams, J. A. Berry, and O. Björkman. 1983b. Photosynthetic characteristics of plants of a Californian cool coastal environment. Oecologia 57:38–42.
Mooney, H. A., and S. L. Gulmon. 1979. Environmental and evolutionary constraints on the photosynthetic characteristics of higher plants. Pp. 316–337 *in* O. T. Solbrig, S. Jain, G. B. Johnson, and P. H. Raven (eds.), Topics in plant population biology. Columbia Uniersity Press, New York.
Munz, P. A. 1975. A California flora, with supplement. University of California Press, Berkeley.
Newman, E. I. 1974. Root and soil water relations. Pp. 363–440 *in* E. W. Carson (ed.), The plant root and its environment. University Press of Virginia, Charlottesville.
– 1976. Water movement through root systems. Phil. Trans. R. Soc. Lond. 273:463–478.
Oosterhuis, D. M. 1983. Resistances to water flow through the soil-plant system. S. Afr. J. Sci. 79:459–465.
Osmond, C. B., K. Winter, and H. Ziegler. 1982. Functional significance of different pathways of CO_2 fixation in photosynthesis. Pp. 479–547 *in* O. L. Lange, P. S. Nobel, C. B. Osmond, and H. Ziegler (eds.), Physiological plant ecology, II, vol. 12B, Encyclopedia of plant physiology, new series. Springer-Verlag, New York.
O'Toole, J. C., J. L. Ozbun, and D. M. Wallace. 1977. Photosynthetic response to water stress in *Phaseolus vulgaris.* Physiol. Plant. 40:111–114.
Passioura, J. B. 1980. The transport of water from soil to shoot in wheat seedlings. J. Exp. Bot. 31:333–345.
– 1982. Water in the soil-plant-atmosphere continuum. Pp. 5–34 *in* O. L. Lange, P. S. Nobel, C. B. Osmond, and H. Ziegler (eds.), Physiological plant ecology, II, vol. 12B, Encyclopedia of plant physiology, new series. Springer-Verlag, New York.
Pearcy, R. W., and J. Ehleringer. 1984. Comparative ecophysiology of C_3 and C_4 plants. Plant Cell Environ. 7:1–13.
Powell, D. B. B. 1978. Regulation of plant water potential by membranes of the endodermis in young corn roots. Plant Cell Environ. 1:69–76.
Raschke, K. 1979. Movements of stomata. Pp. 383–441 *in* W. Hapt and M. E. Feinleib (eds.), Physiology of movements, vol. 7, Encyclopedia of plant physiology, new series. Springer-Verlag, New York.
Sánchez-Díaz, M. F., and H. A. Mooney. 1979. Resistance to water transfer in desert shrubs native to Death Valley, California. Physiol. Plant. 46:139–146.
Sanderson, J. 1983. Water uptake by different regions of the barley root. Pathways of radial flow in relation to development of the endodermis. J. Exp. Bot. 43:240–253.

Sands, R., E. L. Fiscus, and C. P. P. Reid. 1982. Hydraulic properties of pine and bean roots with varying degrees of suberization, vascular differentiation and mycorrhizal infection. Austral. J. Plant Physiol. 9:559–569.

Schulze, E.-D. 1982. Plant life forms and their carbon, water and nutrient relations. Pp. 615–676 *in* O. L. Lange, P. S. Nobel, C. B. Osmond, and H. Ziegler (eds.), Physiological plant ecology, II, vol. 12B, Encyclopedia of plant physiology, new series. Springer-Verlag, New York.

Schulze, E.-D., and A. E. Hall. 1982. Stomatal responses, water loss and CO_2 assimilation rates of plants in constrasting environments. Pp. 181–230 *in* O. L. Lange, P. S. Nobel, C. B. Osmond, and H. Ziegler (eds.), Physiological plant ecology, II, vol. 12B, Encyclopedia of plant physiology, new series. Springer-Verlag, New York.

Schulze, E.-D., N. C. Turner, and G. Glatzel. 1984. Carbon, water and nutrient relations of two mistletoes and their hosts: a hypothesis. Plant Cell Environ. 7:293–299.

Sharkey, T. D. 1984. Transpiration-induced changes in the photosynthetic capacity of leaves. Planta 160:143–150.

Sharkey, T. D., and M. R. Badger. 1982. Effects of water stress on photosynthetic electron transport, photosphorylation and metabolite levels of *Xanthium strumarium* mesophyll cells. Planta 156:199–206.

Sharkey, T. D., K. Imai, G. D. Farquhar, and I. R. Cowan. 1982. A direct confirmation of the standard method of estimating intercellular partial pressure of CO_2. Plant Physiol. 60:657–659.

Taylor, S. E. 1975. Optimal leaf form. Pp. 73–86 *in* D. M. Gates and R. B. Schmerl (eds.), Perspectives in biophysical ecology. Springer-Verlag, New York.

Taylor, H. M., and B. Klepper. 1975. Water uptake by cotton root systems: an examination of assumptions in the single root model. Soil Sci. 120:57–67.

von Caemmerer, S., and G. D. Farquhar. 1981. Some relationships between the biochemistry of photosynthesis and gas exchange of leaves. Planta 153:376–387.

– 1984. Effects of partial defoliation, changes of irradiance during growth, short term water stress and growth at enhanced p(CO_2) on the photosynthetic capacity of leaves of *Phaseolus vulgaris* L. Planta 160:320–329.

Williams, W. E. 1983. Optimal water-use efficiency in a Californian shrub. Plant Cell Environ. 6:145–151.

Wong, S. C., I. R. Cowan, and G. D. Farquhar. 1979. Stomatal conductance correlates with photosynthetic capacity. Nature 282:424–426.

Zimmermann, M. H. 1983. Xylem structure and the ascent of sap. Springer-Verlag, New York.

7 Internal leaf structure: a three-dimensional perspective

DAVID F. PARKHURST

Introduction

For more than a century, the internal structure of leaves and the variations of that structure with environment have intrigued ecologists. This chapter considers the adaptive significance of certain features of internal leaf structure, especially the existence of distinctively different palisade and spongy mesophyll tissues.

Consideration of adaptive significance implies an interest in the results of natural selection. However, because it is often difficult or impossible to mimic natural selection experimentally, I have developed a mathematical model to integrate the effects of structural variation on photosynthesis. In later sections the model will be described, accompanied by a discussion of three-dimensional considerations. Then the model will be applied to the question of the palisade – spongy dichotomy, and later to a series of other questions. Finally, further questions beyond the scope of the model will be posed.

I have often asked friends, both biologists and others, to guess the typical thickness of leaves from common plants like oaks, tomatoes, and the like. Almost all have responded by gauging an imaginary leaf between thumb and index finger and making a tentative guess: "One or two millimeters?" In fact, such estimates are about 10-fold too high; leaf blades, not counting midribs or major veins, are commonly between one-tenth and three-tenths of a millimeter thick. So leaves are very thin, especially in comparison with their overall outline, and it is no wonder that we tend to see them as two-dimensional planar objects.

Contrary to this simple impression, however, the three-dimensional structure of leaves is very important to their function, and that is one theme of this chapter. Its more general theme is to look at the internal structure of leaves (of C_3 plants) in relation to photosynthetic carbon dioxide uptake.

In what follows, I shall unashamedly speak from the platform of the

"Adaptionist Program" (Gould and Lewontin 1979). I still believe (Parkhurst and Loucks 1972) that natural selection *tends* to lead to optimal form and function in plant leaves, even if it does not do so for all features of all organisms. Generally speaking, leaves have one dominant function – to capture energy by photosynthesis – and this simplicity of purpose allows optimality studies to contribute a great deal to our understanding of leaf biology.

Of course, leaves may affect plant fitness in other ways, such as by shading leaves on competing plants, but photosynthetic carbon assimilation seems likely to override other factors in the vast majority of cases. In the remainder of this chapter, then, I shall assume that leaf form is optimal when it maximizes net photosynthesis rate. The complications caused by covariations in transpirational water loss (Parkhurst and Loucks 1972; Givnish and Vermeij 1976) are avoided here by holding boundary-layer and stomatal resistances, and hence water loss, constant.

Palisade and spongy mesophyll

This study began with an attempt to explain why natural selection might have led to the differentiation of the palisade and spongy mesophyll tissue types in leaves. Inspection of leaf cross sections indicates that these two tissues can differ radically in the amounts of air space, of chlorenchyma, and of other cell types per unit volume of tissue; they also differ in cell size and shape, in cell-wall surface area per unit volume, and in other features. There are many ways in which these aspects of mesophyll structure might affect photosynthesis rates; for examples, see Chabot and Chabot (1977), Dornhoff and Shibles (1976), El-Sharkaway and Hesketh (1965), Nobel, Zaragoza, and Smith (1975), Nobel (1976, 1977), Patterson et al. (1977), and Wilson and Cooper (1967).

Figure 7.1 assembles the puzzle into a single picture. The figure shows how certain fundamental variables like mesophyll thickness, porosity, and cell size and shape can act, by affecting diffusion and chemical reactions, to influence the photosynthetic CO_2 uptake rate by a leaf. For example, if all other factors remained equal, a thicker mesophyll would tend to affect CO_2 assimilation in these ways:

1. Assimilation per gram investment of tissue would *decrease* because of internal self-shading (Givnish 1979) and because the average CO_2 molecule would need to diffuse farther through the intercellular air spaces before it reached a cell wall. Many workers ignore this factor entirely.
2. Assimilation per unit leaf surface area would *increase* because there is more cell-wall area, easing liquid-phase transport of CO_2.

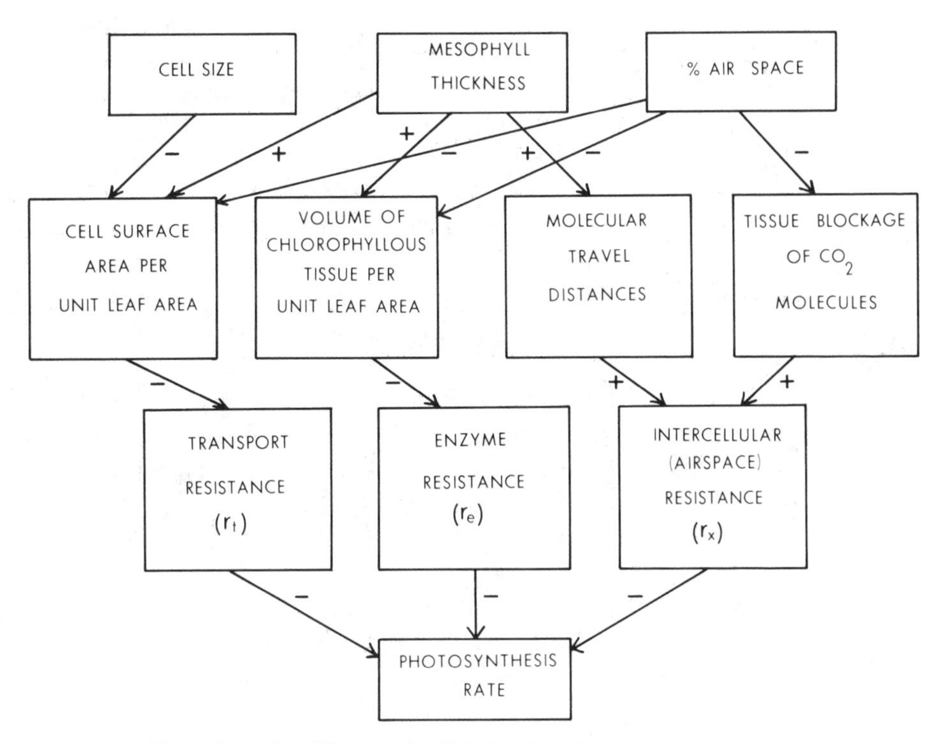

Figure 7.1. Possible causative links between fundamental anatomical variables such as mesophyll thickness, porosity, and cell size (top) and photosynthesis rate (bottom). The intermediate levels show derived variables that help explain the overall effects. The + (or −) on the arrow from factor A to factor B indicates that if all other factors remain constant, an increase in A would lead to an increase (decrease) in factor B.

3. The latter quantity would also increase because there is more "factory" (chloroplasts and enzymes) per unit of leaf surface area.

Some authors (e.g., Nobel et al. 1975) have emphasized single factors such as cell surface area or chlorophyll density per area or volume. Although particular factors of leaf structure will at times be rate-limiting, it is clear that all the factors mentioned are potentially important. The common occurrence of sun leaves having mesophyll differentiated into palisade and spongy tissue types is further evidence that cell-wall area is not, by itself, a sufficient measure of structural adaptation for photosynthesis. If it were sufficient, the leaf would presumably contain only whichever tissue type had the greater surface area for a given carbohydrate investment.

I hold the working hypothesis that in C_3 species, differentiation of the mesophyll into palisade and spongy tissues can be explained largely in terms of promoting CO_2 diffusion (which should in turn increase photosynthesis rates). Specifically, palisade tissue, like a handful of pencils grabbed from a box, allows CO_2 movement through the air spaces between cells mainly in the direction of the axes of the cylindrical cells; horizontal (paradermal) movement is inhibited if the palisade is at all compact. The porosity of a typical spongy tissue, on the other hand, is much more isotropic, or nondirectional. Thus, at least in a hypostomatous leaf, the spongy tissue on the bottom provides for horizontal diffusion and allows palisade cells that are not directly opposite a stomate to be supplied with CO_2. The net effect is that the CO_2 pathway has a shape similar to that of a candelabrum.

Although there may be other evolutionary explanations for the existence of the two cell types, the one stated here seems plausible. From an evolutionary point of view, the existence of the two cell types at least suggests (as have Jones and Slatyer 1972a) that air-space (intercellular) diffusion can limit net photosynthesis; one should not ignore this link in the chain (as some authors have) without careful justification.

Much of the work to be described here represents exploration of the hypothesis that many aspects of internal leaf structure are adaptations for CO_2 diffusion, but other factors such as light scattering will also be considered.

We now turn to the photosynthesis model.

A general mathematical model for CO_2 diffusion inside leaves

An earlier model

The model used for the studies reported in this chapter is a generalization of a more restrictive model developed earlier (Parkhurst 1977a). In that work, a three-dimensional equation for steady-state diffusion with chemical reaction was used to model photosynthetic CO_2 uptake in whole leaves. [A related model, by Rand (1977), uses a one-dimensional approximation.]

In its most general form, the model of Parkhurst (1977a) is

$$D\left[\frac{\partial}{\partial x}\left(p_i\frac{\partial C}{\partial x}\right)+\frac{\partial}{\partial y}\left(p_j\frac{\partial C}{\partial y}\right)+\frac{\partial}{\partial z}\left(p_k\frac{\partial C}{\partial z}\right)\right]=U(x,\,y,\,z,\,C) \quad (7.1)$$

The C in this equation represents the local concentration of CO_2 as it varies from point to point inside the leaf. That variation is of key importance, because reduced C will in general be accompanied by a reduced local

Table 7.1. *Glossary of symbols*[a]

Symbol	Meaning	Units
A_{cw}	Cell-wall area per unit volume	cm^{-1}
C	CO_2 concentration	$g\ cm^{-3}$
C_∞	Ambient CO_2 concentration	$g\ cm^{-3}$
C_ρ, C_z	Concentration gradients, $\partial C/\partial\rho$ and $\partial C/\partial z$	$g\ cm^{-4}$
D	Diffusivity of CO_2 in air	$cm^2\ s^{-1}$
F	Photosynthetic flux density	$\mu mol\ m^{-2}\ s^{-1}$
I	Relative light intensity	
I_0	Light intensity just inside upper epidermis	
L_s	Effective length of stomatal cylinder [equation (7.3)]	cm (or μm)
P	Diffusivity times paradermal porosity/tortuosity	$cm^2\ s^{-1}$
Q	Diffusivity times porosity/tortuosity, normal to leaf surfaces	$cm^2\ s^{-1}$
U	CO_2 uptake function	$g\ cm^{-3}\ s^{-1}$
a	Radius of tissue plug served by one stomate	cm (or μm)
a'	Radius of equivalent stomatal cylinder	cm (or μm)
$f(\rho, z)$	Local CO_2 sink strength, $2/[z_m(r_t + r_e)]$	s^{-1}
$g(\rho, z)$	Local back-production (respiration) of CO_2	$g\ cm^{-3}\ s^{-1}$
k_e	Proportionality constant for r_e function	$s\ cm^{-1}$
k_t	Proportionality constant for r_t function	$s\ cm^{-2}$
p_i, p_j, p_k	Porosity/tortuosity in x, y, and z directions	
r_a	Boundary-layer resistance	$s\ cm^{-1}$
r_e	Enzyme "resistance" to CO_2 assimilation	$s\ cm^{-1}$
r_0	Surface resistance, $r_s + r_a$	$s\ cm^{-1}$
r_s	Stomatal resistance to CO_2 diffusion	$s\ cm^{-1}$
r_t	Transport resistance to CO_2 assimilation (intracellular)	$s\ cm^{-1}$
x, y	Paradermal coordinates	cm (or μm)
z	Coordinate normal to leaf surfaces	cm (or μm)
z_m	Total thickness of mesophyll tissue	cm (or μm)
δ	Density of chlorophyll-containing cells in a particular tissue layer	$cm^3\ cm^{-3}$
λ	Coefficient of anisotropy relating P and Q (i.e., $P = \lambda Q$)	
Γ	CO_2 compensation point [as defined by Jones and Slatyer (1972b)]	$g\ cm^{-3}$
ρ	Radial coordinate in tissue plug	cm (or μm)

[a] Those variables based on leaf surface area are based on the total surface area, or twice the projected area, of the leaf.

photosynthesis rate. Other symbols used in this chapter are defined in Table 7.1.

In its earlier form, the model (7.1) was simplified with the following assumptions:

1. The net CO_2 uptake U at any point in the mesophyll was taken to be

$$U = 2(C - \Gamma)/[z_m(r_t + r_e)] \tag{7.2}$$

2. The effective directional porosities p_i, p_j, and p_k were assumed equal to each other and were taken to be uniform and constant throughout the mesophyll.
3. Γ, r_t, and r_e were also considered uniform and constant through the mesophyll.

With these assumptions, analytic solutions were derived for both one- and three-dimensional geometries, and a number of conclusions were drawn. The solution, which is the concentration profile $C(x, y, z)$ satisfying (7.1), was then used to calculate net photosynthesis by leaves under various conditions.

Generalized model

In this chapter, the model is made considerably more realistic by removing several simplifications and considering cases in which the following hold:

1. The effective porosities p_i, p_j, and p_k need no longer be uniform but may depend on the coordinates x, y, and z. For example, they may differ between palisade and spongy mesophyll, as described in the example that follows, which uses data for *Arbutus menziesii* (Pursh).
2. The directional p's at any given point may differ from one another; for example, in palisade mesophyll tissue, porosity (corrected for tortuosity) tends to be higher in the direction parallel to the long axis of the cylindrical cells than it is in a direction perpendicular to that axis. In other words, the medium may be anisotropic.
3. $U = fC - g$ at any point in the mesophyll, where f is the local sink strength for CO_2, and g is the rate of CO_2 production by respiration. Both f and g may vary with the location in the mesophyll. Thus, the local uptake function U is more general here than in the earlier model. The specific case solved in this chapter is $U = f(C - \Gamma)$, where Γ is assumed uniform throughout the mesophyll; thus, $g = f\Gamma$. However, the solution method developed will work equally well for other g's.

A major point of Parkhurst (1977a), and of this chapter, is that the three-dimensional geometry (*volume* relationships) of CO_2 diffusion within leaves cannot be well approximated by one-dimensional models. However, when there is axial symmetry, it is possible to reduce volume relationships to two *computational* dimensions, by using cylindrical coordinates. If $p_i \neq p_j$, or if axial symmetry is missing for some other reason, the techniques presented here could be applied in three rectangular coordinates, although with more work. Here we assume axial symmetry and define $P = p_i D = p_j D$, and $Q = p_k D$. Hence, P and Q are "effective diffusivities."

To use cylindrical coordinates, suppose a stomate of radius a' serves a cylindrical plug of tissue with radius a, as in Figure 7.2. The present analysis differs from the earlier version by treating the stomate as a circular

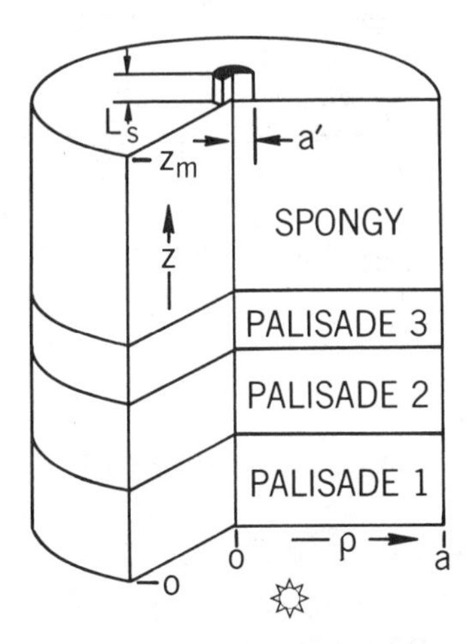

Figure 7.2. Geometric relationships of a typical plug of tissue served by one stomate. Sun symbol indicates direction from which sunlight strikes upper epidermis.

cylinder having an "effective length" L_s such that resistance to diffusion of CO_2 through the cylinder is equal to the measured (or assumed) stomatal resistance plus average boundary-layer resistance of the leaf. This sum, the "surface resistance," is denoted by r_0. The boundary-layer resistance varies from point to point on the leaf surface, increasing with distance from the leading edge (Parkhurst et al. 1968), but because its value is usually much smaller than that of stomatal resistance, it is sufficient to use its average. Technically, CO_2 transfer through the boundary layer is convective rather than purely diffusive. The present model simply adds to the effective length of the equivalent cylindrical stomate an amount sufficient to represent the average boundary-layer resistance.

Figure 7.2 shows the geometrical relationships involved in one "stomatal unit," which represents the average such unit on the leaf. The volumetric region served by a given stomate will never be a true circular cylinder, but a cylinder is an adequate approximation, because diffusion "smears out" the details at the edges. Thus (for a hypostomatous leaf), one chooses values for a, a', and L_s as follows:

1. a is the radius such that the plan area per stomate ($= \pi a^2$) is equal to the plan area of the leaf (one side) divided by the total number of stomates on that one side. That is, let A_1 be the area of the lower surface of a hypostomatous leaf, and let N be the number of stomata on that surface. Then $\pi a^2 = A_1/N$, or $a = (A_1/N\pi)^{1/2}$.
2. a' is the radius such that the ratio of the total stomatal pore area to the plan area of the leaf is equal to the ratio of $\pi a'^2$ to πa^2. Thus, if the total

stomatal pore area is a proportion s_1 of the area A_1, then $\pi a'^2 = s_1 \pi a^2$, or $a' = \sqrt{s_1}$.

3. The diffusional flux (f_1) of CO_2 through one pore will be given by the product of the diffusivity (D), the cross-sectional area of the pore ($\pi a'^2$), and the CO_2 concentration gradient ($\Delta C / L_s$). L_s is the length of the cylinder that simulates the stomatal pore. The total flux through the whole lower surface will be Nf_1 and can also be calculated as $2\Delta C(\pi a^2 N)/r_0$ (because r_0 is defined in terms of total leaf surface area). These relationships allow us to solve for

$$L_s = r_0 D a'^2 / (2a^2) \tag{7.3}$$

In cylindrical coordinates, and given our assumptions, (7.1) becomes

$$D\left[\frac{1}{\rho}\frac{\partial}{\partial \rho}\left(\rho P \frac{\partial C}{\partial \rho}\right) + \frac{\partial}{\partial z}\left(Q \frac{\partial C}{\partial z}\right)\right] = U(\rho, z) \cdot (C - \Gamma) \tag{7.4}$$

where ρ is chosen to denote the radial coordinate, in order to avoid confusion with transfer resistances, which are commonly denoted by lowercase r's. The boundary conditions (for a hypostomatous leaf) are

$$\begin{aligned} &C_\rho(0, z) = C_\rho(a, z) = 0 \\ &C_z(\rho, 0) = 0 \\ &C_z(\rho, z_m) = 0, \qquad a' \le \rho \le a \\ &C(\rho, z_m + L_s) = C_\infty, \qquad 0 \le \rho \le a' \end{aligned} \tag{7.5}$$

where $C_\rho = \partial C / \partial \rho$, and so forth.

These boundary conditions state that $\partial C / \partial \rho = 0$ at $\rho = 0$ (because of the axial symmetry) and at $\rho = a$ (because our representative stomatal unit is considered to be surrounded by other identical units). Also, $\partial C / \partial z = 0$ over the epidermal cells, because we are assuming that CO_2 movement through the epidermis is negligible compared with that through the stomates. (If cuticular CO_2 transport were appreciable, more complex boundary conditions could be used.) Finally, C_∞ is assumed to be the ambient CO_2 concentration at the outside end of the equivalent stomatal cylinder already described.

Example application

To paraphrase Hamming (1973): The purpose of modeling is insight, not numbers. So I turn now to the *use* of the model just described. I shall work with leaves from the sclerophyllous tree, madrone (*Arbutus menziesii* Pursh), first describing how its structural features are included in the model and then how the model can be solved to estimate photosynthesis rates. Then "experiments" are performed by changing the leaf struc-

Table 7.2. *Mesophyll characteristics of an* Arbutus menziesii *leaf*[a]

Tissue layer	Thickness (μm)	Air space	Chlorophyllous tissue	Cell-wall area per volume (cm^{-1})
Palisade 1	64	7.4%	88.4%	990
Palisade 2	60	17.0%	80.3%	980
Palisade 3	41	25.2%	64.4%	663
Spongy	133	39.4%	46.4%	88
Overall	298	26.6%	64.7%	540

[a] The percentages here represent the portions of the volume of each mesophyll layer taken up by air space or by chlorophyllous tissue; hence, they differ from the percentages listed in Table 7.4.
Source: Data from Hays (1976, and personal communication).

ture in the model to see how calculated photosynthesis responds. Finally, the evolutionary implications of the results are discussed.

Hays (1976, and personal communication) measured several features of the internal leaf structure of madrone; some of the measurements are described in Table 7.2, and the leaf cross section is shown in Figure 7.3. For the initial calculations, I have made the following assumptions:

1. Stomatal spacing equals mesophyll thickness (Parkhurst, 1977a).
2. The stomate occupies 1% of the bottom surface area of the tissue plug it serves, when $r_0 = 3$ s cm^{-1} (Parkhurst 1977a). Hence, $a' = 0.1a$.
3. In each tissue layer, Q is calculated as the CO_2 diffusivity D multiplied by the fraction of intercellular air space in that layer. (This neglects any effects of tortuosity in the z, or thickness, dimension.) Thus, Q is larger in the more open spongy tissue than in the denser palisade layers.
4. In the palisade layers, tortuosity and blockage effects are taken to be more restrictive to horizontal diffusion than to vertical, making $P \leq Q$ there. In particular, $P = \lambda Q$ is assumed, where $\lambda = 0.1$, 0.4, 0.7, and 1.0 in the four tissue layers (adaxial palisade to spongy, respectively). Tortuosity is very difficult to measure, and these values have been estimated from visual inspection of sections. It will be shown later that the model is not at all sensitive to the values used.
5. Internal light intensity (I) has the following distribution, relative to 100% (or I_0) entering the adaxial epidermis: (a) $I = 0.95I_0$ at the top of the upper palisade layer; that is, the upper epidermis absorbs 5% of the light. (b) $I = 0.1I_0$ just inside the adaxial epidermis (at the bottom of the spongy layer); that is, the upper epidermis and the mesophyll together absorb 90% of the light. (c) I decays exponentially (Beer's law) through the mesophyll, with the decay coefficient in each layer being proportional to the chlorenchyma cell density in that layer.

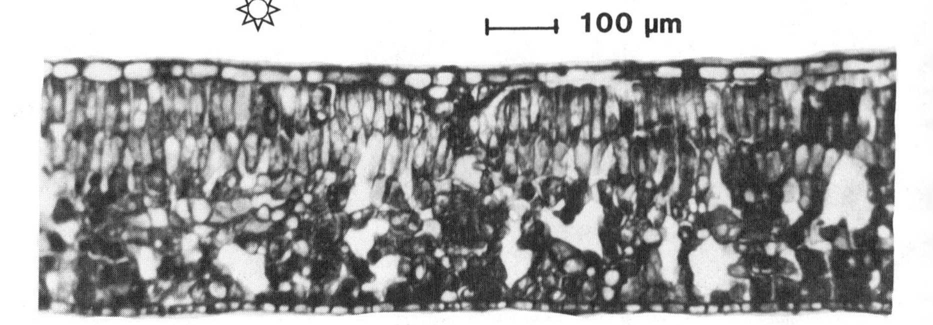

Figure 7.3. Cross section of leaf of *Arbutus menziesii.* Sun symbol indicates upper epidermis. (Photo by Robert L. Hays.)

6. The sink strength at any point in the mesophyll is given by

$$f(\rho, z) = \frac{2}{z_m(r_t + r_e)} \tag{7.6}$$

where r_t and r_e are the local values of transport and enzyme resistances. This function is plotted in Figure 7.4.

7. The transport resistance r_t is made inversely proportional to the local density of chlorenchyma cell surface area. Hence, $r_t = k_t/A_{cw}$, where k_t was adjusted to give r_t a mean value of 1.5 s cm^{-1} through the mesophyll for most runs. This value and that of r_e in the following were obtained by dividing the mesophyll resistance value of 3 s cm^{-1} from Bierhuizen and Slatyer (1964) equally between the two components.

8. $r_e = k_e/(I\delta)$, where δ is the local density of chlorenchyma, and where k_e was similarly adjusted to give r_e a mean value of 1.5 s cm^{-1}.

9. r_0 $(= r_s + r_a) = 3$ s cm^{-1}. (Calculations with $r_0 = 20$ s cm^{-1} will also be described in a later section.)

In other words, the sink strength f is assumed to increase with local light level, with density of chlorenchyma cells (a rough proxy for local enzyme content), and with local cell-wall area per unit volume of tissue. Note that f is a local CO_2 conductance, on a volume basis (its units are s^{-1}).

For the calculations that follow, the "standard conditions" will incorporate all the foregoing assumptions, and in particular, $r_0 = 3$ s cm^{-1}, the presence of a substomatal cavity, exponential decay of light levels, and anisotropic effective diffusivities. There is nothing sacred about all these assumptions, but they capture many of the major variations one might expect to find within a leaf having the structure measured by Hays. They provide an example of the generality of this revised model, and they can easily be modified as more information becomes available or to represent different assumptions.

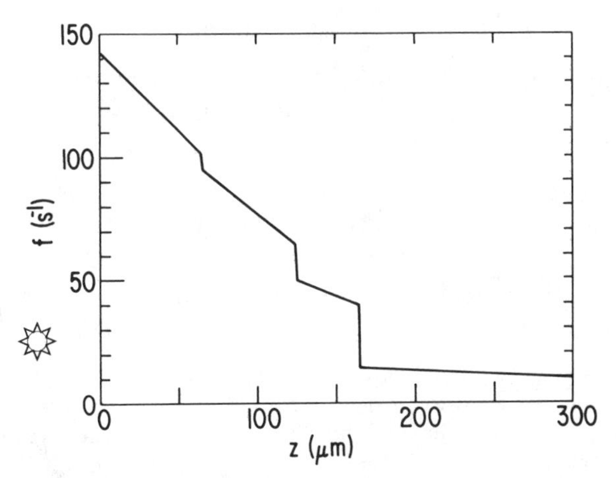

Figure 7.4. Assumed sink-strength function (f) as it varies with depth into the mesophyll (z), measured from the inner surface of the adaxial epidermis. Sun symbol indicates upper epidermis.

Purposes of the model

Because mathematical models can be used in various ways for various purposes, I shall state here the purposes of this one. The first purpose is to serve as a sort of "coat rack" on which to hang various pieces of knowledge from the literature and to see how well they fit together in an integrated whole. To paraphrase Forrester (1968), most workers in any field (such as ecological leaf anatomy) form their own internal mental models. Mathematical modeling forces one's mental model to be laid out concisely and explicitly in a form that is easy for others to criticize and improve on. The formulation stated here is exposed and vulnerable in that way.

Second, the process of making the pieces fit together at a precise mathematical level often allows one to integrate them in other, more general ways. Figure 7.1 was actually derived from the model, rather than the converse.

Third, because a mathematical statement of any part of a model must be very precise, one is forced while modeling to make many decisions. That is, one cannot think vaguely, but must choose specific functional forms (such as Beer's law of decay of light level) and specific values for each parameter (such as $r_0 = 3$ s cm^{-1}). When such decisions are not clear-cut, it is often feasible to try two or more different versions and see how much the results vary. For example, one can run the model first with the Beer's law assumption, and a second time assuming uniform internal light intensity. Such "experiments on the model," or sensitivity tests, will be described later.

Solution by the finite-element method

The earlier model (Parkhurst 1977a) had one advantage over the one described here: It could be solved analytically in terms of Fourier or Bessel series. The current model must be solved numerically, but the gain in generality is worthwhile.

Of the various methods that exist for solving partial differential equations numerically, the finite-element method or FEM (Huebner 1975) was adopted because of its flexibility. The details of the solution will be described elsewhere, but some brief remarks are in order here:

1. The volume of the mesophyll plug is divided into a set of elements, each of which is triangular in cross section and forms a ring concentric to the plug centerline. Elements of this type are used because equation (7.4) contains no angular dependence in any paradermal plane. That is, our representative tissue plug (served by one stomate) is assumed to be radially symmetric, with CO_2 concentration varying primarily in the ρ and z dimensions. Figure 7.5 shows a radial section of the plug (and hence cross sections of all the elements) for the grid of the type used for the majority of calculations reported here. The grid shown comprises 341 elements, defined by 205 nodes (vertices).
2. By means of the Galerkin method (Huebner 1975), the FEM yields a system of n linear algebraic equations in n unknowns that is solved by standard methods. The solution of equation (7.4) by the FEM is the set of CO_2 concentrations C at each of the $n = 205$ nodes.
3. Anywhere away from the nodes, C is determined by a two-dimensional (planar) extension of linear interpolation.
4. Once the nodal C values have been calculated, one can calculate total CO_2 uptake by integrating its local value throughout the volume:

$$\int_V U \, dV = \int_V f(C - \Gamma) \, dV \tag{7.7}$$

CO_2 concentration profiles inside leaves: a three-dimensional phenomenon

Imagine a beaker (with vertical sides) about as deep as its diameter. Fill it with water, and sprinkle a teaspoon of instant iced-tea powder evenly over the surface. Given time, the tea will eventually diffuse evenly through the water. Here the diffusion is one-dimensional; if the water remains still, at any time the concentration of the tea will be uniform in any plane parallel to the water surface. And the average distance a molecule will diffuse will be half the water depth.

Now repeat the experiment with a new beaker of water, but this time

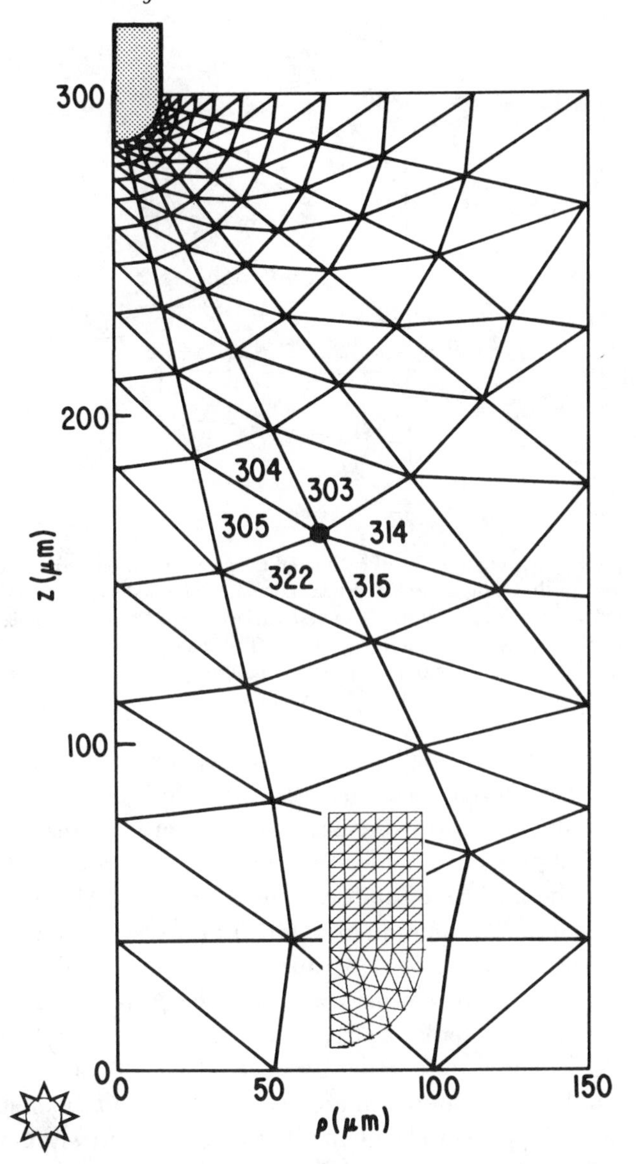

Figure 7.5. Grid of triangular elements used in the FEM. The inset shows the stippled area (in the stomate and substomatal cavity) in greater detail. The heavy dot is at a representative node (number 23), and the numbers surrounding the dot indicate the six elements of which node 23 is a vertex. Sun symbol indicates upper epidermis ($z = 0$).

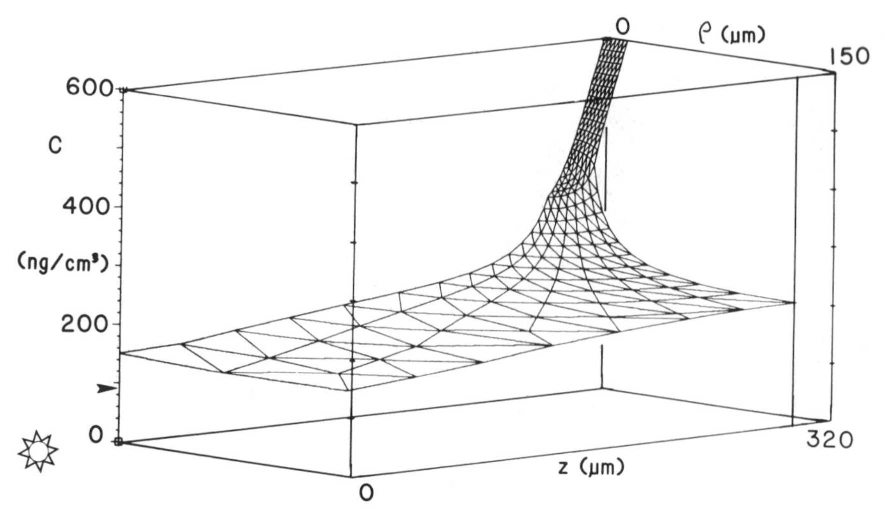

Figure 7.6. Isometric plot of calculated CO_2 concentration within the leaf as a fraction of the ambient value (vertical axis). This plot is shows the result of the "standard run" of the model. The base surface of the "box" is a radial section of the tissue plug, as shown in Figure 7.2. The small arrow represents Γ, the compensation concentration. Sun symbol indicates upper epidermis ($z = 0$).

introduce the powder at a small spot in the very center of the top surface. If surface spreading did not occur (unfortunately, it does in practice), the diffusion would be fully three-dimensional. In theory, the tea will diffuse away from the point of introduction in a hemispherical pattern. If no bulk mixing occurs, the surfaces of uniform concentration will be hemispherical shells, not planes. The amount of water within a distance *d* of the point will increase with the cube of *d*. Because there is more volume in each concentric shell, *the average distance a molecule will diffuse is greater than half the water depth* for this "point-source" case. In fact, for a cylinder with depth equal to diameter, the average distance will be 63% of the diameter.

Now back to leaves (from which the tea was derived in the first place!). Although the leaf appears to be planar, the CO_2 diffusion within it is not. Because each stomate is so small relative to the cross-sectional area of the region of tissue it serves, it acts as a point source, as in the second of our two tea experiments. But why is this important?

The answer lies in the CO_2 concentration profiles which result from three-dimensional diffusion. Figure 7.6 shows the calculated concentration field resulting from the "standard conditions" assumed earlier. The ambient concentration at the outside of the stomate (and boundary layer) is set at 600 ng cm^{-3}, or about 340 ppm (v/v). It drops steeply through the

Table 7.3. *Mean CO_2 concentrations for various surfaces or volumes within the leaf, stated as a fraction of ambient concentration* C_∞

Location	Formula for mean	$\overline{C}/C_\infty$	$\left(\frac{\overline{C}-\Gamma}{C_\infty-\Gamma}\right)$
Inner face of cylindrical pore	$\overline{C}_1 = \frac{2}{a'^2}\int_0^{a'} \rho C(\rho, z_m)\, d\rho$	0.673	0.615
Surface of substomatal cavity	$\overline{C}_2 = \int_0^{\pi/2} C \cos\theta\, d\theta$	0.596	0.525
Lower epidermis (except pore)	$\overline{C}_3 = \frac{2}{(a^2 - a'^2)}\int_{a'}^{a} \rho C(\rho, z_m)\, d\rho$	0.365	0.253
Whole lower epidermis	$\overline{C}_4 = \frac{2}{a^2}\int_0^{a} \rho C(\rho, z_m)\, d\rho$	0.368	0.256
Upper epidermis	$\overline{C}_5 = \frac{2}{a^2}\int_0^{a} \rho C(\rho, 0)\, d\rho$	0.243	0.109
Plug centerline (within mesophyll)	$\overline{C}_6 = \frac{1}{z_m}\int_0^{z_m} C(0, z)\, dz$	0.336	0.219
Outer surface of tissue cylinder	$\overline{C}_7 = \frac{1}{z_m}\int_0^{z_m} C(a, z)\, dz$	0.304	0.181
Entire mesophyll plug	$\overline{C}_8 = \frac{2}{a^2 z_m}\int_0^{z_m}\int_0^{a} \rho C(\rho, z)\, d\rho\, dz$	0.308	0.186

constriction of the stomate, slightly less steeply in the substomatal cavity, and then steeply again as the CO_2 enters the spongy tissue, which is partly constricted by cells and is also taking up CO_2. Finally, the profile flattens out in the bulk of the plug. In the case of the tea, the whole profile would eventually flatten out to a uniform level throughout the cylinder. The leaf, of course, is different, because there is a steady CO_2 uptake in the tissue, and a gradient must be present at all times for this to occur. Note in Figure 7.6 that the minimum CO_2 concentration is about 25% of the ambient value. This minimum occurs along the front edge of the profile, which represents the inner surface of the upper epidermis.

It is interesting to consider some average calculated concentration values at various points in the tissue plug. As listed in Table 7.3, these vary substantially from one place to another in the plug. The ratio $\Delta = (C - \Gamma)/(C_\infty - \Gamma)$ is especially interesting, because it represents the physiologically important concentration *differences* that drive photosynthesis. Note that Δ averages 0.525 over the surface of the substomatal cavity, but only 0.109 at the upper epidermis and 0.186 in the mesophyll plug as a whole.

The difference in mean concentrations between the lower and upper epidermises is not surprising and can be modeled in one-dimensional geometry. The radial variation, which is missed entirely by one-dimensional models, is also large, particularly at the stomatal end of the tissue plug. Note, too, that the overall mean concentration is lower than one might expect, because the outer regions of the cylinder (at large r) contain more volume than do the inner regions (at small r), and C is lowest at large r.

This variability has the same significance to experiments as was discussed in Parkhurst (1977a). That is, if data concerning the water vapor pathway are used to calculate CO_2 concentrations inside the leaf, an average concentration at the wall of the substomatal cavity (C_2 in Table 7.3) will be the result. Clearly (for the modeled leaf) this value of 0.596 is not representative of the whole mesophyll, for which the mean concentration is 0.308, or about half as great. The C value at the upper epidermis, where light is most intense and the tissue is quite dense, is even lower. Thus, results plotted against concentrations calculated in this way (a common technique) may be misleading.

It is instructive to consider what will happen if we neglect, in our calculations, the three-dimensional nature of leaf structure. This is relatively easy to do by modifying the grid of Figure 7.5 in such a way that the stomatal cylinder becomes as wide as the plug it serves ($a' \rightarrow a$). This increases the area available for diffusion by a factor of $(a/a')^2$, or 100 for our standard conditions. To maintain the same value of stomatal resistance, the pore is then assigned a porosity of 1%.

Figure 7.7 shows the concentration profile resulting from this change to one-dimensional geometry. Because the CO_2 will no longer have to spread out in all directions from a nearly point source, it will become more readily available, as shown in the figure. Now the mean Δ values over the lower epidermis, over the upper epidermis, and throughout the mesophyll are roughly 0.41, 0.19, and 0.30, respectively. These are considerably higher than the comparable three-dimensional values of 0.26, 0.11, and 0.19 from Table 7.3.

This "experiment" demonstrates that failure to account for the true three-dimensional nature of CO_2 movement in leaves can lead to substantial errors in calculated concentrations; the differences in the foregoing Δ values are 58%, 73%, and 58%, respectively. The unfortunate part is that the resistance (and conductance) models that have been used so commonly to describe CO_2 uptake are nearly all based on an assumption (usually unstated) of one-dimensional diffusion (Parkhurst 1977a). Three exceptions are models by Laisk, Oja, and Rahi (1970), Sinclair, Goudriaan, and de Wit (1977), and Sinclair and Rand (1979). These models, however, are three-dimensional only at the cellular level; they do not deal with the intercellular diffusion pathway.

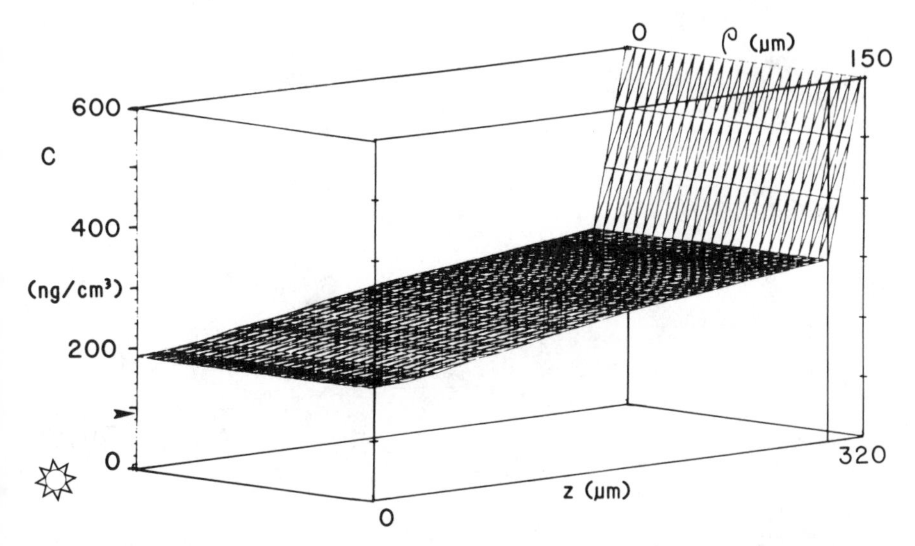

Figure 7.7. Concentration profile calculated with standard conditions, except that the stomatal resistance is "smeared" mathematically over the abaxial surface to yield an artificially one-dimensional situation. Sun symbol indicates upper epidermis ($z = 0$).

When the mean C/C_∞ for the whole plug (0.31, from Table 7.3) is multiplied by an ambient CO_2 concentration of 340 μbar, one obtains an average intercellular concentration of 105 μbar. Farquhar (personal communication) has written that this concentration seems too small to sustain measured CO_2 assimilation rates, given available estimates of carboxylase activity in leaves. Although the general shape of the C profile in Figure 7.6 must be fairly universal, the specific concentration levels do depend on the sizes of the stomatal, transport, and enzyme "resistances." For this reason, Figures 7.8–7.10 were generated to explore how the profiles would change when various resistance components were varied. [Note that the three resistance components are not added algebraically here; rather, they are incorporated in the partial differential equation (7.4) or in its boundary conditions.]

Figure 7.8 shows the result of changing r_0 from 3 to 20 s cm^{-1}. This change was accomplished by using the grid of Figure 7.5, but blocking, mathematically, the outer $(20 - 3)/20$ of the stomatal cylinder by setting the porosity to zero there. As expected, the calculated concentrations are generally lower than in Figure 7.6. They are also more uniform inside the mesophyll, because intercellular CO_2 exchange is now more free in comparison with exchange through the stomate.

As just described, the parametric value (r_0) of stomatal resistance (with

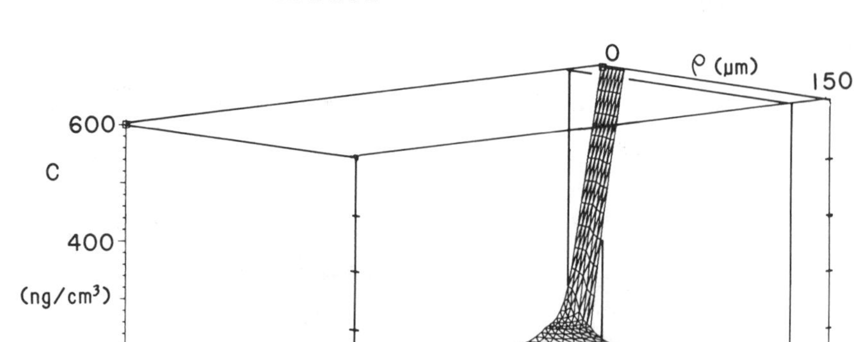

Figure 7.8. Concentration profile when $r_0 = 20$ s cm^{-1}. Sun symbol indicates upper epidermis ($z = 0$).

its small correction for boundary-layer resistance included) was modified by changing only the pore geometry. Those who use resistance models and related conductance models generally assume that stomatal resistances and internal "mesophyll resistances" are independent and simply additive. Parkhurst (1984) has shown that when r_0 is changed from 3 to 20 s cm^{-1}, the value of total "mesophyll resistance" r_m inferred from the resulting photosynthesis rates increases by 46%. That is, r_m, which is *supposed* to depend only on internal geometry and biochemistry, varies when only the stomatal pore diameter, and nothing else, is changed in the model. Thus, resistance and conductance models cannot be expected to apportion variation in photosynthesis correctly to stomatal versus internal causes. See Parkhurst (1984) for further discussion of this issue.

For Figure 7.9, the mean value, through the mesophyll, of enzyme "resistance" r_e is reduced from 1.5 s cm^{-1} to one-tenth that value. It still varies in the same relative way with chlorenchyma density and calculated light level, but its overall level is much smaller. Here, again as expected, the concentration levels are pulled down by the stronger sink, relative to the standard conditions. In Figure 7.10, by contrast, r_e is set to 15 s cm^{-1}, or 10 times its original value, and of course the internal concentrations rise because the mesophyll cannot so readily take up CO_2 at the local level.

It would be highly desirable to have experimental confirmation of CO_2 profiles like those predicted here. Unfortunately, it is not now possible (so

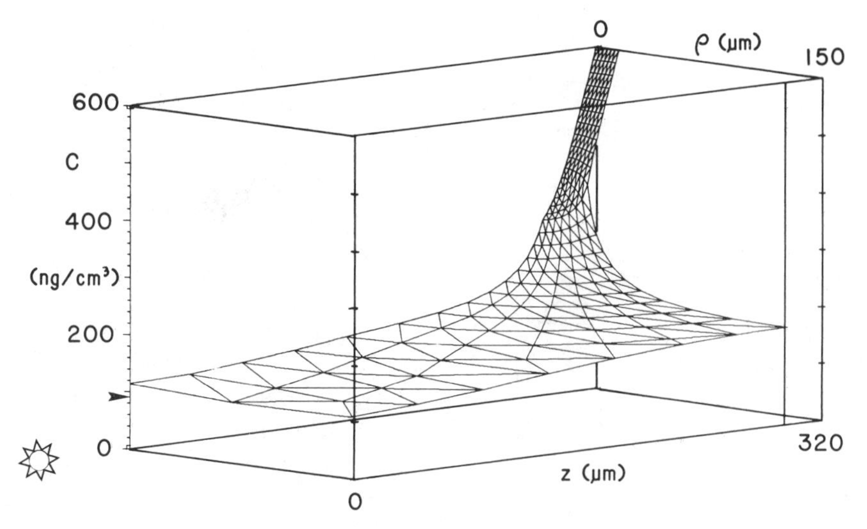

Figure 7.9. Concentration profile when $r_e = 0.15$ s cm^{-1}. Sun symbol indicates upper epidermis ($z = 0$).

far as I know) to measure directly the local CO_2 concentrations at various points inside a leaf (Jarman 1974). (The difficulty, in part, is due to the thinness of leaves referred to earlier.) But neither should we, ostrich-like, ignore potential variations in concentration. Indeed, the inability to make such measurements increases the value of a model like this, which allows the variation at least to be estimated.

Sharkey et al. (1982) measured substomatal concentrations (which they referred to as intercellular ones) for the amphistomatous leaves of *Xanthium* and *Gossypium* and found in the two species that conversions of water vapor resistances to CO_2 resistances gave correct values for CO_2 concentrations near the sites in the leaf from which water evaporates. Also, there was very little variation of CO_2 concentration across the leaves. For an amphistomatous leaf, the effective thickness (distance from a surface bearing stomates) is only half the total mesophyll thickness, and this may partially explain their results. Their leaves may also have been more porous (with more air space) than the madrone leaves modeled in this chapter. In any event, development of methods to measure internal CO_2 concentration profiles might pay off in increased understanding of whole-leaf photosynthesis and is certainly needed to test the present model.

Further explorations seem in order to search for a combination of r_0, r_t, r_e, C levels, and enzyme levels that are mutually consistent. The necessary data need to be derived from a single leaf so that all the components will

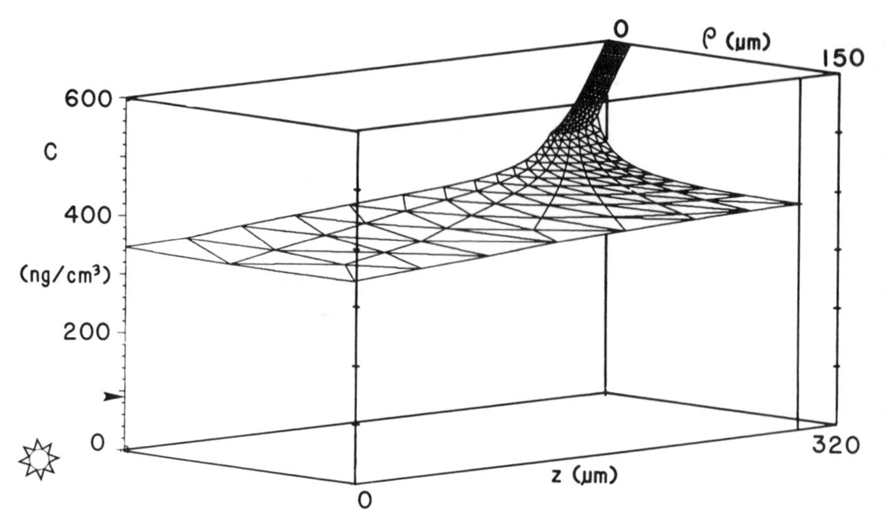

Figure 7.10. Concentration profile when $r_e = 15.0$ s cm^{-1}. Sun symbol indicates upper epidermis ($z = 0$).

represent the same situation. For now, however, the diffusion part of the model is firmly grounded in fundamental physics. The model's shortcomings are most likely to be in r_e, although the values for this parameter have been derived from careful studies by Jones and Slatyer (1972b) and Sinclair et al. (1977). The r_t values are based on those from Laisk et al. (1970).

Calculated photosynthesis rates

Although the internal variation of C is interesting in itself, we are ultimately interested in the resulting CO_2 assimilation rates. First consider local rates $U(\rho, z)$; these are plotted (as a contour diagram) in Figure 7.11. Note the global maximum at $z = 0$, the upper epidermis, where the high sink strength overcomes (or, in another sense, produces) the low C values. There is a much smaller local maximum in U just inside the substomatal cavity, where the abundant CO_2 partially makes up for the low sink strength.

Table 7.4 integrates the U values layer by layer and compares the relative values with the relative "shares" in each layer of several variables that influence CO_2 uptake in the model. At least for the leaf *in silico* (i.e., in computer), no one of the explanatory variables seems to stand out as a predictor of photosynthesis rate.

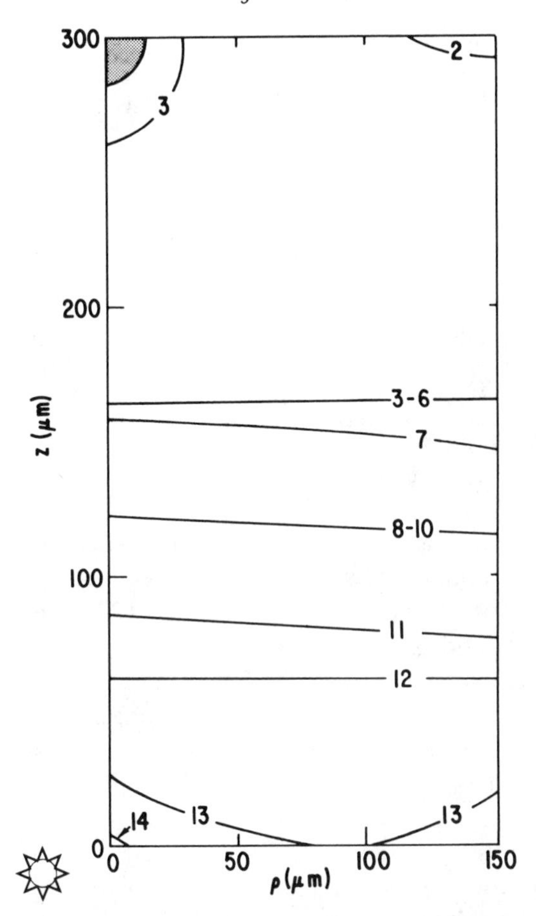

Figure 7.11. Contour plot of the uptake function U (local photosynthetic CO_2 assimilation rate per unit volume of tissue) over a radial section of tissue plug, as shown in Figure 7.5. The U values (e.g., 2–14) have units of g cm^{-3} s^{-1} (per unit ambient concentration C_∞). Sun symbol indicates upper epidermis ($z = 0$).

The calculated local assimilation values vary substantially within the leaf, and, again, experimental confirmation would be welcome. For this question there is *some* information: Mokronosov et al. (1973) fed $^{14}CO_2$ to leaves of several species, then sectioned them paradermally into palisade and spongy tissues. They found greater assimilation rates in the palisade, both absolutely and per unit weight of tissue, and this result would be consistent with Figure 7.11. However, they worked with cut strips of tissue, and CO_2 might have diffused in from the edges.

Outlaw and Fisher (1975) used a similar procedure with *Vicia faba* and determined the amount of ^{14}C incorporated (after a 1-min pulse) per unit

Table 7.4. *Relative photosynthesis rates*[a]

Tissue layer	% PS	% VOL	% CWA	% CTD	% LA	%PS / %VOL	%PS / %CWA	%PS / %CTD	%PS / %LA
Palisade 1	40	21	39	29	54	1.90	1.03	1.38	1.35
Palisade 2	31	20	36	25	25	1.55	0.86	1.24	0.81
Palisade 3	14	14	17	14	9	1.00	0.82	1.00	0.64
Spongy	15	45	7	32	12	0.33	2.14	0.47	0.80

[a] Relative photosynthesis rates (PS) in the four tissue layers, in comparison with relative proportions, associated with each layer, of the total mesophyll volume (VOL), total cell-wall area (CWA), total chlorenchyma cell density (CTD), and light absorption (LA) for the whole leaf. Note that that percentages here have a different basis from those in Table 7.2.

cell volume in the palisade and spongy tissues. The ratio of this quantity (palisade:spongy) declined from about 3 at low light levels to near 1 at high light intensities. Later, Outlaw, Schmuck, and Tolbert (1976) fed $NaH^{14}CO_3$ to isolated palisade and spongy *cells* in aqueous suspension. These authors found that the two cell types (in *V. faba*) were similar in terms of photosynthetic CO_2 fixation, but their results are not directly comparable with the calculations here, because their measurements eliminated the intercellular diffusion aspect of the problem.

These labeling experiments are a beginning toward checking for the variations in local photosynthesis predicted by the model, but they suffer from a major problem. If the labeled CO_2 is fed for a short term (order of seconds), the uptake rate in the chloroplasts far from the stomates will be underestimated (relative to the true steady-state value there). On the other hand, in a longer experiment (order of minutes or hours), variations in label accumulation from one tissue to another could depend as much on differential export rates as on uptake rates. Further developments will be required before such experiments can tell us much about differences in actual photosynthetic carbon uptake between palisade and spongy tissues when both biochemical and diffusional influences are allowed to operate normally. Finally, let us look at overall CO_2 assimilation rates corresponding to the five C profiles previously discussed. For the standard run (Figure 7.6), the CO_2 flux density (F) was 13.9 μmol m^{-2} s^{-1} (with both leaf surfaces as the area basis). For the one-dimensional version (Figure 7.7), F was 22.9 μmol m^{-2} s^{-1}, or 65% higher, even though internal conditions remained the same. This is similar to results obtained with the much simpler model of Parkhurst (1977a). Thus, one-dimensional resistance or conductance models can hardly be expected to represent CO_2 uptake accurately, perhaps not even usefully.

When r_0 increased from 3 to 20 s cm^{-1} (Figure 7.8), F dropped by 71%. For $r_e = 0.15$ s cm^{-1} (Figure 7.9), F increased by 3% over the standard run, for which $r_e = 1.5$; with $r_e = 15$ s cm^{-1} (Figure 7.10), F was reduced to 58% of the standard-run value.

Experiments on the model

We now return to the "experiments." Some of these are sensitivity tests for uncertainties about the model itself, whereas others use the model for its original purpose: to investigate the adaptive nature of leaf structure.

A 2⁴-factorial sensitivity test

In the first experiment (a sensitivity test), four factors were varied in a 2^4-factorial design, a technique borrowed from statistical experimental design (Parkhurst and Loucks 1972; Parkhurst 1978). The particular factors varied (in all 16 possible combinations) were as follows:

1. Whether or not a substomatal cavity is included. The "standard" model assumes that one is present. It is given the shape of a hemisphere of radius 14.9 μm, added on to the inner end of the effective stomatal cylinder. (Hence, its radius is the same as a' when $r_0 = 3$ s cm^{-1}.) As in the stomatal cylinder itself, the values $f = 0$ and $P = Q = D$ are assigned within the cavity. When no cavity is present, spongy tissue is included right up to the base of the stomatal cylinder.
2. Whether the light level is assumed to vary through the tissue as described earlier (standard) or is assigned a uniform value equal to the mean value from the variable case.
3. Whether the anisotropy described earlier is assumed (standard), or, alternatively, whether isotropy ($\lambda = 1$) is assumed throughout the tissue.
4. Whether $r_0 (= r_s + r_a)$ is given the value 3 (standard) or 20 s cm^{-1}. For the latter runs, the stomatal cylinder was given a radius $a'_{20} = \sqrt{3/20}\, a'_3$.

These experiments indicated how sensitive the model is to some of the assumptions on which it is based. For each of the 16 combinations, FEM calculations yielded the nodal CO_2 concentrations, and these, substituted into equation (7.7), allowed calculation of the CO_2 flux densities into a representative stomatal unit. For the standard assumptions (one of the 16 runs) this flux was 13.9 μmol m^{-2} s^{-1}. The mean value over the 16 runs was 9 μmol m^{-2} s^{-1}.

In 2^n-factorial experiments, the "main effect" of a factor A is usually defined as the mean response when A is present less the mean response when A is absent (Davies 1967). For the present experiment, the main effects, here expressed as percentages of the mean photosynthesis rate for all 16 runs, were as follows:

1. Stomatal resistance plus boundary-layer resistance ($r_0 = 20$ or $r_0 = 3$): -111%
2. Substomatal cavity (present or absent): $+11\%$
3. Light variation through the mesophyll (uniform or exponential decay): -0.6%
4. Anisotropy of effective diffusivities ($\lambda = 1$ or $\lambda = 0.1, 0.4, 0.7, 1$): $+0.1\%$

In addition to these main effects, one can calculate various interactions among factors (Davies 1967). In the present case, the only interaction with a magnitude greater than 1% of the mean was between the r_0 resistance level and the presence or absence of the substomatal cavity (-3%). The minus sign indicates that presence of the cavity improves photosynthesis less when r_0 is high than when it is low, or, alternatively, that increasing stomatal resistance reduces F less when the cavity is present than when the cavity is absent.

The large effect of changing the outer resistance r_0 from 3 to 20 s cm^{-1} is not at all surprising, but some of the other effects may be. Removing the cells from a small hemisphere just inside the stomate (just 0.03% of the plug volume) aids CO_2 diffusion enough to increase CO_2 assimilation by 11% on the average. Pickard (1982), using a model of the substomatal cavity itself, found similar advantages. Thus, it appears important to allow for substomatal cavities in any photosynthesis model incorporating geometrical considerations.

The small effect of changing the assumption about the variation of light through the leaf is surprising. Members of seminar audiences have more than once suggested to me that light is scattered and reflected internally in leaves, which would lead to a more uniform distribution than that shown by the results from the absorption model assumed for the standard run. At any rate, given the way light intensity influences the assumed sink strength f in this model, the experiment shows that there is little sensitivity to the way light is distributed, so it does not matter much how the latter is modeled.

One plausible reason for the insensitivity to light distribution is that it may truly not matter much where in the leaf the light is absorbed. If the absorption is concentrated in the upper palisade, then it can be efficiently used, because there is a lot of cell material there to make use of it. On the other hand, if light is evenly absorbed, then more will reach the lower palisade, where it can be efficiently used because CO_2 is plentiful (except at large ρ). To the extent that this explanation is valid for the model, it should also be valid for actual leaves. At any rate, it will be fortunate if whole-leaf photosynthesis really is insensitive to where light is absorbed – it would be difficult to measure light profiles within leaves, just as it is to measure CO_2 concentration profiles.

Finally, the very small effect of anisotropy in horizontal diffusivity is equally encouraging, because it, too, is a very difficult parameter to measure.

Varying the size of the substomatal chamber

The next experiment is a simple one, in principle. Referring to the grid of Figure 7.5, one can see the rings of triangular elements concentric to the point $\rho = 0$, $z = 298\ \mu m$. In the experiment, successive pairs of rings were excised (mathematically) from the tissue, up to the ring that includes the three elements shown in the figure as numbers 303–305. Specifically, for any given run, the sink function was set to zero, and the porosity was set to unity in any elements to be included in the cavity.

It should be noted that *measured* values of stomatal (or "leaf") resistance to water vapor loss will generally contain substomatal component as well as a stomatal component. In this study, the value of r_s represents only the stomatal pore. Changes in the substomatal cavity affect the geometry of CO_2 diffusion, but are independent of the parametric value of r_s or r_0.

Figure 7.12 shows resulting CO_2 assimilation rates versus cavity radii. The amount of chlorenchyma cell volume remaining in the tissue plug is plotted for comparison. When $r_0 = 3\ s\ cm^{-1}$, assimilation reached a maximum at the next to last cavity size, then dropped slightly, and this occurred in spite of the large loss of cell material from the leaf. For $r_0 = 20$, the maximum assimilation rate occurred with the largest cavity modeled.

These results are qualitatively consistent with those of Pickard (1982), whose model showed that "the optimal chamber radius is several times larger than is the pore radius." In fact, it appears that from a diffusional point of view, the chamber should be very large indeed. The actual limiting factor to size in real leaves may be physical strength: If the cavities were too large, there would be little tissue to connect the lower epidermis to the spongy mesophyll! Finally, note that increasing the size of the substomatal chamber decreases average mesophyll thickness. Both Parkhurst (1977a) and Givnish (1979), using different approaches, have previously found optimal leaf thicknesses using theoretical arguments. However, in the present analysis, photosynthesis increases in an absolute sense with increasing chamber size, not just in a relative sense of assimilation per unit of biomass.

As the size of the substomatal cavity increases in these experiments, the

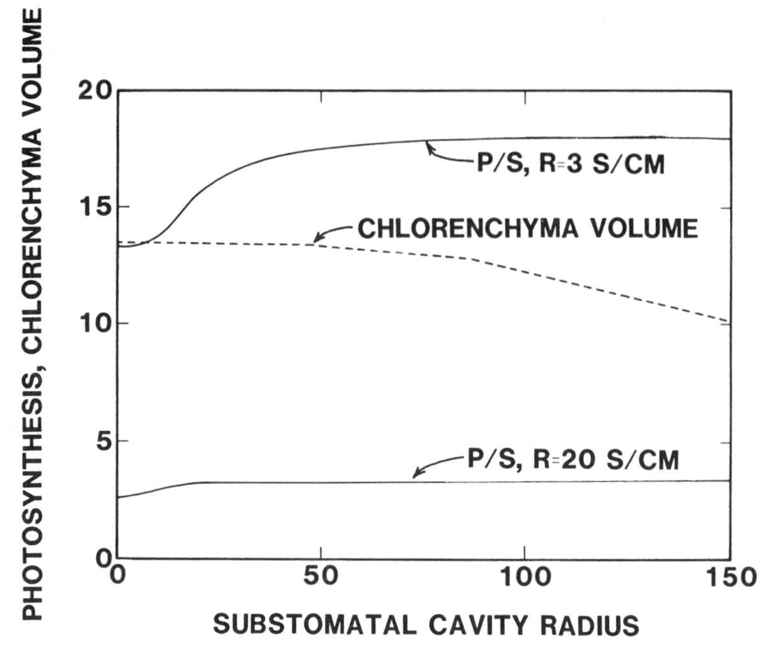

Figure 7.12. Responses of calculated photosynthesis rates to changing sizes of substomatal cavity. The volume of active chlorenchyma cells per tissue plug is also shown.

average diffusion distance for water vapor either will remain constant (if transpired water comes from the guard cells or their immediate neighbors) or will increase (if water evaporates generally from the walls of the substomatal cavity). Hence, the adaptive advantage of a large substomatal cavity attributable to increased carbon fixation will not be compromised by increased transpiration, but if anything will be accompanied by a lower rate of water loss.

Optimal mesophyll structure, or variations on a theme by nature

The next experiment motivated the development of the model in the first place and has as its purpose to help us understand the differentiation of palisade and spongy tissues. The model was built around madrone because of the extensive data set of Hays (1976). Figure 7.13 shows diagrammatically the real madrone leaf with its three layers of palisade tissue and one of spongy.

Now, suppose r_0 is held constant (to keep water loss constant) and total volume of chlorenchymatous cell material is held constant (to keep leaf construction cost constant). If all the chlorenchyma in the mesophyll were

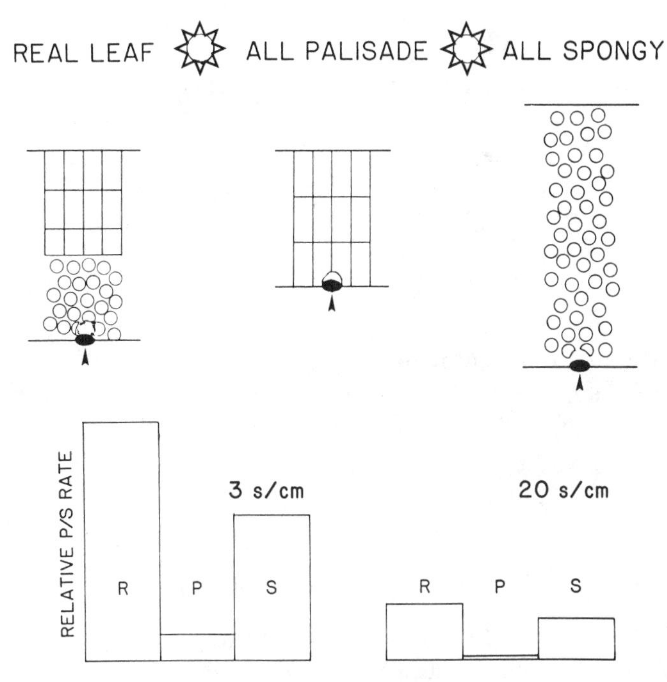

Figure 7.13. Three ways of distributing a given amount of chlorenchyma cell volume in a madrone leaf. The relative calculated photosynthesis rates for the three tissue distributions are also shown, for both low and high values of the surface resistance, r_0. Sun symbols indicate upper surfaces.

turned into cells of the kind found in the uppermost palisade layer, the leaf would be more densely packed, and considerably thinner. It would also be much less open to horizontal diffusion. If all the chlorenchyma were of the spongy cell type, the leaf would have less cell-wall area and more air space and would be thicker. How would these changes affect photosynthesis?

The basic grid was modified for the thinner and thicker leaf types shown, and the sink function was modified to represent the properties of the single cell types throughout the simulated mesophyll. The CO_2 concentration profiles were then calculated, as well as resulting photosynthesis rates. Figure 7.13 shows that for a given expenditure of resources to manufacture chlorenchyma, the plant spends it most efficiently by making a leaf with the structure found in nature; the same resources spent for an all-spongy or all-palisade leaf will not yield as great a return. Parkhurst (1977) also simulated two other tissue distributions: one like the real leaf except with stomates moved to the upper epidermis, and one with spongy tissue under the upper (sunlit) epidermis and palisade at the lower epidermis (the stomatal surface). Those cases are not directly comparable to the present ones because parameter values were different, but neither did as well as the real leaf in that simulation.

My guess is that the actual leaf structure does better than the all-palisade one because of the candelabrum effect. Horizontal diffusion is simply too limited in the all-palisade leaf. The all-spongy leaf must do poorly for different reasons. One possibility is that the spongy tissue is too thick for a hypostomatous leaf, making diffusion distances too great for effective CO_2 distribution. The smaller cell-wall surface area may also make the spongy leaf less efficient.

Although other factors (such as light) may have been involved in the palisade – spongy differentiation, this experiment is at least consistent with the hypothesis that CO_2 diffusion is a major reason for it. On the other hand, amphistomatous leaves sometimes have both types of tissue well developed too, and for them the differentiation is harder to understand.

Other questions about internal structure

The model discussed in this chapter, and the explorations of it, have looked at a number of aspects of internal leaf structure such as those in Figure 7.1. Many interesting questions remain, of course, and the purpose of this section is to raise some of them.

Leaf thickness

Two major questions arise: Why do leaves have the thickness they do? Why does thickness vary with environment as it does? Clearly, there must be some balance between the plant's material investment in a leaf and the income gained from it (Orians and Solbrig 1977). Even without consideration of investment, however, the simpler, uniform-mesophyll version of the model (Parkhurst 1977a) predicted a peak in CO_2 uptake at about 0.1 – 0.2 mm thickness. Much work remains in this area, such as studying leaf thicknesses of particular species in specific environments. Givnish (1979), for example, has provided an explanation for leaves to be thicker in dry and nutrient-poor habitats than elsewhere.

Stomatal spacing and epidermal thickness

The features of leaf structure that contribute most to the three-dimensional nature of CO_2 diffusion are the size and spacing of the stomates. In particular, if CO_2 entered the mesophyll uniformly (as in the calcula-

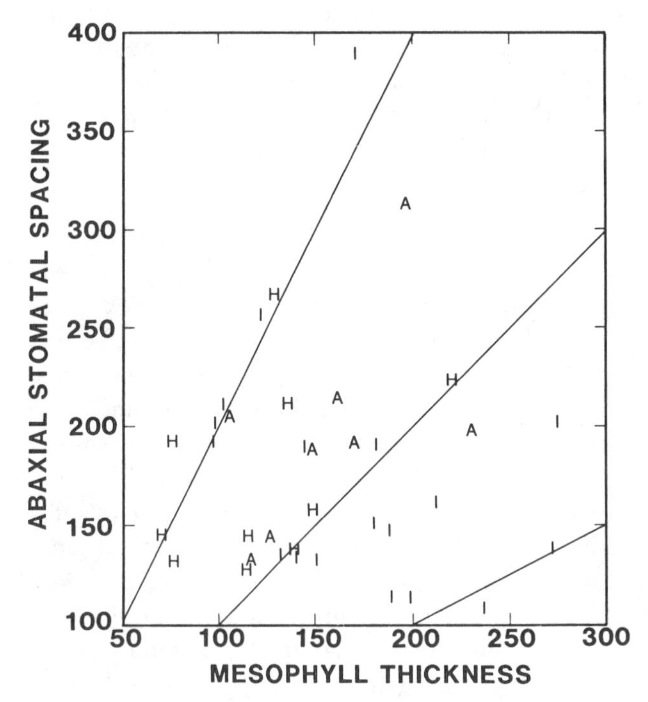

Figure 7.14. Stomatal spacings on lower leaf surface versus mesophyll thickness for 36 individuals (in about 30 species) of Compositae. The three lines represent (from top to bottom) spacing:thickness ratios of 2, 1, and $\frac{1}{2}$ respectively. H = hypostomatous (no stomates on the upper epidermis); A = amphistomatous (upper density/lower density > 0.9); I = intermediate density ratios.

tions for Figure 7.7), then the process would be one-dimensional, and much easier to model. Or if stomates were very closely spaced relative to mesophyll thickness, a one-dimensional approximation might be adequate. However, as stated earlier, stomates quite typically are spaced about as far apart as the mesophyll is thick (Parkhurst 1977a).

Figure 7.14 shows the relationship between mesophyll thickness and stomatal spacing for 36 individuals from some 30 species of herbaceous Compositae from Indiana. The stomatal spacing is calculated from the density (stomates/mm^2) under the assumption (for consistency) that the stomates lie on a triangular grid. In nearly all the cases, the spacing is between $\frac{1}{2}$ and 2 times the mesophyll thickness. My impression is that stomates on woody plants (e.g., oaks) tend to be somewhat more closely spaced than 1:1, but the data are not currently available to say with certainty.

When the uniform-mesophyll model was used to test for optimal spacing, the best spacing was the closest (Parkhurst 1977a, Figure 3b, p. 479). This makes sense, in that close spacing is more like the conditions of Figure 7.7, which had a higher calculated carbon assimilation rate than did the standard case. Finally, epidermal thickness must interact with stomatal spacing and size, because it determines the L_s of equation (7.3).

Light

Even though the 2^4 experiment suggests that the detailed distribution of light absorption within the leaf may not, on the average, be very important to whole-leaf CO_2 assimilation, it is still a phenomenon that needs to be better understood. For example, the vertical cylinders of palisade cells may act as light pipes, channeling light down to the spongy tissue; at the same time, their own chloroplasts could use some of the light before it passes through many cell walls. [Some plant tissues have recently been identified as "optical fibers" by Mandoli and Briggs (1982).] Once the light reaches the spongy cells, it may be reflected back upward by their horizontal cell walls. These effects could occur in addition to diffusional ones, reinforcing the candelabrum effect in making the palisade–spongy dichotomy adaptive. Indeed, light distribution might possibly explain the palisade–spongy differentiation in amphistomatous leaves. Light is certainly scattered by plant cells (Gausman 1977; Latimer 1979), but how this influences CO_2 uptake needs further study.

Other factors

There must be other factors that have influenced the natural selection of leaf structure. It is possible that cytoplasmic streaming has influenced the development of the palisade cells; to the extent that their cytoplasm undergoes cyclosis, this would allow a given chloroplast to drop down for a gulp of CO_2-rich air, then climb back up to bask in the sun. This could be especially important in a hypostomatous leaf, and if it occurs, it will favor development of long palisade cells.

Adaptations for transport of assimilates from the leaf have no doubt contributed to leaf structure. This idea has been better studied in C_4 plants (Hatch and Osmond 1976) than in C_3, but Franceschi and Giaquinta (1983) have recently described a tissue they call "paraveinal mesophyll" in soybeans. Its apparent function is assimilate transport, and it looks remarkably like the layers of nonchlorophyllous cells that occur in the leaves of certain other species of C_3 plants. These cells are common in *Solidago* and its relatives (Anderson and Creech 1975; Parkhurst, unpublished data).

Anatomists frequently refer to them as water-storage cells, but this supposed function makes little sense, given that a leaf will commonly evaporate much more water in an hour than a layer of cells 10 or 20 μm thick can store.

Finally, leaves require physical support and must not be too vulnerable to insect attack; the structure of their photosynthetic tissues may be partly modified to meet these additional requirements.

Conclusions

The purpose of this chapter has been to report progress in attempts to understand the internal structure of C_3 plant leaves. In particular, the hypothesis that many structural features can be explained as adaptations to enhance CO_2 diffusion within the leaf has been explored. The existence of large substomatal cavities was shown to be consistent with this hypothesis, and so was the differentiation of mesophyll into palisade and spongy parenchyma cells. The presence of the two cell types allows the CO_2 to diffuse along a candelabrum-shaped path, and in this way the gas more easily moves into the upper reaches of the palisade, where the cells are densely packed and light is plentiful.

Neither of the foregoing structural features would be necessary, or even helpful, if the diffusion process in the leaf were not *three-dimensional* in important ways. That is, because each stomate occupies only a small proportion of the surface of the region that it serves, it acts essentially like a point source. This means that in successive 1-μm bands of mesophyll away from the pore, the amount of tissue increases more or less with the cube of the distance. This spreading-out effect (which is somewhat like the inverse-square law for the decrease in light intensity away from a light source) tends to inhibit photosynthesis more than would be the case if the whole epidermis transmitted CO_2 uniformly. The latter situation would be one-dimensional, but real leaves are not built that way. To understand leaf structure, I believe it is crucial that we stop using one-dimensional models for photosynthesis. Such models seem to work well for transport processes from the leaf surface on *out*, but not for internal processes.

As a final point, the mathematical model explored here indicates that there are major variations in CO_2 concentrations within the leaf mesophyll. Based as it is on tried and true fundamental physics, the diffusional aspects of the model hardly seem open to question. However, the coupling of the diffusion to the uptake at the cell walls needs further development. To test the model's predictions, we are very much in need of a technique for measuring the variability in CO_2 concentrations inside leaves; however, this does seem a technically difficult problem.

Summary

The hypothesis is posed that many features of internal leaf structure are adaptations to increase the ease of CO_2 diffusion for photosynthesis. A conceptual model is then suggested that integrates the effects on photosynthesis of such variables as amount of cell-wall surface area, cell density, and amount of air space per volume of leaf. These variables become parameters in a partial differential equation describing CO_2 diffusion to the cells, and its uptake by them. Solution by the finite-element method (described briefly) allows this model to be very general, with parameters that vary from point to point in the mesophyll. Thus, the model can account for the different properties of palisade and spongy parenchyma, for example.

The model is then exercised in several experiments, first to calculate the CO_2 profiles within the leaf, and then *in silico* (computer-calculated) photosynthesis rates. The concentration profile derived in the defined "standard run" shows dramatically that CO_2 diffusion is a three-dimensional process. This is so because stomates are small and widely spaced and act as point sources rather than as plane sources.

A 2^4-factorial experiment on the model has shown that stomatal resistance (not surprisingly) has a major effect on carbon assimilation, as does the presence of a substomatal cavity. On the other hand, the way light is distributed within the leaf and the anisotropy of porosity have little effect on calculated assimilation rates. A second experiment calculates photosynthesis rates for substomatal cavities of increasing size and finds an optimum only at a very large size. Structural strength, not diffusional considerations, probably limits the maximum size of these chambers in real leaves.

Another experiment replaces the measured structure of a madrone leaf with a second made of the same amount of tissue, but all palisade cells, and a third of all spongy cells. The photosynthesis rate of the real leaf is higher than that of either of the potential competitors, indicating that the CO_2 diffusion hypothesis is a sufficient explanation (though not necessarily the only one or the correct one) for the existence of both palisade and spongy cell types.

Additional factors such as light piping and cytoplasmic streaming are discussed, but these have not been incorporated into the model. The chapter ends with a plea for experimentalists to develop a means to measure CO_2 concentration from point to point within plant leaves.

Acknowledgments

I thank Robert Hays for the madrone leaf-structure data and for interesting discussions. Karen Chancellor helped with the finite-element computer program, and Tom Givnish made many helpful suggestions to improve the manuscript. Research was supported by NSF grant DEB 77-01205.

References

Anderson, L. C., and J. B. Creech. 1975. Comparative leaf anatomy of *Solidago* and related Asteraceae. Amer. J. Bot. 62:486–493.

Bierhuizen, J. F., and R. O. Slatyer. 1964. Photosynthesis of cotton leaves under a range of environmental conditions in relation to internal and external diffusive resistances. Aust. J. Biol. Sci. 17:348–359.

Chabot, B. F., and J. F. Chabot. 1977. Effects of light and temperature on leaf anatomy and photosynthesis in *Fragaria vesca*. Oecologia 26:363–377.

Davies, O. L. (ed.). 1967. Design and analysis of industrial experiments. Hafner Publishing, New York.

Dornhoff, G. M., and R. Shibles. 1976. Leaf morphology and anatomy in relation to CO_2 exchange rate of soybean leaves. Crop Sci. 16:377–381.

El-Sharkaway, M., and J. Hesketh. 1965. Photosynthesis among species in relation to characteristics of leaf anatomy and CO_2 diffusion resistance. Crop Sci. 5:517–521.

Forrester, J. W. 1968. Principles of systems. Wright-Allen Press, Boston.

Franceschi, V. R., and R. T. Giaquinta. 1983. The paraveinal mesophyll of soybean leaves in relation to assimilate transfer and compartmentation. I. Ultrastructure and histochemistry during vegetative development. Planta 157:411–421.

Gausman, H. W. 1977. Reflectance of leaf components. Remote Sensing of Environment 6:1–9.

Givnish, T. J. 1979. On the adaptive significance of leaf form. Pp. 375–407 *in* O. T. Solbrig, S. Jain, G. B. Johnson, and P. H. Raven (eds.), Topics in plant population biology. Columbia University Press, New York.

Givnish, T. J., and G. J. Vermeij. 1976. Sizes and shapes of liane leaves. Amer. Nat. 110:743–778.

Gould, S. J., and R. C. Lewontin. 1979. Spandrels of San Marco and the Panglossian paradigm – a critique of the adaptationist program. Proc. Roy. Soc., Series B 205:581–598.

Hamming, R. W. 1973. Numerical methods for scientists and engineers, 2nd ed. McGraw-Hill, New York.

Hatch, M. D., and C. B. Osmond. 1976. Compartmentation and transport in C_4 photosynthesis. Pp. 144–184 *in* C. R. Stocking and U. Heber (eds.), Encyclopedia of plant physiology, new series, vol. 3, Transport in plants III. Springer-Verlag, Berlin.

Hays, R. L. 1976. The influence of leaf anatomy on water-use efficiency: a modeling approach. Plant Physiol. Supplement 57:77 (abstract).

Huebner, K. H. 1975. The finite element method for engineers. Wiley, New York.

Jarman, P. D. 1974. The diffusion of carbon dioxide and water vapour through stomata. J. Exp. Bot. 25:927–936.

Jones, H. B., and R. O. Slatyer. 1972a. Effects of intercellular resistances on estimates of the intracellular resistance to CO_2 uptake by plant leaves. Aust. J. Biol. Sci. 25:443–453.

– 1972b. Estimation of the transport and carboxylation components of the intracellular limitation to leaf photosynthesis. Plant Physiol. 50:283–288.

Laisk, A., V. Oja, and M. Rahi. 1970. Diffusion resistance of leaves in connection with their anatomy. Soviet Plant Physiol. 17:31–38.

Latimer, P. 1979. Light scattering vs. microscopy for measuring average cell size and shape. Biophys. J. 27:117–126.

Mandoli, D. F., and W. R. Briggs. 1982. Optical properties of etiolated plant tissues. Proc. Natl. Acad. Sci. USA 79:2902–2906.

Mokronosov, A. T., R. I. Bagoutdinova, E. A. Bubnova, and I. V. Kobeleva. 1973. Photosynthetic metabolism in palisade and spongy tissue of the leaf. Soviet Plant Physiol. 20:1013–1018 (translated from *Fiziologia Rastenii* 20:1191–1197).

Nobel, P. S. 1976. Photosynthesis rates of sun versus shale leaves of *Hyptis emoryi* Torr. Plant Physiol. 58:218–223.

– 1977. Internal leaf area and cellular CO_2 resistance: photosynthetic implications of variations with growth conditions and plant species. Physiol. Plant. 40:137–144.

Nobel, P. S., L. J. Zaragoza, and W. K. Smith. 1975. Relation between mesophyll surface area, photosynthetic rate, and illumination level during development for leaves of *Plectranthus parviflorus* Henckel. Plant Physiol. 55:1067–1070.

Orians, G. H., and O. T. Solbrig. 1977. A cost-income model of leaves and roots with special reference to arid and semiarid areas. Amer. Nat. 111:677–690.

Outlaw, W. H., Jr., and D. B. Fisher. 1975. Compartmentation in *Vicia faba* leaves. III. Photosynthesis in the spongy and palisade parenchyma. Aust. J. Plant Physiol. 2:435–439.

Outlaw, W. H., C. L. Schmuck, and N. E. Tolbert. 1976. Photosynthetic argon metabolism in the palisade parenchyma and spongy parenchyma of *Vicia faba* L. Plant Physiol. 58:186–189.

Parkhurst, D. F. 1977a. A three-dimensional model for CO_2 uptake by continuously distributed mesophyll in leaves. J. Theor. Biol. 67:471–488.

– 1977b. A model for the adaptive significance of xeromorphy in leaves. Bull. Ecol. Soc. Amer. 58:33 (abstract).

– 1978. Adaptive significance of stomatal location on one or both surfaces of leaves. J. Ecol. 66:367–383.

– 1984. Mesophyll resistance to photosynthetic carbon dioxide uptake in leaves: dependence upon stomatal aperture. Can. J. Bot. 62:163–165.

Parkhurst, D. F., P. R. Duncan, D. M. Gates, and F. Kreith. 1968. Wind-tunnel modelling of convection of heat between air and broad leaves of plants. Agric. Meteorol. 5:33–47.

Parkhurst, D. F., and O. L. Loucks. 1972. Optimal leaf size in relation to environment. J. Ecol. 60:505–537.

Patterson, D. T., J. A. Bunce, R. S. Alberte, and E. Van Volkenburgh. 1977. Photosynthesis in relation to leaf characteristics of cotton from controlled and field environments. Plant Physiol. 59:384–387.

Pickard, W. F. 1982. Why is the substomatal chamber as large as it is? Plant Physiol. 69:971–974.

Rand, R. H. 1977. Gaseous diffusion in the leaf interior. Trans. Amer. Soc. Agric. Eng. 20:701–704.

Sharkey, T. D., K. Imai, G. D. Farquhar, and I. R. Cowan. 1982. A direct confirmation of the standard method of estimating intercellular partial pressure of CO_2. Plant Physiol. 69:657–659.

Sinclair, T. R., J. Goudriaan, and C. T. de Wit. 1977. Mesophyll resistance and CO_2 compensation concentration in leaf photosynthesis models. Photosynthetica 11:56–65.

Sinclair, T. R., and R. H. Rand. 1979. Mathematical analysis of cell CO_2 exchange under high CO_2 concentrations. Photosynthetica 13:239–244.

Wilson, D., and J. P. Cooper. 1967. Assimilation of *Lolium* in relation to leaf mesophyll. Nature 214:989–992.

8 Competing root systems: morphology and models of absorption

MARTYN M. CALDWELL AND
JAMES H. RICHARDS

Introduction: the costs of competition

To acquire soil moisture and nutrients, higher plants invest heavily in root systems. In ecosystems such as shortgrass prairie, shrub steppe, and tundra, the proportion of plant biomass below ground exceeds 80%, and a large part of this is invested in fine roots (Table 8.1). Even in ecosystems such as the deciduous forest, where root/shoot ratios are small, the turnover of root systems is high. Thus, the annual root-system production can constitute 60% to 80% of the total net primary production in many biomes (Table 8.1). Belowground respiration represents an additional energy expenditure.

These root-system costs may be escalated because plants usually must compete with their neighbors for soil resources. Analogous to the investment by forest trees in tall stems to compete for light, some of the belowground investment may be solely for the sake of competition. However, it is presently unknown what proportion of the belowground costs are necessary simply to accomplish the purposes of support and absorption of soil nutrients and moisture and what additional costs may be necessary to effectively compete with neighboring plants for these resources.

That plants interfere with one another has been inferred from plant distribution patterns and size–distance correlations of neighboring plants (Yeaton and Cody 1976; Harper 1977; Smith 1979; Bell 1981). Removal experiments in the field have also indicated plant interference in that the water status and aboveground production of plants are improved when neighbors are removed (Pinder 1975; Allen and Forman 1976; Fonteyn and Mahall 1981; Fowler 1981; Robberecht et al. 1983). The ability to recover from severe defoliation is also greatly enhanced by removal of neighboring plants (Mueggler 1972). Certainly, plants acquire more soil resources without neighbors, but they also may do this at a lower unit cost of root-system investment.

Table 8.1. *Standing biomass and production of fine roots: total belowground and aboveground components in vegetation of different ecosystems*

Ecosystem	Standing biomass ($g\ m^{-2}$)			Production ($g\ m^{-2}\ year^{-1}$)		
	Above ground	Below ground	Fine roots	Above ground	Below ground	Fine roots
Arctic tundra[a]	80–130	534–620	375–432	80–130		100
Shortgrass prairie[b]	218	1,173	725	218	618	471
Shrub steppe[c]	416	1,886	1,818	131	435	427
Deciduous forest[d]	14,720	3,742	1,679	609	977	930
Pinus sylvestris[e]	3,102		310	1,126		1,920

[a] Shaver and Billings (1975).
[b] Sims and Singh (1978) (ungrazed shortgrass prairie, Pawnee site, 1971).
[c] Caldwell et al. (1977) (*Atriplex confertifolia* stand).
[d] Reichle et al. (1973) (*Liriodendron tulipifera* forest with stump and large root biomass as part of belowground).
[e] Persson (1983), Ågren et al. (1980) (14-year *Pinus sylvestris* plantation; separate values for stump and large roots not available).

Root systems of different species can interfere with and influence one another by many mechanisms, including allelochemics or alterations of microbial populations associated with neighboring plant roots (Rice 1974; Christie et al. 1978). Although the significance of such processes should not be overlooked, demonstration of these phenomena in the field remains elusive. Competition for soil resources is also difficult to prove unambiguously in a field setting. Yet, much more is known about acquisition of soil resources by roots and the implications of these processes for resource competition than can be surmised for less direct forms of interference. This chapter will focus on aspects of plant interference specifically involving competition for soil resources in the belowground space.

This chapter will first address the implications of physical models of nutrient and moisture absorption between overlapping root systems. The models are set in two-dimensional space; thus, extension of the results of these to three dimensions must be explored. A second purpose of the chapter is to discuss the distribution of root lengths in the belowground space. An example of two tussock grasses will illustrate differences in allocation patterns of root systems. Finally, a few aspects of root-system morphology will be discussed.

Root systems of different species frequently intermingle, and often this overlap of root systems can be extensive (Clements et al. 1929). If soil resources are in limited supply and neighboring plant species have active root systems in the same location in the soil profile, one species may be more effective in acquiring these resources than the other. Yet, little is known about the characteristics of root systems that enable some species to acquire resources more efficaciously than others in a competitive situation. Does a species compete more effectively by virtue of the absorption capacity of its root elements, its total root surface area, its rooting density (length per soil volume), its microbial associations, or its tactics of growth timing? Or, if it competes by a combination of these characteristics, which ones are most important? Does a species acquire more soil resources by intensively controlling belowground space with a system of very fine root elements or by exploiting more belowground space with an extensive root system, or is it a matter of more efficient uptake per unit root element? Though it is not possible to provide answers to such general questions, some implications of root structure and function for the competitive potential of plants can be explored. Physical models of water and nutrient uptake and some empirical information will be used in this exploration.

Perspective on root competition: contribution of models

Cylindrical-flow models and parameter sensitivity

Models for water and solute uptake on the scale of individual roots have been well developed. Most of these models are based on the premise that moisture and solutes move radially toward a root element from a soil cylinder surrounding the root axis and that the root absorption properties are uniform along its length (Gardner 1960; Cowan 1965; Baldwin et al. 1972; Cushman 1979). The generalized equation describing this radial flow is

$$\frac{\partial C}{\partial t} = \frac{1}{r} \frac{\partial}{\partial r} \left(rD \frac{\partial C}{\partial r} + \frac{v_0 a}{b} C \right) \tag{8.1}$$

where C is the concentration of ions in the soil solution or the water content of the soil, D is the effective diffusion coefficient of water or the ion in question, r is the radial distance from the root axis, and t is time. The component of the equation $v_0 aC/b$ applies only to movement of ions by mass flow, where v_0 is the moisture flux rate into the root, a is the root radius, and b is a differential buffer power of the soil for that ion. The diffusion coefficient for water is principally a function of soil moisture

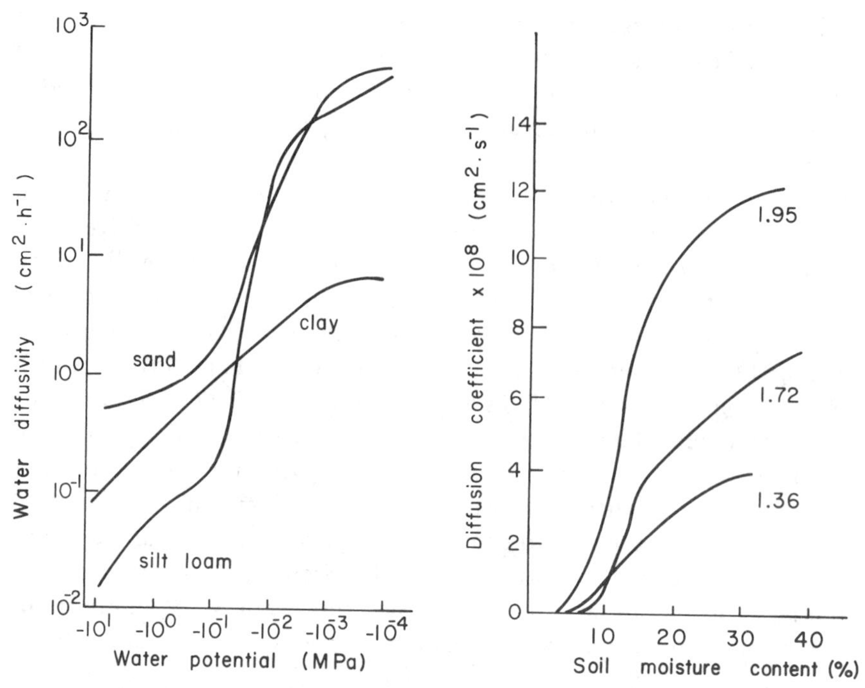

Figure 8.1. Diffusivity of water (left) as a function of soil water potential in soils of differing texture (adapted from McCoy et al. 1984) and diffusivity of rubidium (right) in a sandy loam of differing density (due to compaction) as a function of soil moisture content (adapted from Nye and Tinker 1977). Soil density is indicated for each curve in tons m^{-3}.

potential and soil texture (Lang and Gardner 1970), whereas the diffusion coefficient for soil solutes depends on the diffusion coefficient of the ion in solution and the water content of the soil, as well as other characteristics of the soil that determine its impedance to diffusion (Nye and Tinker 1977). The behaviors of water and ion diffusivities with changes in soil moisture and other characteristics are illustrated in Figure 8.1.

Various numerical and analytical solutions to differential equation (8.1) have been used with slightly different assumptions. Under most conditions, however, the general behaviors of these models are similar (Caldwell 1976; Nye and Tinker 1977). In the solution of equation (8.1) it is necessary to establish the flux rate of water, or solute, from the soil solution at the root surface into the root. This flux per length of root can be variously expressed for mineral nutrients and is often taken as the product of the "root demand coefficient," αa (where α is the root absorbing power, and a is the root radius), and the solute concentration in the soil solution at the root surface (Nye and Tinker 1977). For mineral nutrients, this will reflect

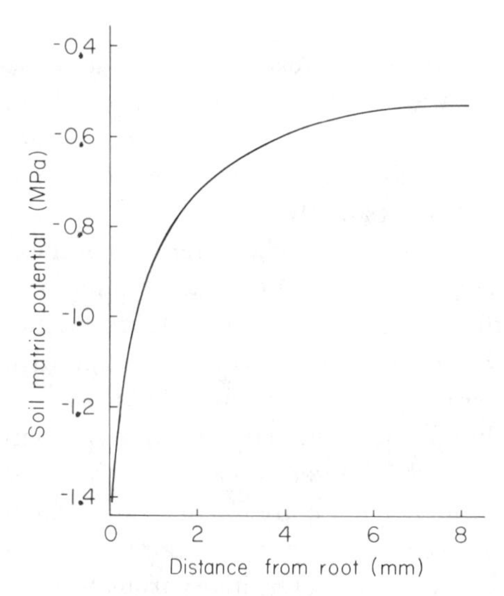

Figure 8.2. Localized depletion of soil moisture near the root surface as indicated by the predicted soil-matrix water-potential gradient using the model of Cowan (1965) (adapted from Caldwell 1976).

the capacity of the root to absorb the specific ion in question; for water flux, the absorption rate will depend on the transpiration of the entire plant and how this water uptake is partitioned among the active roots.

When the root demand coefficient is high and diffusion and mass flow toward the root are slow, radial-flow models predict localized depletion of water or nutrients near the root surface, as indicated in Figure 8.2. This localized depletion is a reflection of the limitation on uptake imposed by diffusion in the soil. These localized depletions can result in further reductions in diffusion to the root caused by the lower water potential in the immediate vicinity of the root (Figure 8.2). The manner in which localized depletions develop has important implications for competition, as will be discussed later. For water, there is no feasible way experimentally to verify the existence of depletion zones at present. Other evidence suggests that these depletion zones do not develop under field conditions until the bulk soil water potential drops below −1.5 MPa, if one assumes that cylindrical-flow models apply (Caldwell 1976); however, as will be discussed later, a different geometry of water flux to the root can result in depletion zones at relatively high (e.g., −0.5 MPa) soil water potentials. For soil nutrients that diffuse slowly in soil, depletion zones are quite probable and have been directly observed by use of phosphorus and sulfur radioisotopes (Bhat and Nye 1973). For nutrients, the depth (reduction in concentration near the root surface) and spread (diameter of influence) of these depletion zones at a point in time are functions of $\overline{\alpha a}/Db$, where $\overline{\alpha a}$ is the average root demand coefficient for the root system. For ions with high diffusivity, such as nitrate, the depletion zones will be shallow and broad, whereas for ions that diffuse slowly, such as phosphorus, the depletion zones will be deep and narrow (Baldwin et al. 1972).

Although an extensive review of the behavioral characteristics of these models does not fall within the scope of this chapter, an example of a sensitivity analysis will be useful to illustrate root and soil characteristics that may be important in nutrient absorption and thus in competitive effectiveness. Silberbush and Barber (1983a) used the model of Cushman (1979) to conduct a sensitivity analysis for potassium uptake by soybean in a silt loam soil (potassium is an ion with an intermediate diffusivity). A portrayal of part of this analysis is contained in Figure 8.3 (left side), which shows the influence of independently changing several parameters from 0.5 to 2.0 times the initial level. Clearly, the most sensitive parameters are those related to root morphology: root growth and root radius. Growth in root length was the most sensitive parameter, because both root length and the soil volume explored increase with growth. This is a linear response, because root surface area increases linearly with length. Root surface area also increases linearly with root radius; however, this is less effective than increasing root length, because the volume of soil explored is not increased, and there are changes in the geometry of radial movement of nutrients toward the root surface as root radius increases. Physiological absorption capacity is less sensitive in the model, except when it is decreased to the extent that it becomes limiting. Buffer power and diffusivity of potassium in the soil are of intermediate sensitivity. The flux rate of water is totally insensitive for potassium uptake, because essentially all uptake is by diffusion, and mass flow of potassium is unimportant.

Because continued root growth is costly and the soil volume that can be explored is usually limited, a more constrained and realistic sensitivity analysis will limit both the amount of biomass invested in roots and the total soil volume explored. Such an analysis was also conducted by Silberbush and Barber (1983a). For a given soil volume, if the same root biomass is distributed in a greater length of thinner root elements, as opposed to a lesser length of thicker roots, the predicted potassium uptake is greatly increased. This is shown in Figure 8.3 (right side). Thus, even with fixed root biomass and soil volume, altering the root structure has a greater effect than physiological characteristics of the roots. A sensitivity analysis for phosphorus uptake indicated similar sensitivity of parameters (Silberbush and Barber 1983b).

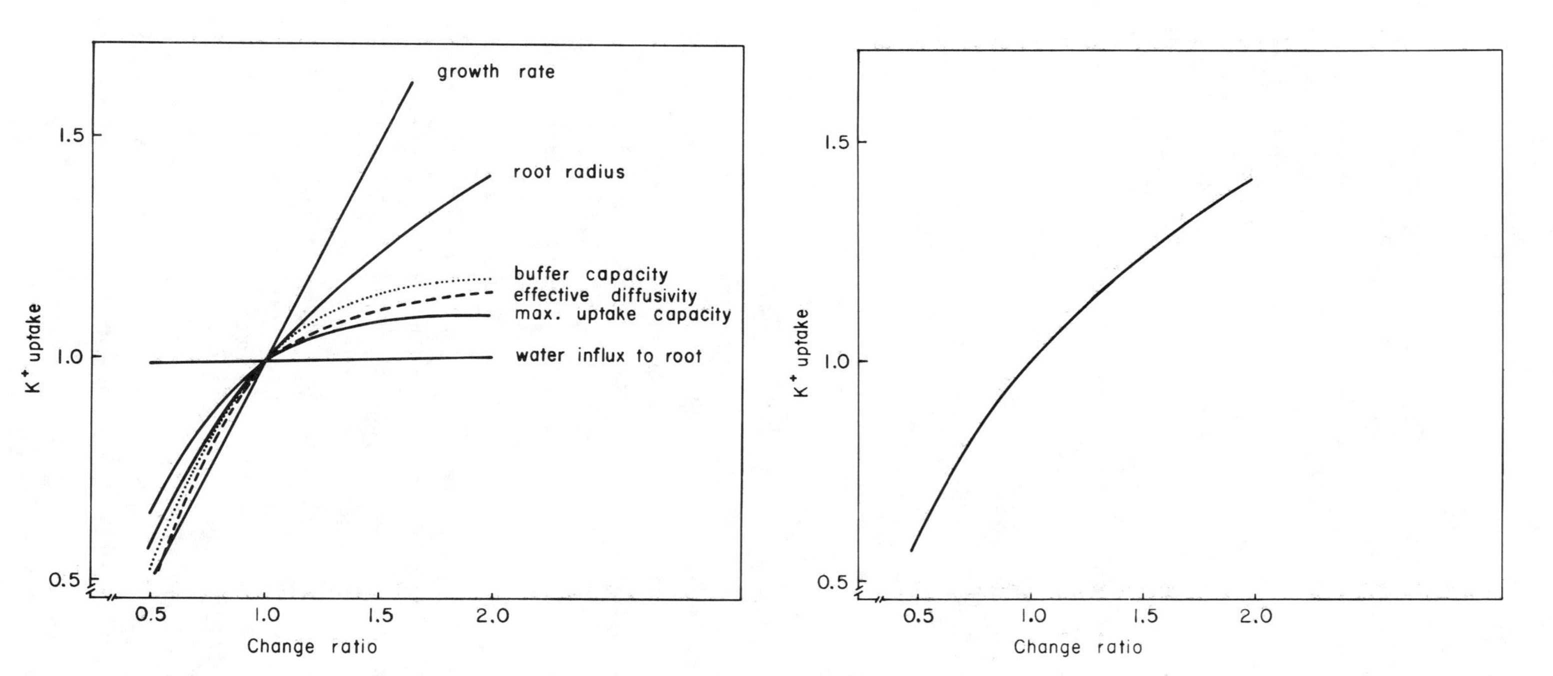

Figure 8.3. Left: Sensitivity analysis of predicted potassium uptake by soybean plants growing in silt loam soil using the model of Cushman (1979). Each parameter was changed independently from 0.5 to 2.0 times the initial value. Initial values were determined experimentally on potted soybean plants growing in a growth chamber (adapted from Silberbush and Barber 1983a). Right: Sensitivity analysis with the same model when total root length is changed between 0.5 and 2.0 times its original value and root thickness is changed concomitantly so that total root biomass and soil volume explored remain constant.

Such analyses indicate that at least for ions of moderate to high $\overline{\alpha a}/Db$, root-system structure appears to be more important than physiological characteristics such as absorption capacity. Empirical evidence also suggests that plants that inhabit soils of low fertility, and presumably are most competitive there, do not have a high nutrient absorption capacity (per unit mass), but instead invest heavily in a large root biomass (Chapin 1980). Thus, both the nutrient-uptake models and empirical evidence suggest that root-system structure is particularly important for efficacious acquisition of nutrients, at least in soils of moderate to low fertility and for ions with low mobility in soil. Similarly, rooting density is a particularly sensitive parameter in models of water uptake from drier soils (Gardner 1960; Cowan 1965).

In addition to the purely theoretical purposes of models of moisture and nutrient uptake, a pragmatic goal has been to investigate plant root parameters that might be of advantage in agricultural situations. For example, fertilizers might be more efficiently applied, and plant varieties might be selected whose root systems can best utilize soil nutrients. These models have not, however, as yet been designed to investigate questions of interspecific competition.

Competition between roots: considerations in two dimensions

The cylindrical-flow models assume that roots are regularly spaced and parallel and that the soil cylinders surrounding each root have a radius of $(\pi L_v)^{-1/2}$, where L_v is the rooting density. If these evenly spaced roots have similar uptake capacities and the soil is uniform, no nutrients or water will flow from one soil cylinder to another, and each root will have sole access to the resources within its cylinder. This is the manner in which root–root competition is often treated in such models (Cowan 1965; Cushman 1979). This will also mean that the concentration of nutrients or soil moisture at the perimeter of each cylinder will represent the bulk soil concentration. Obviously, natural root systems are not composed simply of regularly spaced parallel root elements, and a root may deplete resources within the soil cylinder of a neighboring root. Analytic solutions for the cylindrical model treating irregularly spaced root elements and interactions between roots have not been undertaken.

Baldwin et al. (1972) did employ a two-dimensional electrical analogue to assess interactions between roots (because this was two-dimensional, it necessarily assumed parallel roots, although their distribution could range from regularly spaced to very clumped). This representation gave very reasonable results and led to several conclusions that seem compatible with

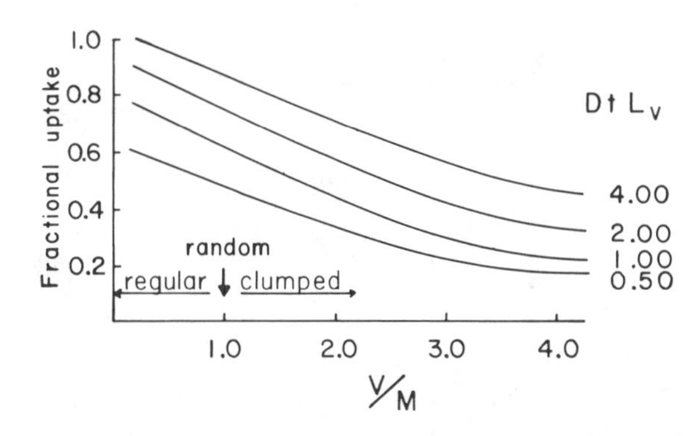

Figure 8.4. Nutrient uptake for root systems with distribution patterns ranging from regular to very clumped. The distribution pattern is expressed as the ratio of the variance, *V*, to the mean, *M*, of interroot distance. The magnitude of uptake is also influenced by the product of diffusivity, *D*, time, *t*, and rooting density, αL_v (adapted from Baldwin et al. 1972).

the cylindrical-flow model predictions. A competitive effect (i.e., when roots influence one another's resources) depends on *D* and time. For regularly spaced roots, a useful generalization is that roots are influencing resources within one another's soil cylinders when the average distance between roots is less than $(Dt)^{1/2}$. The competitive effect also depends on the ratios $\overline{\alpha a}/Db$ and Dt/a^2. The first ratio describes the depth of the depletion zone, and the second ratio the spread of this zone through short periods of time (assuming no renewal of nutrients). A fast-diffusing nutrient such as nitrate, with a low ratio of $\overline{\alpha a}/Db$, will have very broad, but not very deep, depletion zones; that is, the concentrations at the root surface will not tend to be much lower than at the perimeter of the soil cylinder. On the other end of the scale, phosphorus, which has a very high $\overline{\alpha a}/Db$, will have very steep depletion zones, but they will not tend to have much breadth and therefore will not overlap unless root density is very high. An element such as potassium is intermediate. Through time, of course, these depletion zones will broaden, and roots will tend to influence the resources of one another's space.

At the same root density, if roots are irregularly spaced, the degree of competition between roots will increase. This is portrayed in Figure 8.4, from the work of Baldwin et al. (1972). The ratio of the variance of distance between roots to the average distance between roots, *V/M*, is a measure of the regularity with which roots are arranged. A high *V/M* indicates a large degree of aggregation or clumping of root elements. Because more interroot competition exists under these circumstances, the

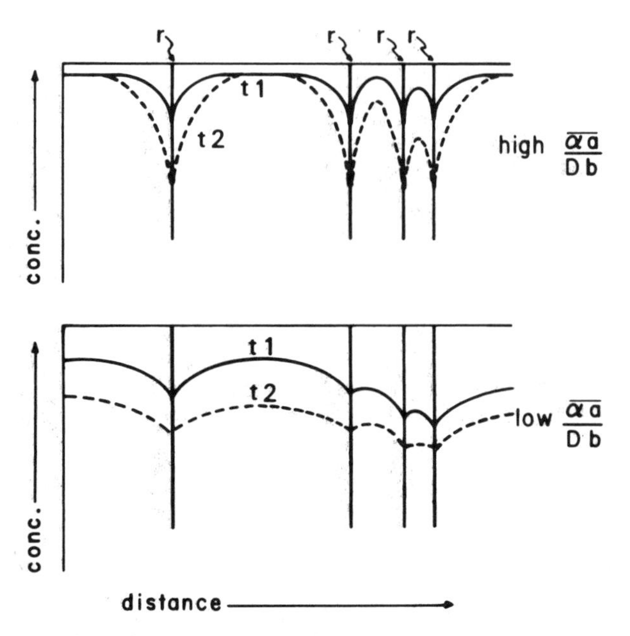

Figure 8.5. Localized depletion zones around parallel roots spaced at irregular intervals. These roots have the same root demand coefficient, $\overline{\alpha a}$ (i.e., they have the same uptake capacity). Concentration profiles are represented for two times. The soil is assumed initially to be uniform, and the roots are not growing. Changes in concentration through time are the result of root absorption and diffusion. For a case of high $\overline{\alpha a}/Db$, there are few effects of neighboring roots on one another until time *t*2. Concentration profiles shown are based on the behavior of nutrient absorption models and portrayal of neighboring root interactions in Nye and Tinker (1977).

total uptake by the root system is reduced. This analysis also suggests that the proportional influence of root density on total root-system uptake is greater for clumped roots than for regularly spaced roots.

When roots are influencing the resources in the soil cylinders of adjacent roots, the bulk soil concentration is changed and should assume some level that depends on the interroot distance, time, and $\overline{\alpha a}/Db$. Depletion zones at two points in time are depicted in Figure 8.5 for irregularly spaced parallel roots. These are envisaged using the behavior of cylindrical-flow models as a basis, but they are not calculated because there are no available analytic solutions for such situations. For a nutrient of moderate to low $\overline{\alpha a}/Db$, the bulk soil concentration (the concentration midway between roots) is quickly changed, and depletion zones are not particularly deep. In contrast, nutrients with high $\overline{\alpha a}/Db$ develop deep depletion zones that do not overlap initially, but with time do begin to overlap and thus alter the

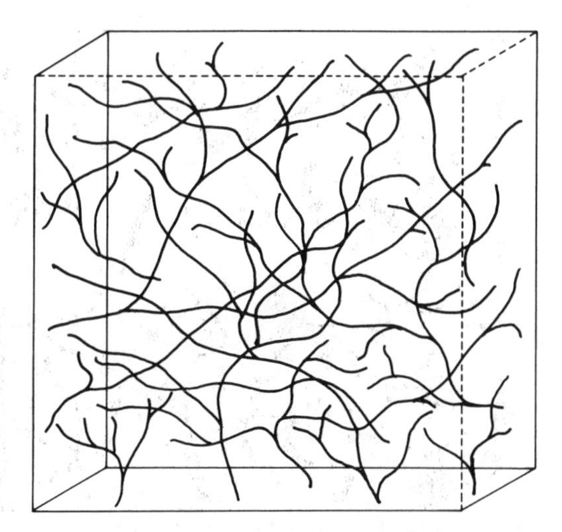

Figure 8.6. Depiction of root distribution in three-dimensional space.

bulk soil concentration. Such scenarios can lead to various predictions about the density of packing of roots in a soil volume and when competition between roots for specific nutrients might ensue.

Three dimensions and other complications

The real situation is necessarily more complex, because roots are distributed in three-dimensional space, the system is dynamic in that roots change their position by growth and senescence, and the location of soil resources can be altered by new wetting fronts within the soil or decay of organic materials. That roots absorb uniformly along their entire length is also an abstraction from reality. Although it is daunting to consider deterministic models to handle three-dimensional space, uneven absorption along the length of a root, and the dynamics through time, one can at least consider a few of the implications of these extensions to reality.

Although three-dimensional root systems (Figure 8.6) hardly approximate the honeycomb arrangement of soil cylinders envisaged by cylindrical-flow models, the extrapolation from two to three dimensions might not be so unreasonable (Baldwin et al. 1972). Marriott (1972) has shown that for straight lines randomly distributed in space, but not necessarily isotropically arranged, the average distance from any point in this space to the nth nearest line will be the same for a plane intersecting this space as for the three-dimensional space.

If a root absorbs unevenly along its length, the conclusions of cylindrical-flow models can be changed considerably. Because water or nutrients

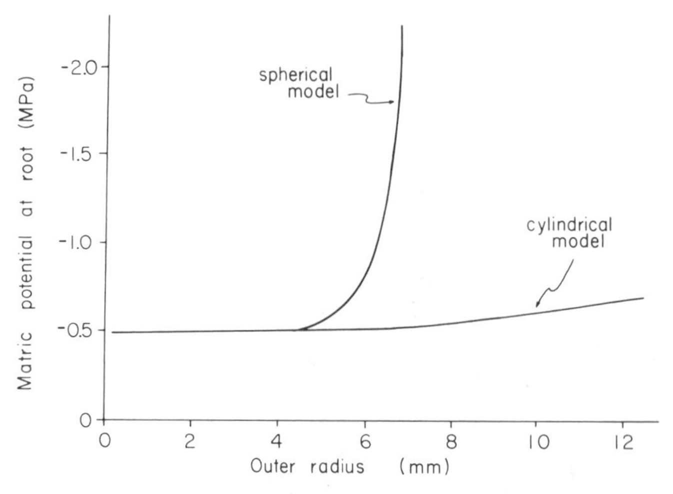

Figure 8.7. Soil-matrix water potential at the root surface as a function of the magnitude of the outer radius of a cylinder or a sphere of soil from which water is absorbed. This assumes a constant rate of water uptake per volume of soil; therefore, as the outer radius increases, rooting density is decreasing, and water uptake per unit length of root is increasing. The bulk soil-matrix potential is −0.5 MPa. In the case of a cylinder, a normal cylindrical-uptake model will apply, and the root is assumed to be absorbing along its length. In the case of the soil sphere, absorption is assumed to take place effectively at a point on the root (adapted from Caldwell 1976).

are absorbed at points rather than along an axis, a spherical rather than a cylindrical model would apply, and depletion zones would develop much more severely for this spherical case (Caldwell 1976). For absorption of water, depletion zones of spherical configuration can develop at much higher water potentials than depletion zones in a cylindrical configuration (Figure 8.7). Of course, a strict spherical model is not likely to apply to root systems; rather, some intermediate situation, as depicted in Figure 8.8, might develop. Not enough is known about the location of absorption along roots under field conditions to conclude that a spherical model is preferable to a cylindrical one at present. Nevertheless, if a root does not absorb evenly along its length, one can surmise from the spherical model that the competitive interactions between roots might be quite different than predicted by the normal cylindrical-flow concept.

Depletion zones become broader through time if nothing else changes. However, the locations of root absorbing zones (principally root tips) change with root growth, and depletion zones recover following cessation of absorption at a particular location. Also, the soil environment is hardly

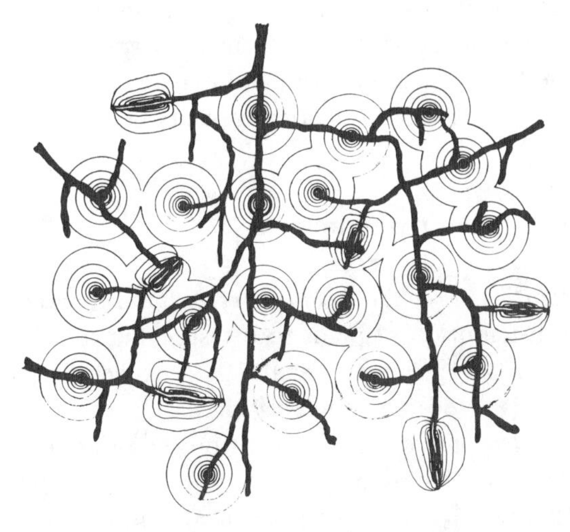

Figure 8.8. Patterns of water uptake indicated by contours of decreasing water potential approaching regions of uptake at individual roots for a perennial plant root system. Most of the root system is suberized. Uptake is considered to occur at widely spaced points where the suberization is disrupted, such as cracks, or at unsuberized root tips, which are shown as white roots. Water flow around cracks is in a spherical configuration. For the unsuberized root tips, the water-uptake pattern is ovoid, with increasing radius toward the basal region of the roots, because the roots are actively growing (from Caldwell 1976).

uniform in time or space. The timing and nature of root growth of perennial plants were discussed in greater detail in an earlier review (Caldwell 1979).

Conclusions from uptake models for competition between roots

A high rooting density and thus thick roots, as well as more evenly spaced roots that minimize competition between root elements of the same plant, would appear to be advantageous for competitive potential. On the other hand, high absorptive capacity of roots may play a lesser role in competitive effectiveness. The severity of competition between roots for resources will depend on soil factors and, in particular, on the diffusivity of nutrient elements and on soil water potential. Interroot distance will play a much larger role for elements of low diffusivity than for those of high diffusivity. The length of time that active absorbing zones remain in the same locations will also increase the probability of competition between roots. The timing and location of root growth are of particular importance in altering the root-system structure as well as the sites of absorption.

Finally, if a species is competing more effectively than another, and if this is the result of soil resource deprivation by the superior species, it is likely that this is accomplished more by the growth and structural characteristics of the root system than by greater absorption capacity per length of root.

Root density and morphometrics

Density and location of roots: an example of two bunchgrasses

Though the distribution of plant root systems in soils has often been described (Cannon 1911; Clements et al. 1929; Kutschera and Lichtenegger 1982; Richards 1986) and there is increasing information on root-system biomass and root production in various ecosystems (Persson 1980; Vogt et al. 1981; McClaugherty et al. 1982), there is a paucity of information on rooting density for nonagricultural plants (Barley 1970; Newman 1974; Kummerow et al. 1978). This is regrettable in view of the overwhelming importance of rooting density in models of moisture and nutrient uptake. Certainly a correlation exists between root biomass and rooting density; however, differences in root-system morphology will alter this correlation considerably. This is illustrated for two *Agropyron* species that are very similar in their growth form, phenology, water relations, and photosynthetic characteristics. These *Agropyron* tussock grasses of the intermountain West have been under intensive study because they differ radically in terms of ability to tolerate defoliation and to compete with the dominant shrub of the region, *Artemisia tridentata* (Caldwell et al. 1981, 1983; Richards 1984).

Root systems of these grasses were excavated to assess the distribution of root length with respect to depth. This was conducted in a field plot where each species of *Agropyron* had been interplanted in a regular matrix mixed with *Artemisia tridentata* such that each tussock grass was surrounded by four *Artemisia* shrubs at a distance of 50 cm. These field plots were established by a series of uniform transplants in 1978, and these root-system excavations were conducted in 1981. The excavations were carried out in part as a monolith beneath the plant and in part by soil cores to sample an extended volume of soil that in total constituted a cylinder 75 cm in diameter and 1 m in depth. After extraction and separation of the apparently living roots, total root length was determined by an optical scanning device (Comair, Inc., Melbourne, Australia).

The rooting density as a function of depth is portrayed for these two species at one point in time in Figure 8.9. Even though the average root biomasses for these were the same (199 g for *Agropyron spicatum* and 192 g

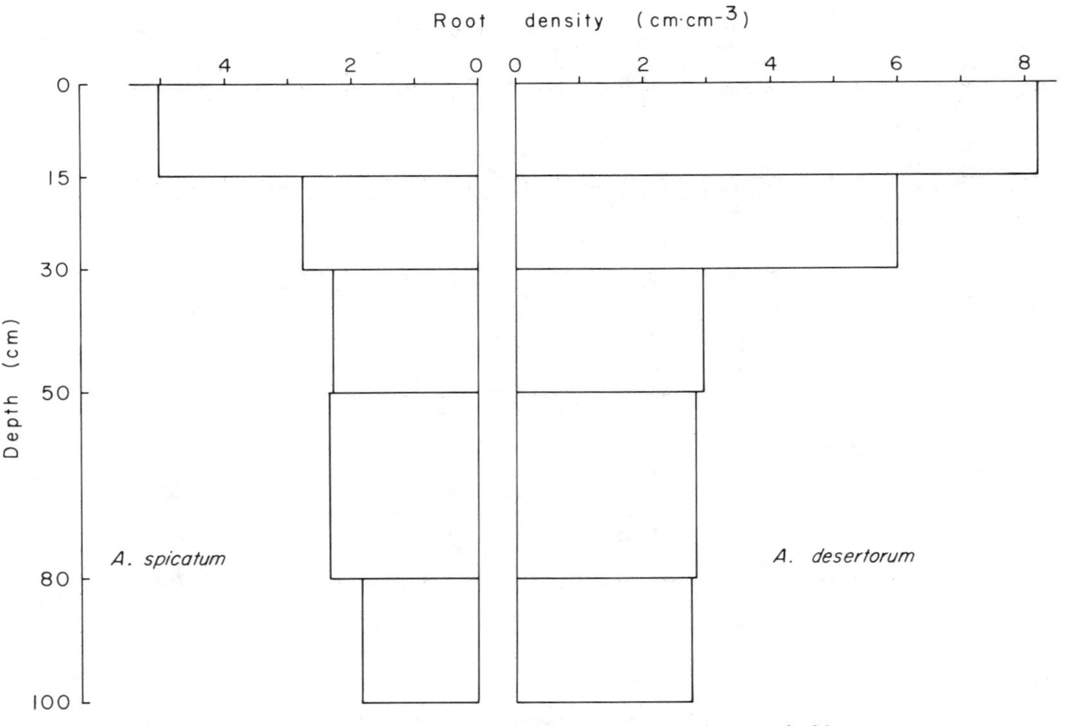

Figure 8.9. Rooting density of two *Agropyron* bunchgrass species on July 20, 1981, at different depths. Each value represents the average rooting density over a 75-cm-diameter cylinder centered about the axis of the bunchgrass plant. These are means of three plants for each species.

Table 8.2. *Mean root diameter and number of branches of* Agropyron *bunchgrasses under field conditions*[a]

Variable	*A. desertorum*	*A. spicatum*
First-order root diameter (mm)	0.18	0.23
Second-order root diameter (mm)	0.26	0.47
Number of first-order roots per cm of second-order root	5.2	3.9

[a] Roots from three depth intervals (0–10, 20–30, 60–70 cm) were sampled uniformly. Species means were significantly different for all characteristics at $p < 0.05$.

for *Agropyron desertorum*), the total root lengths differed by 50% (11.8 km for *A. spicatum* and 18.2 km for *A. desertorum*). The basic branching patterns of these root systems were similar; however, the roots of *A. desertorum* were much thinner, and there was also a greater number of lateral roots, as indicated in Table 8.2. When growing in monocultures, at the same densities, *A. desertorum* can deplete soil moisture more rapidly than *A. spicatum*, especially at greater depths in the profile (Thorgeirsson and Richards, unpublished data). Although this correlation between water extraction and root density is noteworthy, it certainly may not be the sole explanation for why *A. desertorum* is a much more effective competitor with the shrub *Artemisia tridentata* than is *A. spicatum*.

Although the density of rooting is emphasized in models and in this discussion, many other root-system structural characteristics are of consequence. How roots are dispersed, where they are located in the belowground space, their timing of growth, the location and extent of mycorrhizal associations, and root hair characteristics are all of potential importance. Unfortunately, a comprehensive picture of these root-system characteristics under field conditions is seldom available.

Descriptive studies of root-system form have documented the great variability in root distribution that can be exhibited by a given species growing in different habitats and between species growing in a single habitat (Cannon 1911; Clements et al. 1929; Biswell 1935; Huinink 1966; Richards 1986). Certainly species growing together can avoid competition to some extent by having active roots at different depths. Root systems of neighboring plants, however, usually co-occur within the same soil volume and proliferate most where resources are most available (McClaugherty et al. 1982; St. John et al. 1983), leading to considerable overlap of resource exploitation and potential competitive interaction.

When considering competitive relationships, the distribution and degree of overlap of neighboring plant root systems are of particular interest.

An assessment of this has been made for the two *Agropyron* species growing with neighboring *Artemisia tridentata* shrubs described earlier. The assessment of rooting distribution was done in 1983 by the classic profile-wall mapping technique (Boehm 1979). Intersections of roots with a plane established by the profile wall were counted in cells 10 × 10 cm in a 40-cm interval between neighboring plants and to a depth of 1 m. Roots of each neighbor were separately tallied. It was feasible to count only higher-order roots, which for these grasses were primarily second-order roots. [Following the terminology of Fitter (1982), the finest roots that terminate in a meristem are the first-order roots, and the second-, third-, and higher-order roots are those formed by junctions of roots of the preceding order.] Thus, the roots represented in Figure 8.10 usually possessed one lower order of roots and sometimes two. Figure 8.10 is a depiction of the average root distribution patterns from six replicate profile-wall samples for each species pair. The root intersections are portrayed relative to the maximum number of roots found in any particular cell. For these data, this was the cell immediately beneath the *A. spicatum* plant. The relative rooting densities are indicated by the areas of the triangles in the corners of each cell – a triangle for each species. Dominance of the cell by one of the two species is indicated by the color of the cell between the two triangles. Dominance is defined simply as which of the two species had the greater number of root intersections in a particular cell, no matter how dense or sparse the rooting density. (Because the relative rooting density is normalized to the cell under the *A. spicatum* plant, there is no remaining cell area to indicate that *A. spicatum* was dominant in this cell.)

The intermingling of neighboring plant root systems was pronounced – even to the extent that roots of plants were virtually beneath the immediate crown projection of the neighboring plants in some quantity. Virtually every cell had roots of both species represented. By the criterion of dominance, both *Agropyron* bunchgrass species clearly dominated in the 20- to 50-cm depth. For the total profile, *A. desertorum* dominated in 31 of 40 cells with neighboring *Artemisia tridentata,* whereas *A. spicatum* dominated in only 20 of 40 cells with *Artemisia.* As might be expected from the root-density data (Figure 8.9), *A. desertorum* was better represented at depth than *A. spicatum.* In any case, *Artemisia* dominated at the greatest depths, although rooting density for all species was sparse. In the six years following the establishment of uniform-sized transplants, significant differences in crown volume of *A. tridentata* had developed, depending on whether it was interplanted with *A. desertorum* or with *A. spicatum* (J. H. Richards et al., unpublished data). The average crown volume of *Artemisia* interplanted with *A. spicatum* was more than twice the mean volume of the

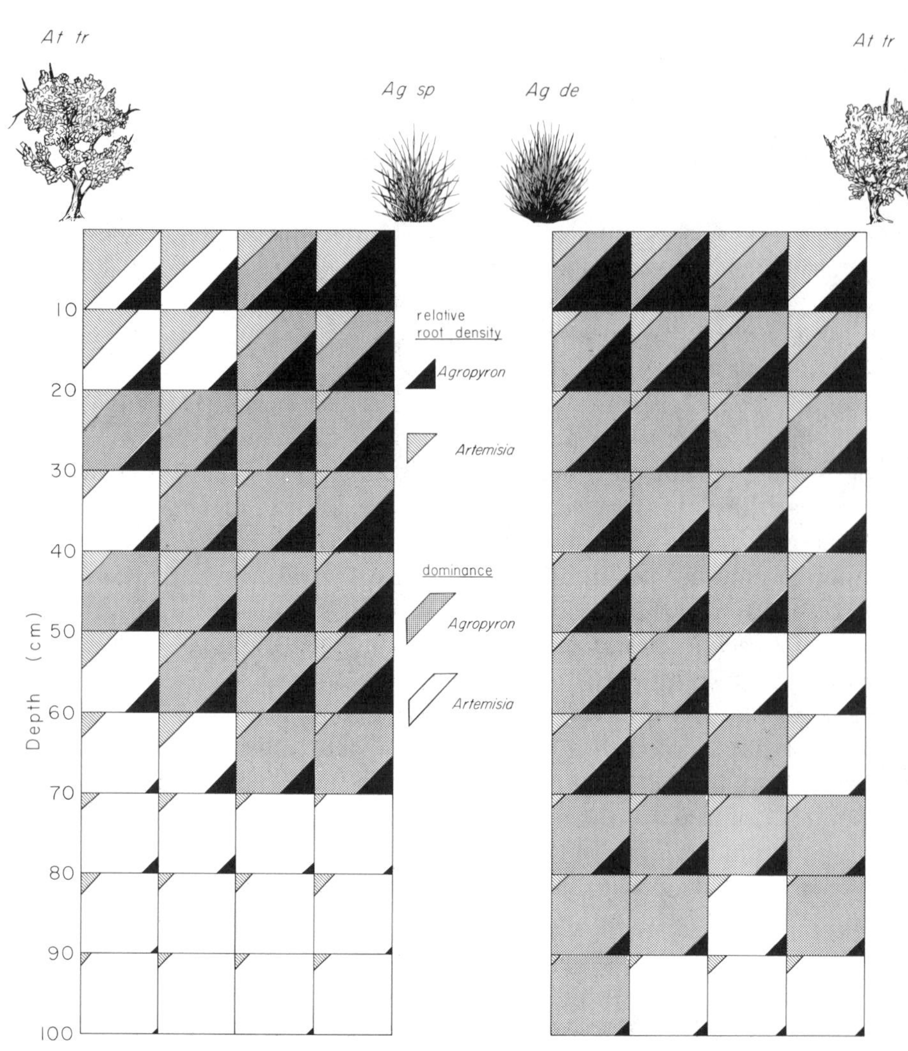

Figure 8.10. Root-system distribution of neighboring *Agropyron* bunchgrasses and *A. tridentata* shrubs. The relative rooting densities of higher-order roots of each species and the dominance in each soil cell are indicated.

Artemisia interplanted with *A. desertorum*. Thus, when compared with *A. spicatum,* the thinner roots of *A. desertorum* would seem to be of advantage both in providing a greater rooting density and in allowing this species to be more dominant in the belowground space in relation to its neighbor. Nevertheless, the crowding success of *A. desertorum* may be related to other factors not yet identified.

Partitioning root biomass into thin root elements, which effects greater rooting density without representing a greater biomass investment, may be of considerable advantage; however, this may have attendant costs. Though data are very sparse, there is some evidence that root systems of species with thin roots experience a greater turnover rate than root systems composed of thicker root elements (Shaver and Billings 1975; Fernandez and Caldwell 1977; Caldwell et al. 1977), which is analogous to the relationship between thickness and turnover rate for leaves (Chapin 1980). Very thin roots may also have a tendency toward greater axial resistance to water transport (Greacen et al. 1976; Fowkes and Landsberg 1981; Sands et al. 1982).

Morphometrics of root systems

The organization of a root system will determine to some extent how effectively the total rooting length is placed within the profile and the location and trajectories of actively absorbing root elements. The dendritic nature of plant root systems suggests that roots have multiple orders of branching. Though branching of plant root systems has received considerable attention, the morphometrics of root systems had been subjected to comparatively little analysis until the recent work of Fitter (1982). He showed that root systems, like rivers in a watershed or bronchioles of lungs, exhibit a general relationship in that the number of branches of each order is a constant proportion of the number of branches in the next order. In this relationship, known as Horton's law of branching, a plot of the log of branches versus branching order yields a straight line of negative slope. The branching ratio, R_b, then is defined as the antilog of the absolute value of this slope. Thus, a larger value of R_b indicates that the number of branches for each branching order decreases more rapidly at successively higher branching orders. Given the same number of branching orders, a larger value of R_b indicates a greater number of branches. Or given the same number of branches on a root system, a larger value of R_b means that these branches are distributed among fewer branching orders.

The range of R_b values for the root systems of different families is great (3 to 12), and this generally overlaps with the range of R_b values for shoot systems of different tree species. However, Fitter found that the median R_b value for root systems was higher than that for tree shoots and also higher than those for branching systems such as the drainage system of a watershed. The advantage of higher branching ratios in root systems is not immediately explicable. However, one would expect the selective forces molding the evolution of root-system structure to be quite different from those directing the architecture of shoot systems. Optimization must involve factors other than considerations of transport within the root system itself.

Assessing root-system morphology with respect to effectiveness in competition for soil resources is hardly a trivial undertaking. Models of soil moisture and nutrient uptake clearly indicate that absorbing root elements should have a high density at the expense of root thickness, provided axial resistances are not too large (Figure 8.3), and they should be as regularly spaced as possible (Figure 8.4). Also, if root tips are the primary locations of absorption, then for a given number of branching orders, a high R_b will be of advantage. Whereas these conclusions are reasonably straightforward, exploration of belowground space and resource acquisition involves much more than the arrangement of soil cylinders. Branching and trajectories of a root system must reach a compromise between directness, regular spacing, total path length for diffusion and transport within the root system, and root biomass investment. Roots should also be sufficiently flexible to be able to proliferate in patches of abundant soil moisture and nutrients. An optimal pattern must, of course, still comply with the constraints dictated by the manner in which a root system grows, whether determinant or indeterminant in pattern. Soils are hardly uniform. Thus, the deployment of roots of varying densities to different parts of the profile involves yet another set of considerations (Fowkes and Landsberg 1981). The analysis is further complicated by changes through time. Root systems continue to grow, points of absorption (new root tips) continue to change locations, and diffusion and mass flow continue to change the soil environment. Roots senesce, die, and are replaced by new growth. Wetting fronts, decay, and other biological activities continually renew and redistribute the soil resources. Consideration of these several factors may curb optimism about generalizations on optimal root systems based on models of single-root absorption from a soil cylinder. Just as one should extrapolate very cautiously from single-leaf behavior to canopy function, so one should not be too expansive in the extension of root-element behavior to that of competing root systems.

Acknowledgments

Some of the information contained in this chapter resulted from research supported by the National Science Foundation (DEB 7907323 and DEB 8207171) and the Utah Agricultural Experiment Station. We gratefully acknowledge assistance with root profile-wall mapping by L. Urness, with root extractions and processing by T. Wilkins, J. Erickson, and L. Jordan, and the drafting of figures by C.

Warner. Review comments by R. Robichaux and P. Groff were very much appreciated.

References

Ågren, G. I., B. Axelsson, J. G. K. Flower-Ellis, S. Linder, H. Persson, H. Staaf, and E. Troeng. 1980. Annual carbon budget for a young Scots pine. Ecol. Bull. Stockholm 32:307–313.

Allen, E. B., and R. T. T. Forman. 1976. Plant species removals and old-field community structure and stability. Ecology 57:1233–1243.

Baldwin, J. P., P. B. Tinker, and P. H. Nye. 1972. Uptake of solutes by multiple root systems from soil. II. The theoretical effects of rooting density and pattern on uptake of nutrients from soil. Plant Soil 63:693–708.

Barley, K. P. 1970. The configuration of the root system in relation to nutrient uptake. Adv. Agron. 22:159–201.

Bell, D. T. 1981. Spatial and size-class patterns in a central Australian spinifex grassland. Aust. J. Bot. 29:321–327.

Bhat, K. K. S., and P. H. Nye. 1973. Diffusion of phosphate to plant roots in soil. I. Quantitative autoradiography of the depletion zone. Plant Soil 38:161–175.

Biswell, H. H. 1935. Effects of environment on the root habits of certain deciduous forest trees. Botanical Gazette 96:676–708.

Boehm, W. 1979. Methods of studying root systems. Springer-Verlag, Heidelberg.

Caldwell, M. M. 1976. Root extension and water absorption. Pp. 63–85 *in* O. L. Lange, L. Kappen, and E.-D. Schulze (eds.), Water and plant life. Springer-Verlag, Heidelberg.

– 1979. Root structure: the considerable cost of belowground function. Pp. 408–427 *in* O. T. Solbrig, S. Jain, G. B. Johnson, and P. H. Raven (eds.), Topics in plant population biology. Columbia University Press, New York.

Caldwell, M. M., T. J. Dean, R. S. Novak, R. S. Dzurec, and J. H. Richards. 1983. Bunchgrass architecture, light interception, and water-use efficiency: assessment by fiber optic point quadrats and gas exchange. Oecologia 59:178–184.

Caldwell, M. M., J. H. Richards, D. A. Johnson, R. S. Nowak, and R. S. Dzurec. 1981. Coping with herbivory: photosynthetic capacity and resource allocation in two semiarid *Agropyron* bunchgrasses. Oecologia 63:14–24.

Caldwell, M. M., R. S. White, R. T. Moore, and L. B. Camp. 1977. Carbon balance, productivity, and water use of cold-winter desert shrub communities dominated by C_3 and C_4 species. Oecologia 29:275–300.

Cannon, W. A. 1911. The root habits of desert plants. Carnegie Inst. Washington publication no. 131.

Chapin, F. S., III. 1980. The mineral nutrition of wild plants. Ann. Rev. Ecol. Syst. 11:233–260.

Christie, P., E. I. Newman, and R. Campbell. 1978. The influence of neighbouring grassland plants on each others' endomycorrhizas and root-surface microorganisms. Soil Biol. Biochem. 10:521–527.

Clements, F. E., J. E. Weaver, and H. C. Hanson. 1929. Plant competition: an analysis of community function. Carnegie Inst. Washington publication no. 398.
Cowan, I. R. 1965. Transport of water in the soil-plant-atmosphere system. J. Appl. Ecol. 2:221–239.
Cushman, J. H. 1979. An analytical solution to solute transport near root surfaces for low initial concentration: I. Equations development. Soil Sci. Soc. Amer. J. 43:1087–1090.
Fernandez, O. A., and M. M. Caldwell. 1977. Phenology and dynamics of root growth of three cool semi-desert shrubs under field conditions. J. Ecol. 63:703–714.
Fitter, A. H. 1982. Morphometric analysis of root systems: application of the technique and influence of soil fertility on root system development in two herbaceous species. Plant Cell Environ. 5:313–322.
Fonteyn, P. J., and B. E. Mahall. 1981. An experimental analysis of structure in a desert plant community. J. Ecol. 69:883–896.
Fowkes, N. D., and J. J. Landsberg. 1981. Optimal root systems in terms of water uptake and movement. Pp. 109–125 *in* D.A. Rose and D. A. Charles-Edwards (eds.), Mathematics and plant physiology. Academic Press, London.
Fowler, N. 1981. Competition and coexistence in a North Carolina grassland. J. Ecol. 69:843–854.
Gardner, W. R. 1960. Dynamic aspects of water availability to plants. Soil Sci. 89:63–73.
Greacen, E. L., P. Ponsana, and K. P. Barley. 1976. Resistance to water flow in the roots of cereals. Pp. 86–100 *in* O. L. Lange, L. Kappen, and E.-D. Schulze (eds.), Water and plant life. Springer-Verlag, Heidelberg.
Harper, J. L. 1977. Population biology of plants. Academic Press, New York.
Huinink, W. A. E. van Donselaar-ten Bokkel. 1966. Structure, root systems and periodicity of savanna plants and vegetations in northern Surinam. Wentia 17:1–162.
Kummerow, J., D. Krause, and W. Jow. 1978. Seasonal changes of fine root density in the southern Californian chaparral. Oecologia 37:201–212.
Kutschera, L., and E. Lichtenegger. 1982. Wurzelatlas mitteleuropaeischer Gruendlandpflanzen. Bd. 1. Monocotyledoneae. Gustav Fischer, Stuttgart.
Lang, A. R. G., and W. R. Gardner. 1970. Limitation to water flux from soils to plants. Agron. J. 62:693–695.
McClaugherty, C. A., J. D. Aber, and J. M. Melillo. 1982. The role of fine roots in the organic matter and nitrogen budgets of two forested ecosystems. Ecology 63:1481–1490.
McCoy, E. L., L. Boersma, M. L. Ungs, and S. Akratanakul. 1984. Toward understanding soil water uptake by plant roots. Soil Science 137:69–77.
Marriott, F. H. C. 1972. Buffon's problems for non-random distribution. Biometrics 28:621–624.
Mueggler, W. F. 1972. Influence of competition on the response of bluebunch wheatgrass to clipping. J. Range Mange. 25:88–92.
Newman, E. I. 1974. Root and soil water relations. Pp. 363–440 *in* E. W. Carson (ed.), The plant root and its environment. University Press of Virginia, Charlottesville.

Nye, P. H., and P. B. Tinker. 1977. Solute movement in the soil-root system. University of California Press, Berkeley.
Persson, H. 1980. Spatial distribution of fine-root growth, mortality and decomposition in a young Scots pine stand in Central Sweden. Oikos 34:77–87.
– 1983. The importance of fine roots in boreal forests. Pp. 595–608 *in* Wurzeloekologie und ihre Nutzanwendung. Bundesanstalt Gumpenstein, Irdning, Austria.
Pinder, J. E., III. 1975. Effects of species removal on an old-field plant community. Ecology 56:747–751.
Reichle, D. E., B. E. Dinger, N. T. Edwards, W. F. Harris, and P. Sollins. 1973. Carbon flow and storage in a forest ecosystem. Pp. 345–365 *in* G. M. Woodwell and E. V. Pecan (eds.), Carbon and the biosphere. U. S. Atomic Energy Commission.
Rice, E. L. 1974. Allelopathy. Academic Press, New York.
Richards, J.H. 1984. Root growth response to defoliation in two *Agropyron* bunchgrasses: field observations with an improved root periscope. Oecologia 64:21–25.
– 1986. Root form and depth distribution in several biomes. *In* D. Carlisle, W. L. Berry, J. R. Watterson and I. R. Kaplan (eds.), Mineral exploration: biological systems and organic matter. Prentice-Hall, Englewood Cliffs, N.J.
Robberecht, R., B. E. Mahall, and P. S. Nobel. 1983. Experimental removal of intraspecific competitors – effects on water relations and productivity of a desert bunchgrass, *Hilaria rigida*. Oecologia 60:21–24.
Sands, R., E. L. Fiscus, and C. P. P. Reid. 1982. Hydraulic properties of pine and bean roots with varying degrees of suberization, vascular differentiation and mycorrhizal infection. Aust. J. Plant Physiol. 9:559–569.
Shaver, G. R., and W. D. Billings. 1975. Root production and root turnover in a wet tundra ecosystem, Barrow, Alaska. Ecology 56:401–409.
Silberbush, M., and S. A. Barber. 1983a. Sensitivity analysis of parameters used in simulating K uptake with a mechanistic mathematical model. Agron. J. 75:851–854.
– 1983b. Sensitivity of simulated phosphorus uptake to parameters used by a mechanistic-mathematical model. Plant Soil 74:93–100.
Sims, P. L., and J. S. Singh. 1978. The structure and function of ten western North American grasslands. II. Intraseasonal dynamics in primary producer compartments. J. Ecol. 66:547–572.
Smith, A. P. 1979. Spacing patterns and crown size variability in an Ecuadorian desert shrub species. Oecologia 40:203–205.
St. John, T. V., D. C. Coleman, and C. P. P. Reid. 1983. Growth and spatial distribution of nutrient-absorbing organs: selective exploitation of heterogeneity. Plant Soil 71:487–493.
Vogt, K. A., R. I. Edmonds, and C. C. Grier. 1981. Seasonal changes in biomass and vertical distribution of mycorrhizal and fibrous-textured conifer fine roots in 23- and 180-year-old subalpine *Abies amabilis* stands. Can. J. For. Res. 11:223–229.
Yeaton, R. I., and M. L. Cody. 1976. Competition and spacing in plant communities: the northern Mohave Desert. J. Ecol. 64:689–696.

9 Belowground costs: hydraulic conductance

EDWIN L. FISCUS

There are three traditionally recognized functions of root systems: anchorage, water absorption, and procurement of mineral nutrients. More recently, production of growth regulators has been recognized as another important root function. In terms of survival advantage, the extremely high degrees of interaction and interdependence of these functions make it impossible to state that any one is more important than any other, or even that such a question really has any meaning. I shall confine my remarks to the water absorption function and arbitrarily treat that function as the sole benefit derived by a plant from its root system. It should be apparent, however, that because the other functions are also dependent on system size, most of the growth analysis will be relevant to them as well. Further, I shall not attempt a comprehensive review of this subject but shall use data primarily from my own laboratory on *Phaseolus vulgaris*, because similarly extensive data are difficult to find.

This chapter will be divided into two sections, the first dealing with root conductance as the benefit, and the second dealing with the cost to the plant of producing that water supply system. Specifically, the ultimate aim of this chapter is to estimate the cost of a unit of hydraulic conductance.

Benefits

Theory of hydraulic conductance

It has long been recognized that the flux of a fluid through a system (roots, pipes, etc.) can be described as the product of a driving force and some sort of conductance. The simplest kind of system is one in which the conductance is constant and the force–flux relationship passes through the origin. Unfortunately, root systems are generally not that simple. The main complicating factor arises because root systems accumulate solutes of various kinds at the expense of metabolic energy, thus creating osmotic gradients within the system. The relationships between

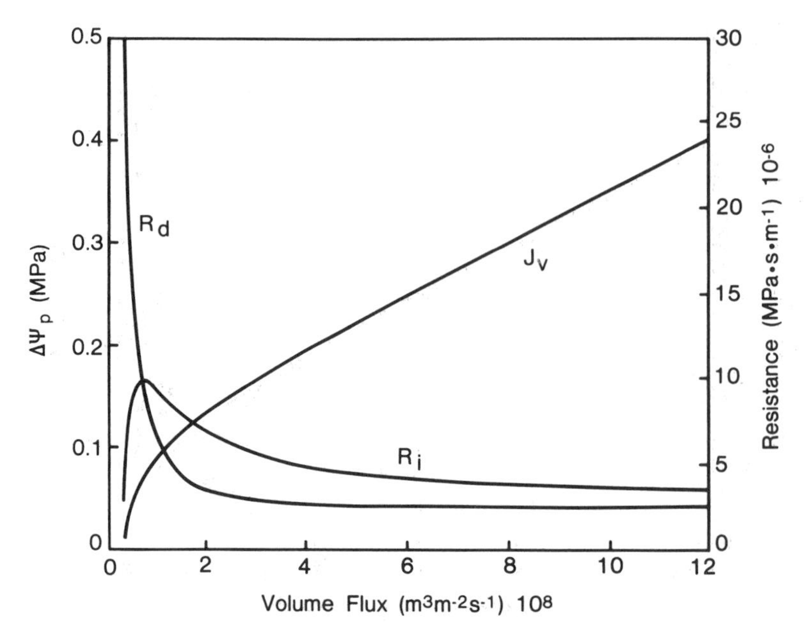

Figure 9.1. Total volume flux and instantaneous and differential resistance functions for a typical *Phaseolus* root system. $L_p = 4 \times 10^{-7}$ m s^{-1} MPa^{-1}; $J_s^* = 1.5 \times 10^{-6}$ mol m^{-2} s^{-1}; $\omega = 1 \times 10^{-6}$ mol m^{-2} s^{-1} MPa^{-1}; $\sigma = 0.97$; $\Pi^o =$ 0.11 MPa.

the osmotic gradient and the hydrostatic gradient, usually the major driving force, are flux-dependent, so that depending on particular circumstances, the two forces may aid or oppose each other. The result of these interactions is a system with characteristics similar to those in Figure 9.1, which shows conductance (not a conductance coefficient) directly related to the flux rate. A great many problems arise in the interpretation of these relationships that are directly traceable to inconsistent and imprecise use of terminology, especially that related to conductance and resistance. For example, in discussing pressure- or tension-driven radial flow through roots, there are many ways in which the term "conductance" can be used in a single flux equation. Throughout the rest of this section I shall attempt to clarify this terminology and then discuss the interpretation of other relevant transport coefficients to define more clearly the water transport characteristics of a *Phaseolus* root system under different circumstances.

Water flow through whole root systems can be described adequately by the model we proposed several years ago (Fiscus 1975). This model was based on the well-known membrane transport relationship

$$J_v = L_p(\Delta P - \sigma\Delta\Pi) \tag{9.1}$$

where J_v is the volume flux in $m^3\ m^{-2}\ s^{-1}$, ΔP is the pressure difference in MPa, $\Delta\Pi$ is the osmotic pressure difference in MPa, σ is the reflection coefficient or the osmotic efficiency, and L_p is the hydraulic conductance coefficient of the system in $m^3\ m^{-2}\ s^{-1}\ MPa^{-1}$. Please note that in order to make comparisons with the previous literature, the numerical value of L_p as given here is 0.1 times the value in $cm^3\ cm^{-2}\ s^{-1}\ bar^{-1}$, a frequently used set of units. Also, J_v in $m^3\ m^{-2}\ s^{-1}$ is 0.01 times the value in $cm^3\ cm^{-2}\ s^{-1}$.

The relationship between xylem tension, included in ΔP, and water absorption is frequently nonlinear (Mees and Weatherley 1957; Lopushinsky 1961, 1964; Kuiper and Kuiper 1974; Fiscus 1975; Fiscus and Kramer 1975; Markhart et al. 1979; Sands et al. 1982). By examining the osmotic component of the driving force, we can discover possible reasons for the nonlinearity (Dalton et al. 1975; Fiscus 1975). If, for convenience, we use the van't Hoff approximation, we can write the osmotic difference in equation (9.1) as

$$\Delta\Pi = RT(C^o - C^i) \tag{9.2}$$

where R is the gas constant, T is the temperature in degrees Kelvin, and the C's are the concentrations of solutes in mol m^{-3} outside the root and inside the xylem, respectively.

An important feature of equation (9.2) is that C^i is an inverse function of J_v and a direct function of J_s, the total solute transport rate, so that

$$C^i = J_s/J_v \tag{9.3}$$

Therefore, C^i, as a component of the driving force, is itself determined by the total volume flux.

We can further discover the roles of various solute transport components by examining the total solute flux relationship

$$J_s = \overline{C}_s(1 - \sigma)J_v + \omega\Delta\Pi + J_s^* \tag{9.4}$$

Equation (9.4) separates the total solute flux into three components, convective, diffusive, and active, each with its own distinctive coefficient.

Functional interpretation of parameters

The relationships in equations (9.1) and (9.4) were developed for and strictly apply to single-membrane systems, which the root is almost certainly not. And although we have repeatedly emphasized the operational interpretation of these relationships, some controversy has arisen from my use of the coefficients. Perhaps it would have been better to have altered them to indicate more clearly their functional nature when applied to roots. Not having changed the notation in the past, I shall take some space here to clarify my use of these coefficients. Throughout, however, remember that the transport of any substance in any system can be described as the sum of the convective, diffusive, and active transport components, whether or not a membrane is involved. So, even if the notation were altered, the flux relationship would resemble equation (9.4) in form.

Convective solute flux. The $\overline{C}_s$ term is usually taken as the average concentration of solutes across the membrane, with the restriction that the difference be small (Katchalsky and Curran 1974). Clearly, applying $\overline{C}_s$ to flow across a root can be very risky, because (1) the root is a multiple-membrane system, (2) the concentration difference may be quite large, and (3) it is only under special circumstances that we can expect to estimate with any degree of reliability the actual concentration difference. For these reasons, we found $\overline{C}_s$ to be inappropriate for describing convective solute fluxes through roots. We concluded that we really need a more functional definition of the convective term, one that will tell us how much of the external solutes are being conveyed to the root xylem as a result of being dissolved in the water that is moving in that direction. Convection in an open system can be described simply as the product of the volume flux and the concentration. However, when the solution crosses a barrier that in any way impedes the passage of one substance relative to the other, such as a semipermeable membrane, then we need a coefficient to describe the degree of retardation or filtration. The traditional reflection coefficient σ, just as it appears in equation (9.4), serves this purpose. But $\overline{C}_s$ will now logically be replaced by C^o (Fiscus 1977), the external solute concentration, because we are interested only in how much of the solute passes the barrier with the water. In this way, we need not worry about the specific features of the barrier (or barriers) involved, but can describe it in terms of a measured reflection coefficient. The first term in equation (9.4) will therefore be altered to read $C^o(1 - \sigma)J_v$. Clearly, when $\sigma = 1$ there is no convection, and when $\sigma = 0$ the solute flux will not be restricted to any degree, an interpretation consistent with the development of the concept of the reflection coefficient, but one that must be kept carefully in proper context.

Diffusive and active solute flux. The diffusive and active components of J_s are also interpreted functionally, that is, as the sum of unknown processes operating somewhere between the exterior of the root and the xylem. J_s^* can be further expanded both by the usual Michaelis–Menten kinetics for isothermal conditions and by enlarging that treatment to span a range of temperatures, because temperature is an important environmental variable affecting not only J_s^* but also L_p (Kramer 1942, 1948; Kuiper 1964; Markhart et al. 1979). However, these enhancements to the J_s^* term are beyond the scope and purpose of this chapter.

The coefficient of solute mobility in the membrane, ω, is in the nature of a diffusion coefficient. As such, it, too, is expected to vary with temperature, both because of the Q_{10} for diffusion and because of temperature-induced alterations in the fluidity of the cell membranes (Markhart et al. 1979; Markhart 1982).

Combined volume and solute flux

Substituting the modified equation (9.3) into equation (9.2) leads to the result (Fiscus 1977)

$$J_v = L_p\left[\Delta P + \frac{\sigma RT(J_s^* - \sigma C^o J_v)}{J_v + \omega RT}\right] \tag{9.5}$$

For values of L_p, σ, J_s^*, and ω determined experimentally (Fiscus 1977), a typical flow curve for a *Phaseolus* root system looks like Figure 9.1. Equation (9.5), from which Figure 9.1 is calculated, predicts several important features of this force–flux relationship. First, the overall flux curve will be nonlinear initially but will approach a limiting slope at high flow rates. This type of relationship has been demonstrated repeatedly for *Phaseolus* (Kuiper and Kuiper 1974; Fiscus 1975, 1977, 1981a), *Glycine* (Newman et al. 1973; Markhart et al. 1979), *Lycopersicon* (Mees and Weatherley 1957; Lopushinsky 1961, 1964), *Zea* and *Helianthus* (Newman et al. 1973; Boyer 1974), *Brassica* (Markhart et al. 1979), and *Pinus* (Sands et al. 1982). The limit of the slope is L_p, actually $1/L_p$ in the case of Figure 9.1. This limit will be discussed more fully later under the heading of the differential resistance. The second prediction is that the intercept of the curve at $\Delta P = 0$ is largely determined by J_s^* and is relatively insensitive to changes in L_p. A third and very important prediction is that extrapolation of the linear portion of the curve back to the ordinate should result in an intercept equal to $\sigma^2\Pi^o$. That this does not usually happen was pointed out (Newman 1976) and forms the basis of the only serious criticism of the model.

Intermediate osmotic compartment. We attempted to address Newman's criticism and decided that there were two possibilities that had to be considered: (1) an external boundary-layer buildup that would increase Π at the root surface above the value in the bulk solution and (2) an intermediate compartment, although not with the characteristics he then proposed, acting in opposition to the flow. We concluded that the former explanation was unsatisfactory because the linear portion of the J_v–ΔP

curve typically continued linearly up to at least 0.7 MPa (unpublished data). If boundary-layer buildup were a problem, we would expect to reach a point of diminishing slope long before that level of force (or flux: $J_v = 20 \times 10^{-8}$ m s^{-1}). The most likely explanation, at this time, involves an intermediate compartment, some of the characteristics of which we discovered while searching for explanations for certain growth regulator effects. We were able to show that a pool of solutes existed in *Phaseolus* roots that could be mobilized by abscisic acid (ABA) treatment (Fiscus 1981a). There appeared, in fact, to be two pools of sequestered solutes (Fiscus 1983), one of which, possibly the cytoplasm, had an ABA mobilization threshold at or below 10^{-10} mol ABA cm^{-2} root surface, and the other, possibly the vacuole, with a threshold of 10^{-7} mol ABA cm^{-2}. The details of these pools are not essential now, but the important feature observed was that when the ABA treatments occurred and the pools of solutes were mobilized, there was a simultaneous increase in volume flux in proportion to the amount by which the expected intercept (just discussed) and the measured intercept differed. Thus, it appeared that the low-threshold pool was acting in opposition to the pressure-induced flow, and when that pool was mobilized, the entire $J_v - \Delta P$ curve was shifted by an amount that made it conform to Newman's intercept test. It appears, then, that there is an intermediate compartment between the root surface and the xylem containing normally nonmobile solutes that are asymmetrically distributed toward the interior of the root. This pool of solutes must be of a relatively stable size, because the intercept does not appear to shift much during the course of an experiment in which ABA is not involved. The implications of this for the model are quite simply that equation (9.5) will require that the bulk external osmotic pressure be replaced by an effective pressure (Fiscus 1977). Also, bearing in mind our discussion of the reflection coefficient, we shall need to use the actual bulk osmotic pressure when calculating that parameter from the limiting internal concentration as J_v approaches infinity (Fiscus 1977). Making these interpretational adjustments to equation (9.5) brings it very closely into line with experiment.

High fluxes and steady states. Finally, there are two very important points relating to the experimental determination of the parameters in equation (9.5): Almost all parameters need to be determined at high fluxes (Fiscus 1975, 1977, 1983; Markhart et al. 1979) and under steady-state conditions. It is only at high fluxes that we can expect to minimize artifacts that are largely due to our inability to determine concentrations deep within the root. Also, it is only at the steady states of volume and solute fluxes that we can minimize other artifacts, particularly hysteresis effects (Kuiper and Kuiper 1974; Fiscus 1977) that are due to loading and unloading of solutes from the tissues adjacent to the conducting pathway. Because the exchange of ions between these tissues is slow, it may take several hours and a total volume turnover of 10 times (Markhart 1982; Fiscus 1983) to bring a root system to the steady state with respect to both fluxes.

Comparison of L_p values. Another question that is frequently raised concerning the model is how our solution-grown root systems compare with those grown in soil. Certainly they differ in appearance, the solution-grown ones being somewhat thinner and of a more uniform diameter. Functionally, however, as nearly as we can tell, with regard to their water conduction characteristics they act very much the same. The evidence for this is currently derived by comparison with the literature and is indirect. Conductance values calculated for soil-grown plants from data in the literature reveal that *Phaseolus, Glycine, Helianthus, Zea,* and *Gossypium* all fall within the same range: 0.8×10^{-7} to 6.1×10^{-7} $m^3\ m^{-2}\ s^{-1}\ MPa^{-1}$ (Fiscus 1983). This is the entire range of values previously observed (Fiscus and Markhart 1979) for solution-grown *Phaseolus.* Another opportunity for comparison comes from a recent article (Fiscus et al. 1983) in which we estimated L_p in a model system designed to explain some peculiarities of whole-plant water transport reported by many authors. We estimated L_p as 4.3×10^{-7} $m^3\ m^{-2}\ s^{-1}\ MPa^{-1}$, a value that with minor refinements will provide a good fit to the whole-plant data (Boyer 1974) on which the estimates are based.

In addition to these comparisons there are theoretical reasons for believing that L_p values determined in a root-system pressure chamber are more realistic than those determined by osmotic methods or at low flux rates. The main reason for this is that the effects of both the intermediate osmotic compartment and any standing gradients in the system may be minimized at high fluxes. The real problem that we are trying to overcome at high fluxes is our ignorance of the axial distribution of the major parameters of the model. However, the experimental approach we use allows us to conclude that the values of L_p measured in the pressure chamber are accurate and relevant to whole-plant water flow models.

Differential resistance

Given that we can determine reliable values of L_p and can show how these values vary as the plant grows (Fiscus and Markhart 1979), we still have the problem of relating L_p to whole-plant water flow or whole-plant or even organ resistance or conductance. We would also like to relate the various plant organ conductances and resistances to each other and to an overall plant conductance or resistance.

In a recent article, we discussed the concepts of instantaneous and differential resistances and conductances (Fiscus 1983) and concluded that the instantaneous concept, based on the Ohm's law analogy, was not of very great utility in dealing with plant water flow. These concepts differ in that the instantaneous resistance (based on a misinterpretation of Ohm's law) is simply the ratio of the force to the corresponding flux, whereas the differential resistance is the slope of the force–flux curve. For a system in which the force–flux relationship is linear and passes through the origin, these two concepts are interchangeable. However, many plant transport systems are not linear in nature, and their force–flux relationships do not pass through the origin. Figure 9.1 provides a useful example of how these concepts compare. The three curves on the graph are the volume flux, J_v, the differential resistance, R_d, and the instantaneous resistance, R_i. It is clear from examining the two resistance curves that experimental interpretation can vary considerably depending on which concept is used. In the one case (R_i), the resistance is seen to increase with the flux to a peak, after which point it decreases, slowly approaching the actual slope of the J_v curve as a limit. Because we have some intuitive notion about the meaning of the word "resistance," we might spend considerable time and effort searching for the cause of this outrageous behavior. In the other case (R_d), however, the resistance is seen to decrease continuously with increasing flow. In this instance, however, we can immediately discover the cause of the resistance change by examining the transport function and its derivative. For example, solving equation (9.5) for ΔP gives us

$$\Delta P = \frac{J_v}{L_p} - \frac{\sigma RT(J_s^* - \sigma C^o J_v)}{J_v + \omega RT} \tag{9.6}$$

which describes the force–flux curve in Figure 9.1. The slope of this curve, the differential resistance, is

$$R_d = \frac{d\Delta P}{dJ_v} = \frac{1}{L_p} + \frac{\sigma RT(J_s^* + \sigma\omega\Pi^o)}{(J_v + \omega RT)^2} \tag{9.7}$$

which is the curve labeled R_d inFigure 9.1. Immediately from equation (9.7) we can see that the hydraulic conductance coefficient L_p is only part of the resistance term, which is seen to decrease toward $1/L_p$ as a limit while the volume flux increases.

At this point it may be useful to recap the various uses of the terms "resistance," "resistivity," "conductance," and "conductivity." We have described briefly the instantaneous and differential resistances, and by the same reasoning there must also be differential and instantaneous conductances. Also, there is the hydraulic conductance coefficient L_p, which when applied to root systems may be modified, depending on which root dimension we want to relate to the water flux. Any of the dimensions may be useful under different conditions, and they are equally correct as long as they are clearly defined. Thus, we can with equal validity have a flow based on unit area, unit length, or even unit volume if we wish. The real problem arises when the concepts are mixed indiscriminantly, such as measuring the instantaneous resistance and then equating it with $1/L_p$, however defined.

Also from equation (9.7) we can draw some inferences about how R_d might vary with changes in all the parameters of the equation. The equation does in fact suggest which parameters might be most profitably manipulated to achieve some desired end. For instance, in the past it was common practice to use natural root-pressure exudation rates ($\Delta P = 0$) to determine L_p values. Notwithstanding the standing gradient effects, it is easy to show (Fiscus 1975, 1977, 1981a) that the volume flux rates in such a system are determined almost entirely by J_s^* and should be very insensitive to changes in L_p over a wide range.

Without belaboring the point any further, we can conclude this section by stating that in addition to providing us with experimentally testable hypotheses, the differential resistance concept also provides a common link that joins transport processes throughout the whole plant (Fiscus et al. 1983) and allows a unified physical treatment of diffusive, convective, and active transport processes in the liquid and vapor phases of the entire system.

Having defined the various kinds of conductances and resistances and briefly indicated their use, we shall now make some estimates concerning the cost to the plant of producing these facilities for water uptake and transport.

Costs

In earlier work we found that the *Phaseolus* root system could be divided into four very distinctive size classes according to external diameter (Fiscus 1981b). The variability in diameters was such that there was practically no overlap between the size classes, and it became a relatively easy, though tedious, task to measure the contribution of each class to the total root surface area. We also found that once the plants had reached a size of about 0.1 m^2 total root area, the proportions of the various size classes remained relatively constant. Because of this constancy, we were able to use average values for plants larger than 0.1 m^2 for many of our calculations and for some purposes will continue to do so here.

Growth model

Growth of the projected leaf area, the area of each root size class, and the total root area can be described by an exponential sigmoid function of the form

$$A = \frac{a}{1 + be^{nt}} \quad (m^2) \tag{9.8}$$

where t is the time in days, a is the maximum area, and b and n are growth coefficients given in Table 9.1. Although we used a simple power function in previous work, it is not generally considered a good growth model because it does not account for maturity and senescence effects. So we replace the power function with equation (9.8) and fit our previous data (Fiscus 1981b) to that function.

Although the root surface area is the geometric parameter we think most relevant to determining rates of water absorption, accounting for the energy necessary to produce that surface area requires some relationship between rates of dry-matter and surface-area accumulation. We can start with the fundamental relationships

$$M = \rho V \quad \text{and} \quad V = \frac{Ad}{4} \tag{9.9}$$

where M is the dry matter in g, ρ is the dry weight (DW) density in g dry matter [cm^3 fresh volume]$^{-1}$, V is the volume in cm^3, A is the surface area in cm^2, and d is the mean diameter for each root class or for the whole root system in cm. Values for ρ and d are taken from previous work (Fiscus 1981b). Combining relationships (9.8) and (9.9), we get for M, in g, as a function of time,

$$M = \frac{\rho d}{4} A = \frac{\rho d a}{4(1 + be^{nt})} \tag{9.10}$$

and for the rate of dry-matter accumulation,

$$\frac{dM}{dt} = -\frac{\rho dabne^{nt}}{4(1 + be^{nt})^2} \tag{9.11}$$

in g DW day^{-1}.

From the relationships (9.9), simple geometry allows us to form the area/dry-matter ratio

$$A/M = 4/\rho d \tag{9.12}$$

Table 9.1. *Geometric dimensions and energy and mass relationships for solution-grown* Phaseolus *root systems*[a]

Size class	Diameter (cm)	A/M ($cm^2\ g^{-1}$)	% of class 1	A/E_b ($cm^2\ kJ^{-1}$)	a	b	n	r^2
1	0.0245	1458	100	91.1	0.560	130	−0.15	0.98
2	0.0540	661	45.3	41.3	0.155[b]	80	−0.13	0.89
3	0.0831	430	27.5	26.9	0.108[b]	30	−0.10	0.92
4	0.1159	308	21.1	19.2	0.128	20	−0.07	0.61
Total	0.0311	1148	78.7	71.7	0.90	85	−0.13	0.97
Leaf area (m^2)					0.79	125	−0.15	0.96

[a] The average energy content of the dry matter is 16.01 kJ [g DW]$^{-1}$. Note that the coefficients a, b, and n give the areas in m^2.
[b] Coefficients b and n were fitted for a value of a picked to be consistent with the size class proportions discovered earlier (Fiscus and Markhart 1979).

which allows us easily to see that the smallest size class makes by far the most efficient use of dry matter in the production of surface area. The figures for each size class and the total are summarized in Table 9.1, where for purposes of comparison the A/M ratios are given as percentages of class 1 roots.

Required energy. To obtain an estimate of the total energy required to build and maintain the root water transport capacity for the plant, we must consider three factors: the energy content of the building blocks (E_b), the energy required to assemble them into cellular structures and to drive cell expansion (growth respiration, R_g), and the energy required to maintain the structures (maintenance respiration, R_m). Several assumptions and extrapolations are necessary to sort out these various components of energy input. As a starting point we shall assume a constant rate of maintenance respiration equal to 319 J [g DW]$^{-1}$ day^{-1} (Penning de Vries 1975a) (equal to 20 mg glucose [g DW]$^{-1}$ day^{-1}). Given R_m and assuming that R_g (J [g DW]$^{-1}$ day^{-1}) is also constant, we can estimate a value for R_g from Figure 9.2. This figure is the oxygen consumption rate for an entire *Phaseolus* root system measured diurnally at approximate hourly intervals. An estimate of the integral over 24 hr divided by time provides us with an average for the total respiration rate for this period, assuming a respiratory quotient of 1. Calculations based on this figure are given in Table 9.2, where the most significant number for our purposes is R_g.

Now we shall turn our attention to the energy content of the assembled structures. To do so requires some assumptions about the proportions of

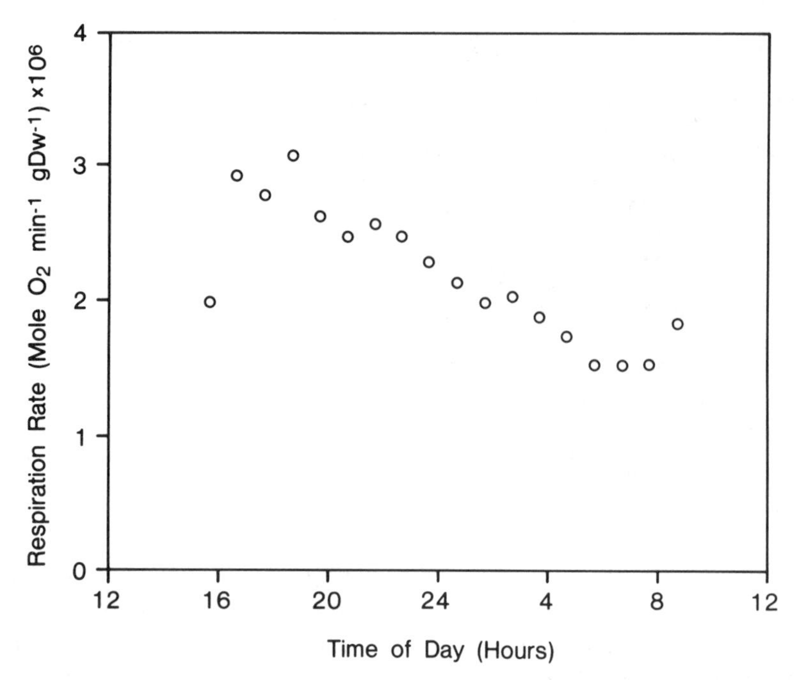

Figure 9.2. Diurnal changes in respiration rate for an entire *Phaseolus* root system. The plant was 18 days old, with a leaf area of 0.14 m^2 at the beginning of the experiment.

proteins, lipids, and carbohydrates in the tissues. Because most plant vegetative tissues are composed of 1% to 5% nitrogen (Bonner and Galston 1952), we assume a value of 3%, and that this is mostly protein, which on average is 16% N. This gives us a protein content of about 19% of the dry weight. Lipids generally compose 5% or less of the dry weight of vegetative organs (Bonner and Galston 1952), and so we also choose an average value of 3% for that component. Mineral ions may compose a further 5% of the dry weight but will be considered to be an integral part of the system and as such not possessed of structural energy in the same sense as we are considering the other components. The remainder of the dry weight (73%) is assumed to be carbohydrate.

Table 9.3 summarizes calculations of the average energy content of the dry matter based on standard bomb-calorimeter combustion values for lipid, protein, and carbohydrate (White et al. 1959) and the foregoing assumptions regarding composition.

Because part of the growth energy finds its way into the structure of the system in the form of chemical bonds, linking carbohydrate and protein subunits, for example, we must somehow account for the relationship

Table 9.2. *Respiratory components and other relevant data for the* Phaseolus *root system used in Figure 9.1*

		Total glucose equivalents (mg)	Energy
Total consumed	2.61×10^{-3} mol O_2	78.4	1,249 J
Dry weight	0.88 g		
dM/dt	0.109 g day^{-1}		
Respiration (maintenance) (R_m)	20 mg glucose [g DW]$^{-1}$ day^{-1}	17.6	319 J [g DW]$^{-1}$ day^{-1}
Respiration (growth) (R_g)	558 mg glucose [g DW]$^{-1}$	60.8	8,888 J [g DW]$^{-1}$

between the energy contents of the building blocks and the final structure and the respiratory energy necessary to assemble the components. To this end we can start by expressing the total energy of the system (E_s) as

$$E_s = E_b + E_a \tag{9.13}$$

where E_a is the energy incorporated into the structure during assembly of the building blocks, and E_b is the energy content of the building blocks.

As a first attempt to separate these components we shall examine the situation with regard to the carbohydrates, because they constitute the largest percentage of the dry matter, and specifically we shall examine cellulose, because that is frequently the largest carbohydrate component present.

Taking the combustion value for glucose as 15.69 kJ g^{-1} and that of cellulose as 17.58 kJ g^{-1} (Crampton and Lloyd 1959), we can see an increase in energy of the cellulose over the glucose building blocks of 1.9 kJ g^{-1} or about 12%. We shall now proceed under the assumption that a similar figure applies to assembly of the lipid and protein components. This may not be an especially good assumption, but the proportions of lipid and protein are relatively small compared with carbohydrate, and so the error may be minor.

Although the 1.9 kJ g^{-1} just calculated for E_a is energy we may measure as part of E_s, in order to separate it from the respiratory energy expenditures from whence it came we must subtract it from E_s and deal only with E_b when further considering rates of energy incorporation into dry matter. That is, E_a is accounted as part of R_g rather than as part of E_s. Leaving E_a as a part of E_s, for accounting purposes, would require that we know the efficiency of incorporation of respiratory energy into dry matter. Al-

Table 9.3. *Assumed composition and energy content for typical vegetative tissues*

Component	Fraction	Combustion value (kJ g^{-1})	Tissue energy content (kJ [g DW]$^{-1}$)
Protein	0.19	22.18	4.21
Lipids	0.03	38.92	1.17
Minerals	0.05		
Carbohydrates	0.73	17.16	12.53
Total	1.00		(E_s) 17.91

though the latter approach to the problem is just as viable, we choose the former as being more convenient at this time.

Carrying out the calculation indicated by equation (9.13) gives us an energy value of 16.01 kJ [g DW]$^{-1}$ for the building blocks (E_b). It is of interest to note that this value is within 2% of the energy value for glucose and within 2% of the average energy value for plant dry matter cited by Gates (1980). This fact may or may not be strictly fortuitous, but in any case this was the number used to calculate A/E_b in Table 9.1.

Direct PAR. Finally, we shall compare the energy required for the establishment and functioning of a root system to the total amount of radiant energy, in the form of direct PAR, available to (incident on) the plant. For simplicity we shall consider only the direct radiation and leave as an exercise for the reader the more refined calculations.

Measurements of the projected canopy area provide us with a good measure of the leaf area exposed to direct radiation. Figure 9.3 is a plot of the relationship between the shadow area of the canopy and the total projected leaf area for a small population of plants covering the range of sizes of interest to us. The linear correlation is quite good, especially for plants larger than 0.1 m^2, so that the area exposed to direct radiation at any time can be calculated by combining the equation for Figure 9.3 and equation (9.8) to yield

$$A_{lD} = k + \frac{ma}{1 + be^{nt}} \qquad (9.14)$$

where A_{lD} is the shadow area (leaf area exposed to direct radiation), k and m are the intercept and slope, respectively, in Figure 9.3, and a and b are as before.

The daily average integral of the light intensity for the relevant green-

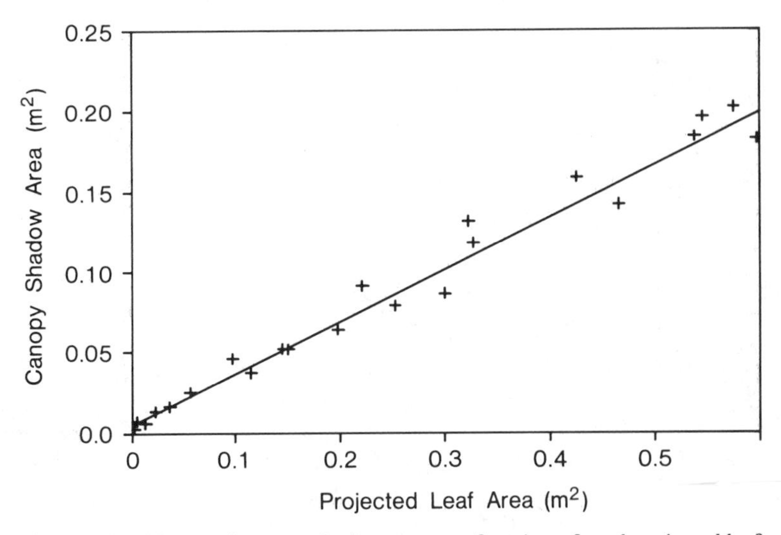

Figure 9.3. Measured canopy shadow area as a function of total projected leaf area used to calculate the direct radiation. $A_{lD} = k + mA_l$; $k = 4.603 \times 10^{-3}$; $m = 3.319 \times 10^{-1}$; A_l is given by equation (9.8).

house growth conditions was 12.4 ± 1.5 mol m^{-2}, about 20% of the outdoor summertime maximum for this location. Using the conversion factor of 4.6 μmol m^{-2} s^{-1} of PAR = 1 watt m^{-2} (Biggs and Hansen 1979) yields an average daily energy incident on the leaves of 2.7 MJ m^{-2}. The total energy incident on the plant in the form of direct PAR in MJ day^{-1} is the product of this figure and the leaf area exposed to direct radiation:

$$\frac{dE_i}{dt} = A_{lD}(2.7 \text{ MJ m}^{-2} \text{ day}^{-1}) \tag{9.15}$$

Efficiency of energy use. The efficiency with which direct radiation is used to form a root system can be defined as the ratio of the rate of energy incident on the leaves (dE_i/dt) to the rate of total energy usage by the root system (dE_T/dt). In this sense, E_T is the total of the energy of the building blocks used to make the system (E_b), the growth energy required to assemble those blocks into roots (R_g), and the energy necessary to maintain the existing system (R_m). We can therefore write for the daily rate of energy use

$$\frac{dE_T}{dt} = E_b\frac{dM}{dt} + R_g\frac{dM}{dt} + R_mM \tag{9.16}$$

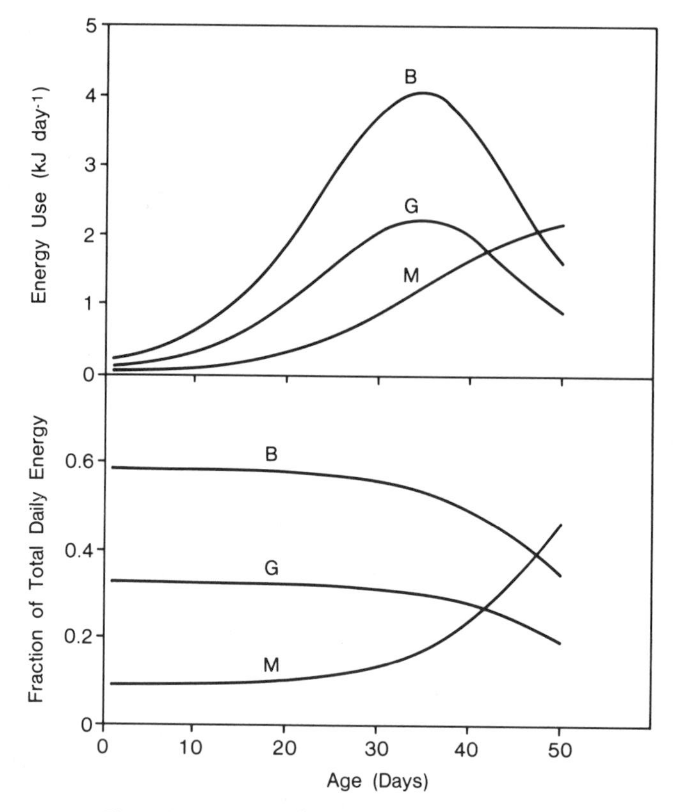

Figure 9.4. Proportions of the energy content and growth and maintenance energy expenditures as functions of plant age as calculated from equation (9.16). B: $E_b\ (dM/dt)$. G: $R_g\ (dM/dt)$. M: R_mM.

Rearrangement and substitution of equation (9.10) for M and equation (9.11) for dM/dt gives the total daily rate of energy consumption by the root system as

$$\frac{dE_T}{dt} = -\frac{\rho dabne^{nt}}{4(1+be^{nt})^2}\left[E_b + R_g - \frac{R_m(1+be^{nt})}{bne^{nt}}\right] \tag{9.17}$$

where R_g and R_m are the growth and maintenance respiratory coefficients given in Table 9.2.

The three components of equation (9.16) were calculated and plotted in Figure 9.4. The first term, dE_b/dt, is the rate of incorporation of new material into the roots and is the product of the rate of dry-matter increase and the average energy content of that dry matter (Table 9.1). Term two is the energy required to assemble the dry matter into root tissue and con-

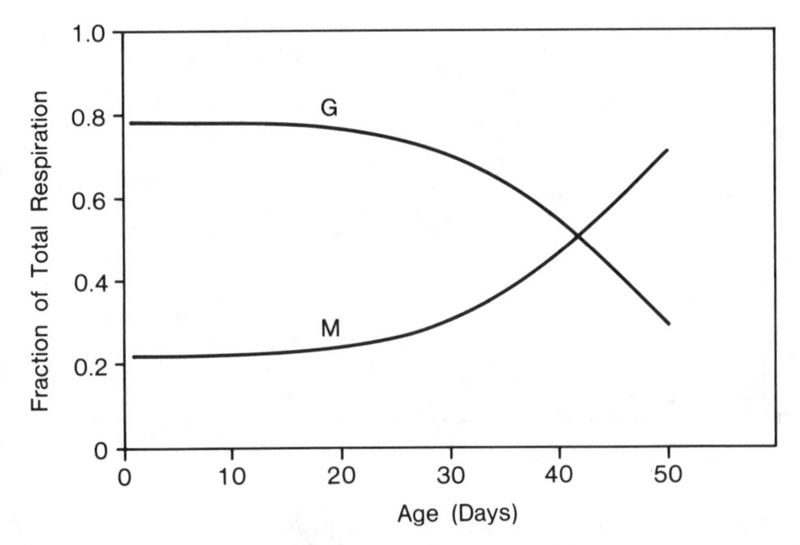

Figure 9.5. Partitioning of total respiratory energy use between growth (G) and maintenance (M).

tains the growth respiratory coefficient and the rate of increase of dry matter. Finally, the third term is the energy required to maintain the existing system and is the product of the maintenance respiratory coefficient and the quantity of dry matter present at any time.

As expected, the energies of the building blocks and the growth component decline as the plant approaches maturity. Maintenance energy, however, continues to increase, approaching a plateau of about 2.5 kJ day^{-1}. As a proportion of the total energy use by the roots, maintenance goes from about 10% during early growth to about 50% at day 50.

The calculated growth and maintenance components are broken out for comparison with each other in Figure 9.5.

Now we can form the efficiency ratio (dE_T/dE_i) simply by dividing equation (9.17) by equation (9.15). Figure 9.6 is a plot of both the components and the efficiency, which is seen to range from about 0.5% to a little over 3%, with an arithmetic mean of 2.2 ± 0.8%.

Cost of conductance

The final point we are now able to address is the question of the energy cost per unit of total root-system conductance. Because the differential conductance (Fiscus 1983) varies with the flow rate, we shall examine only the average hydraulic conductance coefficient (L_R). L_R is the

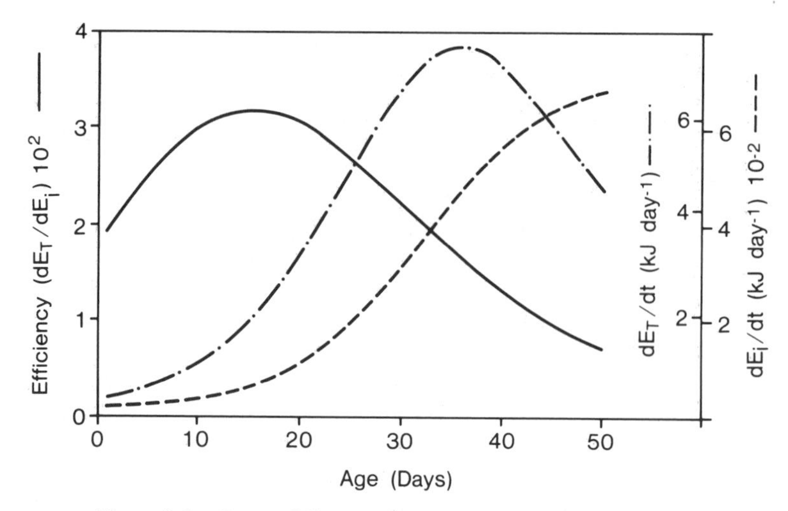

Figure 9.6. Rates of direct-radiation input (dE_i/dt) and total energy usage by the root system (dE_T/dt) and the ratio of the two (efficiency).

product of the hydraulic conductance (L_p) and the root surface area. The units of L_R are $m^3\ s^{-1}\ MPa^{-1}$, and the curve in Figure 9.7 was constructed from previous data (Fiscus and Markhart 1979).

The conductance of the root system changes with time, but at any particular time its value is related to the energy expended to build and maintain the system up until that time. For this reason we have chosen the integral of the daily rate of energy usage (dE_T/dt) as the appropriate figure to compare with L_R. The integral can be formed from equation (9.17) and is

$$E_T = \frac{\rho da}{4(1 + be^{nt})}\left[E_b + R_g + R_m(1 + be^{nt}) \cdot \left(t - \frac{\ln(1 + be^{nt})}{n}\right)\right] + C \qquad \text{(kJ)} \qquad (9.18)$$

where C, the integration constant, is -87.93 kJ. For curiosity's sake, the direct-radiation integral can be formed from equation (9.15),

$$E_i = 2{,}700\left[k + ma\left(t - \frac{\ln(1 + be^{nt})}{n}\right)\right] + C \qquad \text{(kJ)} \qquad (9.19)$$

where C is -22.8 MJ.

The ratios of equations (9.18) and (9.19) to L_R were formed and plotted in Figure 9.8 as functions of plant age. Both functions are seen to cycle, starting out high, because the early conductance is so low, reaching a

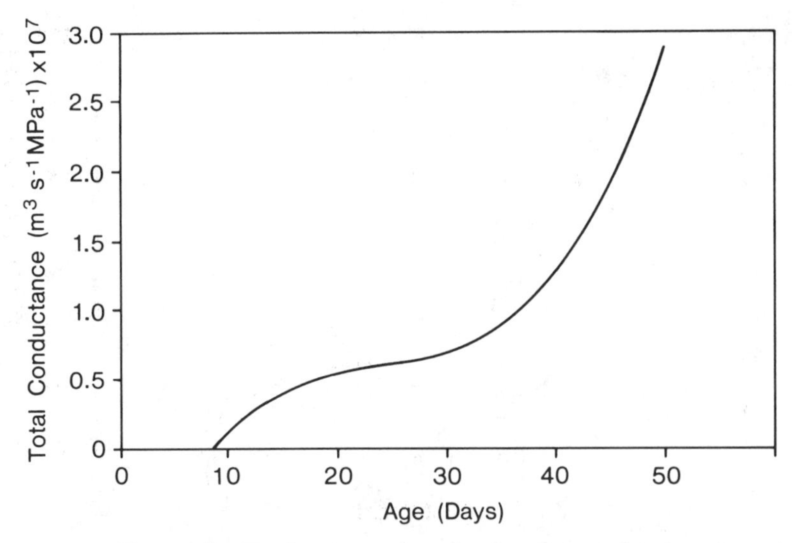

Figure 9.7. Total root-system conductance (L_R) as a function of age. Curve was calculated as $L_R = a + bt + ct^2 + dt^3$; $a = -1.1941 \times 10^{-7}$; $b = 0.1987 \times 10^{-8}$; $c = -7.77 \times 10^{-10}$; $d = 1.0869 \times 10^{-11}$.

minimum and then peaking again. The final roll-off in E_T/L_R results from both the rapid increase in L_R as the plants mature and the decline in the rate of total energy use during the same period. The ratio E_i/L_R follows a similar cycle for substantially the same reasons.

The ratios of the individual integrals to L_R (Figure 9.9) for maintenance, growth, and building materials also cycle in the same way, except that E_M/L_R is somewhat out of phase with the others. This is because the maintenance energy expenditures remain elevated even as the others are falling dramatically. The average values for the integrals in Figures 9.8 and 9.9 are given in Table 9.4 in several different combinations of units that might ease comparisons with the previous literature.

Final comments

The purpose of this somewhat lengthy but simplistic analysis has been to provide some first approximations of the cost to the plant, in terms of energy expenditures, of furnishing the structures necessary to absorb water from the soil. As we have seen repeatedly in this chapter, the energy actually used by the root system and to build the root system is generally only a small fraction of the total direct radiation falling on the leaf system.

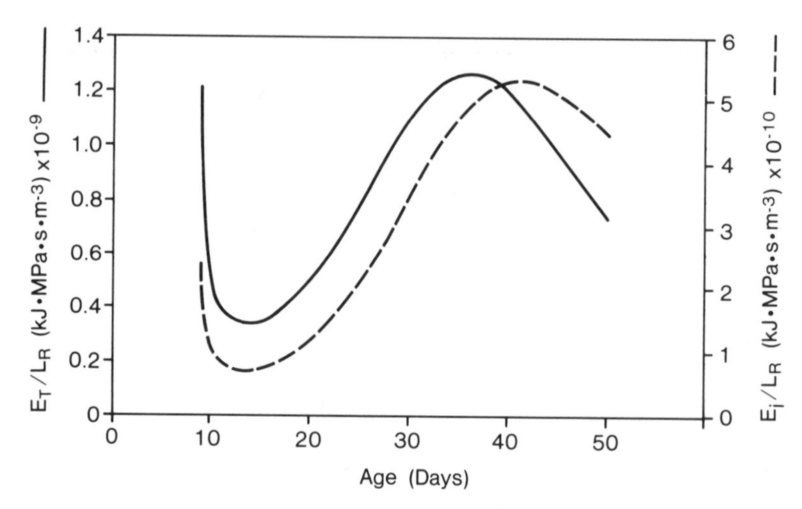

Figure 9.8. Cumulative energy input per unit of conductance. Energy is given both as the total used (E_T) and the total direct radiation (E_i). Means and standard deviations are $E_T/L_R = 8.5 \times 10^8 \pm 3.3 \times 10^8$ and $E_i/L_R = 3.17 \times 10^{10} \pm 178 \times 10^{10}$.

Even if we include the aboveground portions, which in this case constitute the other three-fourths of the plant dry matter (fractional dry weight of roots : stems : leaves = 0.222 : 0.226 : 0.552), the total efficiency, as defined earlier, will still average only about 9%. If we accept the assumption that leaves not receiving direct radiation produce adequate photosynthate to meet the demands of maintenance respiration (Tanner and Sinclair 1983), then the efficiency of use of direct radiation will fall even more.

Our assumptions about root composition may be erroneous, but it seems unlikely that the values will be too far wrong. Direct comparisons are difficult, but previously published compositions of leaf material (Penning de Vries 1975b) give us cause for optimism. The composition given is 25% nitrogenous compounds, 66.5% carbohydrates, 2.5% lipids, and 6% for minerals and lignin. The respective estimates for the root composition given in Table 9.3 are 19%, 73%, 3%, and 5% for minerals. Later Kjeldahl analysis on plants grown under similar conditions showed 3.04%, 2.50%, and 3.96% Kjeldahl N in the roots, stems, and leaves, respectively. These values work out to 19%, 15.6%, and 24.8% protein. The 19% value for roots is exactly what we estimated, and the 24.8% for the leaves is very close to the value given by Penning de Vries. We think that the nearly 6.5% difference between his value for leaf carbohydrate and our estimate for root carbohydrate is very nearly balanced by the almost 6% difference between the root and leaf percentage protein compositions shown by

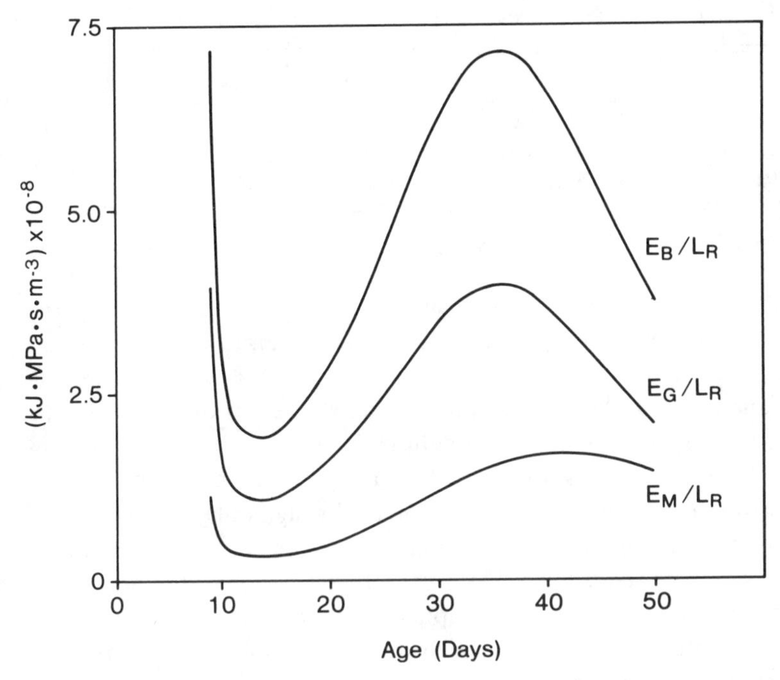

Figure 9.9. Cumulative energy input per unit of conductance broken down into construction materials (E_B), growth energy (E_G), and maintenance energy (E_M). Means and standard deviations are $E_B/L_R = 4.76 \times 10^8 \pm 1.84 \times 10^8$; $E_G/L_R = 2.64 \times 10^8 \pm 1.02 \times 10^8$; $E_M/L_R = 1.08 \times 10^8 \pm 0.53 \times 10^8$.

Kjeldahl analysis. We therefore believe that our estimates of composition, no matter how crudely derived, are reasonable.

Our assumption about the value of R_m may also prove erroneous, but it and the resulting value of R_g determined from experiment are not out of line with those of other workers (Evans 1975). It is more likely, however, that our assumption of their constancy will prove false.

Water and nutrient absorption and transport are inextricable. However, one might just as easily choose to view the root system as though its main function were nutrient uptake and everything else merely a consequence of that function. The energy actually expended to extract nutrients from the soil is included in R_m (Penning de Vries 1975a) and constitutes about 15% of R_m in the case of our plants. Therefore, nutrient uptake constitutes only 1–2% of the total energy used by the root system until growth begins to decline (Figure 9.4). Of course, the apparatus of extraction must exist, and so all the growth-related expenditures should be included in the cost of extracting nutrients from the soil.

Table 9.4. *Average integrals from Figures 9.8 and 9.9 given in various units*[a]

Units	E_T/L_R	E_B/L_R	E_G/L_R	E_M/L_R
kJ MPa s m^{-3}	8.5×10^8	4.76×10^8	2.64×10^8	1.08×10^8
mg glucose MPa s cm^{-3}	5.42×10^4	3.03×10^4	1.68×10^4	6.88×10^3
g DW MPa s cm^{-3}	53.1	29.7	16.5	6.7

[a] Values used for conversions were 15.69 kJ g^{-1} glucose and 16.01 kJ g^{-1} DW.

Considering the quantities of plant growth regulators formed in the roots, the energy expenditures specifically involved in those activities are probably negligible.

The function of anchorage is a much more slippery proposition to analyze. In some respects, anchorage might be viewed as a secondary benefit deriving from the plant's water- and nutrient-extracting activities. Secondary growth necessary for support and anchorage of larger plants may simply be the consequence of generating a water supply system adequate to meet the needs of the foliage. However, the existence of the additional root functions of support, nutrient uptake, and hormone secretion means that our estimate of the direct cost of the hydraulic conductance may be an upper limit for the actual cost, because the costs of the other functions have been included.

In short, no matter how we choose to view the root system functionally, the cost of building and maintaining it will be the same.

References

Biggs, W. W., and M. C. Hansen. 1979. Radiation measurement. Pp. 45–46 *in* Instrumentation for biological and environmental sciences (product catalog). Licor, Inc., Lincoln, Nebr.

Bonner, J., and A. W. Galston. 1952. Principles of plant physiology. W. H. Freeman, San Francisco. Pp. 274–286.

Boyer, J. S. 1974. Water transport in plants: mechanism of apparent changes in resistance during absorption. Planta 117:187–207.

Crampton, E. W., and L. E. Lloyd. 1959. Fundamentals of nutrition. W. H. Freeman, San Francisco.

Dalton, F. N., P. A. C. Raats, and W. R. Gardner. 1975. Simultaneous uptake of water and solutes by plants. Agron. J. 67:334–339.

Evans, L. T. 1975. The physiological basis of crop yield. Pp. 327–355 *in* L. T. Evans (ed.), Crop physiology. Cambridge University Press.

Fiscus, E. L. 1975. The interaction between osmotic- and pressure-induced water flow in plant roots. Plant Physiol. 55:917–922.

- 1977. Determination of hydraulic and osmotic properties of soybean root system. Plant Physiol. 59:1013–1020.

- 1981a. Effects of abscisic acid on the hydraulic conductance of and the total ion transport through *Phaseolus* root systems. Plant Physiol. 68:169–174.

- 1981b. Analysis of the components of area growth of bean root systems. Crop Science 21:909–913.

- 1983.Water transport and balance within the plant: resistance to water flow in roots. Pp. 183–194 *in* H. M. Taylor, W. R. Jordan, and T. R. Sinclair (eds.), Limitations to efficient water use in crop production. ASA, CSSA, SSA, Inc., Madison.

Fiscus, E. L., and P. J. Kramer. 1975. General model for osmotic and pressure-induced flow in plant roots. Proc. Natl. Acad. Sci. USA 72:3114–3118.

Fiscus, E. L., A. Klute, and M. R. Kaufmann. 1983. An interpretation of some whole plant water transport phenomena. Plant Physiol. 71:810–817.

Fiscus, E. L., and A. H. Markhart. 1979. Relationships between root system water transport properties and plant size in *Phaseolus.* Plant Physiol. 64:770–773.

Gates, D. M. 1980. Biophysical ecology. Springer-Verlag, New York.

Katchalsky, A., and P. F. Curran. 1974. Nonequilibrium thermodynamics in biophysics. Harvard University Press, Cambridge, Mass.

Kramer, P. J. 1942. Species differences with respect to water absorption at low soil temperatures. Am. J. Bot. 29:828–832.

- 1948. Root resistance as a cause of decreased water absorption by plants at low temperatures. Plant Physiol. 15:63–79.

Kuiper, F., and P. J. C. Kuiper. 1974. Permeability and self-induction as factors in water transport through bean roots. Plant Physiol. 31:159–162.

Kuiper, P. J. C. 1964. Water uptake of higher plants as affected by root temperature. Meded. Landbouwhogesch. Wageningen 64:1–11.

Lopushinsky, W. 1961. Effect of water movement on salt movement through tomato roots. Nature 192:994–995.

- 1964. Effects of water movement on ion movement into the xylem of tomato roots. Plant Physiol. 39:494–501.

Markhart, A. H. 1982. Penetration of soybean root systems by abscisic acid isomers. Plant Physiol. 69:1350–1352.

Markhart, A. H., E. L. Fiscus, A. W. Naylor, and P. J. Kramer. 1979. Effect of temperature on water and ion transport in soybean and broccoli systems. Plant Physiol. 64:83–87.

Mees, G. C., and P. E. Weatherley. 1957. The mechanism of water absorption by roots. I. Preliminary studies on the effects of hydrostatic pressure gradients. Proc. R. Soc. London, Series B 147:367–380.

Newman, E. I. 1976. Interaction between osmotic and pressure induced water flow in plant roots. Plant Physiol. 57:738–739.

Newman, H. H., G. W. Thurtell, and K. R. Stevenson. 1973. In situ measurements of leaf water potential and resistance to water flow in corn, soybean, and sunflower at several transpiration rates. Can. J. Plant Sci. 54:175–184.

Penning de Vries, F. W. T. 1975a. The cost of maintenance processes in plant cells. Ann. Bot. 39:77–92.

– 1975b. Use of assimilates in higher plants. Pp. 461–475 *in* J. P. Cooper (ed.), Photosynthesis and productivity in different environments. Cambridge University Press.

Sands, R., E. L. Fiscus, and C. P. P. Reid. 1982. Hydraulic properties of pine and bean roots with varying degrees of suberization, vascular differentiation and mycorrhizal infection. Aust. J. Plant Physiol. 9:559–569.

Tanner, C. B., and T. R. Sinclair. 1983. Efficient water use in crop production: research or re-search? P. 13 *in* H. M. Taylor, W. R. Jordan and T. R. Sinclair (eds.), Limitations to efficient water use in crop production. ASA, CSSA, SSA, Inc., Madison.

White, A., P. Handler, E. L. Smith, and D. Stetten. 1959. Principles of biochemistry. McGraw-Hill, New York. P. 299.

10 Economy of symbiotic nitrogen fixation

JOHN S. PATE

Introduction

Effective absorption of the element nitrogen from the external environment is an obviously crucial factor in the performance and survival of plant species in many natural ecosystems, and any symbiotic association between a higher plant and a microorganism that possesses the capacity to fix atmospheric nitrogen gas directly gives that plant a clear potential advantage over other species that depend on the combined organic or inorganic resources of the rooting medium for their nitrogen nutrition. However, a number of conditions have to be met before the competitive advantage of the N_2-fixing component is likely to be realized. First, the productivity of the ecosystem should be limited primarily by the availability of nitrogen. Second, the symbiotic component should be capable of reserving exclusively for its own use the nitrogen it acquires from the atmosphere. Third, any extra requirements of the symbiotic association for other potentially limiting elements should not be so large that acquisition of these will limit the productivity of the said association. And finally, the symbiotic process per se should not place on the host plant such large extra demands for photosynthate as to jeopardize the growth and productivity of the association in comparison with that of counterparts assimilating combined nitrogen.

This chapter examines the bioenergetics of nitrogen fixation in symbioses involving higher plants, concentrating especially on experimental and theoretically based estimates of the likely costs of symbiotic activity at the biochemical, tissue, and whole-organ levels in the general context of whole-plant functioning. The information presented is drawn largely from the extensive literature on symbioses involving agriculturally important legumes, using as illustrative material the costing studies on legumes conducted by the author and his colleagues. This bias toward legumes is expressed also in those parts of the chapter presenting cost–benefit analyses of symbiotic nitrogen fixation in comparison with other forms of assimilation or acquisition of nitrogen.

Theoretical costs of the nitrogenase: hydrogenase functions of N_2-fixing organisms

The logical starting point for discussion of the bioenergetics of nitrogen fixation is to consider solely the biochemical reactions unique to the process, namely, reduction of nitrogen gas to ammonia and reduction of protons to gaseous hydrogen by the enzyme complex nitrogenase in the bacteroids or other microorganismal component of the symbiotic structure. Based on in vitro measurements of the ATP requirements for the functioning of isolated nitrogenase, a commonly accepted minimum cost is 2 mol ATP per electron involved in reduction of one or the other of the foregoing substrates (Phillips 1980; Schubert and Ryle 1980; Pate et al. 1981). Stoichiometric relationships between the primary reactions of nitrogenase and ATP consumption then read as follows:

$$N_2 + 6e^- + 8H^+ + 12ATP \rightarrow 2NH_4^+ + 12ADP + 12P$$
$$2H^+ + 2e^- + 4ATP \rightarrow H_2 + 4ADP + 4P$$

When extending considerations to include the cost of generation of electrons involved in these reduction sequences, it has been customary to assume that glucose is the respiratory substrate and that its complete oxidative phosphorylation will provide as an absolute maximum 38 mol ATP mol^{-1} (6.3 mol ATP mol^{-1} CO_2, with an associated $P/2e^-$ ratio of 3). On this basis, a total of 21 mol ATP or 0.55 mol glucose will be consumed in the reduction of 1 mol N_2, or 7 mol ATP and 0.18 mol glucose in the production of 1 mol H_2. The corresponding outputs of CO_2 will be 3.33 and 1.11 mol mol^{-1} N_2, respectively. It must be emphasized that these are strictly minimum cost estimates and relate exclusively to the operation of nitrogenase, not at all to cell or tissue costs outside the bacteroids. Also, there is now extensive evidence that the efficiency of coupling of glucose oxidation to ATP synthesis is well below the theoretical maximum, especially in bacteria. Moreover, in legume symbiosis, organic acids may be the primary substrates for bacteroid respiration, as suggested by information on the utilization of carbon sources by free-living *Rhizobium* or isolated bacteroids (Laing et al. 1979; Wilcockson and Werner 1979; Rawsthorne et al. 1980; Dixon et al. 1981).

Complications in relation to costing arise in those organisms in which a unidirectional uptake hydrogenase is operative, and the association thereby is capable of effecting energy savings through utilizing some or all of the H_2 evolved by the hydrogenase. This is evident, for example, in a wide range of legume symbioses (Dixon 1967, 1972, 1978; Carter et al. 1978; Lim 1978; Evans et al. 1979, 1980; Phillips 1980), including legumes under field conditions (Conrad and Seiler 1979). Actinorhizal nodules of *Alnus* also exhibit hydrogenase activity (Benson et al. 1979). The hydrogenase is considered to generate, at maximum, 3 mol ATP per 1 mol H_2 oxidized, so that, were all of the H_2 evolved by nitrogenase to be recycled, 43% of the energy initially consumed in H_2 production would be conserved.

Largely because of inherent difficulties in measuring nitrogenase and hydrogenase functionings simultaneously in vivo (Lim 1978; Maier et al. 1978), it still is not clear if the nitrogenase systems of different organisms differ basically from one another in the relative extents to which they allocate electrons to N_2 reduction as opposed to H^+ reduction. Nevertheless, substantial differences do exist between symbioses in the amounts of H_2 generated per 1 mol N_2 fixed, both between organisms and (possibly more important) within an organism or symbiotic association in response to environmental variables such as light or temperature (Bethlenfalvay and Phillips 1979; Evans et al. 1980; Rainbird et al. 1983). There is also the possibility that hydrogen evolved from the nitrogenase might build up to inhibitory levels within the nodule, in which case symbioses possessing hydrogenase might be advantaged in being able to ameliorate or eliminate such inhibition (Dixon et al. 1982).

Until the carbon metabolism and interrelationships of hydrogenase and nitrogenase are better understood in a variety of symbiotic associations, it is probably best to continue to rely on the previously mentioned set of purely hypothetical minimum cost values established in terms of ATP consumption, N_2 fixation, and H^+/H_2 metabolism. For example, the data of Table 10.1, drawn from an earlier study (Pate et al. 1981), provide a set of cost estimates for situations involving 20% to 60% allocation of electrons in nitrogenase to H_2 evolution, as opposed to N_2 fixation, compounded with situations in which 0% to 100% of the resulting H_2 is oxidized by uptake hydrogenase. In each of the cases considered, the ratio of moles of H_2 evolved to moles of N_2 fixed is recorded against the net cost of the nitrogenase : hydrogenase reactions. The latter costs are seen to range from 24.0 to 52.5 mol ATP mol^{-1} N_2 – the minimum value involving only 20% electron allocation to H_2 evolution, with full recovery of the H_2 by the uptake hydrogenase, and the maximum value (52.5 mol ATP mol^{-1} N_2 and 4.5 mol H_2 evolved mol^{-1} N_2 fixed) a 60% allocation of electrons to H_2 evolution and zero hydrogenase activity.

The estimates given in Table 10.1 show how steeply the expenditure of ATP would be expected to rise with increased percentage allocation of electrons to H_2 evolution, and how important economies through recycling of H_2 would become were the proportional electron flow to proton

Table 10.1. *Estimated costs of nitrogen fixation in legume nodules in relation to uptake hydrogenase activity and partitioning of electrons for proton and nitrogen reduction by nitrogenase*

Oxidation of evolved H_2 by uptake hydrogenase (%)	Electron flow of nitrogen to H_2 evolution as opposed to N_2 fixation (%)				
	20	30	40	50	60
0	26.3[a]	30.0	35.0	42.0	52.5
	0.75[b]	1.3	2.0	3.0	4.5
20	25.8	29.2	33.8	40.2	49.8
	0.60	1.0	1.6	2.4	3.6
40	25.4	28.5	32.6	38.4	47.1
	0.45	0.8	1.2	1.8	2.7
60	24.9	27.7	31.4	36.6	44.4
	0.30	0.5	0.8	1.2	1.8
80	24.5	27.0	30.2	34.8	41.2
	0.15	0.3	0.4	0.6	0.9
100	24.0	26.1	29.0	33.0	39.0
	0	0	0	0	0

Note: Operation of nitrogenase is costed as requiring inputs of the equivalent of 21 mol ATP mol^{-1} N_2 fixed and the equivalent of 7 mol ATP mol^{-1} H_2 evolved; uptake hydrogenase is assumed to generate 3 mol ATP mol^{-1} H_2 oxidized (data from Pate et al. 1981).
[a] Cost as mol ATP mol^{-1} N_2 fixed.
[b] Net mol H_2 evolved mol^{-1} N_2 fixed.

reduction to exceed 50%. The approach followed in Table 10.1 also allows one to predict the likely range of costs of nitrogenase : hydrogenase functioning that would apply for a specific ratio of H_2 evolved to N_2 fixed. For example, a ratio of 1 mol H_2 evolved per 1 mol N_2 fixed would carry a cost range of 29 to 42 mol ATP mol^{-1} N_2 fixed, a ratio of 2 : 1 from 35 to 45 ATP/N_2, and a ratio of 3 : 1 from 42 to 48 ATP/N_2.

The recent literature on symbiotic associations provides ample evidence of wide variations between different host : *Rhizobium* partnerships in H_2 evolution during N_2 fixation, with some associations typically generating virtually no H_2 while possessing an active hydrogenase (so-called Hup^+ associations), and others evolving in excess of 3 mol H_2 mol^{-1} H_2 fixed, in some cases with no evidence whatever of hydrogenase activity (Hup^- associations). According to the literature surveyed by Evans et al. (1980), Phillips (1980), and Pate et al. (1981), an electron allocation range of 25% to 60% to H_2 evolution would probably cover the recorded behavior of most, possibly all, symbiotic associations involving legumes, thus giving legume nodules generally a range of theoretical minimum costs (see Table 10.1) of 28.0 to 52.5 mol ATP mol^{-1} N_2 (0.73 – 1.38 mol glucose mol^{-1} N_2 fixed) were hydrogenase nonfunctional or absent, or a range of

25 to 39 mol ATP mol^{-1} N_2 (0.66–1.03 mol glucose mol^{-1} N_2 fixed) were all the evolved H_2 to be recovered by hydrogenase. The potential savings in carbohydrate consumption through possession of a fully active hydrogenase would therefore range from 0.07 to 0.35 ml glucose mol^{-1} N_2. The significance of these values will become apparent later when the overall cost of fixation is considered in terms of whole-plant functioning.

Measurement of energy costs of symbiotic N_2 fixation: case study of the legume nodule

The symbiotic association between legume and *Rhizobium* has long been a favored system for studying the total economy of N_2 fixation, particularly in terms of comparisons of growth and yield of plants relying on N_2 as opposed to combined forms of nitrogen such as NO_3^- or NH_4^+ (Phillips 1980; Rawsthorne et al. 1980; Schubert and Ryle 1980). In this instance, all costs relevant to nitrogen fixation are taken into account: biochemical costs at the level of nitrogenase, cellular costs in assimilating ammonia and transporting fixed N from symbiotic organs to host, and costs at the organ level in the initial growth and maintenance and replacement of symbiotic structures. Alternatively, when considering the costs of using combined N, account must be taken not only of the biochemical costs of assimilation and transport but also of any additional costs of building and maintaining extra root biomass to absorb the N compounds from the soil. Our own approach has been to construct budgets for the utilization of plant photosynthate by nodules from data collected for intact nodulated plants whose shoots and roots are enclosed for measurements of respiratory fluxes of CO_2. In its most sophisticated form, this approach also involves a separate enclosure of the proximal nodulated zone of a root in order to measure the release of CO_2 and H_2 during nodule functioning (Figure 10.1) (Layzell et al. 1979). A typical experiment lasts 10 days, during which the total net fluxes of C as CO_2 by root, nodules, and shoot are compared with total H_2 evolution from nodules and the net gains of carbon and nitrogen in dry matter of plant parts. Plants from the population are also regularly sampled for xylem sap and stem-base phloem sap, so that the C : N weight ratios of solute streams leaving the root or supplied to the nodulated root can be examined. A typical set of data obtained using this approach is summarized in Figures 10.2 and 10.3. The experimental species is white lupine, *Lupinus albus* L. (cv. Ultra), partnered with *Rhizobium* WU425, the resulting association being highly active in H_2 evolution

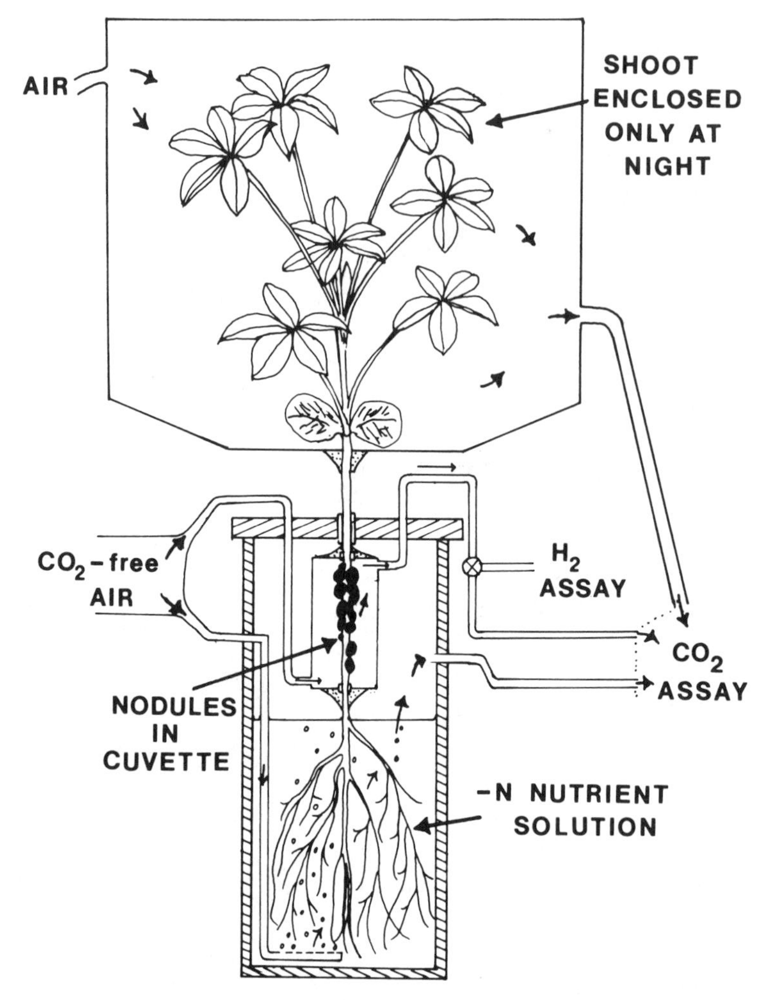

Figure 10.1. Apparatus for continuous measurement of CO_2 fluxes of shoot and root and CO_2 and H_2 evolution by the nodules of an intact legume. See Figures 10.2 and 10.3 for data obtained using this experimental approach.

(2.8 mol H_2 evolved per 1 mol N_2 fixed). Figure 10.2 shows how the carbon of net photosynthate was deployed during the period 51–58 days after sowing, that is, just after the plant had commenced to flower and when the nodule population was maximally active in N_2 fixation. A total of 1,061 mg C was fixed as net photosynthate, and 34.8 mg N was fixed by the nodules during the study interval, giving a weight ratio of assimilatory inputs of C and N of 30 : 1. Nodules received via translocation the equiva-

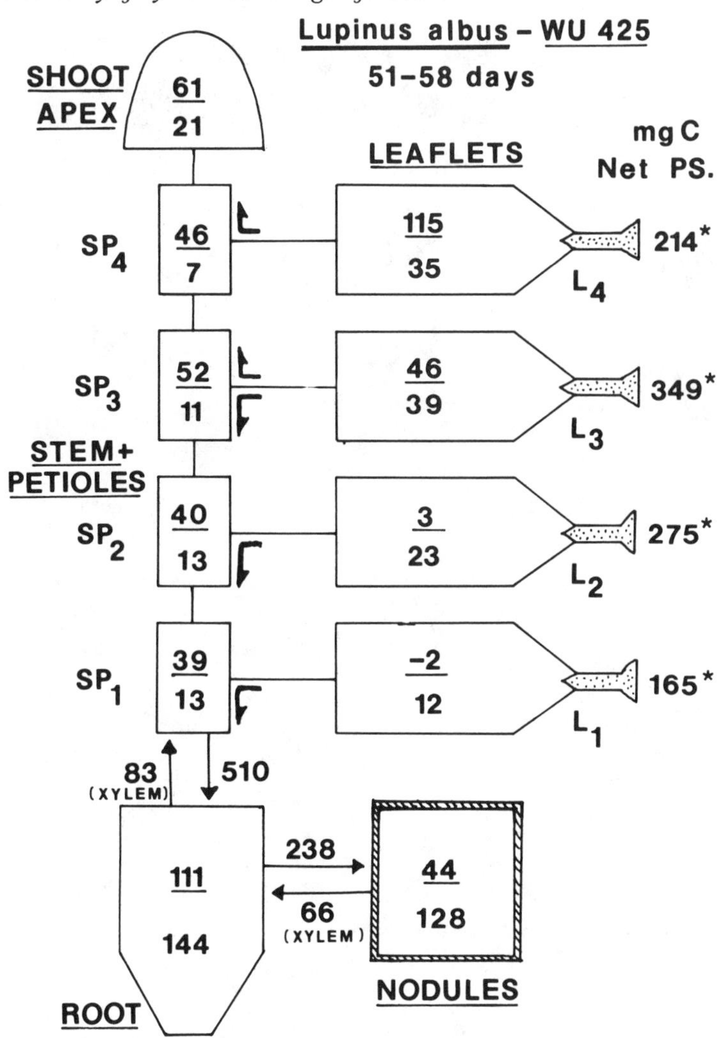

Figure 10.2. Production, partitioning, and consumption of carbon of net photosynthate by 51–58-day N_2-dependent plants of white lupine. The amounts (mg C) of net photosynthate produced by the four strata of leaflets (L_1–L_4) are as shown (asterisks). Nodules fixed 34.8 mg N during the study period. Direction of flow of photosynthate from each leaf stratum is shown, and amounts (mg) of carbon translocated in phloem and exported in xylem from root and nodules are indicated. The top (underlined) number in each box refers to the amount (mg) of carbon incorporated by the plant part into dry matter; the lower number indicates the amount of carbon lost in respiration (data from Pate and Layzell 1981).

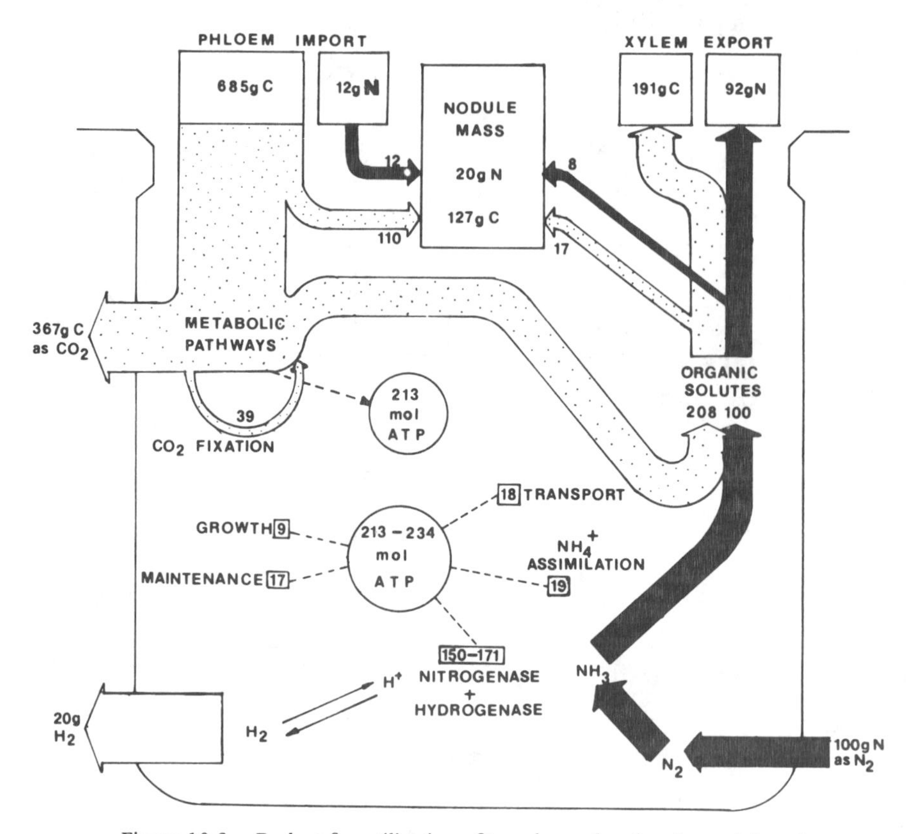

Figure 10.3. Budget for utilization of translocated carbon by nodules of an H_2-evolving association of white lupine with *Rhizobium* WU425. Experimental data on utilization and exchanges of C : N are given relative to fixation of 100 g N by the nodules. Data on H_2 evolution and CO_2 fixation by the nodules are included. Theoretically based costs (dotted lines) for aspects of nodule functioning are given in ATP equivalents (data from Layzell et al. 1979). See Figure 10.2 for relationship between nodule consumption and utilization of photosynthate within the whole plant.

lent of 238 mg C (6.9 mg C mg^{-1} N fixed) or the equivalent of 22.4% of the 1,061 mg C of net photosynthate generated by the host plant.

A detailed breakdown of the nodule's budget for carbon and nitrogen is shown in Figure 10.3. This includes information on H_2 evolution, consumption of C and N in nodule growth, fluxes of C and N in and out of the nodule as determined by C : N weight ratios of xylem and phloem sap, and the net respiratory efflux of CO_2 by the nodules during the experimental period. The supposed extent of CO_2 fixation by the nodules is estimated by assuming a maximum anapleurotic input of carbon from phosphoenolpyruvate (PEP) carboxylase activity in the synthesis of the asparagine, aspartate, and malate exported from nodules in the xylem (Rawsthorne et al. 1980; Pate et al. 1981). The model of functioning predicts that over half (54%) of the translocated carbon is lost in nodule respiration, with 28% returning to the plant attached to fixation products exported via the xylem, and with the remaining 18% being incorporated into nodule dry matter.

Based on an observed ratio of H_2 evolved to N_2 fixed of 2.8 : 1 (mol mol^{-1}), the nitrogenase of the *Lupinus* : WU425 nodule will be expected to consume 150 to 171 mol ATP (Table 10.1), or the equivalent of 70% to 80% of the 213 mol ATP assumed to have been available from oxidation of glucose. Expressed in terms of the total budget of the nodule, 37% to 43% of its acquired translocation product will thus be devoted to providing ATP for nitrogenase : hydrogenase activity. Other fixation-related functions and structures that may have associated respiratory costs are also included in the budget shown in Figure 10.3, including the costs of membrane-mediated transport of NH_4^+ and amino compounds within and out of the nodule and the growth and maintenance components of nodule respiration. All of these items are rated as being small relative to the predicted costs of nitrogenase activity. The reader is referred to two earlier publications (Layzell et al. 1979; Pate et al. 1981) for details on how these estimates were computed.

A number of questions come to mind when evaluations of this kind are made for symbiotic functioning. First, how does percentage consumption of net photosynthate by nodules vary over the life cycle? More important, what is the total investment of net photosynthate in nodules over the whole growth cycle of the plant? The data of Pate et al. (1980) suggest that for every gram of nitrogen fixed by the nodulated root of the *L. albus* : *Rhizobium* WU425 association, an average of 20.2 g C of phloem translocate (the equivalent of 53% of the plant's net photosynthate) is required by the nodulated root. Assuming that the value of 6.9 mg C mg^{-1} N recorded for nodule consumption over the 51–58-day period (Figure 10.3) were to apply for the whole growth period, nodule consumption could account for approximately one-third (34%) of the total translocate acquired by below-ground parts of the plant, or the equivalent of one-sixth (16%) of the total net photosynthate produced by the plant over its life cycle. Then, assuming that nitrogenase : hydrogenase activities were to represent some 37% to 43% of the nodule's budget for translocate (Figure 10.3), these reactions, unique to the fixation process, would be rated as consuming a mere

6% to 7% of the total resource of net photosynthate produced by the plant over its growth cycle. However, although this may not seem to be a large element, one must remember that the same amount of photosynthate, if diverted progressively, say, into increased leaf or root growth, might have greatly increased the plant's growth rate through increased assimilatory capacity.

A second series of questions relates to the likely extent of the differences between legume species in regard to the cost-effectiveness of nodule fixation. Unfortunately, the type of analysis detailed here has been conducted on only a few agriculturally important species, and on only one of these species [cowpea (*Vigna unguiculata*)] in a manner similar to that already discussed for white lupine. The particular cowpea association examined (Figure 10.4) showed lower evolution of H_2 relative to N_2 fixation than in the *L. albus*: WU425 association, and hence a considerable saving in terms of ATP consumption per gram of N_2 fixed and a lesser respiratory output of CO_2 per unit of N_2 fixation. The cowpea nodules also exhibited lower overall consumption of phloem translocate than in lupine because of lesser investments of carbon in nodule growth and export of fixation products. Greater economy in transport was due to the use of ureides (C : N 1 : 1) as principal solutes for export of fixed N in cowpea, as compared with amides (C : N 2–2.5 : 1) in lupine.

Extending the data on cowpea to include consumption of plant net photosynthate by nodules over the growth cycle, it was found that on average, 24.7 g C of net photosynthate was produced for a fixation of 0.79 g N; see the data on C and N economy of the species by Herridge and Pate (1977). Rating nodules as consuming 5.4 g C g^{-1} N_2 fixed (Figure 10.4), the data suggested an overall average consumption of 17.3% of the plant's total net photosynthate by nodules over the growth cycle, or an average allocation of 6.0% to 8.6% of plant photosynthate to the nitrogenase : hydrogenase reaction of the nodules. The cowpea symbiosis thus emerges as remarkably similar to that of lupine in terms of cost-effectiveness in nodule fixation, despite large differences between the two legume species in regard to growth habit, nitrogen metabolism, pattern of translocation, grain yield, and protein content of seed (Pate in press; Pate et al. in press).

A third issue relates to the likely effects on dry-matter gain and yield of fixed N by a legume species were it to be partnered by *Rhizobium* strains lacking in or possessing hydrogenase activity (see discussions by Schubert et al. 1977; Albrecht et al. 1979; Evans et al. 1980; Arima 1981). Again relying almost entirely on cultivated legumes, the evidence so far indicates that H_2-evolving symbioses are not consistently less productive or, in cer-

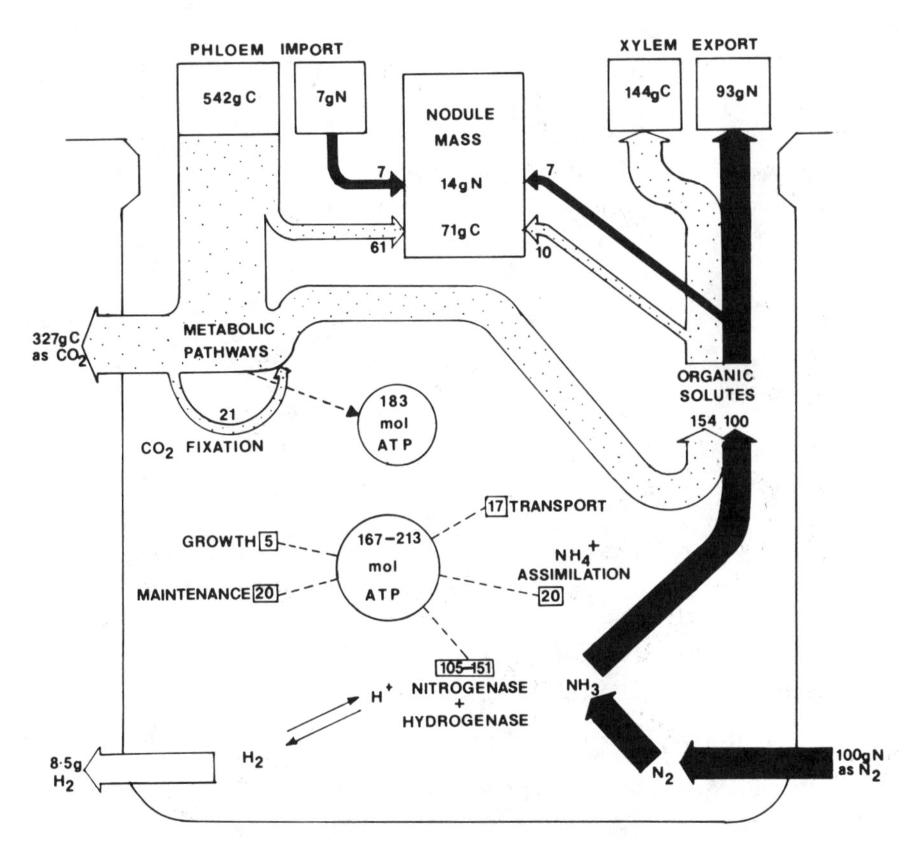

Figure 10.4. Budget for utilization of translocated C by nodules of the *Vigna unguiculata*: *Rhizobium* 176A27 association. Experimentally based data on utilization and transport of C and N are given relative to fixation of 100 g N by the nodules. Data on hydrogen evolution and CO_2 fixation by the nodules are included. Theoretically based costs of nodule functioning (dotted lines) are given in ATP equivalents (derived from Pate et al. 1981).

tain cases, are only marginally less productive than those with little or no H_2 evolution, as borne out by a series of recent comparisons in a variety of legumes (see review by Rainbird et al. 1983). However, distinct differences can be observed in the economy of photosynthate usage by nodules. For example, in a comparison involving a single cultivar of cowpea, low-H_2-evolving nodules showed 16% better economy of carbon usage than did high-H_2-evolving nodules (Rainbird et al. 1983). If, as mentioned earlier for cowpea, nodules were to consume in their growth, maintenance, and functioning some 17% of a plant's net photosynthate, the advantage due to low H_2 evolution would translate to the equivalent of 2.7% of the plant's net photosynthate. A superiority of this small magnitude obviously would be difficult to detect in terms of whole-plant growth and yield, because it might easily be overwhelmed or negated by other differences in economy relating to the symbiotic associations under comparison. However, if, as suggested by Schubert et al. (1978) and Zablotowicz et al. (1980), small differences were to be compounded over time, yield differences of some size might become apparent by the end of a plant's growth cycle.

Cost of symbiotic N_2 fixation versus other forms of assimilation or acquisition of N_2

Nitrate and (in certain circumstances) ammonium are the common sources of nitrogen available to plants from soil or aquatic environments. Nitrate can be assimilated by root or shoot or by a combination of these routes (Pate 1980, 1983; Hunter et al. 1982), with the possibility of shoot-located reduction systems utilizing directly, and essentially at minimal cost, the surplus reductant and ATP generated in photosynthesis (Canvin and Atkins 1974; Schrader and Thomas 1981). The stoichiometry of reduction is likely to be as follows:

$$NO_3^- + 8e^- + OH^- \rightarrow NH_4^+ + 3H_2O$$

If this occurs nonphotosynthetically and is coupled to aerobic respiration proceeding with a $P/2e^-$ ratio of 3, the equivalent of 12 mol ATP will be consumed per 1 mol NH_4^+ produced. Thus, when compared with the suggested cost range of 13 to 26 mol ATP mol^{-1} NH_4^+ for N_2 fixation (Table 10.1), there would appear to be a slight cost saving between respiratory-linked NO_3^- reduction and the most efficient class of nitrogenase : hydrogenase functioning in N_2 fixation, as well as the possibility of considerably greater savings, say, between photosynthetically linked reduction of nitrate and the more expensive forms of H_2-evolving symbiotic systems. Nevertheless, in a root-based assimilation of inorganic N, more root biomass might be involved than in an equivalent N_2-fixing plant, and the costs of this extra tissue would have to be considered.

Whether utilizing N_2, NO_3^-, or NH_4^+, plants autotrophic for N must also bear the cost of assimilating ammonium into organic solutes of N. Theoretically based estimates of such costs demand detailed knowledge of the metabolic pathways involved in synthesis of specific nitrogenous solutes by a species, as well as information on the relative amounts of different solutes it is synthesizing from the ammonia. A recent costing exercise along these lines (Pate et al. 1981) compared the relative amounts of ATP and reductant predicted to be consumed in ammonia assimilation in nodules of three legume species, assuming that the currently accepted metabolic pathways were operating and that maximum advantage accrued from anapleurotic inputs of CO_2 by PEP carboxylase, using CO_2 released in other parts of a synthetic pathway. The high activity of PEP carboxylase in nodules of a wide variety of legumes (e.g., Christeller et al. 1977; Cookson et al. 1980), combined with evidence of $^{14}CO_2$-fixing capacity in nodules and nodulated roots (Laing et al. 1979; Coker and Schubert 1981), indicated that considerable anapleurotic carbon assimilation may accompany ammonia assimilation in these organs, thus effecting considerable savings in terms of the requirement of imported carbon from the host plant. However, other costs of maintaining the associated CO_2-fixing systems would have to be considered before determining whether or not real benefit is involved.

It turns out (Table 10.2) that *Lupinus albus,* an asparagine exporter, exhibits the least cost in terms of ATP equivalents (2.66 mol ATP mol^{-1} NH_4^+), but in having the greatest anapleurotic input of CO_2, it is predicted to have the lowest respiratory quotient (RQ) for ammonium assimilation. The two other species (*Vigna unguiculata* and *Vigna radiata*) utilize ureides as the principal export products, and because of their high activity in synthesis of these compounds, they show greater cost (mung bean, 3.03 mol ATP mol^{-1} NH_4^+; cowpea, 2.75 mol ATP mol^{-1} NH_4^+) than lupine. Disregarding these slight differences between species, the overall costs of NH_4^+ assimilation generally appear to be small relative to those of N_2 or NO_3^- assimilation, suggesting that there will be considerable cost advantages in NH_4^+-utilizers over either N_2-fixers or NO_3^--reducers, at least under conditions in which the two forms of N are equally available in the soil.

Certain plants assimilate nitrogen in an already-assimilated form by engaging in carnivory, in parasitic relationships with other higher plants, or in symbiotic relationships with saprophytic fungi (Pate 1983). In these cases, the principal forms of nitrogen acquired from prey or host are likely to be organic solutes of N (e.g., amides, amino acids, or ureides), in which case the costs of converting N_2 or NO_3^- to NH_4^+ and of converting NH_4^+ to amino compounds will be eliminated. A comparison between the overall cost relationships of these heterotrophs and plants autotrophic for N_2, NO_3^-, and NH_4^+ is outlined diagrammatically in Figure 10.5. The figure includes the ranges of theoretically based costings mentioned earlier for N_2 fixation and respiration-linked or photosynthetic NO_3^- reduction, all costs being strictly intracellular metabolic costs.

Table 10.2A. *Theoretically based costs of assimilating ammonia into exportable organic solutes in nodules of three species of legumes: composition of products exported from nodule in xylem*[a]

	Relative abundance (% molar basis)		
Compound	*Lupinus albus* *Rhizobium* WU425	*Vigna unguiculata* *Rhizobium* 176A27	*Vigna radiata* *Rhizobium* CB756
Ureide(s)[b]	—	37	49.2
Organic acids[c]	7.0	19	25.9
Asparagine	65.7	12	4.6
Glutamine	9.3	18	11.9
Aspartic acid	7.1	1	1.0
Glutamic acid	2.7	—	—
γ-aminobutyric acid	1.9	2	0.6
Serine	1.7	1	—
Alanine	1.5	—	—
Valine	1.4	1	1.0
Histidine	0.8	1	0.4
Minor constituents	0.9	0.8	5.0

[a] Based on analyses of bleeding sap of detached nodules or decapitated nodulated roots.
[b] Allantoin and allantoic acid.
[c] As equivalents of malate.

Table 10.2B. *Net cost of ammonia assimilation and ancilliary carbon metabolism*

	Legume association		
Budget item[a]	*Lupinus albus* *Rhizobium* WU425	*Vigna unguiculata* *Rhizobium* 176A27	*Vigna radiata* *Rhizobium* CB756
Net mol ATP consumed mol^{-1} NH_4^+	2.00	2.69	2.82
Net mol reductant mol^{-1} NH_4^+	0.22	0.02	0.07
Net cost (mol ATP equivalents mol^{-1}/NH_4^+)	2.66	2.75	3.03
Mol CO_2 fixed by nodule mol^{-1} NH_4^+	0.46	0.24	0.20
Predicted RQ (NH_4^+ assimilation)	0.43	1.18	0.95

[a] All items costed on the basis of a net synthesis of organic compounds in proportions indicated in Table 10.2A; reactions of synthesis costed according to the metabolic schemes suggested by Pate and associates (1981). Anapleurotic inputs from PEP carboxylase are maximized and the metabolic pathways balanced such that the only products that accumulate are the exported products listed in Table 10.2A (data from Pate et al. 1981).

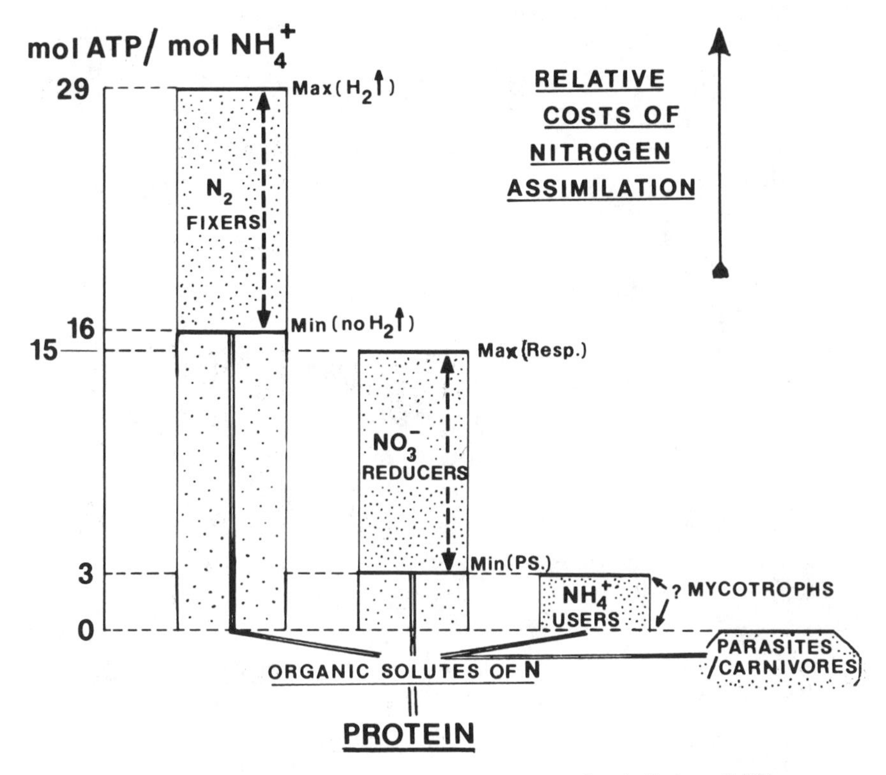

Figure 10.5. Relative theoretically based costs of assimilation of different forms of nitrogen by different trophic classes of plants. Total costs (ATP mol^{-1} NH_4^+) are given only for stages up to the synthesis of organic solutes of N. See text for details of how minimum and maximum costs were determined.

Legumes, and other higher plants with the potential to fix nitrogen symbiotically or to utilize combined forms of N, offer the possibility of making detailed studies of plant growth, photosynthate partitioning, and energy balance when relying on these alternative sources of N (see the studies of Mahon 1977, 1979; Ryle et al. 1979; and Phillips 1980). For comparisons to be meaningful, rates of growth and nitrogen gain by the symbiotic and nonsymbiotic sets of plants must be strictly comparable, and this requires careful matching of the plant's supply of nitrate or ammonium to the rate of fixation of nodulated counterparts. However, even when matched perfectly in terms of N assimilation rate, differences may still exist between the symbiotic and nonsymbiotic plants in features such as shoot and root morphology, leaf area, root : shoot weight ratio, and percentage N in plant organs (Pate et al. 1981).

The objective of several of the foregoing types of studies has been to determine if the allegedly greater cost of N_2 assimilation can be detected as a greater respiratory output by nodulated roots than by NH_4^+- or NO_3^--fed, nonnodulated roots, or as a poorer overall efficiency of conversion of photosynthate to dry matter in N_2-fed plants than in nonsymbiotic plants. Our own experiments in this connection have examined white lupine (Pate et al. 1980), a species that reduces NO_3^- almost entirely in its roots (Atkins, Pate, and Layzell 1979), and cowpea, a species that reduces nitrate principally in its leaves (Atkins et al. 1980)

The results for lupine were that nodulated plants showed less (57%) conversion of C of net photosynthate to dry matter than did NO_3^--fed plants (69%), and this difference was due principally to the belowground parts of nodulated plants consuming a larger share (58%) of net photosynthate than NO_3^--fed plants (50%). Reflecting this, nodulated roots respired 10.2 mg C per mg N assimilated, versus 8.1 mg C per mg N in the roots assimilating nitrate. The experiments on cowpea showed that there were virtually identical efficiencies of conversion of net photosynthate to dry matter in NO_3^-- and N_2-dependent plants, but noticeably greater proportional consumption of net photosynthate in belowground organs of N_2-fed plants (37% of net photosynthate) than in NO_3-fed plants (23–26%). The greater consumption of photosynthate by nodulated roots was only partly due to the requirement for carbon in the synthesis of organic solutes of N exported from nodules, because dry-matter gain and respiration losses were higher in nodulated roots than in NO_3^--fed roots.

A number of other authors have made essentially similar comparisons between NO_3^-- and N_2-fed legumes (see the reviews of Phillips 1980; Schubert and Ryle 1980; and Pate et al. 1981). Some of these estimates were based on study of relative growth rates (e.g., Silsbury 1979), some simply on differences in final yield, still others on detailed estimates of specific components of root and nodule respiration (Mahon 1977, 1979; Ryle et al. 1978, 1979; Dixon et al. 1981; Witty et al. in press). The majority of studies suggested that overall costs are indeed greater for N_2 assimilation than for NO_3^- utilization, but that these relate as much to the extra resources of photosynthate required for growth and maintenance of nodules as to the supposed extra cost specifically involved in the initial fixation of N_2 to ammonia.

A second ancillary class of differences in cost-effectiveness between N_2- and NO_3^--dependent legumes relates to maintenance of ion balance of shoot and root when plants are subsisting on these different forms of nitrogen. Differences in these respects may also extend to species that reduce nitrate principally in their shoots or in their roots. In certain of these situations, considerable costs are likely to be involved if organic acids accumulate to maintain charge balance in transport fluids or tissues of root or shoot. The types of effects involved are discussed, for example, by Raven and Smith (1976), Armstrong and Kirkby (1979), Kirkby and Armstrong (1980), Pate (1980), and Israel and Jackson (1982).

Cost–benefit analysis of symbiotic N_2 fixation under ecological conditions

Though rarely the dominant components of vegetation, nitrogen-fixing associations involving legumes or, more rarely, cycads or nonleguminous angiosperms, are significant components of a wide variety of natural ecosystems across the world, ranging from forest to open grasslands or heathlands and even tundra. Some nitrogen-fixers are slow-growing long-lived perennials (e.g., *Macrozamia, Casuarina,* and certain relatively long-lived species of *Acacia* in Australian forest ecosystems), occurring as subdominant understory elements that show the potential to augment soil nitrogen capital through to the climax stages of a plant community. Other nitrogen-fixers, by contrast, are relatively short-lived, fast-growing "pioneer" species, establishing large populations immediately after environmental disturbance by fire, landslide, drought, glacial retreat, or volcanic activity, and thus having the effect of replenishing the resources of nitrogen within a recovering ecosystem. On the death of these ephemeral nitrogen-fixing associations, the nitrogen and other nutrients of their biomass may become available to other species: for example, *Ceanothus* and *Alnus* play a role in nutrient accretion in recently burned coniferous forests in the western United States, and an essentially similar role is played by a range of short-lived shrub legumes (e.g., *Acacia, Bossiaea*) in various eucalypt forests of Australia (Lamont 1983).

Although quantitative measurements of nitrogen fixation have been made in only a few nitrogen-fixing associations under natural conditions, there is evidence that the annual returns of nitrogen that they effect may be quite substantial, even in ecosystems with extremely low levels of available nutrients. For example, in the nutrient-deficient forest soils of southwestern Australia, fixation rates of 2 to 15 kg N ha^{-1} $year^{-1}$ have been recorded for stands of short-lived legumes establishing after fire (Malajczuk and Grove 1977; Bowen 1981), and the long-lived cycad *Macrozamia riedlei* in symbiosis with blue-green algae is estimated to return 1.4 to 19 kg N ha^{-1} $year^{-1}$ in the same or similar habitats (Halliday and Pate 1976; Grove et al. 1980). In soils of higher nutrient status, returns of nitrogen are likely to be much higher.

It can be concluded from the foregoing that symbiotic N_2 fixation is an effective means of supplying the parent association with nitrogen during its lifetime and ultimately making this nitrogen available to other components. Symbiotic activity is obviously likely to confer maximum advantage to a species under circumstances in which soil nitrogen is limiting but where other essential mineral elements, water, and light are all fully available. In certain forest and heathland communities, the years immediately following fire may well realize such conditions, particularly because of release of nutrients such as phosphorus and trace elements from ash. However, in the immediate period after burning, ammonium and nitrate become transiently available, presumably released from organic matter or organisms killed by the fire. A range of non-N_2-fixing fire ephemerals then germinate and are able to compete on even terms with N-fixing pioneer species. This was evident, for example, in a recent study of growth of stands of *Acacia pulchella* in the nutrient-poor coastal sands of *Banksia* woodlands, near Perth, Western Australia (Monk et al. 1981). In the first year after a fire, annual grasses and other nonfixing fire ephemerals grew faster than the *A. pulchella* and showed rates of uptake of nitrogen as high as or higher than those of the latter species. The *Acacia* seedlings were poorly nodulated in the first year, presumably because of the inhibitory effects of soil nitrogen on nodulation, and C_2H_2 reduction assays suggested a reliance on symbiosis for only 8% of the N that they accumulated during this first growing season. During the second to fourth years, the *Acacia* stands grew into dense thickets, almost totally dominating the understory. By that time, soil N had been depleted, causing the demise, or much reduced growth, of nonfixing species, just as the *Acacia* stands were reaching peak biomass and reproductive achievement. During that period, the plants of *A. pulchella* gained more than two-thirds of their annual return of N from their root nodules, indicating how important symbiotic activity had been to their growth and productivity.

In undisturbed vegetation not recently experiencing fire, perennial N_2-fixing components will survive only if they are able to compete with other species for nonnitrogenous nutrients. Here the symbiotic association may be doubly disadvantaged, in being condemned to a slower rate of growth of roots because of the extra drain on resources of photosynthate for maintenance and seasonal replacement of symbiotic organs, while also possibly requiring greater amounts of certain key nutrients (e.g., P, Co, Mo) than do nonfixing counterparts, in view of the special involvement of these elements in the ATP regenerating system (P) or in the enzymatic process (Co, Mo) of N_2 fixation (e.g., Riley and Dilworth 1982; Lamont 1983). Legumes are also reported to require higher levels of Ca, S, and Zn and to be less tolerant of acid soils than are most nonleguminous, nonfixing species, but it is debatable that effects of this kind relate specifically to nodule functioning, as opposed to more general attributes of the host species.

The necessity in nutrient-impoverished soils for nitrogen-fixing associations to compete effectively for limiting elements such as P and certain trace elements has undoubtedly promoted the evolution within these plants of mechanisms for enhancing nutrient uptake. This is well borne out in the data compiled for Mediterranean-type ecosystems of South Africa and Western Australia by Lamont (1983). Virtually all of the legumes of the Fabaceae and Mimosaceae surveyed within these regions have been shown to possess proteoid roots, VA mycorrhiza, or ectomycorrhizae, in addition to bearing root nodules. These specialized root structures presumably allow the legume component to compete on equal terms with nonlegumes, the latter collectively exhibiting an array of root modifications virtually identical with that of their N_2-fixing counterparts. In any event, mycorrhizal-type symbioses inevitably incur costs to the macrosymbiont, and in associations that are autotrophic for N, these costs must be added to any "extra" costs that the N_2-fixer must bear in relation to growth and maintenance of its symbiotic organs. Unfortunately, little experimental evidence is available on the likely costs to a host plant of association with mycorrhizal fungi. According to Kucey and Paul (1982), approximately 6% of plant photosynthate is consumed in maintaining a VA-type mycorrhizal association in nodulated *Vicia faba*, probably a lesser cost than that concerned with maintaining N_2 fixation in nodules. Mycorrhizal associations (see mycotrophs, Figure 10.5) may also equip non-N_2-fixing species for highly efficient utilization of soil N, as suggested by France and Reid (1982) for tree ectomycorrhiza and by Lamont (1983) for a range of normally unnodulated, but mycorrhizal, legumes in the Caesalpiniaceae. It is not clear which forms of nitrogen (organic N or NH_4^+) are made available to the host plant from mycotrophic associations. Presumably, N_2-fixing associations might also facultatively acquire N heterotrophically if bearing mycorrhiza.

It is difficult to place a cost on the habit of bearing cluster-type "proteoid" roots, because little work has been done on their rates of respiration or, indeed, on the precise nature of their functioning. However, in certain cases under moderately low nutrient conditions, they may collectively represent up to 5% of the dry weight of belowground parts of a plant (unpublished data on the legume *Viminaria juncea*, B. A. Walker and J. S. Pate), and so expenditure in their development as structures may be considerable, let alone in maintenance of their functioning.

It holds widely that the foliage of leguminous and other N-fixing components of ecosystems is, on the average, richer in N (percentage N in dry weight) than that of nonsymbiotic components [e.g., see the literature reviewed by Lamont (1983) and the data of Pate and Dell (1984)] and that differences of this nature can extend to other plant parts, including fleshy roots and seeds (Pate and Layzell 1981). Legumes accordingly rank as important contributors of protein to the diet of a range of grazing vertebrates and phytophagous or seed-predating arthropods, and in response they have developed a virtual armory of mechanisms against such attack. These defense strategies include woodiness and spinescence of leaf and stem and acquisition of a range of toxic qualities, including possession of organic selenium compounds (e.g., *Astragalus*) (Brown and Shrift 1981), fluoroacetate (*Oxylobium, Gastrolobium*) (Main 1981), cyanogens (*Phaseolus, Lotus*), alkaloids (*Genista, Lupinus*), depilatory agents (e.g., mimosine), carcinogens, and neurotoxins (Bell 1980; Pate 1983). In certain cases the toxic principles (e.g., canavanine, homoarginine, pipecolic acid) can accumulate in large amounts, especially in relation to insect attack on seeds, in which case a significant fraction of the nitrogen resource can be employed for purposes of defense (Bell 1980). Once germination occurs, the same nitrogen can be recycled and used for seedling growth.

The success of the foregoing strategies may relate more to a narrowing of the range of predatory agents attacking the species in question than to the conferring of total immunity to attack. For example, this applies particularly to attack on seeds by bruchid beetles, where usually one insect species coevolves with its host legume and in the process acquires immunity against high levels of specific toxic agent(s) (e.g., Rosenthal et al. 1976). Parallel cases of resistance of grazing vertebrates to legume toxins are also known, such as the case cited by Main (1981) of populations of forest-dwelling marsupials showing unusually high tolerance to fluoroacetate when consuming foliage or seeds of certain common understory legumes (e.g., *Gastrolobium, Oxylobium*) possessing this toxin. Introduced European animals (foxes and rabbits), or populations of marsupials not normally exposed to fluoroacetate-containing plant species, proved to be markedly intolerant to the toxin.

In addition to investing significant amounts of nitrogen and resources of other elements in defense of their seeds and seedlings against attack by animals, nitrogen-fixing plant associations incur a significant disadvantage over nonfixing species during their establishment phase, in having to make a substantial investment of cotyledon carbon and nitrogen in forming the symbiotic structures that will eventually confer advantage in terms of fixing atmospheric nitrogen. For example, up to 15% of the mobilizable seed reserve of dry matter of *Macrozamia riedlei* may be consumed in forming the extensive sets of apogeotropic roots, forerunners of the cyanophytic N-fixing coralloid roots of the species. Indeed, seedlings of this cycad may not commence fixation until the second or third year after germination, suggesting that there will be a protracted drain in terms of maintenance respiration of these presymbiotic structures. Seedlings whose apogeotropic roots are removed early in development, if they have access to nitrate, achieve almost twice as much biomass over the following 10 months as those allowed to develop their symbiotic potential under nitrogen-deficient conditions (J. S. Pate, unpublished data).

Penalties during prefixation seedling growth have been widely reported among legumes, particularly in species with small seeds. In *Trifolium* and *Medicago,* for instance, a nonnodulated seedling raised in non-N culture solution produces a heavier and more extensive root system than does a comparable seedling developing nodules (Pate and Dart 1961; Pate 1977). Moreover, if nitrate is applied to the nonnodulated seedling at the time when its nodulated counterpart is commencing to fix N, the nonnodulated seedling will grow much more rapidly. In both small- and large-seeded legumes, a progressive dilution of cotyledon reserves of N during prefixation development may lead to a so-called nitrogen-hunger state, characterized by minimal levels of N in the dry matter of seedling parts, a hiatus in growth, and a pronounced yellowing of cotyledons and first-foliage leaves. This condition may last for several days, but it is quickly reversed once fixed N commences to be exported from the nodules to the shoot. According to ^{15}N-labeling studies on cowpea (J. S. Pate, C. A. Atkins, and I. Matthews, unpublished data), the development of N_2-fixing potential in young nodules is autocatalytic, in that the first nitrogen that is fixed is used almost entirely for synthesis of bacteroid protein. So long as this continues, the nitrogen-hunger state of the seedling cannot be alleviated. No hunger symptons are evident in nodulated or nonnodulated seedlings if combined forms of N such as NH_4^+ or NO_3^- are supplied during seedling development, supporting the view that the condition is primarily one of N deficiency during the critical transitional period between cotyledon nutrition and commencement of N fixation (Pate 1977).

Bearing in mind the aforementioned costs of allocation to developing nodules and the autocatalytic nature of the generation of nodule mass, it is not surprising to find that symbiotically fixing associations have an inherently slow start to their seedling growth (Pate 1977). This is likely to be particularly disadvantageous when the plants in question have very short life spans, that is, when competing with nonfixing members of an ephemeral flora.

A further area in which symbiotically fixing legumes may incur disadvantages involves conditions of severe environmental stress, because root nodules or equivalent N_2-fixing structures frequently show ultrasensitivity to factors such as drought, excessively high or low pH, and frost or heat stress (Pate 1977). Nodulation of legumes, for instance, can be highly seasonal and apparently is restricted to a shorter time scale, say, than is the assimilation of inorganic N. However, because N_2-fixing associations are also facultative users of NO_3^- or NH_4^+ and may exhibit mycorrhizal-type associations, they presumably have recourse to soil N during periods when symbiotic organs are absent or nonfunctional.

Finally, returning to the picture of the relative costs of assimilation or acquisition of nitrogen shown in Figure 10.5, it would be misleading to leave the impression that heterotrophic forms of nutrition, as displayed, say, by carnivorous plants and various types of mycotrophic and parasitic plants, are inherently more cost-effective simply by virtue of their lacking an energy requirement for assimilation of elemental N to the level of amino compounds. Against this apparent advantage must be set what might well be a much larger resource involvement in the construction and maintenance of specialized nutritional apparatus. As examples, southwestern Australian *Drosera* species have been estimated to deploy the equivalent of some 3% to 6% of their net photosynthate in the production of mucilage on their glandular leaves (J. S. Pate, unpublished data), and mature root hemiparasites (e.g., *Olax, Exocarpus*) from the same geographical region are found to possess many hundreds of root haustorial connections per plant, these collectively representing a significant proportion (up to 3%) of root dry weight and exploiting a wide range of hosts over an area of soil up to 50 to 100 times that occupied by the plant's foliar canopy. Root–shoot biomass ratios for certain of these species are therefore high in comparison with fully autotrophic species (K. W. Dixon and J. S. Pate, unpublished data), and presumably there are commensurate penalties in terms of photosynthate allocation for growth and maintenance of these large root systems and their complements of haustoria.

Conclusions and summary

Symbiotic nitrogen fixation, though of obvious benefit to a species in situations in which soil nitrogen is limiting, incurs substantial penalties to a host plant in terms of cost of maintenance and functioning of symbiotic structures and their nitrogenase systems. Where nutrients other than N are also in short supply, a N_2-fixing association will be advantageous only if it is able to acquire these limiting nutrients with the same effectiveness as do competing nonfixing plant species. The extra requirement of N_2-fixing plants for elements such as MO, Co, and P may prove disadvantageous for symbiotic N-fixers under certain circumstances, although this is by no means fully proven.

Costed purely in terms of assimilation of N to the level of synthesis of organic solutes of N, symbiotic N-fixers are likely to have slightly higher levels of expenditure per unit N assimilated than are plants reducing NO_3^- in their roots. Plants utilizing NH_4^+ or reducing NO_3^- photosynthetically under conditions in which provision of ATP and reductant by photosynthesis is nonlimiting will have substantially lesser assimilatory costs than those fixing N or engaging in respiration-linked NO_3^- reduction. Costs of NH_4^+ assimilation to organic solutes of N are small relative to those of reducing NO_3^- or N_2 to NH_4^+, suggesting that carnivorous plants, mycotrophic plants, and hemiparasitic plants acquiring organic forms of N from their prey or hosts are at only a slight cost advantage over those utilizing NH_4^+, but they are at a decided cost advantage over those fixing N_2 or assimilating NO_3^- by respiration.

Against these cost elements identified specifically against N assimilation, one must consider a range of associated costs incurred in relation to each form of assimilation or acquisition of N (e.g., in the growth, maintenance, and seasonal replacement of nodules, in haustoria and mycorrhizal investments, and in the cost of mucilage and enzyme production in carnivores). Little is known of the absolute and relative magnitudes of such costs, let alone of how they might affect the overall resource deployment of a species when relying on a specific type of nitrogen assimilation.

References

Albrecht, S. L., R. J. Maier, F. J. Hanus, S. A. Russell, D. W. Emerich, and H. J. Evans. 1979. Hydrogenase in *Rhizobium japonicum* increases nitrogen fixation by nodulated soybeans. Science 203:1255–1257.

Arima, Y. 1981. Respiration and efficiency of N_2 fixation by nodules formed with a H_2-uptake positive strain of *Rhizobium japonicum.* Soil Sci. Plant Nutr. 27:115–119.

Armstrong, M. J., and E. A. Kirkby. 1979. Estimation of potassium recirculation in tomato plants by comparison of the rates of potassium and calcium accumulation in the tops with their fluxes in the xylem stream. Plant Physiol. 63:978–983.

Atkins, C. A., J. S. Pate, G. J. Griffiths, and S. T. White. 1980. Economy of carbon and nitrogen in nodulated and non-nodulated (NO_3^--grown) cowpea (*Vigna unguiculata* (L.) Walp). Plant Physiol. 66:978–983.

Atkins, C. A., J. S. Pate, and D. B. Layzell. 1979. Assimilation and transport of nitrogen in non-nodulated (NO_3^--grown) *Lupinus albus* L. Plant Physiol. 64:1078–1082.
Bell, E. A. 1980. The non-protein amino acids of higher plants. Endeavour 4:102–107.
Benson, D. R., D. J. Arp, and R. H. Burris. 1979. Cell-free nitrogenase and hydrogenase from actinorhizal root nodules. Science 205:688–689.
Bethlenfalvay, G. J., and D. A. Phillips. 1979. Variation in nitrogenase and hydrogenase activity of Alaska pea root nodules. Plant Physiol. 63:816–820.
Bowen, G. D. 1981. Coping with low nutrients. Pp. 33–64 *in* J. S. Pate and A. J. McComb (eds.), The biology of Australian plants. University of Western Australia Press, Perth.
Brown, T. A., and A. Shrift. 1981. Exclusion of selenium from proteins of selenium-tolerant *Astragalus* species. Plant Physiol. 67:1051–1053.
Canvin, D. T., and Atkins, C. A. 1974. Nitrate, nitrite, and ammonia assimilation by leaves: effect of light, carbon dioxide and oxygen. Planta 116:207–224.
Carter, K. R., N. T. Jennings, J. Hanus, and H. J. Evans. 1978. Hydrogen evolution and uptake of nodules of soybeans inoculated with different strains of *Rhizobium japonicum*. Can. J. Microbiol. 24:307–311.
Christeller, J. T., W. A. Laing, and W. D. Sutton, 1977. Carbon dioxide fixation by lupin root nodules. I. Characterization, association with phosphenolpyruvate carboxylase, and correlation with nitrogen fixation during nodule development. Plant Physiol. 60:47–50.
Coker, G. T., and K. R. Schubert. 1981. Carbon dioxide fixation in soybean roots and nodules. I. Characterization and comparison with N_2 fixation and composition of xylem exudate during early nodule development. Plant Physiol. 67:691–696.
Conrad, R., and W. Seiler. 1979. Field measurements of hydrogen evolution by nitrogen-fixing legumes. Soil Biol. Biochem. 11:689–690.
Cookson, C., H. Hughes, and J. Coombs. 1980. Effects of combined nitrogen on anapleurotic carbon assimilation and bleeding sap composition in *Phaseolus vulgaris* L. Planta 148:338–345.
Dixon, R. O. D. 1967. Hydrogen uptake and exchange by pea root nodules. Ann. Bot. 31:179–188.
– 1972. Hydrogenase in legume root nodule bacteroids: occurrence and properties. Arch. Mikrobiol. 85:193–201.
– 1978. Nitrogenase-hydrogenase interrelationships in rhizobia. Biochim. 60:233–236.
Dixon, R. O. D., Y. M. Berlier, and P. A. Lespinat. 1981. Respiration and nitrogen fixation in nodulated roots of soya bean and pea. Plant Soil 61:135–143.
Dixon, R. O. D., E. A. G. Blunden, and J. W. Searl. 1982. Intercellular space and hydrogen diffusion in pea and lupin root nodules. Plant Sci. Letters 23:109–116.
Evans, H. J., D. W. Emerich, R. J. Maier, F. J. Hanus, and S. A. Russell. 1979. Hydrogen cycling within the nodules of legumes and non-legumes and its

role in nitrogen fixation. *In* J. C. Gordon, C. T. Wheeler, and D. A. Perry (eds.), Symbiotic nitrogen fixation in the management of temperate forests.

Evans, H. J., D. W. Emerich, T. Ruiz-Argueso, R. J. Maier, and S. J. Albrecht. 1980. Hydrogen metabolism in legume-*Rhizobium* symbiosis. Pp. 69–86 *in* W. E. Newton and W. H. Orme-Johnson (eds.), Nitrogen fixation, vol. II. University Park Press, Baltimore.

France, R. C., and C. P. P. Reid. 1982. Interactions of nitrogen and carbon in the physiology of ectomycorrhizae. Can. J. Bot. 61:961–964.

Grove, T. S., A. M. O'Connell, and N. Malajczuk. 1980. Effects of fire on the growth, nutrient content and rate of nitrogen fixation of the cycad *Macrozamia riedlei.* Aust. J. Bot. 28:271–281.

Halliday, J., and J. S. Pate. 1976. Symbiotic nitrogen fixation by coralloid roots of the cycad *Macrozamia riedlei:* physiological characteristics and ecological significance. Aust. J. Plant Physiol. 3:349–358.

Herridge, D. F., and J. S. Pate. 1977. Utilization of net photosynthate for nitrogen fixation and protein production in an annual legume. Plant Physiol. 60:759–764.

Hunter, W. J., C. J. Fahring, S. R. Olsen, and L. K. Porter. 1982. Location of nitrate reduction in different soybean cultivars. Crop Sci. 22:944–948.

Israel, D. W., and W. A. Jackson. 1982. Ion balance, uptake and transport processes in N_2-fixing and nitrate- and urea-dependent soybean plants. Plant Physiol. 69:171–178.

Kirkby, E. A., and M. J. Armstrong. 1980. Nitrate uptake by roots as regulated by nitrate assimilation in the shoot of castor oil plants. Plant Physiol. 65:286–290.

Kucey, R. M. N., and E. A. Paul. 1982. Carbon flow, photosynthesis, and N_2 fixation in mycorrhizal and nodulated faba beans (*Vicia faba* L.). Soil Biol. Biochem. 14:407–412.

Laing, W. A., J. T. Christeller, and W. D. Sutton. 1979. Carbon dioxide fixation by lupin root nodules. II. Studies with ^{14}C-labeled glucose, the pathway of glucose catabolism and the effects of some treatments that inhibit nitrogen fixation. Plant Physiol. 63:450–454.

Lamont, B. 1983. Mechanisms for enhancing nutrient uptake in plants, with particular reference to Mediterranean South Africa and Western Australia. Bot. Rev. 48:597–689.

Layzell, D. B., R. M. Rainbird, C. A. Atkins, and J. S. Pate. 1979. Economy of photosynthate use in N-fixing legume nodules: observations on two contrasting symbioses. Plant Physiol. 64:888–891.

Lim, S. T. 1978. Determination of hydrogenase in free-living cultures of *Rhizobium japonicum* and energy efficiency of soybean nodules. Plant Physiol. 62:609–611.

Mahon, J. D. 1977. Respiration and the energy requirement for nitrogen fixation in nodulated pea roots. Plant Physiol. 60:817–821.

– 1979. Environmental and genotypic effects on the respiration associated with symbiotic nitrogen fixation in peas. Plant Physiol. 63:892–897.

Maier, R. J., N. E. R. Campbell, F. J. Hanus, F. B. Simpson, S. A. Russell, and H. J. Evans. 1978. Expression of hydrogenase activity in free-living *Rhizobium japonicum.* Proc. Natl. Acad. Sci. USA 75:3258–3262.

Main, A. R. 1981. Plants as animal food. Pp. 342–361 *in* J. S. Pate and A. J. McComb (eds.), The biology of Australian plants. University of Western Australia Press, Perth.

Malajczuk, N., and T. Grove. 1977. Legume understorey biomass, nutrient content and nitrogen fixation in eucalypt forests of southwestern Australia. Pp. 36–39 *in* Nutrient cycling in indigenous forest ecosystems. CSIRO, Division of Land Resources and Management, Perth.

Monk, D., J. S. Pate, and W. A. Loneragan. 1981. Biology of *Acacia pulchella* R.Br. with special reference to symbiotic nitrogen fixation. Aust. J. Bot. 29:579–582.

Pate, J. S. 1977. Functional biology of dinitrogen. Pp. 473–517 *in* R. W. F. Hardy and W. S. Silver (eds.), A treatise on dinitrogen fixation, section III, Biology. Wiley, New York.

– 1980. Transport and partitioning of nitrogenous solutes. Ann. Rev. Plant Physiol. 31:313–340.

– 1983. Patterns of nitrogen metabolism in higher plants and their ecological significance. Pp. 225–255 *in* J. A. Lee (ed.), Nitrogen as an ecological factor. British Ecological Society, London.

– 1984. The carbon and nitrogen nutrition of fruit and seed–case studies of selected grain legumes. Pp. 41–82 *in* D. R. Murray (ed.), Seed physiology, Vol. I. Development. Academic Press, New York.

Pate, J. S., C. A. Atkins, and R. M. Rainbird. 1981. Theoretical and experimental costing of nitrogen fixation and related processes in nodules of legumes. Pp. 105–116 *in* A. H. Gibson and W. E. Newton (eds.), Current perspectives in nitrogen fixation. Elsevier, Amsterdam.

Pate, J. S., and P. J. Dart. 1961. Nodulation studies in legumes. IV. The influence of inoculum strain and time of application of ammonium nitrate on symbiotic response. Plant and Soil 15:329–346.

Pate, J. S., and B. Dell. 1984. Economy of mineral nutrients in sandplain species. Pp. 227–258 *in* J. S. Pate and J. S. Beard (eds.), Kwongan: plant life of the sandplain. University of Western Australia Press, Nedlands.

Pate, J. S., and D. B. Layzell. 1981. Carbon and nitrogen partitioning in the whole plant–a thesis based on empirical modelling. Pp. 94–134 *in* J. D. Bewley (ed.), Nitrogen and carbon metabolism. Martinus Nijhoff, The Hague.

Pate, J. S., D. B. Layzell, and C. A. Atkins. 1980. Transport exchanges of carbon, nitrogen and water in the context of whole plant growth and functioning–case history of a nodulated annual legume. Ber. Deutsch. Bot. Ges. 93:243–255.

Pate, J. S., W. W. Williams, and P. Farrington. in press. Lupins. *In* R. J. Summerfield and E. H. Roberts (eds.), Grain legume crops. Granada, London.

Phillips, D. A. 1980. Efficiency of symbiotic nitrogen fixation in legumes. Ann. Rev. Plant Physiol. 31:29–49.

Rainbird, R. M., C. A. Atkins, J. S. Pate, and P. Sanford. 1983. Significance of hydrogen evolution in the carbon and nitrogen economy of nodulated cowpea. Plant Physiol. 71:122–127.

Raven, J. A., and F. A. Smith. 1976. Nitrogen assimilation and transport in vascular plants in relation to intracellular pH regulation. New Phytol. 76:415–431.

Rawsthorne, S., F. R. Minchin, R. J. Summerfield, C. Cookson, and J. Coombs. 1980. Carbon and nitrogen metabolism in legume root nodules. Phytochem. 19:341–355.

Riley, I. T., and M. J. Dilworth. 1982. Cobalt and the contribution of grown and lateral nodules to nitrogen fixation of *Lupinus angustifolius* L. New Phytol. 90:717–721.

Rosenthal, G. A., D. L. Dehlman, and D. H. Janzen. 1976. A novel means for dealing with L-canavanine, a toxic metabolite. Science 192:256–258.

Ryle, G. J. A., C. E. Powell, and A. J. Gordon. 1978. Effect of source of nitrogen on the growth of Fiskeby soya bean: the carbon economy of whole plants. Ann. Bot. 42:637–648.

– 1979. The respiratory costs of nitrogen fixation in soyabean, cowpea and white clover. II. Comparisons of the costs of nitrogen fixation and the utilization of combined nitrogen. J. Exp. Bot. 30:145–153.

Schrader, L. E., and R. J. Thomas. 1981. Nitrate uptake, reduction and transport in the whole plant. Pp. 49–93 *in* J. D. Bewley (ed.), Nitrogen and carbon metabolism. Martinus Nijoff, The Hague.

Schubert, K. R., J. A. Engelke, S. A. Russell, and H. J. Evans. 1977. Hydrogen reactions of nodulated leguminous plants. I. Effect of rhizobial strain and plant age. Plant Physiol. 60:651–654.

Schubert, K. R., N. T. Jennings, and H. J. Evans. 1978. Hydrogen reactions of nodulated leguminous plants. II. Effects of dry matter accumulation and nitrogen fixation. Plant Physiol. 61:398–401.

Schubert, K. R., and G. J. A. Ryle. 1980. The energy requirements for nitrogen fixation in nodulated legumes. Pp. 85–96 *in* R. J. Summerfield and A. H. Bunting (eds.), Advances in legume science. Royal Botanic Gardens, Kew, England.

Silsbury, J. H. 1979. Growth, maintenance and nitrogen fixation of nodulated plants of subterranean clover (*Trifolium subterraneum* L.). Aust. J. Plant Physiol. 6:165–176.

Wilcockson, J., and D. Werner. 1979. Organic acids and prolonged nitrogenase activity by non-growing, free-living *Rhizobium japonicum.* Arch. Microbiol. 122:153–159.

Witty, J. F., F. R. Minchin, and J. E. Sheehy. 1983. Carbon costs of nitrogenase activity in legume root nodules determined using acetylene and oxygen. J. Exp. Bot. 34:951–963.

Zablotowicz, R. M., S. A. Russell, and H. J. Evans. 1980. Effect of the hydrogenase system in *Rhizobium japonicum* on the nitrogen fixation and growth of soybeans at different stages of development. Agron. J. 72:555–559.

11 Ecological patterns in xylem anatomy

PIETER BAAS

Introduction

Wood or secondary xylem provides a woody plant with a complex tissue for sap transport (at relatively low resistance to flow, and often over long distances), for mechanical strength, and for metabolic processes such as storage and mobilization of reserve carbohydrates and lipids. These vital functions can lead to conflicting demands on wood structure for physiological fitness. Consequently, the study of adaptive strategies in xylem evolution is a complex field, bestrewed with speculative hypotheses. Here I shall advance some points of view as a descriptive wood anatomist, fully aware of the limitations of the comparative method. If some of these ideas should lead to more sophisticated ecological analyses of wood anatomical patterns and to experimentation on a larger scale to establish the precise functions of some of the wood anatomical characteristics discussed here, the purpose of this chapter will have been served.

The study of ecological patterns in xylem anatomy is as old as wood anatomy itself. For instance, the discussions of growth-ring variation in ring-porous woods and conifers in relation to climate and of the inconspicuousness of ring boundaries in tropical trees by Antoni van Leeuwenhoek in the seventeenth century showed a profound understanding of the influence of growth rate and rhythm on wood structure and properties (Baas 1982a). Ecological wood anatomy as a means to understand factors governing wood evolution has its roots in the late-nineteenth-century debates on the systematic value of wood anatomical characteristics. In contrast to those who claimed that natural families are characterized by specific wood anatomies, the botanist Vesque held the view that because of its important physiological functions, the secondary xylem could never escape adaptive pressures, and therefore its anatomy would be too plastic to be of great taxonomic value (Vesque 1889).

In the twentieth century, Bailey and Tupper's (1918) brilliant reconstruction of the major trends in xylem evolution, supported and extended by many later studies, provided an inspiring guideline for formulating functionally adaptive hypotheses in attempts to give a causal explanation of these evolutionary trends (Baas 1973, 1976, 1982b; Baas et al. 1983; Van der Graaff and Baas 1974; Carlquist 1975, 1977, 1980; Van den Oever et al. 1981). These and many other studies have revealed weak or strong correlations between various wood anatomical characteristics and ecological factors such as water availability, temperature, and seasonality. Functionally adaptive interpretations of these ecological trends remain highly speculative and have sometimes given rise to controversy, because the precise functional significance of variation in most wood anatomical traits has never been experimentally established.

In this chapter, the salient ecological trends, as far as they are known, will be reviewed for characteristics of vessels, fibers, axial parenchyma, and rays. Interpretation of these trends in terms of adaptive strategies must unfortunately remain tentative and will be balanced with alternative explanations in terms of functionless trends due to correlative restraints. Wood formation and differentiation constitute a complex of closely integrated developmental processes, and an adaptive change in, for instance, vessel diameter or type of perforation may imply a number of correlated changes in other characteristics that are not necessarily of functional significance.

By comparing the anatomical characteristics of fast- and slow-growing species or individuals, I shall also explore possible constraints of wood structure on primary productivity.

Methodology

Ecological patterns in wood anatomy have been studied in a number of fundamentally different ways, and it is perhaps appropriate to discuss briefly the advantages and limitations of the different approaches.

1. Studies of ecological variations within a single species or, preferably, even within clonal material of identical genotypes permit analysis of the phenotypic responses of wood formation and differentiation to environmental factors. Although this type of study is very important for forestry and forest-products research, it has little bearing on evolutionary strategies. The latter should result in the selection of different genotypes, optimally suited for specific ecologies. However, detailed knowledge of phenotypic plasticity remains essential for a critical appraisal of the ecological trends established by the methods to be discussed later, and the degree of plasticity in wood structure – as a heritable trait – may play a role in the evolutionary success of a species.

2. Studies of ecological variations within genetically diverse, but phylogenetically related, material in genera or families with wide ecological

amplitudes are likely to reveal ecological patterns in genetic characteristics [see Carlquist's article on Compositae (Carlquist 1966) and my own work on *Ilex* (Baas 1973)]. The requirements for comparability (i.e., in habit and general wood histology) are usually obeyed in closely related taxa such as the species of one genus. In taxa of higher rank (families and orders), the diversity in habit and wood structure may reduce the comparability. Often, however, closely related taxa are more or less similar in their ecological preferences (taken broadly and not considering microclimatological and edaphic preferences), and the number of genera with truly wide ecological ranges (e.g., ranging from arid to mesic, or from hot tropical to cool temperate) is quite limited. This puts a limit to this approach in ecological wood anatomy.

3. By comparing all woody species in a local flora or vegetation type of known ecology, the incidence of certain wood anatomical characteristics in these floras or vegetation types can reveal trends that may at least partly be due to selection of the fittest wood anatomical types for each category of physical environment (Kanehira 1921; Webber 1936; Carlquist 1975, 1977; Baas 1976; Baas et al. 1983). On the other hand, the comparability of data from remotely related species in a woody flora is restricted, and the floristic composition of a region as determined by its phytogeographical history may influence the trends. The limited range of wood anatomical variations within a single genus or family and the relative incidences of their species thus affect the results (Baas et al. 1983).

In all three methods of analyzing ecological patterns in wood anatomy, great care must be taken to eliminate the well-known effects of radial and axial variations within a single tree or shrub. This is not always possible if one wishes to compare the mature wood of tall trees with, for instance, alpine and arctic or desert shrubs. There simply are no "standard positions" in the trunk of an emergent tree that are equivalent to parts of a multistemmed dwarf shrub. Yet comparisons of the wood anatomies of these extreme habit categories are of functional interest.

All the pitfalls in the methodology of establishing ecological trends can, in my opinion, best be overcome by pursuing the aforementioned approaches together as complementary tools. They gain in value if their results reinforce one another. In cases of conflicting results, more critical and detailed analyses should be carried out.

The evolutionary basis

The major evolutionary trends in the xylem anatomy of angiosperms, as established by Bailey and Tupper (1918) and elaborated by later workers, have been discussed in many studies providing direct or cirumstantial evidence to support them. The Baileyan transformation series was based primarily on comparisons of extant plants, but a convincing starting point for the series was provided by the fossil records of gymnosperms and woody cryptogams. Later paleobotanical studies (Barghoorn 1964; Page 1981; Baas 1982b) strongly supported most of the postulated phylogenetic trends. Moreover, independently developed views on angiosperm phylogeny are in line with the distribution of primitive and advanced wood anatomical characteristics (*sensu* Bailey and Tupper) in extant dicotyledons (Dickison 1975; Stern 1978; Baas 1982b). Here a very brief summary must suffice (Figure 11.1).

Early angiosperms, or at least protoangiosperms, had vesselless xylem* formed by cambium containing very long fusiform initials. This type of wood survives in a small number of extant angiosperms. In the course of early angiosperm diversification, vessels with scalariform perforation plates, resembling scalariformly pitted tracheids, evolved. Subsequently, simply perforated vessels composed of relatively short vessel members (derived from short cambial initials) and alternate-wall pitting evolved. Solitary vessels preceded vessel multiples or clusters. Concomitant with the specialization of the conductive tissue, libriform fibers (i.e., fibers with few and much-reduced bordered or simple pits) evolved via the intermediate phase of fiber-tracheids from the ancestral type, with tracheids (i.e., fiber-like cells with large bordered pits) exclusively for both conduction and mechanical support. Axial parenchyma was scarce and most probably diffuse or diffuse-in-aggregates in the early angiosperms; a common trend of specialization was the development of more abundant, predominantly paratracheal, parenchyma (i.e., associated with the vessels) of the vasicentric, aliform, or confluent-to-banded type. The rays of the early angiosperms were markedly heterocellular (i.e., composed of upright and procumbent cells) and of two distinct sizes; they preceded exclusively uniseriate or exclusively multiseriate rays composed of procumbent cells only.

The question of reversibility or irreversibility of these major evolution-

* Young's contention (1981) that the early angiosperms had vessels and that the vesselless xylem in some extant angiosperms is due to secondary loss through paedomorphosis has been criticized on methodological as well as wood anatomical grounds by several authors. Young's main reason (viz., that the origin of vessels should not have appeared in evolution too many times) is unconvincing, if one considers that other specialized features such as libriform fibers and simple vessel perforations must have arisen countless times within the angiosperms. Thus, there is no reason to assume that scalariform perforations should not have evolved several times in descendants of vesselless ancestors with scalariform pitting in their tracheid walls.

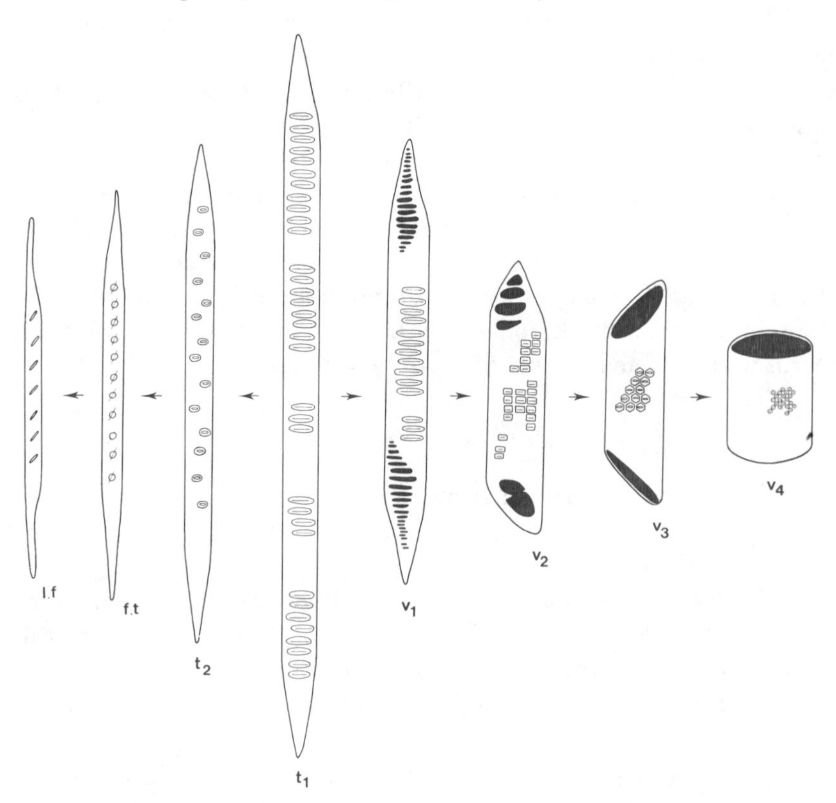

Figure 11.1. Summary of the major evolutionary trends (arrows) of perforate and imperforate axial elements in angiosperm wood; $v_1 - v_4$ = vessel members of progressive specialization levels; $t_1 - t_2$ = tracheids of progressive specialization levels; f.t = fiber-tracheid; l.f = libriform fiber (adapted from Bailey and Tupper 1918).

ary trends has been discussed at some length by Carlquist (1980) and Baas (1973, 1982b). In my opinion, the major qualitative trends for vessel members and fibers (i.e., the phylogenetic series from scalariform to simple perforations and from fiber-tracheids to libriform fibers) can be considered to be largely irreversible. Cambial initial length must have been subjected to a considerable amount of reversibility, although the reversion from short to long initials most probably was never of such magnitude that a specialized taxon gave rise to descendents with fusiform initials as long as those of the primitive angiosperms. For parenchyma and ray specialization, the possibilities for deviation from or reversion of the major trends are likely to have been much greater (Baas 1982b).

In part of the literature on evolutionary trends in wood anatomy, great importance is also attached to characteristics like vessel diameter and vessel distribution (i.e., diffuse versus ring porosity). In phylogenetically primitive woods with exclusively scalariform perforations there are indeed constraints on vessel diameter and distribution (almost always narrow and diffuse), but in moderately to highly advanced woods these characters are probably unstable in an evolutionary sense; in other words, although genetically determined to some extent, the change from the primitive to the specialized condition was and is highly reversible, so that either very wide or very narrow vessels can be typical for phylogenetically specialized taxa.

The foregoing summary of wood anatomical specialization and diversification may serve as background for the following survey of ecological patterns in the xylem anatomy of extant woody plants.

Survey of ecological trends

Vessel elements

Ecological trends in vessel characteristics have been extensively discussed elsewhere (Baas 1976; Baas et al. 1983; Fahn et al. in press). In summary, two partly parallel trends emerge from floristic analyses as well as from studies of anatomical variation within widely distributed genera and families.

1. Within mesic floras or taxa, vessel member length and vessel diameter decrease, and vessel frequency (number per square millimeter) and the incidence of scalariform perforations increase, with increasing latitude or altitude of provenance, that is, in cooler environments. The incidence of spiral (tertiary) thickenings increases from tropical to more temperate, seasonal climates.
2. If mesic floras or taxa are compared with xeric ones, woods from drier habitats appear to have shorter vessel members, and the incidence of scalariform perforation plates is reduced (to nil in desert floras). Vessel diameter may be reduced in part of the arid taxa, but many xeric species show two vessel size classes: numerous, very narrow vessels in addition to wide vessels that may even exceed those of more mesic tropical and temperate taxa in maximum diameter. The incidence of spiral thickenings in extremely arid floras is fairly low (Baas et al. 1983) to moderately high (Webber 1936), but it is very high in the moderately arid Mediterranean or sclerophyllous vegetation (Webber 1936; Baas et al. 1983).

The foregoing statements are quantitatively illustrated in Figures 11.2–11.4 and in Tables 11.1 and 11.2. The separation of tropical–temperate–arctic and mesic–xeric trends is partly artificial. Prevailing temperature regimes and water availability vary together and result in interdependent effects on the water balance of a plant, and consequently they must have

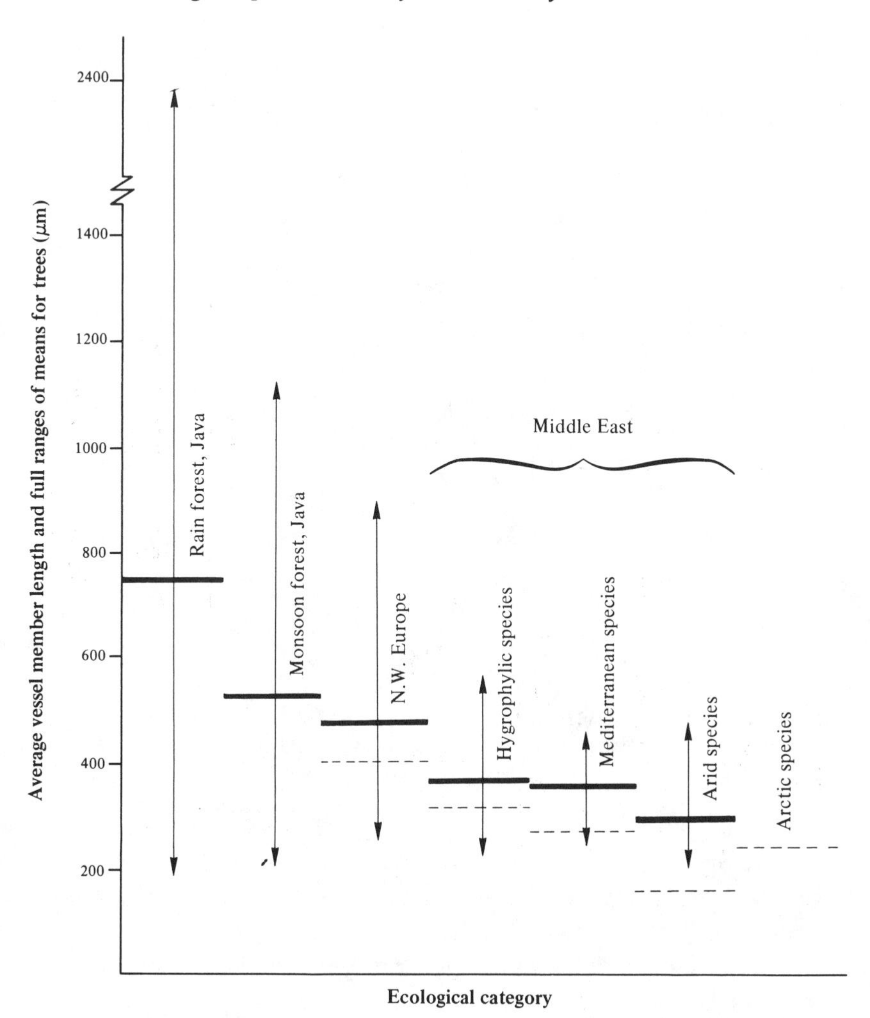

Figure 11.2. Average values for vessel member length in different ecological categories. For trees, the full ranges of average species values and the floristic averages (solid bar) are given. For (erect) shrubs, only floristic averages are given (broken lines) (reproduced from Baas et al. 1983, with permission of the *IAWA Bulletin*).

been interactive in selective pressures responsible for the ecological trends.

On the assumption that primitive characteristics (great vessel member length, scalariform perforations) were retained in environments in which they were of adaptive value when they arose in evolution, the ecological patterns in the xylem anatomy of extant plants point to tropical rain forests

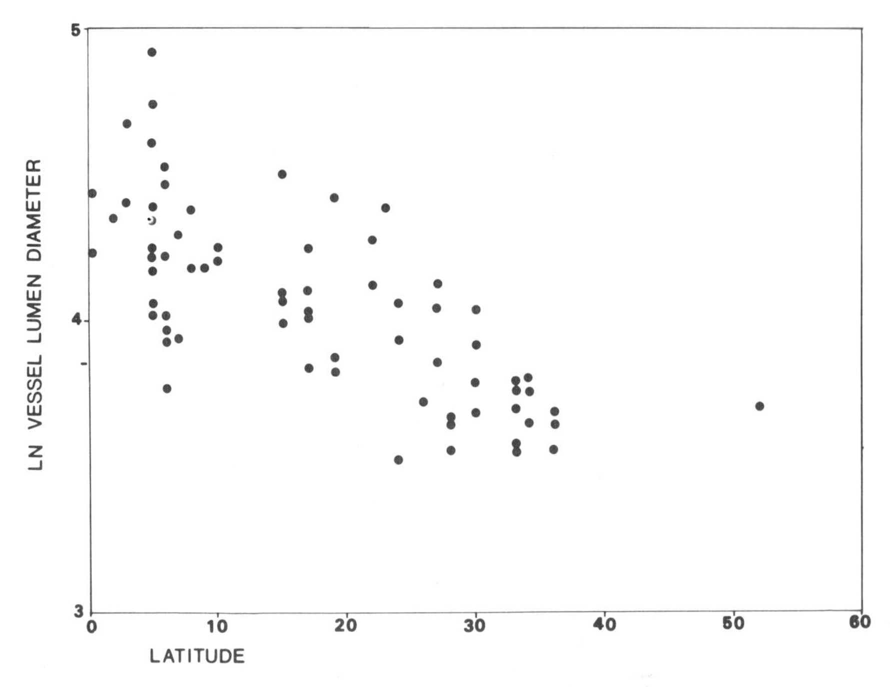

Figure 11.3. Correlation between vessel diameter and latitude of provenance in the genus *Symplocos* (reproduced from Van den Oever et al. 1981, with permission of the *IAWA Bulletin*).

of submontane to montane altitude for the early origin of vesselled angiosperms. In fact, this is still the predominant ecology of most vesselless angiosperms (Carlquist 1975). Radiation into other environments, but also diversification within the rain forest, presumably involved a substantial shortening of the fusiform initials of the cambium, resulting in a wide range of length classes in the mesic forests and more uniformly short initials (and thus vessel members) in arid floras and cool temperate to arctic floras (Figure 11.2). Xeric conditions apparently favored selection for simple perforations and to some extent for two distinct vessel diameter classes. In the cold arctic or alpine environment, the primitive type of vessel perforation could be retained, as well as the more or less uniformly narrow vessels in a diffuse distribution pattern. It is by no means certain that selective pressures of the arid or cool environments were instrumental in the evolutionary changes of the vessel member. It is quite possible that the different vessel member types and size classes had already evolved in the less extreme temperate or tropical forests and that in the extreme habitats wood anatomically suitable types were selected from the available

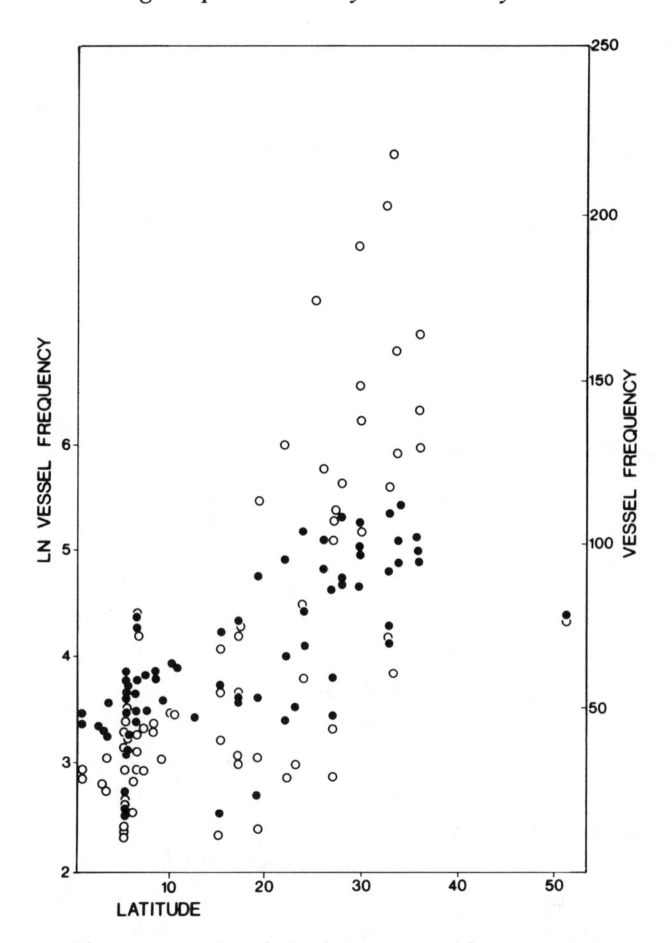

Figure 11.4. Correlation between vessel frequency and latitude or provenance in the genus *Symplocos*. Solid dots: ln-transformed data. Open circles: raw data (reproduced from Van den Oever et al. 1981, with permission of the *IAWA Bulletin*).

types coexisting in the milder biotopes. This is very likely for a characteristic like cambial initial or vessel member length: An analysis of different floras has revealed that arctic or arid floras show the bottom end of the much wider range occurring in the tropical forests (Figure 11.2). The floras in extreme habitats are thus not characterized by "new" or "exclusive" wood anatomical features, but show a much higher incidence of characteristics that are at an extreme end of the variation within less extreme habitats.

In terms of function, the total vessel length distribution is, of course, a more important characteristic than vessel member length, but the avail-

Table 11.1. *Frequencies of occurrence of scalariform perforations and spiral thickenings in trees and shrubs from various climatic zones*[a]

Climatic zone	Percentage of genera with scalariform perforations	Percentage of genera with spiral thickenings on the walls
Tropics		5–15
Rain forests	(0) 5–8	
Monsoon forests	1–5	
Mountain forests	15–33	
Subtropical to warm temperate		
Various regional floras	13–22	38–66
Mountain flora	16–40	
Deserts	0	22(–52[b])
Cool temperate to arctic	23–53	

[a] Data from Baas (1973, 1976), Baas et al. (1983), Van der Graaff and Baas (1974), and Webber (1936).
[b] Data of Webber (1936).

able data are insufficient to deduce any ecological trend. From the strong correlation between diameter classes and total vessel length within temperate woods (Zimmermann 1982) (Figure 11.5), one might expect a general correlation between vessel diameter and total vessel length. By inference one could forecast the tropical monsoon and rain forest floras and certain components of extremely arid floras to have the longest vessels (in part of the arid flora mixed with narrow, short vessels), and the cool temperate and arctic species to have the shortest vessels (Baas et al. 1983) (Figure 11.2). A large number of desert species (viz., those shrubs with numerous, narrow vessels) are also expected to have low values for total vessel lengths.

Fibers

Table 11.3 gives the incidences of taxa with fiber-tracheids in different regional floras or vegetation types. As I suggested previously (Baas 1982b), the low percentage of genera with fiber-tracheids in the flora of Java is probably a reflection of their scarcity in the lowland; tropical mountain floras have a higher incidence of fiber-tracheids. Although further critical analyses are badly needed, the emerging ecological trend of retention of primitive fibers (i.e., fiber-tracheids) in warm to cool temperate floras and their paucity in the tropical lowland flora and in arid habitats has a parallel in the much better established ecological trends for vessel perforation plate type (Table 11.1). Scalariform perforation plates

Table 11.2. *Annual rainfall and vessel member length*[a]

Florula	Approximate annual rainfall (cm)	Average vessel member length (μm)
Southwest Australia		
Karri Forest understory shrubs	150	385
Bog shrubs	150	361
Coastal shrubs	100	342
Sandheath shrubs	65	307
Desert shrubs	22	217
Israel		
Hydrophyllic shrubs (near water)	40–100	308
Mediterranean shrubs	40–100	275
Desert shrubs	3–25	150

[a] Based on data by Carlquist (1977) and Fahn et al. (in press) on florulas in southwest Australia and Israel, respectively.

are, like fiber-tracheids, very rare (to absent) in both the tropical lowland tree flora and in xeric shrubs or treelets. This parallel may be entirely due to the phylogenetically and ontogenetically integrated specialization of different cell types of the secondary xylem, because regardless of ecological conditions, fiber-tracheids tend to be associated with primitive perforation plates (Chalk 1983). As discussed in a subsequent section of this chapter, these parallel trends lead to a paradox if interpreted rigidly in terms of functional adaptation.

Other fiber characteristics like length, wall thickness, and presence of spiral thickenings or septation have not been analyzed for ecological trends on any comprehensive scale. Fiber length depends on cambial length and on the amount of intrusive growth prior to fiber maturation. Ecological trends probably are the same as those for vessel member length. This is borne out by studies within widely distributed genera (Baas 1973; Van der Graaff and Baas 1974; Van den Oever et al. 1981). Floristic analyses for the Middle East (Fahn et al. in press) show a similar parallel. It seems that the relative amount of intrusive growth does not show significant ecological tendencies and that the ecological fiber length trends are mainly governed by the trend for fusiform initial length, which also determines vessel member length.

The presence of spiral thickenings in fibers is to some extent associated with spiral thickenings in vessels, and there is no doubt that subtropical and temperate floras will show a higher incidence of fibers with spirals than the tropical floras. However, precise figures are not available.

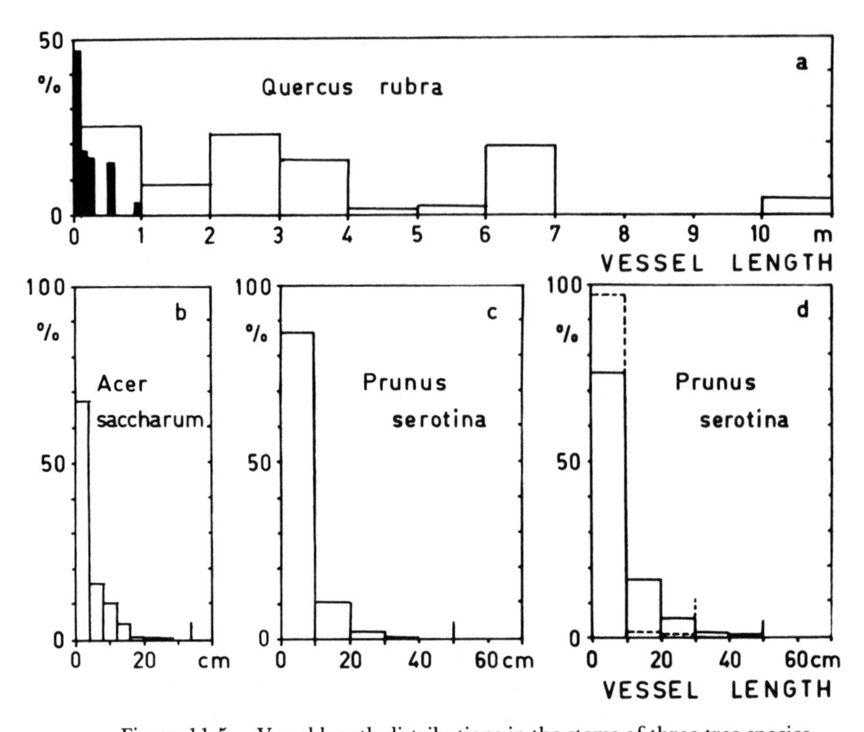

Figure 11.5. Vessel length distributions in the stems of three tree species. Note that the percentage scale is the same in all diagrams (a–d), but the vessel length scale in (a) is different from those in (b–d). (a) Vessel length distribution in *Quercus rubra.* The 1-m bars show the length classes of the wide earlywood vessels; the narrow black bars are the much shorter narrow latewood vessels. (b and c) In *Acer saccharum* and *Prunus serotina* the vessels are not only narrower than in *Quercus* but also much shorter. The longest vessel lengths are indicated by small vertical bars at 34 and 50 cm. There are so few vessels in the longer length classes that the bars often do not exceed the baseline of the graph. (d) Vessel length distribution in *P. serotina* shown separately for relatively wide earlywood vessels (solid lines) and narrower latewood vessels (dashed lines) (reproduced from Zimmermann 1982).

Fiber wall thickness varies tremendously from species to species and also within species or individual trees, often depending on the growth rate fluctuations of the latter. Each biotope potentially harbors species with light as well as heavy woods (i.e., with thin-walled as well as with thick-walled fibers). The woody flora of the Middle East (Fahn et al. in press) shows a high percentage of species with thick-walled fibers (ca. 50%) in both arid and Mediterranean localities; most species growing along streams or lakes have more thin-walled fibers. Data are inadequate to postulate generalized ecological trends for fiber wall thickness. If one considers that the tropical lowland forest yields both extremely heavy and

Table 11.3. *Frequencies of occurrence of fiber-tracheids in trees and shrubs in different regional floras*[a]

Woody flora of	Percentage of genera (or species) with fiber-tracheids
Central Europe	42
Japan	31
New Zealand	34
Middle East	
Arid flora	11
Mediterranean flora	32
Java	18
World (on species basis)	33

[a] Data from Baas (1982b) and Fahn et al. (in press).

very light timbers, it can be anticipated that any ecological trend will be rather weak.

Axial parenchyma

Axial parenchyma tends to be more abundant in tropical species than in temperate ones (Baas 1973, 1982b). Mesic–xeric trends have not been reliably established. Any interpretation of the functional significance of ecological trends in parenchyma distribution and abundance suffers from incomplete knowledge of the incidence of living fibers (nucleated, often septate) in many species, which can take the role of axial parenchyma (Carlquist 1975).

Rays

Ecological trends for ray composition are not very clear (Table 11.4). North temperate floras have the highest incidence of specialized homocellular rays, but it is puzzling that they are very rare in the South temperate flora of New Zealand (ecological trends for other wood traits apply well to New Zealand). Obviously, all types of environments, mesic or arid, as well as hot or cool, accommodate a majority of woody species containing various proportions of upright cells in addition to procumbent cells. This does not imply that all heterocellular rays are relictual or primitive in these floras. Paedomorphosis or juvenilism (i.e., the retention of characteristics of first-formed secondary xylem near the pith, where rays show a high proportion of upright cells) offers a possibility for reversion of the specialization series from heterocellular to homocellular (Carlquist 1962; Baas 1982b), and rays largely composed of upright cells (i.e., of the

Table 11.4. *Frequencies of occurrence of homocellular rays (composed of procumbent cells only) in trees and shrubs in different regional floras*[a]

Woody flora of	Percentage of genera (or species) with homocellular rays
Central Europe	16–25
Japan	31
New Zealand	3
Middle East	
Arid flora	8
Mediterranean flora	21
Various tropical floras	13–22
World (on species basis)	21

[a] Data from Baas (1982b) and Fahn et al. (in press).

juvenilistic type) are especially common in the arid flora of the Middle East in otherwise highly specialized families or genera (Fahn et al. in press).

Ray height shows latitudinal and sometimes also altitudinal trends parallel to those for vessel member and fiber length within widely distributed genera (Baas 1973; Van der Graaff and Baas 1974; Van den Oever et al. 1981): Species of cool climates show lower rays than species from the tropical lowland. Floristic analyses of ray size are not available.

Functional significance of the ecological trends

Vessels: safety and efficiency in the ascent of sap

Trends in vessel diameter, vessel perforation plate type, vessel frequency, vessel member length, total vessel length, vessel wall sculpturing, and fiber type have all been discussed in terms of their input to the safety and efficiency of water transport (Zimmermann and Brown 1971; Zimmermann 1978, 1983; Carlquist 1982a, 1982b; Baas et al. 1983). For efficiency or maximal conductivity, vessel diameter is no doubt the most important variable, because the former is proportional to its fourth power. Minor changes in vessel diameter can thus have a greater effect on conductivity than more substantial changes in vessel frequency. Both diameter and frequency are of importance in the safety of water conduction: Embolisms are less likely to be initiated in very narrow vessels than in wide vessels, and if embolisms do occur in wood with a high vessel frequency, the ascent of sap will not be significantly affected by the incapacity of part of its numerous vessels. Total vessel lengths and lateral connections between individual vessels in the three-dimensional vessel network are, of

Table 11.5. *Maximum vessel diameters (as a measure of efficiency) and incidences of species with two different vessel diameter classes (a combined measure for safety and efficiency) in shrubs from different vegetation types in Israel and adjacent regions*[a]

Vegetation type or soil type	Average maximum tangential vessel diameter (μm)		Percentage of species with two vessel size classes	
Desert shrubs	82		64	
Rock crevices		74		61
Wadis and oases		119		41
Stony soils		49		50
Xerohalophytes		52		97
Hydrohalophytes		71		88
Desert sands		100		65
Mediterranean shrubs	73		36	
Maquis and dwarf shrubs		71		38
Pseudosavanna		130		0
High mountains		48		40
Hydrophyllic shrubs	88		29	
Climbers	127		50	

[a] Data from Fahn et al. (in press).

course, also crucial in the spreading or trapping of embolisms, as well as in resistance to flow per unit volume of sapwood. As explained before, our knowledge of the latter aspects is too limited for a meaningful discussion of ecological trends, let alone their functional significance.

Considering the ecological trends for vessel diameter and frequency, it becomes evident that for mesic floras the tropical lowland forests show the highest average efficiency and lowest degree of safety, and in cool tropical mountain floras as well as temperate floras there is a shift toward a less efficient but safer hydraulic architecture. Alpine and arctic floras show the extreme of this tendency. Within the temperature floras, the ring-porous woods with their narrow outer layer of sapwood provide a special case, with highly efficient but very vulnerable wide earlywood vessels and inefficient but safe narrow latewood vessels (Zimmermann and Brown 1971). If we compare xeric and mesic floras, a complex picture emerges: In arid floras (e.g., deserts and dry steppes) a number of shrubs and small trees have provisions for safe but inefficient sap transport; another category shows a hydraulic architecture suitable for both efficient and safe conduction through the possession of two vessel diameter classes; a third category provides only for high conductive efficiency (Baas et al. 1983). The latter category presumably can dispose of subterranean water. Throughout the comparisons of xeric and mesic woody floras (Baas et al. 1983) (Tables 11.5 and 11.6), the tendency in drier areas for higher conductive effi-

Table 11.6. *Maximum vessel diameters and incidences of species with two different vessel diameter classes in trees from different local floras and ecological categories*[a]

Species	Average maximum tangential vessel diameter (μm)	Percentage of species with two vessel size classes
Desert species in Israel (mainly from wadis and oases)	193	25
Mediterranean species from Israel	109	38
Hydrophyllic species from Israel	168	30
Temperate species from northwest Europe	100	23
Rain forest species from Java	132	3
Monsoon forest species from Java	164	15

[a] Data from Baas et al. (1983).

ciency is at least as strong as the tendency for increased safety. These results are in conflict with previous views (e.g., Carlquist 1975) that drought is mostly associated with devices for optimal safety (viz., numerous and very narrow vessels).

The ecological trend for the incidence of scalariform perforation plates (Table 11.1) calls for a similar explanation: Ecotypes with reputedly high (continuous or periodical) sap conduction (i.e., the tropical lowland rain forest and arid woody floras) have very few species with scalariform perforation plates, whereas vegetations with lower transpiration rates of the tropical montane and temperate to arctic zones have retained a relatively high proportion of taxa with this primitive type of perforation plate. This is, of course, what one would expect if scalariform plates add to the resistance to flow. Petty (1978) presented experimental evidence for resistance to flow at the site of scalariform perforation plates in birch.

For both narrow vessels and scalariform perforation plates in cool temperate regions, functional significance has been claimed in terms of increased safety (viz., prevention or localization of embolism) (Zimmermann 1978, 1982). Whatever the actual role of these attributes may be in the safety of the ascent of sap, their predominance in cool temperate or tropical montane floras does not have to be explained by selective pressure favoring their retention or development. The lack of high demands on the efficiency of the conductive system in these habitats with relatively low transpiration rates may simply be accounted for by the fact that there has been no selective pressure for *elimination* of these characteristics, which were ubiquitous among the early angiosperms. Those primitive woods with numerous narrow and scalariformly perforated vessels qualify as safe but rather inefficient conductors.

Selective pressure for increased hydraulic efficiency is probably sufficient to explain the high incidence of simple perforations and wide vessels in tropical lowland forests and some components of the arid zone floras. Selection for increased safety has perhaps been instrumental only in those arid zone species and ring-porous temperate woods in which two diameter size classes of vessels combine the virtues of safe and efficient transport.

The role of vessel member length in the ascent of sap remains elusive. All attempts to ascribe a functional significance to short vessel members in arid or cool temperate to arctic floras are highly speculative and seem artificial. The presumed superior mechanical properties of vessels built of many short members (Carlquist 1975) to withstand high negative pressures is open to criticism. The more numerous perforation rims hypothesized to provide additional strength might just as well be places of weakness because of the discontinuities in the superimposed wall layers and compound middle lamella. In fact, in wood surfaces exposed for scanning electron microscopy by splitting, intact vessel member ends are often observed because the vessel has split at the interface of the compound middle lamella and secondary wall or at the S_1 and S_2 layers of the latter. Moreover, in a coherent tissue like wood, it is most unlikely that negative pressures will ever result in actual damage to individual cells; both vessels and surrounding cells are probably sufficiently flexible to respond to high tensions without mechanical failure. Such failures should have been noted in wood anatomical studies, especially those with the scanning electron microscope, but they have never been reported, so far as I know. An alternative hypothesis (Carlquist 1982a, 1982b) proposes that vessels built of short members offer more recesses (perforation rims, tails of vessel members) for trapping the embolisms caused by cavitation due to high tensions in the water column in xerophytes or in the top of tall trees, or due to thawing of frozen xylem sap in cool temperate to arctic trees and shrubs. This hypothesis has the merit that it is compatible with the predominance of short vessel members in hot arid as well as cool mesic environments; it should be put to the test by simple experimentation.

Fiber-tracheids also contribute to some extent to the conductive capacity of the secondary xylem (Braun 1970); they must also contribute to the safety of sap ascent by virtue of their high number, narrow lumens, and restricted length. Their relatively high incidence in temperate floras and their scarcity in tropical lowland floras and arid vegetations cannot be explained in terms of selective advantage, because additional conductive capacity and safety are least needed in the temperate environment. The lack of selective pressure to eliminate primitive wood anatomical characteristics associated with scalariform perforations in the temperate to arctic zones provides a more simple and satisfactory explanation.

Concepts of vulnerability and mesomorphy

In the foregoing discussion of ecological trends and their possible significance for efficiency and safety of conduction, the concepts of vulnerability and mesomorphy, as introduced by Carlquist (1977), have not been considered.

The vulnerability index (= mean vessel diameter divided by mean vessel frequency) was said to be indicative of the sensitivity of a species to the risks of embolisms: Low vulnerability values of woods with numerous, narrow vessels would indicate safety; high vulnerability values are typical for efficient water conductors with wide, infrequent vessels.

Mesomorphy (= vulnerability multiplied by mean vessel member length) was said to be a measure of the water availability of the species, with high values being typical for species with a mesic ecology.

The lack of predictive value and the inadequacy of these concepts were demonstrated by Van Vliet (1979) in his study of the wood of *Terminalia* species and by Van den Oever et al. (1981) for *Symplocos.* In response to his critics, Carlquist (1980) suggested that for *Terminalia* the long-lived foliage acts as a buffer, which might explain why here the concepts do not apply so well. Meanwhile, several recent studies have continued to apply the concepts of mesomorphy and vulnerability, regardless of whether the species are deciduous or evergreen.

Apart from their general lack of predictive value noted earlier, there are four other reasons that I find Carlquist's concepts an unfortunate contribution to ecological and functional wood anatomy: (1) Any equation suggesting a linear relationship between functional fitness and vessel diameter ignores the fact that conductivity is not *linearly* related to diameter, but to its fourth power. (2) As discussed earlier, the role of vessel member length in hydraulic architecture remains elusive, and any formula suggesting equality in function or in ecological response of vessel member length and vessel diameter is bound to be misleading. (3) A concept of safety, taking into account vessel diameter, total vessel length, vessel frequency, and interconnections of individual vessels in the three-dimensional vessel network, as adopted in the literature on xylem physiology, is more meaningful

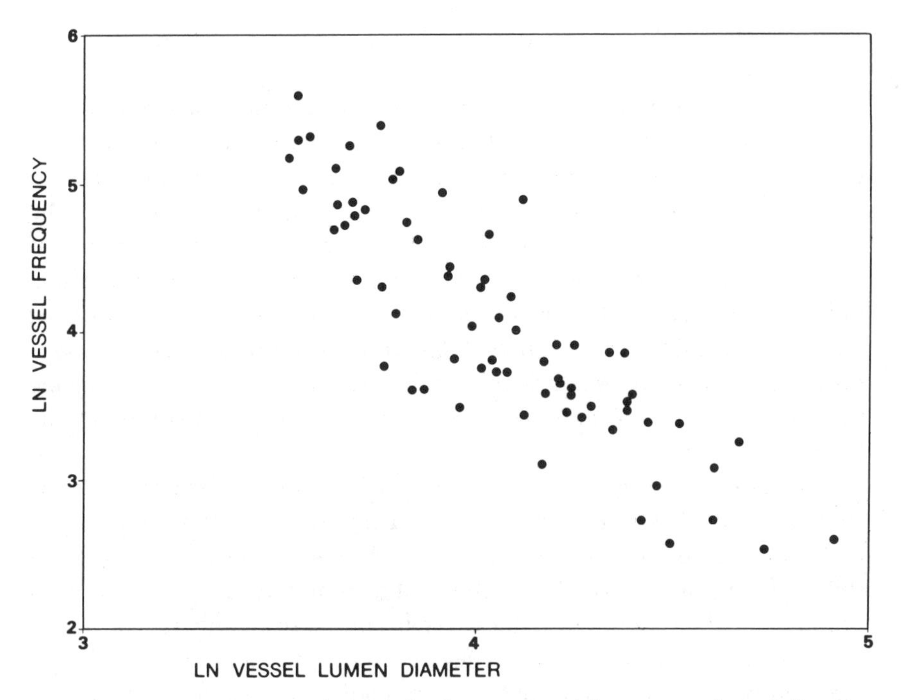

Figure 11.6. Mutual correlation between vessel frequency and vessel diameter in the genus *Symplocos* (reproduced from Van den Oever et al. 1981, with permission of the *IAWA Bulletin*).

than vulnerability and makes the latter concept redundant. (4) Formulas like the one for mesomorphy are, moreover, likely to illustrate the results of correlative phenomena in xylem differentiation rather than to be meaningful in an ecophysiological sense. Baas (1973), Van der Graaff and Baas (1974), and Van den Oever et al. (1981) have shown that within genera, vessel member length tends to be closely related to vessel diameter and inversely related to vessel frequency. For instance, in *Symplocos*, the mutual correlation between these traits is often stronger than the correlation of one of them with latitude or provenance (Van den Oever et al. 1981) (Figure 11.6). An adaptive significance of one or two of these parameters in ecological trends would be reinforced by the other(s) in the formula for mesomorphy, regardless of a mesic or xeric ecology of the species: The correlations referred to earlier were largely established within genera composed of mesic species only, but covering wide altitudinal and latitudinal ranges. Developmentally and phylogenetically determined correlations do not rule out the possibility of coadaptive trends, but should caution us in applying rigid indexes like vulnerability and mesomorphy as ecophysiological parameters.

Fibers and parenchyma: mechanical strength and carbohydrate metabolism

Apart from conduction of xylem sap, mechanical strength and storage and mobilization of metabolites are the main functions wood has to carry out in the living plant. For mechanical strength, fiber wall thickness is probably the most important factor. Tissue proportions (especially the percentage of fibers), fiber length, and the relative amount of overlap at fiber ends are also of relevance. Ecological trends have been established only for fiber length; the trend for tropical lowland species to have the longest fibers makes sense functionally on the assumption that tree size is positively related to demands for mechanical strength, and considering that canopy species and emergents of the tropical rain forest are among the tallest trees.

The high percentage of species with thick fiber walls in desert and steppe species in the Middle East, as mentioned earlier (Fahn et al. in press), is more difficult to account for functionally, unless occasional strong winds should be a real hazard. From the well-documented great variation in density, and thus in fiber wall thickness, in commercial timbers from many types of natural forests, it becomes likely that the amount of cell-wall material in the trunks of large trees is not a limiting factor for the evolutionary success of a species, provided that minimal requirements for strength are met. This conclusion also weakens the functional interpretation of fiber length trends, discussed earlier, because the role of fiber length in strength properties is probably inferior to that of fiber wall thickness. However, future analyses should consider the successional status of species, because there is some evidence that late successional species have denser and mechanically safer wood than early successional species (T. J. Givnish, personal communication).

For storage and mobilization of metabolites, mainly carbohydrates, both axial parenchyma and ray parenchyma play vital roles, as do living fibers. Lack of detailed knowledge of the incidence of the latter tissue category and ray volume percentages in different woody floras precludes a meaningful discussion of poorly established ecological trends for parenchyma abundance. Previously I commented (Baas 1982b) that the abundance of axial parenchyma in many tropical lowland species seems rather pointless in these evergreen trees, which are continuously capable of intensive photosynthesis; in an alternative vein, one might also interpret this abundant parenchyma as highly functional in trees that show flush-like growth and massive flowering and fruiting involving great temporary demands for metabolites. These two opposing alternatives precisely show

how fruitless functional interpretations of ecological trends can become in the absence of adequate physiological information. Both comments will remain equally futile until storage and mobilization of carbohydrates have been studied comprehensively in a large number of species in relation to their phenology.

At the same time, attention should be paid to the role of paratracheal contact parenchyma (i.e., the part of the paratracheal parenchyma immediately bordering on the vessels and provided with numerous pit contacts to the vessels) in supplying the xylem sap with osmotically active sugars. According to Braun (1984) this can result in positive pressure and "osmotic water shifting" in the tree. This may be especially significant throughout the year in tropical rain forests, where transpiration is limited by the high relative humidity of the air. Abundant parenchyma surrounding the contact parenchyma cells would then provide a useful means for storing large amounts of starch that can be mobilized through the high metabolic activity in the contact cells. In temperate regions, osmotic water shifting is mainly operational in spring, when the trees "come into sap" (Braun 1984), that is, before budbreak.

Growth rate, ecological trends, and wood structure

Many of the ecological trends reported earlier are also related to different axial and radial growth rates in the different environments. Especially the low values for vessel member length and fiber length in arid and cool temperate to arctic species are associated with slow growth of the shrubs and small trees of these environments. This raises the question whether or not the ecological trends for element length are not partly due to a phenotypic growth rate response. Baas et al. (1984) found that in dwarfed specimens of 11 species of softwoods and hardwoods, the length-on-age graphs usually show a significant decline toward the periphery, whereas in normal- or fast-growing individuals of the same species, fiber or vessel member length remains at a high level or continues to increase from the pith outward. The relative amount of the reduction in element length in the periphery of dwarfed stems varies from species to species and is not linearly related to the percentage of reduction in growth rate. Yet, the order of magnitude of the phenotypically induced shortening of xylem elements in this case is sometimes similar to the differences found between ecological categories in floristic analyses, as illustrated in Figure 11.2. An adaptive interpretation of the shortening of elements in dwarfed trees is out of place; the phenotypic response should be regarded as a deficiency phenomenon, comparable to certain pathological responses of the cambium ultimately leading to the death of the tree.

The answer to the question whether the ecological trends discussed in this chapter are due to a phenotypic response rather than to selection for more desirable hydraulic or mechanical properties is probably a partial yes and no. Yes, because the association of slow growth with shorter cambial initials is also a general phenomenon within species (Panshin and De Zeeuw 1980). No, because within, for instance, the tropical rain forest flora, all possibilities of extremely short to long vessel members (and consequently cambial initials) are realized (Figure 11.2), without any apparent relationship to tree size or growth rate. Moreover, by extensive progeny tests in forestry research it has been demonstrated that element length is under genetic control to a large extent. Apparently the phenotypic plasticity of element length as related to growth rate and habit is restricted between certain limits (which may be different for each genotype), and the great differences in average length values for the rain forest flora on the one hand and for the arid or arctic floras on the other (Figure 11.2) can never be fully accounted for by phenotypic, growth-rate-related effects.

Wood structure and primary productivity

The theme of this volume (evolutionary constraints on primary productivity) invites the question whether or not different types of wood structure can be limiting for biomass production of individual trees or species. This question can only be tentatively answered by comparing the wood anatomies of some fast- and slow-growing species.

If one considers that among the species most popular in forestry programs for their fast growth there are such different wood structural types as several conifer species (vesselless woods) and moderately to highly specialized hardwoods like species of *Eucalyptus, Gmelina, Anthocephalus,* and *Populus,* it is obvious that the general wood histology, or more specifically the hydraulic architecture, cannot be a serious limiting factor in the primary productivity of these species. The foregoing hardwood examples are diverse for fiber type (libriform or fiber-tracheids), ray size and type (uniseriate or multiseriate; homocellular or heterocellular), and parenchyma distribution (apotracheal or paratracheal). The only wood anatomical characteristic that is only rarely associated with great tree size, and presumably also with fast growth, is the scalariform perforation plate. In the tropical lowland rain forest, scalariform perforations are almost entirely restricted to small trees or shrubs of the substage or undergrowth; in the temperate floras, scalariform perforation plates also tend to be restricted to small trees and shrubs (see the lists of genera in Baas 1976). However, to conclude that scalariform perforation plates constitute a constraint on growth rate and primary productivity would still be rather far-fetched. The generally narrow diameter of vessels with scalariform perforations (a correlative constraint) may be a more important limiting factor in conductive efficiency and consequently in growth rate and ultimate tree size than the type of vessel perforation plate. Minimal requirements for conductive efficiency are easily met by sufficiently wide vessels, and in wood with simple perforations this character is probably not under evolutionary constraint. However, it is significant that vesselless conifers apparently achieve the same efficiency with their more voluminous sapwood zone of narrow, imperforate elements only.

Whitmore (1975) listed a number of fast- and slow-growing tropical rain forest species that all develop into large trees. By comparing their wood anatomies, I found no salient differences between the groups of slow- and fast-growing species: Both were wood anatomically diverse and included species with thin- or thick-walled fibers (light or heavy timbers). All species had fairly wide vessels with simple perforations. Other studies, summarized by Panshin and De Zeeuw (1980), have often indicated that fast growth is associated with low density (thin-walled fibers). Usually this is due to growth conditions rather than to genetically controlled relationships between growth rate and density. Within species, moreover, it is possible to select for genotypes that combine high density and fast growth. Genetically fixed low density is thus no prerequisite for high growth rates, and by the same token thick-walled fibers do not create an evolutionary constraint on radial or axial growth rate or on primary productivity in woody plants.

Conclusions

The ecological trends of certain wood anatomical characteristics surveyed in this chapter can partly be interpreted as the result of adaptive evolution. This is especially true for replacement of the scalariform perforation plate by the simple perforation and for the adjustments in vessel diameter to comply with periodic or continuous demands for efficient xylem sap conduction in xeric woody plants and in tall rain forest trees in the tropics.

Some of the ecological trends (e.g., fiber-tracheids are more common in temperate regions than in the tropics) may be due to correlative restraints because of their general association with scalariform perforation plates that have not been eliminated in cool mesic floras to the same extent as in the tropical lowland or arid zone floras.

Other salient trends, notably the distinctive shortening of cambial initials and consequently of vessel members and fibers in arid and cool temperate to arctic vegetations, remain elusive. In part, but not entirely, the latter trends may be due to low growth rate and small plant size in these extreme habitats.

In descriptive ecological wood anatomy there is still much scope for refinement and expansion of correlations between wood structure and environmental factors. In interpretative ecological wood anatomy, especially where attempts are made at functional explanations of the observed trends, the possibilities for meaningful contributions will very soon be exhausted if they cannot be enriched with comprehensive information from comparative tree physiology, which should consider experimental results in relation to wood structural diversity. In this respect, the work of Zimmermann (1983) is a great step forward, and the challenging ideas in his book deserve to be pursued in future experimental research programs.

References

Baas, P. 1973. The wood anatomy of *Ilex* (Aquifoliaceae) and its ecological and phylogenetic significance. Blumea 21:193–258.

– 1976. Some functional and adaptive aspects of vessel member morphology. Pp. 157–181 *in* P. Baas, A. J. Bolton, and D. M. Catling (eds.), Wood structure in biological and technological research. Leiden Botanical Series no. 3.

– 1982a. Leeuwenhoek's contributions to wood anatomy and his ideas on sap transport in plants. Pp. 79–107 *in* L. C. Palm and H. A. M. Snelders (eds.), Antoni van Leeuwenhoek 1632–1723. Editions Rodopi, Amsterdam.

– 1982b. Systematic, phylogenetic, and ecological wood anatomy. History and perspectives. Pp. 23–58 *in* P. Baas (ed.), new perspectives in wood anatomy. Nijhoff/Junk, The Hague.

Baas, P., Lee Chenglee, Zhang Xinying, Cui Keming, and Deng Yuefen. 1984. Some effects of dwarf growth on wood structure. IAWA Bull., n.s. 5:45–63.

Baas, P., E. Werker, and A. Fahn. 1983. Some ecological trends in vessel characters. IAWA Bull., n.s. 4:141–159.

Bailey, I. W., and W. W. Tupper. 1918. Size variation in tracheary cells. I. A comparison between the secondary xylems of vascular cryptogams, gymnosperms and angiosperms. Proc. Amer. Arts Sci. 54:149–204.

Barghoorn, E. S. 1964. Evolution of cambium in geological time. Pp. 3–17 *in* M. H. Zimmermann (ed.), The formation of wood in forest trees. Academic Press, New York.

Braun, H. J. 1970. Funktionelle Histologie der sekundären Sprossachse. I. Das Holz. Encyclopedia of Plant Anatomy IX (1). Borntraeger, Berlin.

– 1983. Zur Dynamik des Wassertransportes in Bäumen. Ber. Deutsch. Bot. Ges. 26:29–47.

– 1984. The significance of the accessory tissues of the hydrosystem for osmotic water shifting as the second principle of water ascent, with some thoughts concerning the evolution of trees. IAWA Bull., n.s. 5:275–294.

Carlquist, S. 1962. A theory of paedomorphosis in dicotyledonous woods. Phytomorphology 12:30–45.

– 1966. Wood anatomy of Compositae: a summary, with comments on factors controlling wood evolution. Aliso 6:25–44.

– 1975. Ecological strategies of xylem evolution. University of California Press, Berkeley.

– 1977. Ecological factors in wood evolution, a floristic approach. Amer. J. Bot. 64:887–896.

– 1980. Further concepts in ecological wood anatomy, with comments on recent work in wood anatomy and evolution. Aliso 9:499–553.

– 1982a. Wood anatomy of Daphniphyllaceae: ecological and phylogenetic considerations, review of pittosporalean families. Brittonia 34:252–266.

– 1982b. Wood anatomy of *Illicium* (Illiciaceae). Phylogenetic, ecological and functional interpretations. Amer. J. Bot. 69:1587–1598.

– 1983. Wood anatomy of *Bubbia* (Winteraceae) with comments on origin of vessels in dicotyledons. Amer. J. Bot. 70:578–590.

Chalk, L. 1983. Fibres. Pp. 28–38, *in* C. R. Metcalfe and L. Chalk (eds.), Anatomy of the dicotyledons. II. Wood structure and conclusion of the general introduction. Clarendon Press, Oxford.

Dickison, W. C. 1975. The bases of angiosperm phylogeny: vegetative anatomy. Ann. Missouri Bot. Gard. 62:590–620.

Fahn, A., E. Werker, and P. Baas. In press. Wood anatomy and identification of trees and shrubs from Israel and adjacent regions. Israel Academy of Sciences, Jerusalem.

Kanehira, R. 1921. Anatomical characters and identification of Formosan woods, with critical remarks from the climatic point of view. Bureau of Productive Industries, Taihoku.

Page, V. M. 1981. Dicotyledonous wood from the Upper Cretaceous of California. III. Conclusions. J. Arn. Arbor. 62:437–455.

Panshin, A. J., and C. De Zeeuw. 1980. Textbook of wood technology, 4th edition. McGraw-Hill, New York.

Petty, J. A. 1978. Fluid flow through the vessels of birch wood. J. Exp. Bot. 29:1463–1469.

Stern, W. L. 1978. A retrospective view of comparative anatomy, phylogeny, and plant taxonomy. IAWA Bull. 2–3:33–39.

Van den Oever, L., P. Baas, and M. Zandee. 1981. Comparative wood anatomy of *Symplocos* and latitude and altitude of provenance. IAWA Bull., n.s. 2:3–24.

Van der Graaff, N. A., and P. Baas. 1974. Wood anatomical variation in relation to latitude and altitude. Blumea 22:101–121.

Van Vliet, G. J. C. M. 1979. Wood anatomy of the Combretaceae. Blumea 25:141–223.

Vesque, J. 1889. De l'emploi des caractères anatomiques dans la classification des végétaux. Bull. Soc. Bot. France 36:41–87.

Webber, I. E. 1936. The wood of sclerophyllous and desert shrubs and desert plants of California. Amer. J. Bot. 23:181–188.

Whitmore, T. C. 1975. Tropical rain forests of the Far East. Clarendon Press, Oxford.

Young, D. A. 1981. Are the angiosperms primitively vesselless? Syst. Bot. 6:313–330.

Zimmermann, M. H. 1978. Structural requirements for optimal water conduction in tree stems. Pp. 517–532 *in* P. B. Tomlinson and M. H. Zimmermann (eds.), Tropical trees as living systems. Cambridge University Press.

– 1982. Functional xylem anatomy of angiosperm trees. Pp. 59–70 *in* P. Baas (ed.), New perspectives in wood anatomy. Nijhoff/Junk, The Hague.

– 1983. Xylem structure and the ascent of sap. Springer series in wood science 1. Springer, Berlin.

Zimmermann, M. H., and C. L. Brown. 1971. Trees. Structure and function. Springer, Berlin.

12 Turgor maintenance in Hawaiian *Dubautia* species: the role of variation in tissue osmotic and elastic properties

ROBERT H. ROBICHAUX,
KENT E. HOLSINGER,
AND SUZANNE R. MORSE

Introduction

Insufficient moisture availability is a major factor limiting terrestrial plant productivity on a worldwide basis (Fischer and Turner 1978; Turner and Kramer 1980; Boyer 1982; Hanson and Hitz 1982). In agricultural communities, worldwide losses in yield as a result of water deficits probably exceed the losses from all other causes combined (Kramer 1980). In natural communities, water deficits appear to play a central role in determining the distribution and abundance of many plant species (Lange et al. 1976; Jones et al. 1981).

A wide variety of growth-related processes are affected by water deficits, with cell expansion being one of the most sensitive (Hsiao et al. 1976; Bradford and Hsiao 1982). For many of these processes, changes in turgor pressure may be the principal means by which small changes in plant water status are transduced into changes in metabolism (Hsiao et al. 1976; Zimmermann 1978). Rates of leaf expansion in sorghum and sunflower, for example, are markedly influenced by small changes in tissue turgor pressure (Hsiao et al. 1976; Takami et al. 1981, 1982).

This turgor dependence suggests that plant growth under conditions of low moisture availability may be enhanced by mechanisms promoting the maintenance of high turgor pressures as tissue water content decreases (Turner and Jones 1980; Jones et al. 1981). Two mechanisms that promote turgor maintenance involve changes in tissue osmotic and elastic properties (Tyree and Jarvis 1982). First, a decrease in the tissue osmotic potential at full hydration causes an increase in the maximal value of tissue turgor pressure. Second, a decrease in the tissue elastic modulus causes a decrease in the magnitude of the change in turgor pressure for a given fractional change in tissue water content.

In this chapter we shall examine the nature and significance of variation in these tissue osmotic and elastic properties in several Hawaiian species of *Dubautia* (Compositae). We begin by analyzing the relationship between variation in these properties and diurnal turgor maintenance. This analysis enables us to evaluate the potential functional significance of this variation. We then examine the extent to which these properties vary (1) among individuals of one species growing in different moisture regimes, (2) among two species and their natural hybrid, and (3) among six species that differ in habitat and diploid chromosome number. Of particular interest to us are the genetic and evolutionary implications of these results. Finally, we evaluate three hypotheses concerning the mechanistic basis of the variation in tissue elastic properties. These hypotheses encompass the possible effects of cell size, cell-wall composition, and apoplasmic water loss.

Relationship to diurnal turgor maintenance

The relationship between variation in tissue osmotic and elastic properties and diurnal turgor maintenance can be illustrated with two *Dubautia* species that grow sympatrically at approximately 2,000 m elevation on the slopes of Mauna Loa in Hawaii. The ecological distributions of the two species at this site of sympatry are very unusual: *Dubautia scabra* is restricted to a 1935 lava flow, whereas *Dubautia ciliolata* is restricted to an older, prehistoric lava flow (Carr and Kyhos 1981; Robichaux 1984). The two lava flows differ not only in age but also in physical structure (Robichaux 1984). Because the two lava flows form a mosaic at this site, individuals of the two species commonly grow within a few meters of one another.

Associated with this habitat difference is a difference in the tissue water deficits experienced by the two species. During a typical summer day, tissue water potentials in *D. ciliolata* are significantly lower than in *D. scabra* (Figure 12.1). At midday, the difference in the water potentials of these two species commonly exceeds 0.45 MPa (megapascals).

The two species also exhibit significant differences in their tissue osmotic and elastic properties. This is apparent in Figure 12.2, where tissue turgor pressure is plotted as a function of tissue water content. Differences in the *y* intercepts of these two curves reflect differences in tissue osmotic properties, and differences in their slopes reflect differences in tissue elastic properties. Relative to *D. scabra, D. ciliolata* exhibits a lower tissue osmotic potential at full hydration and a lower tissue elastic modulus

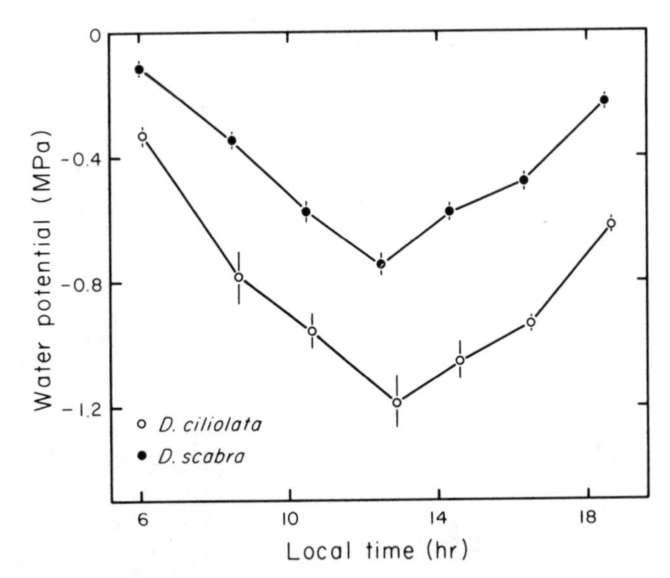

Figure 12.1. Diurnal tissue water potentials for *Dubautia ciliolata* and *D. scabra* on 4 August 1982 near Puu Huluhulu, Saddle Road, Hawaii (Robichaux 1984). Vertical lines indicate ± 1 SD for five replicates of *D. ciliolata* between 0830 and 1230 hr and five replicates of *D. scabra* at 1230 hr. At other times, vertical lines indicate the range of variation for duplicates.

near full hydration (Robichaux 1984). These differences have pronounced effects on the relative abilities of the two species to maintain high and positive turgor pressures as tissue water content decreases. At a relative water content of 0.93, for example, the value of tissue turgor pressure is 0.77 MPa in *D. ciliolata* and only 0.06 MPa in *D. scabra*. In addition, the relative water content at which turgor pressure reaches zero is significantly lower in *D. ciliolata* (0.75) than in *D. scabra* (0.88).

These differences in tissue osmotic and elastic properties appear to have a marked influence on diurnal turgor maintenance. From the data in Figure 12.2 it is possible to derive relationships between tissue water potential and tissue turgor pressure for the two species (Figure 12.3). By combining the information in Figure 12.3 with that in Figure 12.1, one can then estimate the diurnal values of turgor pressure experienced by *D. ciliolata* and *D. scabra* during a typical summer day (Robichaux 1984). As illustrated in Figure 12.4, the two species experience very similar diurnal turgor pressures, which contrasts markedly with the large differences in their diurnal water potentials. Turgor pressures in both species decline from a maximal value of approximately 0.80 MPa in the early morning to a minimal value of 0.14 MPa at midday. Turgor pressures then increase in

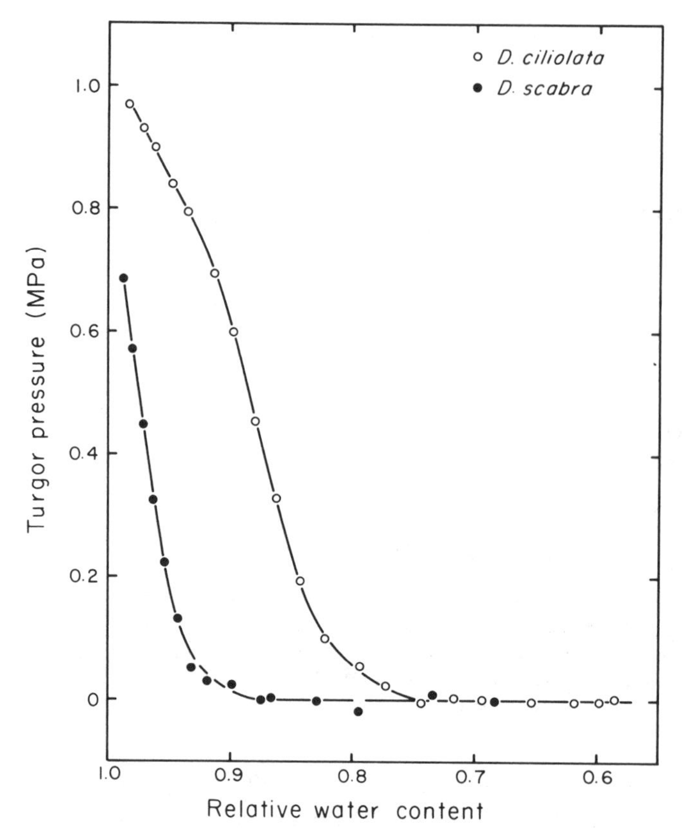

Figure 12.2. Relationships between tissue turgor pressure and tissue water content for *Dubautia ciliolata* and *D. scabra*. These relationships were obtained for branches with 6–10 mature leaves.

both species during the afternoon and reach a value of approximately 0.60 MPa by early evening.

It is also possible to calculate diurnal turgor pressures for a hypothetical individual that exhibits tissue osmotic and elastic properties similar to those of *D. scabra* but experiences diurnal water potentials similar to those of *D. ciliolata* (Robichaux 1984). In such a hypothetical individual, turgor pressure declines rapidly from a maximal value of approximately 0.50 MPa in the early morning to 0 MPa at 1030 hr. Turgor pressure remains at 0 MPa until 1630 hr, then increases to 0.25 MPa by early evening. Hence, turgor pressures in such a hypothetical individual are significantly lower than the actual turgor pressures experienced by *D. ciliolata* during a typical summer day (Robichaux 1984).

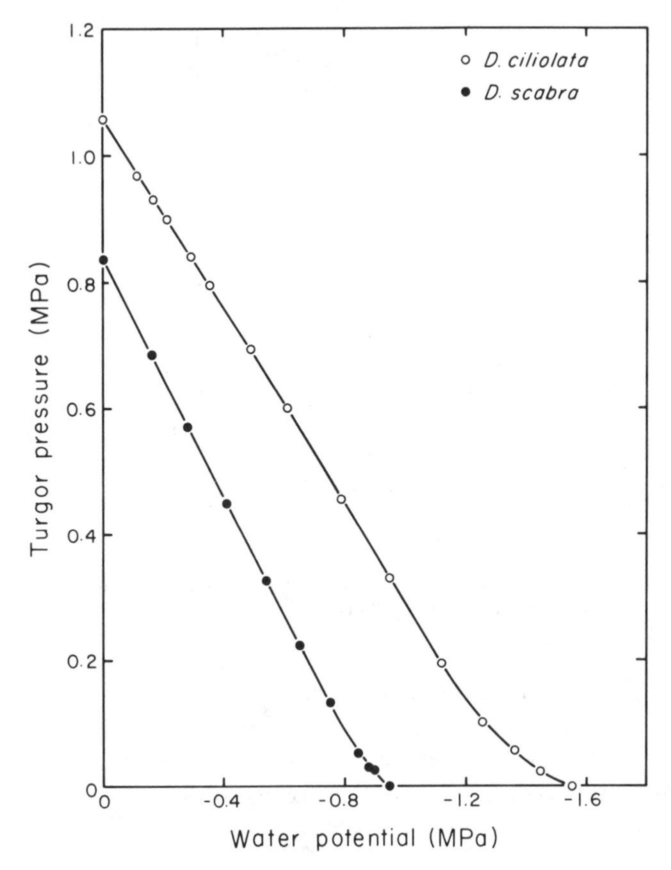

Figure 12.3. Relationships between tissue turgor pressure and tissue water potential for *Dubautia ciliolata* and *D. scabra* (Robichaux 1984).

In terms of diurnal turgor maintenance, these calculations suggest that *D. ciliolata* is able to tolerate more severe tissue water deficits than *D. scabra*. The importance of diurnal turgor maintenance stems from the fact that a variety of physiological processes, such as leaf expansion and stomatal opening, appear to depend intimately on tissue turgor pressure (Hsiao et al. 1976; Turner and Jones 1980). Acevedo et al. (1979), for example, have demonstrated that diurnal turgor maintenance by osmotic adjustment is an important mechanism promoting high rates of leaf expansion in sorghum and maize plants growing under field conditions. Turgor maintenance by osmotic adjustment also appears to be an important mechanism promoting high stomatal conductances at low leaf water potentials (Turner et al. 1978; Ludlow 1980; Turner and Jones 1980). High stoma-

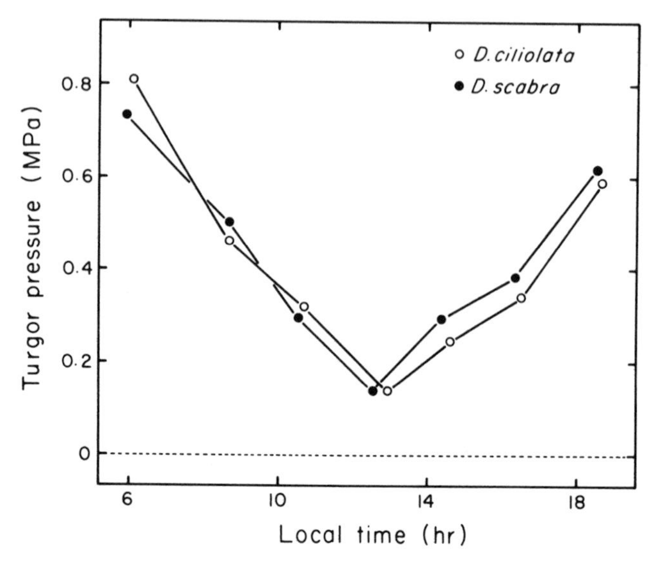

Figure 12.4. Estimated diurnal turgor pressures for *Dubautia ciliolata* and *D. scabra* on 4 August 1982 (Robichaux 1984).

tal conductances, in turn, appear to promote high photosynthetic rates at low leaf water potentials (Jones and Rawson 1979; Turner and Jones 1980; Ehleringer 1983). Further field observations and laboratory experiments should allow us to examine the possibility that turgor maintenance serves these functions in the *Dubautia* species as well.

Phenotypic variation within one species

An important question arises from these results: To what extent are the differences in tissue osmotic and elastic properties genetically based? In other words, if the two *Dubautia* species were growing in similar moisture environments, then to what extent would the differences in their tissue osmotic and elastic properties disappear? In order to answer these questions, it is first necessary to examine the degree of phenotypic variation that exists in these properties.

In *D. ciliolata*, the degree of phenotypic variation appears to be relatively limited (Robichaux 1984). This is illustrated in Figure 12.5, where tissue turgor pressure is plotted as a function of tissue water content for plants growing under well-watered conditions in a glasshouse and under natural conditions in the field. Although the tissue water potentials experienced by the two groups of plants differ significantly (Robichaux 1984), their

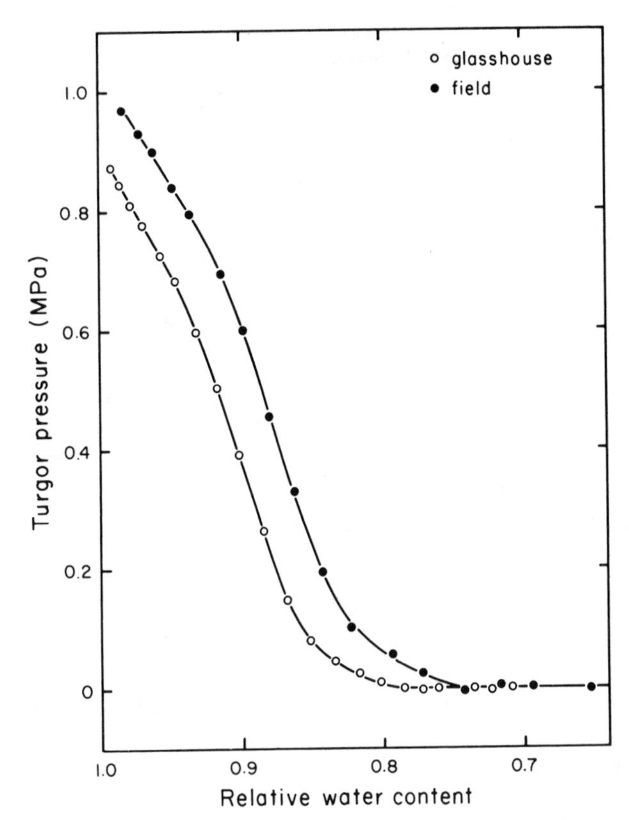

Figure 12.5. Relationships between tissue turgor pressure and tissue water content for plants of *Dubautia ciliolata* growing in the glasshouse and in the field (Robichaux 1984).

tissue osmotic and elastic properties are similar. The principal difference between them is a 0.1 MPa shift in the tissue osmotic potential at full hydration, which results in a lower maximal value of turgor pressure in the glasshouse plants (Table 12.1, Figure 12.5). The tissue elastic modulus near full hydration is also slightly higher in the glasshouse plants, reflecting the fact that the initial rate at which turgor pressure declines with decreasing tissue water content is slightly higher in these plants (Table 12.1, Figure 12.5). However, the overall shapes of the two curves are strikingly similar. In particular, the maximal rate at which turgor pressure declines with decreasing tissue water content is virtually identical in the two groups of plants. The latter similarity is reflected in the similar maximal elastic moduli of plants growing under the two conditions (Table 12.1).

Table 12.1. *Tissue osmotic potential at full hydration* (π_i), *tissue elastic modulus near full hydration* (E_i), *maximal tissue elastic modulus* (E_m), *and sample size* (n) *for plants of* Dubautia ciliolata *growing in the glasshouse and in the field.*[a]

Parameter[b]	Glasshouse	Field
π_i (MPa)	−0.98 (0.01)[c]	−1.08 (0.04)
E_i (MPa)	3.50 (0.34)	2.22 (0.48)
E_m (MPa)	5.17 (0.49)	5.19 (0.44)
n	8	6

[a] Data from Robichaux (1984).
[b] The differences in π_i and E_i are statistically significant at $P < 0.001$.
[c] Standard deviations are given in parentheses.

The limited nature of this phenotypic variation in *D. ciliolata* becomes particularly evident when the tissue osmotic and elastic properties of these two groups of plants are compared with those of *D. scabra* growing in the field (Robichaux 1984). In the latter plants, the tissue osmotic potential at full hydration is −0.81 ± 0.04 MPa, and the tissue elastic modulus near full hydration is 10.23 ± 1.75 MPa. These values are significantly higher than those for both groups of *D. ciliolata* plants (Table 12.1). These differences have a marked effect on the magnitude of tissue turgor pressure at moderate tissue water contents. As mentioned previously, the value of tissue turgor pressure at a relative water content of 0.93 is only 0.06 MPa in plants of *D. scabra* growing in the field. In plants of *D. ciliolata* growing in the glasshouse and in the field, the corresponding values are 0.59 MPa and 0.77 MPa, respectively. Hence, relative to *D. scabra,* plants of *D. ciliolata* growing under both conditions exhibit a large capacity for maintaining high turgor pressures as tissue water content decreases.

The tissue water deficits experienced by *D. ciliolata* in the glasshouse are very similar to those experienced by *D. scabra* in the field (Robichaux 1984). Because large differences in the tissue osmotic and elastic properties of the two species persist under these conditions, the differences are probably genetically based to a significant extent.

Variation among two species and their hybrid

Analyzing the genetic basis of these differences in tissue osmotic and elastic properties will require controlled production of F_2 hybrids and backcrosses. Although such crosses are not yet available, it has been possi-

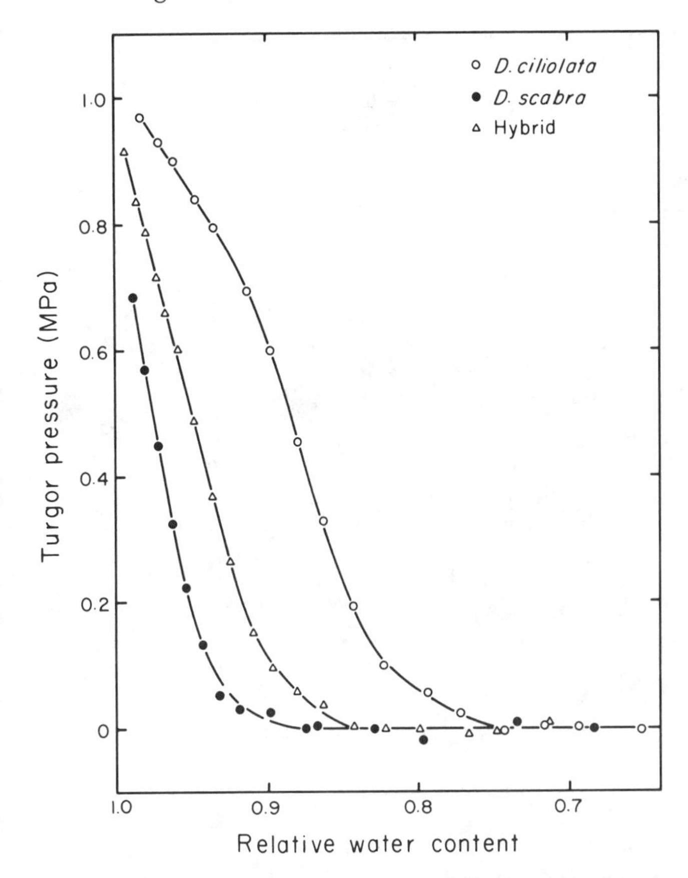

Figure 12.6. Relationships between tissue turgor pressure and tissue water content for *Dubautia ciliolata, D. scabra,* and their natural hybrid (Robichaux 1984).

ble to analyze the tissue osmotic and elastic properties of the natural hybrid between *Dubautia ciliolata* and *D. scabra* (Robichaux 1984). These naturally occurring individuals are probably F_1 hybrids, although the possibility exists that they represent recombinants beyond the F_1 generation (Carr and Kyhos 1981).

The tissue osmotic and elastic properties of the natural hybrid are intermediate between those of *D. ciliolata* and *D. scabra* (Figure 12.6). As a result, at all relative water contents above 0.85, turgor pressures in the hybrid are intermediate between those of the two parental species. At a relative water content of 0.93, for example, the value of tissue turgor pressure in the hybrid is 0.31 MPa, which is midway between the values given previously for *D. ciliolata* and *D. scabra.* In addition, zero turgor

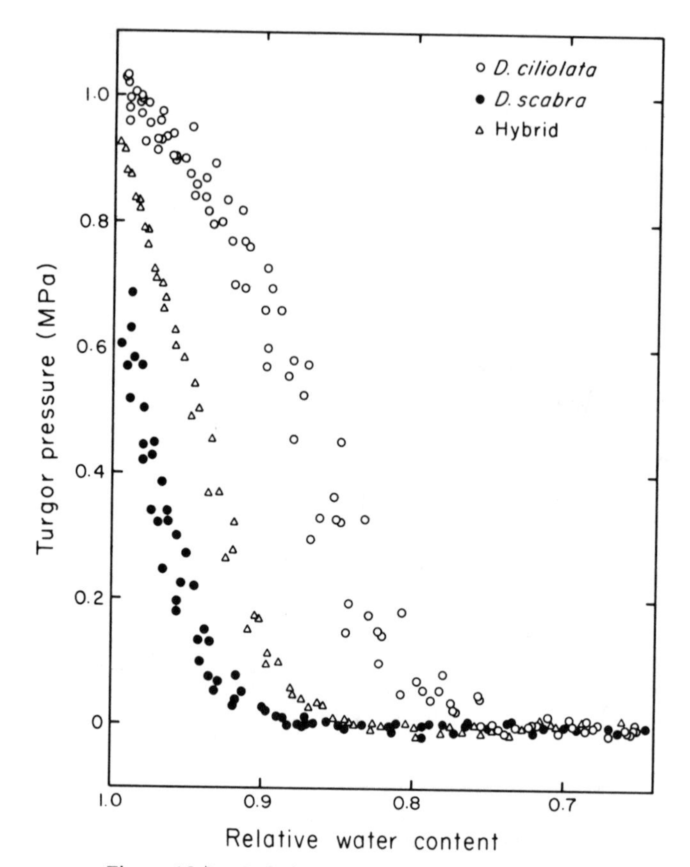

Figure 12.7. Relationships between tissue turgor pressure and tissue water content for six individuals of *Dubautia ciliolata*, five individuals of *D. scabra*, and three individuals of their natural hybrid.

pressure is reached at an intermediate relative water content in the hybrid in comparison with the two parental species.

The response curves plotted in Figure 12.6 represent single branches for each of the two parental species and the hybrid. However, the intermediate nature of the hybrid response curve is characteristic of all of the hybrid individuals that we have examined (Figure 12.7). Indeed, the range of variation among individual response curves is quite small, not only in the hybrid but also in the two parental species.

The turgor dependence of the elastic modulus in the hybrid is also intermediate relative to that in *D. ciliolata* and *D. scabra* (Figure 12.8). The elastic modulus in *D. scabra* increases linearly with turgor pressure to a maximal value of approximately 10 MPa. The elastic modulus in *D. ciliolata*, in contrast, increases to a maximal value of approximately 5 MPa at

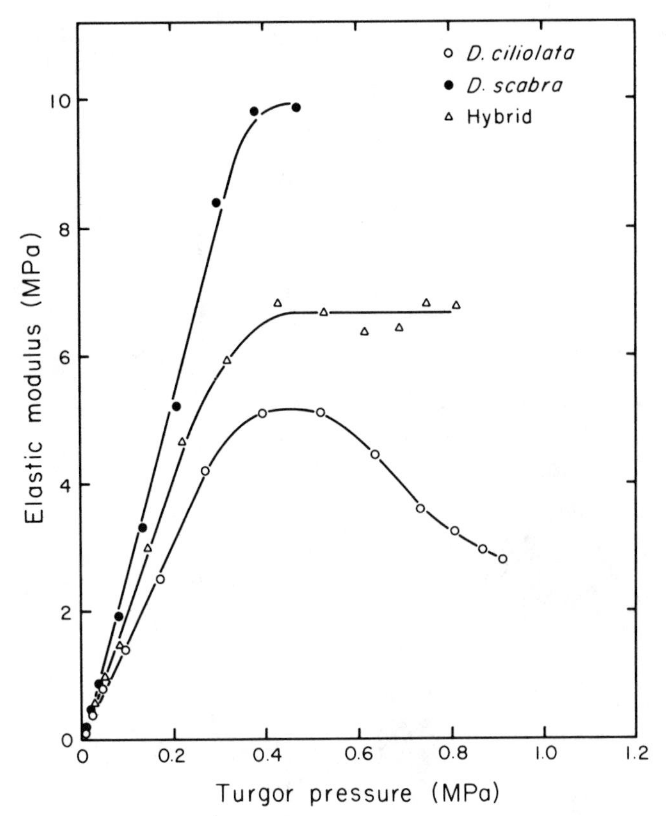

Figure 12.8. Relationships between tissue elastic modulus and tissue turgor pressure for *Dubautia ciliolata, D. scabra,* and their natural hybrid (Robichaux 1984).

intermediate turgor pressures, then decreases to a value of 2–3 MPa at high turgor pressures. In the hybrid, the elastic modulus increases to a maximal value of approximately 7 MPa at intermediate turgor pressures, then remains essentially constant until maximal turgor pressures are reached. The initial rate at which the elastic modulus increases with turgor pressure is highest in *D. scabra,* lowest in *D. ciliolata,* and intermediate in the hybrid.

Variation among six additional species

Given the large differences between *D. ciliolata, D. scabra,* and the hybrid, the question arises as to how tissue osmotic and elastic properties vary among the other 19 species of *Dubautia.* These additional species

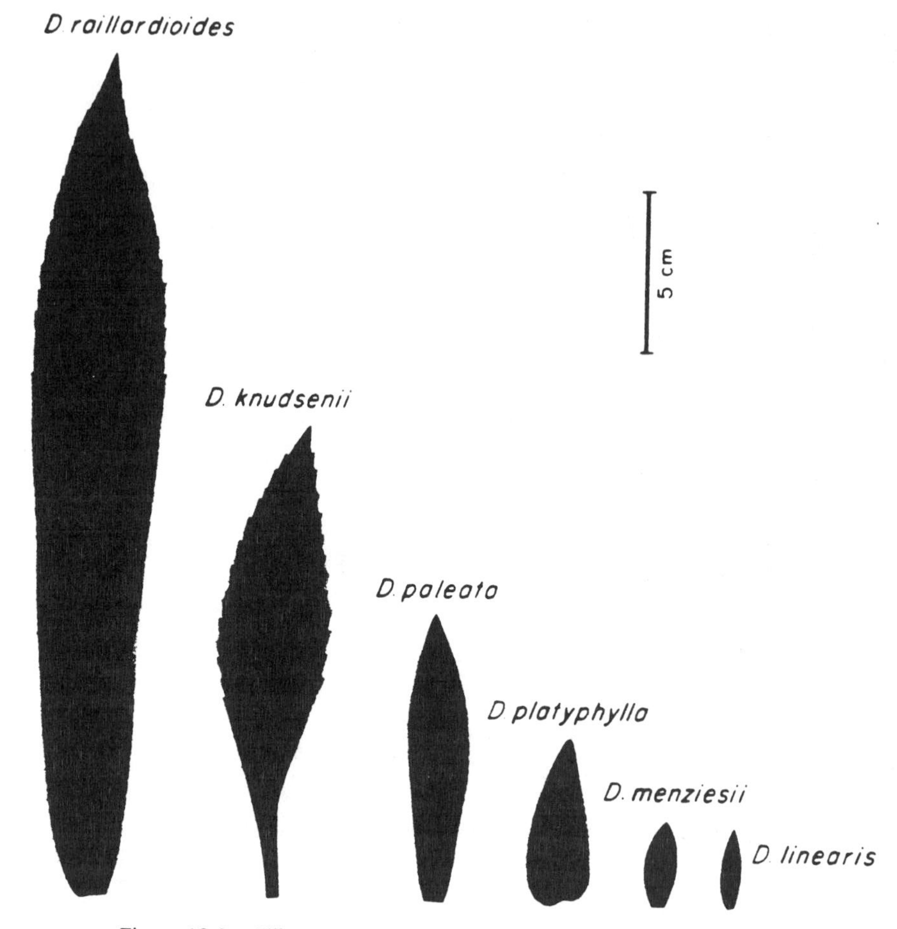

Figure 12.9. Silhouettes of mature leaves of six *Dubautia* species.

grow in habitats as varied as exposed cinder, dry scrub, dry woodland, mesic forest, wet forest, and bog (Carlquist 1980; Carr and Kyhos 1981). These habitats constitute an unusually wide array of moisture environments, with annual rainfall ranging from less than 400 mm/year in the dry scrub habitat to more than 12,300 mm/year in the wet forest and bog habitats. Indeed, the latter habitats are among the wettest terrestrial environments on earth (Grosvenor 1966).

We have recently analyzed the tissue osmotic and elastic properties of six of these species, the mature leaves of which are illustrated in Figure 12.9. *Dubautia raillardioides, D. knudsenii,* and *D. paleata* grow in wet forest, mesic forest, and bog habitats, respectively, on the island of Kauai. *Dubautia platyphylla, D. menziesii,* and *D. linearis,* in contrast, grow in a variety of

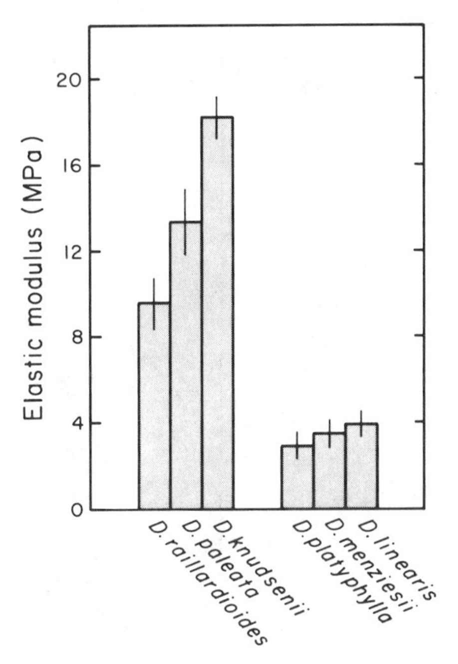

Figure 12.10. Tissue elastic moduli near full hydration for six *Dubautia* species. The vertical line at the top of each bar indicates ±1 SD for four (*D. raillardioides* and *D. linearis*), five (*D. platyphylla* and *D. menziesii*), or six replicates (*D. paleata* and *D. knudsenii*).

dry scrub habitats on the islands of Maui and Hawaii, with the habitat of *D. platyphylla* being the most mesic of the three. These species differ not only in habitat but also in diploid chromosome number. *Dubautia raillardioides, D. knudsenii,* and *D. paleata* have 14 pairs of chromosomes in their diploid complement, whereas *D. platyphylla, D. menziesii,* and *D. linearis* have 13 pairs of chromosomes (Carr 1978; Carr and Kyhos 1981).

Significant degrees of variation exist in the tissue osmotic and elastic properties of these six species. Tissue osmotic potentials at full hydration range from −0.86 MPa in the wet forest *D. raillardioides* to −1.22 MPa in the dry scrub *D. linearis,* and tissue elastic moduli near full hydration range from 2.9 MPa in the dry scrub *D. platyphylla* to 18.2 MPa in the mesic forest *D. knudsenii* (Robichaux and Canfield 1985). The three 13-paired species from dry habitats tend to exhibit lower osmotic potentials and lower elastic moduli than the three 14-paired species from mesic to wet habitats, with the differences in elastic moduli being particularly striking (Figure 12.10).

As in the case of the 13-paired *D. ciliolata* and the 14-paired *D. scabra,*

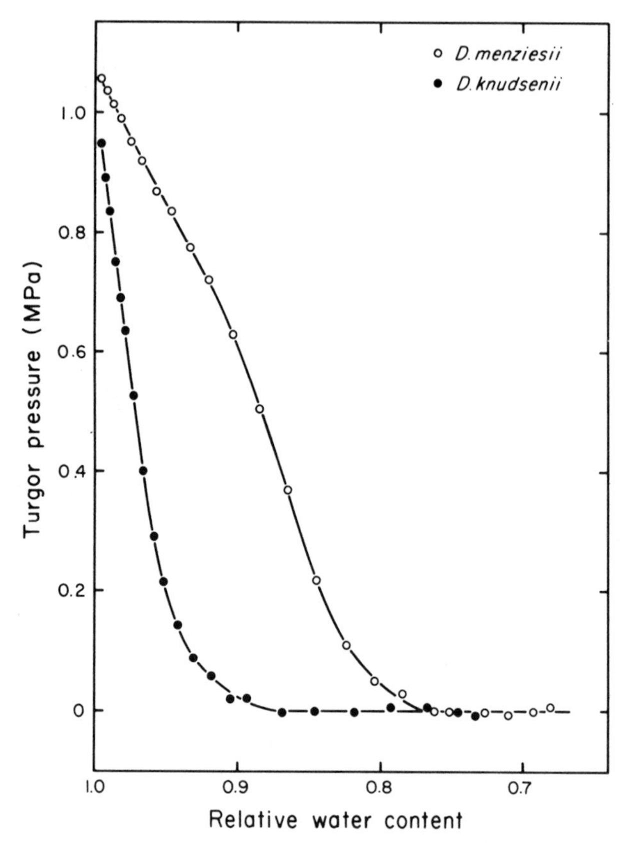

Figure 12.11. Relationships between tissue turgor pressure and tissue water content for *Dubautia menziesii* and *D. knudsenii* (Robichaux and Canfield 1985).

these large differences in elastic moduli have marked effects on the relative abilities of these species to maintain high turgor pressures as tissue water content decreases (Robichaux and Canfield 1985). This is illustrated in Figure 12.11 for the 13-paired *D. menziesii* and the 14-paired *D. knudsenii.* These two species exhibit a relatively small difference in their tissue osmotic potentials at full hydration, with the osmotic potential in *D. menziesii* (-1.12 ± 0.04 MPa) being slightly lower than that in *D. knudsenii* (-1.07 ± 0.05 MPa). However, they exhibit a very large difference in their tissue elastic moduli near full hydration (Figure 12.10). As a consequence, while the maximal values of turgor pressure are approximately the same in the two species, the initial rates at which turgor pressure declines with decreasing tissue water content are markedly different (Figure 12.11). The effect of this latter difference is striking. At a relative water content of 0.93, the value of tissue turgor pressure is 0.76 MPa in *D. menziesii* and only 0.08 MPa in *D. knudsenii.*

The turgor dependence of the elastic modulus in the three 13-paired species also differs from that in the three 14-paired species (Robichaux and Canfield 1985). In the latter species, the elastic modulus increases steadily with turgor pressure, such that its maximal value is reached near full hydration. In the 13-paired species, in contrast, the elastic modulus increases to a maximal value of approximately 5–6 MPa at intermediate turgor pressures, then decreases to a significantly lower value at high turgor pressures. As a result, there appears to be a pronounced qualitative difference in the tissue elastic properties of the two groups of species. This qualitative difference is similar to that discussed earlier for the 13-paired *D. ciliolata* and the 14-paired *D. scabra.*

These results indicate that the evolutionary diversification of the Hawaiian *Dubautia* species has been accompanied by a significant degree of change at the physiological level (Robichaux and Canfield 1985). At least one major evolutionary event, the aneuploid reduction in diploid chromosome number (Carr and Kyhos 1981), may have been accompanied by major shifts in tissue osmotic and elastic properties. The latter physiological shifts may well have enabled the derived 13-paired species to exploit significantly drier habitats than those occupied by the ancestral 14-paired species.

Mechanistic basis of the variation in tissue elastic properties

Although the mechanistic basis of the variation in tissue elastic properties is unknown, preliminary evidence suggests that at least three factors may be involved. These three factors are cell size, cell-wall composition, and apoplasmic water loss.

Influence of cell size

The difference in the tissue elastic properties of *D. ciliolata* and *D. scabra* is correlated with a large difference in the sizes of their mature leaf cells (Table 12.2) (Robichaux and Morse, unpublished data). Cell lengths and widths in the upper epidermis and cell lengths in the palisade parenchyma are significantly smaller in *D. ciliolata* than in *D. scabra.* Small but significant differences between the two species are also present in the widths of their palisade parenchyma cells and the lengths and widths of their spongy mesophyll cells. In each case, cell dimensions in *D. ciliolata* are

Table 12.2. *Cell dimensions in recently mature leaves of* Dubautia ciliolata, D. scabra, *and their natural hybrid*[a]

Tissue	*D. ciliolata* (μm)	*D. scabra* (μm)	Hybrid (μm)
Upper epidermis			
Cell length	26.4 (4.1)[b]	89.6 (16.3)	33.5 (6.6)
Cell width	33.5 (8.8)	63.3 (10.5)	44.4 (11.1)
Sample size	39	54	51
Palisade parenchyma			
Cell length	41.7 (8.3)	73.8 (16.7)	48.9 (12.7)
Cell width	15.9 (2.9)	17.1* (3.0)	17.5* (3.2)
Sample size	84	108	100
Spongy mesophyll			
Cell length	59.2* (13.9)	64.9 (14.4)	60.2* (14.6)
Cell width	43.9* (11.4)	49.0 (11.3)	45.8* (11.2)
Sample size	80	126	96
Lower epidermis			
Cell length	23.8* (4.4)	22.6* (4.7)	22.0* (4.4)
Cell width	34.6* (8.8)	28.0 (7.5)	34.2* (8.4)
Sample size	42	48	51

[a] All paired comparisons in a given row are significantly different ($P < 0.05$), except those with asterisks. Leaves were fixed in the field in a formalin : propionic acid : alcohol mixture and were sectioned with a Lab-Line Instruments Corp. model 1225 plant microtome. Maximal lengths and widths for a randomly chosen sample of cells in the central portion of each leaf were determined with an ocular micrometer calibrated against a stage micrometer at ×200 power. Leaves were sampled from at least three individuals of each taxon.
[b] Standard deviation.

smaller than in *D. scabra*. Cell dimensions in *D. ciliolata* exceed those in *D. scabra* only in the case of the widths of the lower epidermal cells. Because the upper epidermis and palisade parenchyma constitute 35–40% of the cross-sectional area of the leaf in each species, a large fraction of the cells in mature leaves of *D. ciliolata* is significantly smaller than in mature leaves of *D. scabra*.

This difference in average cell size appears to exist not only for mature leaves but also for entire branches (Robichaux and Morse, unpublished data). Average cell size in branches of *D. ciliolata* and *D. scabra* can be estimated with the formula: (weight of symplasmic water in the tissue at full hydration) ÷ (tissue dry weight). For the branches used in the measurements of tissue elastic properties, this ratio is 2.26 ± 0.24 g/g in *D. ciliolata* and 4.72 ± 0.54 g/g in *D. scabra*. In theory, this difference could reflect a difference in apoplasmic fractions rather than a difference in average cell sizes (Cutler et al. 1977; Tyree and Jarvis 1982). However, the apoplasmic fractions of *D. ciliolata* and *D. scabra* are not significantly different (Robichaux 1984). Hence, the average cell size in branches of *D. ciliolata* appears to be significantly smaller than that in branches of *D. scabra.*

This difference in average cell size may contribute to the difference in the tissue elastic properties of the two species. Cellular elastic moduli have been shown to vary with cell size in a variety of algal and higher plant cells, such that lower maximal elastic moduli are associated with smaller cells (Steudle et al. 1977; Husken et al. 1978; Zimmermann and Steudle 1980; Steudle et al. 1983). Thus, the lower maximal tissue elastic modulus in *D. ciliolata* may be related in part to its smaller average cell size.

The hybrid between *D. ciliolata* and *D. scabra* exhibits intermediate tissue elastic properties. If cell size actually influences these properties, then the hybrid should also possess cells of intermediate size. At least for cells in the upper epidermis and palisade parenchyma, this appears to be the case (Table 12.2).

Influence of cell-wall composition

The difference in the tissue elastic properties of the *Dubautia* species also appears to be correlated with a difference in the chemical composition of their cell walls. According to Carlquist (1959), leaves of *Dubautia* species from dry habitats accumulate large amounts of pectin in their cell walls relative to leaves of species from mesic to wet habitats. It is conceivable that an increase in the ratio of pectin to cellulose in the cell wall could result in an increase in wall elasticity. Thus, the lower tissue elastic moduli in the dry-habitat *Dubautia* species may be related in part to their greater pectic accumulations. (A lower elastic modulus indicates greater wall elasticity.) However, a more detailed analysis of the chemical composition of the cell walls of these species will be necessary before this hypothesis can be evaluated critically.

Influence of apoplasmic water loss

Whereas cell size and cell-wall composition may influence the actual value of the tissue elastic modulus in the *Dubautia* species, apoplasmic water loss may influence the calculated value of the elastic modulus (Robichaux and Holsinger, unpublished data). The weight-averaged bulk tissue elastic modulus, E, is defined as the change in tissue turgor pressure, P, for a given fractional change in the weight of symplasmic water in the tissue, W_s (Tyree and Jarvis 1982). Hence,

$$E = \frac{dP}{dW_s} W_s \qquad (12.1)$$

If the weight of apoplasmic water in the tissue, W_a, remains constant as W_s changes, then

$$E = \frac{dP}{dR}(R - R_a) \tag{12.2}$$

where R is the tissue relative water content and R_a is the relative water content of the tissue apoplasm, or the apoplasmic fraction. Equation (12.2) was used in calculating E for all of the *Dubautia* species (Robichaux 1984; Robichaux and Canfield 1985). However, if W_a does not remain constant as W_s changes, then

$$E = \frac{dP}{dR}(R - R_a)\left(1 + \frac{dW_a}{dW_s}\right) \tag{12.3}$$

(see Appendix I). Under this condition, the calculated value of E in the *Dubautia* species would differ from the actual value of E.

The assumption that W_a remains constant as W_s changes is a fundamental assumption of the pressure-chamber technique (Tyree and Karamanos 1981; Tyree and Jarvis 1982). Although we have no reason to suspect that this assumption was violated during our measurements, we have asked the question: Could apoplasmic water loss account for the differences in the tissue elastic properties of the *Dubautia* species?

We have addressed this question theoretically by examining the way in which dW_a/dW_s would have to vary as a function of R in order for the turgor dependence of E in a dry-habitat species to match that in a mesic-habitat species (see Appendix II). We have assumed in this analysis that no apoplasmic water is lost in the latter species (i.e., $dW_a/dW_s = 0$).

For simplicity, we have chosen to compare two species for which the turgor dependence of E can be calculated over the same range of turgor pressures (Figure 12.12). In the dry-habitat species, *D. ciliolata*, E increases to a maximal value of approximately 5 MPa at intermediate values of P, then decreases to a value of approximately 3 MPa at high values of P. In the mesic-habitat species, *D. plantaginea*, E increases steadily with P to a maximal value of approximately 14 MPa. When P reaches 0.9–1.0 MPa, the value of E in *D. plantaginea* exceeds that in *D. ciliolata* by a factor of 5 (Robichaux and Canfield 1985).

In order for the turgor dependence of E in *D. ciliolata* to equal that in *D. plantaginea* (Figure 12.13), dW_a/dW_s would have to vary as a function of R in the manner illustrated in Figure 12.14. The maximal value of dW_a/dW_s would be near full hydration, where the greatest difference exists in the calculated values of E in the two species (Figure 12.12). Near full hydration, a given amount of symplasmic water loss in *D. ciliolata* would be

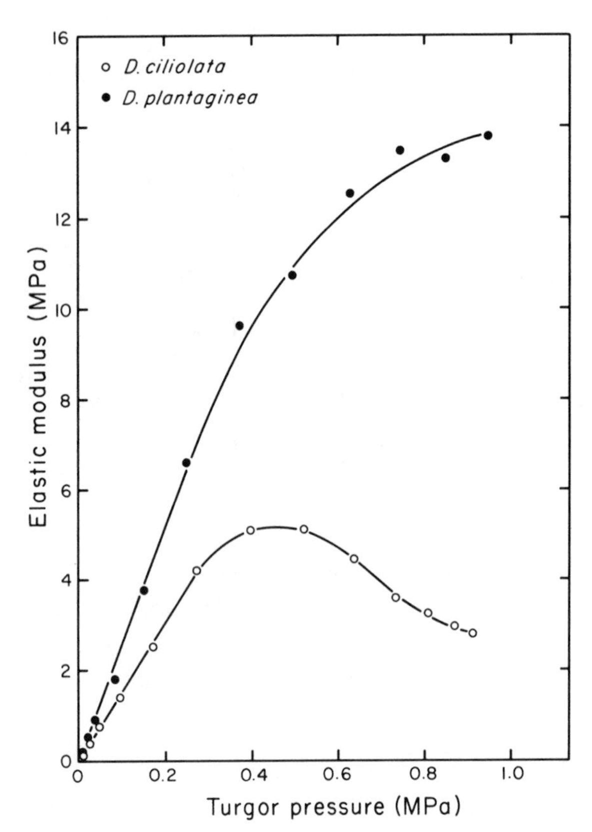

Figure 12.12. Relationships between tissue elastic modulus and tissue turgor pressure for *Dubautia ciliolata* and *D. plantaginea* (Robichaux and Canfield 1985).

accompanied by three to four times as much apoplasmic water loss. As R declined, the value of dW_a/dW_s would decline steadily and would eventually reach zero at the same value of R at which P reached zero in *D. ciliolata* (0.75).

From the information in Figure 12.14 it is also possible to calculate the total amount of apoplasmic water that would have to be lost in *D. ciliolata*. As R declined from 1.0 to 0.75, the total change in R_a would be 0.13. For the individual of *D. ciliolata* illustrated in Figure 12.12, R_a would decline from 0.31 to 0.18. [The latter value of 0.18 was obtained from linear extrapolation of the $1/\psi$-versus-R relationship to $1/\psi = 0$ (Robichaux 1984).] Thus, 50% of the water loss would have to be from the apoplasm.

The individual of *D. plantaginea* illustrated in Figure 12.12 exhibits tissue elastic properties that are close to the average for the *Dubautia* species from mesic to wet habitats. For example, the maximal value of the

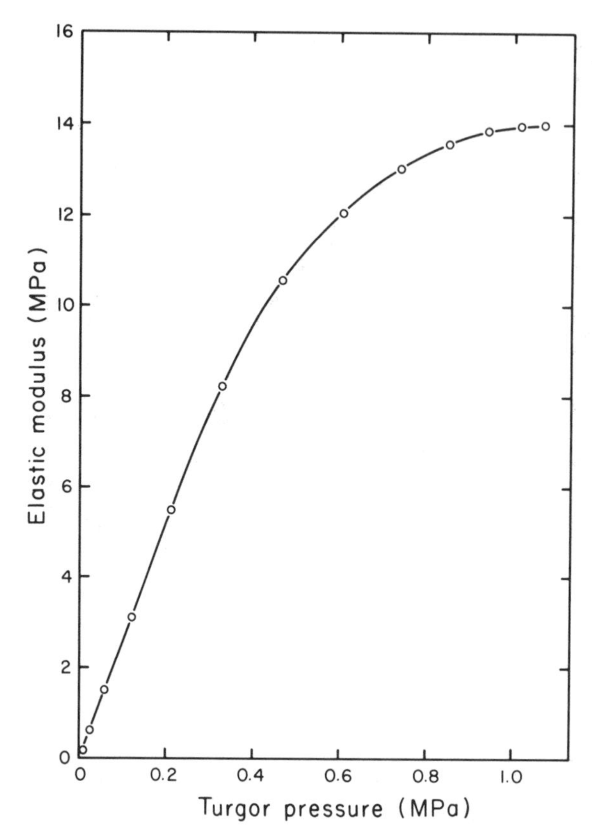

Figure 12.13. Calculated relationship between tissue elastic modulus and tissue turgor pressure for *Dubautia ciliolata* (Robichaux and Holsinger, unpublished data).

elastic modulus (E_m) in the former individual is 13.8 MPa, and the mean value of E_m among the latter species is 13.2 MPa (Robichaux 1984; Robichaux and Canfield 1985). Similarly, the individual of *D. ciliolata* illustrated in Figure 12.12 exhibits tissue elastic properties that are close to the average for the species from dry habitats. The elastic modulus near full hydration (E_i) and E_m in the former individual are 2.8 MPa and 5.1 MPa, respectively. The corresponding mean values of E_i and E_m among the latter species are 3.2 MPa and 5.4 MPa, respectively (Robichaux 1984; Robichaux and Canfield 1985). Hence, the results obtained for the two individuals are representative of the results obtained for the two groups of species.

These calculations imply that a large amount of apoplasmic water would have to be lost in *D. ciliolata*. In addition, most of this water would have to be lost at values of R above 0.93 and hence at values of ψ above

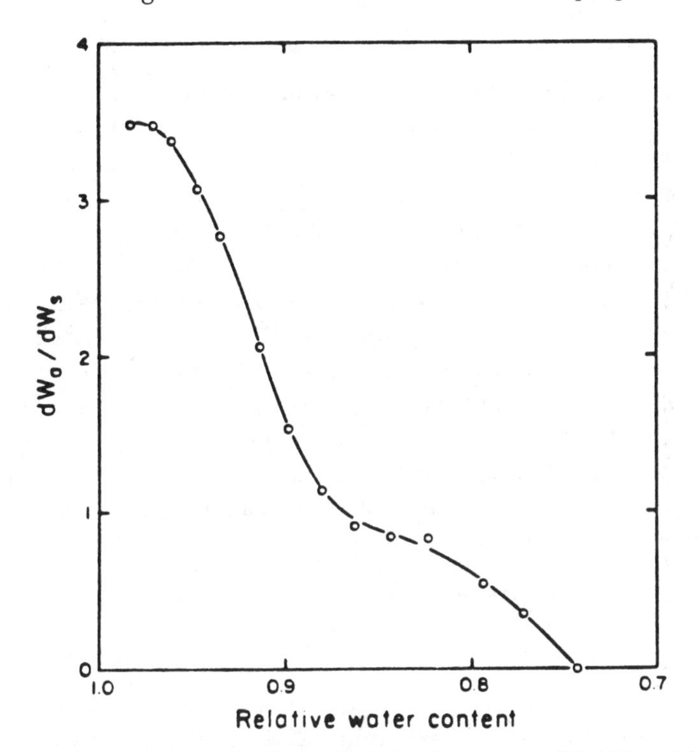

Figure 12.14. Calculated relationship between dW_a/dW_s and tissue water content for *Dubautia ciliolata* (Robichaux and Holsinger, unpublished data).

−0.38 MPa. At present, we cannot envision a mechanism by which this might occur, particularly given that such a mechanism cannot be present in the mesic- to wet-habitat species. For example, it is unlikely that differential drainage of xylem vessels is involved, because vessel elements have significantly larger diameters, and thus are more likely to drain, in the *Dubautia* species from mesic to wet habitats than in the species from dry habitats (Carlquist 1958, 1974). In addition, drainage of xylem vessels appears to be minimized in the pressure-chamber technique (Tyree and Karamanos 1981). As a result, we suggest that it is very unlikely that such a large amount of apoplasmic water loss actually occurs in *D. ciliolata* and the other dry-habitat species. Such a suggestion is consistent with the arguments of Tyree and Karamanos (1981), who provided several lines of evidence in support of the assumption that W_a remains constant as W_s changes. However, at least we now have a specific theoretical prediction to test with the *Dubautia* species once the experimental methods for analyzing apoplasmic water loss become available.

Conclusion

We have concentrated in this chapter on mechanisms promoting turgor maintenance in higher plants. We have shown that a decrease in the tissue osmotic potential at full hydration or a decrease in the tissue elastic modulus will have a marked effect on the degree to which high turgor pressures are maintained as tissue water content decreases.

In conclusion, we emphasize that plant growth under conditions of low moisture availability may be enhanced by a variety of other mechanisms as well. For example, characteristics that increase the amount of water extracted from the soil or that decrease the amount of water lost from the leaves will enable plants to maintain higher tissue water contents (Jones et al. 1981). These higher tissue water contents will result, in turn, in higher tissue turgor pressures. By examining these additional characteristics in the *Dubautia* species, we hope eventually to obtain an integrated view of the mechanisms by which they have adapted to dry environments.

Acknowledgments

This research was supported by NSF grant DEB-8206411 to R.H.R. and by a Miller postdoctoral research fellowship to K.E.H. We thank G. Carr, D. Kyhos, J. Canfield, and L. Stemmermann for many stimulating discussions of Hawaiian evolutionary biology. We also thank M. Feldman for kindly reviewing the appendixes.

Appendix I: An expression for the bulk tissue elastic modulus when water is lost from the apoplasm

We define R, the relative water content, and R_a, the relative water content of the apoplasm, as

$$R = \frac{W_s + W_a}{W_s^0 + W_a^0} \qquad (12.I.1a)$$

$$R_a = \frac{W_a}{W_s^0 + W_a^0} \qquad (12.I.1b)$$

where W_s is the weight of symplasmic water, W_a is the weight of apoplasmic water, W_s^0 is the weight of symplasmic water at full hydration, and W_a^0 is the weight of apoplasmic water at full hydration. Let R_a^* be the value of R_a when all symplasmic water has been lost (i.e., when $W_s = 0$).

We assume that there is some function, $f(W_s)$, that specifies W_a uniquely for a given value of W_s; that is, there is one and only one weight of apoplasmic water corresponding to a given weight of symplasmic water. In particular, we assume that $W_a^0 = f(W_s^0)$; that is, no water is lost from the apoplasm when it is saturated unless some water has already been lost from the symplasm or the symplasm loses water simultaneously.

Concerning this function, we further assume that it is continuous and differentiable and that $dW_a/dW_s \geq 0$; that is, as water is lost from the symplasm, it is not added to the apoplasm.

From (12.I.1a) it is apparent that

$$W_s = R(W_s^0 + W_a^0) - W_a \tag{12.I.2}$$

Let

$$g(W_s) = R(W_s^0 + W_a^0) - W_a - W_s \tag{12.I.3}$$

For a given R, let $\hat{W}_s$ be a root of $g(W_s) = 0$. We seek to show that this root, which obviously corresponds to W_s in (12.I.2), is unique.

For $R = 1$, $g(W_s^0) = 0$; so $\hat{W}_s = W_s^0$ is a root of $g(W_s) = 0$. Now consider $g(W_s)$ for $W_s < W_s^0$:

$$\begin{aligned} g(W_s) &= (W_s^0 + W_a^0) - [f(W_s) + W_s] \\ &= (W_s^0 - W_s) + [W_a^0 - f(W_s)] \end{aligned} \tag{12.I.4}$$

For any $W_s < W_s^0$, the right side of (12.I.4) is obviously positive, because $dW_a/dW_s \geq 0$ implies that $f(W_s) \leq W_a^0$. Thus, $\hat{W}_s = W_s^0$ is the only root of $g(W_s) = 0$ for $R = 1$.

For $R = R_a^*$, $g(0) = 0$; so $\hat{W}_s = 0$ is a root of $g(W_s) = 0$. Now consider $g(W_s)$ for $W_s > 0$:

$$\begin{aligned} g(W_s) &= R_a^*(W_s^0 + W_a^0) - [f(W_s) + W_s) \\ &= -W_s + [f(0) - f(W_s)] \end{aligned} \tag{12.I.5}$$

For any $W_s > 0$, the right side of (12.I.5) is obviously negative, because $dW_a/dW_s \geq 0$ implies that $f(0) \leq f(W_s)$. Thus, $\hat{W}_s = 0$ is the only root of $g(W_s) = 0$ for $R = R_a^*$.

Consider now a given R, $R_a^* < R < 1$.

$$g(W_s^0) = R(W_s^0 + W_a^0) - (W_a^0 + W_s^0) = (R - 1)(W_s^0 + W_a^0) \tag{12.I.6a}$$

$$\begin{aligned} g(0) &= R(W_s^0 + W_a^0) - W_a = R(W_s^0 + W_a^0) - R_a^*(W_s^0 + W_a^0) \\ &= (R - R_a^*)(W_s^0 + W_a^0) \end{aligned} \tag{12.I.6b}$$

$$\frac{dg}{dW_s} = -\frac{dW_a}{dW_s} - 1 \tag{12.I.6c}$$

For any $R < 1$, the right side of (12.I.6a) is obviously negative. Similarly, for any $R > R_a^*$, the right side of (12.I.6b) is obviously positive. Finally, the right side of (12.I.6c) is obviously negative, because $dW_a/dW_s \geq 0$ by assumption. But if $g(W_s^0) < 0$, $g(0) > 0$, and $dg/dW_s < 0$, then there is one and only one value of W_s for which $g(W_s) = 0$. Thus, $0 < \hat{W}_s < W_s^0$, and $\hat{W}_s$ is unique.

Because R varies continuously in the interval $[R_a^*, 1]$, and because we have shown that for a given R in this interval there is one and only one value of W_s corresponding to it, we can specify a function, $h(R)$, defined on this interval that maps $R \in [R_a^*, 1]$ into $W_s \in [0, W_s^0]$.

We know that turgor pressure, P, is a function of W_s. Because we also know that under the restrictions noted earlier, W_s is solely a function of R, a simple application of the chain rule yields

$$\frac{dP}{dR} = \frac{dP}{dW_s}\frac{dW_s}{dR} \tag{12.I.7}$$

But

$$\left(\frac{dW_s}{dR}\right)^{-1} = \frac{dR}{dW_s} \tag{12.I.8}$$

So

$$\frac{dP}{dW_s} = \frac{dP}{dR}\frac{dR}{dW_s} \tag{12.I.9}$$

The bulk tissue elastic modulus, E, is defined as

$$E = \frac{dP}{dW_s} W_s \tag{12.I.10}$$

(Tyree and Jarvis 1982). Substituting from (12.I.9) yields

$$E = \frac{dP}{dR}\frac{dR}{dW_s} W_s \tag{12.I.11}$$

Substituting into (12.I.11) the expressions from (12.I.1a) and (12.I.2) and evaluating yields

$$\begin{aligned} E &= \frac{dP}{dR}\frac{1}{W_s^0 + W_a^0}\left(1 + \frac{dW_a}{dW_s}\right)[R(W_s^0 + W_a^0) - W_a] \\ &= \frac{dP}{dR}(R - R_a)\left(1 + \frac{dW_a}{dW_s}\right) \end{aligned} \tag{12.I.12}$$

Appendix II: Estimating the necessary loss of apoplasmic water to make two sets of observations equivalent

For a given value of P, let E be the elastic modulus that is to be matched and E^* be the elastic modulus that is observed. From the results in Appendix I we know that

$$E = E^* \left(1 + \frac{dW_a}{dW_s}\right) \tag{12.II.1}$$

Thus

$$\frac{dW_a}{dW_s} = \frac{E}{E^*} - 1 \tag{12.II.2}$$

From (12.II.2) we obtain an initial estimate of dW_a/dW_s. Given this initial estimate, we then have to correct for the fact that the linear extrapolation of the $1/\psi$-versus-R relationship to $R = 1$ is no longer valid (Tyree and Jarvis 1982; see Figure 1 in Robichaux 1984). A nonlinear extrapolation results in new estimates of the tissue osmotic potential, π, as a function of R. New estimates of π result, in turn, in new estimates of P as a function of R. A new P-versus-R relationship then changes the calculated value of E^*, which changes the value of dW_a/dW_s in (12.II.2).

We use the following procedure to obtain new estimates of π as a function of R. Suppose that at one point we know π, R_a, and R. Let these values be π_1, R_{a1}, and R_1. We seek π_2 and R_{a2} given R_2. For this purpose, we define the relative water content of the symplasm, R_s, as

$$R_s = \frac{W_s}{W_s^0 + W_a^0} \tag{12.II.3}$$

Notice that $R_s = R - R_a$. In the limit as R_2 approaches R_1, the difference between R_2 and R_1, dR, can be written as

$$dR = \frac{\partial R}{\partial W_s} dW_s + \frac{\partial R}{\partial W_a} dW_a \tag{12.II.4}$$

Given the foregoing definitions, (12.II.4) reduces to

$$dR = \frac{1}{W_s^0 + W_a^0} (dW_s + dW_a) \tag{12.II.5}$$

Noticing that $dW_a = (dW_a/dW_s)\, dW_s$ and substituting into (12.II.5) yields

$$dR = \frac{dW_s}{W_s^0 + W_a^0}\left(1 + \frac{dW_a}{dW_s}\right) \tag{12.II.6}$$

In discrete form this can be written as

$$\begin{aligned} R_2 - R_1 &= \frac{W_{s2} - W_{s1}}{W_s^0 + W_a^0}\left(1 + \frac{dW_a}{dW_s}\right) \\ &= (R_{s2} - R_{s1})\left(1 + \frac{dW_a}{dW_s}\right) \end{aligned} \tag{12.II.7}$$

provided that the second- and higher-order terms in the Taylor-series expansion of R_2 about R_1 are negligible. Thus,

$$R_{s2} = \frac{R_2 - R_1}{1 + dW_a/dW_s} + R_{s1} \tag{12.II.8a}$$

and

$$R_{a2} = R - R_{s2} \tag{12.II.8b}$$

To find π_2, we first recall that

$$\pi_1 W_{s1} = \pi_2 W_{s2} \tag{12.II.9}$$

(Tyree and Jarvis 1982). Dividing both sides of (12.II.9) by $(W_s^0 + W_a^0)$ and rearranging yields

$$\pi_2 = \pi_1 \frac{R_{s1}}{R_{s2}} \tag{12.II.10}$$

By evaluating (12.II.8a), (12.II.8b), and (12.II.10) successively, starting at the value of R at which P reaches zero and working backward to $R = 1$, it is possible to obtain new estimates of π as a function of R.

New estimates of P as a function of R are then obtained, and a new estimate of E^* is calculated as

$$E^* = \frac{dP}{dR}(R - R_a)\left(1 + \frac{dW_a}{dW_s}\right) \tag{12.II.11}$$

where the values of R_a and dW_a/dW_s used are those corresponding to the value of R used. This new estimate of E^* is then used to calculate a new estimate of dW_a/dW_s. If $(dW_a/dW_s)'$ is the new estimate and dW_a/dW_s is the previous estimate, then

$$\left(\frac{dW_a}{dW_s}\right)' = \frac{E}{E^*} - 1 + \frac{dW_a}{dW_s} \tag{12.II.12}$$

This new estimate of dW_a/dW_s is then used to reevaluate R_s, R_a, π, P, and

E^* in the manner just described. This process is repeated until a satisfactory approximation to E is obtained at each value of P.

References

Acevedo, E., E. Fereres, T. C. Hsiao, and D. W. Henderson. 1979. Diurnal growth trends, water potential, and osmotic adjustment of maize and sorghum leaves in the field. Plant Physiol. 64:476–480.

Boyer, J. S. 1982. Plant productivity and environment. Science 218:443–448.

Bradford, K. J., and T. C. Hsiao. 1982. Physiological responses to moderate water stress. Pp. 263–324 *in* O. L. Lange, P. S. Nobel, C. B. Osmond, and H. Ziegler (eds.), Encyclopedia of plant physiology, new series, vol. 12B, Springer-Verlag, Berlin.

Carlquist, S. 1958. Wood anatomy of Heliantheae (Compositae). Tropical Woods 108:1–30.

– 1959. Studies on Madinae: anatomy, cytology, and evolutionary relationships. Aliso 4:171–236.

– 1974. Island biology. Columbia University Press, New York.

– 1980. Hawaii: a natural history. Pacific Tropical Botanical Garden, Lawai.

Carr, G. D. 1978. Chromosome numbers of Hawaiian flowering plants and the significance of cytology in selected taxa. Amer. J. Bot. 65:236–242.

Carr, G. D., and D. W. Kyhos. 1981. Adaptive radiation in the Hawaiian silversword alliance (Compositae-Madiinae). I. Cytogenetics of spontaneous hybrids. Evolution 35:543–556.

Cutler, J. M., D. W. Rains, and R. S. Loomis. 1977. The importance of cell size in the water relations of plants. Physiol. Plant. 40:255–260.

Ehleringer, J. 1983. Ecophysiology of *Amaranthus palmeri,* a Sonoran Desert summer annual. Oecologia 57:107–112.

Fischer, R. A., and N. C. Turner. 1978. Plant productivity in the arid and semiarid zones. Ann. Rev. Plant Physiol. 29:277–317.

Grosvenor, M. B. (ed.). 1966. National Geographic atlas of the world. National Geographic Society, Washington, D.C.

Hanson, A. D., and W. D. Hitz. 1982. Metabolic responses of mesophytes to plant water deficits. Ann. Rev. Plant Physiol. 33:163–203.

Hsiao, T. C., E. Acevedo, E. Fereres, and D. W. Henderson. 1976. Water stress, growth, and osmotic adjustment. Phil. Trans. R. Soc. London, Series B 273:479–500.

Husken, D., E. Steudle, and U. Zimmermann. 1978. Pressure probe technique for measuring water relations of cells in higher plants. Plant Physiol. 61:158–163.

Jones, M. M., and H. M. Rawson. 1979. Influence of rate of development of leaf water deficits upon photosynthesis, leaf conductance, water use efficiency, and osmotic potential in sorghum. Physiol. Plant. 45:103–111.

Jones, M. M., N. C. Turner, and C. B. Osmond. 1981. Mechanisms of drought resistance. Pp.15–37. *in* L. G. Paleg and D. Aspinall (eds.), The physiology and biochemistry of drought resistance in plants. Academic Press, New York.

Kramer, P. J. 1980. Drought, stress, and the origin of adaptations. Pp. 7–20 *in* N. C. Turner and P. J. Kramer (eds.), Adaptation of plants to water and high temperature stress. Wiley, New York.

Lange, O. L., L. Kappen, and E.-D. Schulze (eds.). 1976. Water and plant life: problems and modern approaches. Springer-Verlag, Berlin.

Ludlow, M. M. 1980. Adaptive significance of stomatal responses to water stress. Pp. 123–138 *in* N. C. Turner and P. J. Kramer (eds.), Adaptation of plants to water and high temperature stress. Wiley, New York.

Robichaux, R. H. 1984. Variation in the tissue water relations of two sympatric Hawaiian *Dubautia* species and their natural hybrid. Oecologia 65:75–81.

Robichaux, R. H., and J. E. Canfield. 1985. Tissue elastic properties of eight Hawaiian *Dubautia* species that differ in habitat and diploid chromosome number. Oecologia 66:77–80.

Steudle, E., H. Ziegler, and U. Zimmermann. 1983. Water relations of the epidermal bladder cells of *Oxalis carnosa* Molina. Planta 159:38–45.

Steudle, E., U. Zimmermann, and U. Luttge. 1977. Effect of turgor pressure and cell size on the wall elasticity of plant cells. Plant Physiol. 59:285–289.

Takami, S., N. C. Turner, and H. M. Rawson. 1981. Leaf expansion of four sunflower (*Helianthus annuus* L.) cultivars in relation to water deficits. I. Patterns during plant development. Plant Cell Env. 4:399–407.

– 1982. Leaf expansion in four sunflower (*Helianthus annuus* L.) cultivars in relation to water deficits. II. Diurnal patterns during stress and recovery. Plant Cell Env. 5:279–286.

Turner, N. C., J. E. Begg, and M. L. Tonnet. 1978. Osmotic adjustment of sorghum and sunflower crops in response to water deficits and its influence on the water potential at which stomata close. Aust. J. Plant Physiol. 5:597–608.

Turner, N. C., and M. M. Jones. 1980. Turgor maintenance by osmotic adjustment: a review and evaluation. Pp. 87–103 *in* N. C. Turner and P. J. Kramer (eds.), Adaptation of plants to water and high temperature stress. Wiley, New York.

Turner, N. C., and P. J. Kramer (eds.). 1980. Adaptation of plants to water and high temperature stress. Wiley, New York.

Tyree, M. T., and P. G. Jarvis. 1982. Water in tissues and cells. Pp. 35–77 *in* O. L. Lange, P. S. Nobel, C. B. Osmond, and H. Ziegler (eds.), Encyclopedia of plant physiology, new series, vol. 12B. Springer-Verlag, Berlin.

Tyree, M. T., and A. J. Karamanos. 1981. Water stress as an ecological factor. Pp. 237–261 *in* J. Grace, E. D. Ford, and P. G. Jarvis (eds.), Plants and their atmospheric environment. Blackwell, Oxford.

Zimmermann, U. 1978. Physics of turgor- and osmoregulation. Ann. Rev. Plant Physiol. 29:121–148.

Zimmermann, U., and E. Steudle. 1980. Fundamental water relations parameters. Pp. 113–130 *in* R. M. Spanswick, W. J. Lucas, and J. Dainty (eds.), Plant membrane transport: current conceptual issues. Elsevier/North Holland, Amsterdam.

13 Adaptations for water and thermal balance in Andean giant rosette plants

FREDERICK MEINZER AND
GUILLERMO GOLDSTEIN

The vegetation of high mountain areas in or near equatorial regions has long been noted for its convergence in physiognomy and growth forms (Troll 1958; Hedberg 1964). Perhaps the most conspicuous of these growth forms is the so-called giant rosette characterized by a dense mass of spirally arranged leaves with little internode elongation. The apical bud is typically massive and contains a large bank of unexpanded leaves. This form has arisen in members of the genus *Espeletia* from the tropical Andes of northern South America, in species of *Senecio* and *Lobelia* from equatorial Africa, and in *Argyroxiphium* from the Hawaiian volcanoes (Figure 13.1). There are other examples, but the South American and African mountains contain the greatest numbers of species with the greatest geographical extent.

This great morphological convergence in geographically isolated regions suggests that the giant rosette form represents an adaptive solution to the special selective pressures that prevail in high-altitude tropical environments. Several workers have stressed the singularities of tropical high-altitude climates in which diurnal temperature variations are larger than seasonal ones, and freezing temperatures are frequent and can occur any night of the year (Hedberg 1964; Coe 1967; Troll 1968; Monasterio and Reyes 1980). Other features commonly mentioned are the high radiant-energy input and consistently low air temperatures, although more attention has been given to low nighttime thermal-energy input than to low daytime temperatures. Plants growing in cold tropical environments lacking temperature seasonality are not subjected to the same behavioral constraints as those growing in cold temperate climates. For example, giant rosette plants would be expected to exhibit a more or less constant level of physiological activity throughout the year rather than relying on dormancy as a mechanism for withstanding periods of low temperatures.

Little is known of the physiology of Andean and African giant rosette plants, but several morphological features have been singled out and hy-

Figure 13.1. Convergence in the giant rosette growth form in geographically isolated tropical alpine regions: (a) *Espeletia lutescens* at 4,200 m in the Venezuelan Andes. (b) *Puya raimondii* at 3,800 m in the Peruvian Andes. (c) *Senecio keniodendron* at 4,200 m on Mt. Kenya, Africa. (d) *Lobelia telekii* at 4,200 m on Mt. Kenya. (e) *Argyroxiphium sandwicense* in Haleakala Crater, Hawaii. (Photos b–d courtesy of Alan P. Smith; photo e courtesy of The National Park Service.)

potheses ventured concerning their adaptive value. Hedberg's observations (1964) are an excellent source of testable hypotheses concerning adaptations in African giant rosette species. Despite some initial field and laboratory data available for Andean giant rosette species (e.g., Smith 1974, 1979; Larcher 1975; Baruch and Smith 1979), the precise adaptive significance of most aspects of giant rosette morphology is not entirely clear. A given morphological feature may enhance chances for survival in the presence of an environmental pressure such as low temperature, but in order for our understanding of its adaptive value to be complete, its "hidden costs" in terms of constraints imposed on various aspects of plant functioning and behavior must be analyzed.

Our objective here is to examine some of the consequences of three prominent morphological features in several Andean giant rosette species: (1) the voluminous central pith found in the stems of both caulescent and acaulescent species, (2) the thick layer of marcesent leaves surrounding the stem in caulescent species, and (3) leaf pubescence. These features are thought to influence water, thermal, and carbon balance. Water economy should be of paramount importance in the high-altitude tropics, because water availability is limited on a daily basis by low temperatures and may be limited seasonally by low soil moisture levels. We analyze the significance of within- and between-species differences in pith volume in terms of the pith's impact on hydraulic capacitance and the avoidance of transitory early morning water stress. Regulation of thermal balance is also of interest, because tropical high-altitude environments present the unusual combination of high radiant-energy inputs and low thermal-energy inputs. Our analysis includes an evaluation of some of the adaptive benefits provided by each feature and some of its costs in terms of behavioral and functional constraints. We also discuss the ways in which marcesent leaves and leaf pubescence may influence water balance, carbon economy, and growth through their effects on stem and leaf temperatures, respectively. As an aid to understanding benefits, as well as functional and behavioral constraints, we have used field and laboratory measurements to develop predictive models describing plant responses under simulated environmental conditions. Based on an evaluation of their consequences, we have made predictions as to how these morphological features should vary along environmental gradients, and where possible we have tested the predictions with measurements.

Stem pith

In the Andes, as well as Africa, giant rosette plants exhibit varying degrees of caulescence, ranging from sessile, acaulescent forms to forms with stems several meters long (Figure 13.2). Stems of caulescent giant rosette species show a striking similarity in the appearance of their central parenchymatous pith surrounded by a cylinder of well-developed secondary xylem. Hedberg (1964) has suggested that the pith of the African caulescent giant rosettes may act as a water source during periods of cold

Figure 13.2. Morphological variation in *Espeletia*, particularly with respect to caulescence and therefore pith volume per unit leaf are (PV/LA). Acaulescent or nearly acaulescent species have the lowest PV/LA values. (a) *E. lutescens* at

temperature and low water availability. For this function to be realized, however, relatively good hydraulic connections between the pith and stem xylem would have to exist.

Most discussions of water transfer through the soil–plant–atmosphere continuum have considered the soil as the only source of water, even for meeting short-term demands. Only recently has the importance of plant tissues as important sources of water been emphasized (Jarvis 1975; Powell and Thorpe 1977; Waring and Running 1978). Two categories of internal water reservoirs have been distinguished (Hinckley et al. 1978): (1) elastic tissues such as fruits, buds, and foliage that undergo dimensional changes when water is exchanged with the transpirational stream; (2) inelastic tissues such as sapwood that do not undergo dimensional changes. In trees, stem sapwood seems to account for most of the water available from internal sources (Jarvis 1975; Hinckley et al. 1978). Quantitative estimates of capacitance, the change in water content per unit change in water potential, can be obtained only if both the volume and moisture-releasing properties of the tissue under study are known (Powell and Thorpe 1977; Edwards and Jarvis 1982). The proportion of tissue water that is thermodynamically available to the leaves can be determined with a pressure–volume curve (Tyree and Hammel 1972), the slope of which corresponds to the tissue capacitance. In the discussion that follows, we shall use capacitance in a more general sense, one that reflects the ability of the plant to dampen fluctuations in leaf water potential through reliance on internal water reservoirs rather than soil water.

Hydraulic connections and relative capacitance

The pith water potential of *Espeletia timotensis*, a caulescent species common at an elevation of 4,200 m, shows diurnal changes that parallel those in the leaves. In individuals deprived of water, the pith water potential declines relatively rapidly and recovers rapidly on rewatering (Figure 13.3). The pith thus has relatively good hydraulic connections with the rest of the plant and is capable of at least partially compensating daily transpirational losses.

The *Espeletia* species that grow in the Andean paramos differ not only in pith volume but also in the ratio between transpiring surface and the volume of the water reservoir (Figure 13.2). We used pith volume per unit leaf area (PV/LA) as a measure of relative water storage capacity for seven

Caption to Figure 13.2 (*cont.*)
4,200 m partially sectioned longitudinally to show pith and thick layer of marcesent leaves. (b) *E. moritziana* at 4,200 m. (c) *E. schultzii* at 3,550 m. (d) *E. floccosa* at 3,550 m. (e) *E. atropurpurea* at 3,100 m. (f) *E. schultzii* at 2,600 m.

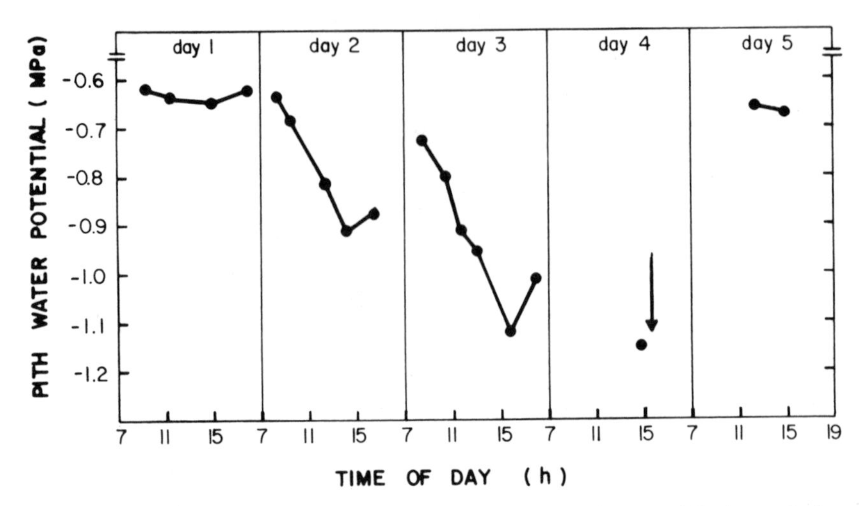

Figure 13.3. Fluctuation in pith water potential of a potted *E. timotensis* plant deprived of irrigation from day 1 to 4. The arrow indicates the time of rewatering. Two ceramic cup psychrometers previously installed inside the pith were used to monitor the changes in water potential (from Goldstein and Meinzer 1983).

Espeletia species growing in three paramos that represent different points along an environmental gradient (Table 13.1). Relative water storage capacity is better expressed by PV/LA than its inverse, because the total leaf area of the rosettes remains essentially constant once the plants reach a certain height. Intraspecific and interspecific differences in water storage capacity are thus due primarily to differences in stem length and therefore pith volume. Species from colder and drier paramos tend to have higher relative capacitance than species from warmer, wetter sites (Table 13.1).

Available water

The pith can fully recover from exposure to water potentials corresponding to the turgor loss point (water potential corresponding to zero turgor) of the leaves (Goldstein and Meinzer 1983), although these potentials are lower than the turgor loss point of the pith itself (Goldstein et al. 1984). The amount or fraction of available water in the pith can be estimated from the change in pith relative water content between saturation (full turgor) and a limit water potential corresponding to the turgor loss point of the leaves. This is obtained from the curve describing the relationship between water potential and water content for pith tissue (Goldstein et al. 1984). The leaf turgor loss point is lower in species with lower PV/LA values, and for this reason their pith contains a greater

Table 13.1. *Summary of site characteristics where* Espeletia *species were studied to estimate the period of time during which the water removed from the pith could replace the water transpired*[a]

Paramo site and elevation (m)	Annual precipitation (mm)	Mean temperature (°C)	Species	PV/LA (cm^3 cm^{-2})	ΔM (g)	T (g h^{-1})	Hours of T
Piedras Blancas 4,200	798	2.8	*E. lutescens*	0.105	176	70.7	2.5
			E. moritziana	0.057	57	39.7	1.4
			E. spicata	0.056	160	81.9	2.0
Mucubaji, 3,600	969	5.4	*E. schultzii*	0.047	99	95.2	1.0
			E. floccosa	0.013	27	44.8	0.6
Batallon, 3,100	1,213	9.3	*E. marcana*	0.038	86	55.3	1.6
			E. atropurpurea	0.018	9	16.3	0.6

[a] Hours of transpiration were calculated by dividing the mass of available water in the pith (ΔM) by the transpiration rate (T) of the entire rosette. Pith volume per unit leaf area (PV/LA) is a measure of relative capacitance.

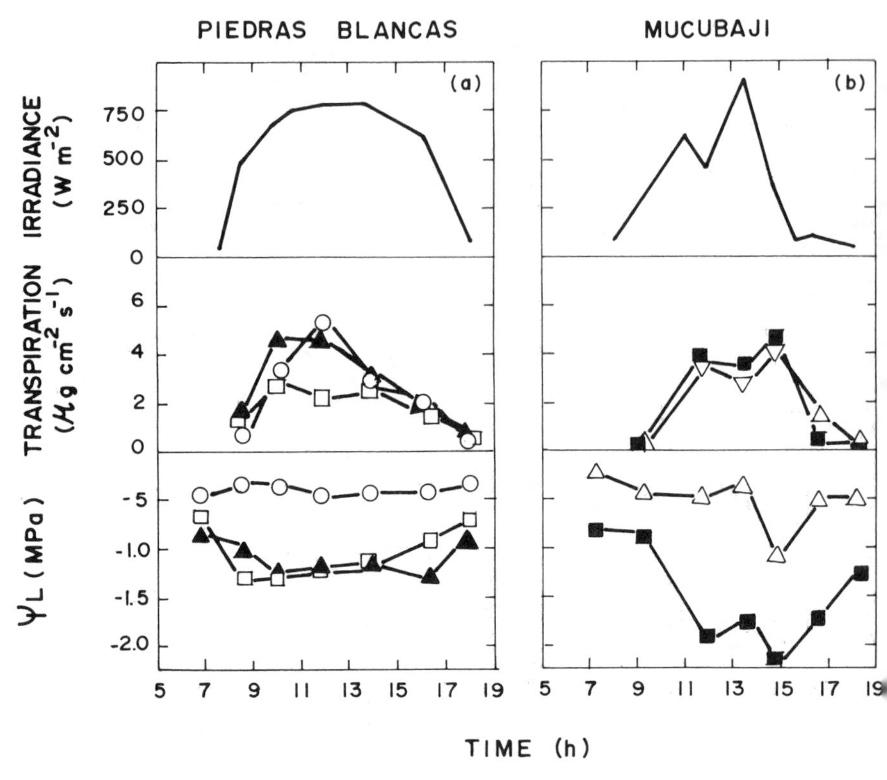

Figure 13.4. Daily courses of global radiation, transpiration, and leaf water potential (ψ_L) during the dry season. In Piedras Blancas paramo (4,200 m), the species were *E. lutescens* (circles), *E. moritziana* (triangles), and *E. spicata* (squares), and in Mucubaji paramo (3,550 m), *E. schultzii* (triangles) and *E. floccosa* (squares) (from Goldstein et al. 1984a).

proportion of available water (Table 13.1). However, estimates of the amount of time that water from the pith can fully replace transpirational losses show that relative storage (PV/LA) exerts more influence than the fraction of available water. In terms of replacement of transpirational losses, pith water storage capacity in the high-elevation species is greater than most of the water storage values reported for elastic reservoirs (Jarvis, 1975; Hinckley et al. 1981), but falls within the range of values of sapwood storage calculated for conifers (Richards 1973; Waring and Running 1978; Running 1980).

Relative capacitance and water relations in the field

Differences in relative capacitance are reflected in patterns of regulation of water balance in the field. The species that have the highest storage capacity in a given paramo either do not exhibit pronounced diurnal changes in leaf water potential or exhibit a morning drop in leaf water potential that lags with respect to that of the other species (Figure 13.4). The leaf water potential of a caulescent species such as *E. lutescens* may change no more than 0.15 MPa even during a day with relatively high evaporative demand. Under similar conditions, an acaulescent species such as *E. floccosa* may experience a 1.0-MPa or larger fluctuation in leaf water potential (Figure 13.4). From the standpoint of water economy, caulescent giant rosettes might be considered as functionally similar to succulent plants. The resemblance is only superficial, however, not only because species of *Espeletia* have high stomatal conductance, unlike most succulents, but also because the pith serves more to meet morning transpirational needs than to store water for prolonged drought periods.

The influence of pith water storage on water economy can be evaluated in a more quantitative manner by determining the relationship between PV/LA and the effective hydraulic resistance between the soil and leaves. The relative drop in leaf water potential per unit of transpiration, a measure of total liquid-flow resistance, is much lower in species with greater relative capacitance (Figure 13.5a). That is, in tall caulescent plants, greater transpiration rates can be sustained for a given drop in leaf water potential. As might be expected, turgor loss occurs at lower water potentials in leaves of species with higher effective hydraulic resistances (lower PV/LA, Figure 13.5b). In this manner, lower water potentials could be sustained in leaves of species with lower relative capacitance without the occurrence of stomatal closure due to turgor loss. The relationship in Figure 13.5b may be extendable to other paramo growth forms. For instance, the leaves of the evergreen shrub *Hypericum laricifolium* (PV/LA $\cong$ 0) lose turgor between -2.0 and 2.2 MPa during the dry season.

Using the data available for seven *Espeletia* species, both linear and asymptotic functions are highly significant in describing the relationship between PV/LA and both hydraulic resistance and leaf turgor loss point (Figure 13.5). The asymptotic functions, which predict that hydraulic resistance and leaf turgor loss point eventually become independent of pith capacitance, may have a higher mechanistic reality. It is reasonable to expect that at some point the benefits provided by increased pith volume diminish in relation to the limitation on rate of water movement from pith to xylem. At this point, hydraulic resistance and leaf turgor loss point reach their minimum and maximum values, respectively.

Intraspecific relationships

In the caulescent species that grow at high elevations, relative capacitance is expected to vary strongly through the life of each individual. In *E. lutescens,* relative capacitance initially increases slowly as both rosette

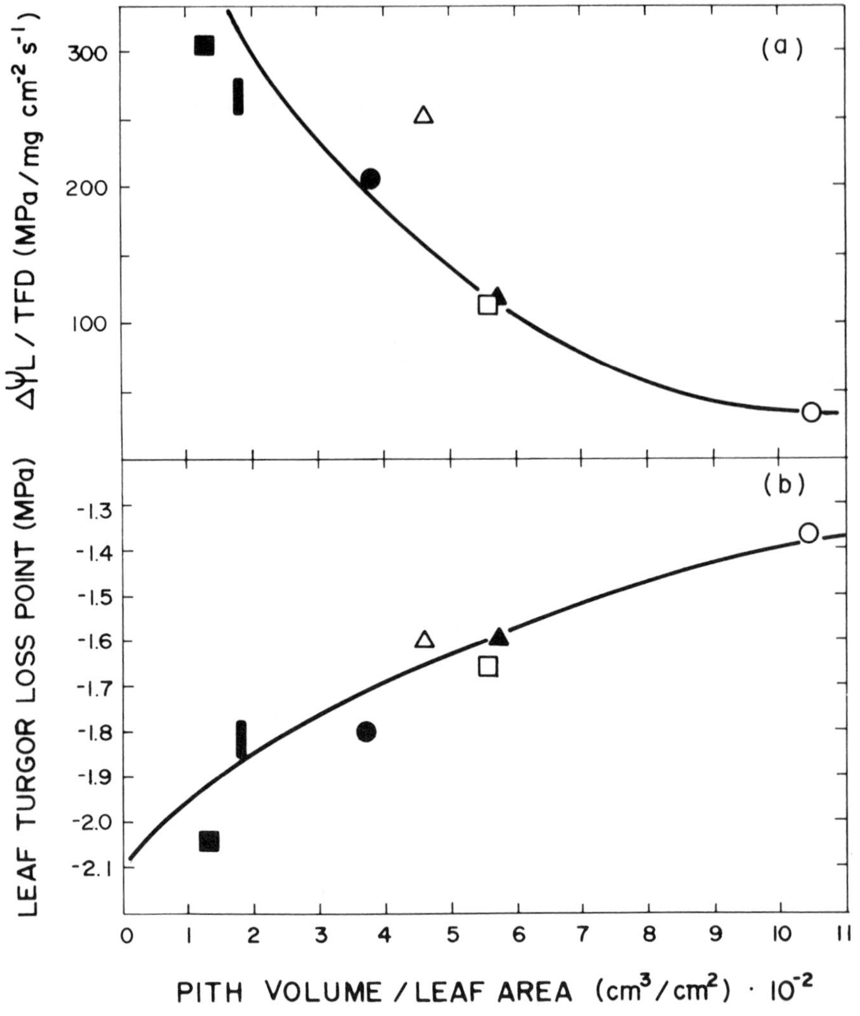

Figure 13.5. (a) Relationship between hydraulic resistance ($\Delta\psi_L$/TFD) and PV/LA in adult individuals of *E. lutescens* (open circle), *E. moritziana* (filled triangle), *E. spicata* (open square), *E. marcana* (filled circle), *E. schultzii* (open triangle), *E. atropurpurea* (bar), and *E. floccosa* (filled square). (b) Leaf turgor loss point determined from pressure–volume curves in relation to PV/LA. Species and symbols as in part (a) (from Goldstein et al. 1984).

area and stem length are increasing (Figure 13.6a). As the rosette area begins to stabilize, a rapid increase in relative capacitance with increasing height occurs. The importance of these changes in relative capacitance is reflected in patterns of water balance in individuals of different heights (Figure 13.6b). During the dry season, minimum leaf water potentials of

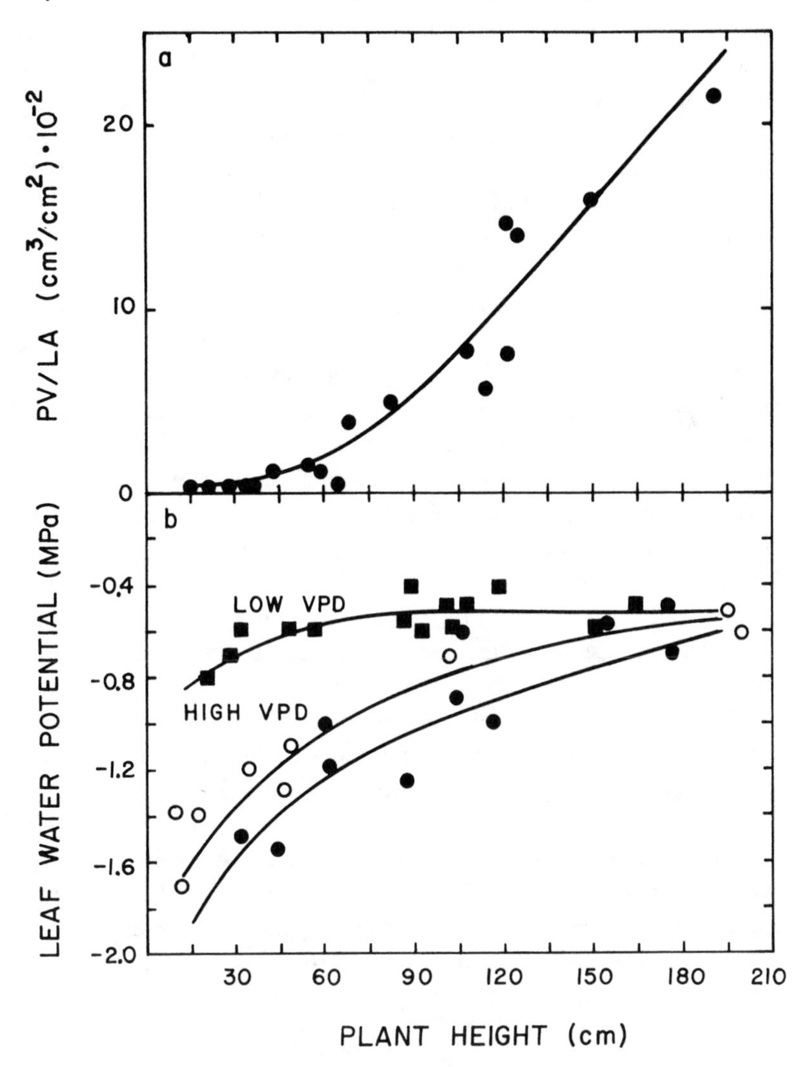

Figure 13.6. (a) Pith volume per unit leaf area (PV/LA) in relation to plant height for individuals of an *E. lutescens* population at 4,200 m. (b) Minimum leaf water potential in relation to plant height on days with low and high leaf-to-air vapor-pressure deficit (VPD) during the dry season (from Goldstein et al. 1985).

taller, adult individuals remain well above the turgor loss point, whereas minimum water potentials are lower in shorter plants, and wilting may occur (Goldstein et al. 1985). Because stomatal opening in *E. lutescens* is rather sensitive to variations in leaf water potential, it is likely that carbon dioxide uptake will be restricted in younger individuals (Meinzer, Goldstein, and Hinckley, unpublished observations).

Analyses of pressure–volume curves show that turgor loss points in leaves of individuals of different ages are nearly identical, indicating that *E. lutescens* has little capacity to adjust osmotically. The leaves begin to suffer irreversible damage at incipient plasmolysis (Goldstein et al. 1985), suggesting a tradeoff between a stable water balance provided by high relative capacitance and resulting low dehydration tolerance. This could represent an important constraint when water availability is limited by low soil moisture in addition to low temperatures. In particularly dry years, irreversible wilting in younger individuals of *E. lutescens* is fairly common.

It is interesting that whereas the typical trend in temperate-zone mountains is a reduction in plant height with increasing elevation, species of *Espeletia* above timberline tend to be taller at higher altitudes. Several explanations have been offered for this apparent paradox, among them increased adult longevity, escape from ground-level freezing stress, and escape from frost-related drought stress (Larcher 1975; Smith 1980). We suggest that the caulescent habit in the genus *Espeletia* plays an important role in water economy, especially in higher, colder habitats. The height of giant rosette plants could be at least partially related to maximization of the effectiveness of water storage. From a hydraulic standpoint, a long, narrow cylindrical reservoir that maximizes the area of contact between xylem and pith is more efficient than a short wide one. At higher elevations, water uptake may be more greatly impeded by low root membrane permeability when soil temperatures are near or below freezing (Kaufmann 1977; Running and Reid 1980; Goldstein 1981). The stems of high-elevation *Espeletia* species are insulated by marcesent leaves that prevent freezing of the pith (Goldstein and Meinzer 1983). With an insulated pith water reservoir, both the distance and resistance between water source and atmospheric sink are minimized during periods of low temperature-limited water availability.

Marcesent leaves

Another conspicuous feature shared by both Andean and African caulescent giant rosette species is a thick cylinder of dead, persistent leaves adhering to and surrounding the stem. In some high-elevation African species these marcesent leaves are absent, and the stem is surrounded by a thick layer of cork (Hedberg 1964). There is no doubt that marcesent leaves provide considerable temperature insulation. In giant rosette plants they prevent nighttime stem temperatures from falling below 0°C even when environmental temperatures are several degrees below zero (Hedberg 1964; Coe 1967; Smith 1979). In other growth forms, such as the palm *Washingtonia filifera,* marcesent leaf cylinders provide thermal insulation against heat from fires (Vogl and McHargue 1966). In the Andean species *E. schultzii* growing at 3,600 m, this insulating effect is important for survivial, as shown by increased mortality among individuals that had been experimentaly deprived of their marcescent leaves (Smith 1979). In more extreme paramo sites, marcesent leaf removal may induce 100% mortality (Goldstein and Meinzer 1983).

The functional and behavioral costs associated with the presence of marcesent leaves may not be readily apparent. An understanding of the specific mechanisms through which marcesent leaf removal induces mortality is important for evaluating these costs. Most recent hypotheses concerning the adaptive value of the marcesent leaf cylinder suggest that in its absence, frozen stem tissue would make water temporarily unavailable for replacing morning transpirational losses, resulting in leaf wilting and death (Hedberg 1964; Smith 1974). This proposed mechanism is essentially the same as the freeze-desiccation hypothesis often invoked to explain the altitudinal limit for tree growth in temperate regions (Wardle 1974; Tranquillini 1979).

Possible costs

Because the marcesent leaf layer dampens diurnal temperature fluctuations, exposed stems will reach higher maximum temperatures than insulated ones. In contrast to most species for which data are available, water flow resistance through the stem of the caulescent *E. timotensis* is highly temperature-dependent even when freezing has not occurred (Figure 13.7). The exponential increase in hydraulic resistance at low temperatures cannot be attributed to the increased viscosity of water. Nearly identical patterns of hydraulic resistance increases at low temperatures have been reported for roots (Kaufmann 1975; Running and Reid 1980). In roots, the changes in resistance are presumably due to low-temperature effects on the permeability of the endodermal cell membranes. Diffusion of water across the living membranes of the pith cells and into the xylem could account for much of the low-temperature effect on hydraulic resistance in the stem of *E. timotensis.* The preferential pathway for water movement from pith to leaves probably would not be a high-resistance, vertical one through the pith, but rather a lower-resistance, lateral pathway from pith to stem xylem, where numerous pith rays penetrate the xylem. Other factors may be involved in the resistance changes, but field and laboratory observations suggest that pith cell membranes are the most important source of hydraulic resistance variations (Figure 13.3 and 13.5).

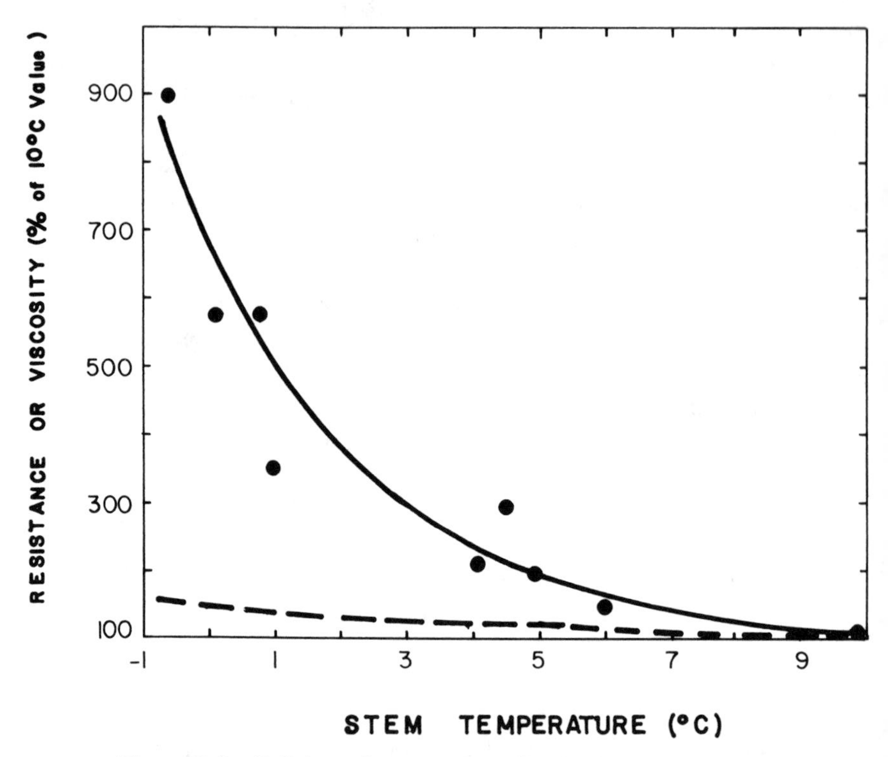

Figure 13.7. Relative resistance to water flow inside the stem of *E. timotensis* and relative viscosity of water (dash line) from 10°C to −1°C stem temperature (from Goldstein and Meinzer 1983).

Maximum stem core temperatures for *E. timotensis* growing at 4,200 m are much lower than maximum air temperature and remain below about 4°C for much of the day. An exposed stem will attain considerably higher daytime temperatures (Figure 13.8). Thus, at least during the day, marcesent leaves reduce the efficiency of the pith as a water reservoir because of their effect on hydraulic resistance (Figure 13.7). Nevertheless, the positioning of the pith immediately below the rosette and in contact with the xylem probably results in lower overall hydraulic resistance than in a bare-stemmed pithless plant relying on soil water to meet short-term demands.

Marcesent leaves and survival

Enhanced chances of survival due to avoidance of direct stem freezing damage compensate for the disadvantage of lower daytime stem temperatures and reduced potential efficiency of water transport from pith to xylem. Studies carried out with *E. timotensis* at 4,200 m have fur-

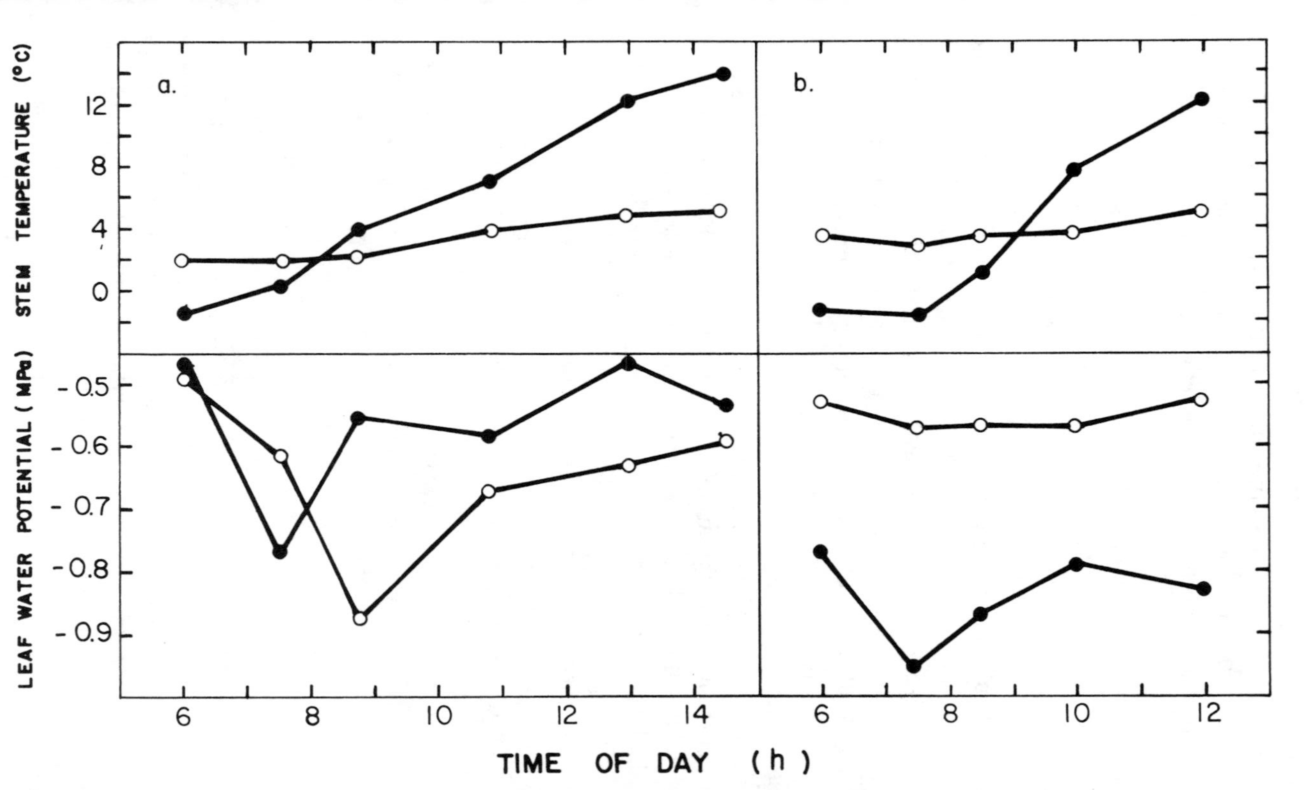

Figure 13.8. Courses of leaf water potential and stem temperatures for intact (open circles) and stripped (filled circles) *E. timotensis* plants at 4,200 m. Marcesent leaves were removed on January 14. (a) January 15; (b) January 29 (from Goldstein and Meinzer 1983).

nished information concerning the nature of the protection provided by marcesent leaves. Stem temperatures increased more rapidly in the morning in plants stripped of their marcesent leaves, and maximum stem temperatures were greater than in intact plants (Figure 13.8). On the day following marcesent leaf removal, the altered stem temperature pattern actually improved water balance, probably because of its effects on hydraulic resistance (Figure 13.8a). In a second observation two weeks later, symptoms of damage were evident. The leaves were wilted, and both predawn and subsequent values of leaf water potential in the stripped plants were lower than on the first day of observation (Figure 13.8b). Hydraulic resistance on this day was four times greater in the stripped plants. All individuals subjected to removal of marcesent leaves died within two months. Our observations do not support previous hypotheses that marcesent leaf removal in giant rosette plants causes death due to frozen stems making water temporarily unavailable for replacing morning transpirational losses. However, if periods of frequent frosts coincide with periods of infrequent precipitation, nightly recharge of the pith in stripped plants could be inhibited by freezing stem temperatures and low soil moisture availability. This could affect plant water balance on a long-term basis and result in higher apparent hydraulic resistances.

Laboratory measurements suggest that there are at least two other low-temperature-related mechanisms that could produce permanent effects on water balance in stripped plants: (1) embolisms in the xylem vessels, which can occur at temperatures slightly below 0°C when water movement is slow, such as at night; (2) direct freezing injury to the pith tissue, which would affect its water exchange capacity (Goldstein and Meinzer 1983). In the case of embolism, the nearly daily cycles of freezing and thawing in higher paramos will make it difficult for air bubbles to redissolve in time to make xylem elements functional for daily transpirational demands. Tissue death occurs at about −5°C in the pith of *E. timotensis*. Absolute minima in its habitat are below −5°C, and exposed stems experiencing high radiative heat loss would be even cooler. Given the physiological constraints associated with functioning of xylem elements and development of frost resistance in pith tissue, the net adaptive value of the marcesent leaf cylinder seems clear.

Leaf pubescence

The adaptive significance of leaf pubescence in plants has been well studied in only a limited number of situations. It is now clear that leaf pubescence of many species from hot, arid habitats results in reduced absorption of solar radiation, reduced leaf temperature, and thus lower transpirational losses (Smith and Nobel 1977; Ehleringer and Björkman 1978; Ehleringer and Mooney 1978; Ehleringer et al. 1981) (see Chapter 2). It has been suggested that the consequences of leaf pubescence in Andean and African giant rosettes are similar to those described for arid-zone species (Hedberg 1964; Baruch and Smith 1979). However, environmental pressures resulting in the evolution of leaf pubescence may only be superficially similar in hot, arid habitats and higher tropical mountains. Although both habitats are characterized by high incoming solar radiation, the patterns of moisture availability and air temperature are quite different. During most of the year in Andean paramos, low temperatures, rather than lack of precipitation, limit water availability. Air temperatures are consistently low, with no distinct favorable season for growth.

The properties of the leaf pubescence itself must also be considered when making predictions concerning its consequences. For example, the thickness of the pubescent layer affects not only leaf spectral properties but also the thickness of the boundary layer of air next to the leaf and thus heat and mass transfers (Smith and Nobel 1977). The relative importance of pubescence effects on radiation absorption versus boundary-layer resistance can be evaluated with energy-balance equations and experimental manipulations.

Characteristics

There are large interspecific and intraspecific variations in both the quantity and characteristics of leaf hairs in Andean giant rosette species. Leaf hairs in *E. timotensis,* a species that occurs between elevations of 3,900 and 4,500 m, are alive, and the thickness of the hair layer ranges from 2 mm in older leaves to as much as 3.5 mm in expanding leaves. These thicknesses are about 10 times greater than those reported for leaf hair layers of plants from desert habitats (Smith and Nobel 1977; Ehleringer and Mooney 1978). If a 2- to 3-mm-thick hair layer represented an effective boundary layer, its effects on heat and mass transfers would be considerable.

Leaf absorptance to solar radiation in the 400–700-nm waveband is about 0.69 for intact *E. timotensis* leaves and 0.84 for leaves with their pubescence removed. These correspond to total short-wave (400–3,000 nm) absorptances of 0.38 and 0.49, respectively (Ehleringer et al. 1981). The value for intact leaves is not particularly low, especially when the thickness of the hair layer is taken into account. Leaf absorptances (400–3,000 nm) of arid-zone plants can be as low as 0.15 (Ehleringer et al. 1981).

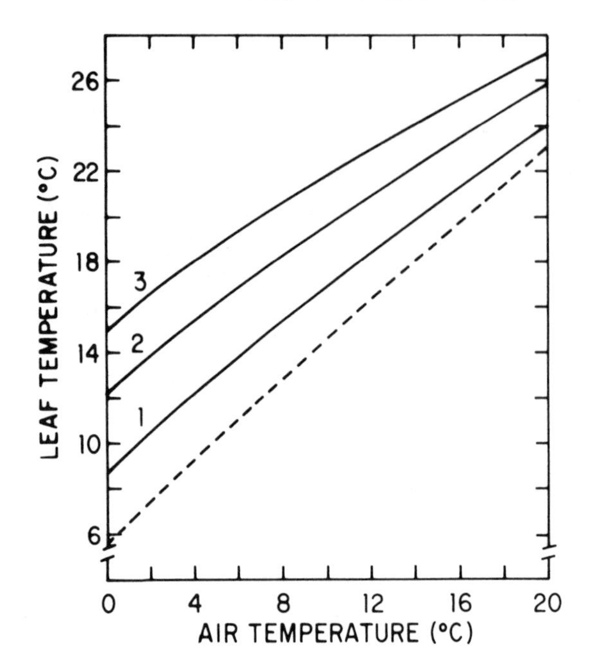

Figure 13.9. Predicted leaf temperature versus air temperature for *E. timotensis* leaves with 0 (dash line), 1, 2, and 3 mm of pubescence. Incident solar radiation was held at 800 W m^{-1}, wind speed at 2 m s^{-1}, and stomatal resistance at 150 s m^{-1} for the simulation (from Meinzer and Goldstein 1984).

If the leaf hairs are living, they often do not greatly reduce absorptance to solar radiation (Gausman and Cardenas 1969; Wuenscher 1970).

Coupling of leaf temperature to environmental variables

Energy-balance equations indicate that leaf pubescence in *E. timotensis* exerts its principal influence on leaf temperature through increased boundary-layer resistance to heat transfer, rather than through reduced absorptance to solar radiation (Meinzer and Goldstein 1984). Under clear-day conditions at 4,200 m, the temperature of a pubescent leaf will be higher than that of a nonpubescent leaf, in spite of the larger amount of solar radiation absorbed by the latter (Figure 13.9). The leaf hairs thus decrease coupling between leaf and air temperature, with the effect being most pronounced at prevailing habitat temperatures.

Because leaf hairs in *E. timotensis* significantly impede convective and evaporative heat losses without greatly reducing absorption of solar radiation, they increase coupling between leaf temperature and incident solar radiation (Figure 13.10). The same mechanism that reduces convective heat loss when solar radiation is intense should cause reduced convective

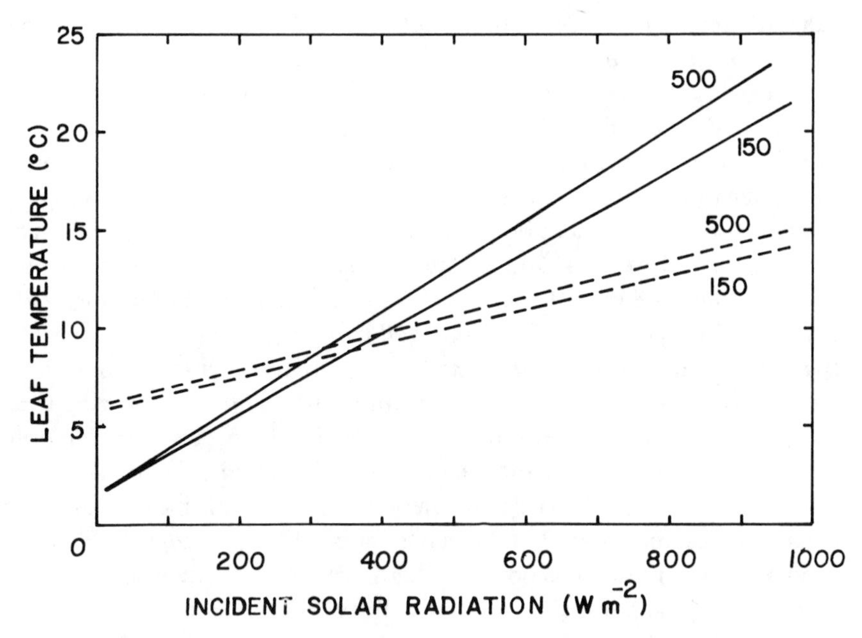

Figure 13.10. Predicted leaf temperature versus incident solar radiation for *E. timotensis* leaves with 0 (dash lines) and 2 mm (solid lines) of pubescence and stomatal resistances of 150 and 500 s m^{-1}. Air temperature was held at 8 °C and wind speed at 2 m s^{-1} for the simulation.

heat gain when incident radiation is low. For this reasons, during the night or at low levels of solar radiation, *E. timotensis* leaves should be cooler than nonpubescent leaves of comparable dimension and stomatal aperture (Figure 13.10). The extent to which leaf pubescence may inhibit long-wave emission to the night sky is not known, but temperature measurements in the field for intact and partially shaved leaves have confirmed that leaves with more pubescence are cooler at night (Meinzer and Goldstein 1984). Another interesting prediction of energy-balance equations is that in leaves with thick pubescence, stomatal movements affect the degree of coupling between leaf temperature and incident radiation (slopes of lines in Figure 13.10). The effect increases with increasing thickness of pubescence. Coupling increases as stomata close in leaves with a thick boundary layer, because decreases in latent heat loss are not readily compensated by increases in convective heat loss.

Pubescence-induced temperature increases have previously been demonstrated for leaves (e.g., Wuenscher 1970) and other structures such as inflorescences (Krog 1955). There are several reasons why this feature could be of adaptive value in high-altitude tropical environments where radiant-energy inputs are high but thermal-energy inputs are low throughout the year. Earlier hypotheses that the adaptive value of leaf pubescence in Andean and African giant rosette plants must lie in reduced radiation absorption, reduced leaf temperature, and reduced transpiration (Hedberg 1964; Baruch and Smith 1979) did not take into account that while radiation in these habitats may indeed be intense, prevailing air temperatures tend to be suboptimal for processes such as translocation of assimilates, leaf growth, and possibly photosynthesis. References to intense heating of exposed rocks in these environments should not imply that this will occur in a leaf far removed from the soil boundary layer. In the high paramos where *E. timotensis* occurs, environmental temperatures frequently may be suboptimal for translocation and leaf growth. Leaf temperatures in *E. timotensis* and other giant rosette species are commonly 5–15°C above air temperature (Smith 1974; Meinzer and Goldstein 1984; Meinzer et al. 1984). Because a typical Q_{10} for translocation and leaf expansion may be about 2, pubescence-induced temperature increases of even a few degrees could be highly significant, especially in an environment lacking temperature seasonality. Initial measurements suggest that leaf growth rates in other giant rosette species of *Espeletia* from high paramos are relatively rapid considering the low mean air temperature. For example, *E. spicata* produces about 360 leaves per year (C. Estrada, unpublished observations), and *E. lutescens,* a species with leaves about the same size as those of *E. timotensis,* produces about 120 leaves per year (M. Monasterio, personal communication). Annual leaf production in *E. lutescens* represents about 370 g dry matter per square meter ground area, comparable to that for temperate forests and much greater than that reported for tree forms at treeline in temperate-zone mountains (Tranquillini 1979). Annual leaf biomass production for the African giant rosette *Senecio keniodendron* growing at 4,100 m was estimated as 166 g m^{-2} (Beck et al. 1980). Another factor to be taken into account when evaluating the importance of assimilate translocation is the high proportion of living nonphotosynthetic tissue present in the caulescent giant rosette growth form. The massive apical bud contains a large bank of unexpanded leaves, and the woody stem contains a large volume of living pith cells.

It has also been suggested that leaf pubescence in giant rosette species may result in higher nighttime leaf temperatures due to reduced radiative heat loss (Hedberg 1964). This effect would be expected to be minor because only massive plant organs with high heat capacities show considerable delay in reaching a steady-state radiation balance. In fact, energy-balance simulations and field data indicate that exposed pubescent leaves are actually cooler at night because of reduced convective heat transfer from

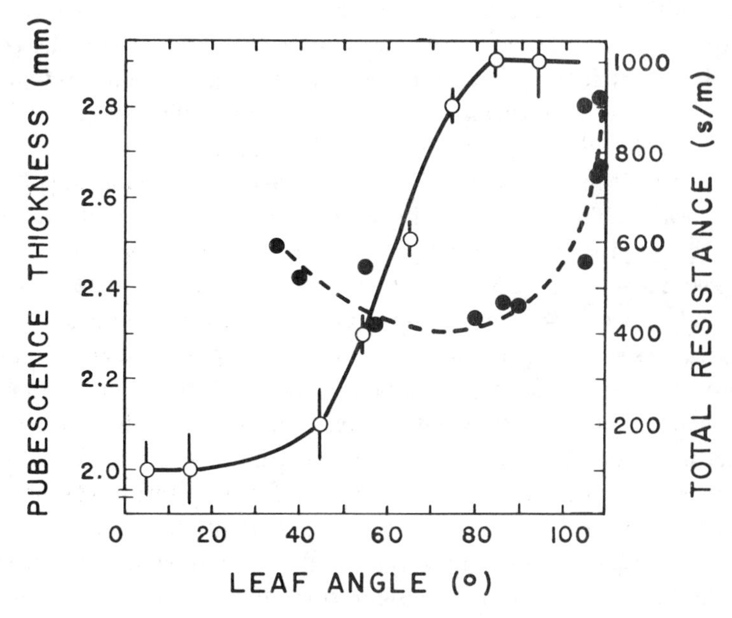

Figure 13.11. Pubescence thickness (open circles) and total diffusive resistance (filled circles) in relation to leaf angle (leaf age) in an *E. timotensis* rosette. Angle is 0 degrees for a horizontal leaf (from Meinzer and Goldstein, 1984).

air to leaf. However, in massive apical buds that have experienced daytime heating, leaf nyctinasty and the presence of successive layers of leaves do have insulating effects by reducing heat conduction and convection from one leaf layer to the next (Smith 1974).

Leaf pubescence and stomatal behavior

Energy-balance models predict that at a given level of incident radiation, reductions in stomatal opening will produce significant temperature increases in *E. timotensis* leaves (Figure 13.10). This effect will be greatest in leaves with the thickest hair layer. Increasing the leaf angle in the rosette from nearly horizontal senescent leaves to vertical expanding leaves corresponds to a gradient of increasing thickness of pubescence (Figure 13.11). These differences in leaf angle also result in a gradient of incident radiation. Measurements of total diffusive resistance show that in nonsenescent leaves, total resistance increases with increasing leaf angle and decreasing leaf age (Figure 13.11). The major component of the increase is stomatal resistance, not boundary-layer resistance. Temperature measurements for *E. schultzii* (Smith 1974; Larcher 1975) suggest that the apical bud and expanding leaves are exposed to heavy daytime radiation loads due to reflected short-wave and emitted long-wave radiation from surrounding leaves. Stomatal resistance in expanding leaves of most plants is higher than that of recently expanded leaves, but in *Espeletia* leaves with thick pubescence this behavior enhances coupling between incident radiation and leaf temperature. In view of the low prevailing air temperatures, this probably will favor more rapid leaf expansion.

Increasing the thickness of the hair layer exerts two opposing effects on transpiration. The increased diffusive resistance will tend to decrease transpiration. On the other hand, transpiration will tend to increase because of increased leaf temperature and thus an increased leaf-to-air vapor-pressure difference. The relative importances of these two effects depend on stomatal resistance, boundary-layer thickness, and the leaf temperature range over which transpiration is being measured (Smith and Geller 1980; Geller and Smith 1982). Stomatal resistance of *E. timotensis* is such that under most conditions the leaf temperature increase induced by 2 – 3 mm of pubescence will not result in increased transpiration (Meinzer and Goldstein 1984). This may have important adaptive consequences. Even when absolute soil moisture levels are high, water movement to the rosette may be limited by low stem temperatures that inhibit passage of water from the stem pith to the stem xylem (Goldstein and Meinzer 1983). If transpiration were to increase under these conditions of high flow resistance, increased leaf water deficits would result (Goldstein et al. 1984).

Morphological adaptation and constraints on functioning and behavior

From the preceding discussion of some of the consequences of stem pith, marcesent leaves, and leaf pubescence, it should be apparent that specific predictions can be made as to how these features should vary along environmental gradients. We can also make predictions concerning environmental extremes at which these features will no longer possess any adaptive value or will even be detrimental. We have used quantitative field and laboratory data to make such predictions, some of which can be easily tested with additional measurements.

Andean giant rosette plants represent a nearly ideal system for undertaking such an exercise. They are all of the same general growth form and are closely related, as manifested by the frequent formation of hybrids. They also occur in a wide variety of habitats within the vegetation type known as paramo. Some species such as *E. schultzii* have altitudinal ranges that span almost the entire paramo zone.

Table 13.2. *Summary of morphological and physical characteristics of five.* E. schultzii *populations occurring along an elevation gradient*

Elevation (m)	Pith volume (cm^3)	Leaf area ($cm^2 \cdot 10^4$)	PV/LA ($cm^3\ cm^{-2}$)	Pubescence thickness (mm)	Leaf absorptance 400–700 nm
2,600	89	1.16	0.008	1.1	0.78
3,100	336	0.96	0.039	1.6	0.74
3,550	424	0.68	0.063	2.1	0.70
3,850	702	0.64	0.116	2.3	0.67
4,200	873	0.48	0.179	2.6	0.69

Stem pith and marcesent leaves

If stem pith in giant rosettes is indeed an adaptation to low temperature-limited water availability, then pith volume or more important, relative capacitance (PV/LA) should increase with increasing altitude. This is the case for *E. schultzii,* which occurs from about 2,600 to 4,200 m (Table 13.2). This altitude span represents a drop in mean temperature from 13°C to 2.8°C and corresponds to an abrupt gradient in increasing frequency of night frosts (Monasterio and Reyes 1980). Although there is a reduction in total leaf area of the rosette with increasing elevation, the greater relative capacitance in the higher sites is attained mostly through an increase in stem height.

The presence of marcesent leaves results in a tradeoff between frost protection for sensitive stem tissues and maintenance of an acceptable hydraulic resistance. We attempted to evaluate the relative advantages and disadvantages of different thicknesses of marcesent leaves along an environmental gradient (Figure 13.12). Minimum stem temperatures along a gradient of increasing altitude were predicted for plants with different amounts of insulation. Relative hydraulic resistance was also estimated for the period between sunrise and midday, when the soil is still cold but stomata are open. As expected, the initial increments in thickness of the insulation should have the greatest effect in preventing stem temperature from falling below 0°C (Figure 13.12a). Because the marcesent leaf layer is typically 15 to 30 cm thick, stems should be protected from freezing at elevations above the approximate 4,500-m limit for caulescent giant rosettes. However, above about 4,000 m, relative hydraulic resistance is predicted to increase sharply for all thicknesses of insulation, suggesting rapidly increasing costs in terms of reduced efficiency of water transport to the rosette (Figure 13.12b). This can partially explain why the

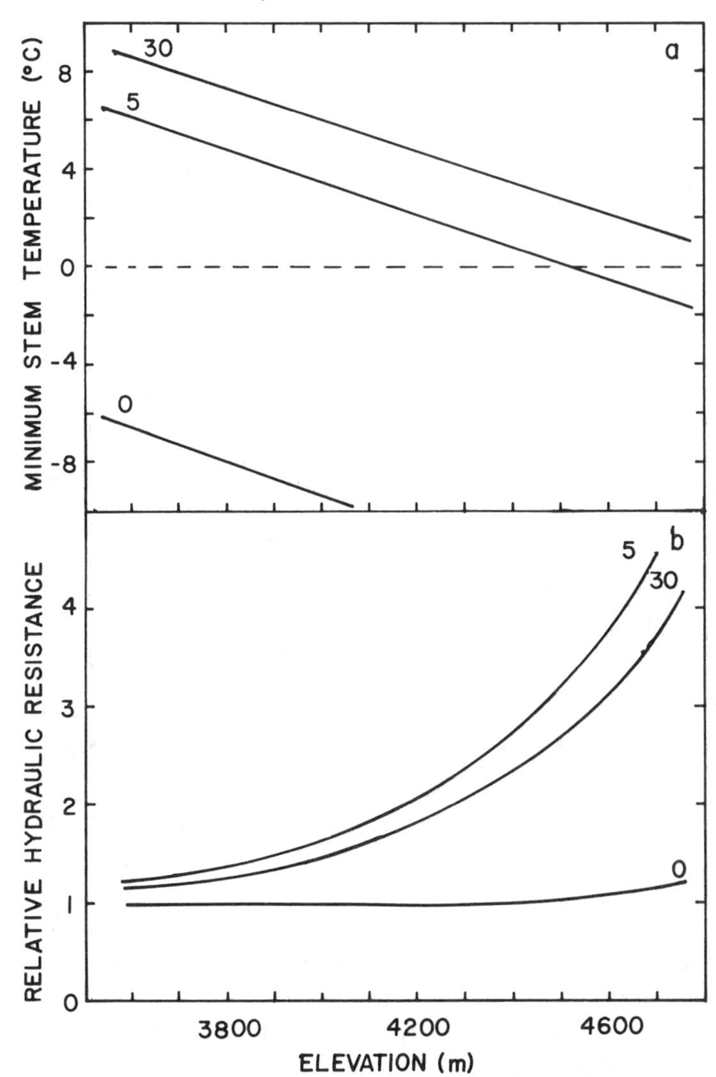

Figure 13.12. (a) Predicted minimum stem temperature of *E. timotensis* in relation to elevation for different lengths (cm) of marcesent leaves. Stem temperatures were predicted with an energy-balance model that simulated diurnal courses of air temperature and incident radiation. (b) Predicted average relative hydraulic resistances between 0800 and 1200 hr for different lengths of marcescent leaves. Relative resistance = 1 at 10°C stem temperature (Figure 13.7) and was calculated from hourly predictions of stem temperature.

upper altitude limit for frost protection exceeds the upper limit for growth. During periods of unusually low temperatures, the insulation isolines in Figure 13.12 should be displaced toward lower elevations.

The relationship between hydraulic resistance and stem temperature can be used to better understand the selective pressures favoring higher relative capacitance for maintenance of water balance at high altitudes. Because leaf water potential is a function of both transpiration rate and hydraulic resistance, giant rosette plants with a higher relative capacitance will have a higher leaf water potential at a given transpiration rate (Figure 13.13a). If stem temperature were to drop, the same relative capacitance would be less effective, and leaf water potential could approach the turgor loss point (Figure 13.13b). High relative capacitance in colder sites prevents excessive water potential drop and permits stomata to remain open for CO_2 assimilation during the early part of the day, when light levels are favorable for photosynthesis, but stem and soil temperatures are unfavorable for water uptake and transport.

Leaf pubescence

Pubescence thickness should increase along an elevational gradient if its adaptive value is related to increases in leaf temperature. The thickness of the leaf hair layer in *E. schultzii* more than doubles along an altitude gradient from 2,600 to 4,200 m (Table 13.2). The accompanying decrease in absorptance to solar radiation is not very great considering the amount of increase in thickness of the hair layer. The possible significance of these changes was evaluated with a simulation model (Figure 13.14). Under midday conditions, the predicted leaf-to-air temperature difference should always be greater for intact leaves versus leaves without pubescence. For glabrous leaves, the degree of coupling between leaf and air temperatures should remain essentially constant as maximum air temperature decreases with increasing elevation. For intact leaves, the influence of the increase in pubescence thickness should be great enough to decrease coupling between leaf and air temperatures as elevation increases. In the presence of low thermal-energy inputs, it would appear that species such as *E. timotensis* and *E. schultzii* are capable of increased efficiency of solar energy use to satisfy their requirements for thermal regulation. Information concerning the temperature dependence of photosynthesis would be necessary for evaluating the consequences of pubescence for carbon gain.

It is interesting to speculate concerning other possible mechanisms that could result in leaf temperature increases in giant rosette plants of the high-altitude tropics. An increase in stomatal resistance, by decreasing latent heat loss, would tend to raise leaf temperature, but computer simu-

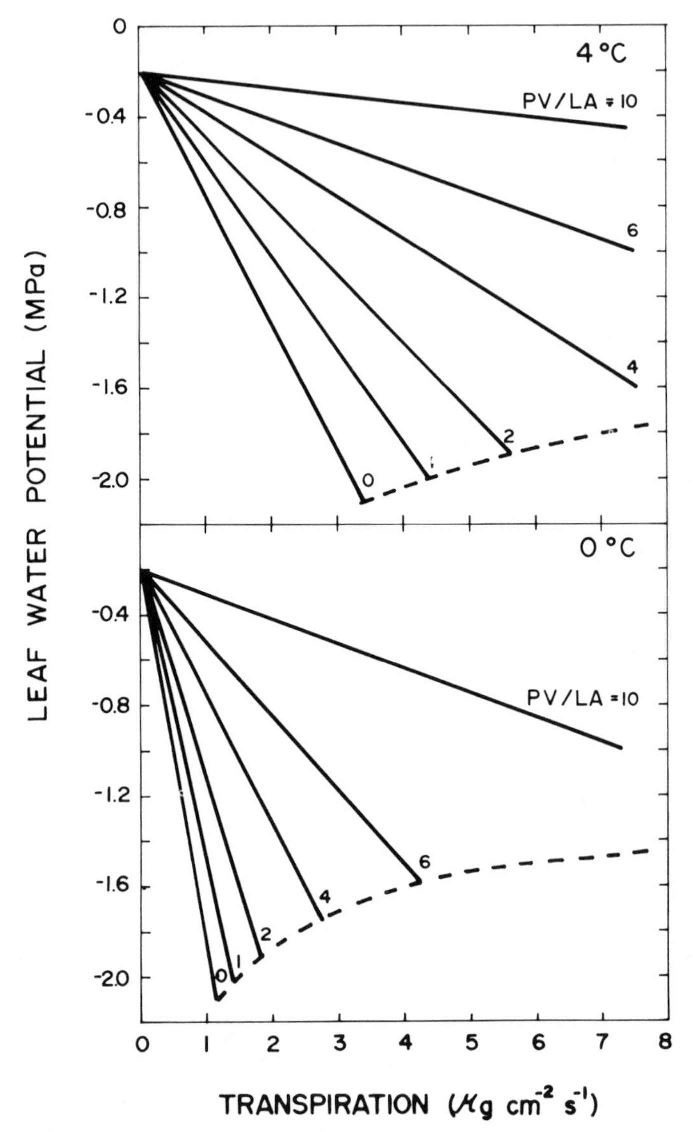

Figure 13.13. Predicted relationship between leaf water potential and transpiration for stem temperatures of 4°C and 0°C and different values of relative water storage capacity (PV/LA) (see Figure 13.5). The dash line defines the boundary at which leaf turgor loss would occur for given transpiration rate and PV/LA.

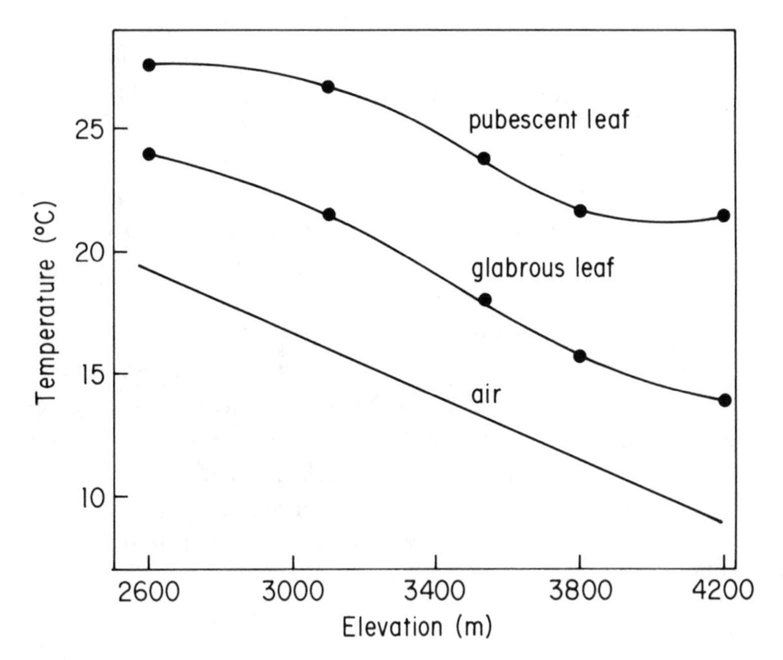

Figure 13.14. Typical maximum air temperatures and predicted temperatures of intact and glabrous *E. schultzii* leaves along an elevation gradient. Dimensions, pubescence thickness, and absorptance to solar radiation measured for leaves of five populations occurring at 2,600, 3,100, 3,600, 3,800, and 4,200 m were used in an energy-balance model to predict leaf temperature under standard midday conditions. Pubescence was removed to determine absorptance to solar radiation of glabrous leaves.

lations predict that this should not be a very effective mechanism in nonpubescent leaves the same size as those of *E. timotensis* and *E. schultzii*. High stomatal resistance should also severely restrict carbon assimilation. Thick leaf pubescence, on the other hand, permits leaf temperature increases with relatively open stomata. The additional boundary-layer thickness would be expected to have a much smaller relative effect on CO_2 absorption than on transpiration because of the probable higher internal resistance to CO_2 transfer and much lower internal resistance to water transfer. Increased leaf width should also increase boundary-layer thickness and thus leaf temperature. However, the thickness of this variable aerodynamic boundary layer is very sensitive to changes in wind speed. The amount of increase in leaf dimension needed to produce the same effect on leaf temperature as a fixed 2-mm layer of pubescence can be calculated. At a constant wind speed of 2 m s^{-1}, and with stomatal resistance and other environmental conditions equal, the nonpubescent leaf would have to be about 35 cm wide. None of the high-altitude species of *Espeletia* without leaf hairs have leaves approaching 35 cm wide, but some of the African species of *Senecio* studied by Hedberg (1964) have leaves around 15 cm to more than 20 cm wide, with margins rolled downward and pubescence on the underside of the leaf, but not on the upper surface.

It should be stressed that the environmental pressures acting on *Espeletia* species in the high paramo zone and on pubescent plants from warm deserts are quite different. Typical maximum air temperatures in warm desert habitats can exceed 40°C, at least 30°C above maximum air temperatures in high paramos. Water availability can be limited in both habitats, but it is apparent that leaf temperature variations at the higher desert temperatures will have a much greater effect on potential rates of transpiration because of the much steeper slope of the saturation vapor-pressure curve near 40°C. The consequences of leaf pubescence in some *Espeletia* species and in desert plants are similar in the sense that in both cases the pubescence represents a mechanism by which prevailing air temperatures can be avoided.

Conclusions

In tropical high mountain climates where diurnal temperature fluctuations are much larger than seasonal ones, low-temperature limitations on water availability and growth have led to a series of adaptations in giant rosette plants that tend to uncouple the plant from its temperature environment. Insulating marcescent leaves, for example, constitute a finely tuned adaptation effective only in habitats in which freezing temperatures last only a few hours. A large volume of stem pith just below the rosette uncouples the plant from the inhibitory effects of low soil temperatures on water absorption. Thick leaf pubescence uncouples rosette temperature from low daytime air temperatures and permits a certain degree of thermoregulation by coupling leaf temperatures more closely to incident radiation.

We have outlined some of the more apparent costs of these morphological adaptations to the unique pattern of low-temperature stress in the high-altitude tropics. Rather than having a neutral value, some of these morphological features represent evolutionary courses that may impose serious constraints on functioning and survival outside of the environment in which they occur. High relative capacitance, for instance, while conferring a great advantage under cold, moist conditions, is associated with high susceptibility to wilting and dehydration damage when the soil is dry

(Goldstein et al. 1985). Under warmer conditions, possession of a large volume of nonphotosynthetic living pith tissue can result in excessively high maintenance costs.

Although many predictions can be made, other adaptive consequences and costs of stem pith, marcesent leaves, and leaf pubescence remain to be evaluated. In the case of leaf pubescence, it would be of interest to evaluate its possible costs in terms of additional resistance to CO_2 diffusion and interception of photosynthetically active radiation. In leaves of some acaulescent *Espeletia* species (e.g., *E. floccosa*) having a thin (0.2 mm) reflective hair layer, it is likely that leaf pubescence has consequences quite different from those mentioned here for *E. timotensis* and *E. schultzii.*

Maintenance costs for the large respiring stem pith can be appreciable, but this potential cost can be partially reduced by low stem temperatures. It is also necessary to evaluate the importance of sources of capacitance other than the pith, because pith water alone cannot account for the stability of daily courses of leaf water potential observed in some species. Multiple sources of stored water and declines in hydraulic resistance during the day would both tend to dampen diurnal fluctuations in leaf water potential.

The giant rosette form represents a general adaptive solution to the environmental pressures of the high-altitude tropics. Rough comparisons of individual morphological traits in species from different tropical mountain regions suggest that measurable degrees of variation occur in adaptive solutions to similar environmental pressures. An approach such as that used here with *Espeletia* can be employed to generate and test hypotheses concerning the adaptive values and costs of specific morphological features in giant rosette species from other tropical mountain regions.

Acknowledgment

This chapter is based in part on research supported by CDCH grant C-180-81 from the Universidad de los Andes.

References

Baruch, Z., and A. P. Smith. 1979. Morphological and physiological correlates of niche breadth in two species of *Espeletia* (Compositae) in the Venezuelan Andes. Oecologia 38:71–82.

Beck, E., R. Schiebe, M. Senser, and W. Müller. 1980. Estimation of leaf and stem growth of unbranched *Senecio keniodendron* trees. Flora 170:68–76.

Coe, J. M. 1967. The ecology of the alpine zone of Mount Kenya. Dr. W. Junk, The Hague.

Edwards, W. R. N., and P. G. Jarvis. 1982. Relations between water content, potential and permeability in stems of conifers. Plant Cell Environ. 5:271–277.

Ehleringer, J., and O. Björkman. 1978. Pubescence and leaf spectral characteristics in a desert shrub, *Encelia farinosa*. Oecologia 36:151–162.

Ehleringer, J., and H. A. Mooney. 1978. Leaf hairs: effects on physiological activity and adaptive value to a desert shrub. Oecologia 37:183–200.

Ehleringer, J., H. A. Mooney, S. L. Gulmon, and P. W. Rundel. 1981. Parallel evolution of leaf pubescence in *Encelia* in coastal deserts of North and South America. Oecologia 49:38–41.

Gausman, H. W., and R. Cardenas. 1969. Effect of leaf pubescence of *Cynura aurantiaca* on light reflectance. Bot. Gaz. 130:158–162.

Geller, G. N., and W. K. Smith. 1982. Influence of leaf size, orientation and arrangement on temperature and transpiration in three high-elevation large-leafed herbs. Oecologia 53:227–234.

Goldstein, G. 1981. Ecophysiological and demographic studies of white spruce (*Picea glauca* (Moench) Voss) at treeline in the central Brooks Range of Alaska. Ph.D. dissertation, University of Washington, Seattle.

Goldstein, G., and F. C. Meinzer. 1983. Influence of insulating dead leaves and low temperatures on water balance in an Andean giant rosette plant. Plant Cell Environ. 6:649–656.

Goldstein, G., F. C. Meinzer, and M. Monasterio. 1984. The role of capacitance in the water balance of Andean giant rosette species. Plant Cell Environ. 7:179–186.

– 1985. Physiological and physical factors in relation to size-dependent mortality in *Espeletia lutescens*. J. Ecol. (in press).

Hedberg, O. 1964. Features of afroalpine plant ecology. Acta Phytogeogr. Suecia 49:1–144.

Hinckley, T. M., J. P. Lassoie, and S. W. Running. 1978. Temporal and spatial variations in the water status of forest trees. Forest Science Monograph 20. Society of American Foresters, Bethesda, Md.

Hinckley, T. M., R. O. Teskey, F. Duhme, and H. Richter. 1981. Temperate hardwood forests. Pp. 153–208 *in* T. T. Kozlowski (ed.), Water deficits and plant growth, vol. VI. Academic Press, New York.

Jarvis, P. G. 1975. Water transfer in plants. Pp. 369–394 *in* D. A. de Vries and N. K. Van Alfen (eds.), Heat and mass transfer in the environment of vegetation. Script Book Co., Washington, D.C.

Kaufmann, M. R. 1975. Leaf water stress in Engelmann spruce. Plant Physiol. 56:841–844.

– 1977. Soil temperature and drying cycle effects on water relations of *Pinus radiata*. Can. J. Bot. 55:2413–2418.

Krog, J. 1955. Notes on temperature measurements indicative of special organization in arctic and sub-arctic plants for utilization of radiated heat from the sun. Physiol. Plant. 8:836–839.

Larcher, W. 1975. Pflanzenökologische Beobactungen in der Paramostufe der venezolansichen Anden. Anzeiger math.-naturw. Klasse der Österreich. Akad. Wiss. 11:194–213.

Meinzer, F. C., and G. Goldstein. 1985. Leaf pubescence and some of its consequences in an Andean giant rosette plant. Ecology 66:512–520.
Meinzer, F. C., G. Goldstein, and P. W. Rundel. 1985. Morphological changes along an altitude gradient and their consequences for an Andean giant rosette plant. Oecologia 65:278–283.
Monasterio, M., and S. Reyes. 1980. Diversidad ambiental y variación de la vegetación in los páramos de los Andes venezolanos. Pp 47–91 *in* M. Monasterio (ed.), Estudios ecológicos en los páramos andinos. Ediciones de la Universidad de los Andes, Mérida, Venezuela.
Powell, D. B. B., and M. R. Thorpe. 1977. Dynamic aspects of plant-water relations. Pp. 259–279 *in* J. J. Landsberg and C. V. Cutting (eds.), Environmental effects on crop physiology. Academic Press, London.
Richards, G. P. 1973. Some aspects of the water relations of Sitka spruce. PhD dissertation, University of Aberdeen, Scotland.
Running, S. W. 1980. Relating plant capacitance to the water relations of *Pinus contorta.* For. Ecol. management 2:237–252.
Running, S. W., and P. C. Reid. 1980. Soil temperature influences of *Pinus contorta* seedlings. Plant Physiol. 65:635–640.
Smith, A. P. 1974. Bud temperature in relation to nyctinastic leaf movement in an Andean giant rosette plant. Biotropica 6:263–266.
– 1979. The function of dead leaves in *Espeletia schultzii* (Compositae), an Andean giant rosette plant. Biotropica 11:43–47.
– 1980. The paradox of plant height in an Andean giant rosette species. J. Ecol. 68:63–73.
Smith, W. K., and P. S. Nobel. 1977. Influences of seasonal changes in leaf morphology on water-use efficiency for three desert broadleaf shrubs. Ecology 58:1033–1043.
Smith, W. K., and G. N. Geller. 1980. Leaf and environmental parameters influencing transpiration: theory and field measurements. Oecologia 46:308–313.
Tranquillini, W. 1979. Physiological ecology of the alpine timberline. Springer-Verlag, New York.
Troll, C. 1958. Zur Physiognomik der Tropengewächse. Jber. Ges. Fr. Ford. rhein. Friedrich-Wilhelms-Univ. Bonn E.V. 1958:1–75.
– 1968. The cordilleras of the tropical Americas. Aspects of climate, phytogeography and agrarian ecology. Pp. 13–56 *in* C. Troll (ed.), Geo-ecology of the mountain regions of the tropical Americas. UNESCO, New York.
Tyree, M. T., and T. H. Hammel. 1972. The measurement of the turgor pressure and the water relations of plants by the pressure-bomb technique. J. Exp. Bot. 23:267–282.
Vogl, R. J., and L. T. McHargue. 1966. Vegetation of California fan palm oases on the San Andreas fault. Ecology 47:532–540.
Wardle, P. 1974. Alpine timberlines. Pp. 371–402 *in* J. D. Ives and R. G. Barry (eds.), Arctic and alpine environments. Methuen, London.
Waring, R. H., and S. W. Running. 1978. Sapwood water storage; its contribution to transpiration and effect upon water conductance though the stems of old growth Douglas-fir. Plant Cell Environ. 1:131–140.
Wuenscher, J. E. 1970. The effect of leaf hairs of *Verbascum thapsus* on leaf energy exchange. New Phytol. 68:65–73.

Part II

Economics of support

Many leaf and crown traits involve unavoidable tradeoffs between potential carbon gain and allocation to unproductive tissue required to ensure mechanical stability, increase height growth, reduce self-shading, and/or provide an adequate supply of water and nutrients to the canopy. These tradeoffs create an *economics of support,* linking the photosynthetic benefits associated with different leaf or crown geometries to the structural costs of supporting and supplying those crown arrangements. As with the economics of gas exchange, we generally expect natural selection to favor plants whose form tends to maximize the difference between these benefits and associated structure costs.

Figure II.1 summarizes prominent features of the economics of support in a constant environment. Among plants having foliage with comparable gas-exchange properties, the relative rate of whole-plant carbon gain can be influenced by differences in leaf form, arrangement, and branching pattern. These latter traits affect photosynthesis per unit leaf mass through effects on (1) leaf overlap and self-shading within crowns, (2) rate of height growth, (3) mechanical stability, (4) shading by or of other plants, and (5) leaf microclimate and light interception. Such traits also influence whole-plant carbon gain by affecting (6) allocation to foliage versus relatively unproductive support and transport tissue, and (7) allocation to veins and petioles versus more permanent elements of a plant's support skeleton, with consequent effects on the recurring costs of construction and maintenance of structural tissue.

Three important conclusions can be drawn from the interactions portrayed in Figure II.1 and their extension to time-varying environments:

1. As with the economics of gas exchange, most traits that affect the balance of photosynthetic benefits and structural costs do so through a multiplicity of different pathways, interacting with the effects of several other traits. Branching pattern, for example, affects whole-plant carbon gain through effects on leaf overlap, height growth, pattern of horizontal spread, and allocation to productive versus unproductive tissue. The ben-

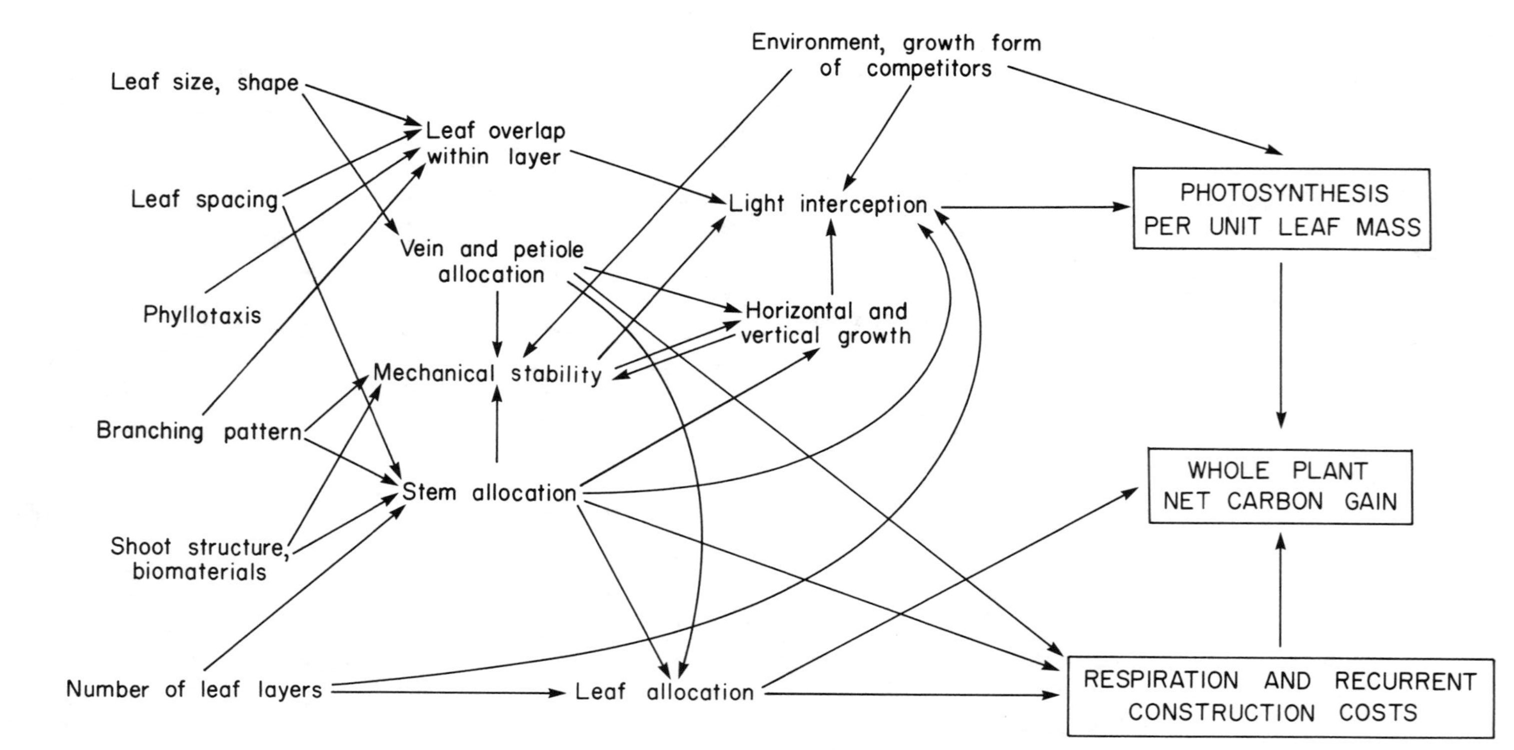

Figure II.1. Summary of interactions among selected plant traits involved in the economics of support.

efits associated with a particular branching pattern, furthermore, depend on leaf shape, size, and phyllotaxis and may also vary with plant size and the growth patterns adopted by competitors (see Chapter 16). Adaptations affecting the economics of support should thus be integrated and co-adapted with each other, and the advantage they yield should be context-specific.

2. The need for mechanical stability plays a central role in the economics of support. It imposes a constraint on the *minimum* amount of tissue required to support different leaf or crown forms, with a cascading series of implications for the structural costs associated with different growth forms. Furthermore, selection on competitors with similar growth forms should reduce support allocation to the minimum compatible with structural integrity. It is thus extremely important to understand the physical constraints on the mechanical stability of self-supporting biological structures.

For example, what are the constraints on stem diameter in trees or other terrestrial plants? McMahon (1973) showed that the most likely mode of stem failure is not compressive failure (rupture under a plant's own weight) but elastic toppling. Elastic toppling occurs when a stem fails to generate enough force, through stem flexure and elasticity of stem material, to resist the torque caused by a slight deviation from the vertical. If stem tissue accounts for most of a plant's mass, stem diameter should scale like the $\frac{3}{2}$ power of stem height to avoid elastic toppling and achieve a given height with minimum support tissue (Greenhill 1881). Indeed, the trees surveyed by McMahon (1973) obey this scaling law and have diameters at a given height roughly twice those that would lead to elastic toppling. Theoretical refinements to account for the effects of buttresses and crown mass have been given by Henwood (1973), McMahon and Kronauer (1976), King and Loucks (1978), and Givnish (1982).

A consideration of such biomechanical constraints can lead to more complex models for optimal plant form and growth. King (1981), for instance, demonstrates that in self-thinning stands of aspen – in which height growth is at a premium – the observed allocation to trunk versus crown is quite close to the allocation that would maximize height growth and still ensure mechanical stability. Different environments are likely to present different structural challenges and thus present different biomechanical constraints on optimal plant form (Wainwright et al. 1976; Grace and Russell 1977; Patterson 1985). Water is dense and incompressible, for example, so that in aquatic habitats tensile failure is likely to be more important than elastic toppling or compressive failure, with consequent implications for the forms of various sessile organisms (see Chapter 18).

Finally, in some situations structural *failure* may be advantageous, if a plant's growth form incorporates strategically placed weak points that break and reduce drag under heavy loads, allowing survival of infrequent, unusual stresses at low cost.

3. Many traits that affect the economics of support also affect the economics of gas exchange. Consider plants that arrange foliage with similar gas-exchange properties in different crown geometries. Horn (1971) compared net photosyntheses for leaves packed in a single layer versus those scattered in many layers and showed that multilayered crowns are more productive at high light intensities, and monolayered crowns are more productive at low light intensities. Furthermore, leaf carbon gain in multilayered crowns should be maximized if leaf layers are added until the shaded, lowermost leaves are operating at the light compensation point, at which leaf photosynthesis balances leaf respiration. Thus, based purely on the economics of gas exchange, the optimal number of leaf layers – and consequent ability to compete or survive under different light conditions – should be set by ambient light intensity and leaf compensation point.

But there are additional complications. For a leaf to yield a net gain, it must balance not only instantaneous leaf respiration, but also nighttime respiration and the amortized costs of leaf construction. Root costs and photosynthetic opportunity costs arising from retranslocation of nutrients between leaves must also be covered. In addition, leaves must balance the amortized costs of constructing and maintaining the structural material needed to support them (see Chapter 14, Givnish 1984). This creates a connection between the economics of support and the economics of gas exchange. Plants with shorter or less costly support skeletons should support more leaf layers under given light conditions, have a lower whole-plant light compensation point, and be able to survive at a lower light intensity, water availability, and/or soil fertility than plants with taller or more costly skeletons. Similar connections between the economics of support and of gas exchange arise in the study of any trait that affects both gas exchange and the efficiency of crown support.

The six chapters that follow discuss adaptation and functional constraints on support structures and their implications for light capture and competitive interactions in plants ranging from trees and forest herbs to marine algae.

Raven (Chapter 14) provides an exceptionally broad perspective on the constraints and structural costs associated with three fundamental growth forms: plankton, benthic algae, and terrestrial vascular plants. First, the effective light compensation point of whole plants is calculated as a function of the relative allocation to nonproductive, structural material in these three growth forms. This analysis helps specify the range of light intensities over which autotrophy is a feasible means of energy capture. Only plankton or short attached plants with inexpensive support skeletons should survive in the most densely shaded habitats.

Second, Raven discusses the potential significance of salient features of each growth form. For plankton, the importance of cell size and motility for carbon gain and context-specific competitive ability is analyzed. For attached algae, the potential constraints associated with unicellularity, nutrient absorption, long-distance transport, and resistance to mechanical stresses are discussed. Constraints on water transport, self-support, and photosynthesis in terrestrial plants are also considered. Using this theoretical framework, Raven concludes with an insightful review of the evolutionary history of photosynthetic organisms.

Fisher (Chapter 15) evaluates the significance of branching patterns and angles in determining crown form, leaf overlap, and efficiency of light capture in trees. The position and orientation of leaf-bearing axes depend on primary and secondary patterns of shoot organization and growth, phenology, and shoot demography. Spatial models of branching patterns have now been devised that account for many aspects of crown form with relatively few rules and branching parameters. Computer simulations for the geometrically simple tree *Terminalia* show that the observed branching angles, relative branch lengths, and number of branches per crown tier are close to the values that would optimize leaf packing and minimize self-shading. An important refinement of this approach will be to incorporate the costs of branch construction associated with different crown forms, as Borchert and Tomlinson (1984) have done for *Tabebuia* in a highly simplified fashion. This will not be easy: In plants of indeterminate growth, should the cost of a branch be weighed against the benefits of foliage it supports now or foliage it will help support in the future?

Givnish (Chapter 16) analyzes the structural costs and competitive benefits associated with four aspects of crown geometry in forest herbs: crown height, branching pattern, crown shape, and leaf shape. Many forest herbs have determinate aboveground shoots, so that the conceptual difficulties associated with indeterminate growth in woody plants can be avoided, and parallel questions about crown form can be more easily addressed. Case studies are presented on the adaptive significance of (1) leaf height, (2) branched versus unbranched canopies, branching angle, and branching height in *Podophyllum peltatum,* (3) the arching growth habit in *Polygonatum biflorum* and *Smilacina racemosa,* and (4) leaf shape in *Viola.* All are based on analyses of how the amount of mechanical tissue required to support a given amount of leaf tissue varies with plant size and geometry. In the first study, these mechanical costs are balanced against the photosynthetic benefits associated with differences in leaf height to predict optimal leaf height, using a game-theoretic model. In the remaining three studies, data on the allometry of support tissue are used to analyze which canopy geometry would minimize the cost of supporting a given amount of photosynthetic tissue at a given height. Each of these studies shows that the advantage conferred by a given biomechanical adaptation depends on a plant's competitive context, as well as physical constraints.

Schulze, Küppers, and Matyssek (Chapter 17) present a synthetic discussion of the effects of branching pattern and carbon gain on the growth of woody species. First, productivities in *Fagus* and *Picea* are related to differences in leaf phenology, photosynthetic rate, and growth habit. Second, the competitive abilities of species involved in different phases of hedgerow succession are related to the marginal structural cost of height growth and horizontal spread in different branching patterns. Third, differences in stem allocation between *Larix* species are related to differences in transpiration rate and the additional constraint on woody plant stems imposed by possible differences in the conductive requirements for sapwood. Finally, the authors assess the relative roles of water and nutrient relations in determining relative rates of growth in xylem-tapping mistletoes and their hosts.

Koehl (Chapter 18) provides an excellent overview of the peculiar biomechanical constraints facing attached aquatic plants. Terrestrial plants must support their own weight and resist the mechanical drag exerted by moving air. For aquatic plants, the second constraint is far more important, because water is incompressible and many times denser than air. Koehl presents a detailed engineering analysis of how the form of macroalgae and the mechanical properties of their tissues affect their exposure to drag-induced stresses, uptake of nutrients, and deformation and breakage in moving water. Tradeoffs between mechanical sturdiness and productive capacity frequently appear. For example, seaweeds exposed to strong currents and waves tend to have narrow, thick flat blades, which reduce form drag and permit survival in exposed habitats at the expense of greater internal shading and higher allocation to nonproductive tissue. Species from more protected sites often have thinner, broader blades with ruffled edges. Ruffles and other blade protuberances increase turbulence and reduce the thickness of the blade boundary layer, thus increasing uptake of carbon dioxide and mineral nutrients in relatively still waters, but at the cost of increased drag and mortality in wave-swept sites. Algal biomaterials that resist drag-induced stresses are very tough, being either stiff and strong or very stretchy and weak. Seaweeds with weak, extensible tissue are expected wherever high flow forces are of short duration, relative to the time required to stretch them to breaking length. As Koehl notes, algae offer a rich field for quantitative experimental studies on plant functional morphology and optimal behavior (see also Littler and Littler 1980; Lubchenco and Cubit 1980; Steneck 1982).

Hay (Chapter 19) explores aspects of algal functional morphology related to light capture. He presents a model for crown structure, blade shape, and blade optical density to account for species distribution along spatial and temporal gradients of light intensity, based largely on Horn's (1971) monolayer – multilayer analysis for trees. Monolayered species with broad, optically dense blades should grow more slowly than multilayers under high light levels, and predominante under low light levels. Multilayered species with narrow blades or broad but optically translucent blades should grow rapidly under intense light, but grow slowly or die in dense shade. These predictions are largely confirmed by comparative data on the depth zonation of different growth forms and by experimental studies on species differences in growth and abundance at a tropical site with a storm-induced seasonality in light intensity. At the latter site, maintenance of species richness apparently occurs because long periods of storm-induced turbidity and low light availability cause selective mortality of the species that are capable of growing most rapidly when light is intense. The ecological distributions of monolayered and multilayered species in marine habitats differ from those seen in the forests studied by Horn (1971), in that frequent disturbance and wave action prevent monolayers from casting complete shade and thus inducing a successional shift from light-loving multilayers to shade-adapted monolayers that supplant pioneers that cannot reproduce in their own shade. In shallow marine habitats, multilayers appear to be competitively superior, with monolayers being prevalent only in deeper or more turbid areas that are photosynthetically marginal for multilayers.

References

Borchert, R., and P. B. Tomlinson. 1984. Architecture and crown geometry in *Tabebuia rosea* (Bignoniaceae). Amer. J. Bot. 71:958–969.

Givnish, T. J. 1982. On the adaptive significance of leaf height in forest herbs. Amer. Nat. 120:353–381.

– 1984. Leaf and canopy adaptations in tropical forests. Pp. 51–84 *in* E. Medina, H. A. Mooney, and C. Vásquez-Yánes (eds.), Physiological ecology of plants of the wet tropics. Dr. Junk, The Hague.

Grace, J., and G. Russell. 1977. The effect of wind on grasses. III. Influence of continuous drought or wind on anatomy and water relations in *Festuca arundinacea* Schreb. J. Exp. Bot. 28:368–378.

Greenhill, A. G. 1881. Determination of the greatest height consistent with stability that a vertical pole or mast can be made, and of the greatest height to which a tree of given proportions can grow. Proc. Camb. Phil. Soc. 4:65–73.

Henwood, K. 1973. A structural model of forces in buttressed tropical rain forest trees. Biotrop. 5:83–93.

Horn, H. S. 1971. The adaptive geometry of trees. Princeton University Press, Princeton, N.J.

King, D. 1981. Tree dimensions: maximizing the rate of height growth in dense stands. Oecol. 51:351–356.

King, D., and O. L. Loucks. 1978. The theory of tree bole and branch form. Rad. Envir. Biophys. 15:141–165.

Littler, M. M., and D. S. Littler. 1980. The evolution of thallus form and survival strategies in benthic marine macroalgae: field and laboratory tests of a functional-form model. Amer. Nat. 116:25–44.

Lubchenco, J., and J. Cubit. 1980. Heteromorphic life histories of certain marine algae as adaptations to variations in herbivory. Ecol. 61:676–687.

McMahon, T. A. 1973. Size and shape in biology. Science 179:1201–1204.

McMahon, T. A., and R. E. Kronauer. 1976. Tree structures: deducing the principle of mechanical design. J. Theor. Biol. 59:443–466.

Patterson, M. R. 1985. The effect of mechanical loading on growth in sunflower (*Helianthus annuus* L.) seedlings. J. Exp. Bot. (in press).

Steneck, R. S. 1982. A limpet-coralline algae association: adaptations and defenses between a selective herbivore and its prey. Ecol. 63:507–522.

Wainwright, S. A., W. D. Biggs, J. D. Currey, and J. M. Gosline. 1976. Mechanical design in organisms. Arnold Press, London.

14 Evolution of plant life forms

JOHN A. RAVEN

Introduction

We follow Clements (1920) and Schulze (1982) in defining the "life form" of a plant as a morphological feature independent of environment, and the "growth form" as the direct and quantitative response made by a plant to different habitats and conditions. Growth forms are, accordingly, the phenotypes we immediately observe, whereas life forms are abstractions that can be deduced from a consideration of the range of growth forms of a given genotype. This usage grades into the use of the term "life forms" to indicate ecologically important structural and phenological groupings of genotypes (Raunkaier 1934). Both the life form and the growth form of a plant can be construed as contributing to the evolutionary fitness of the plant. However, the criteria of fitness rightly proposed by Osmond et al. (1980), involving competition between genotypes, have rarely been applied in the analysis of life forms. Thus, most of the evidence we shall consider is anecdotal, especially when fossil taxa are considered. However, it is believed that some selective advantages can be relatively unambiguously assigned in comparisons of widely differing life forms.

The plan to be followed in this chapter involves a discussion of classifications of life forms, followed by an analysis of constraints on the minimum photon flux densities that different photolithotrophic organisms require for maintenance and for growth. This analysis computes maintenance and growth energy costs per cell or per square meter of habitat for comparison with energy input as fractional light absorption and efficiency of energy transformation to yield the minimum incident photon flux density. The next sections are devoted to a discussion of extant life forms of photolithotrophs, followed by an attempt to analyze the evolution of these life forms in the context of the fossil record. The final section summarizes these discussions and points to areas in which further work is required.

In considering the energetics of plant life forms in relation to evolution, it is important to consider not only the *efficiency* of a process but also the *capital costs* of the catalytic and structural machinery required for the process and the *safety* and *controllability* of the process (Raven 1981a).

The *efficiency* of a process is important in the straightforward thermodynamic sense of useful work done per unit energy input and in a wider sense of growth per unit of any resource *used up* by the plant during its growth; the prime example here is transpired water (i.e. water-use efficiency). *Efficiency,* then, is a ratio that relates plant performance to the *flux* of energy or water.

Capital costs relate to the rate at which a resource is manipulated by the plant per unit of that resource (or some other) that is committed to the catalysts and structures needed for the manipulation. Although it is fashionable to refer, for example, to nitrogen-use efficiency, this is not an efficiency in the sense of energy-use efficiency or water-use efficiency; it has time-related units (e.g., mol C fixed mol^{-1} leaf N s^{-1}). Capital costs of plant growth may be significant in cost–benefit analyses at both low and high resource input levels. At low resource input levels (e.g., low N availability), a high "N-use efficiency" might be selectively favored. Equally, a high N-use efficiency might also be favored when the availability of N, and other resources, is high, in that a high specific growth rate can be achieved only if the specific reaction rates of individual catalyzed processes are high. Because these two specific rates are related by the concentration of catalysts in the organism, we see that a generally low rate of catalyzed reactions demands a higher catalyst content, and hence higher biomass, per unit growth, and thus a lower specific growth rate.

Whereas *efficiency* and *capital costs* are quantifiable in cost–benefit analyses, controllability and safety are less readily quantified and tend to be thought of as the selective rationale of components that do not obviously contribute to maximizing efficiency or minimizing capital costs. Some aspects of chemical defense from herbivory, a safety-related process, are discussed by Chew and Rodman (1979) and Swain and Cooper-Driver (1981).

Classification of life forms

Most life-form classifications involve only terrestrial plants (e.g., Raunkaier 1934; Box 1981; Schulze 1982), although Luther (1949), Feldman (1966), and Den Hartog and Segal (1964) have produced schemes for aquatic plants. Because this chapter deals with both aquatic *and* terrestrial plants, the basic categorization adopted here (Table 14.1) has as its major trichotomy *planophytes* (phototrophs suspended in a fluid medium), *haptophytes* (phototrophs attached to solid particles that are large relative to the size of the plant), and *rhizophytes* (phototrophs with a portion embedded in solid particles that are small relative to the size of the plant). Within the planophytes we deal almost entirely with the *planktophytes* (microplanophytes). This scheme relegates all terrestrial vascular plants to a subdivision of the rhizophytes (Raven 1981b). It is within these terrestrial rhizophytes that such important life-form dichotomies as woody/herbaceous and sclerophyll/mesophyll are to be found; other important terrestrial dichotomies (e.g., evergreen/deciduous) have their parallels in aquatic plants (Table 14.1).

The significance of the planktophyte-haptophyte-rhizophyte trichotomy is to be seen not only in morphological and evolutionary considerations but also in relation to physiology and energetics, which will be the central focus of this chapter. From the point of view of the evolution of energetic processes, we note that the trichotomy represents a series in which the fraction of the plant body that is photosynthetic generally decreases in the order planktophytes > haptophytes > rhizophytes. Similarly, the fraction of plasmalemma area that is involved in the influx of dissolved solutes (HPO_4^{2-}, NO_3^-, Fe^{2+}, etc.) decreases in the order planktophytes > haptophytes > rhizophytes. Only in rhizophytes is there a part of the plant that is not illuminated; only in (most) terrestrial rhizophytes is this part of the plant the only one that has access to liquid water and dissolved solutes. Only terrestrial rhizophytes can (as case of the vascular plants) be homoiohydric rather than poikilohydric (Walter 1955) (Table 14.2). Poikilohydric plants (or parts of plants) have water contents dictated largely by environmental water potential. Homoiohydric plants have various structures and regulatory mechanisms that permit these plants to stay hydrated for hours to years when the external water supply is restricted. The homoiohydric condition involves the occurrence of a non-photosynthetic fraction of the plant body that is greater than that implicit in other (poikilohydric) life forms (see later section on benthic algae). The implications of the fraction of non-green tissue for the incident photon flux density that is required for photolithotrophic growth will be considered in the next section.

We note that the "r and K" scheme of MacArthur and Wilson (1967) and the "R, C, and S" scheme of Grime (1979) can be superimposed on the scheme shown in Table 14.1, as has been done by Raven (1981b, 1984a). Such a superimposition will not be attempted here.

Table 14.1. *A scheme of life forms for photolithotrophs*[a]

Category	Subcategory	Submerged aquatic examples	Amphibious/intertidal/ interfacial examples	Terrestrial examples
Planophytes: plants not attached to the substrate	Planktophytes: microscopic planophytes	All are short-lived microphytes; many members of the cyanobacteria and Chlorophyta; most members of the Dinophyta, Bacillariophyta, Prymnesiophyta, Chrysophyta, Cryptophyta; includes many symbioses with protozoans (Anderson et al. 1983)	None	None; no photosynthetically active aerial plankton
	Pleustophytes: macroscopic planophytes			
	Benthic pleustophytes: plants resting on substrate	Aegegrophilic algae and cyanobacteria *Cladophora aegegrophila, Nostoc puncriforme*; Many foraminifera symbiotic with dinophyte algae; *Convoluta* symbiotic with *Platymonas; Elysia* symbiotic with *Codium* chloroplasts	*Ascophyllum nodosum* ecad *Mackei* (derived from haptophytic populations) *Fucus vesiculosus, Macroeystis pyrifera* (Gerard and Kirkman 1984)	Some terrestrial angiosperms, bromeliads (*Tillandsia* spp.) some lichens (e.g., *Chondriopsis*) (Rogers 1971)
	Mesopleustophytes: plants between substrate and	*Lemna trisulca, Utricularia vulgaris, Ceratophyllum*	None	None

	surface of overlying field	*demersum* (angiosperms), *Riccia fluitans* (hepatic derived from haptophytic populations), *Sargassum* spp., *Enteromorpha* spp.		
	Acropleustophytes: float on surface of water; at least some part of plant permanently exposed to air	*Eichornia, Pistia, Hydrocharis mossus–ranae, Stratiotes alloides* (sinks in winter), *Lemna minor, Azolla filiculoides, Salvinia natans* (heterosporous ferns)	None	None
Haptophytes: microphytes or macrophytes, *attached to* substrate particles that are *large* relative to the plant		Many submerged microalgae; most submerged macroalgae; all submerged lichens, most submerged bryophytes (includes epilithic and epiphytic plants); a few submerged flowering plants (freshwater Podostemaceae, marine *Phyllospadix*); most submerged aquatic symbioses of coelenterates, porifera, and *Tridacna* with microalgae	Most intertidal algae, all intertidal lichens (includes epiphytic and epilithic plants)	Epilithic and epiphytic algae, lichens, and bryophytes
Rhizophytes: macrophytes, with rhizoids or roots *in* a substrate with particles *small* relative to the plant		Many submerged macroalgae in the Chlorophyta (Charophyceae: Characaceae;	Some algae: *Nitella terrestris, Protosiphon, Botrydium, Fritschiella* (Iyengar 1932); many	Most soil-inhabiting terrestrial bryophytes; almost all terrestrial vascular plant sporo-

Table 14.1. (*cont.*)

Category	Subcategory	Submerged aquatic examples	Amphibious/intertidal/ interfacial examples	Terrestrial examples
		Chlorophyceae: *Protosiphon*; Ulvaphyceae: *Caulerpa* spp.), Tribophyta (*Vaucheria: Botrydium*); most submerged vascular plants (lycopsid *Isoetes;* angiosperms in freshwater and seawater)	vascular plants (float-leaved or emergent, including mangroves, and those that can complete their life cycle either emerged or submerged – *Littorella uniflora*)	phytes (except, e.g., some bromeliads) and their free-living gametophytes

[a] Data from Boysen-Jensen (1932), Den Hartog and Segal (1964), Feldman (1966), Luther (1949), Raven (1981b, 1984a), Raven and Richardson (1984a), and Schulze (1982).

Table 14.2. *Relationships of poikilohydry and homoiohydry to desiccation tolerance*[a]

	Desiccation-resistant	Desiccation-intolerant
Poikilohydric	Many intertidal algae, terrestrial algae, lichens, bryophyte and pteridophyte gametophytes	Many subtidal (permanently submerged) algae; some terrestrial bryophyte and pteridophyte gametophytes; most submerged vascular plants
Homoiohydric	A few ("resurrection plants") terrestrial vascular plants (0.65% of pteridophytes; 0.1% of monocotyledons; 0.01% of dicotyledons)	Most terrestrial vascular plants (99.35% of pteridophytes; 100% of gymnosperms; 99.9% of monocotyledons; 99.99% of dicotyledons)

[a] "Poikilohydric" has been used rather loosely [e.g., by Raven (1977)] to mean both plants that lack the homoiohydric system for regulating water loss and plants that are resistant to desiccation. Walter (1955) used "poikilohydric" to indicate just the first of these attributes. Because not all poikilohydric plants are desiccation-tolerant, and some homoiohydric plants are desiccation-resistant, it would seem best to *not* use "poikilohydric" to mean desiccation-tolerant and homoiohydric to imply desiccation-intolerant. The scheme here shows the four possibilities (Hsiao 1973; Page 1979; Bewley and Krochko 1982; Raven 1977, 1984b).

Photon flux density requirements for photolithotrophs of various life forms

We can relate the incident photon flux density to the growth rate of the phototroph by equation (14.1), modified from Penning de Vries (1972):

$$\mu = JAP\phi DFR - M \qquad (14.1)$$

where μ is the relative growth rate (moles C assimilated per mole plant C per second), J is the incident photon flux density (mol photon m^{-2} s^{-1}), A is the fractional absorption of incident photons, P is a factor relating moles C in photosynthetic organs to area exposed to incident photons (m^2 mol^{-1} C), ϕ is the quantum yield of photosynthesis (moles C fixed per mole photon absorbed) (see Appendix I). The term $JAP\phi$ gives the rate of photosynthesis in moles C fixed per mole C in the light period; this photosynthetic rate is corrected for any photorespiratory or carbon-pumping costs and losses, but *not* for any losses via "dark" respiratory processes. D, the fraction of the 24-hr light–dark cycle for which the mean photon flux density J occurs, averages the C gain over the full 24 hr. F is the fraction of the photosynthate C that is incorporated into plant material (i.e., the fraction that is *not* lost in dark respiration). R is the fraction of the plant C that is associated with the photosynthetic organs. The factor DFR converts the photosynthetic C assimilation rate per unit C in photosynthetic organs into the rate of accumulation of C per unit C in the whole plant. Before this latter can be equated with μ, it must be corrected for M, the maintenance respiration rate (moles C used per mole C in the plant per second).

This equation can be used to compute the minimum photon flux density for *maintenance* of existing plant material by setting $\mu = 0$ and putting the term $F = 1$ (no growth means no growth-associated respiratory losses); thus (Penning de Vries 1975; Loehle 1983),

$$J_{\text{maint}} = \frac{M}{AP\phi DR} \tag{14.2}$$

We shall use equations (14.1) and (14.2) to compute growth rates and maintenance light requirements for planktophytes, haptophytes, and aquatic and terrestrial rhizophytes. We note that the maintenance light requirement, as defined here, is not the "ecological compensation point" of Givnish (1982) in that it does not include constructional costs or effects of herbivory.

Clearly, the computed photon flux densities are only very approximate. For example, the assumed constancy of $F = 0.7$ assumes a tradeoff between "costly" (e.g., protein, especially with NO_3^- as N source, and lignin) and "cheap"(e.g., polysaccharide) plant constituents between protein-rich microalgae and lignin-rich trees [see Penning de Vries et al. (1974) for synthesis costs of various major organic components of plants]. Further, to the extent that M is a function of the quantity and rate of turnover of proteins in the biomass, it is unlikely to be constant. Finally, the equations do not explicitly include the light saturation of growth.

For a planktophyte, we illustrate the use of equations (14.1) and (14.2) by reference to the 10-μm-diameter spherical model cell of Raven (1982, 1984a). The data used by Raven (1982) are $A = 0.137$, $P = 15$ m^2 projected cell area per mole cell C, $F = 0.7, R = 1.0, M = 10^{-8}$ s^{-1}. The value of ϕ (0.1) that we use here is lower than that used by Raven (1982) (see Appendix I); then $J_{\text{maint}} = 10^{-7}$ mol photon m^{-2} s^{-1} [equation (14.2)]. We note that the water-surface photon flux density is unlikely to exceed $2 \cdot 10{-}3$ mol photon m^{-2} s^{-1} (i.e., maintenance uses only $2 \cdot 10^{-4}$ of the maximum photon flux density.

Turning to growth, the maximum photon flux density of $2 \cdot 10^{-3}$ mol photon m^{-2} s^{-1} could give a specific growth rate μ of $1.4 \cdot 10^{-4}$ s^{-1}, corresponding to a generation time of 2.9 hr [equation (14.1)]. The major constraint that limits *observed* algal generation times at 20°C to values of about 8 hr (Eppley 1972; Banse 1982) (i.e., $\mu = 5 \cdot 10^{-3}$ s^{-1} for cells with

Table 14.3. *Minimum photon flux densities at which growth of microphytic photolithotrophs is observed*

Organism	Minimum photon flux density at which growth is observed (μmol m^{-2} s^{-1})	References
Amphidinium carterae (eukaryote, Dinophyta)[a]	$\leqslant 1$	Richardson and Fogg (1982), Hersey and Swift (1976)
Ditylum brightwellii (eukaryote, Bacillariophyceae)[a]	$\leqslant 4$	Perry et al. (1981)
Scenedesmus protuberans (eukaryote, Chlorophyta)[a]	>1	Gons (1977)
Oscillatoria spp. (prokaryote, Cyanobacteria)[a]	$\geqslant 0.25$	Van Liere and Mur (1979)
Prosthecochloris (prokaryote, Chlorobineae)[b]	>0.2	Biebl and Pfennig (1978)
Chlorobium (prokaryote, Chlorobineae)[b]	~ 0.2	Biebl and Pfennig (1978)
Chromatium (prokaryote, Rhodospirillineae)[b]	~ 0.8	Biebl and Pfennig (1978)
Rhodoseudomonas (prokaryote, Rhodospirillineae)[b]	~ 0.8	Biebl and Pfennig (1978)

[a] Uses H_2O as donor in photosynthesis and evolves O_2.
[b] Does not use H_2O as donor in photosynthesis or evolve O_2.

diameters of 5–10 μm) is a decrease in the value of ϕ. Raven (1984a) showed that this is related to problems of fitting sufficient catalysts into the cells [granted their observed specific reaction rates and M_r (relative molecular mass values)] to handle all absorbed light energy.

Returning to the low-light end of the scale, we find that, putting $\mu = 10^{-6}$ (generation time = 7 days), the required value of J is $1.4 \cdot 10^{-5}$ mol photon m^{-2} s^{-1}. Reference to Table 14.3 shows that growth of microphytes can occur at photon flux densities an order of magnitude lower; although the growth in Table 14.3 was often at specific growth rates lower than 10^{-6} s^{-1}, it was not an order of magnitude lower. Assuming that the cells were exposed to *solar* radiation of $1.4 \cdot 10^{-5}$ mol photon m^{-2} s^{-1}, that could increase μ by up to fourfold. The only possibility of maintaining μ [equation (14.1)] at decreased vectorial J values is by increasing A; P is fixed, ϕ is already as low as is plausible, D can be increased only two-fold, and F cannot be increased to more than 0.9. A is determined (at constant cell volume) largely by the pigment concentration in the cells. The concentration of chlorophyll in the cells for which the A value of 0.137 was computed was 16 g kg^{-1} organic weight. Because each gram of chlorophyll is associated with at least 4 g of protein in light-harvesting complexes, 16 g chlorophyll per kilogram organic weight involves at least 64 g protein per kilogram organic weight. Doubling the pigment content increases the protein committed to pigment-protein complexes to 128 g, and another doubling increases it to 256 g, thus approaching *half* of the protein content of the cells (600 g per kilogram organic weight). Quadrupling of the pigment content would not lead to a quadrupling of A (Kirk 1983; Raven 1984a); at most, A could be 0.5, leading (if other values are held constant) to $\mu = 10^{-6}\ s^{-1}$ at $J = 5 \cdot 10^{-6}$ mol photon $m^{-2}\ s^{-1}$. We note that a further increase in A would, at best, decrease J only to $2.5 \cdot 10^{-6}$ mol photon $m^{-2}\ s^{-1}$; such a doubling of A would involve essentially *all* of the cell protein being devoted to pigment-protein complexes.

In terms of costs and benefits, growth at $10^{-6}\ s^{-1}$ with a photon flux density of $1.4 \cdot 10^{-5}$ mol photon $m^{-2}\ s^{-1}$ requires some 64 g protein and 16 g chlorophyll per kilogram organic weight. A 10% increase in A would involve an *approximately* 10% increase in pigment-protein content [because $(\Delta A/A)/(\Delta \text{pigment}/\text{pigment}) \simeq 1$ at $A = 0.137$], which would increase the cost of synthesis of a cell by 1.2% (i.e., substantially less than the 10% increase in light absorption), so that μ could increase to $1.086 \cdot 10^{-6}\ s^{-1}$. Viewed solely from the point of view of added pigment, it is not until $(\Delta A/A)/(\Delta \text{pigment}/\text{pigment})$ falls to about 0.12 that adding pigment-protein would cost more than the extra light absorbed could provide *at the quoted value* of J. However, as we have already seen, at these high A values (~ 0.8), essentially *all* of the cell protein would be in pigment-protein complexes, so that other factors would obviously intervene.

A point that will be taken up in a later section concerns the significance of cell size in planktophytes in relation to photon absorption. Use of the analytical methods of Kirk (1983) shows that decreasing cell size *decreases* A at a given pigment concentration per unit cell volume but that the efficiency of photon absorption per mole pigment *increases,* as does the value of C [equation (14.1)]. These effects mean that μ can be maintained at $10^{-6}\ s^{-1}$ at $J \simeq 1\ \mu$mol photon $m^{-2}\ s^{-1}$ as cell size decreases (Table 14.3).

Turning to macrophytes, the simplest of these have very little non-green tissue; examples are aquatic haptophytic algae such as *Ulva, Enteromorpha,* and *Porphyra,* juveniles of many more complex macroalgae, and the aquatic or terrestrial gametophytes (haptophytes or rhizophytes) of many phaeophytes, bryophytes, and homosporous pteridophytes. Values for the variables in equations (14.1) and (14.2) for algae such as *Ulva, Enteromorpha,* and *Porphyra* are $A = 0.60$ (Ramus 1978; Ramus and Rosenberg

Table 14.4. *Minimum photon flux densities at which growth of poikilohydric, terrestrial, rhizophytic photolithotrophs is observed; all the examples are the prothalli of leptosporangiate ferns*

Organism (criterion)	Minimum photon flux density at which growth is observed (μmol m^{-2} s^{-1} for 12 hr per day)	References
Dryopteris filix-mas (gametophytes; cell division)	1	Mohr (1965)
Onoclea sensibilis (gametophytes; cell division; dry weight?)	5	Miller and Miller (1961)
Cibotium (gametophytes)	1.5	Friend (1975)

1980; B. A. Osborne, unpublished data), $P = 2.0$ m^2 per mole cell C (Seybold and Egle 1938; Niell 1976; Ramus 1978), $\phi = 0.1, D = 0.5, F = 0.7, R = 1.0$, and $M = 10^{-8}$ s^{-1}. From equation (14.2), we deduce J_{maint} of $1.7 \cdot 10^{-7}$ mol photon m^{-2} s^{-1}; by comparison with our model planktophyte, the rather higher J_{maint} means that the higher A does not quite offset the lower P of the 70-μm-thick thallus (Ramus 1978). Growth at 10^{-6} s^{-1} [easily achieved by, for example, *Porphyra* (Jackson 1980)] would require a photon flux density of at least $2.4 \cdot 10^{-5}$ mol photon m^{-2} s^{-1}. This is a reasonable value provided the thallus area index is ≤ 1.0 (Raven et al. 1979).

The chlorophyll content equivalent to absorption of 0.6 of incident photons is, in green algae, equivalent to some 9 g chlorophyll per kilogram organic weight (Seybold and Eagle 1938; Ramus 1978), that is, some 45 g pigment-protein complex per kilogram organic weight. We note that slow growth of morphologically rather similar (rhizophytic) fern prothalli can occur at 1–5 μmol photon m^{-2} s^{-1} (Table 14.4).

For larger submerged plants (e.g., the haptophyte *Laminaria*), the values of variables in equations (14.1) and (14.2) are $A = 0.95$ (Lüning 1979), $P = 0.20$ m^2 [mol c]$^{-1}$ (Seybold and Egle 1938; Larkum 1972), $\phi = 0.10, D = 0.5, F = 0.7, R = 0.33$ (Kremer 1980), and $M = 10^{-8}$ s^{-1}. The value of P assumes a thallus area index of 1.0, with 250 g dry weight per square meter of substratum.

Using equation (14.2), we find J_{maint} of $3.15 \cdot 10^{-6}$ mol photon m^{-2} s^{-1}. For a doubling time of 6 months (Kain 1979), that is, a specific growth rate of $4.4 \cdot 10^{-8}$ s^{-1}, we find a mean photon flux density requirement of $2.4 \cdot 10^{-5}$ mol photon m^{-2} s^{-1}. This is reasonable in terms of the values in Table 14.5. A thallus area index value of 1 corresponds to light-limited

Table 14.5. *Minimum photon flux densities at which growth of poikilohydric, aquatic, haptophytic, macrophytes and photolithotrophs is observed*

Organism (criterion)	Minimum photon flux density at which growth is observed (μmol m^{-2} s^{-1} for 12 hr per day)	References
Lithothamnion spp. (Rhodophyta) (from natural habitat)	$\leqslant$1 (assumes surface photon flux density is 2,000 μmol. m^{-2} s^{-1})	Lüning and Dring (1979)
	~0.4 (from yearly total photo-synthetically active radiation assuming mean 12 hr day^{-1})	
Gametophytes of Laminariales (Phaeophyta) (cell division)	0.25	Lüning and Neushul (1978)
Laminaria hyperborea (Phaeophyta) (mature sporophytes; from natural limit of kelp growth)	$\leqslant$10 (assumes surface photon flux density is 2,000 μmol m^{-2} s^{-1}	
	~4.5 (from yearly total photosynthetically active radiation assuming mean 12 hr day^{-1})	Lüning and Dring (1979)
Laminaria spp. sporophytes (Phaeophyta) (from natural limit under *Macrocystis* canopy)	$\leqslant$12	Neushul (1971a)
Macrocystis spp. juvenile sporophytes (Phaeophyta) (from natural limit under *Macrocystis* canopy)	$\leqslant$16	Neushul (1971a)
Cystoseira spp. sporophyte (Phaeophyta) (from natural limit under *Macrocystis* canopy)	$\leqslant$6	Neushul (1971a)
Desmarestia spp. sporophytes (Phaeophyta) (from natural limit under *Macrocystis* canopy)	$\leqslant$10	Neushul (1971a)
Eisenia spp. sporophytes (Phaeophyta) (from natural limit under *Macrocystis* canopy)	$\leqslant$12	Neushul (1971a)
Various benthic algae under *Zostera marina* bed (from natural habitat)	$\leqslant$80	Dennison et al. (1981)

growth at depth; at higher photon flux densities, thallus area indices (again for one side of the thallus) of 10 have been found. These can be regarded as having $A = 1.0$ and $P = 0.02$ m^2 per mole C, so that J_{maint} for dense *Laminaria* beds is $3.0 \cdot 10^{-5}$ mol photon m^{-2} s^{-1}, and J_{growth} for a generation time of $4.4 \cdot 10^{-8}$ s^{-1} is $2.3 \cdot 10^{-4}$ mol photon m^{-2} s^{-1}. This again accords with ecological data (Kain 1979; Lüning 1979).

The A value of 0.95 for a *Laminaria* thallus involves a chlorophyll + fucoxanthin content of the thallus of some 1,000 μmol m^{-2}, that is, about 3 g pigment per kilogram organic weight, or 15 g pigment-protein complex per kilogram organic weight (Anderson 1967; Raven et al. 1979; Raven 1984a).

For a small terrestrial herb (e.g., an understory plant in a forest), the parameters in equations (14.1) and (14.2) might have the values (Evans 1972) $A = 0.90$ (leaf area index = 1), $P = 0.48$ (0.02 m^2 leaf area (one side) per gram leaf dry weight), $\phi = 0.10$, $D = 0.5$, $F = 0.7$, $R = 0.5$, $M = 10^{-8}$ s^{-1}. J_{maint} is then $9.2 \cdot 10^{-7}$ mol photon m^{-2} s^{-1}, and for a 6-month doubling time (i.e., a specific growth rate of $4.4 \cdot 10^{-8}$ s^{-1}), $J_{growth} = 7.1 \cdot 10^{-6}$ mol photon m^{-2} s^{-1}; again, these figures are in accord with data in Table 14.6.

Finally, we look at terrestrial forest, with 10 kg DW m^{-2} (100 metric tons DW per hectare). Values for the parameters (Ovington 1962; Kira 1975; Grier and Logan 1977; Farnum et al. 1983) are $A = 0.95$ (leaf area index = 3), $P = 0.16$ (one-third of leaf area of 0.48 m^2 (one side of leaf) per mole leaf C), $\phi = 0.10$, $D = 0.5$, $F = 0.7$, $R = 0.06$ (for living material; live/dead = 0.25), and $M = 10^{-8}$ s^{-1}. J_{maint} is then $2.19 \cdot 10^{-5}$ mol photon m^{-2} s^{-1}, and J_{growth} for a doubling time of 10 years for live + dead biomass (i.e., a specific growth rate of $3.17 \cdot 10^{-9}$ s^{-1}) is (allowing for maintenance of *live* material only) $7.1 \cdot 10^{-5}$ mol photon m^{-2} s^{-1}. We note that total respiration (growth + maintenance) in this example equals 0.55 of gross photosynthesis (=JAPϕ); in 80-m-tall *Pseudotsuga menziesii* this fraction can apparently reach 0.9 or so as a result of the very large quantity of live tissue to be maintained, and a low growth rate (Grier and Logan 1977) (see later section on benthic algae). Understory trees (Table 14.6) can grow with less than 20 μmol photon m^{-2} s^{-1}.

The computed values, based on the simple model of equations (14.1) and (14.2), show reasonable agreement with the measured values for the minimum photon flux density for growth under laboratory and field conditions: The minimum photon flux densities under which growth was *observed* to occur generally fall between the values *computed* for the maintenance requirement and those *computed* for slow growth ("slow" defined in

Table 14.6. *Minimum photon flux densities at which growth of homoiohydric, terrestrial, vascular rhizophytic photolithotrophs is observed*

Organism (criterion)	Minimum photon flux density at which growth is observed (μmol m^{-2} s^{-1} for 12 hr per day)	References
Lowland tropical rain forest, Lamington National Park: Queensland shrubs, herbs, below canopy (from natural habitat)	$\leqslant 5$	Björkman and Ludlow (1972)
Shrubs, herbs in lowland tropical forest (from natural habitat)	$\leqslant 10$ (assumes surface photon flux density is 2,000 μmol m^{-2} s^{-1})	Huber (1978)
Shrubs, herbs in tropical cloud forest (from natural habitat)	$\leqslant 26$ (assumes surface photon flux density is 2,000 μmol m^{-2} s^{-1})	Huber (1978)
Understory trees (C_3 *Claoxylon sandwicence*, C_4 *Euphorbia forbesii*) (in Hawaiian tropical rain forest)	$\leqslant 20$	Pearcy (1983), Pearcy and Calkin (1983)
Seedlings of *Acmena smithii* (rain-forest tree of eastern Australia)	$\leqslant 20$ (growth at ~1% of full daylight; author's extrapolation to zero growth at 0.18% full daylight not used here)	Ashton and Franckenberg (1976)
Pteris cretica (sporophytes)	$\leqslant 20$ (minimum photon flux density at which growth was tested)	Hariri and Priouhl (1978)

the context of observed growth rates for each category of plants) (Tables 14.3–14.6).

In the context of our categories of planktophyte, haptophyte, and rhizophyte, we can see that the photon flux densities needed for maintenance and for slow growth increase in this order: microalgae and microscopic, poikilohydric stages of macrophytes; macrophytic poikilohydric plants with little non-green tissue (haptophytes *and* rhizophytes); more differentiated poikilohydric haptophytes and rhizophytes, and the smaller homoiohydric rhizophytes; trees.

Starting from a canopy provided by large evergreen trees with ~2 mmol photon $m^{-2}\,s^{-1}$ (400–700 nm) incident on it, at least 0.95 of the incident photons are absorbed by the canopy, giving a *maximum* photon flux density of 100 μmol photon $m^{-2}\,s^{-1}$ under the canopy (Horn 1971). In many tropical evergreen forests the values are in the range 5–26 μmol photon $m^{-2}\,s^{-1}$ (Table 14.3); that is, absorption by plants >2 m high is ~99%. The work of Pearcy (1983) and Pearcy and Calkin (1983) shows that woody plants (understory trees including the C_3 *Claoxylon sandwicence* and the C_4 *Euphorbia forbesii*) can grow at mean photon flux densities of 20 μmol photon $m^{-2}\,s^{-1}$. Vascular plants (homoiohydric) of lower stature can clearly grow at mean photon flux densities of ~5 μmol photon $m^{-2}\,s^{-1}$, and poikilohydric plants (soil algae; fern and moss gametophytes) can still grow after a further twofold to fivefold light attenuation by the vascular plants to 1–2.5 μmol photon $m^{-2}\,s^{-1}$.

This phenomenon of stratification in high-biomass communities is mimicked by marine kelp forests (see later section on benthic algae). Here all of the components are poikilohydric haptophytes, but there is still a hierarchy of minimum photon flux densities for maintenance and for growth that is related to the ratio of photosynthetic to nonphotosynthetic biomass; this ratio is lower for mature specimens of canopy-forming species than for understory species.

Regeneration of the canopy dominants would clearly be very difficult if the mature canopy dominants attenuated the light by another order of magnitude (i.e., to ~0.1% rather than 1% of that incident at the top of the canopy), even if gaps in the canopy caused by death of a canopy dominant were taken into account (Horn 1971; Grubb 1977; Kirkman 1981). A possible reason for this apparent limitation of light attenuation by the upper layers of canopy may be found in considerations of the effectiveness in photosynthesis of hypothetical additional layers of canopy. Such layer(s) of photosynthetic structures may not be below their *individual* light compensation points, assuming that the organism has sufficient phenotypic plasticity to produce photosynthetic organs capable of carrying out net photosynthesis at 2,000 μmol photon $m^{-2} s^{-1}$ (at the top of the canopy) and at 5 μmol photon $m^{-2} s^{-1}$ (at the bottom of the canopy of the mature organism), which would be appropriate to juveniles of the species. Such variation over a 400-fold range of photon flux densities is seen, for example, in *Laminaria* (Table 14.5) and *Cibotium* (Tables 14.4 and 14.6) (Friend 1975). However, it may well be that the resources tied up in this layer of photosynthetic structures (e.g., N, P, K, mobilizable organic C) could be better employed (from the viewpoint of natural selection) elsewhere in the plant (e.g., in producing an increment of height of the canopy, and thus increasing carbon gain, and an increment of depth or density of rooting, or in reproduction) than in promoting photosynthesis that could provide for well under half of the organism's overall maintenance requirement (Givnish 1984). Self-pruning (often involving abscission) of individuals (Addicott 1982) and "self-thinning" in terms of density of individuals (Schiel and Choat 1980; Pickard 1983; Kirkman 1984) may be involved here.

By contrast, the phototrophs of smaller stature (including juveniles of the dominants) have much smaller growth maintenance requirements per unit of light-absorbing machinery than do the mature dominants and can grow on the "leavings" of the taller plants.

In brief, the taller a plant becomes in its (selective) competition for light, the more light it needs to support its preexisting biomass and to achieve unit height growth (with implications for structural and transporting elements). In many cases (see later sections on planktophytes and benthic algae), the high photon flux density requirement of even the lowermost photosynthetic elements of such plants means that structural adjustments to the plant and the community permit sufficient light to penetrate to permit other life forms of photolithotrophs to survive beneath the main canopy.

General discussions of adaptations to various photon flux densities, including problems of photoinhibition at high photon flux densities and, for terrestrial plants, compromises related to thermal and water relations, can be found in Boardman (1977), Cowan (1977), Cowan and Farquhar (1977), Björkman (1981), Raven (1984a), and Richardson et al. (1983).

Plant construction types

Planktophytes

The "ur-phototroph" is a unicellular alga; such organisms in the marine plankton contribute about 25% of global primary production (Raven and Richardson 1984a). A major characteristic of planktophytes is that they are *small;* maximum cell dimensions range from <1 μm for the eukaryotic and cyanobacteria picoplankton to >100 μm in, for example, some centric diatoms. It is a mistake to dismiss phytoplankters as being merely small; within a single class, the Bacillariophyceae (diatoms), the range of cell volumes (Sournia 1982) is as great as the range of body volumes in terrestrial mammals (K. Richardson, personal communication). However, all planktophytes *are* small by the standards of the largest haptophytes (the 50-m-long *Macrocystis*) or rhizophytes (the 100-m-high *Sequoia* or *Eucalyptus*).

Colinvaux (1980) suggested that the relative absence of larger phototrophs in the open ocean [an exception being the mesopleustophyte specimens of *Sargassum* species (Table 14.1)] is a function of the difficulties of returning propagules to the appropriate water body. Pleustophytic *Sargassum* recruits drifting material (which was originally haptophytic) into the gyre of the Sargasso Sea. Colinvaux (1980) argued that the nutrient supply to a drifting pleustophyte (with gas bladders) need not be a problem. However, most analyses based on concentrations of available N and P in oligotrophic oceanic waters and the achievable relative motion between organisms and the surrounding water suggest that planktophyte-size phototrophs are at a considerable advantage (Vogel 1981).

Granted that planktophytes are small, we can inquire as to the influence of their size on functional attributes; in this discussion, *shape* will also be an important consideration (Sournia 1982).

There is a widespread tendency for metabolic rates per unit biomass (e.g., specific growth rate, specific respiration rate) to decrease in larger organisms of a particular ecological category such as terrestrial homoiotherm vertebrates (Peters 1983). A common exponent for the decrease is -0.25 (i.e., metabolic rate/biomass declines as biomass$^{-0.25}$). This decline is greater than can be explained by a "surface law," which would suggest, for organisms of similar shapes but different sizes, an exponent of -0.33 (Banse 1982; Peters 1983). Such a -0.25 law is likely (but not well quantified) for the maximum specific growth rates of terrestrial vascular plants of various sizes; such a law does apply to their specific respiration rates (Banse 1982). It is, accordingly, of interest that the exponent for the light- and nutrient-saturated specific growth rates of two classes of marine planktophytes (the Bacillariophyceae and the Dinophyceae) is apparently substantially less than -0.25 (i.e., -0.10 to -0.15) (Banse 1982). This means that the biomass dependence of the specific growth rate is relatively small, in contrast to the behavior of microchemoorganotrophs from the same environment with a similar range of body sizes that have a larger, more typical, biomass dependence of the specific growth rate (Banse 1982). The chemoorganotrophs generally have higher specific growth rates than do phototrophs of similar size (except at the high end of the range), and among the phototrophs, diatoms of a given size grow faster than do dinoflagellates, with other planktophytes intermediate between these two classes. The ecological relevance of these findings is somewhat obscure, in that the biomass dependence of resource-limited growth rates may be more relevant [but see Goldman et al. (1979)].

Another important size-dependent function of planktophytes is fractional photon absorption (Kirk 1983). At a given pigment concentration per unit cell volume, fractional absorption decreases with cell size. Taking the maximum fraction of the biomass that can be taken up by pigment-protein complexes as 0.5 of the organic dry weight, an analysis according to Kirk (1983) shows that cells containing equimolar chlorophyll and phycobilin chromophores (e.g., cyanobacteria, cryptophytes) with 40 g chlorophyll per kilogram organic dry weight have fractional absorptions of incident photons that are strongly dependent on cell size. At the chlorophyll *a* absorption maximum at 435 nm (appropriate to the blue light that penetrates deepest in oceanic waters), a 28.8-μm-radius spherical cell or colony would absorb 0.95 of incident photons, a 4-μm-radius cell would have a fractional absorption of 0.70, and a 0.4-μm-radius picoplankton cell would absorb only about 0.15 of the incident photons; in all cases we deal with a vectorial light field (J. Raven, in preparation).

However, things look very different when we consider C-specific absorption of light as a function of cell size. If the cells or colonies considered earlier had C contents equivalent to one-eighth of their wet weight, 10 μmol photon $m^{-2}\ s^{-1}$ of 435-nm light would give a photon absorption rate of $9.5 \cdot 10^{-5}$ mol photon per mole C per second for 26.8-μm-radius cells or colonies; for the 4-μm-radius cells, $5 \cdot 10^{-4}$ mol photon per mole C per second are absorbed; for the 0.4-μm-radius cells, the photons absorbed per second are $1.07 \cdot 10^{-3}$ mol photon per mole C. We use equation (14.1) and these values for a 28.8-μm-radius cell or colony: $J = 10^{-5}$ mol photon $m^{-2}\ s^{-1}$, $A = 0.95$, $P = 10\ m^2$ per mole C in the cells, $\phi = 0.1$, $D = 0.5$, $F = 0.7$, $R = 1.0$, $M = 10^{-8}$. We find $\mu = 3.3 \cdot 10^{-6}\ s^{-1}$. For 4-$\mu$m-radius cells, with $A = 0.70$ and $P = 71$, and other values as before, we find $\mu = 17 \cdot 10^{-6}\ s^{-1}$. For 0.4-$\mu$m-radius cells, with $A = 0.15$ and $P = 710$, and other values as before, we find $\mu = 37 \cdot 10^{-6}\ s^{-1}$. Accordingly, we find that there is the potential for more than a 10-fold increase in potential specific growth rate at rate-limiting photon flux densities as the cell (or colony) radius is decreased 72-fold.

At lower pigment contents (1 g chlorophyll per kilogram organic weight), where the differences in fractional absorptance (on a cell projected-area basis) are larger (0.55 for the 28.8-μm-radius cell or colony, 0.34 for the 4-μm-radius cells, and 0.035 for the 0.4-μm-radius cell) and the differences in computed μ values are much smaller (i.e., $1.9 \cdot 10^{-6}$ s^{-1} for the largest cell or colony, $8.3 \cdot 10^{-6}$ s^{-1} for the 4-μm-radius cell, and $8.6 \cdot 10^{-6}$ s^{-1} for the 0.4-μm-radius cell) (J. Raven, in preparation), these computations show that in the range of pigment concentrations and cell sizes in which self-shading and "package effects" are small, the light-limited specific growth rates *are independent of cell size.* However, as pigment content and/or cell size increase, the potential light-limited specific growth rate is higher for smaller cells. We note that the cost–benefit analysis of increasing the pigment content of a light-limited planktophyte cell, as presented earlier, involved cells in which $(\Delta A/A)/(\Delta\text{pigment}/\text{pigment})$ was linear (i.e., those in which there is a negligible computed size dependence of light-limited specific reaction rate).

Considering the data in Table 14.3, there is a tendency for the lowest photon flux densities at which growth can be detected to occur for the smallest of the cells tested.

Further considerations of photon absorption involve the effects of changes in cell shape (Kirk 1975a, 1975b, 1976) and the effect of the presence of a vacuole (Raven 1984a) in a cell with a constant quantity of pigment per cell and constant cytoplasmic volume per cell. Both a nonspherical shape and the presence of a vacuole can help to offset the "package effect" that reduces the effective specific absorption coefficient of pigment molecules. Because, at a given pigment concentration in the cytoplasm, the package effect increases with cytoplasmic volume, it is of interest that cell shapes that depart most strongly from nearly isodiametric spheroids and cylinders, and the presence of vacuoles occupying a large fraction of the cell volume, are commonest in cells large enough to suffer a substantial package effect if they were spherical and nonvacuolate (Sournia 1982). However, Raven (1982, 1984a) has shown that for turgid-walled cells, increased wall construction costs can largely offset increased photon absorption resulting from shape change or vacuolation.

A further effect of cell size and shape and the degree of vacuolation of planktophytes that relates to light handling is the area of plasmalemma per unit cytoplasmic volume. Raven (1984d, 1986) has shown that for spherical cells of radius <0.5 μm, a generation time of 3 hr at light saturation could be achieved if *all* of the photon-harvesting and ATP- and NADPH-generating reactions were to be located in the plasmalemma, granted reasonable sizes and specific reaction rates of the catalysts involved (Raven 1980a, 1980b, 1984a). Such a course could not be followed in eukaryotic picoplankton, where photosynthesis "light reactions" are confined to chloroplast-located thylakoids. However, such a limitation is not present in cyanobacteria (e.g., *Gloeobacter*, in which phototrophic growth occurs with solely plasmalemma-located photoreactions) (Stanier and Cohen-Bazire 1977). If such organisms occur in the picoplankton, they might not be readily distinguished from chemoorganotrophic plankton by electron microscopic studies, as opposed to fluorescence studies, of natural samples of plankton (Johnson and Sieburth 1979, 1982; Waterbury et al. 1979; Platt et al. 1983; Li et al. 1983).

Nutrient acquisition is another aspect of planktophyte biology that can be related to cell size and shape (Raven 1980a; Sournia 1982). Small size, nonisodiametric shapes, and extensive vacuolation *all* increase the plasmalemma area per unit cytoplasm, increasing the area of plasmalemma into which nutrient porters can be fitted and which is exposed to the bulk, nutrient-supplying medium. These potential adaptations to low-nutrient-concentration environments must be viewed in the context of their effects on movement of the organisms relative to the bulk medium. All of the tricks by which the plasmalemma area per unit cytoplasm can be increased will tend to *reduce* the sinking rate (other factors being equal) and thus reduce the Munk and Riley (1952) effects of sinking on reduction in unstirred-layer thickness and hence in increasing potential nutrient flux to the cell surface. However, we note that the effects of movement of cells relative to the medium in enhancing nutrient uptake may have been overestimated by Munk and Riley (1952) [see Vogel (1981)].

Sinking of cells that are denser than water ultimately removes them from the euphotic zone. The great majority of planktophyte cells are denser than the surrounding medium, the exceptions being gas-vacuolate cyanobacteria and some marine members of the Dinophyceae with vacuolar solutes of low density (Raven 1984a). Water movements in mixed waters are generally several orders of magnitude higher than the sinking rate of planktophytes (up to a few meters per day for "healthy" cells).

Motile planktophytes (dinophytes and members of most other algal classes, with the exception of diatoms) can usually move at velocities substantially in excess of their intrinsic sinking rate; dinophytes can move at up to 1 mm s^{-1} (Sournia 1982). However, such velocities of movement do not permit maintenance of their positions relative to the water surface in mixed waters, where vertical water movement can occur at 10 mm s^{-1} (Sournia 1982). Dinophytes can, and do, show vertical migration in stratified waters, after occupying a station nearer the surface in the day than their lower nocturnal station. There is a surface-high gradient of photon flux density and, usually, a surface-low gradient of nutrient availability in stratified waters. Raven and Richardson (1984b) carried out a cost–

benefit analysis of migration and showed that the energy costs of migrating into higher-light environments in the day are more than offset by the extra energy gained at the higher station; furthermore, the energetic gain is not negated in contribution to a higher growth rate by decreased time spent in the high-nutrient environments.

In attempting to integrate this resource-acquisition and potential resource-saturated growth-rate data into ecological "preferences" of the different planktophytes, we can distinguish the nonmotile diatoms, which are more characteristic of mixed water, from the motile dinoflagellates, which are often dominant in stratified water. The analysis of growth as a function of photon flux density that was carried out by Richardson et al. (1983) showed that the Dinophyceae and Bacillariophyceae had similar low minimum photon flux densities at which growth could be detected, but the Bacillariophyceae had substantially higher light requirements to saturate growth and for the onset of photoinhibition of growth. This finding that there is a greater range of photon flux densities between compensation and saturation in diatoms correlates well with the higher light-saturated specific growth rate that characterizes diatoms (Banse 1982). Furthermore, the capacity to use, and tolerate, a greater range of photon flux densities in diatoms than in dinoflagellates correlates with the frequent dominance of diatoms in mixed waters, where vertical movement relative to the surface exposes cells to wide ranges of photon flux densities over time scales of minutes (Sournia 1982). Dinoflagellates of stratified waters can, by contrast, "choose" the mean photon flux density of their environment and tolerate a smaller range of photon flux densities between compensation of growth and photoinhibition of growth.

Aside from this generalization, it is difficult to find any *general* correlation between morphology of planktophytes and sun or shade adaptation. Sournia (1982) pointed out that the planktophyte "shade flora" that preferentially inhibits the lower levels of the euphotic layer is taxonomically and morphologically diverse; there is no single morphological solution to problems of life at low photon flux densities on the part of planktophytes. This conclusion is not in disagreement with the earlier conclusions as to the utility of small cell size, markedly nonspherical shape, or a high degree of vacuolation in maximizing the light interception capacity of cells on a unit pigment basis.

The morphology of planktophytes is also *relatively* poorly correlated with nutrient availability in their environment. The work of Margalef (1978), as summarized by Sournia (1982), suggests that small, rounded nonmotile cells with a high potential growth rate dominate relatively nutrient-rich and turbulent waters, whereas larger cells, often of a nonrounded shape, and motile, with a lower potential growth rate, dominate less turbulent nutrient-poor waters. This can be related to an ecological succession from the mixed, relatively nutrient-rich waters of the spring bloom to the less turbulent, low-nutrient waters of midsummer in temperate waters (Sournia 1982).

A final and very important correlate of morphology is susceptibility to grazing. This is a safety factor, as described earlier, and, as always, there is a tendency to attribute to defense against grazing those characteristics that cannot be accounted for in terms of the efficiency or rate of resource handling. Both extremes of size can be advantageous in decreasing losses due to grazing (Banse 1982; Sournia 1982).

The reference to safety in relation to grazing brings us to the other two desiderata (efficiency and capital cost) mentioned earlier. In relation to cell size, the maximum (resource-saturated) growth rate is clearly a function of the quantity of catalysts and structures that can be packaged into the cell and of the maximum specific reaction rate of the catalysts; that is, it is a function of capital costs in the sense that the smaller is the investment in catalytic or structural machinery required to achieve a certain rate of substrate transformation, the greater is the potential specific growth rate of the cells. We find that there are *relatively* restricted ranges of catalytic capacities and molecular sizes among examples of homologous catalysts from different organisms (Raven 1984a), so that it is not easy to attribute differences in maximum growth rate to, for example, differences in the mass of catalyst needed to produce a certain rate of conversion.

A major example of *analogous* catalysts in planktophytes are the light-harvesting pigment-protein complexes of the three major pigment groups: the chlorophyll $a + b$ complexes of the chlorophytes, the chlorophyll $a + c$ + carotenol complexes of the chromophytes, and the phycobilins of the phycobiliphytes (see later section on evolution of life forms). Here there are substantial (twofold or more) differences in the quantities of protein required to provide a light-harvesting complex with a given light-absorbing capacity at the wavelengths of peak absorption of the respective complexes; the phycobilins cost more than the other complexes (Raven 1984a). Even though this difference relates directly to an efficiency term (the quantity of complex needed to permit a cell to absorb a certain fraction of incident photons), it may also be significant at light saturation, where a certain minimum quantity of light-harvesting machinery is still needed. It is not clear how this factor might be related to differences in maximum specific growth rate between planktophytes.

In general, it is not easy to relate differences in maximum growth rate of planktophytes to the quantity of catalytic machinery or to its potential activity.

Turning to resource-limited growth, we have already seen that the efficiencies of particular catalysts of energy metabolism, whether of light harvesting or energy transformation, do not vary much *on a molar basis.* However, it would seem that the energy costs of light harvesting are higher in large, nonvacuolate spherical cells than in small spherical cells, vacuolate cells, cells with markedly nonspherical shapes, or cells that use phycobilins as their light-harvesting complexes. The effects are, at the moment, poorly quantified, although they are susceptible to cost–benefit analysis.

It would seem that planktophytes are amenable to much more cost–benefit analysis than has been carried out to date.

Benthic algae

The benthic algae range in size from unicells of similar dimensions to small planktophytes up to the giant kelps *(Macrocystis)* some 50 m long (Bold and Wynne 1977). This size range involves a decrease in surface area per cubic meter of cell volume from $600 \cdot 10^3$ m^2 for a 5-μm-radius cell to $2 \cdot 10^3$ m^2 for a macrophyte with a mean thickness of a laminar thallus of 1 mm. This decrease in area per unit volume of *only* 300-fold for a 10^4-fold increase in maximum dimension is related to transition from essentially isodiametric unicells to elongated planar macrophytes. For comparison, wet biomass increases from some 0.5 ng for a 5-μm-radius cell to tens of kilograms for a *Macrocystis* plant. The larger of the benthic algae can be either rhizophytes or haptophytes; the smaller (<100 μm) can only be haptophytes (Table 14.1).

The benthic, as opposed to the planophytic, habit has a number of potential advantages in shallow water (i.e., water in which light penetration permits the growth of photolithotrophs attached to, or embedded in, the substratum). One advantage, at least in comparison with planophytes of mixed waters, is the relatively constant position of the organism in the light field, despite tidal fluctuations in the water level in the sea. Another advantage of the benthic habit is improved nutrient supply, at least to macrophytes. Haptophytes often live in regimes of relatively high water velocity that reduce the unstirred-layer thickness and hence increase the potential flux of solutes to the surface of the organism from the bulk medium (Raven 1981b, 1984a). For rhizophytes, which frequently live in regimes of lower water velocity, there is the frequently realized possibility of obtaining nutrients from the sediment *via* rhizoids (or, in higher plants, roots); the sediment receives organic detritus courtesy of gravity and lower water velocities, and microorganisms mineralize this detritus, yielding much higher concentrations of available N and P than is the case for the bulk water medium (Raven 1981b, 1984a).

The significance of the large sizes of benthic algae that live in relatively undisturbed, relatively resource-rich environments is presumably related to competition for light with other organisms at the same trophic level (Givnish 1982) and to the release of propagules into the bulk water, or at least into the boundary layer around the plant, rather than into the much thicker boundary layer over the substrate (Neushul 1972). The longevity of these large benthic forms means that in habitats with seasonal availability of nutrients, the organism's storage capacity can be used to enhance yearly productivity (although storage has an energy cost). Raven (1984a) has discussed the storage capacity for organic C (and thus energy) and N in macroalgae such as *Laminaria.* The storage capacity is sufficient for about one doubling of biomass while maintaining the nutrient concentration above the critical level, below which the growth rate cannot be maintained at its maximum value. In *Laminaria,* this storage of carbon + nutrients helps to maintain growth over a substantial fraction of the year despite seasonal variations in cool temperate environments in inputs of energy (higher photon availability in summer) and nitrogen (greater availability in winter) (Mann 1982; Raven 1984a). However, it must be noted that the large phaeophyte algae are also successful in less seasonally variable warm temperate environments where resource storage is not at such a premium (Raven 1984a).

Special cases of haptophytic macroalgae are the symbioses between "haptozooic" coelenterates, sponges, and bivalves and various microalgae (Table 14.1). Many of these symbioses can be regarded as photolithotrophic as far as energy and carbon are concerned, with a major role for their phagotrophic capacity in supplying particulate N and P in an environment that is poor in dissolved N and P (Raven 1981b). Quantitatively, the most significant of these invertebrate-based haptophytes are the corals, which, like many of the marine symbioses of this type, are essentially confined to waters above 18 °C. Corals and kelps are thus allopatric, because the kelps cannot complete their life cycle above 18 °C.

The capacity to obtain particulate N and P is probably a major determinant of the success of corals in nutrient-poor surface waters in the tropics. Indeed, recent work (Kinsey and Davies 1979) suggests that the absence of corals from upwelling regions may not be due so much to the lower water temperature as to the higher dissolved available N and P levels. While dissolved N and P are readily taken up by symbiotic corals and stimulate the growth of the organic components even at low (by temperate winter surface-water standards) concentrations, formation of the $CaCO_3$ skeleton is *inhibited,* probably as a result of inorganic P in seawater acting as a crystal poison (Kinsey and Davies 1979). Accordingly, it would seem that corals *require* a low level of dissolved inorganic P in their medium at all times; by contrast, kelps need a higher inorganic P (and N) concentration for at least a part of the year, a condition fulfilled in temperate and polar waters in winter and in upwelling areas all year round.

Another possible advantage of the haptophytes whose symbiotic phototrophs are generally acquired anew after each act of sexual reproduction [although there is evidence for transovarian continuity of *Symbiodinium* in some corals (Law and Lewis 1983)] over those whose photosynthetic capacity arose from a symbiosis hundreds if not thousands of years ago (i.e., the endosymbiotic origin of plastids) is that of selection of particular *Symbiodinium* genotypes by a given coral genotype depending on the habitat. An obvious possibility here is adaptation to the photon flux density prevailing at a given depth. Although Dustan (1979) in his transplant experiments, has not conclusively proved that this occurs, Falkowski and Dubinsky (1981) have not definitely disproved this possibility (Chang and Trench 1982).

A final point related to the haptophytic symbioses involves modular growth. The growth of macrophytes, like that of sessile, colonial animals, is typically indeterminate, with the capacity to add new "modules" as genetics and environment determine (Hallé et al. 1978; Harper and Bell 1979; Larwood and Rosen 1979; Niklas 1982; Tomlinson 1982). This principle applies to algae as well as to tracheophytes (Tomlinson 1982). It has been suggested (Harper and Bell 1979; Tomlinson 1982) that the regenerative capacity of modular, nonmotile organisms is analogous to behavioral responses in motile organisms in terms of adaptation to predation (and competition). It is of interest that one of the major microalga-invertebrate symbioses is between *Symbiodinium* (a dinophyte) and the modular, colonial corals. Thus, a phototrophic macroorganism has microphototrophs associated with a "preformed" modular heterotrophic framework. This assertion oversimplifies what is clearly a coevolutionary process, but it does emphasize the difference between the symbiotic and the microalga-to-macroalga progression as a means of producing aquatic macrophytes (Raven 1984a). However, it must be remembered that important phototropic symbioses also occur between *Symbiodinium* and nonmodular invertebrates such as *Tridacna* (Horn 1971; Raven 1981b).

This section examines the various constructional types among benthic algae and attempts to point out cases in which cost–benefit analyses could be employed if more data were available.

The large benthic algae are differentiated into growth, photosynthetic, storage, attachment, and reproductive regions; there are also spatial variations in nutrient absorption capacity (Raven 1984a). Differentiation requires some differential gene activation in different parts of the organism and intraplant transport of nutrients, metabolites, and (probably) morphogens (developmental signals). As we shall see, these two (possibly conflicting) requirements for *separation* of gene products and *exchange* of low-molecular-weight solutes have been met in different ways in different macroalgae; in some cases the exact nature of the mechanism is less than clear (Raven 1976b, 1981b, 1984a).

Even a superficial acquaintance with the macroalgae shows that the production of large differentiated phototrophs at the algal level involves more constructional types than are involved in the homoiohydric sporophytes of vascular plants. The macroalgae are broadly distinguishable into those in which the organism is acellular (giant-celled) and those in which the organism is multicellular (although some organisms are "mixtures," e.g., Charales). Within both the acellular and multicellular organisms can be seen essentially filamentous construction, that is, with a transection of the plant revealing only a main axial cell, with or without branches, with "primary" (produced at the time of branch inception) connections between branches, and the main axis occurring only at the point of inception of the branches. Within the multicellular lines, filaments occur in the simpler members of the Chlorophyta, Phaeophyta, and Rhodophyta; acellular filaments are found only in the Chlorophyta and Tribophyta (formerly Xanthophyta).

More complex modes of construction include pseudoparenchymatous and parenchymatous structures. The pseudoparenchymatous mode of construction is essentially an elaboration of branched filaments (i.e., involving a *surface* subdivision of the thallus during growth) (Graham 1982). Pseudoparenchymatous construction is found in all of the more complex Rhodophyta and many of the larger Phaeophyta in the "cellular" lines, and in many of the ulvaphycean Chlorophyta in the "acellular" lines. Parenchymatous growth involves *internal* subdivision as well as surface subdivision of the plant body into cells (Graham 1982) and is mainly the province of the Phaeophyta among the algae, but it also occurs in the Chlorophyta; this type of construction is, of course, limited to cellular organisms. The sporophytes of vascular plants are also (mainly) parenchymatous in construction, as will be discussed in a later section.

It has been suggested that the pseudoparenchymatous mode of construction has two main deficiencies (e.g., Corner 1964), one related to cohesion within the plant body, the other to communication and transport within the plant body. The *external* mode of ramification of the thallus can, it is argued, lead to the apposition within the thallus of cells (in multicellular organisms) or parts of a cell (in acellular organisms) whose walls do not share covalent bonds and whose cytoplasm does not have intraplasmalemma links (plasmodesmata in multicellular organisms) established during growth. Any such structural and communication links between adjacent cells must be formed secondarily; the secondary pit connections of many Rhodophyta are examples of (probable) communication links formed after the spatial relationships of cells within the thallus have been established. By contrast, the three-dimensional continuity of wall bonding and of plasmodesmatal (symplastic) connections in a parenchymatous structure is laid down at the time of cell division (i.e., during the primary morphogenetic events). However, it is difficult to find good examples of deficiencies of structure or communication in pseudoparenchymatous organisms relative to parenchymatous organisms from the same habitat and with similar external morphology. The "superiority" of parenchymatous construction can, apparently, be demonstrated only by pointing to the fact that the largest and most differentiated algae, as well as all homoiohydric vascular plant sporophytes, are parenchymatous.

The various modes of construction are not mutually exclusive. Thus, Phaeophytes in the Fucales and Laminariales have their primarily parenchymatous thalli "stuffed" with filamentous, pseudoparenchymatous medullary tissue. Further, many multicellular plants have coenocytic, giant-celled portions recalling the acellular mode of construction. Examples of giant cells within a multicellular plant body include the internodal cells of the Characeae (Chlorophyta: Charophyceae), the giant cells of *Griffithsia* (Rhodophyta) and the "solenocytes" involved in long-distance transport in *Saccorhiza* (Phaeophyta) (Raven 1976b, 1984a; Emerson, Buggeln, and Bal 1982). The corticate species of *Chara* (Characeae), and *Saccorhiza,* thus include parenchymatous and pseudoparenchymatous construction, with small and giant cells, in the same thallus. The likely functional significance of the giant-celled habit will be discussed later.

The location of elongation growth in the thallus is also a variable at the algae level of organization. Growth is uniformly *apical* in algal rhizoids, regardless of whether the algae are rhizophytes or haptophytes; in the latter case the rhizoids are functioning as a holdfast (Table 14.1). Apical growth in rhizoids can be rationalized in terms of adhesion of the rhizoids to the substrate in their attachment role, with only the apex able to grow in length without leading to lateral buckling. The parts of the plant that are in the bulk water phase are free from this restriction. They are also free from a structural restriction on intercalary growth (i.e., growth in a defined region other than the apex) in terrestrial plants; the similar densities of water and of submerged plants means that mechanical support in a gravitational field is less of a problem for an extending region, which is necessarily plastic, than is the case for plant parts in air (Corner 1964). In the nonrhizoidal parts of differentiated macroalgae, growth is essentially apical in Chlorophyta, Rhodophyta, and many Phaeophyta, but intercalary in some Phaeophyta (e.g., the Laminariales). In most cases, some extension occurs basipetal to an apical growth region and acropetal to an intercalary growth region.

Raven (1976b, 1981b, 1984a) and Raven et al. (1979) have attempted to interpret the acellular/multicellular distinction in macroalgae in terms of intraplant transport processes. The giant-celled condition permits long-distance transport of solutes by mass flow in streaming cytoplasm (driven by the actomyosin ATPase system), with no (acellular Ulvaphyceae, Tribophyceae) or few (characean, charaphycean) interruptions in the form of plasmodesmata. In small-celled multicellular algae, the occurrence of frequent arrays of plasmodesmata in the transport pathway means that symplastic transport cannot provide an adequate flux of metabolites and nutrients when the distances involved exceed a few millimeters or tens of millimeters because of the concentration decrement that occurs at each plasmodesmatal array, providing the driving force for the diffusive symplastic flux.

The limitations on symplastic transport involving *diffusion* through plasmodesmata can be illustrated by reference to a cylindrical organism with symplastic axial flux of solute of 100 μmol s^{-1} per square meter cross-sectional area of the organism. Data from Robards and Clarkson (1976) suggest that with $2 \cdot 10^4$ plasmodesmatal arrays per meter of path length, and assuming perfect mixing in the path between plasmodesmatal arrays, the flux of 100 μmol m^{-2} s^{-1} requires a concentration drop of at least 40 mol m^{-3} for each millimeter of path length. The ultimate constraint on this concentration drop is the total osmolarity of the symplast. Even in marine algae, a concentration difference for a single solute in excess of 400 mol m^{-3} seems unlikely (Raven 1984a), giving a maximum path length of 10 mm if the axial symplastic flux is to be 100 μmol m^{-2} s^{-1}. If the solute is a 6-C product of photosynthesis such as mannitol, the axial flux is 600 μmol C m^{-2} s^{-1}, which is similar to the area-based rate of macroalgae photosynthesis (Raven 1981b, 1984a). Accordingly, the axial symplastic transport over 10 mm can cope with photosynthate only from a surface area of plant equal to the cross-sectional area of the organism, with no capacity for transport of photosynthate from lateral photosynthetic areas.

In three orders (Fucales, Laminariales, Desmarestiales) of the Phaeophyta, the larger representatives have specialized phloem-like tissues in which mass flow of solution through enlarged plasmodesmata (sieve pores) can occur. At least in the Laminariales, such transport occurs over distances of meters (Lobban 1978). The "streaming" option, like the "phloem" option, permits the long-distance transport implicit in differentiation, but provides (at least in acellular plants) no *obvious* means of localizing the products of differential gene activation, which is presumably a prerequisite of differentiation. Acellular algae, and algae with a major structural contribution from giant cells, do not produce thalli as large as do the largest of the macroalgae made of small cells (Bold and Wynne 1977; Raven 1981b, 1984a). Despite possible structural deficiencies relative to small-celled macrophytes, the acellular and other coenocytic forms can occupy environments characterized by quite high energies of water movements (Raven 1981b, 1984a; Raven et al. 1979).

An important distinction (Table 14.1) is that between rhizophytes and haptophytes. *All* Rhodophyta and Phaeophyta are haptophytes, as are most chlorophycean and many charophycean and ulvaphycean members of the Chlorophyta. The classes Charophyceae and Ulvaphyceae in the Chlorophyta, as well as the Tribophyta, have significant fractions of rhizophyte members. Haptophytes take up photons, inorganic carbon, and other nutrients over the whole plant surface. Rhizophytes can absorb photons only in their nonrhizoid parts, and most inorganic carbon is taken up by the nonrhizoid "shoot." The rhizoids in the relatively nutrient-rich sediments (freshwater Characeae in the Charophyceae, and marine ulvaphyceans such as *Caulerpa* and some species of *Halimeda*) may have a major role in the uptake of N and P sources (Raven 1981b, 1984a). The shoot of the rhizophyte may accordingly specialize in photosynthetic acquisition of energy and carbon, whereas the haptophyte shoot is also involved in acquisition of N, P, Fe, and so forth. That the morphology of the haptophyte shoot that is appropriate for photosynthesis may not always be optimal for acquisition of N, P, Fe, and so forth, is seen in the production of colorless hairs on the thalli of many haptophytes. The extent of hair production is enhanced when the supplies of N, P, Fe, and so forth, are low relative to those of photons or inorganic carbon, either because of low concentrations in the bulk medium or because of low velocities of water flow over the thallus (Raven 1981b, 1984a). Hairs can project through the bulk unstirred layer around the thallus and thus enhance the availability of nutrients to the plant as a whole, provided concentrative transport of the nutrients occurs at the hair plasmalemma (Raven 1984c).

Large macroalgae in environments that provide sufficient nutrients produce a closed canopy with several millimoles of photosynthetic pigment per square meter of habitat and a thallus-area index (total thallus area) of 10–50 (Raven 1981b, 1984a). The pigment content is sufficient to absorb 0.95–0.99 of the incident photons and is at least as high as that found in many terrestrial communities (Neushul 1971a, 1971b; Raven 1981b, 1984a). The thallus-area index of 20 – 50 seems to be high relative to that found in many terrestrial vascular plant communities (e.g., Schulze 1982), but it must be borne in mind that the *internal* amplification of the area available for gas exchange in terrestrial tracheophyte leaves (e.g., Nobel 1977) should be considered in comparing terrestrial and aquatic plants.

The rationale for this approach lies in the general finding that rates of light-saturated photosynthesis in both terrestrial and aquatic plants are restricted by transport processes in the aqueous phase plus subsequent biochemical reactions. This assertion must be true for aquatic plants; in C_3 vascular land plants it is clear that at least two-thirds of the rate limitation is associated with reactions other than diffusion in the gas phase (Raven 1970; Smith and Walker 1980; Farquhar, O'Leary, and Berry 1982; Farquhar and von Caemmerer 1982; Raven et al. 1982). Quantifications of this aqueous-phase transport limitation plus biochemical limitation, and comparisons between plants, are best achieved by reference to the area of plant surface across which the aqueous-phase transport to the carboxylase occurs (Raven 1970, 1984a).

When comparisons between plants are carried out in terms of the area across which aqueous-phase transport takes place (Table 14.7), it is found that the thallus-area index of submerged algae and of submerged or terrestrial mosses, the leaf-area index of submerged aquatic vascular plants, and the "mesophyll-cell-area index" of terrestrial vascular plants come more closely into alignment (Nobel 1977; Proctor 1979; Harvey 1980; Raven 1981b, 1984a; Schulze 1982).

The extent to which the high pigment content of submerged plant communities is related to the lower spatial stability of light-absorbing elements in most plants of aquatic environments, because of the effects of water movements on compliant plants, is not clear. Horn (1971) [cf. Lobban (1978) and Raven (1981b)] pointed out that *rigid* corals are perhaps the nearest aquatic phototroph analogues to trees as far as spatial fixity of light-absorbing elements is concerned. However, the areal pigment density of corals is similar to that of most other benthic plants (Raven 1981b).

Closed canopies of large, dominant algae can be contrasted with the niches occupied by the smaller algae. Ephermeral or opportunist algae (e.g., the ulvaphycean *Ulva* and *Enteromorpha* and the rhodophytan *Porphyra*) generally have higher specific growth rates [μ in equations (14.1) and 14.2)] than the dominants, shorter mean life spans, and less complex thalli, with little differentiation and (often) no symplastic transport. It is of interest that the microscopic, long-lived *Conchocelis* phase in the life cycle

Table 14.7. *Ratio of area of plant surface over which aqueous-phase transport of inorganic carbon occurs to area of habitat for plants of various life forms from aquatic and terrestrial environments*

Plant	Area (m^2) for aqueous-phase inorganic carbon transport per square meter of habitat	References
Spirulina spp. (freshwater cyanobacterial planktophyte)	480[a,b]	Talling et al. (1973)
Chara hispida (freshwater algal rhizophyte)	41	Raven (1984a)
Caulerpa spp. (marine algal rhizophyte)	16	Raven (1984a)
Laminaria spp. (marine algal haptophyte)	16	Raven (1984a)
Posidonia oceanica (marine angiosperm rhizophyte)	14	Raven (1984a)
Vallisneria denserulata (freshwater angiosperm rhizophyte)	19	Raven (1984a)
Tortula intermedia (terrestrial moss rhizophyte)	12[c]	Proctor (1979)
Mnium hornum (terrestrial moss rhizophyte)	36[c]	Proctor (1979)
Pseudoscleropodium purum (terrestrial moss rhizophyte)	40–50[c]	Proctor (1979)
Mesophytic terrestrial angiosperm rhizophytes	45 (–120)[d]	Nobel (1983)

[a] Computed from measured chlorophyll per square meter of habitat (300 mg), assuming cells are 1-μm-diameter cylinders with a density of 1,000 kg m^{-3} and 2.5 g chlorophyll per kilogram wet weight (Raven 1984a).
[b] Data apply to Ethiopean soda lake.
[c] Assuming that the quoted leaf-area indices refer to *only one side* of the moss leaf.
[d] Assuming that the quoted values of mesophyll cell surface area per unit leaf (one side) apply to a canopy with a leaf-area index of 3 [3 m^2 of leaf area (one side) per square meter of habitat area].

of the genus *Porphyra* is more differentiated than the macroscopic *Porphyra* phase. The extent of resource storage in ephemeral algae appears to be less than in other macroalgae (Raven 1984a), in accord with their short life spans, which may not encompass large variations in the availability of different resources (e.g., photons and nitrogen). Ephemeral algae are common in temporary habitats but cannot be construed as an *obligatory* stage in the colonization of a habitat (Horn 1971; Connell 1972; Connell and Slatyer 1977).

The long-lived "shade" algae growing below canopies of dominant algae or at the lower depth limit of algae growth are often of small stature

(Neushul 1971a, 1971b; Foster 1975a, 1975b). Their longevity and lack of resources for replacement of lost ("disturbed") parts make these plants at once "apparent" to biophages and intolerant of biophagy. The extent of mechanical and chemical protection against biophagy in these shade algae is poorly documented. However, the encrusting growth habit of long-lived [many decades (Johansen 1981)] *Lithothamnion* and *Lithophyllum* species (Rhodophyta), often on the undersides of rocks, serves to protect them from biophagy (Steneck 1983; Littler et al. 1983; Littler and Kaukker 1984). Such a growth habit exacerbates resource deprivation, in that plants on the undersides of rocks are shaded, and encrusting plants are within the thick boundary layer of the rock that they are encrusting (Raven 1981b, 1984a).

The macroalgae, together with secondarily aquatic Bryophyta and Tracheophyta, can be contrasted with the terrestrial Bryophyta and, especially, the homoiohydric Tracheophyta. The aquatic environment imposes different structural requirements, and alters nutrient and photon availability, relative to the terrestrial environment. The structural requirements of aquatic plants as compared with terrestrial plants run to a need for tension-resistant elements rather than compression-resistant elements because of the lower density differences between plant and water than between plant and air and the larger drag forces imposed by water than by air moving at their respective "normal" velocities (see Chapter 18). Nutrient availability differences refer especially to carbon dioxide supply, with the D_{CO_2} in air some 10^4 times the D_{CO_2} in aqueous solution. Photon availability is restricted by the intrinsic photon absorption properties of water and of substances dissolved therein. The aquatic environment, however, does not have the insatiable thirst of the unsaturated atmosphere. Whereas saline and hypersaline environments may cause problems for volume, turgor, and osmotic regulation of aquatic plants, they do not extract a price of a quantity of water lost per unit C on energy fixed, as is the case for terrestrial plants (Givnish and Vermeij 1976; Givnish 1979, 1984). The unique combination of features of the sporophytes of terrestrial vascular plants (xylem, intercellular air spaces in parenchymatous tissue, cuticle, stomata) is directly related to the unavoidable loss of water when photons and carbon dioxide are absorbed by a subaerial plant (Raven 1977, 1984a).

However, similarities between aquatic and terrestrial plants must not be ignored. The basic problems of *acquiring* environmental resources and *defending* them from biophagy are very similar. In particular, the advantages of large size in relation to competition for photons and in dispersal of propagules are similar in aquatic and terrestrial plants. The similarities are particularly evident when terrestrial plants (which are all rhizophytes) are compared with rhizophytic aquatic plants (Table 14.1), because both have polarity in the acquisition of photons and (generally) carbon by shoots and of other nutrients predominantly by roots or rhizoids.

Terrestrial vascular plants

These plants are, in some respects, less structurally diverse than are the macroalgae. They are all essentially multicellular, parenchymatous rhizophytic organisms with only small coenocytic incursions [e.g., certain laticifers (Fahn 1979)]. Furthermore, even among the *macro*algae there are substantial differences in degrees of differentiation between ephemeral-ruderal and canopy-dominant strategies; such differences are absent in vascular plants. *All* terrestrial tracheophytes have the same differentiation of conducting tissue and of the components of the "homoiohydric gas-exchange system" (except stomata in the amphibious *Stylites*). Almost all have recognizable roots, stems, and leaves; exceptions include the rootless *Psilotum* and *Tmesipteris* and leafless epiphytic orchids (Table 14.1).

The occurrence of the homoiohydric water-conduction and gas-exchange systems in the terrestrial vascular plants is related to the obligatory water loss to the atmosphere during net carbon dioxide fixation. In many circumstances, regulation of the stomatal aperture in response to light, gas-phase CO_2 and H_2O concentrations, and plant water potential results in maximization of carbon gained per unit water lost (Cowan 1977, 1982; Cowan and Farquhar 1977; Wong et al. 1979). The occurrence of this homoiohydric system is *generally* negatively correlated with tolerance of desiccation of the vegetative sporophytic plant body: If the homoiohydric system fails to keep the plant hydrated in the face of limited soil moisture availability and/or a large evaporative demand from the atmosphere, then the plant dies. This statement applies, as far as present knowledge goes, to 99.99% of dicotyledon, 99.9% of monocotyledon, 100% of gymnosperm, and 99.35% of pteridophyte species, although most of these plants have the capacity to produce a poikilohydric, desiccation-resistant (Table 14.2) phase in the life cycle (Raven 1977, 1984b; Bewley and Krochko 1982. The exceptions (0.65% of pteridophyte, 0.1% of monocotyledon, and 0.01% of dicotyledon species) are the "resurrection plants" (Page 1979; Bewley and Krochko 1982), generally of low stature (<1 m high, the tallest attaining 1.7 m), that share the tolerance of vegetative desiccation and height limitation of many of the poikilohydric nontraceophytes such as algae, lichens, and bryophytes, the tallest of which (≤1 m) is the moss *Dawsonia* from damp forest habitats. Such resurrection plants incur whatever costs are involved in preventing and/or repairing damage caused by desiccation, as well as the costs of producing and maintaining the homoiohydric water-conduction and gas-exchange systems. Energy costing for these two systems is not in a very advanced state (Raven 1984b). We reiterate that the occurrence of the homoiohydric system involves certain minimal energy and resource allocations to this system, with a consequent increase in the minimum photon flux density required for growth of an organism of a particular size.

Other general structural features of terrestrial vascular plants can be profitably compared with those of aquatic macrophytes, such as the interrelated topics of the location of extension growth and the problems of mechanical support of bulky subaerial structures.

Terrestrial vascular plants are characterized by exclusively apical growth of belowground organs, whereas intercalary extension (subject to constraints related to mechanical support of weak growing tissue) is common in aboveground organs. That this is related to the mechanical properties of the medium rather than the morphological (Goethenian) category to which the organ belongs is seen by comparing the exclusively apical growth of subterranean stems with the apical *plus* intercalary growth of many aerial roots (Gill and Tomlinson 1975). The elongation rates for exclusively tip-growing entities (underground stems, roots, root hairs, fungal hyphae) are limited to less than 1 $\mu m\, s^{-1}$ because of the restricted length over which elongation can occur, whereas aerial organs can elongate at more than 1 $\mu m\, s^{-1}$ (Bennett-Clark and Ball 1951; Corner 1966; Trinci 1973; Gill and Tomlinson 1975; Scott Russell 1977). Intercalary elongation of aerial organs has important repercussions for canopy development (Corner 1964; Stebbins 1976).

Mechanical support of the bulk of the body for the larger terrestrial tracheophytes involves dead, lignified cells instead of the hydrostatic skeleton of turgid living cells that suffices in herbs and in many leaves of trees (Wainwright 1970). A *rigid,* compression-resistant structural element is required if erect subaerial plant organs are to reach heights of more than a few hundred millimeters (Wainwright 1970). Plants of such small stature require, if the potential for photosynthesis of their shoots (even if they lack internalization of gas-exchange surfaces, as in the earliest vascular plant *Cooksonia*) is to be realized, a xylem system if their transpiratory water loss is to be made good without recourse to external (ectohydric) water transport. A quantitative argument concerning the utility of xylem (the endohydric mechanism) in such small plants has been presented by Raven (1977, 1984b).

Wainwright (1970) and Raven (1977, 1984b) have argued that the earliest role for rigid, compression-resistant lignin was as a means of resisting implosion of the water-conducting xylem elements in plants of small stature that did not need lignin for support of the plant against gravitational forces and wind drag. The mechanism of transpiratory water movement is generally held to involve *tension* in the water column in the xylem elements, which results in a *compressive* force within the cell walls to which the water in transit adheres (Pickard 1981). Failure of the xylem system can occur by compressive buckling of the walls of the conducting element or by embolism as a result of failure of *cohesion* within the water column or of its *adhesion* to the walls of the elements.

These requirements for reliability must be set, in terms of selective forces, against the requirements for efficiency of transport. The capital cost of xylem transporting a water flux F $m^3 s^{-1}$ with a given pressure gradient $\Delta\psi$ Pa per meter depends on the radius of the conducting elements (Poiseuille's law) and on the :hickness of their walls. Because F increases as the fourth power of the radius of the conducting elements of $\Delta\psi$ m^{-1} is held constant, the cost of xylem-element synthesis is minimized by making xylem elements of large radius with the thinnest cell walls compatible with avoiding buckling under tension. For a given mean xylem water tension, wall thickness must be proportional to xylem-element radius in order to provide a given safety factor in preventing compressional buckling. These two requirements of maximizing water flux and minimizing the risk of buckling under compression with minimal capital investment are served by large-radius elements. However, failure by embolism is increased at large xylem-element radius as well as by increased tension in the xylem. The requirements of xylem in terms of efficiency of water transport and prevention of buckling and embolism in plants in different environments have been quantified by Givnish and Raven (in preparation).

The lignified walls involved in mechanical support of the plant as a whole are, according to the thesis developed earlier, derived from the implosion-resisting walls of water-conducting elements. In most extant trees, mechanical support is related to secondary thickening – an increase in diameter of a nonelongating stem or root. This apparently permits *inter alia* a plant to have a mechanically near-optimal geometry at all stages of its growth. It also permits replacement of conducting elements of xylem and phloem. The vascular cambium requires, because it increases the circumference of the shoot, provision (*via* a phellogen) for protection of the surface of the shoot from water loss and biophage attack. However, we note that many successful extant trees have very limited secondary thickening or, in most palms (Arecaceae), none at all. The columnar stems of palms may be mechanically less efficient throughout the life of the plant (i.e., an overprovision of strength in the short stout trunk of the young palm), although increased cell size and degree of lignification with age may help here (T. J. Givnish, personal communication). Furthermore, there is no apparent provision for replacement of conducting elements of both the cavitation-prone xylem and the enucleate sieve elements of the phloem. Nevertheless, palms can live for centuries and attain heights of 60–70 m (Corner 1966; Tomlinson 1979). These achievements are substantial in comparison with the longevities (few in excess of 1,000 years) and heights (maximum of ~ 100 m) of dicotyledonous and coniferous trees with "conventional" secondary thickening.

In terms of the energetics of tree construction, an empirical approach is clearly possible in that meaningful estimates can be made for the costs of lignin and polysaccharide synthesis (Penning de Vries et al. 1974; Chung and Barnes 1977; Lawton 1984) and the costs of synthesis and maintenance of a living cell in a given tree. However, prediction of optimal tree forms is a less than exact science (Borchert and Slade 1981; Honda et al. 1981; King 1981), so that the minimum energy cost of supporting (mechanically, and with plumbing) a given area of leaves in a monolayer or multilayer (Horn 1971) is not easy. Particularly, it is not possible to analyze the relative energetics of different tree life forms; 23 are recognized by Hallé et al. (1978). It would *seem* that trees 80 m high (*Pseudotsuga menziesii*) are living on a very tight energy budget, with plant respiration taking over 90% of yearly gross photosynthesis (Grier and Logan 1977; Farnum et al. 1983). However, it is significant that estimates of gross primary productivity of more than 200 tons dry weight per hectare per year (Grier and Logan 1977) imply, with 50% C in dry matter and a quantum requirement for gross photosynthesis of 10 mol photon per mole C, a mean *absorbed* photon flux density of 525 μmol photon m^{-2} s^{-1} over the entire (daylight) year. Incomplete absorption, or lower quantum yield, implies a higher mean *incident* photon flux density. It would appear that the quoted gross productivity is barely possible at the latitude (44° N) of growth of the *P. menziesii* community. It is not certain that the high computed dark respiration of Grier and Logan (1977) is needed in terms of growth and maintenance respiration, granted the measured net productivity and living biomass (Kirk 1975a; Farnum et al. 1983).

We have already considered the importance of self-pruning of large photolithotrophs in relation to optimal distribution of photosynthetic structures and the loss of photosynthetic structures and their supporting members whose contribution to net carbon and energy balance is negative, or, if positive, does not selectively justify the allocation of recyclable N, P, K, and C to them. Addicott (1982) has produced a scholarly review of abscission and related topics.

Our view of the tree in this section accordingly is that of a demand for height to get light in competition with other organisms of similar life form. Increased height involves complications of structural support, of water supply to high leaves, and of phloem transport. It is not currently possible to produce even an approximate costing of the energetic factors that might contribute to limiting plant height to not much in excess of 100 m. Schulze (1982) discussed reasons for the inability of trees to dominate vegetation, in many areas of natural vegetation, and presented well-argued accounts of the ''aridity'' and "altitude" treelines. In stable habitats, we find shrubs and herbaceous perennials replacing trees as the dominants as water becomes limiting. *Energetically* important is the longevity (and the cost of synthesis of structures and catalysts that permit a given rate of C fixation and higher constructional costs) of photosynthetic organs, with evergreens generally having lower photosynthetic capacity, per unit leaf area, but longer leaf life, than deciduous plants. Evergreens may be favored in nutrient-poor environments, and deciduous plants favored in environments with a hot dry season during which even a thick waxy cuticle and tightly closed stomata might permit too much water loss (Chabot and Hicks 1982; Cowan 1982).

Turning to the plants that can grow beneath closed canopies, we have already seen that the photons that the large dominant plants find it selectively disadvantageous to harvest can provide a photon flux density lower in the canopy that can support net growth of tracheophytes of lower stature, as well as bryophytes and algae. These photons are the resource most limiting to the growth of these plants; the tracheophytes and bryophytes are generally perennial.

Finally, we come to annuals. We have already noted a significant difference between algae and tracheophytes, in that annual-ephemeral benthic algae commonly have a qualitatively simpler structure than dominant algae. This is not the case for ephemeral terrestrial tracheophytes in comparison with dominant terrestrial tracheophytes. We note (Table 14.1) that most ephemeral tracheophytes are angiosperms. They occur in disturbed habitats and are generally replaced (unless another disturbance event intervenes) by slower-growing, longer-lived plants.

Evolution of life forms: the fossil record

The preceding sections of this chapter have dealt with the structure and function of extant plants. The remainder of the chapter deals

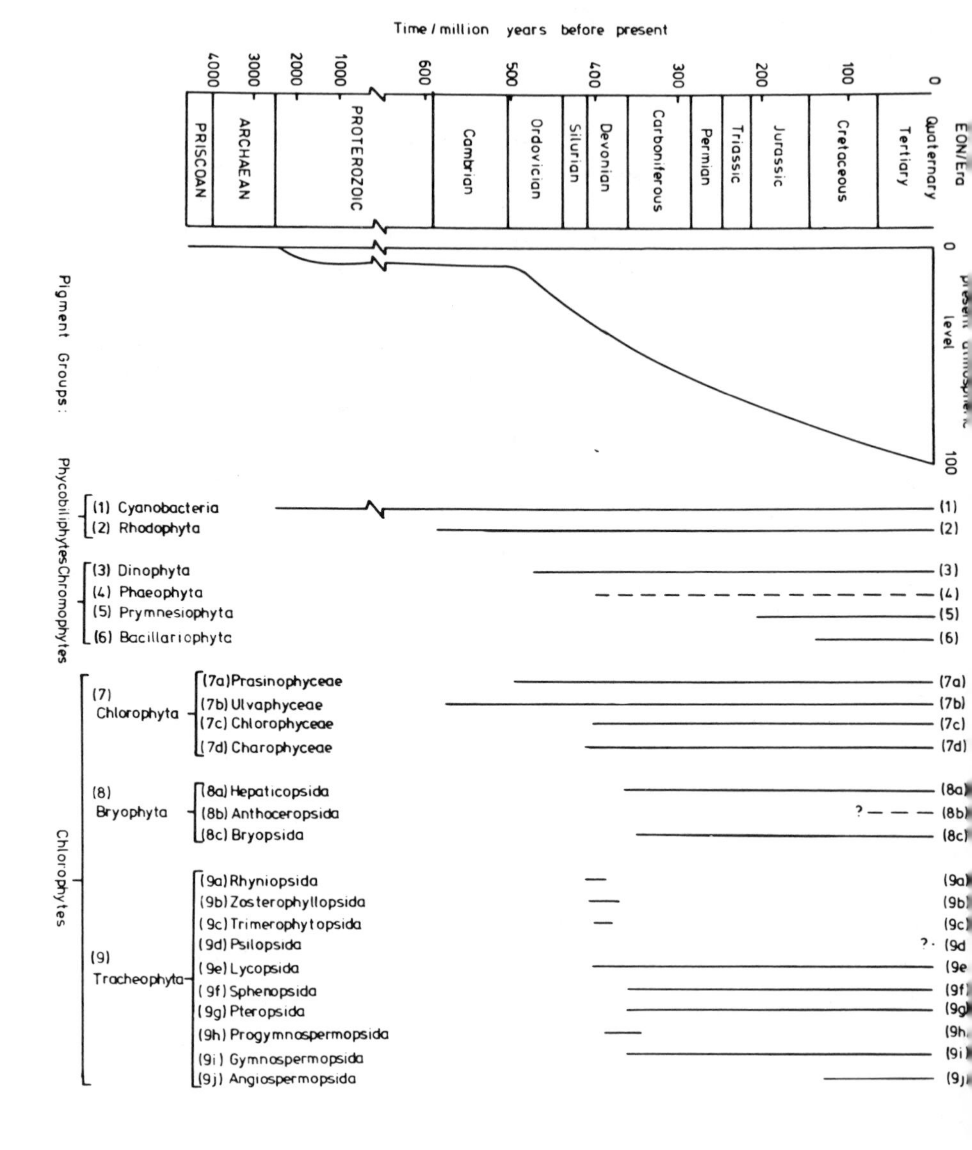

Figure 14.1. Occurrences of major taxa of phototrophs, and the O_2 content of the atmosphere, as functions of time. Data on fossil occurrences of major taxa, and their taxonomy, from Harland et al. (1967), Emberger (1969), Knoll and Rothwell (1981), Miller (1982), and Taylor (1981); O_2 content of the paleoatmosphere from Hart (1978) (on the understanding that many other interpretations are possible!). It is probable that eukaryotic phototrophs, of unknown affinities, occurred in the Proterozoic.

with what the fossil record can tell us about the way in which life forms of plants evolved and their relationships to environments at various times and to the constraints perceived for extant plants in the preceding sections.

Accepting that the presence of significant amounts of free O_2 on earth is a result of photosynthetic activity, the earliest O_2-evolving organisms occurred 2 – 3 billion years ago, as indicated by "banded-iron" deposits. The organisms responsible for this O_2 evolution were cyanobacteria, occurring as planktophytes, or as benthic haptophytes in the case of stromatolite-formers (Margulis 1981). Because much of the O_2 produced by these marine organisms would have been used in inorganic oxidations (e.g., $Fe^{2+} \rightarrow Fe^{3+}$) and in respiratory oxidation of photosynthate, the level of free O_2 in the atmosphere would have been low (Figure 14.1). Correspondingly, the O_3 level, and thus the capacity to absorb ultraviolet (UV)

Caption to Figure 14.1 (*cont.*). The distribution of the major taxa mentioned among the planktophytes, haptophytes, and rhizophytes is as follows:

	Represented as		
Taxon	Plankto-phyte	Hapto-phyte	Rhizo-phyte
(1) Cyanobacteria	+	+	–
(2) Rhodophyta	–	+	–
(3) Dinophyta	+	+	–
(4) Phaeophyta	–	+	–
(5) Prymnesiophyta	+	+	–
(6) Bacillariophyta	+	+	–
(7) Chlorophyta			
(7a) Prasinophyceae	+	+	–
(7b) Ulvaphyceae	–	+	+
(7c) Chlorophyceae	+	+	+
(7d) Charophyceae	–	+	+
(8) Bryophyta			
(8a) Hepaticopsida	–	+	+
(8b) Anthoceropsida	–	–	+
(8c) Bryopsida	–	+	+
(9) Tracheophyta			
(9a) Rhyniopsida	–	–	+
(9b) Zosterophyllopsida	–	–	+
(9c) Trimerophytopsida	–	–	+
(9d) Psilopsida	–	–	+
(9e) Lycopsida	–	–	+
(9f) Sphenopsida	–	–	+
(9g) Pteropsida	–	–	+
(9h) Progymnospermopsida	–	–	+
(9j) Gymnospermopsida	–	–	+
(9k) Angiospermopsida	–	+	+

radiation of wavelengths 250–400 nm, would also have been much lower than at present (Caldwell 1982; Stolarski 1982). Because the sun's overall luminosity probably was not much lower $2 \cdot 10^9$ years ago than it is now (Newman 1980; Wigley and Brimblecombe 1981), and the UV component probably was greater than it is now (Canuto et al. 1982), the UV photon flux density at the surface of the earth (or sea) would have been very substantially greater than it is now.

The discussion by Calkins and Thódardóttir (1982) and Worrest (1982) on the relative absorption of UV radiation (wavelength 280–320 nm, absorbed by, and damaging to, proteins and nucleic acids) and of photosynthetically active (400–700 nm) radiation by marine waters suggest that a higher ratio of UV to photosynthetically active radiation at the sea surface $2 \cdot 10^9$ years ago may have forced phototrophs to live in habitats with a low photon flux density of photosynthetically active radiation if they were not to be exposed to intolerable UV levels. The attenuation of radiation could have occurred in planktophytes deep (> 10 m) in a stratified water column, where gas vacuoles may have been useful in maintaining their position (Calkins and Thórdardóttir 1982; Worrest 1982), or in haptophytes within sediments or stromatolites; see Smith (1982) for data on the wavelength dependence of attenuation of radiation by soils.

This indication that early O_2-evolvers were "shade plants" has some important repercussions (Anderson 1983; Richardson et al. 1983). One is that genotypic shade adaptation probably was the primitive condition for O_2-evolvers, with sun adaptation a derived condition that could be capitalized on only when the UV/visible ratio of photon flux densities had decreased sufficiently because of O_2 (and hence O_3) accumulation. It is likely, however, that many extant shade phototrophs are evolutionarily derived from sun phototrophs (see the discussion of terrestrial vascular plants that follows) (Richardson et al. 1983). Another important repercussion is that limitation of early O_2-evolvers to habitats of low photon flux density would have limited their potential productivity, which in turn would have imposed an upper limit on the rate of O_2 buildup in the hydrosphere and atmosphere; this upper limit would be further reduced by O_2 consumption in inorganic oxidations and in heterotrophic reoxidation of photosynthetically reduced carbon. This negative feedback of low atmospheric O_2 levels on the possibility of generating *more* O_2 also has repercussions for the evolution of more complex constructional types of phototrophs, and thus of many life forms (Table 14.1). The argument here is that put forward earlier: that larger organisms with more nonphotosynthetic tissue require higher photon flux densities for growth. Not until productivity levels increased would there be selection for tall forms with marked competitive advantages in productive, closed benthic communities (cf. Givnish 1982). Furthermore, unless a gas-phase distribution system for O_2 was present in the tissues of such plants, their innermost living tissues could not carry out O_2-dependent respiration at low external O_2 concentrations. Finally, at high UV/visible ratios of photon flux densities, the limitation on the maximum photon flux density for growth probably would have been determined by UV flux density rather than by visible photon flux density. Accordingly, for perhaps one-half to one-third of the total time for which O_2-evolvers have existed, there would have been little selection pressure related to the countering of photoinhibition by visible light [see review by Osmond (1981)]. This could have allowed photoinhibition-sensitive processes to have become entrenched in the photosynthetic apparatus, whence they may have been difficult to dislodge when plants were able to occur in high visible photon flux density environments as the ratio of UV to visible photon flux density decreased. There could be an analogy here with Rubisco, whose ubiquitous oxygenase function (Lorimer 1981) presumably was of little selective disadvantage when it was operating at very low O_2 tensions in Precambrian cyanobacteria. The possible involvement of the oxygenase activity of Rubisco in preventing photoinhibition under some circumstances (Osmond 1981) makes the analogy worth further exploration.

The foregoing discussion of problems with the occurrence of macrophytes at low O_2 tensions and high UV fluxes presupposes the occurrence of eukaryotic phototrophs, because prokaryotes are not known to occur as macrophytes. Evidence for the occurrence of eukaryotic microalgae in the Precambrian remains rather weak. The evidence that microfossils represent eukaryotes if they have internal structures that would represent nucleoli or thylakoids, or if they exceed a certain size, can be challenged. The internal-structure evidence is weakened by experiments showing that cyanobacteria subjected to artificial-fossilization procedures in the laboratory can exhibit such structures (Taylor 1981), and the occurrence today of *Prochloron,* an O_2-evolving prokaryote with cells up to 30 μm in diameter (Margulis 1981), negates some of the "big = eukaryote" argument (Larkin and Strohl 1983). Adopting a more positive approach, there is good evidence for unicellular members of the Chlorophyta (*Tasmanites, Botryococus*) from the Ordovician onward, of Dinophyta from the Silurian onward, of Prymnesiophyta (coccolithophlorids, etc.) from the Jurassic onward, and of Bacillariophyceae (diatoms) from the Cretaceous (marine) and Miocene (freshwater) (Taylor 1981) (Figure 14.1).

Thus, the evidence for eukaryotic planktophytes *assignable to extant divisions or classes* prior to the Ordovician is not strong. However, it is clear that by the Silurian, all three of the main pigment types of O_2-evolvers were established. The chlorophyll-*a*/phycobilin group was presumably represented in the oldest cyanobacteria and, as rhodophytan macrophytes, from the Cambrian onward. The chlorophyll-*a*/chlorophyll-*b* group was represented by the Ordovician *Tasmanites* (probably in the class Prasionophyceae of the Chlorophyta), as well as by macrophytic ulvaphyceans (Chlorophyta) from the Cambrian onward, and macrophytic characeans (Charophycease: Chlorophyta) from the Silurian onward. The chlorophyll-*a*/chlorophyll-*c* group was represented by the Dinophyta from the Silurian onward. We note that there is good fossil evidence for macrophytes in the chlorophyll-*a*/phycobilin series and in the chlorophyll-*a*/chlorophyll-*b* series before there is fossil evidence of the occurrence of unicellular *eukaryotes* from these pigment groups. Because there are many more characters available to the paleobotanist in macrophytes than in microphytes, we should not take this to mean that there were *not* eukaryotic unicellular ancestors of the macrophytes.

Turning to the algal macrophytes themselves, most of the fossils that can be assigned to families or orders in extant classes are of calcified algae. In the Rhodophyta, the family Solenoporaceae extended from the Cambrian to the Tertiary; the extant family Corallinaceae is also well represented in the fossil record, with the extant genus *Lithothamnion* first occurring in the Jurassic (Taylor 1981; Johansen 1981). In the Chlorophyta, the (ulvaphycean) receptaculids and cyclocninoids extended from the Ordovician to the Devonian and Silurian, respectively, and the extant Dascyclydales (Ulvaphyceae) were first found in the Cambrian. The (charophycean) Charales extend from the upper Silurian to the present. With the exception of the mainly freshwater Charales, these calcified algae are all marine. Noncalcified examples of the Rhodophyta, Ulvaphyceae, and Charophyceae are only found more rarely; very striking examples of characean noncalcified fossils are the *Palaeonitella* rhizoids in the Lower Devonian Rhynie chert (Edwards and Lyon 1983); this is probably the earliest direct proof of the rhizophyte nature of any macroalga. Thus, marine macroalgae are known to have existed since the Cambrian, and freshwater macroalgae since the Silurian. Whereas extant chlorophytan and rhodophytan marine macroalgae occur worldwide, the marine macroalgae belonging to the chlorophyll-*a*/chlorophyll-*c* group (i.e., the Phaeophyta) are restricted to cooler waters (Van Den Hoek 1982; Gaines and Lubchenko 1982). The macrophototrophs of warmer waters with chlorophyll-*a*/chlorophyll-*c* are the symbiotic hermatypic corals. Corals were certainly present in the early Paleozoid, as was the division of algae (Dinophyta, as hystrichospheres) that provides the symbionts (*Symbiodinium*) for the extant symbiotic corals. However, it is not clear that the earliest corals were symbiotic; the earliest corals with massive exoskeletons, at present indicative of symbiosis, are from the Triassic, with critical evidence for symbiosis dating from the mid-Jurassic (Law and Lewis 1983). As for the dominant algae of present-day cooler rocky shores (i.e., the large Phaeophyta), these are not calcified, and so the relative absence of fossils is perhaps not surprising. Various fossil genera [e.g., *Prototaxites (=Nematothallus), Foerstia (=Protosalvinia),* both from Lower Devonian strata] have been referred to the Phaeophyta, although this assignment is not definite (Taylor 1981).

If we attempt to summarize the fossil evidence on benthic macroalgae growth forms, it would appear that potential canopy-formers have been present in marine habitats since the Cambrian (Elliott 1978; Gaines and Lubchenko 1982; Van Den Hoek 1982) and in freshwater habitats since the Silurian. In the marine habitat, the earliest evidence suggests that the algae were haptophytes [as are most, but not all, extant calcified marine algae (Hillis-Colinvaux 1980)]; scanty evidence for the rhizophytic habit comes from later, noncalcified fossils (Emberger 1969). Evidence for crustose, shade-tolerant, and grazing-resistant algae comes from the fossils of *Lithothamnion* (Jurassic) and related (earlier) genera. Ephemeral algae would not be expected to fossilize well; it is of interest that the perennial *Conchocelis* phase of the ephemerals *Porphyra* and *Bangia* occurred, as *Palaeoconchocelis,* in the Silurian (Campbell 1980).

As with extant rhodophyte and calcified chlorophyte macroalgae, the constructional types represented in the fossil record are multicellular filamentous pseudoparenchymatous (Rhodophyta) and coenocytic filamentous pseudoparenchymatous (Chlorophyta). Parenchymatous construction is seen only in the dubiously phaeophycean *Foerstia.*

For freshwater benthic algae, the major rhizophytes in the fossil record (Silurian onward) and today are the Charales. Other freshwater benthic algae are poorly represented in the fossil record; what records are available are for the Chlorophyta, mainly Chlorophyceae, with charophyceans of smaller stature than the Charales possibly represented by the Devonian *Parka,* which may be analogous to the extant *Coleochaete* (Taylor 1981; Graham 1982). There seems to be no fossil record of common macrophytic haptophytes such as the ulvaphycean *Cladophora* and the rhodophytes *Batrachospermum* and *Lemanea* (Raven et al. 1982).

Paleontological data do not permit us to estimate the extent of canopy cover (thallus-area index) achieved by the canopy-forming macroalgae; what data are available suggest that their habitats were similar to those of extant analogues (Taylor 1981). An important point concerning the rhizophytic algae is the extent of competition from rhizophytic submerged vascular plants. For freshwater rhizophytic algae (Charales), the major competition would have been from submerged angiosperms, which did not arise until the Cretaceous or later (Hutchinson 1975; Raven 1981b). Completely submerged nonangiosperm vascular rhizophytes are today represented by species of the lycopsids *Isoetes* and *Stylites* (Isoetaceae). The fossil record of the Isoetaceae is best documented from the Cretaceous onward, although the family could have been present as early as the Triassic (Taylor 1981); Isoetaceae and Characeae are found growing together today (Spence 1967), although their life forms are dissimilar. Fossils of the Isoetaceae do not seem to indicate the occurrence of stomata, which in *Isoetes* indicates a terrestrial rather than an aquatic habitat; *Stylites* never seems to produce stomata, even when growing with its leaves in air (Keeley et al. 1984).

In the marine habitat, there seem to have been no vascular competitors for rhizophytic algae prior to the seagrasses in the late Cretaceous onward (Den Hartog 1970; Domning 1981).

Competitors of haptophytic freshwater macroalgae are, today, the submerged bryophytes. Bryophytes have existed since the Devonian; it is not clear whether or not they were aquatic. Marine haptophytic algae today have very few seagrass competitors; these would not have appeared before the late Cretaceous at the earliest (Raven 1981b, 1984a).

Turning to a consideration of the *terrestrial* vascular plants from which the *submerged* vascular plants presumably were derived, their origin from algae ancestors continues to perplex paleobotanists. It should be recognized that there probably was a poikilohydric land flora at the algae level of organization for many tens of millions of years prior to the first known *vascular* plant (i.e., vascularized specimens of *Cooksonia*) in the late Silurian (Edwards et al. 1983). This prevascular flora would have had to contend with high UV photon flux densities and hence would have had to live in environments with low UV radiation and low photosynthetically active radiation, as well as low O_2 content of the atmosphere, all of which (as in the aquatic algae considered earlier) would have militated against the production of tall plants. The terrestrial (soil-dwelling) algae would also have had to contend with desiccation; both the desiccation and the UV problems would have been exacerbated for these early soil algae, relative to many of their extant counterparts (Metting 1981), by the absence of a canopy of vascular plants above them.

These considerations of UV photon flux density and low oxygen concentration cannot, however, be used to place an upper limit on the size of the first *terrestrial* ancestors of higher plants. Thus, by the time that the first vascular plants are found (Upper Silurian), the UV photon flux density must have been low enough, relative to the visible photon density, and the oxygen concentration high enough to permit the occurrence of terrestrial macrophytes; there is ample fossil evidence for such nonvascular plants in the Middle and Upper Silurian (Taylor 1981; Miller 1982). However, these nonvascular plants are of uncertain taxonomic affinity; it is clear that the algal ancestors of vascular plants were charophycean algae (Stewart and Mattox 1975; Raven 1977, 1984b). Church (1919), Corner (1964), and Jonker (1981) favor a large "transmigrant" (i.e., a complex plant that had evolved in an aquatic environment). At the other extreme is the suggestion by Stebbins and Hill (1980) that the soil-dwelling, charophycean, ancestor of the vascular plants was unicellular. Other authors (Fritsch 1945; Raven 1977) favor a middle view, with a transmigrant a few millimeters in maximum dimension, such as a charophycean analogue of the extant chlorophycean *Fritschiella,* or an organism like the extant charophycean *Coleochaete* (cf. the Devonian *Parka* and *Pachytheca*) (Banks 1970; Taylor 1981; Graham 1982).

In view of the predominantly parenchymatous construction of the sporophytes of vascular plants, it is clear that the immediate macrophytic ancestor of the vascular plants must have been of parenchymatous construction. We note that pseudoparenchymatous terrestrial phototrophs (i.e., the lichens) are all poikilohydric, although they have the intercellular gas spaces on which the homoiohydric condition in vascular plants is, in part, predicated (Raven 1977). There are also analogues of other important components of homoiohydric plants (cuticle, stomata, xylem) in fungi. The upper surfaces of many lichenized fungal thalli are water-repellent and may also have a low water permeability (Snelgar et al. 1981a). The pores in the surfaces of terrestrial lichens that provide a gaseous pathway from the bulk atmosphere to the internal atmosphere from which the phycobionts obtain their CO_2 (Snelgar et al. 1981b) do not show changes in aperture of the stomatal type (i.e., related to turgor changes in the cells bordering the pore that are brought about by processes other than "passive" changes related to the water status of the plant). However, there seems to be no reason why a pseudoparenchymatous thallus should not be able to produce a stomatal mechanism. Although some nonlichenized nematode-trapping fungi exhibit rapid closure responses of their traps (Cooke 1977), the analogy with a stomatal mechanism is superficial only in terms of the likely mechanism of (irreversible) closure of the nematode trap and the signals to which it responds. The final homoiohydric attribute (i.e., the xylem) has a fungal analogue in the water-conducting apoplastic "vessels" of certain mycorrhizal rhizomorphs (Duddridge et al. 1980), although the extent to which these elements function under tension (as does the xylem) is not clear. These cases, albeit from a range of (often nonlichenized) fungi, show that the major attributes of homoiohydric vascular plants have analogues in pseudoparenchymatous fungal structures.

It is important to note that whereas pseudoparenchymatous construction apparently *could* produce the essential attributes of homoiohydric plants, at least two of these attributes (xylem and stomata) require that the plant have some cross-walls (i.e., it is not an acellular structure). Intercellular gas spaces certainly occur in aquatic acellular photosynthetic pseudoparenchymatous algae (Ramus 1978; Ramus and Rosenberg 1980), and there is no reason that pores connecting these spaces to the outside atmosphere, or a cuticle over the plant surface, could not occur in acellular terrestrial pseudoparenchymatous plants. However, the production of a lysigenous water-conducting system (i.e., one produced by programmed cell death) clearly requires that the organism concerned have *at least* two cells: one whose death produces the "xylem," and another whose continued life is the *raison d'être* of the xylem. A schizogenous "xylem" (*inter*cellular) could be produced in an acellular pseudoparenchymatous plant, but rendering it implosion-proof might prove difficult (Raven 1983). Furthermore, the turgor differences between the guard cells and the surrunding cells that are required for the operation of stomata could not occur unless the guard cells were isolated from the rest of the plant. For these reasons, the acellular poikilohydric algae found on many damp soils (e.g., the tribophycean *Botrydium* and the chlorophycean *Protosiphon*) are *not* potential ancestors of homoiohydric plants (Raven 1977, 1981b).

It is perhaps necessary to reiterate at this point that the size of the transmigrant cannot be deduced from consideration of the selective forces favoring an increase in size. We have already seen that selection in favor of large size, related to competition for light and the release of fluid-dispersed propagules outside the boundary layer present over the substratum, could occur on land as well as under water.

The vascular plants (division Tracheophyta) probably are monophyletic (Knoll and Rothwell 1981). The earliest known [~406 million years ago (m.y.a.)] vascular plant is *Cooksonia* (Rhyniopsida) (Banks 1975, 1980; Chaloner and Sheerin 1979); earlier (Lower Silurian) specimens have not been unequivocally shown to have had xylem (Edwards et al. 1983). *Cooksonia* had erect, dichotomizing axes with terminal sporangia; it had a cuticle, but no stomata or (probably) intercellular air spaces. The next class of the Tracheophyta that is known from the fossil record is the Zosterophyllopsida (e.g., *Zosterophyllum,* 395 m.y.a.), which "introduced" lateral sporangia (Banks 1980). *Zosterophyllum* is of particular importance in the evolution of the homoiohydric condition in that it was the first fossil tracheophyte to have stomata (Lele and Walton 1961; Walton 1964). Although it has not been established (from the compression fossils available) that *Zosterophyllum* had intercellular gas spaces, it is highly likely that such spaces were present (Raven 1977, 1984b). It is thus very likely that the second oldest vascular plant known had the complete homoiohydric apparatus of xylem, cuticle, stomata, and intercellular gas spaces (Raven 1977, 1984b). Raven (1977, 1984b) pointed out that the sequence observed in the fossil record [i.e., pretracheophyte parenchymatous plant with cuticle; cuticularized plant with xylem (e.g., *Cooksonia*); plant with cuticle, xylem, stomata, and (probably) intercellular gas spaces (e.g., *Zosterophyllum*)] is, from the ecophysiological point of view, the most reasonable order of acquisition of these attributes. A major quantitative point in the argument (Appendix 4 of Raven 1984b) is that xylem is needed to cope with transpirational water losses from *Cooksonia*-size plants that lack internalization of gas-exchange surfaces.

The homoiohydric attributes were certainly all present in the vascular plants found in deposits from 370–380 m.y.a. These organisms were in the Rhyniopsida (e.g., *Rhynia*), Lycopsida (possibly derived from Zosterophyllopsida: *Asteroxylon*), and Trimerophytopsida (e.g., *Psilophyton*, probably derived from Rhynopsida). These plants were more complex than *Cooksonia* with its smooth, dichotomizing axes bearing terminal sporangia (Banks 1980; Chaloner and Sheerin 1979). Accordingly, some or all of these Middle Devonian plants exhibited (1) unequal branching, (2) "enations" (e.g., spines) on the axis, (3) microphylls, some bearing adaxial sporangia, on lycopods, (4) rhizomes (horizontal underground axes), or (5) the distinction between sterile and fertile branches in trimeophytes. However, they lacked roots (other than possible adventitious roots). Niklas (1982) pointed out that the various constructional types discernible in these plants can be interpreted in terms of four processes and that "reiteration" (i.e., a modular growth pattern) is an important part of their development. Earlier, Niklas (1978) showed that different morphometric relationships occurred in these plants, with implications for the economy of their construction with respect to mechanical requirements.

It is likely that these plants of the late Lower Devonian (mid-Siegeman-Emsian) could have formed a fairly dense photosynthetic cover of relatively branched organisms up to 1 m high. We are probably dealing here with perennial, evergreen herbs; it is unlikely that their life cycle (both sporophyte and gametophyte generations, e.g., "spore" to "spore") could have been completed within a year (cf. the relatively advanced nature of extant annual pteridophytes, e.g., *Ceratopteris* species) (Stein 1971; Loyal and Chopra 1977). The tendency (implicit in most mechanisms of fossilization) for fossils to come from aquatic habitats means that we do not have much data on how far these early Devonian plants had gone in capitalizing on their homoiohydry in dry environments. Furthermore, we have little knowledge of how the vegetation could have stratified into two parts: a canopy and shade-adapted understory plants.

Secondary thickening, as indicated by the occurrence of secondary xylem (and hence, presumably, of vascular cambium and secondary phloem), first occurred 370–365 m.y.a. (Banks 1980; Chaloner and Sheerin 1979). This innovation (Barghoorn 1964) provided the means by which most fossil plants and extant land plants achieved substantial height (tens of meters), although tree ferns and palms have achieved large structure without recourse to either vascular cambium or phellogen (cork cambium). We note that periderm has been observed, as a wound response, in *Psilophyton,* which lacked secondary thickening (Banks 1981). Periderm became "institutionalized" as a concomitant of the more abundant secondary thickening, with increased axis circumference, some 365–359 m.y.a. The abundant secondary xylem, with readily observable secondary phloem, and periderm, opened the way to the arborescent habit, which, indeed, characterized many plants that were extant 365–359 m.y.a. and in later strata. These plants of 370–359 m.y.a. belonged to the Lypopsida, Sphenopsida, Pteropsida, and Progymnospermopsida (Banks 1980; Knoll and Rothwell 1981); the last three probably were derived from the trimerophytes and have been characterized as "megaphyllous," a term that found little favor in a recent critical review of stelar morphology, at least insofar as it relates to the occurrence of "leaf gaps" (Beck et al. 1982). At all events, the leaves of these classes probably were modified planate, webbed derivatives of some of the branch systems found in plants living 370–359 m.y.a., whereas those of lycopsids probably were never more complex than an unbranched branch of limited (determinate) growth.

With *Archaeopteris* (359–349 m.y.a.), we find a plant of truly tree proportions, perhaps 30 m high. This progymnospermopsid had abundant secondary xylem, acting in mechanical support [cf. the ~0.1 m possible with a "hydrostatic" skeleton (Wainwright 1970)] as well as in water and nutrient conduction; it had a well-developed primary root system, almost completely planate megaphyllous leaves, and marked heterospory (but not quite a gymnosperm). Abscission of lower megaphylls probably was important in controlling the morphology of the canopy of the plants (Addicott 1982). Abscission in the vascular plants may have been important in dehiscence of sporangia (370 m.y.a. the earliest vascular plants had no obvious method of spore release other than herbivore damage to the sporangial wall) before it became significant in vegetative shaping of the plants and, even later, in seed abscission when the seed became the unit of dispersal. Advanced heterospory and the arborescent habit, with vegetative abscission and secondary thickening, occurred in lycopsids and sphenopsids as well as in progymnospermopsids.

At the end of the Devonian (349 m.y.a.) and onward, we find true seeds and the true gymnospermous habit in the Gymnospermopsida (derived from the Progymnospermopsida). The end of the Devonian is a convenient point at which to take stock of life forms that had been produced in the first 60 million years of tracheophyte evolution (i.e., about one-seventh of the total time tracheophytes had been in existence). Four evolutionary series had attained the status of trees: the sphenopsids, the lycopsids, the gymnosperms (and progymnospermopsids), and the pteropsids. The latter were somewhat isolated in their relative lack of secondary thickening and of heterospory. Even within the progymnosperm-gymnosperm line there were considerable differences in the extent to which the secondary xylem was a major supporting element; the secondary xylem apparently was never as important in arborescent lycopods and sphenopsids as it was in the pyconoxylic (conifer-like secondary xylem anatomy) members of the progymnosperms (e.g., *Archaeopteris*) and gymnosperms. Periderm and persistent leaf bases were important supporting/protective elements to different extents in the different lines. These Upper Devonian trees were generally less branched than are many modern trees; this may have restricted the efficiency of light capture (per unit energy investment in the tree's supporting tissue); an effective leaf (or leaflet) mosaic would have been less easy to devise, even granted the capacity to abscise photosynthetic structures at a variety of levels (leaflets, whole compound leaves, simple leaves, branches; we note that the distinction between these morphological entities was only just becoming definite in the late Devonian).

At all events, these arborescent entities from four phyletic lines were probably capable of forming a closed canopy at some meters or tens of meters elevation. Below this canopy would be juveniles of the canopy species, and smaller vascular and nonvascular plants. To the extent that smaller plants have a lower maintenance requirement per unit light-harvesting apparatus than do large plants, they have lower light compensation points. For reasons mentioned earlier, it is likely that homoiohydric plants have a higher light compensation point for a plant of a given biomass than do poikilohydric plants. This fact is capitalized on in extant homosporous pteridophytes with photolithotrophic gametophytes, where the young sporophyte is, initially, at least partly parasitic on the gametophyte. Here the surplus photosynthate of a poikilohydric phototroph (the gametophyte) supports the early growth of the homoiohydric sporophyte. This was probably the situation in Devonian pteropsids. Other pteridophyte gametophytes (e.g., *Lycopodium*) are chemoorganotrophic, using soil organic C via parasitism on their mycorrhizal symbionts. Here the energy substrates that the developing sporophyte obtains from the gametophyte are obtained not by photolithotrophic activity on the part of the gametophyte but from organic material in litter provided, ultimately, by previous photolithotrophic activity by other members of the plant community. These two options for energy supply to the developing sporophyte *via* sources other than the spore food reserves are forced on the homosporous pteridophyte by virtue of small spore size. The large size of megaspores (heterospory was a fairly early innovation) permits heterosporous parent plants to provide more food for female gametophytes, and thus juvenile sporophytes. The culmination of this trend is for the fertilized megaspore (embryo) in its megasporangium (i.e., the *seed*) to be the unit of dispersal; here, in gymnosperms, the placental state has been achieved (Harper et al. 1970), in that commitment of most of the food reserves from the parent sporophyte to the new sporophyte generation takes place after fertilization. This admits of less waste than the provisioning of a megaspore that is capable of producing fertilizable ova in gametophyte archegonia only after fertilization. We note that there is a placental relationship between fertilized ova and gametophytes in bryophytes, pteridophytes, rhodophytes, and *Coleochaete* (Graham and Wilcox 1983); only in spermatophytes does the placental relationship extend to a sporophyte-sporophyte transfer of nutrients. Finally, in many orchids, very small seeds are chemoorganotrophic *via* mycotrophy for their early growth, recalling the situation with *Lycopodium* prothalli. It is likely that the two generations of a tree fern, between them, can efficiently use almost the entire range of photon flux densities that terrestrial plants can use for growth (e.g., Friend 1975; cf. Keddy 1981). Certainly a Devonian "mycotrophic seedling" possibility cannot be ruled out, although the large sizes (Harper et al. 1970) of many early seeds speak against this.

The sizes of the largest Upper Devonian terrestrial plants (Chaloner and Sheerin 1979) suggest that the atmospheric oxygen concentration could not have been much lower than present levels; otherwise, respiration in bulky boles and roots would have been difficult, bearing in mind the probable maintenance respiration rate of living cells in these tissues and the length of the diffusion path for oxygen. This is consistent with evidence (fossil charcoal) for fires in the Devonian (Cope and Chaloner 1980, 1981; Clark and Russell 1981). A further point related to atmospheric oxygen levels concerns the synthesis and breakdown of lignin. Synthesis and biological breakdown of lignin both involve enzyme systems with relatively low oxygen affinities (Gross 1980; Dagley 1976a, 1976b); although these low affinities could possibly have been offset by increased amounts of enzymes, such a course would not have been very feasible for lignin-degrading organisms living on a substrate of very low N content (Mattson 1980). D. H. Lewis (personal communication) has related lignification and atmospheric oxygen tensions to the evolution of basidiomycetes, the major group of wood-rotting fungi in the Devonian-Carboniferous. We note that burning, an alternative to microbial activity for lignin degradation, also requires relatively high oxygen tensions.

The Carboniferous period (~345–280 m.y.a.) saw a further development of arborescent lycopsids, sphenopsids, pteropsids, and gymnosperms. These trees represented a wider phyletic range than do extant trees, in that the present-day angiosperms are closer to the gymnosperms than were the arborescent lycopsids or sphenopsids. It is, then, perhaps not surprising that a relatively large fraction of the 23 models that Hallé and associates (1978) recognize as describing the range of constructional types in extant trees were present in the Carboniferous closed tropical forest. Tomlinson (1982) pointed out that all 23 of the models are represented by extant dicotyledonous trees; 8 of the 23 are found in extant monocotyledonous trees, 5 of the 23 in present-day conifers, and 2 of the 23 in the living cycads. The diversity of models is substantially greater in tropical than in temperate communities. Hallé et al. (1978, pp. 263–268) analyzed Carboniferous trees and found that it is likely that 9 of the 23 extant models are represented in the Carboniferous tropical forests. Although some dubiety attaches to this attribution of 9 of the models to Carboniferous trees, because ontogenetic data are needed for firm assignment of some of the models, it is of interest that some 0.4 of the extant models probably were present in the Carboniferous forest. It is likely that few additional models occurred until the origin of Angiospermopsida in the Cretaceous.

We accordingly envisage a Carboniferous forest with a substantial range of constructional types of trees belonging to four main classes. These trees were large, up to 40 m high, about half the height of the tallest extant trees (100 m, *Sequoia, Sequoiadendron, Eucalyptus*), probably limited by the structural, transport, and maintenance considerations discussed earlier. Phillips and Dimichelle (1981) analyzed the quantitative composition of the Carboniferous forest and suggested that an arborescent understory occurred in many places in the forest. Furthermore, there is evidence for the occurrence of herbaceous lycopsids and sphenopsids allied to the ex-

Table 14.8. *Qualitative and quantitative data from the fossil record relevant to the functioning of terrestrial tracheophytes*

Datum	Potential utility in interpreting function of Paleozoic plants	Conclusions as to functioning of Paleozoic plants	References
Size of photosynthetic organ		?	
Internal anatomy of photosynthetic organ	Qualitative: indicating which pathway of photosynthesis (C_3, C_4, CAM) was operative	Uniformly C_3	Raven (1984b)
	Quantitative: ratio of internal cell area exposed to intercellular gas spaces to external leaf area; indicator of maximum rate of photosynthesis that can be achieved on a leaf-area basis at a given intercellular-space CO_2 concentration (based on maximum rates attained in extant plants on the basis of internal exposed area)	Within the same range as extant terrestrial plants	Analysis of published micrographs (Raven 1977)
Cuticular thickness	Gas permeability of plant surface when stomata are closed; however, main barrier to movement of H_2O, CO_2, and O_2 is the wax layer, which does not fossilize well	The cuticular carrier of the wax layer was present, but no quantitative deduction as to permeability of the cuticle plus (putative) wax layer is possible	Raven (1984b)
Stomatal frequency and dimensions	Gas permeability of plant surface when stomata are open; hence, maximum rate of CO_2 influx or H_2O efflux with specified concentration differences	Stomatal size within extant range; frequency at or below bottom end of extant range in earliest stoma-bearing plants; within extant range for later (Carboniferous) plants; maximum gas permeability of plant surface lower than ex-	Meidner and Mansfield (1968) (for extant frequencies, dimensions, and method of computation); fossil data from Chaloner and Collinson (1975), Hueber (1983), Stubblefield and Banks (1978)

		tant in early earliest stoma-bearing plants, in the extant range for later plants	
Carbon-isotope ratio of organic material	Discrimination between C_3 (large discrimination against the ^{13}C present in atmospheric CO_2) and C_4 or CAM metabolism (small discrimination against ^{13}C); if CO_2 diffusion is the main limit on the rate of photosynthesis, all pathways will yield small discriminations against ^{13}C	All data show that Paleozoic terrestrial photosynthesis was C_3; the larger discrimination also, of course, shows that CO_2 diffusion was not the main factor limiting the rate of photosynthesis; *if* atmospheric CO_2 concentration was similar to that present today, this would also suggest that stomatal strategies, and hence potential water-use efficiencies (moles C fixed in photosynthesis per mole H_2O transpired), were similar to those operative today in C_3 plants	Troughton (1971), Smith (1976), Pflug (1984) (for past and present carbon-isotope ratios in organic and inorganic material); O'Leary (1981), Farquhar (1980, 1983), Farquhar et al. (1982) (for theory of discrimination by diffusion and carboxylation reactions in C_3, C_4, and CAM photosynthesis; relationship of carbon-isotope ratios to potential water-use efficiency; implications of changed atmospheric CO_2 levels); Raven (1984b, 1984d) (paleobotanical interpretations, including estimates of past CO_2 levels and their effects)
Xylem anatomy	Computation of specific conductivity ($m^2\ s^{-1}\ Pa^{-1}$) of xylem to water, from application of Poiseuille's equation to xylem elements, with correction for discrepancy between prediction and observation found with extant xylem occasioned by presence of (pitted) cross-walls, etc.	Values of specific conductivity of xylem are within the range of extant plants (mainly pteridophytes and gymnosperms), which lack vessels in their xylem; very long tracheids (15 mm) in the Carboniferous liane *Sphenophyllum;* however, cost–benefit analyses for fossil plants are rendered difficult by absence of data on ratio of leaf area to xylem transectional area, and on length of pathway from	Analysis of published micrographs for dimensions [few data on length of tracheids (Cichan and Taylor 1984)]; Woodhouse and Nobel (1982), Zimmermann (1983) (for data on conductivity of extant xylem and deviations from ideal Poiseuille behavior); Raven (1977, 1984b) (for paleobotanical interpretations)

Table 14.8. (*cont.*)

Datum	Potential utility in interpreting function of Paleozoic plants	Conclusions as to functioning of Paleozoic plants	References
		water-absorbing zone underground to transpiring surface, in fossil plants	
Morphometric relationships	Analysis reveals apparent "engineering specifications" that underlay selection of morphology in evolving Paleozoic plants	There were numerous different mechanical "patterns"	Niklas (1978, 1982), Niklas and O'Rourke (1982)
Mean distance between minor veins in photosynthetic organs	Analysis of extant plants suggests wider spacing corresponds to lower photosynthetic capacity on an external-area-of-organ basis	Paleozoic plants had a similar (or, in early Devonian plants, a lower) capacity for photosynthesis on an external-area basis as do extant C_3 plants of similar life forms	Analysis of published micrographs of sections of fossil plants; analysis of modern plants by Raven (1984a) (cf. Gibson 1982)

tant *Lycopodium (Lycopodites)* and *Equisetum (Equisetites). Sphenophyllum* probably represented a liane type of plant. Evidence from fossilized tree stumps *in situ* suggests a relatively dense forest (Pickard 1983). The forest (despite the occurrence of abscission of leaves, etc.) probably was evergreen (Addicott 1982; Chabot and Hicks 1982).

The picture of the Carboniferous forest that has emerged is that of the occurrence of canopy-forming and shade-tolerating perennials in this (relatively) stable environment (Grime 1979; Howe and Smallwood 1982). The canopy-formers were the trees; beneath them were the shade-adapted plants of lower stature (together with juveniles of the trees). What of the ruderals or ephemerals? Presumably there were temporary habitats in the Carboniferous in which rapidly growing plants with short life cycles could occur before destruction of the habitat, or before their replacement in a stable habitat by longer-lived plants (without any obligately successional connotations) (Connell and Slatyer 1977). However, there may well have been a paucity of tracheophytes that could have exploited these habitats in the way that extant ephemerals such as the angiosperm *Arabidopsis* could have done, with its four to six weeks from seed to seed.

The only extant nonangiosperm tracheophytes that are able to complete their life cycles (spore to spore) in eight weeks are leptosporangiate forms, *Ceratopteris* species (Stein 1971; Loyal and Chopra 1977). It may be of significance that the adaptive radiation of the leptosporangiate ferns approximately paralleled (Jurassic-Cretaceous) that of the angiosperms (Cretaceous) (Taylor 1981). The mechanistic determinants of minimum life-cycle lengths in tracheophytes are poorly defined. Trees (polycarpic) begin to reproduce at ~0.1 (dicotyledons) or 0.05 (conifers) of their parental life spans (Harper and White 1974; Calow 1978). These minimal generation times of >10 years are certainly not the minimum times that the Angiospermopsida can achieve (i.e., 4–6 weeks in *Arabidopsis*, ~0.01 of the tree life cycle). The gymnosperms, however, may have a *mechanistic* constraint on the minimal length of their life cycles, with a three-year period in many conifers between initiation of a megasporangiate strobilus and the release of viable seed; perhaps the relatively earlier onset of reproduction in coniferous trees is related to this (Cavalier-Smith 1978).

Comparing extant plants with fossil plants at the pteridophyte level of organization, we have already seen that herbaceous pteridophytes today are only rarely annuals and that the leptosporangiate ferns (to which these potential annual organisms belong) arose only in the Jurassic. Accordingly, it is entirely possible that no tracheophytes prior to the Jurassic were ruderals or ephemerals with life cycles occupying less than a year.

Overall, it would seem that terrestrial plants in the Carboniferous had gone a long way toward achieving the range of life forms, with the corresponding energy strategies, that we find in extant terrestrial plants (Table 14.8). The absence of vessels in fossil pre-Cretaceous tracheophytes is significant in relation to the possible evolutionary significance of vessels in relation to habitat and leaf longevity (Chabot and Hicks 1982; cf. Cavalier-Smith 1978).

Conclusions

We have seen that the life forms of plants can be related, in a general fashion, to the minimum photon flux density at which phototrophic growth can occur, with larger plants (both submerged and terrestrial) having a higher minimum photon flux density at which growth can occur. Among vascular land plants, the diversification into trees, shrubs, and (probably perennial) herbs occurred relatively rapidly (by the end of the Devonian) in at least three major evolutionary lines. The lycopsids and sphenopids lost their arborescent representatives at the end of the Paleozoic, and the Gymnospermopsida were the arborescent phototrophs in the Triassic and Jurassic, with pteropsids, sphenopids, and lycopsids as the herbaceous flora. With the origin of the Angiospermopsida (from glossopterid gymnosperms?) (Retallack and Dilcher 1981; Melville 1983) in the Cretaceous, a rather wider range of constructional types became available within a single class than had been found in all of the other tracheophyte classes (Hallé et al. 1978). Approximately in parallel with the adaptive radiation of the angiosperms, the leptosporangiate ferns enjoyed a rapid diversification.

It appears that the angiosperms had few *vegetative* features that were not shared by some other tracheophyte classes. The angiosperms were, however, the only class to have capitalized (in an evolutionary sense) on *vessels* in xylem, as opposed to their sporadic occurrence in other tracheophyte classes. Zimmermann (1983) and Baas (see Chapter 11) discuss water transport in xylem, including the costs of xylem construction in the context of efficiency and safety.

Acknowledgments

Drs. T. J. Givnish and R. H. Robichaux have been most generous with their time and expertise in commenting on the first draft of this chapter.

Appendix I: The quantum requirement of net CO_2 fixation

Justification of 13 mol photon *absorbed* per mole CO_2 fixed in C_3 photosynthesis is given by Farquhar and von Caemmerer (1982). NADPH for required net fixation of 1 mol CO_2 is $(2 + 2v_0/v_c)$ mol NADPH, and the ATP requirement is $(3 + 3v_o/v_c)$ mol ATP, where v_c is achieved rate of $RuBP_c$ activity, and v_o is achieved rate of $RuBP_o$ activity. In C_3 terrestrial vascular plants in air, $v_o/v_c = 0.25$, so that the fixation of 1 mol CO_2 requires 2.5 mol NADPH and 4.75 mol ATP. If the ATP/NADPH ratio in noncyclic photophosphorylation is $\frac{4}{3}$ ($H^+/e^- = 2$, $H^+/ATP = 3$), then 10 mol photon used in noncyclic photophosphorylation ($photon/e^- = 2$) would yield 2.5 mol NADPH and 3.33 mol ATP. The remaining 1.417 mol ATP mol^{-1} CO_2 could be produced by cyclic photophosphorylation, with an ATP/e^- of $\frac{2}{3}$ ($H^+/e^- = 2$, $H^+/ATP = 3$, $photon/e^- = 1$), using 2.13 mol photon mol^{-1} CO_2 fixed, giving an overall quantum requirement of 12.13 mol photon mol^{-1} CO_2 fixed. Rounding up to 13 encompasses a nominal 7% of losses in excitation energy transmission from antenna pigment to reaction centers, H^+ leakage through thylakoid membranes, and so forth.

The noncyclic scheme proposed earlier, with its H^+/e^- stoichiometry of 2, does not involve a proton-motive Q cycle, whereas the cyclic scheme, also with an H^+/e^- stoichiometry of 2, does use a proton-motive Q cycle. If a proton-motive Q cycle *is* used in noncyclic electron transport and H^+ pumping, then the provision of 2.5 mol NADPH, using 10 mol photon, also generates 5.0 mol ATP ($H^+/e^- = 3$, $H^+/ATP = 3$, $photon/e^- = 2$); that is, the net fixation of 1 mol of CO_2 uses 10 absorbed moles photon with a surplus of 0.25 mol ATP. Some data from algae (see following paragraphs) require that this latter stoichiometry, with a noncyclic proton-motive Q cycle, is operative (Raven 1976a, 1980a, 1984a). A facultative operation of the proton-motive Q cycle in noncyclic redox reactions is a distinct possibility (Rathenow and Rumberg 1980).

Photosynthesis that involves CO_2 supply to Rubisco other than by diffusion from bulk air in a terrestrial C_3 plant may have a different quantum requirement. Terrestrial C_4 and CAM plants may have essentially complete suppression of $RuBP_o$ activity (i.e., $v_o = 0$), so that net CO_2 fixation via the PCRC needs 3 mol ATP and 2 mol NADPH per mole CO_2. However, the auxiliary C_3-C_4 cycles that increase the $[CO_2]/[O_2]$, and hence v_c/v_o, at the site of Rubisco activity require ATP input, thus increasing the quantum requirement from the 8 $photons/CO_2$ ($H^+/e^- = 3$ in noncyclic photophosphorylation) or 8.5 photons/CO_2 ($H^+/e^- = 2$ in noncyclic photophosphorylation) for the basic PCRC operation to values similar to those found for terrestrial C_3 plants in air. Farquhar (1983) discussed the quantum requirements of the various types of C_4 photosynthesis in relation to leakage of CO_2 from bundle sheath cells.

For submerged aquatic C_3 plants with diffusive CO_2 supply for air-equilibrated solution, the unstirred-layer effects will decrease v_c/v_o relative to what would be found in a terrestrial C_3 plant with similar Rubisco kinetics, thus increasing the quantum requirement predicted from the Farquhar and von Caemmerer (1982) equations to more than that predicted for terrestrial plants. The various CO_2-concentrating mechanisms in many aquatic plants (Raven 1980a, 1984a) and some terrestrial lichens (Raven 1980a; Snelgar and Green 1980; Green and Snelgar 1981; Coxson et al. 1982; Bauer 1984) that increase the [CO_2] and [CO_2]/[O_2] at the site of Rubisco, and thus increase v_c/v_o, reduce the quantum requirement for the PCRC + PCOC, but add an (unknown) requirement for ATP, and thus photons, to drive the (leaky) CO_2-accumulating mechanisms. We note that some algal growth data at very high bulk-phase [CO_2]/[O_2] ratios require that the PCRC be driven with 8 mol photon per mole CO_2 fixed (i.e., require a noncyclic proton-motive Q cycle) (Raven 1976b, 1980a, 1984a).

All of the data discussed here (except the algae at high CO_2) pertain to present-day atmospheric [CO_2] and [O_2]. A lower [O_2] in the Paleozoic would have reduced the quantum requirement of net carbon fixation by the (C_3) terrestrial plants then present, as would a higher [CO_2]; a lower [CO_2] would have increased the quantum requirement. We note that the computations in the text are more influenced by the *known* variations with growth of factors determining the minimum photon flux density for growth ($\geq 2\times$ variations in fractional photon absorption, $\geq 10^3\times$ variations in specific growth rate; $5\times$ variation in the fraction of the living carbon in the plant that is present in photosynthetic cells; $5\times$ variation in the fraction of organic carbon in the plant that is present in living cells); the ranges of variation in minimum quantum requirements that occur today, or might have been expected in the past ($\leq 2\times$ variation; 10–20 photons per mole C fixed in photosynthesis), are small. Thus, likely variations in quantum requirements for photosynthesis would not upset the arguments in the text; however, within a given life form, variations in quantum requirements could be significant in relation to competition. We note that whereas C_3 terrestrial tracheophytes have the greatest range of life forms, a wide range is also found in CAM and C_4 plants (Osmond et al. 1982; Schulze 1982), although both CAM and C_4 plants are poorly represented in shade environments on land.

In terms of the translation of photosynthate (at the level of carbohydrate) into plant material, we have assumed in the text that 30% of the photosynthate must be respired in carrying out the growth-related carbon skeleton interconversions, NO_3^- and SO_4^{2-} reduction, and active nutrient transport processes that are required for growth (Raven 1976a, 1984a). This means that the effective quantum requirement for *growth* is increased from 10–13 mol photon absorbed per mole CO_2 fixed in photosynthesis to 14.2–18.6 absorbed moles photons per mole CO_2 incorporated during plant growth. It is significant that these values are, for a given plant composition, little altered if direct use of photoproduced cofactors (ATP and reductant) in processes other than CO_2 fixation occurs in photosynthesizing cells (Raven 1984a). Such a direct use of cofactors *increases* the apparent quantum requirement of the initial CO_2 fixation (because photoproduced cofactors are diverted to other energy-requiring processes) but *decreases* the fraction of the C fixed that is subsequently respired to regenerate ATP and reductant (Raven 1984a).

References

Addicott, A. B. 1982. Abscission. University of California Press, Berkeley.

Anderson, J. M. 1983. Chlorophyll-protein complexes of a *Codium* species including a light-harvesting siphonoxanthin-chlorophyll *a/b* complex, an evolutionary relic of some Chlorophyta. Biochim. Biophys. Acta 724:370–380.

Anderson, M. C. 1967. Photon flux, chlorophyll content and photosynthesis under natural conditions. Ecology 48:1050–1053.

Anderson, O. R., N. R. Swanberg, and P. Bennett. 1983. Fine structure of yellow-green symbionts (Prymnesida) in solitary Radiolaria and their comparison with similar Acantharian symbionts. J. Protozool. 30:718–722.

Ashton, D. M., and J. Franckenberg. 1976. Ecological studies of *Acmaena smithii* (Pair). Merrill and Perriy with special reference to Wilson's Promentary. Austr. J. Bot. 24:453–487.

Banks, H. P. 1970. Evolution and plants of the past. Macmillan, London.

– 1975. The oldest land plants: a note of caution. Rev. Palaeobot. Palynol. 20:13–25.

– 1980. Floral assemblages in the Siluro-Devonian. Pp. 1–24 *in* T. N. Dilcher and T. N. Thomas (eds.), Biostatigraphy of fossil plants. Dowden, Hutchinson and Ross, Stroudsberg, Pa.

– 1981. Peridermal activity (wound repair) in an early Devonian (Emsian) Trimerophyte from the Gaspé Peninsula, Canada. The Palaeobotanist 28–29:20–25.

Banse, K. 1982. Cell volumes, maximal growth rates of unicellular algae and ciliates, and the role of ciliates in the marine pelagial. Limnol. Oceanogr. 27:1059–1071.

Barghoorn, E. S. 1964. Evolution of the cambium in geologic time. Pp. 3–18 *in* M. H. Zimmermann (ed.), The formation of wood in forest trees. Academic Press, New York.

Bauer, H. 1984. Net photosynthetic CO_2 compensation concentrations of some lichens. Z. Pflanzenphysiol. 114:45–50.

Beck, C. B., R. Schmid, and G. W. Rothwell. 1982. Stelar morphology and the primary vascular system of seed plants. Bot. Rev. 48:691–815.

Bennett-Clark, T. A., and N. G. Ball. 1951. The diageotropic behaviour of rhizomes. J. Exp. Bot. 2:169–203.

Bewley, J. D., and J. E. Krochko. 1982. Dessication-tolerance. Pp. 325–378 *in* O. L. Lange, P. S. Nobel, C. B. Osmond, and H. Ziegler (eds.), Physiological plant ecology, II, Water relations and carbon assimilation, vol. 12B, Encyclopedia of plant physiology, new series. Springer-Verlag, Berlin.

Biebl, H., and N. Pfennig, 1978. Growth yields of green sulfur bacteria in mixed cultures with sulfur and sulfate reducing bacteria. Arch. Microbiol. 117:9–16.

Björkman, O. 1981. Responses to different quantum flux densities. Pp. 57–107 *in* O. L. Lange, P. S. Nobel, C. B. Osmond, and H. Ziegler (eds.), Physiological plant ecology, I, Responses to the physical environment, vol. 12A, Encyclopedia of plant physiology, new series. Springer-Verlag, Berlin.

Björkman, O., and M. M. Ludlow. 1972. Characterisation of the light climate on the floor of a Queensland rainforest. Carnegie Inst. Washington Yearbook 71:85–94.

Boardman, N. K. 1977. Comparative photosynthesis of sun and shade plants. Annu. Rev. Plant Physiol. 28:355–377.

Bold, M. C., and M. J. Wynne, 1977. Introduction to the algae: structure and reproduction. Prentice-Hall, Englewood Cliffs, N.J.

Borchert, J. H., and N. A. Slade. 1981. Bifurcation ratios and adaptive geometry of trees. Bot. Gaz. 142:394–401.

Box, E. O. 1981. Macroevolution and plant life forms: an introduction to predictive modelling in phytogeography. Dr. Junk, The Hague.

Boysen-Jensen, P. 1932. Die Stoffproduktion der Pflanzen. Fisher, Jena.

Caldwell, M. M. 1982. Solar U. V. radiation as a selective force in the evolution of terrestrial plant life. Pp. 663–675 *in* J. Calkins (ed.), The role of solar ultraviolet radiation in marine ecosystems. Plenum Press, New York.

Calkins, J., and T. Thórdardóttir. 1982. Penetration of solar UV-B into waters off Iceland. Pp. 309–319 *in* J. Calkins (ed.), The role of solar ultraviolet radiation in marine ecosystems. Plenum Press, New York.

Calow, P. 1978. Life cycles. Chapman & Hall, London..

Campbell, S. E. 1980. *Palaeconchocelis stramachii,* a carbonate boring microfossil from the Upper Silurian of Poland (425 million years old): implications for the evolution of the Bangiaceae (Rhodophyta). Phycologia 19:25–36.

Canuto, V. M., J. S. Levine, T. R. Augustsson, and C. L. Imhoff. 1982. Ultraviolet radiation from the young sun and oxygen and ozone levels in the prebiological atmosphere. Nature 296:816–820.

Cavalier-Smith, T. 1978. Nuclear volume control by nucleoskeletal DNA, selection for cell volume and cell growth rate, and the solution of the DNA C-value paradox. J. Cell Sci. 34:247–278.

Chabot, B. F., and D. J. Hicks. 1982. The ecology of leaf life spans. Annu. Rev. Ecol. Syst. 13:229–259.

Chaloner, W. G., and M. E. Collinson. 1975. Application of SEM to a Sigillarian impression fossil. Rev. Palaeobot. Palynol. 20:85–101.

Chaloner, W. G., and A. Sheerin. 1979. Devonian macrofloras. Pp. 145–161 *in* The Devonian system: special papers in palaeontology 23. The Palaeontological Society, London.

Chang, S. S., and R. K. Trench. 1982. Peridinin – chlorophyll *a* proteins from the symbiotic dinoflagellate *Symbiodinium (= Gymnodinium) microadriaticum* Freudenthal. Proc. Roy. Soc. London B 215:191–210.

Chew, F. S., and J. E. Rodman, 1979. Plant resources for chemical defence. Pp. 271–307 *in* G. A. Rosenthal and F. H. Jansen (eds.), Herbivores: their interaction with secondary plant metabolites. Academic Press, New York.

Chung, H.-H., and R. L. Barnes. 1977. Photosynthate allocation in *Pinus taeda.* I. Substrate requirements for synthesis of shoot biomass. Can. J. For. Sci. 7:106–111.

Church, A. H. 1919. Thalassiophyta and the subaerial transmigration. Oxford Botanical Memoir Number 3. Clarendon Press, Oxford.

Cichan, M. A., and T. N. A. Taylor. 1984. A method for determining tracheid lengths in petrified wood by analysis of cross sections. Ann. Bot. 53:219–226.

Clark, F. R. S., and D. A. Russell. 1981. Fossil charcoal and the palaeoatmosphere. Nature 290:428.

Clements, F. E. 1920. Plant indicators. The relation of plant communities to process and practice. Carnegie Institution of Washington Publication 290.

Colinvaux, P. 1980. Why big fierce animals are rare. Penguin Books, Harmondsworth.

Connell, J. H. 1972. Community interactions on marine rocky intertidal shores. Annu. Rev. Ecol. Syst. 3: 169–192.

Connell, J. H., and Slatyer, R. O. 1977. Mechanisms of succession in natural communities and their role in community stability and organisation. Am. Nat. 111:1119–1144.

Cooke, R. 1977. The biology of symbiotic fungi. Wiley, London.

Cope, M. J., and W. G. Chaloner. 1980. Fossil charcoal as evidence of past atmosphere composition. Nature 283:647–649.

– 1981. Fossil charcoal and the palaeoatmosphere. Nature 290:428.

Corner, E. J. H. 1964. The life of plants. Weidenfeld and Nicolson, London.

– 1966. The natural history of palms. Weidenfeld and Nicolson, London.

Cowan, I. R. 1977. Stomatal behaviour and environment. Adv. Bot. Res. 4:117–228.

– 1982. Regulation of water use in relation to carbon gain in higher plants. Pp. 589–613 *in* O. L. Lange, P. S. Nobel, C. B. Osmond, and H. Ziegler (eds.), Physiological plant ecology, II, Water relations and carbon assimilation, vol. 12B, Encyclopedia of plant physiology, new series. Springer-Verlag, Berlin.

Cowan, I. R., and G. D. Farquhar. 1977. Stomatal function in relation to leaf metabolism and environment. Soc. Exp. Biol. Symp. 31:471–505.

Coxson, D. S., G. P. Harris, and K. A. Kershaw. 1982. Physiological-environmental interactions in lichens. XV. Contrasting gas exchange patterns between a lichenised and non-lichenised terrestrial *Nostoc* cyanophyte. New Phytol. 92:561–572.

Dagley, S. 1976a. A new biochemical perspective for undergraduate biologists. Part I. Biochem. Ed. 4:4–8.

– 1976b. A new biochemical perspective for undergraduate biologists. Part II. Biochem. Ed. 4:25–28.

Den Hartog, C. 1970. The sea-grasses of the world. North-Holland, Amsterdam.

Den Hartog, C., and C. Segal. 1964. A new classification of water plant communities. Acta Bot. Neerl. 13:367–393.

Dennison, W. C., D. Mauzerall, and R. S. Alberte. 1981. Photosynthetic responses of *Zostera manina* (Eelgrass) to *in situ* manipulations of light. Biol. Bull. 161:311–312.

Domning, D. P. 1981. Sea cows and sea grasses. Palaeobiol. 7:417–420.

Duddridge, J. A., A. Halibari, and D. J. Read. 1980. Structure and function of mycorrhizal rhizomorphs with particular reference to their role in water transport. Nature 287:834–836.

Dustan, P. 1979. Distribution of zooxanthellae and photosynthetic chloroplast pigments of the reef-building coral *Montastrea annularis* Ellis and Scholander in relation to depth of a West Indian coral reef. Bull. Mar. Sci. 29:79–95.

Edwards, D., J. Feenan, and D. G. Smith. 1983. A late Wenlock flora from Country Tipperary, Ireland. Bot. J. Linn. Soc. 86:19–36.

Edwards, D. S., and A. G. Lyon. 1983. Algae from the Rhynie chert. Bot. J. Linn. Soc. 86:37–55.

Elliott, G. F. 1978. Ecological significance of post-palaeozoic green calcereous algae. Geol. Mag. 115:437–442.

Emberger, L. 1969. Les plantes fossiles dans leurs rapports avec les vegetaux vivants (2nd ed.). Masson et Cie, Paris.

Emerson, C. J., R. G. Buggeln, and K. Bal. 1982. Translocation in *Saccorhiza dermatodea* (Laminariales, Phaeophyceae): anatomy and physiology. Can. J. Bot. 60:2164–2184.

Eppley, R. W. 1972. Temperature and phytoplankton growth in the sea. Fishery Bull. 70:1063–1085.

Evans, G. C. 1972. The quantitative analysis of plant growth. Blackwell Scientific, Oxford.

Fahn, A. 1979. Secretory tissues in plants. Academic Press, London.

Falkowski, P. G., and Z. Dubinsky. 1981. Light-shade adaptation of *Styllophora pistilata,* a hermatypic coral from the Gulf of Eilat. Nature 289:172–174.

Farnum, P., R. Timmis, and J. L. Kulp. 1983. Biotechnology of forest yield. Science 219:694–702.

Farquhar, G. D. 1980. Carbon isotope discrimination by plants: effects of carbon dioxide concentration and temperature *via* the ratio of intercellular and atmospheric CO_2 concentrations. Pp. 105–110 *in* G. I. Pearman (ed.), Carbon dioxide and climate: Australian research. Australian Academy of Sciences, Canberra.

– 1983. On the nature of carbon isotope discrimination in C_3 species. Austr. J. Plant Physiol. 10:205–236.

Farquhar, G. D., and S. von Caemmerer. 1982. Modelling of photosynthetic responses to environmental conditions. Pp. 615–676 *in* O. L. Lange, P. S. Nobel, C. B. Osmond, and H. Ziegler (eds.), Physiological plant ecology, II, Water relations and carbon assimilation, vol. 12B, Encyclopedia of plant physiology, new series. Springer-Verlag, Berlin.

Farquhar, G. C., M. H. O'Leary, and J. A. Berry. 1982. On the relationship between carbon isotope discrimination and intracellular carbon dioxide concentration in leaves. Austr. J. Plant Physiol. 9:121–137.

Feldmann, J. 1966. Les types biologiques des cryptogrames non vasculaire. Les types biologiques d'algues marines benthiques. Mem. Soc. Bot. France, pp. 45–60.

Foster, M. S. 1975a. Algal succession in a *Macrocystis pyrifera* forest. Mar. Biol. 32:313–329.

– 1975b. Regulation of algal community development in a *Macrocystis pyrifera* forest. Mar. Biol. 32:331–342.

Friend, D. J. C. 1975. Adaptation and adjustment of photosynthetic characteristics of gametophytes and sporophytes of Hawaiian tree-ferns *(Cibotium glaucum)* grown at different irradiances. Photosynthetica 9:157–164.

Fritsch, F. E. 1945. Studies on the comparative morphology of the algae. IV. Algae and archegoniate plants. Ann. Bot. 9:1–30.

Gaines, S. D., and J. Lubchenko. 1982. A unified approach to marine plant-herbivore interactions. II. Biogeography. Annu. Rev. Ecol. Syst. 13:111–138.

Gerard, V. A., and H. Kirkman. 1984. Ecological observations on a branched, loose-lying form of *Macrocystis pyrifera* (L) C. Agardh. in New Zealand. Botanica Marina 27:105–109.

Gibson, A. C. 1982. The anatomy of succulence. Pp. 1–17 *in* I. P. Ting and M. Gibbs (eds.), Crassulacean acid metabolism. American Society of Plant Physiologists, Rockville, Md.

Gill, A. M., and P. B. Tomlinson. 1975. Aerial roots: an array of forms and functions. Pp. 237–275 *in* J. C. Torrey and D. C. Clarkson (eds.), The development and function of roots. Academic Press, New York. Pp. 237–275.

Givnish, T. J. 1979. On the adaptive significance of leaf form. Pp. 375–407 *in* O. T. Solberg, S. Jain, G. B. Johnson, and P. H. Raven (eds.), Topics in plant population biology. Columbia University Press, New York.

– 1982. On the adaptive significance of leaf height in forest herbs. Am. Nat. 120:353–381.

– 1984. Leaf and canopy adaptations in tropical forests. Pp. 51–84 *in* E. Medina, H. Mooney, and C. Vásquez-Yánez (eds.), Physiological ecology of plants of the wet tropics. W. Junk, The Hague.

Givnish, T. J., and G. J. Vermeij. 1976. Sizes and shapes of lianae leaves. Am. Nat. 110:743–778.

Goldman, J. C., J. J. McCarthy, and D. G. Peavey. Growth rate influence on the chemical composition of phytoplankton in oceanic waters. Nature 279:210–215.

Gons, H. J. 1977. On the light-limited growth of *Scendesmus protuberans* Fritsch. PhD thesis, University of Amsterdam.

Graham, L. E. 1982. The occurrence, evolution and phylogenetic significance of parenchyma in *Coleochaete* Brieb (Chlorophyta). Am. J. Bot. 69:447–454.

Graham, L. E., and L. W. Wilcox. 1983. The occurrence and phylogenetic significance of putative placental transfer cells in the green alga *Coleochaete.* Am. J. Bot. 70:113–120.

Green, T. G. A., and W. P. Snelgar. 1981. Carbon dioxide exchange in lichens: relationship between net photosynthetic rate and CO_2 concentration. Plant Physiol. 68:199–201.

Grier, G. C., and R. S. Logan. 1977. Old-growth *Pseudotsuga menziesii* communities of a western Oregon watershed: biomass distribution and production budgets. Ecol. Monogr. 47:373–400.

Grime, J. P. 1979. Plant strategies and vegetation processes. Wiley, Chichester.

Gross, G. G. 1980. The biochemistry of lignification. Adv. Bot. Res. 8:26–63.

Grubb, P. J. 1977. The maintenance of species-richness in plant communities: the importance of the regeneration niche. Biol. Rev. 52:107–145.

Hallé, F., R. A. A. Oldemann, and P. B. Tomlinson. 1978. Tropical trees and forests: an architectural analysis. Springer-Verlag, Heidelberg.

Hariri, M, and J. L. Priouhl. 1978. Light-induced adaptive responses under greenhouse and controlled conditions in the fern *Pteris cretica* var *auvardii.* II. Photosynthetic capacities. Physiol. Plant. 42:97–102.

Harland, W. B., et al. 1967. The fossil record: a symposium with documentation. Geological Society of London.

Harper, J. L., and A. D. Bell. 1979. The population dynamics of growth form in organisms with modular construction. Pp. 29–52 *in* R. M. Anderson, D. B. Turner, and L. R. Taylor (eds.), Population dynamics. Blackwell Scientific, Oxford.

Harper, J. L., P. H. Lovell, and K. G. Moore. 1970. The shapes and sizes of seeds. Annu. Rev. Ecol. Syst. 1:327–356.

Harper, J. L., and J. White, 1974. The demography of plants. Annu. Rev. Ecol. Syst. 5:419–463.

Hart, M. H. 1978. The evolution of the atmosphere of the earth. Icarus 33:23–39.

Harvey, G. W. 1980. Photosynthetic performance of isolated cells from sun and shade plants. Carnegie Inst. Washington Yearbook 79:160–164.

Hersey, D. L., and E. Swift. 1976. Nitrate reductase activity of *Amphidinium carterae* and *Cachoninina niei* (Dinophycease) in batch culture: diel periodicity and effects of light intensity and ammonia. J. Phycol. 12:36–44.

Hillis-Colinvaux, L. 1980. Ecology and taxonomy of *Halimeda.* Adv. Mar. Biol. 17:1–327.

Honda, H., P. B. Tomlinson, and J. B. Fisher. 1981. Computer simulation of branch interaction and regulation by unequal flow rates in botanical trees. Am. J. Bot. 68:569–585.

Horn, H. S. 1971. The adaptive geometry of trees. Princeton University Press, Princeton, N.J.

Howe, H. F., and J. Smallwood. 1982. Ecology of seed dispersal. Annu. Rev. Ecol. Syst. 13:201–228.

Hsiao, T. C. 1973. Plant responses to water stress. Ann. Rev. Plant Physiol. 24:519–570.

Huber, O. 1978. Light compensation point of vascular plants of a tropical cloud forest and an ecological interpretation. Photosynthetica 12:382–390.

Hueber, F. M. 1983. A new species of *Baragwanathia* from the Sextant formation (Emsian) Northern Ontario, Canada. Bot. J. Linn. Soc. 86:57–79.

Hutchinson, G. E. 1975. A treatise on limnology. vol. III. Limnological botany. Wiley, New York.

Iyengar, M. O. P. 1932. *Fritschiella,* a new terrestrial member of the Chaetophoracea. New Phytol. 31:329–335.

– 1958. *Nitella terrestris* sp. nov., a terrestrial chlorophyte from South India. Bull. Bot. Soc. Bengal 12:85–90.

Jackson, G. A. 1980. Marine biomass production through seaweed aquaculture. Pp. 31–58 *in* A. San Pietro (ed.), Biochemical and photosynthetic aspects of energy production. Academic Press, New York.

Johansen, H. W. 1981. Coralline algae, a first synthesis. CRC Press, Boca Baton, Fla.

Johnson, P. W., and J. H. Sieburth. Chroococcoid cyanobacteria in the sea: a ubiquitous and diverse phototrophic biomass. Limnol. Oceanogr. 24:928–935.

– 1982. In situ morphology occurrence of eukaryotic phototrophs of bacterial size in the picoplankton of estuarine and oceanic waters. J. Phycol. 18:318–327.

Jonker, F. P. 1981. The questionable origin of early land plants from algae. The Palaeobotanist 28–29:423–426.

Kain, J. M. 1979. A view of the genus *Laminaria.* Oceanogr. Mar. Biol. Annu. Rev. 17:101–161.

Keddy, P. A. 1981. Why gametophytes and sporophytes are different. Am. Nat. 118:452–454.

Keeley, J. E., C. B. Osmond, and J. A. Raven. 1984. *Stylites,* a vascular land plant without stomata, absorbs CO_2 via its roots. Nature 310:694–695.

King, D. 1981. Tree dimensions: maximising the rate of height growth in dense stands. Oecologia 51:351–356.

Kinsey, D. W., and P. J. Davies. 1979. Effects of elevated nitrogen and phosphorus on coral reef growth. Limnol. Oceanogr. 24:935–940.

Kira, T. 1975. Primary production of forests. Pp. 3–40 *in* J. P. Cooper (ed.), Photosynthesis and productivity in different environments. Cambridge University Press.

Kirk, J. T. O. 1975a. A theoretical analysis of the contribution of algal cells to the attenuation of light within natural waters. I. General treatment. New Phytol. 75:11–20.

– 1975b. A theoretical analysis of the contribution of algal cells to the attenuation of light in natural waters. II. Spherical cells. New Phytol. 75:21–36.

– 1976. A theoretical analysis of the contribution of algal cells to the attenuation of light within natural waters. III. Cylindrical and spheroidal cells. New Phytol. 77:341–358.

– 1983. Light and photosynthesis in aquatic ecosystems. Cambridge University Press.

Kirkman, H. 1981. The first year in the life history and the survival of the juvenile marine macrophyte, *Ecklonia radiata* (Turn) J. Agardh. J. Exp. Mar. Biol. Ecol. 55:243–253.

– 1984. Standing crop and production of *Ecklonia radiata* (C. Ag.) J. Agardh. J. Exp. Mar. Biol. Ecol. 76:119–130.

Knoll, A., and G. Rothwell. 1981. Palaeobotany: perspectives in 1980. Palaeobiology 7:7–35.

Kremer, B. P. 1980. Transversal profiles of carbon assimilation in the fronds of three *Laminaria* species. Mar. Biol. 59:95–103.

Larkin, J. M., and W. R. Strohl. 1983. *Beggiatoa, Thiothrix* and *Thioplaca.* Annu. Rev. Microbiol. 37:341–367.

Larkum, A. W. D. 1972. Frond structure and growth in *Laminaria hyperborea.* J. Mar. Biol. Assoc. U.K. 52:405–408.

Larwood, G., and P. B. Rosen (eds.). 1979. Biology and systematics of colonial animals. Academic Press, London.

Law, R., and D. H. Lewis. 1983. Biotic environment and the maintenance of sex – some evidence from mutualistic symbioses. Biol. J. Linn. Soc. 20:249–276.

Lawton, R. O. 1984. Ecological constraints on wood density in a tropical montane rain forest. Am. J. Bot. 71:261–267.

Lele, K. H., and J. Walton. 1961. Contributions to the knowledge of *Zosterophyllum myretonianum* Penhallow from the lower old red sandstone of Angus. Trans. Roy. Soc. Edinb. 54:469–476.

Li, W. K. W., D. V. Subba Rao, W. G. Harrison, J. C. Smith, J. J. Cullen, B. Irwin, and T. Platt. 1983. Autotrophic picoplankton in the tropical ocean. Science 219:292–295.

Littler, M. M., and B. J. Kaukker. 1984. Heterotrichy and survival strategies in the red alga *Corallina officinalis* L. Bot. Mar. 27:37–44.

Littler, M. M., D. S., Littler, and P. R. Taylor. 1983. Evolutionary strategies in a tropical barrier reef ecosytem: functional-form groups of marine macroalgae. J. Phycol. 19:229–237.

Lobban, C. S. 1978. The growth and death of the *Macrocystis* sporophyte (Phaeophyceae, Laminariales). Phycologia 17:196–212.

Loehle, C. 1983. Growth and maintenance respiration: a reconciliation of Thornley's view and the traditional view. Ann. Bot. 51:741–748.

Lorimer, G. H. 1981. The carboxylation and oxygenation of ribulose 1,5-bisphosphate: the primary events in photosynthesis and photorespiration. Annu. Rev. Plant Physiol. 32:349–383.

Loyal, D. S., and A. K. Chopra. 1977. In vitro life cycle, regeneration and apospony in *Ceratopteris pteridioides.* Current Science 46:39–43.

Lüning, K. 1979.Growth strategies of three Laminaria species (Phaeophyceae). Mar. Ecol. Progr. Series 1:195–207.

Lüning, K., and M. J. Dring. 1979. Continuous underwater light measurements near Helgoland (North Sea) and its significance for characteristic light limits in the sublittoral region. Helg. Wiss. Meeres. 32:403–424.

Lüning, K., and M. Neushul. 1978. Light and temperature demands for growth and reproduction of laminanian gametophytes in south and central California. Mar. Biol. 45:297–309.

Luther, H. 1949. Vorschlag zu einer ökologischen Grundeinleitung der Hydrophyten. Acta. Bot. Fenn. 44:1–15.

MacArthur, R. H., and O. E. Wilson. 1967. The theory of island biogeography. Princeton University Press, Princeton, N.J.

Mann, K. H. 1982. Ecology of coastal waters. A systems approach. Blackwell Scientific, Oxford.
Margalef, R. 1978. Life-forms of phytoplankton as survival alternatives in an unstable environment. Oceanolog. Acta 1:493–509.
Margulis, L. 1981. Symbiosis in cell evolution. W. H. Freeman, San Francisco.
Mattson, W. J., Jr. 1980. Herbivory in relation to plant nitrogen content. Annu. Rev. Ecol. Syst. 11:119–161.
Meidner. H., and T. A. Mansfield, 1968. Physiology of stomata. McGraw-Hill, London.
Melville, R.1983. Glossopteridae, Angiospermidae and the evidence for angiosperm origin. Bot. J. Linn. Soc. 86:279–323.
Metting, B. 1981. The systematics and ecology of soil algae. Bot. Rev. 47:195–312.
Miller, H. A. 1982. Bryophyte evolution and geography. Biol. J. Linn. Soc. 18:145–196.
Miller, J. H., and P. M. Miller. 1961. The effect of different light conditions and sucrose on the growth and development of the gametophyte of the fern, *Onoclea sensibilis.* Am. J. Bot. 48:154–159.
Mohr, H. 1965. Die Steuerung der Entwicklung durch Licht am Biespel der Farngametophyten. Ber. Deut. Bot. Ges. 78:54–68.
Munk, W. H, and G. A. Riley. 1952. Absorption of nutrients by aquatic plants. J. Mar. Res. 11:215–240.
Neushul, M. 1971a. Submarine illumination in *Macrocystis* beds. Beiheft zur Nova Hedwigia 32:241–254.
– 1971b. The kelp community of seaweeds. Beiheft zur Nova Hedwigia 32:265–267.
– 1972. Functional interpretation of benthic algal morphology. Pp. 45–74 *in* I. A. Abbott and M. Korogi (eds.), Contributions to the systematics of benthic marine algae of the North Pacific. Japanese Society of Phycologists, Kobe.
Newman, M. J. 1980. The evolution of the solar constant. Origins of Life 10:105–110.
Niell, F. X. 1976. C:N ratios in some marine macrophytes and its possible ecological significance. Bot. Mar. 19:347–350.
Niklas, K. J. 1978. Morphometric relationships and rates of evolution among Palaeozoic vascular plants. Evolutionary Biol. 11:509–545.
– 1982. Computer simulations of early land plant branching morphologies: canalisation of patterns during evolution? Palaeobiology 8:196–210.
Niklas, K. J., and T. D. O'Rourke. 1982. Growth patterns of plants that maximise vertical growth and minimize internal stresses. Am. J. Bot. 69:1367–1374.
Nobel, P. S. 1977. Internal leaf area and cellular CO_2 resistance: photosynthetic implications of variations with growth conditions and plant species. Physiol. Plant. 40:137–144.
– 1983. Biophysical plant physiology and ecology. W. H. Freeman, San Francisco.
O'Leary, M. H. 1981. Carbon isotope fractionation in plants. Phytochem. 20:553–568.

Osmond, C. B. 1981. Photorespiration and photoinhibition. Some implications for the energetics of photosynthesis. Biochim. Biophys. Acta 639:77–98.
Osmond, C. B., O. Björkman, and D. J. Anderson. 1980. Physiological processes in plant ecology: toward a synthesis with *Atriplex*. Springer-Verlag, Berlin.
Osmond, C. B., K. Winter, and M. Ziegler. 1982. Functional significance of different pathways of CO_2 fixation in photosynthesis. Pp. 479–548 *in* O. L. Lange, P. S. Nobel, C. B. Osmond, and M. Ziegler (eds.), Physiological plant ecology, II, Water relations and carbon assimilation, vol. 12B, Encyclopedia of plant physiology, new series. Springer-Verlag, Berlin.
Ovington, J. D. 1962. Quantitative ecology and the woodland ecosystem concept. Adv. Ecol. Res. 1:103–192.
Page, C. N. 1979. Experimental aspects of fern ecology. Pp. 551–589 *in* A. F. Dyer (ed.), The experimental biology of ferns. Academic Press, London.
Pearcy, R. W. 1983. The light environment and growth of C_3 and C_4 tree species in the understory of a Hawaiian forest. Oecologia 58:19–25.
Pearcy, R. W., and H. W. Calkin. 1983. Carbon dioxide exchange of C_3 and C_4 tree species in the understory of a Hawaiian forest. Oecologia 58:26–32.
Penning de Vries, F. W. T. 1972. Respiration and growth. Pp. 327–347 *in* A. R. Rees, K. E. Cockshull, D. W. Hand, and R. G. Hurd (eds.), Crop processes in controlled environments. Academic Press, London.
– 1975. The cost of maintenance processes in plant cells. Ann. Bot. 39:77–92.
Penning de Vries, F. W. T., A. H. M. Brunsting, and H. H. Van Laar. 1974. Products, requirements and efficiency of biosynthesis: a quantitative approach. J. Theoret. Biol. 45:339–377.
Perry, M. J., M. C. Larsen, and R. S. Alberte. Photoadaptation in marine phytoplankton: responses of the photosynthetic unit. Mar. Biol. 62:91–101.
Peters, R. H. 1983. The ecological implications of body size. Cambridge University Press.
Pflug, H. D. 1984. Early geological record and the origin of life. Naturwiss. 71:63–68.
Phillips, T. L., and W. A. Dimichelle. 1981. Palaeoecology of Middle Pennsylvanian age coal swamps in Southern Illinois – Herrin Coal member at Sahara mine No. 6. Pp. 231–284 *in* K. J. Niklas (ed.), Palaeobotany, palaeoecology and evolution, vol. 1. Praeger, New York.
Pickard, W. F. 1981. The ascent of sap in plants. Progr. Biophys. Mol. Biol. 37:181–229.
– 1983. Three interpretations of the self-thinning rule. Ann. Bot. 51:749–758.
Platt, T., D. V. Subba Rao, and B. Irwin. 1983. Photosynthesis of picoplankton in the oligotrophic ocean. Nature 300:702–704.
Proctor, M. C. F. 1979. Structure and eco-physiological adaptation in bryophytes. Pp. 479–509 *in* G. C. S. Clarke and J. G. Duckett (eds.), Bryophyte systematics. Academic Press, London.
Ramus, J. 1978. Seaweed anatomy and photosynthetic performance: the ecological significance of light guides, heterogenous absorption and multiple scatter. J. Phycol. 14:352–362.
Ramus, J., and G. Rosenberg. 1980. Diurnal photosynthetic performance of seaweeds measured under natural conditions. Mar. Biol. 56:21–28.

Rathenow, M., and B. Rumberg. 1980. Stoichiometries of proton translocation during photosynthesis. Ber. Bunsenges. Physik. Chem. 84:1059–1062.
Raunkaier, C. 1934. The life forms of plants and statistical plant geography. Clarendon Press, Oxford.
Raven, J. A. 1970. Exogenous inorganic carbon sources in plant photosynthesis. Biol. Rev. 45:167–221.
– 1976a. Division of labour between chloroplasts and cytoplasm. Pp. 403–443 *in* J. Barber (ed.), The intact chloroplast. Elsevier, Amsterdam.
– 1976b. Transport in algal cells. Pp. 129–188 *in* U. Lüttge and M. G. Pitman (eds.), vol. 11A, Encyclopedia of plant physiology, new series. Springer-Verlag, Berlin.
– 1977. The evolution of vascular land plants in relation to supracellular transport processes. Adv. Bot. Res. 5:153–219.
– 1980a. Nutrient transport in micro-algae. Adv. Microb. Physiol. 27:47–226.
– 1980b. Chloroplasts of eukaryotic microorganisms. Symp. Soc. Gen. Microbiol. 30:181–205.
– 1981a. Introduction to metabolic control. Pp. 3–27 *in* D. A. Rose and D. A. Charles-Edwards (eds.), Mathematics and plant physiology. Academic Press, London.
– 1981b. Nutritional strategies of submerged benthic plants: the acquisition of C, N and P by rhizophytes and haptophytes. New Phytol. 88:1–30.
– 1982. The energetics of freshwater algae: energy requirements for biosynthesis and volume regulation. New Phytol. 92:1–20.
– 1983. Phytophages of xylem and phloem: a comparison of animal and plant sap-feeders. Adv. Ecol. Res. 13:135–234.
– 1984a. Energetics and transport in aquatic plants. A. R. Liss, New York.
– 1984b. Physiological correlates of the morphology of early vascular plants. Bot. J. Linn. Soc. 88:105–126.
– 1984c. Energy transmission along algal and cyanobacterial membranes. Pp. 289–296 *in* W. J. Cram, K. Janacek, R. Rybova and K. Sigler (eds.), Proceedings of international symposium on membrane transport in plants. Academia, Praha.
– 1984d. A cost–benefit analysis of photon absorption by photosynthetic unicells. New Phytol. 98:593–625.
– 1985. The comparative physiology of adaptation of plants and arthropods to life on land. Phil. Trans. Roy. Soc. Lond. 309:273–288.
– 1986. Physiological consequences of extremely small size for autotrophic organisms in the sea. Can. Bull. Fish Aquat. Soc. (in press).
Raven, J. A., J. Beardall, and H. Griffiths. 1982. Inorganic carbon sources for *Lemanea, Cladophora* and *Ranunculus* in a fast-flowing stream: measurements of gas exchange and of carbon isotope ratio and their ecological implications. Oecologia 53:68–78.
Raven, J. A., and K. Richardson, 1984. Dinophyte flagella: a cost–benefit analysis. New Phytol. 98:259–276.
– 1986. Photosynthesis in marine environments. *In* N. R. Baker and S. P. Long (eds.), Photosynthesis in specific environments. Elsevier, Amsterdam (in press).

Raven, J. A., F. A., Smith, and S. M. Glidewell. 1979. Photosynthetic capacities and biological strategies of giant-celled and small-celled macro-algae. New Phytol. 83:299–309.

Retallack, G., and D. L. Dilcher. 1981. Arguments for a glossopterid ancestry of angiosperms. Paleobiology 7:54–67.

Richardson, K., J. Beardall, and J. A. Raven. 1983. Adaptation of unicellular algae to irradiance: an analysis of strategies. New Phytol. 93:157–191.

Richardson, K., and G. E. Fogg. 1982. The role of dissolved organic material in the nutrition and survival of marine dinoflagellates. Phycologia 21:17–26.

Robards, A. W., and D. T. Clarkson. 1976. The role of plasmodesmata in the transport of water and nutrients across roots. Pp. 181–202 *in* B. E. S. Gunning and A. W. Robards (eds.), Intercellular communication in plants: studies on plasmodesmata. Springer-Verlag, Berlin.

Rogers, R. W. 1971. Distribution of the lichen *Chondriopsis semiviridis* in relation to its heat and drought tolerance. New Phytol. 70:1069–1077.

Schiel, D. R., and J. H. Choat. 1980. Effect of density on monospecific stands of marine algae. Nature 285:324–326.

Schulze, E.-D. 1982. Plant life forms and their carbon, water and nutrient relations. Pp. 615–676 *in* O. L. Lange, P. S. Nobel, C. B. Osmond, and H. Ziegler (eds.), Physiological plant ecology, II, Water relations and carbon assimilation, vol. 12B, Encyclopedia of plant physiology, new series. Springer-Verlag, Berlin.

Scott Russell, R. 1977. Plant root systems: their functions and interaction with the soil. McGraw-Hill, London.

Seybold, A., and K. Eagle, 1938. Quantitative Untersuchungen über die Chlorophyll and Carotinoide der Meeresalgen. Jahrb. wiss. Bot. 84:50–80.

Smith, B. N. 1976. Evolution of C_4 photosynthesis in response to changes in carbon and oxygen concentrations in the atmosphere through time. Biosystems 8:24–32.

Smith, F. A., and N. A. Walker. 1980. Photosynthesis by aquatic plants: effects of unstirred layers in relation to CO_2 and HCO_3^- and to carbon isotope discrimination. New Phytol. 86:245–259.

Smith, H. 1982. Light quality, photoperception, and plant strategy. Annu. Rev. Plant Physiol. 38:481–518.

Snelgar, W. P., and T. G. A. Green. 1980. Carbon dioxide exchange in lichens: low carbon dioxide compensation levels and lack of apparent photorespiratory activity in some lichens. The Bryologist 83:505–507.

Snelgar, W. P., T. G. A. Green, and A. L. Wilkens. 1981a. Carbon dioxide exchange in lichens: resistance to CO_2 uptake at different thallus water contents. New Phytol. 88:353–361.

Snelgar, W. P., T. G. A. Green, and C. K. Beltz. 1981b. Carbon dioxide exchange in lichens: estimation of internal thallus CO_2 resistances. Physiol. Plant. 5:417–422.

Sournia, A. 1982. Form and function in marine phytoplankton. Biol. Rev. 57:347–394.

Spence, D. H. N. 1967. Factors controlling the distribution of freshwater macrophytes with particular reference to the lochs of Scotland. J. Ecol. 55:147–170.

Stanier, R. Y., and G. Cohen-Bazire. 1977. Phototrophic prokaryotes: the cyanobacteria. Annu. Rev. Microbiol. 31:225–274.

Stebbins, G. L. 1976. Seeds, seedlings and the origin of angiosperms. Pp. 300–311 *in* C. B. Beck (ed.), Origin and early evolution of angiosperms. Columbia University Press, New York.

Stebbins, G. L., and G. J. C. Hill. 1980. Did multicellular plants invade the land? Am. Nat. 115:342–353.

Stein, D. B. 1971. Gibberillin-induced fertility in the fern *Ceratapteris thalictroides* (L) Braagn. Plant Physiol. 48:416–418.

Steneck, R. S. 1983. Escalating herbivory and resulting adaptive trends in calcareous algal crusts. Palaeobiology 9:44–61.

Stewart, K. D., and K. R. Mattox. 1975. Comparative cytology, evolution and classification of the green algae with some consideration of other organisms with chlorophylls *a* and *b*. Bot. Rev. 41:104–135.

Stolarski, R. S. 1982. Atmospheric evolution and UV-B radiation. Pp. 677–684 *in* J. Calkins (ed.), The role of solar ultraviolet radiation in marine ecosystems. Plenum Press, New York.

Stubblefield, S., and H. P. Banks. 1978. The cuticles of *Drepanophycus spinaeformis*, a long-ranging Devonian lycopod from New York and eastern Canada. Am. J. Bot. 65:110–118.

Swain, T., and G. Cooper-Davis. 1981. Biochemical evolution of early land plants. Pp. 103–134 *in* K. J. Niklas (ed.), Palaeobotany, palaeoecology and evolution, vol. 1. Praeger, New York.

Talling, J. F., R. B. Wood, M. V. Prosser, and R. M. Baxter. 1973. The upper limit of photosynthetic productivity by phytoplankton: evidence from Ethiopian soda lakes. Freshwater Biol. 3:53–76.

Taylor, T. N. 1981. Palaeobotany. An introduction to fossil plant biology. McGraw-Hill, New York.

Tomlinson, P. B. 1979. Systematics and ecology of the Palmae. Annu. Rev. Ecol. Syst. 10:85–107.

– 1982. Chance and design in the construction of plants. Acta Biotheoretica 31A:162–183.

Trinci, A. P. J. 1973. Growth of wild type and spreading colonial mutants of *Neurospora crassa* in batch culture and on agar medium. Arch. Microbiol. 91:113–126.

Troughton, J. H. 1971. Aspects of the evolution in the photosynthetic carboxylation reaction in plants. Pp. 124–129 *in* M. D. Hatch, C. B. Osmond, and R. O. Slatyer (eds.), Photosynthesis and photorespiration. Academic Press, New York.

Van Den Hoek, C. 1982. The distribution of benthic marine algae in relation to temperature regulation of their life histories. Biol. J. Linn. Soc. 18:81–144.

Van Liere, L., and L. R. Mur. 1979. Growth kinetics of *Oscillatoria agardhii* Gomont in continuous culture, limited in its growth by light energy. J. Gen. Microbiol. 115:153–160.

Vogel, S. 1981. Life in moving fluids: the physical biology of flow. Willard Grant Press, Boston.

Wainwright, S. A. 1970. Design in hydraulic organisms. Naturwiss. 57:321–330.

Walter, H. 1955. The water economy and the hydrature of plants. Annu. Rev. Plant Physiol. 6:239–252.
Walton, J. 1964. On the morphology of *Zosterophyllum* and some other Devonian plants. Phytomorphol. 14:155–160.
Waterbury, J. B., S. W. Watson, R. R. L. Guillard, and L. E. Brand. 1979. Widespread occurrence of a unicellular, marine, planktonic cyanobacterium. Nature 277:293–294.
Wigley, T. M. L., and P. Brimblecombe. 1981. Carbon dioxide, ammonia and the origin of life. Nature 291:213–215.
Wong, S. C., I. R. Cowan, and G. D. Farquhar. 1979. Stomatal conductance correlates with photosynthetic capacity. Nature 282:424–426.
Woodhouse, R. M., and P. S. Nobel. 1982. Stipe anatomy, water potentials and xylem conductances in seven species of ferns (Filicopsida). Am. J. Bot. 69:135–140.
Worrest, R. C. 1982. Review of literature concerning the impact of UV-B radiation upon marine organisms. Pp. 309–319 *in* J. Calkins (ed.), The role of solar ultraviolet radiation in marine ecosystems. Plenum Press, New York.
Zimmerman, M. H. 1983. Xylem structure and the ascent of sap. Springer, Berlin.

15 Branching patterns and angles in trees

JACK B. FISHER

Introduction

A fundamental constraint on the geometry of plant canopies, especially in the crowns of woody plants, involves the branching pattern, orientation, and demography of leaf-bearing axes. Branching complexity ranges from plants with a single axis to two-dimensional rhizome systems and culminates in large trees with many orders of branching in three-dimensional space. The branching pattern of the shoot has a direct effect on the spatial distribution of photosynthetic surfaces by forming the plant skeleton, which is composed of twigs, branches, and trunk. It thus directly affects energy capture, water loss, mechanical support, wind resistance, and presumably competitive ability.

In this chapter we shall broadly review the structural basis of branching pattern and orientation and note the relationships between axis positions and leaf surfaces. The problem of variability caused by phenotypic plasticity in an analysis of tree form is stressed, especially as it relates to simulations of branching patterns. Lastly, we shall examine the approaches of optimization studies and ecological correlations with regard to the adaptive significance of observed branching patterns and crown forms.

Structural basis

The geometry of the stem is determined by the direction of growth of the seedling apical bud, and later by the outgrowth and direction of lateral buds. The behavior of the shoot meristems continues to determine the geometry of successive orders of branches in the shoot. The form of the tree crown is in great part a consequence of the primary organization within the apical bud and within those lateral buds that are released at various times.

Primary orientation

The fundamental direction of growth of the shoot apex and the region of elongation behind it establishes the initial inclination of the axis. In broad terms, growing shoots are usually either erect and positively gravitropic (i.e., orthotropic) or basically horizontal and diagravitropic (i.e., plagiotropic). Later, tissue lignification and thickening growth add strength to the axis. Lateral direction or azimuthal angle is most commonly established by bud position, as noted later. The branch angles of newly released buds are apparently affected by hormone levels, although this effect may be closely related to later changes in axis position (Williams and Billingsley 1970; Jankiewicz and Stecki 1976).

Primary pattern

The branches of most seed plants originate from lateral buds. Bud positions around the parent axis are determined by the phyllotaxis, the arrangement of leaves, with which each bud is associated. Thus, azimuthal angle (compass direction) and possibly inclination, depending on the gravitropic response of the bud and the orientation of the parent axis, are established. In addition, the outgrowth of lateral buds may be either rhythmic or continuous with respect to growth of the parent axis. Lateral buds may grow out immediately, termed sylleptic branching, or may first undergo a period of rest after initiation, termed proleptic branching (Hallé et al. 1978). Their outgrowth along the axis may also be closely correlated to distance from the parent apex, which indicates the degree of apical dominance. The later growth of lateral branches may be under the apical control of the parent apex (Zimmermann and Brown 1971). Often bud outgrowth or rest is related to the position of the bud on the parent axis with respect to gravity. Differences in the relative levels of plant hormones and nutrients between upper and lower sides have been shown to be the mechanism for bud release and the type of shoot that develops (Fisher 1972). The interplay between any number of these factors usually results in a regular pattern of branching characteristic for a given species. In addition, there is a close empirical correlation between increasing degree of branching and decreasing size of leaves in both evergreen and deciduous trees (White 1983a, 1983b). Both compound leaves and large simple leaves tend to be borne on thick, sparsely branched axes.

Branch dimorphism

Some trees have a single shoot axis, such as coconut, or many branching axes that are similar in form and growth behavior, such as dracaena, tree euphorbia, *Miconia,* and *Didymopanax.* However, many plants have two or more types of axes distinguished by their primary orientation, symmetry, or form. Such dimorphisms establish differences between leader axes, which are usually radially symmetrical, and lateral branch axes, which are usually dorsiventrally symmetrical. Among temperate trees such as birch and maple, but seldom in tropical species, long and short shoots occur in which only short shoots bear leaves in the older, interior parts of the crown. In most lateral branches there are differences in vigor among twigs such that one predictably dominates in growth to establish the main branch axis. This results from differences in initial vigor or from long and short shoot production. A well-defined main axis is established, commonly in a regular, alternating zigzag pattern. Small-scale differences in branch form and growth behavior can have large-scale effects on crown form, as will be shown.

Later changes

Branching patterns may change with time when self-pruning of twigs is caused by die-back or abscission and results in an increase in interbranch distances. Conversely, the outgrowth of lateral buds can decrease interbranch distances with time. Secondary thickening growth of an axis often obscures the fundamental branching pattern, such as the sympodial construction of the straight trunks of *Alstonia* or *Hura* (see Figures 15.5E and 15.5F on p. 498).

Orientation also can be age-dependent, as shown in Figure 15.1. Increased loading may change the primary orientation of an axis. Secondary thickening of a trunk can push the branch outward and cause the crotch angle to approach 90° as in some conifers (Jankiewicz and Stecki 1976). Woody axes can reorient as a normal part of crown development in some species (Figures 15.1 – 15.4). Such later changes in axis position are often associated with the formation of specialized "reaction wood" (Mergen 1958; Fisher and Stevenson 1981), which in dicots increases the longitudinal stress within the axis to a point at which bending occurs. As a result, axes may be pulled down (Figure 15.1C) or pulled up (Figures 15.2A and 15.3A) after primary elongation has ended. Reaction-wood formation is also a response that brings axes back to their original positions after they have been displaced because of land slips, wind throws, or ice loading, and that counteracts the increased mechanical stresses within the axes (Scurfield 1973; Wilson and Archer 1977).

Tree architecture

Botanists have long been classifying the diversities of plant form and branching patterns found in nature. Earlier attempts at cataloging life forms and crown shapes have culminated in recent efforts to relate general

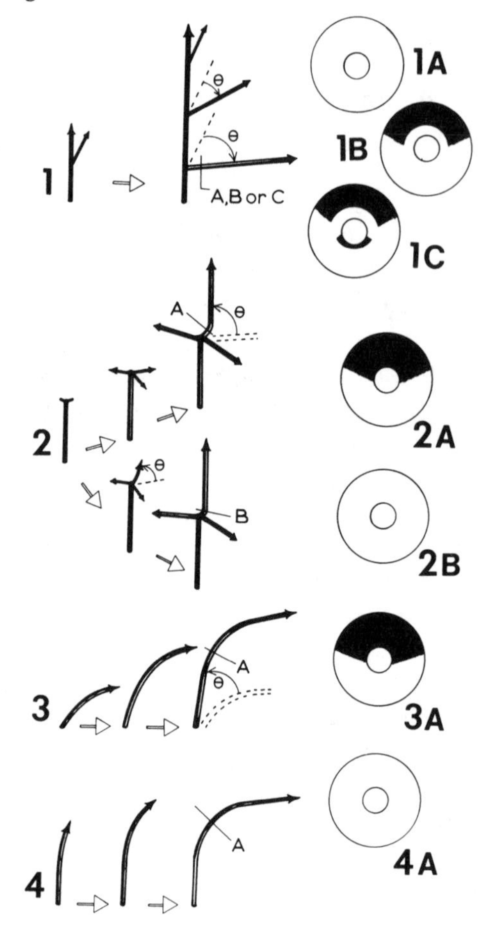

Figures 15.1–15.4. Diagrams of the normal reorientation of axes that occur in dicotyledonous tree architecture. Axis movement is indicated at two or three time intervals, with change in orientation noted by branch-angle change (θ) with respect to direction of gravity. Transverse sections of axes have reaction wood (RW) shaded, if present. Figure 15.1. Lateral branch angle (θ) increases; found in Aubréville's, Massart's, Rauh's, Roux's and other models (according to Hallé et al. 1978) with monopodial leaders; A, no RW (*Dryobalanops, Cordia*); B, RW only on upper side (*Cornus, Eucalyptus*); C, RW first on lower side and later on upper side (*Terminalia, Hevea*). Figure 15.2. Lateral branch angle (θ) decreases with respect to gravity; found in Koriba's model; A, older lateral branch becomes erect, with RW on upper side at base (*Hura*); B, young lateral branch with no RW at base grows upward sooner than in Figure 15.2A (*Alstonia*). Figure 15.3A. Leader axis angle (θ) decreases; found in Troll's model; RW usually found on upper side (*Ulmus, Brownea*). Figure 15.4A. Leader axis angle increases by a change in direction of primary growth rather than a reorientation (as in Figure 15.3A); found in Mangenot's model; RW absent (*Oxydendrum*) (from Fisher and Stevenson 1981).

patterns of crown geometry to the ecology of trees (Horn 1971; Brunig 1976). The most comprehensive study and classification of growth patterns and forms of trees was carried out by Hallé et al. (1978). Their scheme of 23 tree architectural models is based on features of growth from seed to reproductive adult. These features are illustrated in 10 of their models shown in Figure 15.5: (1) Shoots can be branched (Figure 15.5A–J) or unbranched as in a tree fern or coconut palm. (2) The vegetative and reproductive behaviors of buds can produce either monopodial (Figure 15.5B) or sympodial (Figure 15.5E,F) axes. (3) Trunk and lateral branches can be clearly differentiated (Figure 15.5B) or not (Figure 15.5C). (4) The direction of growth of axes can be orthotropic (Figure 15.5B, trunk) or plagiotropic (Figure 15.5B, branches) or can change over time (Figure 15.5G,I). (5) Branching can be continuous (Figure 15.5B) or rhythmic (Figure 15.5H). Architectural models can be thought of as clearly defined, dynamic blueprints of development for a species. Some trees express their models in striking simplicity, such as palms and conifers, whereas others repeat the plan of the model many times during crown development. Such architectural reiteration is normal for many species but can obscure the basic model (de Castro e Santos 1980). In this chapter, only the pattern and orientation of branches will be stressed; a wider analysis of architecture must be deferred. Although a full discussion of this approach cannot be given here, any worker dealing with questions of tree form and function should refer to the original, fully documented account by Hallé et al. (1978) and to further discussions by Fisher (1984) and Givnish (1984). A parallel analysis of architecture in marine algae is offered by Hay in Chapter 19.

Relationship between axis position and leaf surface

Factors determining leaf position

Although the stem axes establish the basic skeleton that supports the crown, the actual distribution of leaf surfaces in the crown is determined by many factors, including (1) the positions of the leaf-bearing branch axes, which usually are the distal units of a branch or trunk, (2) the longitudinal distribution of leaves along the axis, which in turn depends on internode lengths and relative leaf birth/death rates, (3) the presence of long and short shoots, because there are differences between the life spans of leaf-producing meristems (e.g., only short shoots in beech or birch continue to produce leaves from the branch periphery after one growing season), (4) phyllotaxis, which determines the basic dorsiventral versus

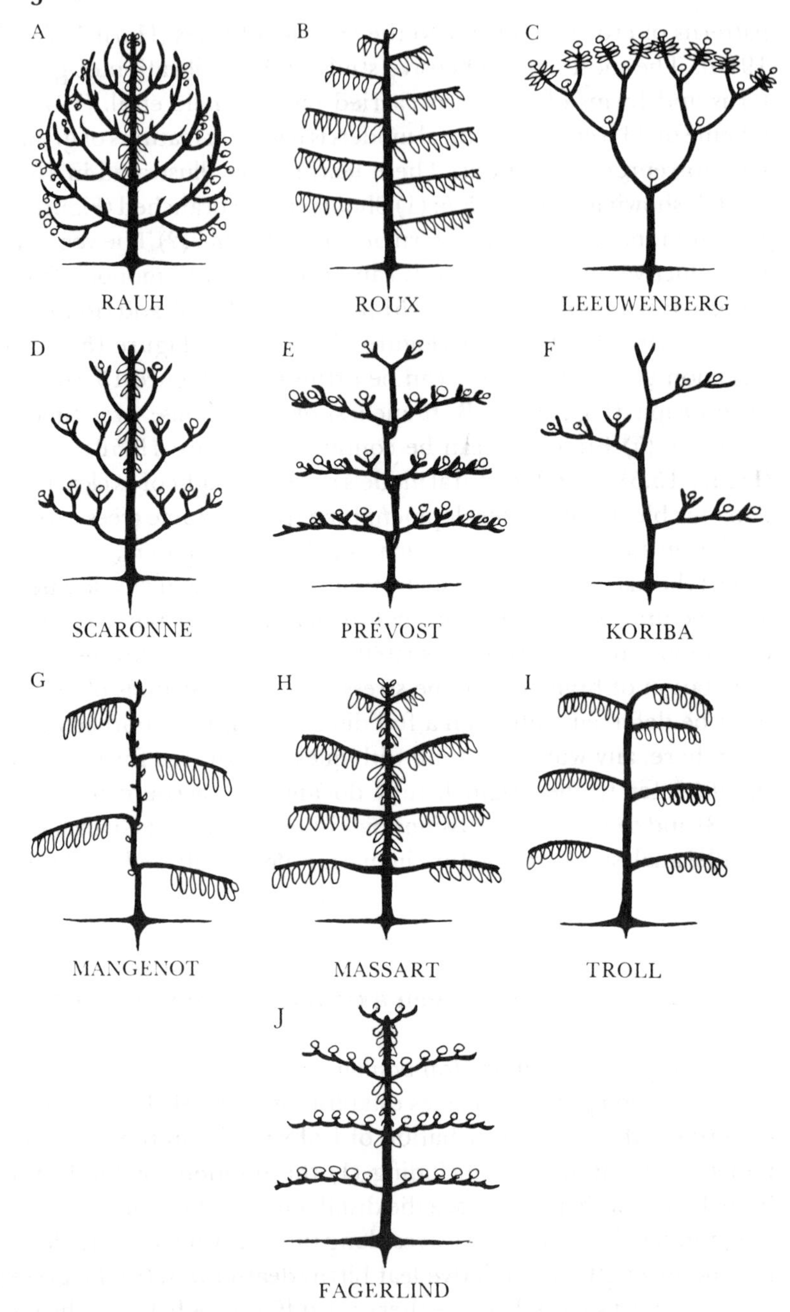

Figure 15.5. Principal architectural models occurring in the pioneer vegetation in French Guiana; 85% of the trees followed six models: A, B, C, D, E, and G (from Foresta 1983, after Hallé et al. 1978).

radial symmetry of leaf attachment onto the axis and also the actual orientation of leaves in many trees, and (5) secondary reorientation of the leaf blade caused by internode twisting, petiole bending, or pulvinus movement that can alter the initial phyllotactic symmetry of the axis (e.g., presence of dorsiventral symmetry on the basically radially symmetrical decussate branch of maple, or differences in blade orientation between sun leaves and shade leaves). Any of these five factors can modify the geometry of the crown and, as a result, affect energy capture by the tree. A more detailed examination of the structural and adaptive relationships between leaf form, leaf arrangement, and canopy structure than can be covered here is presented by Givnish (1984).

Effects of branching on leaf surface

Axis position has widely varying effects on the presentation of photosynthetic surfaces because of the factors listed earlier. As an example, we shall examine trees that follow a single architectural model (Aubréville's model diagrammed in Figure 15.6) but differ in leaf arrangement. In many *Terminalia* species (Combretaceae), leaves are congested in a tight rosette at the erect, short-shoot end of each branch unit because there is one long basal internode (Figure 15.6A). Thus, the foliage is borne in a horizontal layer of closely packed rosettes in *Terminalia nitida* (Figure 15.7A). In *Bucida spinosa* (Combretaceae), the leaf length is shorter relative to the mostly leafless branch axis, so that the rosettes are widely spaced in the foliage layer (Figure 15.7B). On the other hand, leaves are distributed more evenly along the branch-unit axis in *Manilkara* (Sapotaceae), *Pachira* (Bombacaceae), and *Sassafras* (Lauraceae), so that discrete clusters are uncommon (Figure 15.7C). Thus, leaf distribution along an axis may vary while axis position is unchanged.

The branches of *Terminalia catappa* are very planar, but those of *Manilkara* are not, so that *Manilkara* leaves are not distributed in a single, well-defined layer. Small differences in branch-unit orientation with respect to gravity result in noticeable differences in leaf area density (ratio of total leaf area within a unit volume) in the crowns of these two species (Fisher and Hibbs 1982, Figure 2). The longer life span for the leaf-producing apices of the old branch units in *Terminalia* results in a greater leaf area density within its crown interior than in *Manilkara*.

Leaf presentations can be quite different in trees with very similar axis geometries. Although species of *Cordia* (Ehretiaceae) have regularly forked branches (Figure 15.8) that are spatially similar to those of *Terminalia,* they display a different architectural model. Because the apex of the branch unit aborts, there is no distal short shoot that continues to produce

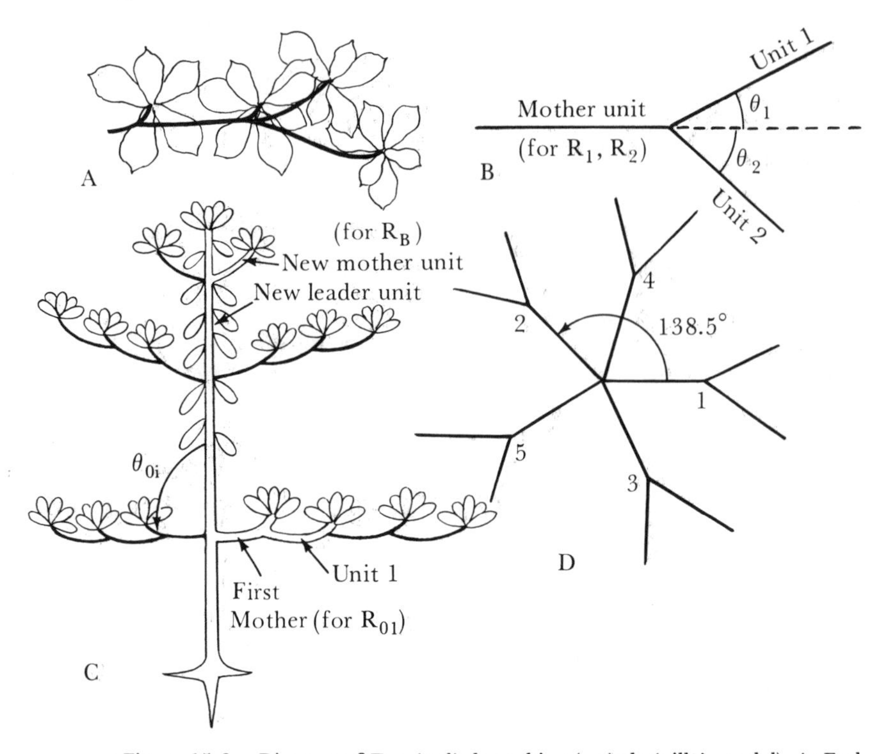

Figure 15.6. Diagram of *Terminalia* branching (= Aubréville's model). A: End of lateral branch showing mother unit and two new daughter units 1 and 2; viewed from below and to one side. B: Same part of lateral branch, but viewed from overhead, showing branching angles, θ_1 and θ_2, and branch length ratios, R_1 (= unit-1/mother) and R_2 (= unit-2/mother). C: Small tree viewed from side showing increasing lateral branch angles (θ_{0i}) as branch tiers age, and the branch length ratios of the first forking in a lateral branch, R_{01} (=first-unit-1/first-mother) and R_{02} (= first-unit-2/first-mother), and the ratio R_B (= new-first-mother/subtending-new-leader). D: View of young branch tier viewed overhead showing the branch sequence by age (1 = first initiated) and the angle of branch divergence (= 138.5°). The foregoing variables are used in Table 15.1.

a rosette of leaves as in *Terminalia*. In addition, leaves are distributed more uniformly along branch-unit axes in *Cordia*. However, there are also small-scale differences among *Cordia* species. Leaves are clustered in pseudowhorls in *Cordia nodosa* (Figure 15.8A), with the upper leaves being

Figure 15.7. Parts of the lateral branches of species displaying *Terminalia* branching pattern but with variations in relative size and placement of leaves; viewed directly overhead. A: *Terminalia nitens* Presl. B: *Bucida spinosa* Jennings (from Fisher and Honda 1979). C: *Manilkara* species. See Figure 15.6 for diagram of branching pattern.

A
B
C

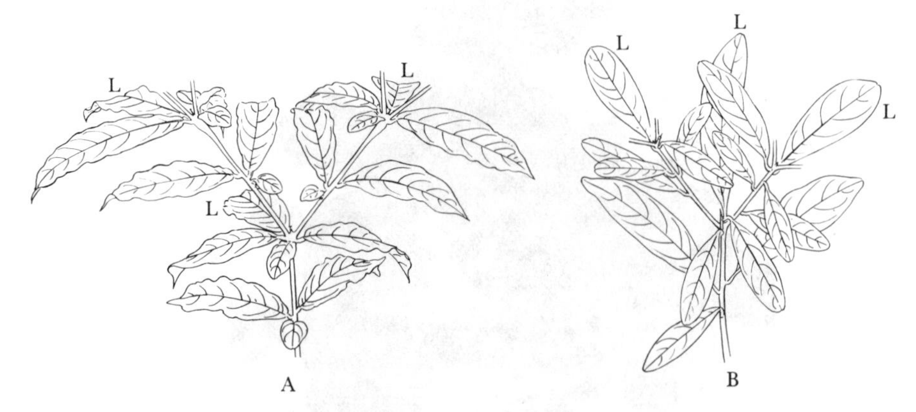

Figure 15.8. Parts of the lateral branches of two *Cordia* species showing the same bifurcating pattern but with different arrangements of leaves; viewed directly overhead. Trees were understory trees in French Guiana. A: *Cordia nodosa* Lam. B: *Cordia nitida* Vahl. L = lowest distal leaf of a unit.

reduced in size. Leaves are spread more uniformly along the axes in *Cordia nitida* (Figure 15.8B). In addition, the lowest distal leaf of the branch unit (L in Figure 15.8) always points forward in *C. nitida,* filling the space between the two new daughter units.

In some trees, the branches themselves may compose the main photosynthetic surface of the crown, as with the fleshy expanded stem axes of tree euphorbs and cacti. In species of Cupressaceae, the tiered dorsiventral branches bear scale leaves, so that the branch axes can be considered the functional photosynthetic area, rather than individual leaves.

Horizontal layers of foliage can be constructed in several ways. The development of several adjacent lateral buds on the trunk forms a tier of branches, each one growing horizontally. The leaves may be borne as radially symmetrical rosettes in the plane (as in Figure 15.7A,B), or they may form a dorsiventral surface along each branch (as in elm, beech, *Annona,* and many other species), or they may be appressed to the branch axes (as in Cypressaceae). Another phenomenon of tree growth results in a layered leaf surface from nonhorizontal axes. The vertical growth of twigs along an oblique branch axis can be so correlated that all leaf-bearing distal twigs remain at the same horizontal level. This is illustrated by a lateral branch of *Neea* (Nyctaginaceae), a midsized tree native to the coasts of Central America (Figure 15.9). The branch units terminate in an inflorescence, so that leaves are carried only by newly produced units. The basal region of an oblique lateral branch is shown in Figure 15.9. The branch is

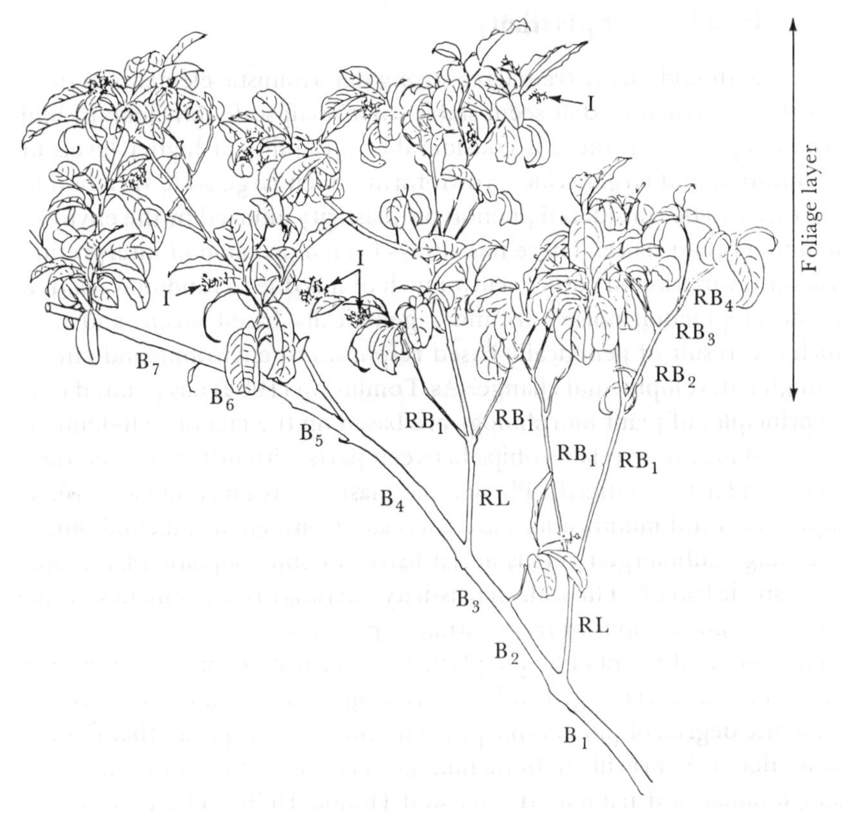

Figure 15.9. Part of a lateral branch of *Neea amplifolia* Donn. showing oblique main axis and horizontal foliage layer; viewed from the side. This small tree is native to coastal Costa Rica. Each branch unit terminates with an inflorescence (I) and produces two new units (B). The main branch axis is composed of a series of units (B_1, B_2, . . .); reiteration leaders (RL) developed at the base of the branch establish a series of new oblique major axes (RB_1, RB_2, . . .) of reiteration branches.

composed of units $B_1 - B_5$, which have long since lost their leaves. As the branch ages, buds are released to produce a series of vertical units (RL) that behave as diminutive leaders, each producing two to three lateral branches composed of branch units ($RB_1 - RB_3$). The growth of axes, which are like small leaders and repeat the architectural model, are termed reiterations (Hallé et al. 1978). The growth of the distal horizontal units of the original branch (not shown in Figure 15.9) and the vertical reiterations are in such harmony along the branch that a horizontal foliage layer is formed and maintained. In this case, both branching pattern and correlative growth determine the configuration of the leaf surface.

Problems of plasticity

Although many trees have strong deterministic elements in their growth and structure that seem stable on a small scale (phyllotaxis, bud form, twig growth), there is considerable variability at higher levels of organization and larger scale (crown form). Such large-scale variation in structure can be considered phenotypic plasticity induced by the environment in a broad sense. There has always been a problem of interpreting structural variation in nature. How much of it is due to random variation or "developmental noise" around a genetically based mean, and how much is a result of genetically based responses to environmental effects through a developmental change? As Tomlinson (1982) has pointed out, the principles of plant morphology are based on the fact of well-defined, structural means or relationships between parts with little variation (i.e., small standard deviations). Phenotypic plasticity is more obvious when major structural modifications are correlated with environmental differences (e.g., submerged versus aerial leaves of some aquatic plants; sun versus shade leaves). The scale of plasticity can range from branches within a crown to populations of trees within a species.

The potential for phenotypic plasticity of crown geometry is revealed when clonal material is grown in contrasting environments. We shall examine the degree of phenotypic plasticity shown by a species that follows the model of Aubréville, a branching pattern found in a wide variety of plant families and habitats (Fisher and Honda 1979). The erect trunk periodically produces a tier of five horizontal lateral branches (Figure 15.6C,D). Each branch consists of a series of well-defined branch units (Figure 15.6A). The unit is initially horizontal and then turns up into an erect, leaf-bearing short shoot. At its distal end, the branch unit normally produces two new horizontal units that continue to enlarge the branch (Figure 15.7). Shoot cuttings taken from a single tree of *Terminalia muelleri* Benth. were rooted and grown in pots under light shade in Miami, Florida, until they were about 1 m tall. Five pots were moved into full sun, and five were moved into an adjacent shade house that had approximately 90% shade. After 10 months, the plants were observed and parameters recorded for regions of each plant that were produced before (= old) and during (= new) the 10-month treatment were recorded. The means of these parameters under sun and shade conditions are given in Table 15.1.

The angles of new daughter branch units in the horizontal plane (θ_1 and θ_2) were not significantly different in the two environments (Table 15.1). The absolute lengths of mother and daughter branch units (both the first and the later-produced units on a lateral branch complex) were different

Table 15.1. *Parameters of crown geometry for a clone of* Terminalia muelleri *Benth. grown in pots in full sun or in 90% shade for 10 months (means from 5 plants per treatment)*

Parameter[a]	Sun	Shade	Difference[b]
Branching angle			
θ_1 (degrees)	33.0	31.9	NS
θ_2 (degrees)	39.4	43.9	NS
First unit on branch			
First mother length (cm)	16.6	22.1	**
Unit-1 length (cm)	12.5	16.7	**
Unit-2 length (cm)	11.1	16.1	**
Ratio unit-1/first mother (R_{01})	0.76	0.77	NS
Ratio unit-2/first mother (R_{02})	0.68	0.74	NS
Subsequent branch units			
Mother length (cm)	14.1	17.4	**
Unit-1 length (cm)	10.1	13.3	**
Unit-2 length (cm)	9.5	12.4	*
Ratio unit-1/mother (R_1)	0.72	0.77	NS
Ratio unit-2/mother (R_2)	0.61	0.70	NS
New leader and mother units			
New leader-unit length (cm)	23.1	75.8	**
New mother length (cm)	15.8	21.8	**
Ratio new-mother/new-leader (R_B)	0.73	0.30	**
Branch angle (θ_{0i})[c]			
New branches (degrees)	63.1	75.8	*
Old branches (degrees)	73.9	89.3	**
Leaves per cluster			
Old units	8.48	6.19	**
New units	11.32	5.88	**
Leaf angle[c]			
Above plane of branch (degrees)	54.6	81.2	**
Below plane of branch (degrees)	112.7	99.3	NS

[a] Symbols for parameters are illustrated in Figure 15.6 and are the same as those used by Fisher and Honda (1977).
[b] Statistical significance of the difference determined by a *t*-test: ** = 0.99 level; * = 0.95 level; NS = not significantly different.
[c] Angles measured from vertical with horizontal = 90°.

under the two conditions. However, the ratios of daughter to mother units (R_1 and R_2) were not different. Thus, although branch units were longer in the shade than in the sun, the proportional relationships of sequential units remained constant. Both the lengths of the subjacent leader units and the ratios of the subjacent leader units to first lateral branch units of the new branch (R_B) were different. Although both the new leader and branch units were longer in the shade, the leaders were elongated more relative to the lateral branches. The lateral branches were more nearly horizontal (θ_{0i} closer to 90°) in the shade than in the sun. Finally, there were more leaves per cluster (= per branch unit) in the sun, and these leaves were held closer to the vertical than those in the shade. Leaves were separated into two groups, those above and those below the plane of the branch complex, to facilitate measurement and comparisons.

We concluded that branching angles and relative branch-unit lengths were not plastic under the two environmental conditions of this experiment. Absolute unit lengths, orientation of lateral branch complexes, leaf number per cluster, and leaf angles all showed phenotypic plasticity in this particular clone.

Several examples of obvious plasticity in features of branching, orientation, growth, and ultimately crown shape can be cited for trees. The amount of branching, as measured by the bifurcation ratio, which is the ratio of the numbers of branches in successive orders of branching in the shoot, was found to vary within one crown (Borchert and Slade 1981; Steingraeber 1982) and within the same species growing in different habitats (Steingraeber et al. 1979; Pickett and Kempf 1980; Kempf and Pickett 1981; Boojh and Ramakrishnan 1982; Veres and Pickett 1982). Branches tended to be long and more nearly horizontal in the lower shaded region of the crown of *Quercus* (Pickett and Kempf 1980).

The amount and direction of sunlight have major effects on crown shape and branching pattern. Unidirectional side lighting results in a crown that is distorted by extreme one-sided growth (Fisher and Hibbs 1982), the riverbank effect of Hallé et al. (1978). The spreading crown of an isolated, open-grown tree is an unnatural situation for most species and represents the extreme case of the riverbank effect. Branching and growth in height may be strongly reduced in long-lived, suppressed saplings that are later released after light gaps form in the forest canopy. Kohyama (1980) documented differences in form and growth of *Abies* seedlings under extremes of exposure. Studies of *Populus* clones demonstrated genetic differences in the responses of branch angles to crowding, in the form of a decrease in the first-order branch angles; in effect, the trees "reached" for the light (Nelson et al. 1981).

Other examples of plasticity in trees can be cited. Physical factors such as wind stress (Lawton 1982) greatly affect elongation and thickening of branches, as do water stress and the deforming effects of snow and ice loads. Displacement of axes of many species results in the production of specialized reaction wood that is associated with the reorientation of trunks and branches back to their original positions (Wilson and Archer 1977). However, reorientation of the leaning axes of some soft-wood species may occur without reaction-wood formation (Fisher and Mueller 1983).

Finally, a tree's ability to develop architectural reiteration, in which the sapling pattern of growth repeats itself as a consequence of injury or aging of the original shoot, can have a significant effect on crown form and may ensure survival after trauma. Variations in architectural plasticity in tropical trees were discussed by Fisher and Hibbs (1982) and de Castro e Santos (1980).

Simulations of tree branching

In the previous sections of this chapter, the structural elements and their variability, which establish the tree crown, were reviewed. Using these mechanisms of pattern formation, we now turn to analysis of the branching patterns themselves. Simulations of branching have clarified the control mechanisms and geometrical rules of branching in several trees. Simulations have also allowed theoretical studies of optimal branching patterns and angles in trees.

The complex spatial configuration of branches in a tree was first successfully simulated by Honda (1971) in a theoretical computer study. He demonstrated that simple but realistic tree shapes could be produced using only two parameters: the branching angle of daughter branch to mother branch (in which asymmetry of the two daughter branch angles is possible) the relative lengths of successive branches (causing a geometrical decrease in length at higher orders of branching). His work showed that overall tree crown shapes characteristic of different species can be determined with relatively little information. A small and seemingly insignificant change in one parameter can yield a significant change in crown form after many iterations or branch orders. He did not investigate the possible causes of differences in branch length or angle.

The original model was later refined and modified to simulate a number of different real trees in a series of studies by Honda and associates and has served as a basis for optimization studies (Fisher and Honda 1977, 1979; Honda and Fisher 1978, 1979; Honda et al. 1981, 1982; Borchert and Honda 1984). Rules for branching, geometrical ratios of the parts, limitations on the amount of bifurcation, and real quantitative values were added to permit simulation of branching patterns and geometries that are very similar to those of real trees. However, before reviewing the optimi-

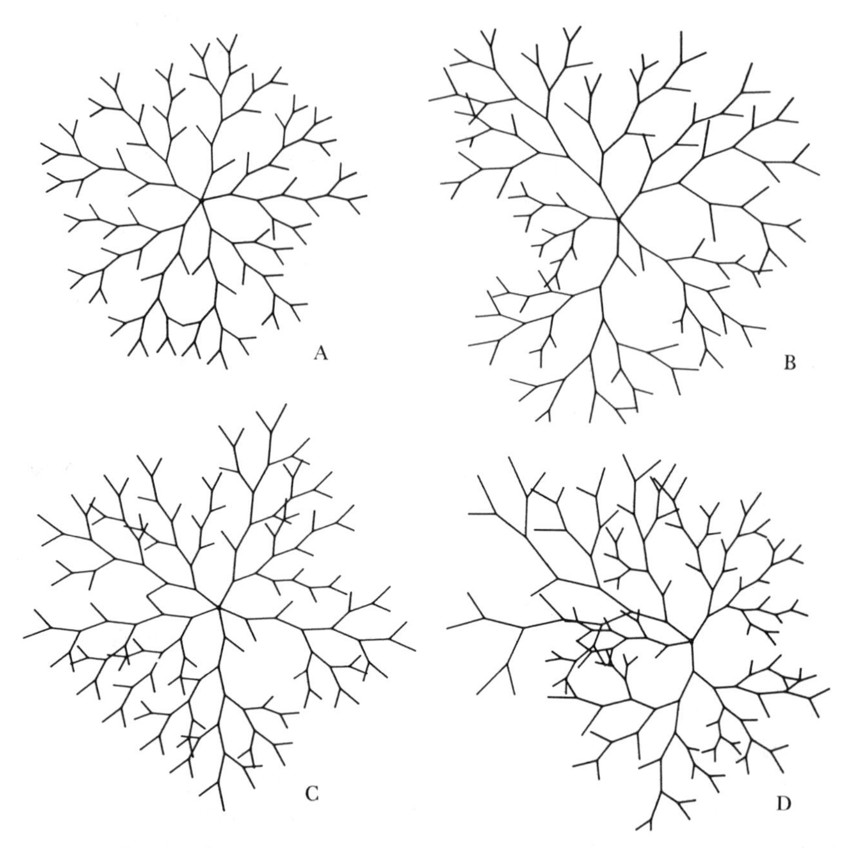

Figure 15.10. Computer-drawn simulations of branch tiers of *Terminalia catappa* viewed overhead, as shown in Figure 15.6D. The trunk is the central circle. A: Deterministic simulation using real means for branching angles and branch length ratios. B: Stochastic simulation using real means and standard deviations for branching angles and branch length ratios. C: Deterministic simulation using optimal values. D: Stochastic simulation using optimal means and real standard deviations. Branching in A–D was limited by branch interaction (Honda et al. 1981), and values used were taken from Fisher and Honda (1979) for A and B, Honda and Fisher (1978, 1979) for C and D. See Figure 15.6 for definition of branching angles (θ_1 and θ_2) and branch length ratios (R_1 and R_2).

zation studies based on these models, we shall briefly review some of the assumptions made in tree modeling.

Models of branching and tree growth have been critically reviewed by Waller and Steingraeber (1985), and some points they have raised can be illustrated with simulations of *Terminalia*. Quantitative spatial models that can deal with problems of adaptive geometry, such as those by Honda and associates cited earlier, can be deterministic or stochastic.

Deterministic models have no random variation of parameters, and all simulations are identical, as in simulations of a branch tier of *Terminalia* (Figure 15.10A,B). The values used for branching angles and branch-unit lengths are constant and are based either on empirical averages (Figure 15.10A) or on theoretical optimal averages (Figure 15.10B).

Alternatively, stochastic models feature random variations of parameters; this makes each simulation unique and thus more natural. Two representative stochastic simulations are shown for branching angles and unit lengths based on empirical (Figure 15.10C) and theoretical means (Figure 15.10D), with observed standard deviations used in both. Rhizome systems, trees, and root systems have been simulated using the known probabilities of branching events, orientations, and/or lengths (Cochrane and Ford 1978; Bell et al. 1979; Harper and Bell 1979; Niklas 1982; Henderson et al. 1983; Reffye 1983; Remphrey et al. 1983; Bell 1984; Niklas and Kerchner 1984; Remphrey and Powell 1984a). However, although stochastic models produce a population of realistic simulations, the introduction of such observed variations into the model may not be useful if mechanistic or evolutionary causes are of interest to the researcher. For example, the mean branch-unit length used in Figure 15.10A is considered to be representative of one population of units for a tree. In fact, the original observations were pooled from several overlapping populations of units showing plastic responses to differing environments. Branch units are shorter when they develop in full sun, during water stress, or on senescing branches. Units are longer on vigorous suckers, within the shaded canopy, or under favorable water and temperature conditions.

Maillette (1982) has shown that there are two populations of buds in the crown of *Betula* that are distinct in their branching behaviors. Similarly, Hatta (1980) found different frequencies in the patterns of shoot formation in the upper, middle and lower crown regions of *Cornus*. The amount of branching also varies within different parts of the crown in *Acer* (Steingraeber 1982) and *Quercus* (Kempf and Pickett 1981). In *Populus*, the frequencies of long versus short shoots vary with crown position (Isebrands and Nelson 1982). In *Larix*, features of bud contents and bud growth fall into distinct classes or populations based on crown position and shoot type (Remphrey and Powell 1984b).

A possible approach to more biologically realistic stochastic simulations would be to use conditional probabilities or, more specifically, Markov processes (Drake 1967), in which the initial distribution and the transitional probability would need to be determined at each bifurcation or node in the tree model. In this type of stochastic model, the present state determines the probability of the next state. Thus, the probability distribution of branch orientation will be quite different in a branch unit developing from a nearly horizontal mother unit versus one developing from a rare, downwardly inclined unit. This phenomenon can be seen in nature when the second case is almost always followed by an upwardly directed branch unit. This orientation brings the branch back to the genetically determined angle by the gravitropic behavior of an elongating unit or the formation of reaction wood in an older unit. Clearly, a great effort would be required to collect the empirical data necessary for determining the transition probabilities for a number of states of each parameter.

Returning now to the basic types of models, we find that both deterministic and stochastic models can be either stationary or nonstationary with respect to the rules of growth. It is important to decide if growth rates or probabilities of branching remain constant (i.e., stationary) or if they change with time (i.e., nonstationary). In nature, trees may initially show stationarity during early vigorous growth of a leader or a lateral branch, but eventually growth and branching frequencies decline, as documented in a variety of trees (Fisher and Honda 1977; Borchert and Honda 1984; White 1984).

The usefulness of these various types of models depends on the purpose of a simulation. Deterministic models can test the spatial effects of varying the parameters and mechanisms that underlie static branching, and in optimization studies. Nonstationary models reflect environmental or density-dependent effects that change with time, size, or position and thus limit growth (Honda et al. 1982; Borchert and Honda 1984). Stochastic models are appropriate for population studies, analysis of real-tree data, and species comparisons, as well as for predictive simulations of one or more realistic trees (Burk et al. 1983; Reffye 1983).

Adaptive significance

Having established the mechanisms of development and the structural elements of tree branching, we come to the task of evaluating the functional significance, if any, of these features. How does a particular feature affect a presumed function? Direct experimentation by changing a structure and observing the effect on function is difficult, at best, or impossible with three branches. Two other approaches are possible: simulation modeling (which in effect is a theoretical experiment) and observing the correlations between tree structure and environment (which might indicate cost–benefit or cause–effect relationships between structure and function).

Optimization studies

To say that a structure is optimal implies that it is the "best" or the most efficient for its environment or its particular function. Many studies have been published relating observed stem structure, orientation, or size to mechanical properties (McMahon and Kronauer 1976; Yamakoshi et al. 1976; King and Loucks 1978; King 1981), to generalized crown shape (Paltridge 1973; Niklas and Kerchner 1984), or to heat exchange (Nobel 1978; Woodhouse et al. 1980, 1983). It has been found that the empirical data closely fit the theoretical values considered optimal for carrying out a function (e.g., stem size to support a load, or stem orientation to maximize interception of winter sun). However, the practice of citing the closeness of fit between features in real plants and predicted values as evidence for optimal evolutionary adaptation has been criticized as circular and Panglossian (Gould and Lewontin 1979). A strong defense for the adaptive optimal argument was offered by Horn (1979), who stressed the usefulness of this approach in comparing different species and different environments. In this way, one can predict what feature is best for a given environment. Analyses of leaf shape and size with respect to adaptive features have been especially productive, as reviewed most recently by Givnish (1984). The collaborative studies of Honda, Tomlinson, and Fisher can serve as examples of the strengths and weaknesses of such an approach to branching patterns.

Our model system has been the tropical almond, *Terminalia catappa*, which is ideally suited for simulation studies because of its geometrical simplicity, which has already been described and illustrated (Figure 15.6). The tight clustering of leaves on the branch units as distal rosettes (Figure 15.7A,B) allows an approximation of leaf surface by discs that are aligned in a plane like the branch axes themselves (Fisher and Honda 1979).

The native environment of *T. catappa* is the coastal open vegetation of Southeast Asia. Because it tends to grow in fully exposed habitats near the equator, direct midday sunlight from the zenith is a realistic ecological feature, and thus light interception and possible casting of deep shade below the crown are ecologically significant characteristics of the branch geometry. The deterministic model assumes that axis forking is nonstationary and declines in frequency away from the trunk and that branch-unit length decreases in a stationary manner as a constant ratio. The two branching angles, θ_1 and θ_2 (Figure 15.6B), can be equal or unequal but are constant for a given simulation. The values used for the basic unmodified simulation are observed means (Figure 15.10A). Total leaf area and effective area, the unshaded or projected area, for simulated lateral branch

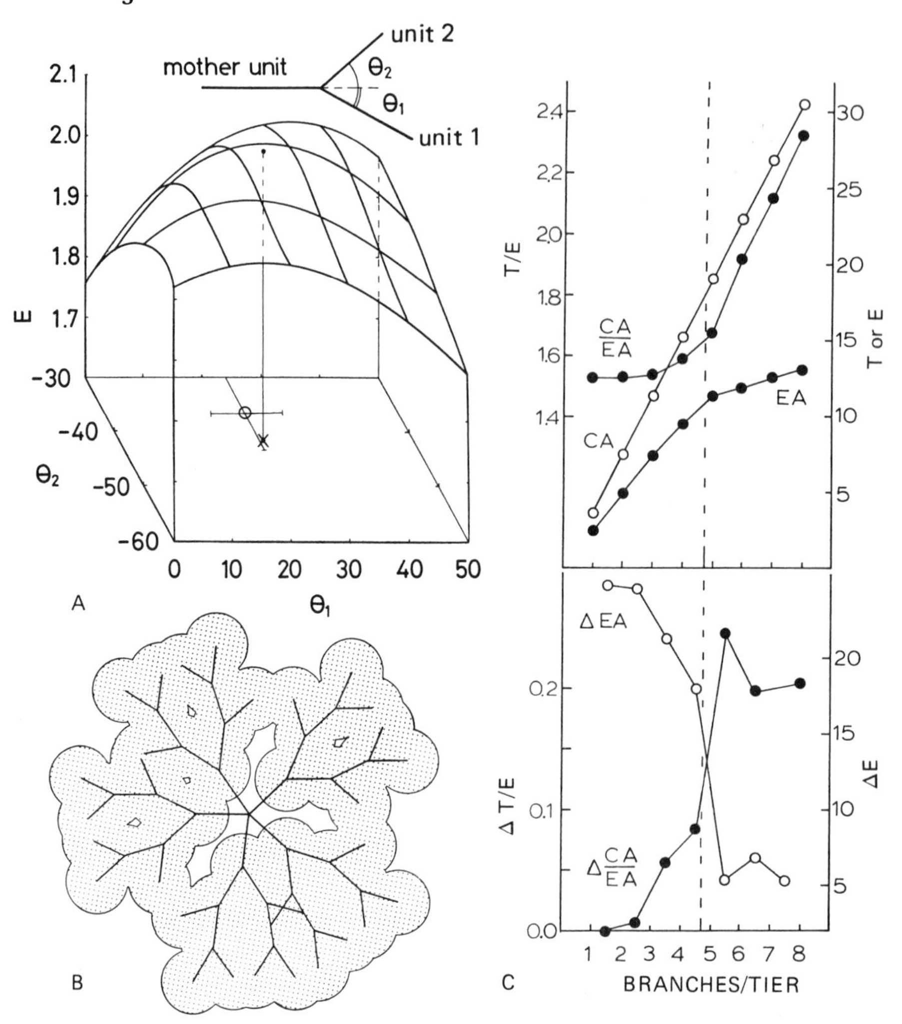

Figure 15.11. A: Effective leaf area (E) in computer simulation of a branch tier of *Terminalia* plotted against branching angles, θ_1 and θ_2. Theoretical optimal angles, ×; real average angles, ○, with bars indicating ±SD. B: Simulated branch tier using optimal angles; effective leaf area is shown (A and B after Honda and Fisher 1978). C: Relationship between number of branches in a tier and total leaf area (T), effective leaf area (E), and ratio T/E from computer simulations of *Terminalia*. Values for the slopes of E and T/E are plotted in the lower graph. Real average number of branches = 4.73, indicated by broken line (from Honda and Fisher 1979).

tiers were calculated by summing the disc approximations and their projections (Figure 15.11B), respectively. When both branching angles were varied, the maximum effective area was produced by a set of asymmetric angles (Figure 15.11A,B). These angles were not significantly different from the mean angles of real trees (Honda and Fisher 1978). It was concluded that, all else being equal, the observed angles are optimal for packing leaf clusters in a horizontal plane. Similarly, the observed unequal lengths of the two branch units at a branch fork were close to theoretical values that gave a minimum variance for the effective area of individual leaf clusters (i.e., the most nearly homogeneous distribution of leaf surfaces in a tier) (Honda and Fisher 1979). The commonly observed asymmetries of branching angles and branch units in trees are related to efficiency of packing and mechanical stability.

One might hypothesize that branching angles optimize the distribution of leaf surfaces for each newly produced branch unit; new branch units would grow out at an angle that would maximize effective leaf area. Such a sequential optimization was carried out with the *Terminalia* model for three orders of asymmetrical branching. Very unreal branching angles were produced with increasing orders of branching (Fisher and Honda 1979, Table 5). Thus, the hypotheses that new branch units grow out to produce optimal filling of the space immediately available can be rejected. In real *Terminalia* trees, the branching angles appear to be strongly determined, with only moderate variations due to local shading. This is reasonable, because the final pattern of the entire tier affects the fitness of the individual tree, rather than the position of a new branch unit within that tier. The benefits of light interception, maximum photosynthesis, and shading of competitors must be balanced with the costs of heat load, transpiration, wind drag, and overall mechanical stability to form an integrated whole.

The number of branches in a tier also seems to reflect efficient leaf display. Because a lateral branch arises from a bud associated with each leaf, the position of the branch on the trunk mirrors the position of the leaf in the phyllotactic spiral. In *Terminalia,* the buds of usually five adjacent leaves grow out to form the five-branched branch tier (Figure 15.6D). Because of phyllotaxis, successive branches have divergence angles of 138.5°. To study the significance of branch number ($\bar{x} = 4.73$), simulations were run in which only branch number was varied. When the slopes of the total and effective leaf areas, which increase as additional branches are added to the tier, are plotted against branch number (Figure 15.11C), we see marked changes between five branches and six branches; at this point the amount of increased effective area, (i.e., its slope) shows its

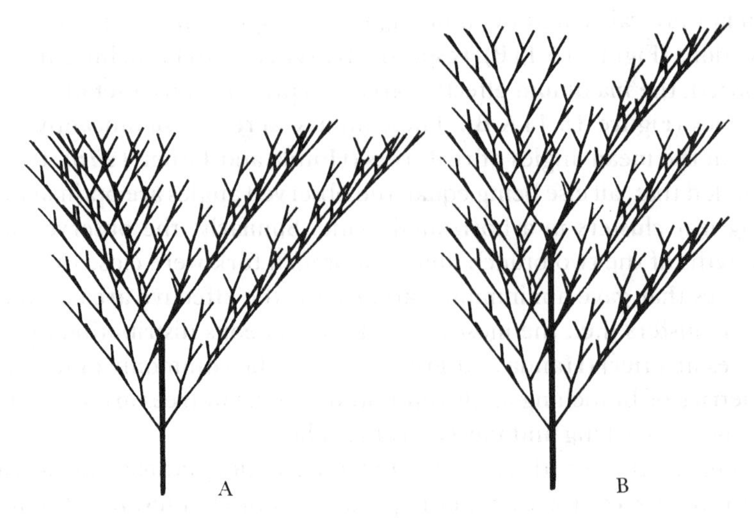

Figure 15.12. Computer simulations of wide (A) and narrow (B) crowns of *Tabebuia* from which data for Table 15.2 were obtained. Only half the crown is plotted in side view (from Borchert and Tomlinson 1984).

largest drop, and the proportion of total area to effective area, which is an index of intratier shading, shows its largest rise. Thus, five appears to be the optimal number of branches per tier in terms of a large effective leaf area with minimum self-shading.

The branch divergence angle of 138.5° in *Terminalia* is close to the ideal angle of 144° that would result in five evenly spaced branches. To my knowledge, only one species has been reported to have such an arrangement. *Anisophyllea* (Rhizophoraceae) has a 2/5 phyllotactic arrangement on the leader, which is a 144° branch divergence, resulting in adjacent horizontal branches being 74° apart (Vincent and Tomlinson 1983). In contrast, *Aralia* (Araliaceae) also has a 2/5 phyllotaxy, but only two or three branches develop below the terminal inflorescence (White 1984). The fact that tiers of five branches occur in many tiered species (e.g., *Cordia* species, other *Terminalia* species, *Cornus alternifolia*) may be a consequence of optimal packing, with divergence angles of branches that are near the ideal Fibonacci angle of 137.5° that is directly related to the optimal packing of leaves on the leader (Leigh 1972).

In the foregoing studies, only one variable was changed, and all others were held constant. From this procedure, the optimal parameter value for displaying leaves efficiently was determined for a set of angles, branch lengths, or tier sizes. It was assumed that all other things remained equal. Although such determinations are useful in understanding an adaptive

Table 15.2. *Comparison of simulated crown characteristics of wide and narrow* Tabebuia *crowns after 21 growth increments, as shown in Figure 15.12*[a]

	Units		
	Wide crown	Narrow crown	Narrow/wide (%)
Leaf area	14,592	11,520	78.9
Projected crown	12,059	4,063	33.6
Weighted sum of branches	3,667	2,028	55.3
Cost	4.53	3.41	75.2
Leaf-area index	1.21	2.84	234

[a] From Borchert and Tomlinson (1984).

function, they must be interpreted in the context of the entire organism. Because branching angles and relative branch lengths both vary in nature, and presumably both undergo natural selection together, an artificial combining of their optimal values produces unrealistic simulations of branch tiers. In real trees, any change in size, angle, or development of a branch is bound to affect other aspects of tree form, because all parts are developed in harmony.

The efficiency of branching geometry in *Tabebuia,* a tree with a bifurcating branch system, was examined by Borchert and Tomlinson (1984) using a spatial deterministic model. It included a realistic regulation of branching frequency by hypothetical flux distributions within the branch system, as presented by Borchert and Honda (1984). In their simulations, Borchert and Tomlinson compared the increase in leaf area of the crown with an index of the support cost per unit leaf area. They necessarily used quite crude estimates of total and projected leaf areas. Cost was viewed as branch biomass, which was approximated in relative terms by the weighted sum of branch units, the number of terminal branch units supplied by every lower-order branch unit. They then simulated wide and narrow crowns, as defined by the branching order at which reiteration or trunk extension began (Figure 15.12), and calculated the estimated values for leaf area, cost, and leaf area index (LAI) during simulated growth (Table 15.2). A wide crown has more leaf area, more branches, and greater cost than a narrow crown at the same age. However, a narrow crown has a very much larger LAI, the total leaf area divided by the projected leaf area on the ground. Thus, the narrow crown has a larger LAI/cost ratio, an indication of greater efficiency in presenting leaf surfaces over a given area of ground. Tall, narrow crowns composed of many tiers would be more efficient than wide, open crowns. Such crowns occur in crowded, forest-grown trees, similar to those of *Terminalia* (Fisher and Hibbs 1982). However, open-grown trees of *Tabebuia* produce wide crowns, an observation that indicates that the simple model and assumptions for efficiency of support of leaf surfaces are not applicable to these trees or growing conditions. Nevertheless, such a theoretical approach is a promising beginning toward understanding the adaptive significance of crown form.

Niklas and Kerchner (1984) have recently determined the optimal solid shapes for relative photosynthetic efficiency, and they have analogues in the flattened thalli of some bryophytes and vascular plants. The model assumes that the surface exposed to direct sun integrated over a day is directly related to the photosynthetic efficiency of the whole plant. Further simulations using a wide range of values for branching angle, rotation angle, and probability of branching in the model gave a variety of theoretical tree shapes. The optimal shapes for photosynthetic efficiency and total moment arm, which relates to mechanical stability, tended to be similar to the shapes of many fossil and living lower vascular plants. Such morphological trends as overtopping (= asymmetric branching angles) and plantation found in the evolution of early vascular plants were broadly interpreted as trends in optimizing the display of photosynthetic tissue (i.e., the display of the axes themselves, because they lack leaves) or in optimizing the balance between photosynthetic efficiency and mechanical stability. This approach seems reasonable from a structural view in these leafless plants, which have only cylindrical stem axes as their photosynthetic organs. However, the presence of leaves on axes greatly complicates the geometry and would make this approach inappropiate for most trees.

Ecological correlations

A second approach to understanding the adaptive significance of branching patterns is to establish correlations between structural characteristics and ecological features such as habitat or successional status.

Bifurcation ratio (R_b), an index of the degree of branching from one order of branching to the next, initially held the promise of a powerfully objective, quantitative tool for comparing tree structures. In an early but limited survey, Whitney (1976) found that R_b was related to successional status. He suggested that high R_b values are associated with early successional species, and low values with late successional species. Unfortunately, the later and more detailed studies showing that R_b varies within a given crown (Borchert and Slade 1981; Steingraeber 1982) and within a given species growing in differing habitats (Steingraeber et al. 1979; Pickett and Kempf 1980; Kempf and Pickett 1981; Boojh and Ramakrishnan 1982; Veres and Pickett 1982) limit the usefulness of this index. In addition, the important aspect of crown geometry is ignored by a measure of the amount of branching in which there is no spatial component.

On a more complex and larger scale, differences in branching patterns and growth were observed by Boojh and Ramakrishnan (1982) for early and late successional trees. Early trees tended to have vertical, indeterminate, and sylleptic (i.e., immediate outgrowth of lateral branches) growth, whereas late successional species had more horizontal, determinate, and proleptic (i.e., delayed outgrowth of lateral branches) growth. White (1983a) found a complex correlation between successional status and branching density, because large leaf size and reduced branching are correlated to each other and to shade intolerance. Givnish (1984, and earlier studies cited therein) has interpreted the occurrence of large-leaved and little-branched early successional trees in cost–benefit models in which large-leaf area or rapid-extension growth is enhanced and the investment in support tissues is reduced by production of light-wooded twigs or "throwaway" branches (= compound leaves).

A difficult but potentially productive area of research involves delineation of those features of architecture or crown geometry that are common to, and presumably convergent in, habitats with similar environmental variables, especially light levels. The correlations between climate and life forms (Böcher 1977) can be extended to the parameters of branching and patterns of foliage distribution. This has been attempted in a preliminary way by Brunig (1976). His different tree ideotypes combine features of leaf and branch distribution that relate to sun and wind exposure. The various ideotypes intercept light in different ways throughout the day; he speculated that this had adaptive value in particular habitats or successional status. In high-moisture habitats, crown shapes should shift from cylindrical, self-shading forms toward flat-topped, fully exposed forms, a trend that Givnish (1984) found in species of deciduous trees in central United States.

In an effort to clarify the utility of architectural models (i.e., the deterministic growth plan rather than the final form of a mature tree), Foresta (1983) distinguished between the architectural spectrum of a flora (% of total species) and the realized architectural spectrum (% of total trees). The former gives a index of floristic diversity and will include many types of habitats. The latter indicates the architectural composition of a selected population or community and has greater ecological significance. He noted the architectural models found in all woody trees present 3.5 years after the clear cutting of a rain forest in French Guiana. He found that 14 species accounted for 86% of the early successional trees and together used only six architectural models, as shown in Figure 15.5. In fact, 71% of all pioneer trees were included in only two models, those of Rauh and Roux (Figure 15.5A,B). The same two models also predominate in the pioneer trees of West Africa and Borneo (Foresta 1983). In the uncut primary rain forest, 85% of trees followed one model, Troll's model (Figure 15.5I), which perhaps mirrors the dominance of legumes in that forest.

Whether a correlation between branching structure and habitat is due to deterministic plans or phenotypic plasticity, such correlations do suggest a common structurally adaptive feature. A widespread feature of planar foliage surfaces, the monolayers of Horn (1971), is a case in point. There is a convergence of different tree architectures that can produce the same overall foliage presentation because of the number of factors affecting leaf position in the tree crown. Thus, tiered crowns or pagoda trees can be produced by several architectural models (e.g., *Terminalia,* Aubréville's model; *Cordia,* Prévost's model; *Fagarea,* Fagerlind's model; *Araucaria,* Massart's model). Conversely, a given model can give rise quite different crown forms; for example, Aubréville's model can produce a solid crown (old *Terminalia catappa*) or a hollow-shell crown (*Manilkara*), and Troll's model can produce the differing crown forms of *Fagus, Ulmus,* and *Acacia* (Fisher and Hibbs 1982).

Conclusions

Branching patterns and angles in woody plants play important roles in establishing both crown form and leaf position. The position and orientation of shoot axes are determined by primary organization and growth and often by later secondary growth. However, many other factors of structure and phenology affect the placement of photosynthetic surfaces within a plant. Similar crown forms can develop by a convergence of different tree architectures; conversely, a given branching pattern or architecture can give rise to different crown shapes.

Spatial models of branching patterns have been devised that account for complex tree form with relatively simple rules and few parameters of branching. In geometrically simple trees, like *Terminalia,* that have regular clustering of leaves at the tips of branch units, deterministic computer simulations have shown that the observed parameters of branching (asymmetrical angles, relative branch lengths, and branch number per tier) are close to theoretical values that would optimize leaf packing. However, optimization studies must be interpreted with caution, because individual parameters are not selected for in isolation. In addition, most species show varying degrees of phenotypic plasticity in their branching parameters. A tree's capacity to respond opportunistically to environmental changes may well be more adaptively significant than its inherited or deterministic form.

Future research in this area should emphasize (1) the extent to which the structural parameters of branching are phenotypically plastic, which can be studied by use of contrasting environments such as sun/shade experiments, and (2) the correlations between native habitat and the parameters of the woody skeleton. Such studies are made difficult and lengthy by both the size and slow growth of trees, but the economic and biological significance of trees makes such an undertaking worthwhile.

Acknowledgments

I thank Carol Weeks and John C. Comfort for assistance with programming and computer graphics and the Computer Laboratory of Florida International University, Miami, for use of facilities. Advance copies of manuscripts were kindly provided by R. Borchert, H. Honda, K. Niklas, D. A. Steingraeber, and D. M. Waller. Figures 15.6, 15.8, and 15.9 were drawn by P. Fawcett. I deeply appreciate the careful and thought-provoking review by T. J. Givnish and an introduction to conditional probability provided by D. F. Parkhurst. This work was supported in part by the National Science Foundation (grant DEB 79-14635).

References

Bell, A. D. 1984. Dynamic morphology: a contribution to plant population ecology. Pp. 48–65 *in* R. Dirzo and J. Sarukhán (eds.), Perspectives on plant population ecology. Sinauer Associates, Sunderland, Mass.

Bell, A. D., D. Roberts, and A. Smith. 1979. Branching patterns: the simulation of plant architecture. J. Theor. Biol. 81:351–375.

Böcher, T. W. 1977. Convergence as an evolutionary process. Bot. J. Linn. Soc. 75:1–19.

Boojh, R., and P. S. Ramakrishnan. 1982. Growth strategy of trees related to successional status. I. Architecture and extension growth. Forest Ecol. Manag. 4:359–374.

Borchert, R., and H. Honda. 1984. Control of development in the bifurcating branch system of *Tabebuia rosea:* a computer simulation. Bot. Gaz. 145:184–195.

Borchert, R., and N. A. Slade. 1981. Bifurcation ratios and the adaptive geometry of trees. Bot. Gaz. 142:394–401.

Borchert, R., and P. B. Tomlinson. 1984. Architecture and crown geometry in *Tabebuia rosea* (Bignoniaceae). Amer. J. Bot. 71:958–969.

Brunig, E. F. 1976. Tree forms in relation to environmental conditions: an ecological viewpoint. Pp. 139–156 *in* M. G. R. Cannell and F. T. Last (eds.), Tree physiology and yield improvement. Academic Press, London.

Burk, T. E., N. D. Nelson, and J. G. Isebrands. 1983. Crown architecture of short-rotation, intensively cultured *Populus*. III. A model of first-order branch architecture. Can. J. For. Res. 13:1107–1116.

Cochrane, L. A., and E. D. Ford. 1978. Growth of a Sitka spruce plantation: analysis and stochastic description of the development of the branching structure. J. Appl. Ecol. 15:227–244.

de Castro e Santos, A. 1980. Essai de classification des arbres tropicaux selon leur capacité de réitération. Biotropica 12:187–194.

Drake, A. W. 1967. Fundamentals of applied probability theory. McGraw-Hill, New York.

Fisher, J. B. 1972. Control of shoot-rhizome dimorphism in the woody monocotyledon, *Cordyline* (Agavaceae). Amer. J. Bot. 59:1000–1010.

– 1984. Tree architecture: relationships between structure and function. Pp. 541–589 *in* R. A. White and W. C. Dickison (eds.), Contemporary problems in plant anatomy. Academic Press, Orlando.

Fisher, J. B., and D. E. Hibbs. 1982. Plasticity of tree architecture: specific and ecological variations found in Aubréville's model. Amer. J. Bot. 69:690–702.

Fisher, J. B., and H. Honda. 1977. Computer simulation of branching pattern and geometry in *Terminalia* (Combretaceae), a tropical tree. Bot. Gaz. 138:377–384.

– 1979. Branch geometry and effective leaf area: a study of *Terminalia*-branching pattern. I. Theoretical trees. II. Survey of real trees. Amer. J. Bot. 66:633–644, 645–655.

Fisher, J. B., and R. J. Mueller. 1983. Reaction anatomy and reorientation in leaning stems of balsa (*Ochroma*) and papaya (*Carica*). Can. J. Bot. 61:880–887.

Fisher, J. B., and J. W. Stevenson. 1981. Occurrence of reaction wood in branches of dicotyledons and its role in tree architecture. Bot. Gaz. 142:82–95.

Foresta, H. de. 1983. Le spectre architectural: application à l'étude des relations entre architecture des arbres et écologie forestière. Bull. Mus. Nat. Hist. Nat. Paris, 4 sér., sect. B, Adansonia 5:295–302.

Givnish, T. J. 1984. Leaf and canopy adaptations in tropical forests. Pp. 51–84 *in* E. Medina, H. A. Mooney, and C. Vásquez-Yánez (eds.), Physiological ecology of plants of the wet tropics. Dr. Junk, The Hague.

Gould, S. J., and R. C. Lewontin. 1979. The spandrels of San Marco and the Panglossian paradigm: a critique of the adaptationist programme. Proc. Roy. Soc. Lond. B 205:581–598.

Hallé, F., R. A. A. Oldemann, and P. B. Tomlinson. 1978. Tropical trees and forests: an architectural analysis. Springer-Verlag, Berlin.

Harper, J. L., and A. D. Bell. 1979. The population dynamics of growth form in organisms with modular construction. Pp. 29–52 *in* R. M. Anderson, B. D. Turner, and L. R. Taylor (eds.), Population dynamics. Blackwell, London.

Hatta, H. 1980. Studies in the crown formation of *Cornus kousa*. I. Shoot elongation and branching pattern. Bull. Nat. Sci. Mus. (Tokyo) ser. B (Bot.) 6:65–76.

Henderson, R., E. D. Ford, and E. Renshaw. 1983. Morphology of the structural root system of Sitka spruce. 2. Computer simulation of rooting patterns. Forestry 56:137–153.

Honda, H. 1971. Description of the form of trees by the parameters of the tree-like body: effects of the branching angle and the branch length in the shape of the tree-like body. J. Theor. Biol. 31:331–338.
Honda, H., and J. B. Fisher. 1978. Tree branch angle: maximizing effective leaf area. Science 199:888–890.
– 1979. Ratio of tree branch lengths: the equitable distribution of leaf clusters on branches. Proc. Natl. Acad. Sci. USA 76:3875–3879.
Honda, H., P. B. Tomlinson, and J. B. Fisher. 1981. Computer simulation of branch interaction and regulation by unequal flow rates in botanical trees. Amer. J. Bot. 68:569–585.
– 1982. Two geometrical models of branching in botanical trees. Ann. Bot. 49:1–11.
Horn, J. S. 1971. The adaptive geometry of trees. Princeton University Press, Princeton, N.J.
– 1979. Adaptation from the perspective of optimality. Pp. 48–61 *in* O. T. Solbrig et al. (eds.), Topics in plant population biology. Columbia University Press, New York.
Isebrands, J. G., and N. D. Nelson. 1982. Crown architecture of short-rotation, intensively cultured *Populus.* II. Branch morphology and distribution of leaves within the crown of *Populus* 'Tristis' as related to biomass production. Can. J. For. Res. 12:853–864.
Jankiewicz, L. S., and Z. J. Stecki. 1976. Some mechanisms responsible for differences in tree form. Pp. 157–172 *in* M. G. R. Cannell and F. T. Last (eds.), Tree physiology and yield improvement. Academic Press, London.
Kempf, J. S., and S. T. A. Pickett. 1981. The role of branch length and angle in branching pattern of forest shrubs along a successional gradient. New Phytol. 88:111–116.
King, D. 1981. Tree dimensions: maximizing the rate of height growth in dense stands. Oecologia 51:351–356.
King, D., and O. L. Loucks. 1978. The theory of tree bole and branch form. Rad. Evniron. Biophys. 15:141–165.
Kohyama, T. 1980. Growth pattern in *Abies mariesii* saplings under conditions of open-growth and suppression. Bot. Mag. Tokyo 93:13–24.
Lawton, R. O. 1982. Wind stress and elfin stature in a montane rain forest tree: an adaptive explanation. Amer. J. Bot. 69:1224–1230.
Leigh, E. G., Jr. 1972. The golden section and spiral leaf-arrangement. Trans. Conn. Acad. Arts Sci. 44:163–176.
Maillette, L. 1982. Structural dynamics of silver birch. I. The fate of buds. II. A matrix model of the bud population. J. Appl. Ecol. 19:203–238.
McMahon, T. A., and R. E. Kronauer. 1976. Tree structure: deducing the principle of mechanical design. J. Theor. Biol. 59:443–466.
Mergen, F. 1958. Distribution of reaction wood in eastern hemlock as a function of its terminal growth. Forest Sci. 4:98–109.
Nelson, N. D., T. Burk, and J. G. Isebrands. 1981. Crown architecture of short-rotation, intensively cultured *Populus.* I. Effects of clone and spacing on first-order branch characteristics. Can. J. For. Res. 11:73–81.

Niklas, K. 1982. Computer simulations of early land plant branching morphologies: canalization of patterns during evolution? Paleobiology 8:196–210.

Niklas, K., and V. Kerchner. 1984. Mechanical and photosynthetic constraints on the evolution of plant shape. Paleobiology 10:79–101.

Nobel, P. S. 1978. Surface temperatures of cacti – influences of environmental and morphological factors. Ecology 59:986–996.

Paltridge, G. W. 1973. On the shape of trees. J. Theor. Biol. 38:111–137.

Pickett, S. T. A., and J. S. Kempf. 1980. Branching patterns in forest shrubs and understory trees in relation to habitat. New Phytol. 86:219–228.

Reffye, P. de. 1983. Modèle mathématique aléatoire et simulation de la croissance et de l'architecture du caféier Robusta. IV. Programmation sur micro-ordinateur du tracé en trois dimensions de l'architecture d'un arbre. Application au caféier. Café Cacao Thé 27:3–20.

Remphrey, W. R., B. R. Neal, and T. A. Steeves. 1983. The morphology and growth of *Arctostaphylos uva-ursi* (L.) Spreng. (bearberry): an architectural model simulating colonizing growth. Can. J. Bot. 61:2451–2458.

Remphrey, W. R., and G. R. Powell. 1984a. Crown architecture of *Larix laricina* saplings: quantitative analysis and modelling of (nonsylleptic) order 1 branching in relation to development of the main stem. Can. J. Bot. 62:1904–1915.

– 1984b. Crown architecture of *Larix laricina* saplings: shoot preformation and neoformation and their relationships to shoot vigor. Can. J. Bot. 62:2181–2192.

Scurfield, G. 1973. Reaction wood: its structure and function. Science 179:647–655.

Steingraeber, D. A. 1982. Phenotypic plasticity of branching pattern in sugar maple (*Acer saccharum*). Amer. J. Bot. 69:638–640.

Steingraeber, D. A., L. J. Kascht, and D. H. Franck. 1979. Variation of shoot morphology and bifurcation ratio in sugar maple (*Acer saccharum*) saplings. Amer. J. Bot. 66:441–445.

Tomlinson, P. B. 1982. Chance and design in the construction of plants. Pp. 162–183 *in* R. Sattler (ed.), Axioms and principles of plant construction. M. Nijhoff/W. Junk, The Hague.

Veres, J. S., and S. T. A. Pickett. 1982. Branching patterns of *Lindera benzoin* beneath gaps and closed canopies. New Phytol. 91:767–772.

Vincent, J. R., and P. B. Tomlinson. 1983. Architecture and phyllotaxis of *Anisophyllea disticha* (Rhizophoraceae). Gardens Bull. 36:3–18.

Waller, D. M., and Steingraeber, D. A. 1985. Branching and modular growth: theoretical models and empirical patterns. *In* R. Cook, L. Buss, and J. B. C. Jackson (eds.), Biology of clonal organisms. Yale University Press, New Haven, Conn.

White, P. S. 1983a. Corner's rules in eastern deciduous trees: Allometry and its implications for the adaptive architecture of trees. Bull. Torrey Bot. Club 110:203–212.

– 1983b. Evidence that temperate east North American evergreen woody plants follow Corner's rules. New Phytol. 95:139–145.

– 1984. The architecture of devil's walking stick, *Aralia spinosa* (Araliaceae). J. Arnold Arbor. 65:403–418.
Whitney, G. G. 1976. The bifurcation ratio as an indicator of adaptive strategy in woody plant species. Bull. Torrey Bot. Club 103:67–72.
Williams, M. W., and H. D. Billingsley. 1970. Increasing the number and crotch angles of primary branches of apple trees with cytokinins and gibberellic acid. J. Amer. Soc. Hort. Sci. 95:649–651.
Wilson, B. F., and R. R. Archer. 1977. Reaction wood: induction and mechanical action. Ann. Rev. Plant Physiol. 28:23–43.
Woodhouse, R. M., J. G. Williams, and P. S. Nobel. 1980. Leaf orientation, radiation interception, and nocturnal acidity increases by the CAM plant *Agave deserti* (Agavaceae). Amer. J. Bot. 67:1179–1185.
– 1983. Simulation of plant temperature and water loss by the desert succulent, *Agave deserti*. Oecologia 57:291–297.
Yamakoshi, K., et al. 1976. Optimality in mechanical properties of branching structure in trees (in Japanese with English abstract). Iyo-denshi To Seitai-kogaku 14:296–302.
Zimmermann, M. H., and C. L. Brown. 1971. Trees: structure and function. Springer-Verlag, New York.

16 Biomechanical constraints on crown geometry in forest herbs

THOMAS J. GIVNISH

Forest herbs show remarkable variations in leaf height, mode of leaf arrangement, and leaf shape at a given effective leaf size. These traits affect the amount of unproductive tissue needed to support and supply a canopy with given photosynthetic characteristics. Such traits are unlikely to affect photosynthesis directly by changing internal determinants of gas exchange, but they can do so indirectly by changing leaf environment. Selection on such traits may thus involve energetic tradeoffs between the photosynthetic benefits and mechanical costs associated with different crown geometries. Competition should favor plants whose crown geometry maximizes the difference between these benefits and costs in a given environment, and thus maximizes their net rate of energy return (Givnish 1979, 1982, 1984).

Here I shall present four case studies of biomechanical adaptation in forest herbs. These focus on the adaptive significance of (1) leaf height, (2) branched versus unbranched canopies, branching angle, and branching height in *Podophyllum,* (3) the arching growth habit in *Polygonatum* and *Smilacina,* and (4) leaf shape in *Viola.* In each case I shall examine how the amount of mechanical tissue required to support a given amount of leaf tissue varies with plant size and canopy geometry. In the first case, these mechanical costs are balanced against the photosynthetic benefits associated with differences in leaf height to predict optimal leaf height. In the latter three cases, data on the allometry of support tissue are used to analyze which canopy geometry would minimize the cost of supporting a given amount of photosynthetic tissue at a given height, and hence be favored by natural selection. An important conclusion of all these studies is that the biomechanically optimal canopy geometry is context-dependent and varies with plant size and/or the competitive environment.

Forest herbs possess many advantages for a study of mechanical constraints on plant growth form. First, they show considerable variation in canopy geometry. Second, analysis of the tradeoffs underlying differences in branching pattern and crown shape in herbs may cast light on analogous but less easily studied differences in tree form (Horn 1971; Givnish 1979, 1984; Honda et al. 1981; King 1981; White 1983, 1984; Borchert and Tomlinson 1984; Fisher 1984). Herbs are physically easier to study with regard to resource allocation than woody plants, by virtue of their smaller size. More important, many forest herbs are conceptually easier to study because they have a determinate pattern of growth, with aboveground shoots springing preformed from winter buds (Foerste 1884; Muller 1978; Kana 1982; Kawano et al. 1983). Determinate growth greatly simplifies the analysis of mechanical costs by removing two difficult questions: (1) Should the cost of a given structure be weighed against what that structure supports currently, or what it will support in the future? (2) How quickly should the cost of a structure be amortized relative to its lifetime and the plant's rate of growth?

I shall first consider the energetic tradeoffs associated with leaf height in forest herbs, drawing heavily on earlier work (Givnish 1982). Material for the remaining studies on branching in *Podophyllum,* canopy form in *Polygonatum* and *Smilacina,* and leaf shape in *Viola* is presented here for the first time.

Adaptive significance of leaf height in forest herbs

Competition for light should be an important selective pressure on leaf height in understory herbs, and it involves a tradeoff between mechanical costs and photosynthetic benefits (Figure 16.1). The proportion of an herb's resources that annually goes into leaves rather than stems should decrease with leaf height, reflecting the disproportionate increase in support tissue needed to ensure mechanical stability. This tends to reduce the growth rate of a plant and, acting alone, would favor plants that arrange their leaves flat on the ground (Givnish 1982). However, competition for light favors increased leaf height to prevent overtopping.

If a plant is much above the leaf height of its competitors, it should photosynthesize at some maximum rate that depends, in a complicated way, on environmental conditions. If it is much below the height of its competitors, it will photosynthesize at some minimum rate, depending on these same complex conditions and (in a very simple way) on the density of competing foliage. For if competing foliage is sparse, it is unlikely that a plant will be next to and thus under a competitor; the expected rate of photosynthesis per unit leaf biomass, averaged over many individuals of the same height, will thus depend only weakly on relative height (Figure 16.1). However, if the density of herbaceous foliage is high, a plant will

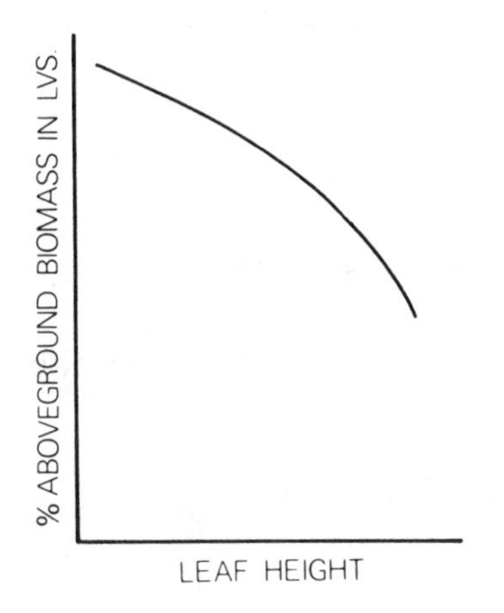

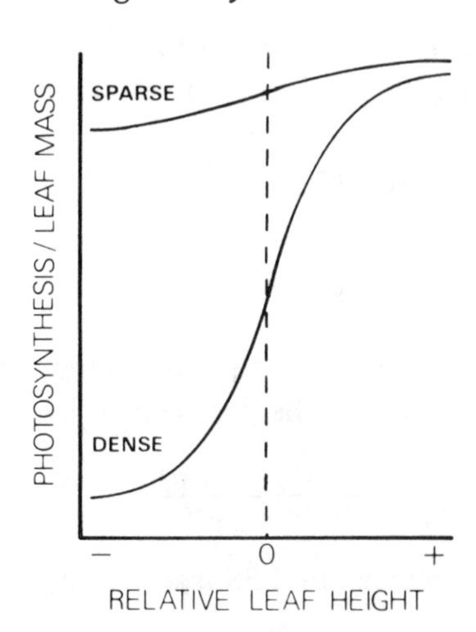

Figure 16.1. Economic tradeoffs associated with the evolution of leaf height in herbs. Left: Taller plants must divert more resources to support tissue to remain mechanically stable, and so should display a lower proportional allocation to foliage. Right: Balanced against this structural cost of greater leaf height is the expected photosynthetic advantage, averaged over many shoots, of holding leaves higher than those of a competitor. This advantage should be small in areas with sparse herbaceous cover, and larger where cover is denser (from Givnish 1982).

likely be next to and thus under a competitor, so that the expected rate of photosynthesis will increase sharply with relative leaf height (Figure 16.1).

Givnish (1982) suggests that these tradeoffs imply an evolutionary game, in which plants with similar photosynthetic responses and growing together under shared conditions should evolve taller leaves until the expected photosynthetic gain each makes by being slightly taller than an opponent is just balanced by the structural cost of a decreased proportion of energy allocated to leaves. Hence, where ecological conditions harshly limit total herbaceous cover, there is little photosynthetic advantage to an increment in relative leaf height, and the structural cost of absolute leaf height favors short plants. In areas where higher resource levels favor dense herbaceous cover, a height increment confers a larger advantage and favors taller plants at evolutionary equilibrium.

In the simplest case, involving competition between two morphs with average heights h_1 and h_2, the preceding conclusions can be stated more precisely. Let $f(h)$ be the proportion of aboveground biomass in photosynthetic tissue as a function of leaf height h, and let $g(h_i - h_j)$ be the photosynthetic rate per unit leaf biomass of the ith morph as a function of the difference between its height and that of its competitor. The height at which neither competitor can gain a net advantage by further increments of leaf height is then given by the solution of

$$\frac{\partial}{\partial h_1}[f(h_1) \cdot g(h_1 - h_2)]|_{h_1=h_2} = 0 = f'(h_1) \cdot g(0) + f(h_1) \cdot g'(0) \qquad (16.1a)$$

$$\frac{\partial}{\partial h_2}[f(h_2) \cdot g(h_2 - h_1)]|_{h_1=h_2} = 0 = f'(h_2) \cdot g(0) + f(h_2) \cdot g'(0) \qquad (16.1b)$$

This implies that the evolutionarily stable strategy (ESS) (Maynard Smith 1983) for leaf height is given by

$$-\frac{f'(h^*)}{f(h^*)} = \frac{g'(0)}{g(0)} \qquad (16.2)$$

where h^* is the ESS leaf height.

Tests of the model

Functional relation of leaf height to herbaceous cover

Givnish (1982) classified each of 72 autotrophic species occurring on an oak-woods/mesic-forest/floodplain-forest transect in northern Virginia into height classes, based on average maximum leaf height in the range 0–5 cm, 5–10 cm, 10–20 cm, 20–40 cm, 40–80 cm, and 80–160 cm. Certain dimorphic species (e.g., *Geum virginianum, Stellaria pubera*) produce leaves of different heights at different seasons, and the winter/spring and summer phases of these plants received different height scores. During 1976–1977, coverage by each species was measured on 13 census dates in ten 1-m^2 permanent plots at nine sampling sites along the transect. In most cases, the average cover in which a species finds itself was calculated using the average of the maximum herbaceous cover achieved yearly at the nine transect sites, weighted by the maximum coverage of the focal species at each site. Spring ephemerals and winter annuals reach peak cover before total herbaceous cover peaks, and so ambient cover for these groups was calculated using the census date for which coverage by these groups was at a maximum. Similarly, ambient cover for the winter phase of dimorphic species and for wintergreen species was calculated using January coverages.

Maximum leaf height is significantly correlated with the average density of herbaceous foliage in which a species or seasonal morph finds itself (Figure 16.2), with $r^2 = 0.271$ and $p < 0.001$ for 74 d.f. for a log-linear regression. Scatter about the regression is large, but this may be partly due to the rather poor estimate of ambient cover given by a 90-m^2 sample. If

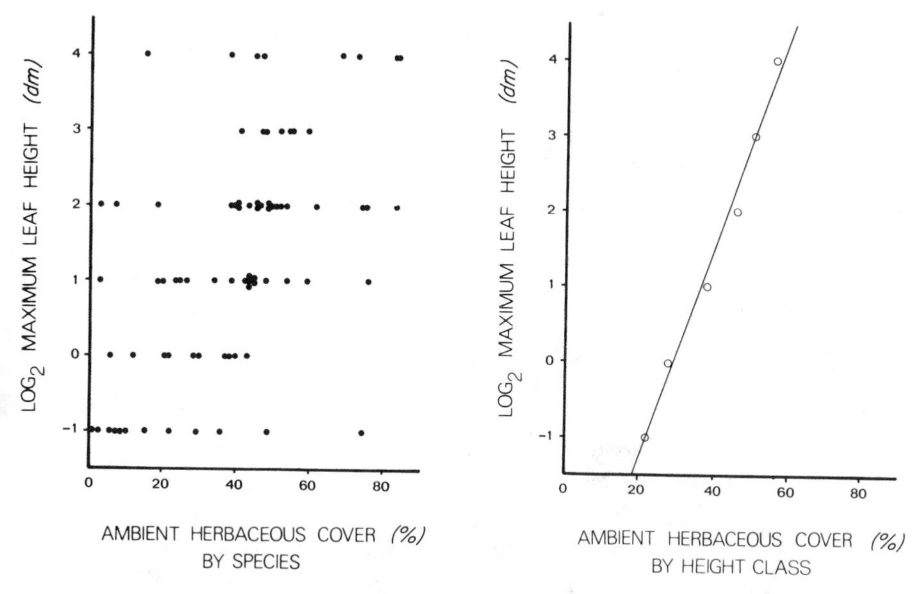

Figure 16.2. Left: Maximum leaf height of species or seasonal morphs versus ambient herbaceous cover. Maximum leaf height is taken as the upper limit of the size class into which a form falls. Right: Maximum leaf height versus average ambient herbaceous cover for all species or seasonal morphs falling into a given height class. Line is the least-mean-squares regression relating the logarithm (base 2) of maximum leaf height to herbaceous cover, $y = -4.03 + 0.138x$ (from Givnish 1982).

one smooths over many of the events leading to chance differences in coverage between sites by calculating the average ambient cover for all species within a height class, a much stronger relation appears between maximum leaf height and herbaceous cover (Figure 16.2), with $r^2 = 0.986$ and $p < 0.001$ for 4 d.f. The least-mean-squares regression line between $\log_2$(maximum leaf height in dm) and average ambient cover is $y = -4.03 + 0.138x$, indicating that leaf height roughly doubles with each 7% increase in coverage.

Quantitative agreement with theory

To determine if this pattern corresponds to the quantitative predictions of the leaf height model, we need to know how the proportions $f(h)$ of leaf tissue versus support tissue vary with absolute leaf height and how the expected rate of photosynthesis $g(\Delta h)$ varies with relative leaf height.

To address the first question, Givnish (1982) presented data on the biomass ratio of photosynthetic tissue to total support tissue and photosynthetic tissue in 21 species of forest herbs. Among the 18 species studied that reach peak coverage before midsummer, there is a linear decrease in the proportion of dry matter allocated to photosynthetic tissue as a function of leaf height h (cm):

$$f(h) = 0.832 - 0.0082 \cdot h \tag{16.3}$$

The three species studied that reach peak coverage after midsummer require less support tissue than do early summer species to support a given leaf mass at a given height. This appears to result, in part, from late summer species scattering their leaves over a broad vertical interval rather than in an umbrella-like arrangement, and thus having narrow crowns and relatively inexpensive horizontal lever arms compared with early summer species. This economy, however, is achieved at the expense of greater self-shading; not surprisingly, late summer species tend to inhabit sites that are better lit in summer than those occupied by early summer species (Givnish et al. 1986).

Givnish (1982) presented a simple model for the expected photosynthetic rate $g(h_1, h_2)$ of a morph in terms of its mean leaf height h_1 and that of its competitor h_2, the proportional variation σ of leaf height about these means, the relative rates of photosynthesis P_{max} and P_{min} in unshaded and shaded leaves, and proportion of herbaceous cover c. Leaves of each morph i were assumed to be distributed uniformly over the vertical interval $h_i(1 \pm \sigma)$; σ was estimated as the ratio of standard deviation to mean leaf height in five plots spanning the range of leaf heights found along the Virginia transect, and a mean value of $\sigma = 0.57 \pm 0.16$ was obtained. Substituting equation (16.3) and the photosynthetic model into equation (16.2), Givnish (1982) obtained

$$h^* = \frac{-\frac{a}{b} \cdot c^2 \cdot (1 - P_{min}/P_{max})}{2\sigma + (1 - \sigma) \cdot c^2 \cdot (1 - P_{min}/P_{max})} \tag{16.4}$$

where $a = 0.832$ and $b = -0.0082$ are the intercept and slope of $f(h)$. Given that the illumination levels on the Virginia transect are close to the compensation point (Givnish et al. 1986), it seems reasonable to assume that shaded leaves respire at approximately the same rate as unshaded leaves photosynthesize; in other words, $P_{min}/P_{max} \simeq -1$. The predicted ESS leaf height resulting from this assumption closely fits the observed relation of leaf height to herb cover at low and intermediate coverages (Figure 16.3).

The model's underestimation of leaf height at high coverages may be more apparent than real. Most of the tallest plants on the Virginia transect,

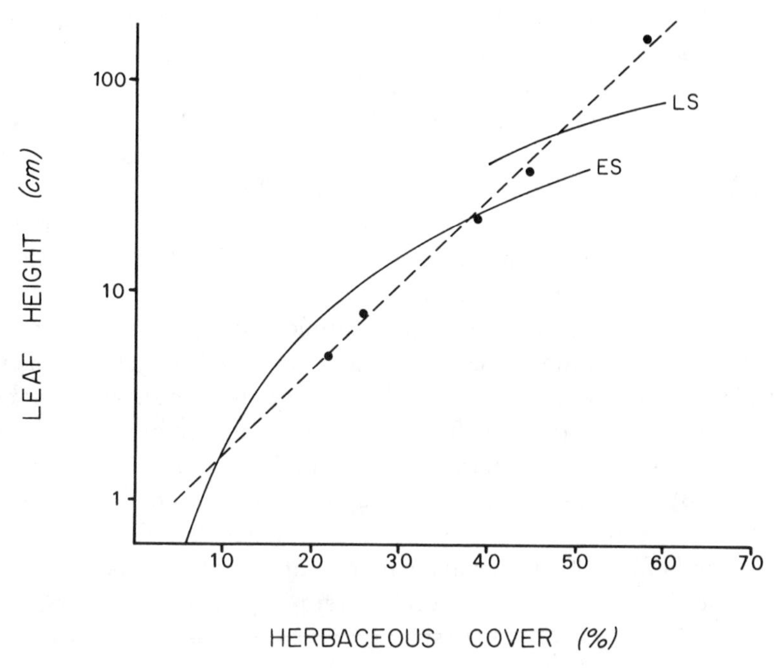

Figure 16.3. Predicted (solid line) and observed (dots, dashed line) relationships of leaf height to herbaceous cover. The ES curve is the predicted relationship between leaf height and cover based on the allometry of umbrella-like, early summer species. The LS curve is the predicted relationship based on allometry of erect, late summer species, which predominate in the taller height classes and denser habitats.

which tend to occur in areas of high coverage, are late summer species that show a less steep decline in $f(h)$ with leaf height (Givnish 1982). A late summer regression forced to have the same intercept as that for early summer species has a slope of $b \simeq -0.005$; substitution of this value results in a 60% multiplicative increase in the ESS leaf heights predicted by equation (16.4) and a much closer agreement with the observed relationship at high coverages (Figure 16.3).

Effects of leaf phenology

Leaf temperature could also be an important selective force on leaf height in winter-active herbs. Winter photosynthesis is likely to be limited not by light but by temperature, and one would expect winter-active species to place their leaves in the warmest microclimate in the winter forest, the sunlit forest floor. Indeed, most evergreens, wintergreens, winter annuals, and the winter phases of dimorphic species native to the northeastern United States place their leaves at or near ground level (Givnish 1982).

If an evergreen herb places its leaves close to the ground to enhance winter photosynthesis, it cannot compete successfully in dense summer-herb communities. This could be one of several factors, in addition to advantages in nutrient conservation and differences in growth rate (Monk 1966; Chabot and Hicks 1982), that restrict evergreen herbs to dry, acid woods that support small amounts of herb cover (Givnish 1982). An exception perhaps proving this rule is Christmas fern, *Polystichum acrostichoides,* the only "evergreen" (really dimorphic) species occurring in high-density herb communities on the Virginia transect. *Polystichum* has the best of summer and winter worlds. In spring and summer, it emerges and holds its fronds erect, thereby increasing its interception of light by about half over what it would experience at ground level in midsummer (Givnish et al. 1986). In winter, its fronds collapse on the ground, leading to elevated leaf temperatures. Givnish et al. (1986) calculated the probable photosynthetic benefit of this behavior by monitoring the temperature of sunlit fronds in winter in their natural winter position, and then experimentally lifting the fronds to their summer posture. The observed increases of roughly 4°C in leaf temperature, combined with measurements of thermal photosynthetic responses, show that *Polystichum* can enhance its photosynthesis by about 22% in this way. This discussion suggests that dimorphic or wintergreen herbs do not face the same competitive constraints as do evergreen herbs and that they should replace evergreens in communities with moderate to high summer coverage (Givnish et al. 1986).

The game-theoretic approach used here makes an important methodological contribution. Most recent analyses of adaptations in plant form (e.g., Orians and Solbrig 1977; Mooney and Gulmon 1979; Givnish 1984) have assumed that natural selection favors traits that tend to maximize a plant's net rate of growth. The growth rates considered have been potential growth rates in the absence of explicit competitors; the hope has been that realized growth rates will parallel these potential rates. Leaf height is an example of a trait in which the strategy that maximizes growth in the absence of competitors is precisely the strategy that does most poorly in their presence. Growth associated with a particular leaf height depends strongly on the competitive context and is negatively correlated with growth in the absence of competition. Only a game-theoretic approach, which analyzes the success of morphs in particular competitive contexts and details how the success of such morphs can change the competitive context, can satisfactorily account for evolutionary patterns in such a trait.

Figure 16.4. Gross morphology of asexual (left) and sexual (right) aerial shoots of *Podophyllum peltatum.*

Economy of branching in *Podophyllum*

Podophyllum peltatum (Berberidaceae) is a common herb in moist deciduous forests of eastern North America, with leaves active in spring and early summer. Its simple growth form permits us to pose several interesting questions regarding the significance of branching in plant canopies. Each vegetative unit, or ramet, is a rhizome bearing one or more aerial shoots at separate nodes (Holms 1899; Ernst 1964; Sohn and Policansky 1977). Each asexual aerial shoot has a single vertical stem surmounted by a symmetric, deeply lobed, centrally attached peltate leaf (Figure 16.4). The stem of each sexual shoot, by contrast, commonly branches where the single flower is inserted, and each of the two branches bears an asymmetric, peripherally peltate leaf. Rare morphological variants are known in which sexual shoots have zero to three leaves, and in which the reproductive structures are inserted above or below the branching of the leaf stems (Foerste 1884; Holms 1899; Harris 1909).

Podophyllum thus presents a model system with which to study the adaptive geometry of branching in plant canopies. This section is organized around three central questions: (1) Why should sexual shoots branch and bear two leaves rather than one? (2) Given that sexual shoots branch, what should be the angle of branching between the leaf stems? (3) What should be the height at which the leaf stems branch?

Sohn and Policansky (1977) have presented evidence that sexual *Podophyllum* shoots require more leaf mass than asexual shoots in order to replace the energy allocated to reproduction. Sexual shoots that produce fruit show the same rhizomatous growth as vegetative shoots, whereas sexual shoots that do not produce fruit show greater rhizomatous growth and higher rates of survival and future reproduction. This suggests that sexual shoots should have more leaf mass than asexual shoots to achieve the same growth, but it does not explain why this greater leaf mass should be packaged in two leaves rather than one.

One possible reason is biomechanical. As leaf mass and area increase, so must the length of the major veins supporting each lobe. The mass of such veins must increase at a high power of their length in order to maintain mechanical stability in a leaf of constant thickness and mass density (see later section on biomechanical principles). Thus, above a critical leaf mass, it may be less costly for a *Podophyllum* shoot to branch and produce two smaller leaves with shorter and hence much less massive veins, even though branching requires a greater allocation to stem tissue above the branchpoint.

To test this idea quantitatively, we need to know (1) how vein mass scales with lobe length and mass, (2) how vein mass scales with leaf mass in sexual and asexual shoots, (3) how stem mass per unit length varies with the leaf mass it supports, and (4) how the length of the stem above the branchpoint varies with leaf mass. In evaluating the costs and benefits of different branching patterns, we shall compare plants supporting the same dry mass of lamina at a given height and ask which branching pattern minimizes the total dry mass of structural-support tissue in stems and major veins. This approach avoids the messy problem of weighing the relative costs of dry mass in foliage and support tissue and assumes only that – other things being equal – plants of comparable laminar mass and height but minimal allocation to relatively unproductive support tissue will be favored by natural selection. A similar approach can be used to predict the optimal angle of branching in sexual shoots (based on tradeoffs between branch length and vein costs of leaf asymmetry) as well as the optimal height of branching (based on tradeoffs between branch length and the relative costs per unit length of the main stem and branches).

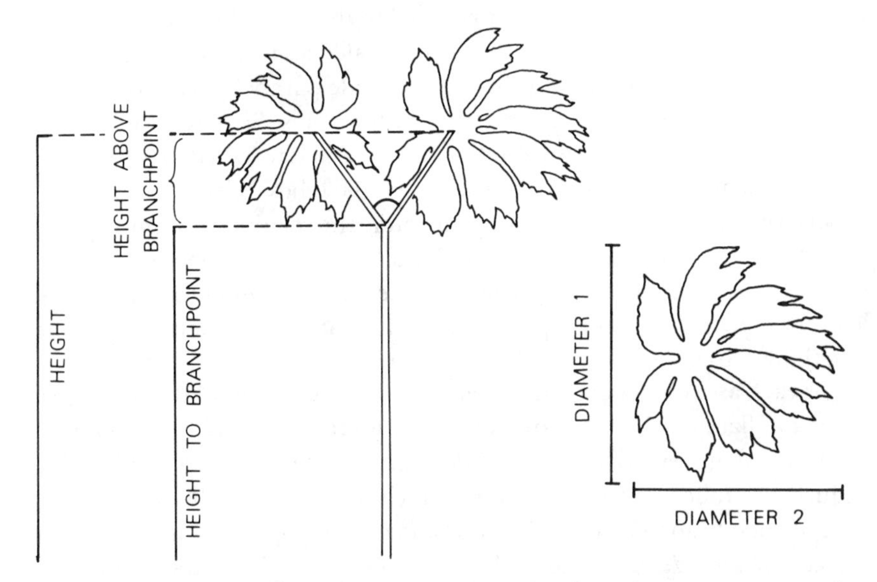

Figure 16.5. Schematic summary of length and angular measurements made on sexual *Podophyllum* shoots.

Methods

In May 1982, 23 asexual and 27 sexual shoots were sampled from a *Podophyllum* population at the Case Estates of the Arnold Arboretum in Weston, Massachusetts. An attempt was made to sample the entire range of shoot sizes present. The leaf diameter and height of each asexual shoot were measured in the field to the nearest mm. Similar measurements were made for sexual shoots: the maximum and minimum diameters of each leaf were noted, and leaf height was measured as the mean height of the points where the branches join their respective leaf blades. In addition, the length of each stem branch above the branchpoint or crotch was measured, as was the vertical height of the leaves above the junction. Finally, the branching angle subtended by the petioles was measured to the nearest degree. Figure 16.5 schematically summarizes the linear and angular measurements made.

Shoots were transported immediately to Cambridge in plastic bags on ice to maintain turgor. In the laboratory, all veins thicker than 0.5 mm were excised carefully from the rest of the lamina. Vein diameter was assessed tactilely and visually under a dissecting microscope, using a semicircular 0.5-mm notch cut into a razor-blade edge. One person measured all veins and placed a cut across each at the threshold diameter, while others carefully excised all veins proximal to the cuts. Microscopic examination of vein cross sections at the cuts revealed a mean diameter of 0.55 mm. We tried to avoid slicing through the pale parenchyma of the major veins and/or leaving lamina adhering to the veins. Structural veins and lamina thus separated were oven-dried at 70°C and weighed to the nearest 0.01 mg using an analytical balance. Individual lobes from several asexual shoots were analyzed separately to determine the relationships among lobe length, lamina mass, and vein mass. For all shoots, each lobe was isolated from its neighbors by cutting from the intervening sinuses to the point of leaf attachment, and then photocopied for later measurement of lobe length and area.

For each asexual shoot, two consecutive 5-cm segments of the stem were cut immediately below the point of petiole attachment and, together with the remaining stem, oven-dried and weighed individually. A similar procedure was followed for each stem branch of the sexual shoots, with the base of each branch being defined by a razor cut coplanar with the upper surface of the adjacent branch at the branchpoint. The stem below the branchpoint, and any reproductive structures present, were also dried and weighed. Length and mass data are summarized in Tables 16.1 and 16.2 for asexual and sexual shoots, respectively. Allometric relationships between variables of the form $y = bx^a$, expected from biomechanical principles (see next section), were determined by fitting log-transformed data to the equation $\ln y = a \ln x + \ln b$, using principal-axis regressions (Sokal and Rohlf 1981).

Biomechanical principles

The deep sinuses between the lobes of a *Podophyllum* leaf make each lobe largely independent of the others for mechanical support. For lobes of a given shape, biomechanical considerations suggest that vein dry mass should scale like the cube of lobe length and the $\frac{3}{2}$ power of lobe laminar mass (Givnish 1979, 1984). This can be seen as follows: If we assume that the midrib of a lobe provides essentially all longitudinal support for the lobe, that it is equally resistant to the bending moments applied to it at different points, and that its mass is small relative to that of the lamina, then the diameter $w(x, T)$ of the midrib at a distance x from the base of a lobe of total length T must satisfy

$$w(x, T)^3 = 2\beta \int_x^T f(z, T) \cdot (z - x) \cdot dz \qquad (16.5)$$

Table 16.1. *Selected length and mass measurements for asexual shoots of* Podophyllum peltatum

	Length (cm)		Mass (g)			
Shoot	Height	Leaf diameter	Lamina	Vein	Stem	Stem mass density (g/cm)[a]
A1	10.8	9.3	0.09632	0.00392	0.02865	0.002679
A2	10.9	11.3	0.15625	0.01222	0.04534	0.004248
A3	14.4	10.5	0.15574	0.01109	0.06834	0.004100
A4	15.6	15.5	0.25543	0.01995	0.09196	0.005131
A5	15.9	11.5	0.17698	0.01251	0.06990	0.003747
A6	17.0	13.6	0.31711	0.02697	0.13643	0.006606
A7	17.3	12.4	0.15897	0.01588	0.07847	0.003618
A8	20.0	14.5	0.33738	0.03093	0.15447	0.006210
A9	21.0	17.0	0.43853	0.04097	0.21101	0.008530
A10	21.3	18.0	0.43944	0.05151	0.19134	0.007246
A11	21.6	16.2	0.43128	0.05434	0.21048	0.007543
A12	22.8	18.0	0.46512	0.05192	0.24278	0.008164
A13	23.0	21.5	0.64987	0.09650	0.30850	0.010408
A14	24.5	19.0	0.64045	0.09165	0.35012	0.011304
A15	25.0	23.5	0.90640	0.15883	0.48709	0.014340
A16	26.0	19.0	0.52980	0.05842	0.28477	0.008700
A17	29.5	21.3	0.86205	0.14200	0.51838	0.013284
A18	32.2	25.8	1.15619	0.21037	0.75216	0.017364
A19	36.8	28.5	1.95193	0.38492	1.44024	0.030035
A20	37.8	34.0	2.33195	0.46982	1.34727	0.027622
A21	43.0	37.5	2.77961	0.71729	2.03618	0.034605
A22	44.0	32.8	1.8125	0.34477	1.27366	0.024241
A23	44.0	36.3	1.96692	0.48593	1.76344	0.034505

[a] Density measurements based on mean oven-dry mass of the two uppermost, adjacent 5-cm segments of stem.

where $f(z, T)$ is the half-width of the lobe at a distance z from its base, $z = x$ is the length of the lever arm between the mass at z and the cross section at x, and β is a constant involving the mass density of fresh lamina, the shape of the midrib cross section, and Young's modulus for vein tissue (Givnish 1976, 1978, 1979). For lobes of a given shape, $f(z, T) = T \cdot f(z/T, 1)$, so that equation (16.5) can be rewritten as

$$w(x, T)^3 = 2\beta \int_x^T T \cdot f(z/T, 1) \cdot T \cdot (z/T - x/T) \cdot dz \qquad (16.6)$$

or

$$w(x, T)^3 = 2\beta T^3 \int_{x'}^1 f(z', 1) \cdot (z' - x') \cdot dz' \qquad (16.7)$$

Table 16.2. *Selected length and mass measurements for sexual shoots of* Podophyllum peltatum [*data for each member of a shoot's leaf pair are tabulated on separate rows for the larger (a) and smaller (b) leaves*]

	Length (cm)				Mass (g)					
Shoot	Height	Height of branchpoint	Leaf diameter[a]	Branch length	Lamina	Vein	Branch	Stem	Branch density (g/cm)[b]	Branching angle(°)
Sla	20.0	11.4	15.2	9.0	0.67507	0.07976	0.11502	0.35398	0.01175	45
b			13.0	9.2	0.41114	0.04467	0.08389		0.00978	
S2a	23.8	14.4	16.8	10.0	0.64212	0.11087	0.14131	0.48612	0.01413	55
b			14.5	9.0	0.52190	0.07084	0.10410		0.01059	
S3a	30.0	17.4	18.8	14.0	0.57125	0.08955	0.14294	0.40857	0.00989	45
b			14.9	13.0	0.29931	0.03937	0.07744		0.00554	
S4a	30.2	21.6	15.5	10.0	0.59767	0.09487	0.12994	0.58559	0.01299	47
b			14.5	7.5	0.48787	0.06806	0.09478		0.01235	
S5a	31.1	21.1	18.4	10.0	0.75051	0.14325	0.15289	0.61607	0.01529	58
b			16.2	12.6	0.38651	0.05471	0.11491		0.00886	
S6a	32.0	20.5	20.8	12.4	0.94450	0.18841	0.18655	0.83961	0.01436	50
b			21.6	12.4	0.94001	0.18812	0.21226		0.01716	
S7a	32.3	21.7	20.8	11.0	1.00653	0.23174	0.25277	0.89082	0.02088	20
b			15.8	10.1	0.34545	0.05127	0.10047		0.00985	
S8a	35.8	25.5	16.5	11.4	0.62270	0.08876	0.12821	0.69426	0.01128	68
b			15.9	12.0	0.40943	0.05469	0.09384		0.00734	
S9a	37.2	27.0	18.7	10.2	0.83791	0.13471	0.16472	0.87020	0.01516	60
b			17.4	11.2	0.57247	0.09666	0.13944		0.01203	
S10a	37.2	24.0	24.6	14.5	1.33412	0.24098	0.30546	1.27514	0.02016	53
b			23.0	14.5	1.11139	0.19788	0.26291		0.01742	
S11a	37.3	22.3	24.0	16.2	1.16596	0.21340	0.32509	0.95691	0.01804	46
b			19.1	15.0	0.62753	0.10110	0.20777		0.01221	
S12a	38.3	24.1	18.7	15.2	0.58490	0.11164	0.17902	0.72752	0.01121	34
b			16.9	14.7	0.46342	0.07157	0.15690		0.01009	
S13a	38.5	24.0	23.8	17.6	0.91586	0.17669	0.27206	0.87469	0.01445	74
b			21.5	18.1	0.62734	0.10932	0.20323		0.01046	

S14a	39.8	27.3	23.6	14.6	1.07847	0.20644	0.25256	0.97209	0.01678	57
b			18.6	13.2	0.50495	0.08183	0.14526		0.01084	
S15a	40.0	26.4	21.6	14.6	2.03965	0.44386	0.70522	1.94209	0.03012	45
b			17.1	14.5	1.33323	0.23233	0.46484		0.02068	
S16a	40.2	24.2	25.3	16.8	1.57371	0.34687	0.43679	1.47229	0.02489	43
b			23.2	16.8	0.98732	0.18966	0.27533		0.01566	
S17a	40.3	30.1	21.8	10.2	0.87536	0.16101	0.16575	0.69426	0.01584	60
b			17.1	11.5	0.56006	0.08341	0.12178		0.01128	
S18a	40.6	32.0	22.9	9.7	1.14433	0.26277	0.20958	1.37756	0.01120	60
b			18.3	10.4	0.79044	0.12872	0.14383		0.01438	
S19a	41.5	24.9	24.6	15.9	1.94263	0.41219	0.43731	1.61316	0.02662	30
b			19.4	17.1	0.70721	0.14754	0.33582		0.01849	
S20a	42.8	29.4	23.9	14.2	1.24089	0.20881	0.28982	1.32493	0.01965	45
b			20.4	14.6	0.93074	0.14369	0.24653		0.01589	
S21a	43.5	29.0	29.3	16.3	2.04610	0.53336	0.53400	2.04948	0.03281	58
b			24.9	15.5	1.28574	0.29714	0.33404		0.02167	
S22a	44.2	26.7	29.8	19.0	2.06515	0.51219	0.62092	2.01881	0.02980	51
b			27.9	19.0	1.45862	0.31905	0.45354		0.02303	
S23a	44.5	33.3	24.0	12.6	1.44670	0.26563	0.29846	1.71914	0.02316	60
b			21.6	13.8	0.92522	0.16575	0.23020		0.01698	
S24a	45.3	26.1	24.8	20.0	1.55771	0.34681	0.53948	1.47968	0.02471	50
b			23.2	20.1	0.92548	0.18304	0.36761		0.01685	
S25a	46.3	26.7	29.4	21.5	1.08989	0.17649	0.26306	1.02947	0.01848	48
b			24.4	21.3	0.55952	0.07771	0.16225		0.01068	
S26a	50.0	35.0	24.7	16.2	1.11267	0.22635	0.30621	1.44050	0.01792	50
b			20.4	16.4	0.68611	0.12438	0.20220		0.01211	
S27a	51.2	36.3	29.5	15.0	1.81392	0.45828	0.45521	2.56336	0.03046	45
b			26.8	16.0	1.36983	0.34154	0.42416		0.02552	

[a] Average of maximum and minimum diameters of each leaf.
[b] Density measurements based on mean oven-dry mass of the two uppermost, adjacent 5-cm segments of branch in most cases. Density based on uppermost 5-cm segment only in branches with length <10 cm.

where $x' = x/T$, $z' = z/T$, and $dz' = z/T$. Thus, total vein mass should be given by the following weighted integral of vein cross-sectional area:

$$\frac{\pi\sigma}{4}\int_0^T w(x, T)^2$$

$$= T^2 \cdot \frac{\pi\sigma}{4} \cdot (2\beta)^{2/3} \int_0^T \left(\int_{x/T}^1 f(z', 1)(z' - x')\, dz'\right)^{2/3} dx \quad (16.8)$$

$$= T^3 \cdot \frac{\pi\sigma}{4} \cdot (2\beta)^{2/3} \int_0^1 \left(\int_{x'}^1 f(z', 1)(z' - x')\, dz'\right)^{2/3} dx' \quad (16.9)$$

where σ is the density per unit fresh volume of dry mass in vein tissue. Because the integral on the left-hand side of equation (16.6) is constant for all lobes of a given shape, we conclude that total vein mass should scale like the cube of lobe length T. Because lobe laminar mass should scale like T^2 for lobes of a given shape ($\int_0^T f(x, T)\, dx = T^2 \cdot \int_0^1 f(x', 1)\, dx'$), this implies that vein mass should scale like the $\frac{3}{2}$ power of lobe laminar mass. I shall show that *Podophyllum* lobes and leaves follow these biomechanical rules and shall analyze the implications these have for the relative economy of different leaf arrangements, after first briefly describing size differences between sexual and asexual shoots.

Description of sexual and asexual shoots

The average laminar mass of asexual shoots is 0.827 ± 0.786 g dry mass, less than half that for sexual shoots, 1.922 ± 0.796 g; the difference in laminar mass is highly significant ($p < 0.001$ for a two-tailed t-test with 48 degrees of freedom). The laminar masses of asexual shoots sampled ranged from roughly 0.096 g to 2.780 g; those of sexual shoots ranged from roughly 0.871 g to 3.524 g (Table 16.2). The transition between one- and two-leaved canopies takes place at a laminar mass roughly midway between the respective mean masses, at about 1.156 g to 1.352 g. Fully 78% of all sexual shoots have laminar masses greater than this range of values, whereas 78% of asexual shoots have laminar masses lower than this value. For convenience, I shall designate 1.2 g as the transition laminar mass.

The number of leaf lobes increases with roughly the logarithm of laminar mass in asexual shoots ($N = 0.939 \cdot \ln L + 6.235$, $r^2 = 0.673$, and $P < 0.001$ for 21 degrees of freedom), suggesting that as total laminar mass increases, the mass and width of individual lobes increase relatively slowly. Leaf height increases as a power of laminar mass in both asexual ($\ln H = 0.412 \cdot \ln L + 3.393$, $r^2 = 0.938$, and $p < 0.001$ for 21 degrees of

Table 16.3. *Principal-axis regressions for* Podophyllum peltatum *relating vein mass* (V) *to lobe mass* (L_l), *lobe length* (T), *and laminar mass of asexual leaves* (L_{asex}), *sexual leaves* ($L_{sex,1}$), *and sexual canopies* ($L_{sex,2}$)[a]

y	x	a	b	r
V_l	T	3.097	1.237×10^{-3}	0.985[b]
V_l	L_l	1.536	0.1216	0.992[b]
L_l	T	2.035	0.0102	0.976[b]
V_{asex}	L_{asex}	1.464	0.1622	0.995[b]
$V_{sex,1}$	$L_{sex,1}$	1.355	0.1882	0.984[c]
$V_{sex,2}$	$L_{sex,2}$	1.378	0.1466	0.980[c]

[a] Data are fit to the logarithmically transformed power-law equation $\ln y = a \ln x + \ln b$.
[b] $P < 0.001$ for 21 d.f.
[c] $P < 0.001$ for 25 d.f.

freedom) and sexual shoots ($\ln H = 0.340 \cdot \ln L + 3.431$, $r^2 = 0.435$, and $p < 0.001$ for 25 degrees of freedom). The mean branching angle between the petioles of sexual shoots is $50.8 \pm 11.3°$, yielding an average deviation of individual petioles from the vertical of 25.6°. The mean leaf height of sexual shoots is 38.29 ± 7.29 cm, and the mean height of the branchpoint is 25.27 ± 5.71 cm.

Economy of branching versus not branching

Vein costs. Table 16.3 summarizes the allometric relationships between vein mass and lobe mass, lobe length, and laminar mass for sexual and asexual shoots, based on principal-axis regressions. The tightness of these relationships is very high, with $r > 0.975$ in every case. Vein mass increases with the 3.097 power of lobe length and the 1.536 power of lobe laminar mass, compared with the expected values of a equal to 3.0 and 1.5 based on biomechanical principles. Lobe mass, as expected, increases with roughly the square of lobe length, with a equal to 2.035.

The vein mass V_{asex} of individual asexual leaves increases at a slightly lower rate with leaf laminar mass than with lobe laminar mass, $V_{asex} = 0.1622 \cdot L^{1.464}$ (Table 16.3, Figure 16.6). This appears to reflect the increasing numbers of lobes in more massive leaves and suggests that lobe width and hence effective leaf size (Taylor 1975; Givnish and Vermeij 1976) increase rather slowly with total laminar mass and area. The total vein mass $V_{sex,2}$ of a pair of sexual leaves is, as expected, less than that of a single asexual leaf of the same mass over almost the entire range of leaf

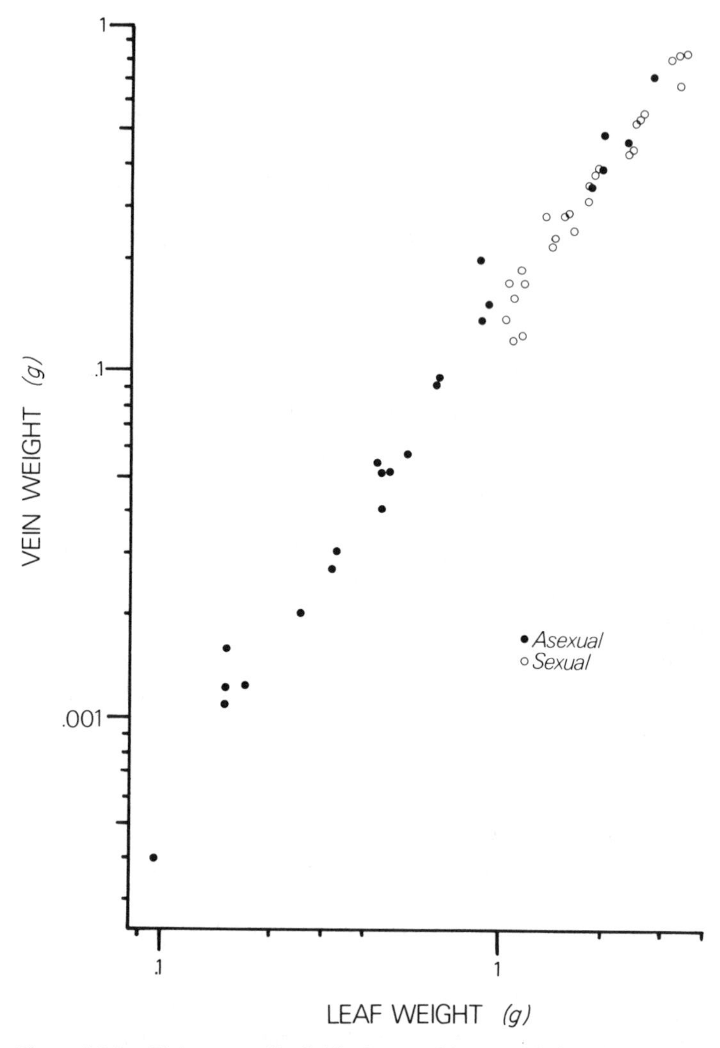

Figure 16.6. Vein mass of individual asexual leaves (filled circles) and sexual canopies (open circles) as a function of total laminar mass. Log-log plot illustrates the roughly $\frac{3}{2}$-power law relating vein and laminar masses, as well as the 15 to 20% lower vein mass of two-leaved sexual canopies versus hypothetical one-leaved asexual canopies of the same laminar mass.

masses observed in the field: $V_{\mathrm{sex},2} = 0.1466 \cdot L^{1.378}$. The relative additional mass of veins in a single asexual leaf, compared with a pair of sexual leaves, is given by

$$\frac{V_{\mathrm{asex}} - V_{\mathrm{sex},2}}{V_{\mathrm{sex},2}} = 1.1337 \cdot L^{0.086} - 1 \qquad (16.10)$$

This implies that the mass of veins in a single asexual leaf is about 15.2% more than that in a pair of sexual leaves for canopies near the transition laminar mass of 1.2 g and that the additional vein mass will increase to about 20.3% at a laminar mass of 2.0 g. The additional mass of veins in asexual, one-leaved canopies compared with sexual, two-leaved canopies can be considered a *cost* of one-leaved canopies versus two-leaved canopies, whether expressed in relative or absolute terms.

The difference in vein mass between asexual and sexual canopies of the same laminar mass results from three factors: (1) leaf number, (2) inequality of leaf sizes in two-leaved canopies, and (3) asymmetry of individual leaves in two-leaved canopies. Our measurements permit an assessment of the relative importance of each of these factors in determining the overall mechanical efficiency of different leaf and canopy configurations. The vein costs associated with these three factors can be determined as follows.

Costs of one- versus two-leaved canopies. Given the allometric relationships of vein mass to laminar mass in asexual leaves, we can calculate the hypothetical benefit gained by constructing two equal, symmetric, asexual-type leaves rather than one leaf of the same laminar mass (Appendix I). This procedure isolates the effect of leaf number per se from the effects of leaf inequality and asymmetry on the vein cost of one-leaved canopies. The vein mass V_{I} of a single asexual leaf is $0.1622 \cdot L^{1.464}$ (Table 16.3). Similarly, the vein mass V_{II} of two asexual leaves, each with mass $L/2$, can be calculated to be $0.1179 \cdot L^{1.464}$ (Appendix I). Thus, the additional vein cost of a single asexual leaf, compared two similar leaves having the same total mass, is 37.8% in relative terms. In absolute terms, this additional cost will amount to roughly 169 mg in a shoot with a laminar mass of 2 g.

Cost of inequality of leaf size in two-leaved canopies. Actual pairs of sexual leaves are more costly than would be expected from a comparison of single and paired asexual leaves, with the additional cost of single asexual leaves being only about 15.2% near the transition laminar mass, rather than 37.8% [equation 16.10)] (Appendix I). Part of the reason for this is that the two leaves of sexual shoots are unequal in size, with the laminar mass of the larger leaflet being 1.634 ± 0.433 (N = 27) times that of the smaller. Insofar as vein mass increases at a greater than linear rate with leaf mass, two leaves of unequal masses will have a greater total vein mass than two equal leaves of the same total laminar mass. We can estimate the cost of leaf inequality by comparing the total vein mass of two equal asexual leaves of a given laminar mass with that of two asexual leaves whose laminar masses have a ratio of 1.634 to 1. This calculation, given in Appendix II, shows that the relative cost of unequal versus equal leaves is 1.4%.

Cost of leaf asymmetry. Because vein mass increases at a high power of lobe length, asymmetric leaves with unequal lobes should require more total vein mass than a radially symmetric leaf having the same laminar mass. This cost of leaf asymmetry can be estimated in four ways. First, direct comparison of the allometries of individual sexual and asexual leaves indicates that the relative cost of the latter is $1.160 \cdot L^{-0.109} - 1$. This amounts to roughly 13.6% for a laminar mass of 1.2 g, and 7.6% for a laminar mass of 2.0 g.

Second, a more mechanistic analysis of the cost of leaf asymmetry can be based on allometry of leaf lobes (Appendix III). In this procedure, the lobe length of a symmetric leaf having the same laminar mass as an asymmetric leaf is calculated in terms of the mean lobe length $\bar{x} = \Sigma x_i/n$ of the asymmetric leaf and the proportional variance about that mean, $V_p = \mathrm{Var}(x_i)/\bar{x}^2$ (Appendix III). Then, the vein masses of symmetric and asymmetric leaves are estimated using the power-law relationship of vein mass to lobe length and laminar mass. The calculated cost of leaf asymmetry is a function of the allometric constants relating lobe length to vein mass and laminar mass, and of leaf asymmetry, as measured by V_p, and is given by

$$\frac{V_{\mathrm{asym}} - V_{\mathrm{sym}}}{V_{\mathrm{sym}}} = \frac{2.698 \cdot V_p}{1 + 0.550 \cdot V_p} \tag{16.11}$$

(Appendix III). The mean value of V_p measured on actual sexual leaves is 0.0121, yielding an expected cost of leaf asymmetry of roughly 3.2%. The mean value of V_p in asexual leaves is 0.0018, yielding an expected vein cost of roughly 0.5% over a hypothetical, perfectly symmetric leaf.

A third approach to the cost of leaf asymmetry is to compare a pair of equal asexual leaves to a pair of equal sexual leaves with the same total laminar mass (Table 16.3). The fractional cost of asymmetry derived in this way is, again, $1.160 \cdot L^{-0.109} - 1$. Finally, the cost of asymmetry can be estimated by comparing a pair of sexual leaves with a pair of asexual leaves whose masses are in the same ratio of 1.634 to 1. Calculations show that the fractional cost of leaf asymmetry derived in this way is $1.223 \cdot L^{-0.086} - 1$ (Figure 16.7). This amounts to 20.4% for leaves with a laminar mass of 1.2 g, and 15.2% for leaves with a laminar mass of 2.0 g.

The incremental costs of leaf arrangements that differ in any combination of leaf number, leaf equality, and leaf asymmetry can be similarly calculated, based on the allometry of leaf lobes, asexual leaves, and sexual

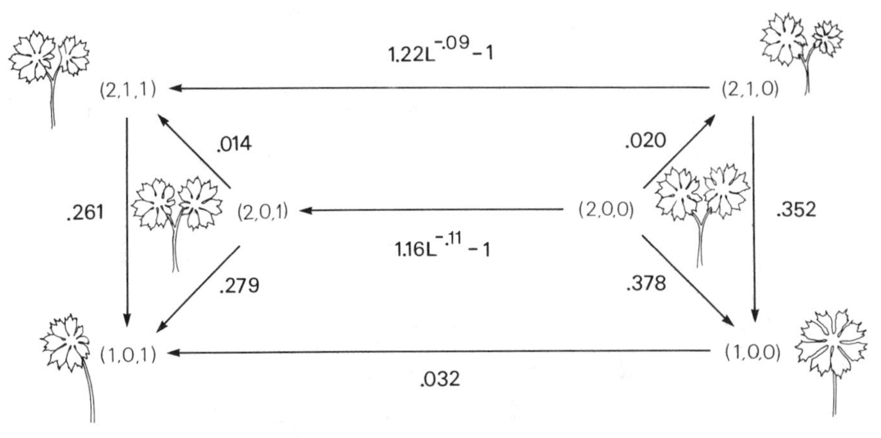

Figure 16.7. Proportional vein costs associated with transitions between hypothetical *Podophyllum* canopies differing in leaf number, leaf equality, and individual leaf symmetry. The ordered triplets specify coordinates in the space determined by the six possible combinations of these characters (1st coordinate: 1 = 1 leaf, 2 = 2 leaves; 2nd coordinate: 0 = leaf equality, 1 = leaf inequality; 3rd coordinate: 0 = leaf symmetry, 1 = leaf asymmetry).

leaves. Six hypothetical leaf arrangements, based on all possible combinations of leaf number, equality, and symmetry, can be defined (2 of $8 = 2^3$ possible states are degenerate because leaf equality is undefined when only one leaf is present). Furthermore, these six states define a group of 18 transformations between states that differ in only one characteristic. The relative vein costs associated with the nine transformations from more efficient to less efficient leaf arrangements are given in Figure 16.7. The arrows in Figure 16.7 point from the more efficient to the less efficient arrangement, and the adjacent number gives the relative increment in vein mass associated with the transformation. The total relative increment in vein cost along any chain of arrows pointing in the same direction is given by the product $\Pi(1 + z_{ij})$, reduced by 1, where z_{ij} is the fractional cost associated with the transformation from state i to state j. If two arrows point in opposite directions from some intermediate state k, then the fractional cost of the transition from state i to state j is given by $(1 + z_{kj})/(1 + z_{ki}) - 1$. The product of transformations $(1 + z_{ij})$ around any loop in Figure 16.7 should equal 1, representing the identity mapping of one leaf arrangement to itself. This provides a useful check of the compatibility of cost estimates obtained from the allometry of leaf lobes, asexual leaves, and sexual leaves. The products of the transformations around the loops at either end of Figure 16.7 balance precisely, reflecting their derivation entirely from the allometry of sexual (left-hand loop) and asexual (right-hand loop) leaves. The upper loop across the width of the figure also balances with good precision and is based on independent measures of sexual and asexual leaves. However, other loops across the width of Figure 16.7 do not completely balance if the cost of asymmetry based on lobe allometry (see Appendix III) is used. The observed products around the loop range from $0.90 \cdot L^{-0.086}$ to $0.96 \cdot L^{-0.109}$, leading to a discrepancy of 5.9% to 11.4% at a laminar mass of 1.2 g.

Conclusions regarding the cost of veins in different leaf arrangements can be summarized as follows: First, the total absolute difference in grams dry matter between the vein mass of a pair of sexual leaves and an asexual leaf of the same total laminar mass L is given by

$$\Delta V = 0.1622 \cdot L^{1.464} - 0.1466 \cdot L^{1.378} \qquad (16.12)$$

Second, pairs of leaves have less costly veins than a single leaf with the same laminar mass, equal leaves have less costly veins than unequal leaves of the same total laminar mass, and symmetric leaves have less costly veins than asymmetric leaves of the same laminar mass. Each of these patterns is readily understood, both qualitatively and quantitatively, in terms of biomechanics and the observed allometry of leaf lobes, asexual leaves, and sexual leaves. Third, a symmetric peltate leaf requires far less vein mass than a lobe-shaped, more or less lanceolate leaf with the same laminar mass, as predicted by Givnish and Vermeij (1976), given the greater length of lever arms in the latter. Vein mass in a lobe-shaped leaf will equal $0.4242 \cdot L^{1.536}$, compared with $0.1622 \cdot L^{1.464}$ in a symmetric peltate leaf. This yields a fractional cost of $2.615 \cdot L^{0.072} - 1$ and amounts to a relative cost of 165.0% and an absolute cost of 350 mg for a 1.2-g leaf and a relative cost of 174.9% and an absolute cost of 783 mg for a 2.0-g leaf.

Stem costs. We shall now analyze the stem costs associated with one- and two-leaved *Podophyllum* canopies. For now, we consider only the dry mass associated with stems above the reproductive structures – this entails the vertical interval above the crotch or branchpoint in actual sexual plants, and the same height interval in hypothetical sexual plants with a single, symmetric, asexual leaf of the same total laminar mass. The reason for this approach is straightforward. The stem of sexual shoots below the reproductive structures must bear a much greater load than a corresponding portion of the stem of asexual shoots with the same laminar mass and height. The fresh mass of the peduncle and nearly ripe fruit averages roughly 13.1 g, based on a mean dry mass of 1.31 ± 0.45 g ($N = 8$) in this study and an average ratio of dry to fresh fruit mass of 0.10 obtained by Sohn and Policansky (1977). If we use the same ratio of dry to fresh mass for laminar tissue, the mean fresh mass of 13.1 g for the fruit and peduncle will compare with a mean fresh mass of 19.2 g for the lamina of sexual shoots, increasing substantially the stresses experienced by the stem below the reproductive structures. By focusing on the costs of the stem above the crotch, we can avoid many of the complications involved in considering the support of the reproductive structures and concentrate on the stem costs of one- versus two-leaved canopies per se. Furthermore, to a first approximation, one- and two-leaved canopies having the same laminar mass and height above the crotch should require the same stem mass below the crotch to support the foliage and reproductive structures.

The difference in stem mass required to support one- and two-leaved canopies depends on four factors: (1) the mass per unit length of a stem supporting an asexual leaf, (2) the mass per unit length of the stem branch supporting each sexual leaf, (3) the height of the leaves above the branch-point in actual sexual plants, as a guide to the length of the petiole required to support the foliage at a comparable height in hypothetical one-leaved sexual shoots, and (4) the actual lengths of the stem branches supporting leaves in sexual plants. In each case, these four factors should be expressed as functions of laminar mass for functional reasons and to make the calculated net stem cost ΔS of two-leaved versus one-leaved canopies comparable to the previously obtained net vein benefit ΔV of two-leaved canopies [equation (16.12)]. When $\Delta S < \Delta V$, the dry-mass advantage of sexual shoots exceeds their incremental stem cost, and natural selection should favor two-leaved canopies. Similarly, when $\Delta S > \Delta V$, the vein-cost advantage of sexual shoots is outweighed by their higher net stem costs, and natural selection should favor one-leaved canopies. Because ΔS and ΔV are functions of laminar mass, the preceding inequalities define a critical laminar mass at which selection should favor a shift from one- to two-leaved canopies.

Based on the uppermost 10 cm of stem tissue, the mass per unit length (g/cm) of stems supporting asexual leaves is a roughly linear function of laminar mass ($y = 0.0127 \cdot L + 0.0023$, $r^2 = 0.953$, and $p < 0.001$ for 21 degrees of freedom). This is somewhat lower than the linear density of stem branches supporting sexual leaves of the same mass ($y = 0.0132 \cdot L + 0.0030$, $r^2 = 0.929$, and $p < 0.001$ for 25 degrees of freedom). The greater density of sexual-leaf stems is expected, given that they are inclined somewhat from the vertical and bear asymmetric leaf loads; sexual stems must thus bear continuous horizontal bending moments, unlike vertical asexual stems, which mainly must withstand elastic toppling *sensu* McMahon (1973).

We can model the net stem cost ΔS of two- versus one-leaved canopies based on the mass per unit length of the stem required to support an asexual leaf of a given laminar mass, the total laminar mass of the 27 sexual shoots studied, their actual stem mass, and the height of their foliage above the branchpoint. First, the linear density of an erect stem supporting a single hypothetical leaf with the same total laminar mass as each of the sexual shoots is calculated, using the equations developed in the preceding paragraph. Second, to obtain the total stem mass associated with a hypothetical asexual leaf of the same laminar mass and height above the crotch as a given sexual shoot, we multiply the stem-mass density by the corresponding height of foliage above the crotch. Third, the stem mass associated with a hypothetical one-leaved canopy is subtracted from the stem mass of the corresponding, actual two-leaved canopy. Finally, this difference in stem mass is related to laminar mass using a principal-axis regression, excluding two extreme outliers. This difference, the net absolute stem cost of two- versus one-leaved canopies, is given by the equation

$$\Delta S = 0.0659 \cdot L + 0.0124 \quad (16.13)$$

($r^2 = 0.759, p < 0.01$ for 23 degrees of freedom). The net vein benefit of two-leaved canopies is, as shown previously, given by equation (16.12). These cost–benefit curves are plotted in Figure 16.8. As discussed, two-leaved canopies should be favored when their marginal benefits exceed their marginal costs, that is, when $\Delta V > \Delta S$. As can be seen from Figure 16.8, two-leaved canopies are indeed favored at large leaf masses, and one-leaved canopies at small leaf masses, supporting the original qualitative argument. However, the critical laminar mass predicted for the transition between the morphs (approximately 4.6 g) exceeds considerably both the observed transition mass (1.2 g) and the largest laminar mass seen in this study (3.5 g).

Three explanations might be advanced to account for this discrepancy. First, dry mass may be an inappropriate measure of the effective costs of vein and stem tissues (Harper 1977; Lovett-Doust and Harper 1980). This is always a possibility in this kind of study, and it is true that the nitrogen content of stem tissue ($\simeq 1.0\%$) is lower than that of vein tissue ($\simeq 1.7\%$). If nitrogen were a more appropriate measure of cost, this difference in concentration would lower the relative cost of stem tissue and move the expected transition between one- and two-leaved canopies in the proper direction (Figure 16.8). However, I believe that stem and vein tissues, as operationally defined here, are as comparable as is ever likely to occur, and that the bulk dry mass required to create mechanical strength – not some unidentified trace element – is an appropriate measure of the effective costs of veins and stems.

Second , we have undoubtedly underestimated the stem mass required

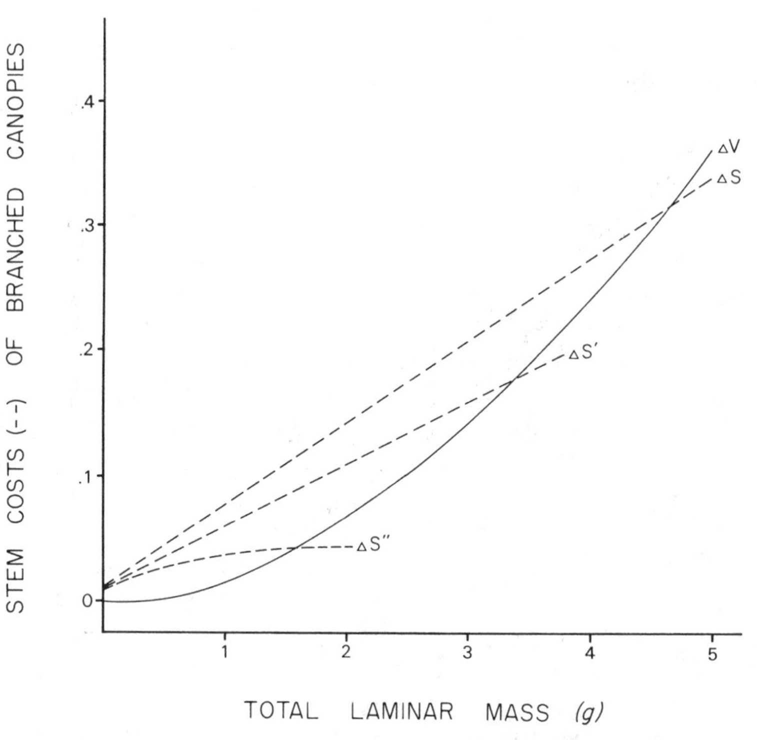

Figure 16.8. Marginal vein benefits (ΔV) and stem costs (ΔS, $\Delta S'$, $\Delta S''$) of two- versus one-leaved *Podophyllum* canopies as functions of total laminar mass. A shift from one- to two-leaved canopies is energetically favored at or above the laminar mass where the marginal benefits of two-leaved canopies exceed their marginal costs. The ΔS cost curve represents the unmodified model; the $\Delta S'$ curve incorporates the need to counterbalance reproductive structures in one-leaved canopies versus two-leaved canopies with unequal leaves; the $\Delta S''$ curve also incorporates the effect of greater variance in the height–laminar-mass relationship in sexual versus asexual shoots. All three models predict a shift toward two-leaved shoots at high laminar masses.

to support a hypothetical one-leaved canopy by assuming that its stem would be vertical. A completely vertical shoot would be seriously unbalanced by the torque exerted by the flower and fruit, which are held off to one side of the stem. To counter this imbalance (and thus avoid costly reinforcement of the stem below the reproductive structures to withstand horizontal bending moments), a one-leaved shoot should incline its leaf away from the flower/fruit and thus counterbalance it. Fruits have a mean fresh mass of about 13.1 g, as described earlier, with a centroid suspended roughly 2 to 4 cm from the main stem. To counterbalance the torque thus created, a single asexual leaf of about 1.9 g dry laminar mass (mean for sexual shoots) and 19.0 g fresh laminar mass should horizontally displace its centroid roughly 2.1 cm from the vertical. This would not increase by more than a few millimeters the petiole length required to hold the leaf at the prescribed height, but it would expose the leaf stem itself to bending moments and would require stronger, denser support tissue. Substituting the stem-density/laminar-mass relationship for sexual leaves with inclined petioles into the model for ΔS, we obtain a lower $\Delta S'$ (Figure 16.8). As shown, this lower marginal cost of two-leaved canopies lowers the expected transition to two-leaved canopies to 3.3 g dry laminar mass. This is within the range of total laminar masses seen in sexual shoots, but it is still considerably above the actual transition between morphs at roughly 1.2 g.

Third, we may have underestimated the stem mass required to support a hypothetical one-leaved canopy by assuming that its leaf stem has a length equal to the vertical distance between the crotch and foliage of the corresponding sexual shoot. In asexual plants, there is a strong correlation between leaf height and laminar mass ($\ln H = 0.412 \cdot \ln L + 3.393$, $r^2 = 0.938$). There is a similar correlation in sexual plants ($\ln H = 0.340 \cdot \ln L + 3.431$, $r^2 = 0.435$). Over the entire range of laminar masses seen in actual sexual shoots, these regressions yield essentially identical leaf heights for sexual and asexual shoots with the same laminar mass. However, the variance about the regression line is almost 10 times greater for sexual shoots than for asexual shoots. It thus appears that the growth form of sexual shoots may afford greater flexibility in leaf height. Flexibility above the expected leaf height should be more important than that below in terms of the competitive edge it may confer in sites with dense leaf cover (see earlier section on leaf height). If we assume that the height of hypothetical, one-leaved sexual shoots must equal the mean height of actual sexual shoots that exceed the expected height in order to be competitively more equivalent to two-leaved shoots, then the stem mass associated with this additional height should be subtracted in analyzing the net stem cost of two- versus one-leaved canopies. The additional height needed in one-leaved shoots is $\Delta H = 2.52 \cdot L^{0.412}$, based on a model assuming the same power-law exponent relating leaf height to laminar mass as in asexual shoots. This additional height, when multiplied by the stem density corresponding to a given laminar mass on an inclined leaf stem, yields the added stem cost of one-leaved canopies. This results in a new marginal stem cost $\Delta S''$ for two-leaved canopies and a predicted transition from unbranched to branched canopies at about 1.6 g laminar mass, reasonably close to the observed transition mass of 1.2 g (Figure 16.8).

This remaining discrepancy in the branching model does not occur in the following models for optimal branching angle and branching height, in which there is dramatically close agreement between the observed patterns and the predictions of simple, straightforward models.

Optimal branching angle

Sexual shoots of *Podophyllum peltatum* have an average branching angle of $50.4 \pm 11.2°$, based on direct measurements, implying a mean divergence of branches from the vertical of $25.2 \pm 5.6°$. Leaf diameters average 20.89 ± 4.41 cm. The mean branch length is 14.02 ± 3.39 cm, and the mean height of the leaves above the branchpoint is 12.83 ± 3.01. This implies a mean divergence angle from the vertical, based on indirect measurements, of $\cos^{-1}(12.83/14.02) = 23.8°$.

Given that sexual shoots branch, the optimal branching angle between the stem branches may be determined by a tradeoff between the following two factors. First, as the branching angle and divergence of the branches from the vertical increase, the length of the branches required to hold the leaves at a fixed height above the ground should increase. Second, this increase in stem costs should be partly offset by a decrease in vein costs, because as leaves of fixed laminar mass are held farther apart, they can each be more nearly symmetric without overlapping. At some optimal angle, the benefits of decreased vein costs should just balance increased stem costs and thus minimize total support costs above the branchpoint.

These costs and benefits can be quantified as follows: Consider an average asexual shoot, with branch length of 14.02 cm, foliage height above the branchpoint of 12.83 cm, and mean leaf radius (roughly equal to lobe length) of $20.89/2 \simeq 10.45$ cm. Branch length is minimized if each branch is vertical, in which case branch length is 12.83 cm. If θ is the angle of a branch's divergence from the vertical, then the branch length λ required to hold a leaf a vertical distance of 12.83 cm above the crotch is $\lambda = 12.83/\cos\theta$. Thus, the net increase in branch length accompanying a divergence angle of θ is

$$\Delta\lambda = 12.83 \cdot (1/\cos\theta - 1) \tag{16.14}$$

The branch itself must withstand horizontal bending moments regardless of the angle of divergence, either because the leaf it supports must be considerably asymmetric at low divergence angles to avoid leaf overlap and self-shading or because the branch is partly horizontal at greater divergence angles. Let us therefore assume that the mass per unit length for such branches bears the same relation to laminar mass as do the values for actual sexual stem branches. The error introduced by this assumption

Figure 16.9. Schematic representation of parameters in model for optimal branching angle in *Podophyllum*. Branching height is held constant as branching angle varies; as branching angle increases, branch length and hence mass increase while leaf symmetry and hence vein mass decrease.

should be vanishingly small for angles near the actual divergence angle. If we make this assumption, the additional branch mass associated with a given divergence angle θ and a single leaf of laminar mass L is

$$\Delta S = (0.0132 \cdot L + 0.0030) \cdot 12.83 \cdot (1/\cos\theta - 1) \qquad (16.15)$$

The additional cost of veins associated with leaf asymmetry can be calculated directly if four reasonable assumptions are made, based on the form of actual shoots. We assume that a sexual leaf has eight lobes that are spaced at regular angular intervals about the point of petiole attachment. We assume further that the shortest lobes are arranged next to each other on either side of a line connecting the points of petiole attachment of the two leaves and that the length of these lobes is adjusted so as to be maximal without leading to overlap with the corresponding lobes of the adjacent leaf (Figure 16.9). The length x_{min} of the shortest lobe under these conditions is $x_{min} = z(\theta)/\cos(360/16)$, where $z(\theta) = 12.83 \cdot \tan\theta$ is half the horizontal distance between the attachment points of the two leaves (Figure 16.9). Finally, assume that the length of each pair of lobes beyond the first is just $1 + n\zeta$ times the length of the shortest lobe, where n is the number of lobes beyond the shortest pair and ζ is adjusted to result in a laminar mass equal to that of a symmetric leaf with eight lobes each 10.45 cm long. The additional vein cost associated with each angle of divergence can then be

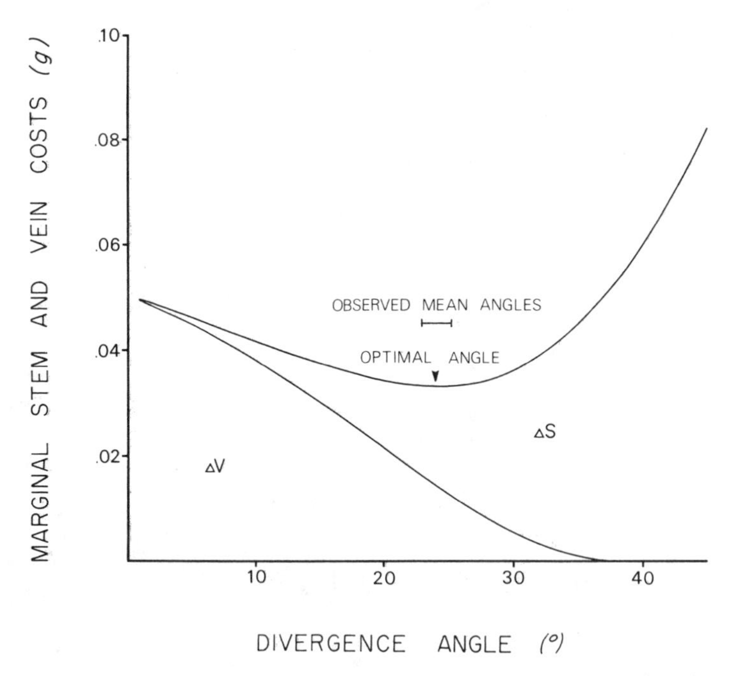

Figure 16.10. Sum of marginal vein (ΔV) and stem (ΔS) costs as functions of divergence angle in sexual *Podophyllum* shoots. Optimal divergence angle occurs at 24.0°, where the sum of marginal costs is minimized, compared with observed mean divergence angles of 23.8° to 25.2°.

calculated by subtracting the vein mass of a packed asymmetric leaf from that of a symmetric leaf with the same laminar mass. The sum of the marginal vein and stem cost is plotted as a function of divergence angle θ in Figure 16.10. Note that the marginal cost of veins beyond approximately 37.0° is zero, because symmetric leaves can be packed without overlap at divergence angles greater than that. As can be seen, marginal vein costs decrease and marginal stem costs increase as a function of divergence angle. Their sum is minimized at $\theta = 24.0°$, within 0.2 to 1.2° of either empirical measurement of the actual divergence angle. The model thus predicts rather precisely the branching angle that the canopies of sexual shoots display and helps account for the seemingly inefficient asymmetry of sexual leaves in terms of the economy of leaf packing. Although successful, the model only approximates reality because it does not take into account leaf inequality in sexual shoots, possible differences in the divergence from the vertical of the branches bearing the larger and smaller leaves, and the (mechanically efficient) nonlinear, upwardly curving shape of the petioles. Nevertheless, the model probably captures the most im-

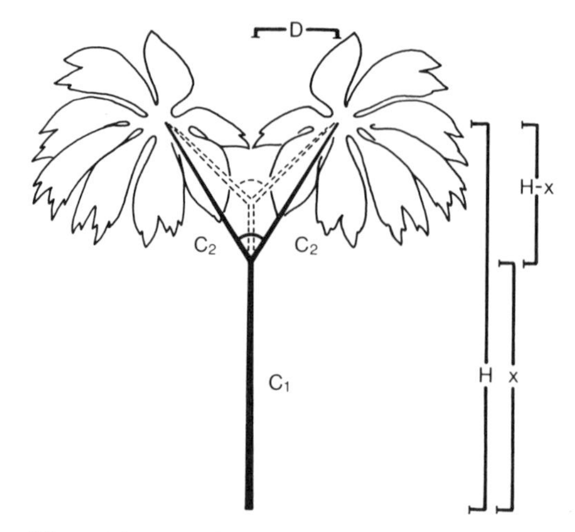

Figure 16.11. Schematic representation of parameters in model for optimal branching in *Podophyllum*. Distance between leaves is held constant as branching height and angle vary; as branching height increases, the length and mass of the unbranched stem increase while the length and mass of the stem branches decrease.

portant energetic tradeoffs that determine branching angle once branching height is fixed, and it provides a close approach between observed and predicted branching geometries.

Optimal branching height

A more general model for optimal branching geometry would allow both branching height and angle to vary as total support mass is minimized. In the previous model, we held constant the height of branching and considered the branch costs and vein benefits of different branching angles. Here we fix the distance between the two leaves of a sexual canopy, at the value observed in nature and apparently determined by the tradeoffs described in the previous section, and let branching height and angle vary.

Consider the stem costs associated with the idealized shoot geometry shown in Figure 16.11. Here, a sexual shoot of height H and separation $2D$ between the bases of two leaves of equal laminar masses branches symmetrically at height x. The length of the unbranched stem is thus x, whereas the length of each of the stem branches is $y = [D^2 + (H - x)^2]^{1/2}$. Thus, if c_1 is the mass per unit length of the unbranched stem and c_2 is the mass per unit length of the stem branches, then the total mass S of the stem assembly is

$$S = c_1 \cdot x + 2c_2 \cdot y(x) \qquad (16.16)$$

provided c_1 and c_2 are constant. The optimal branching height $\hat{x}$ can be derived by setting the derivative of equation (16.16) with respect to x equal to zero, yielding

$$0 = c_1 - 2c_2 \cdot (H - x)/[D^2 + (H - x)^2]^{1/2} \tag{16.17}$$

or

$$\cos \hat{\theta} = \frac{c_1}{2c_2} \tag{16.18}$$

where $\hat{\theta}$ is the angle of divergence of each branch from the vertical. The optimal branching height is then given by

$$\hat{x} = H - D/\tan \hat{\theta} \tag{16.19}$$

(Figure 16.11). If, as is likely, c_1 and c_2 vary with position, then total stem mass is given by

$$S = \int_0^x c_1(z) \cdot dz + 2 \cdot \int_0^{y(x)} c_2(z) \cdot dz \tag{16.20}$$

Taking the derivative of this equation with respect to x and setting it equal to zero, we again obtain the optimality condition specified by equation (16.18). Thus, optimal branching angle and hence height are specified by the local costs of the branched and unbranched sections of stem near the branchpoint. Equation (16.18) is a familiar result of pipeline theory in economics, where the aim is to connect a supply point with two or more sinks at minimum cost; an analogue was developed for branching in leaf veins by Howland (1962). In general, the cost per unit length of a "pipe" can be a function of the properties (e.g., oil demand, laminar mass) of the points it connects, its length, and its spatial position. Finding a pipeline network of minimal cost for a given set of sources and sinks is equivalent to the Steiner problem in mathematics, which currently has no general solution. Equation (16.18) describes the solution of the simplest form of the pipeline problem, connecting one source with two sinks by means of pipes whose cost does not vary with position.

A potential complication in applying equation (16.18) is that the cost c_1 of the stem below the branchpoint may depend on the fruit mass it supports. However, this presents no problem when the predicted branchpoint lies at or above the insertion of the reproductive structures. In this case, the branching angle and hence height depend only on the costs per unit length of the unbranched stem and branches near the branchpoint, and these should depend only on the laminar mass being supported. For a typical sexual shoot with mean total laminar mass of 1.922 g, the mass per unit length of the stem branches is

$$c_2 = 0.0132 \cdot (1.922/2) + 0.0030 = 0.01588 \qquad (16.21)$$

based on the allometry of sexual shoots and their branches, as described earlier. The mass per unit length of the stem below the branchpoint, based on the allometry of the lower 5 cm measured on asexual shoots, is

$$c_1 = 0.0136 \cdot 1.922 + 0.0015 = 0.02764 \qquad (16.22)$$

Thus, the optimal angle of branching is $\hat{\theta} = \cos^{-1} \cdot (\frac{1}{2} \cdot 0.02746/0.01588) = 28.2°$. This is within 3.0° of the mean divergence angle of 25.2 ± 5.6° and does not differ significantly from it. The optimal branching height $\hat{x}$ can be derived from $\hat{\theta}$, using equation (16.19) and values for total shoot height H and leaf separation $2D$. The mean height of sexual shoots is $H = 38.29$ cm. The parameter D can be estimated from the mean branch length λ and height h_L of foliage above the crotch in sexual shoots, using the formula $D = (\lambda^2 - h_L^2)^{1/2}$. The mean branch length of sexual shoots is $\lambda = 14.02$ cm, and the mean height of the foliage above the branchpoint is $h_L = 12.83$ cm, yielding $D = 5.65$ cm. Substituting the values of D, H, and λ into equation (16.19), we obtain the optimal branching height $\hat{x} = 27.75$ cm, which clearly lies within the 95% confidence interval about the expected branching height of 25.28 cm for laminar mass of 1.922 g, calculated as 25.28 ± 10.22 cm, based on a linear least-mean-squares regression (Sokal and Rohlf 1981) relating branchpoint height $h_c = H - h_L$ to laminar mass L ($h_c = 3.6112 \cdot L + 18.3320$, $r^2 = 0.504$, and $P < 0.001$ for 25 degrees of freedom; principal-axis regression $h_c = 14.0284 \cdot L - 1.6940$).

Thus, the preceding model accurately predicts the branching height and angle actually seen in sexual *Podophyllum* shoots. Because the predicted branching height of 27.75 cm is greater than the height assumed for insertion of the reproductive structures at the actual mean of 25.28 cm, the assumption made at the beginning of the preceding paragraph is valid. Why did we not use the density of the unbranched section of actual sexual stems in determining c_1, rather than constructing an estimate for c_1 from the allometry of asexual shoots? The reason is that the density of sexual stems below the branchpoint presumably is a response to the stresses imposed not only by the leaves but also by the reproductive structures it supports. If we had estimated c_1 from the density of sexual shoots below the branchpoint, we would, in effect, have assumed that as branchpoint height increases, the height of the reproductive structures will correspondingly increase and so remain inserted at the branchpoint. Such an assumption overlooks the tradeoffs between the energetic cost of holding flowers and fruit at a given height and the reproductive benefits of holding these organs at various heights above the ground and below the leaves. To avoid such complications, we held the height of the reproductive organs constant at their observed mean height of insertion in sexual shoots.

The fact that the model predicts that reproductive structures should be inserted roughly 2.5 cm below the branchpoint probably reflects slight inaccuracies in the estimations of c_1 and c_2, as well as the use of linear rather than power-law regressions to obtain the relation of stem density to laminar mass. Hence, the disparity should not be taken too seriously. More important is the fact that the reproductive structures and leaves share the same support structure over much of that system's height. Such sharing of support is expected if both leaves and reproductive structures are selected to be held well above ground and if the mass per unit length of a stem increases at a lower than proportional rate with the mass it supports. The latter can occur if there is a linear relation of stem density to leading mass with a positive intercept (as in the case of the regressions for *Podophyllum*) or if there is a power-law relationship of stem density to loading mass with exponent less than 1 [as expected in general (Givnish 1984) and also seen in asexual *Podophyllum* shoots ($y = 0.0157 \cdot L^{0.789}$, $r^2 = 0.985$)]. Thus, it is not surprising that almost all terrestrial herbs in eastern North America that hold both leaves and reproductive structures well above ground level bear both on the same stem. The few cases in which both leaves and flowers are borne well above ground level on separate stems advertise situations of great evolutionary interest, in that some unexpected (perhaps biotic) selective forces may help favor an otherwise inefficient support system. An interesting system for studying this problem occurs in North American species of *Aralia* (Araliaceae). *Aralia nudicaulis* is a common dioecious forest herb bearing leaves and reproductive structures on separate aerial shoots roughly 20 to 40 cm tall (Fernald 1950). *Aralia hispida* and *A. racemosa* bear flowers and leaves together on herbaceous, somewhat taller stems. It is difficult to understand what advantage *A. nudicaulis* might obtain by bearing reproductive structures and leaves on separate stems of roughly similar phenologies, when close relatives [and less closely related species of convergent aboveground morphology (e.g., *Actaea rubra, Panax quinquefolius*)] bear them together.

Competitive advantage of the arching growth form on slopes

Several understory herbs in eastern North American arrange their leaves in a distichous array along an arching stem. Prominent among these are species of the genera *Disporum, Polygonatum, Smilacina, Strep-*

topus, and *Uvularia* in the Liliaceae (Figure 16.12). Such a leaf arrangement should be, and is, mechanically less efficient than an umbrella-like arrangement that supports leaves at the same height on a vertical stem with radial horizontal lever arms. How do species with this growth form compete? It could be argued that *Polygonatum, Smilacina,* and related genera are simply "flattened lilies" adapted to light capture in dimly lit understories and that their mechanically inefficient leaf arrangement is the closest approach to an efficient "umbrella" consistent with phylogenetic constraints. However, the forest herbs that compete with such liliaceous leaners are unaware of the latter's ancestry, and the question remains as to how plants in these genera manage to coexist with mechanically more efficient forms.

A curious feature of arching herbs is that many orient strongly downslope (Figure 16.12), apparently in the direction of greatest light availability. The radially asymmetric canopies of *Polygonatum* and related genera may therefore simply be an adaptation for light interception in the asymmetric light fields that occur on steep slopes. Orientation downslope may also imply that species with arching stems are mechanically more, not less, efficient on sufficiently steep slopes than umbrella-like species.

Consider an arching shoot with foliar centroid a vertical distance H and a horizontal distance L from its base (Figure 16.13). Such a leaning shoot would be less efficient than an umbrella-like shoot of equal height in holding leaves at the same height and position above a horizontal surface. On a slope, however, an umbrella-like shoot would have to be taller in order to hold leaves in the same position as a leaning shoot rooted uphill of it (Figure 16.13). In particular, if the inclination of the slope is θ, then an umbrella-like shoot would require a stem of height $H + L \tan \theta$ in order to hold leaves in the same position as a leaning stem of height H. Thus, as slope inclination θ increases, at some point the height of equivalent umbrella-like shoots must be so great that they will be mechanically less efficient than leaning shoots.

If the fractional allocation $f(H)$ of aboveground vegetative biomass to leaves is given by $\xi(H)$ for umbrella-like herbs, and by $\kappa(H)$ for leaning herbs, then the critical slope inclination θ_{crit}, above which leaners should be favored, is given by

$$\xi(H + L \tan \theta_{crit}) = \kappa(H) \tag{16.23}$$

or

$$\theta_{crit} = \tan^{-1}\left(\frac{\xi^{-1}[\kappa(H)] - H}{L}\right) \tag{16.24}$$

$$= \tan^{-1}(\Delta H/L) \tag{16.25}$$

Figure 16.12. Examples of forest herbs that display leaves along an arching stem. (A) Orientation downslope of *Smilacina racemosa* near Porter's Flat. Growth forms of (B) *Polygonatum biflorum,* (C) *Smilacina racemosa,* (D) *Disporum lanuginosum,* and (E) *Uvalaria sessilifolia.*

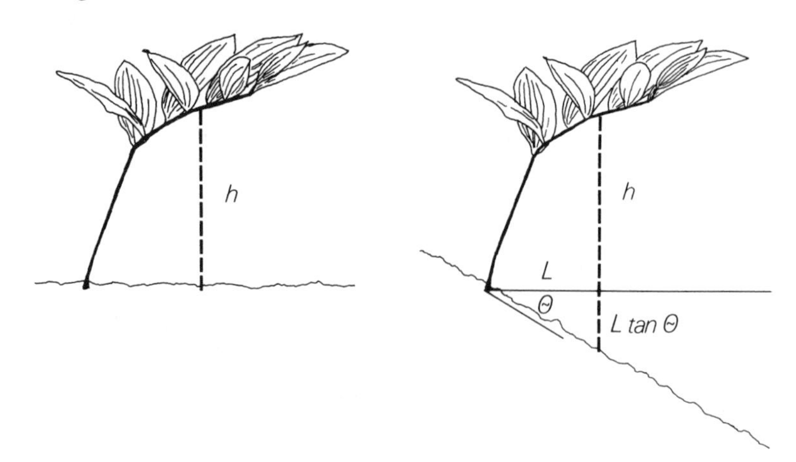

Figure 16.13. Height advantage gained by arching herbs compared with umbrella-like competitors on slopes versus relatively horizontal sites. Arching herbs with leaf centroids a vertical distance H and horizontal distance L from the stem base can hold their leaves in the same position as umbrella-like herbs rooted downslope of height $H + L \tan \theta$, where θ is slope inclination.

where $\Delta H = L \tan \theta_{crit}$ is the additional height required to reduce the foliar allocation of an umbrella-like shoot to that of a leaning shoot of a given height. It should thus be possible to calculate the critical slope inclination θ_{crit}, above which leaning shoots become more efficient than umbrella-like herbs and should tend to supplant them, using the allometry of leaning $[\kappa(H)]$ and umbrella-like $[\xi(H)]$ shoots and the shape (H, L) of leaning shoots. The resulting predictions can be tested against the actual distribution of dominance by leaners as a function of slope inclination.

Methods

In April 1981, we measured the compass orientations of the aerial shoots of *Disporum lanuginosum, Polygonatum biflorum,* and *Smilacina racemosa,* of the heliotropic flowers of *Trillium grandiflorum* and *Viola rostrata,* and of the slopes supporting these plant populations at 13 sites in virgin and slightly disturbed deciduous forest at 850–950 m elevation near Porter's Flat, Great Smoky Mountains National Park. The vegetation of this superb cove forest has been described by Whittaker (1956), Bratton (1976), and Hicks (1980). The compass orientations of *Disporum, Polygonatum,* and *Smilacina* were taken as the azimuth of the principal or only leaf-bearing axis for each shoot visible within 3 m downslope of an observer walking a 10- or 15-m transect. Each slope sampled for plant orientation had an inclination of 25° or more, as determined by clinometer.

The compass orientation of each slope was measured at both ends of each sample transect and averaged to give the characteristic slope orientation. The orientations of 10 to 25 *Trillium* and *Viola* flowers were measured on certain slopes to provide an independent index of the direction of greatest average light availability; flowers of these species appear strongly heliotropic, based on their orientations near clearings and roadsides. The presence or absence of large canopy gaps uphill of each transect was noted.

In May 1982, the distributions of *Disporum, Polygonatum,* and *Smilacina* were tallied at 10-m intervals along a 1-km transect along the trail through the virgin section of the Porter's Flat forest. At each station, coverage by each leaning taxon was estimated in ten 0.5-m^2 plots placed at a fixed distance of 1 to 5 m upslope and downslope of the trail. Leaf coverage was estimated as being 0, 5, 10, 20, 40, 60, 80, 100, or 120% of plot area. Mean coverages were calculated separately for the upslope and downslope blocks of plots at each station. The slope inclination of each block was measured by clinometer to the nearest degree. Plots on boulders or under rhododendron thickets or hemlock trees were excluded from the final analysis because coverage on such sites by most herbs, and particularly *Disporum, Polygonatum,* and *Smilacina,* was nearly zero. The characteristic vertical and horizontal dimensions H and L of *Polygonatum biflorum* and *Smilacina racemosa* were measured in the field on 10 individuals of each species. The proportion of aboveground vegetative biomass in foliage was measured using the methods of Givnish (1982).

Results

Shoots of *Disporum, Polygonatum,* and *Smilacina* show a strong orientation downhill on slopes having an inclination of 25° or more (Figure 16.14). This orientation appears to be based on the direction of greatest light availability, rather than gravity, based on the orientation upslope of arching shoots and heliotropic flowers on transects with a large upslope gap in the canopy (Figure 16.14). Orientation is rather precise, with the mean populational standard deviation for shoot azimuth on slopes lacking gaps equal to 15.2 ± 5.4° ($N = 21$). Orientation is less precise on slopes with canopy gaps, with a mean standard deviation of 44.5 ± 14.8° ($N = 4$).

Calculations of θ_{crit} for *Polygonatum* and *Smilacina* are summarized in Table 16.4. *Polygonatum* has a mean height H of 23.9 cm, horizontal dimension L of 15.3 cm, and a proportion κ of 0.606 of aboveground vegetative biomass in foliage. An umbrella-like shoot of comparable height would be expected to have a greater proportion ξ of 0.636 of aboveground vegetative biomass in foliage, based on the relationship $\xi(H) = 0.832 - 0.0082 \cdot H$ documented by Givnish (1982). Similarly, *Smilacina* has a

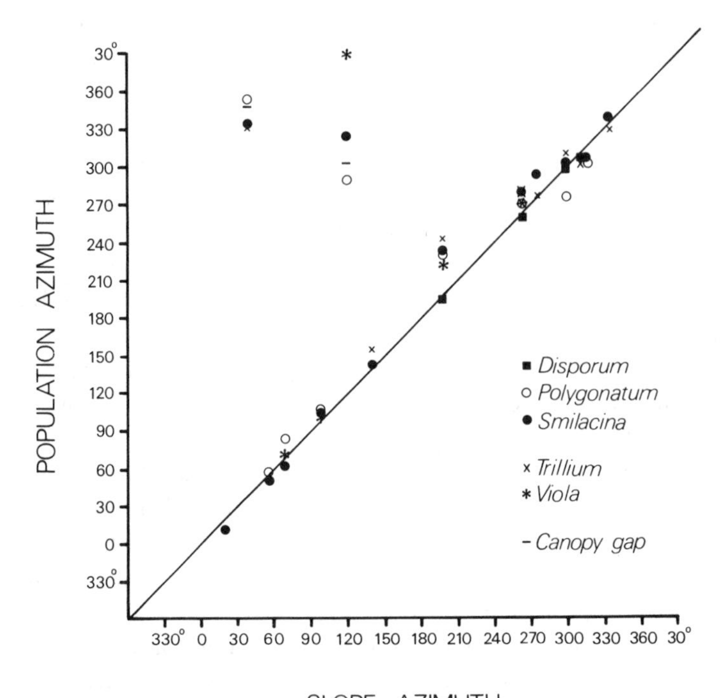

Figure 16.14. Mean compass orientations of arching herbs and heliotropic flowers as functions of slope orientation, Great Smoky Mountains National Park. Note the close approach of plant and slope orientations under continuous canopies, as well as the plant orientation toward canopy gaps under discontinuous canopies.

mean vertical height of 22.4 cm, horizontal dimension of 10.6 cm, and foliar allocation of $\kappa = 0.621$, compared with an expected allocation $\xi = 0.648$. Thus, as expected, these arching herbs are somewhat less efficient than umbrella-like herbs of comparable height.

The critical slope inclination can be calculated by asking either (1) at which angle leaners and umbrellas would be competitively equivalent or (2) at which angle leaners would have a significant advantage at the 95% confidence level. In either case, the procedure used is the same and is based on equation (16.20). First, we calculate the height H' of an umbrella-like shoot that would have the same foliar allocation as an arching shoot of given height H, $H' = \xi^{-1}[\kappa(H)]$. In case (1) we employ the actual dependence $\xi(H)$ of foliar allocation to leaf height in umbrella-like herbs; in case (2) we augment the actual $\xi(H)$ by the 95% confidence interval about the regression. Second, we determine the additional height of an

Table 16.4. *Calculation of critical slope angle for* Polygonatum biflorum *and* Smilacina racemosa

Data	*Polygonatum*	*Smilacina*
Height (H)	23.9 cm	22.4 cm
Length (L)	15.3 cm	10.6 cm
Observed fractional allocation to leaves [$\kappa(H)$]	0.606	0.621
Case 1: Competitive equality		
Expected fractional allocation to leaves in umbrella-like herbs [$\xi(H)$]	0.636	0.648
Expected leaf height in umbrealla-like herbs with same allocation {$H_1 = \xi^{-1}[\kappa(H)]$}	27.6 cm	25.8 cm
Difference in height between species and umbrella-like herb with same foliar allocation ($\Delta H = H - H_1$)	3.7 cm	3.4 cm
Critical slope angle $\left[\theta_{crit} = \tan^{-1}\left(\frac{\Delta H}{L}\right)\right]$	13.6°	17.7°
Case 2: Significant competitive advantage		
Expected fractional allocation to leaves in umbrella-like herbs + 95% confidence interval [$\xi^*(H)$]	0.668	0.675
Expected leaf height in umbrella-like herbs with same allocation {$H_2 = \xi^{*-1}[\kappa(H)]$}	31.2 cm	29.3 cm
Difference in height between species and umbrella-like herb with same foliar allocation ($\Delta H = H - H_2$)	7.3 cm	6.9 cm
Critical slope angle $\left[\theta_{crit} = \tan^{-1}\left(\frac{\Delta H}{L}\right)\right]$	25.5°	23.6°

equivalent umbrella-like shoot, $\Delta H = H' - H$. Finally, we determine θ_{crit} by using equation (16.21), $\theta_{crit} = \tan^{-1}(\Delta H/L)$.

The critical slope inclinations for *Polygonatum biflorum* are 13.6% for competitive equality and 25.5° for significant superiority. The corresponding inclinations for *Smilacina racemosa* are 17.7° for equality and 28.6° for significant superiority over umbrella-like herbs. Thus, dominance by these species should begin to increase on slopes between roughly 15° and 25° and should increase with increasing slope inclination. In addition, there may be a tendency for coverage by *Smilacina* to increase relative to that by *Polygonatum* on slopes of intermediate inclination, given *Smilacina*'s greater disadvantage relative to umbrella-like herbs at lower slopes and its disadvantage relative to *Polygonatum* on the steepest slopes as a result of its lower length/height (L/H) ratio.

These predictions are largely supported by the distributional data. With one exception, all blocks with slope inclination less than 20% have zero coverage by *Polygonatum, Smilacina,* or the somewhat similar *Disporum* (Figure 16.15). The average coverage by such leaners and the proportion

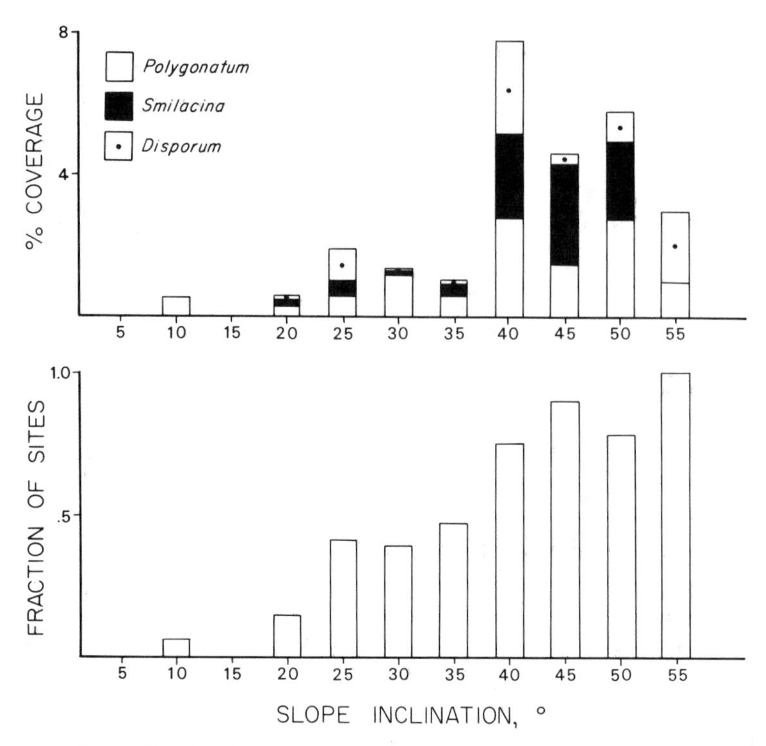

Figure 16.15. Average coverages and percentages of sample blocks occupied by the arching herbs *Disporum lanuginosum, Polygonatum biflorum,* and *Smilacina racemosa* at Porter's Flat.

of blocks in which they occur increase sharply with slope inclination above 20°. The proportion of plots occupied by leaners increases roughly linearly with slope inclination between 20° and 55°, whereas average coverage increases rapidly between 35° and 40° and then decreases somewhat. Although coverage by other herbs was not measured in all blocks, the observed trends in coverage and presence of leaners are not artifacts of any tendency for total herb cover to increase on steeper slopes; if anything, the greatest density of total coverage occurs on relatively level plots at Porter's Flat. Finally, as expected, coverage by *Smilacina* relative to that by *Polygonatum* peaks at intermediate slope inclinations, reaching maximum values between 40° and 50° (Figure 16.15).

These observations support the quantitative prediction that herbs with arching shoots that orient downslope should have a context-specific mechanical advantage over umbrella-like herbs on slopes steeper than a critical angle. This is not to say, however, that *Polygonatum, Smilacina,* and other similar herbs should be excluded from all level sites. They might successfully occupy such sites as fugitive species that disperse rapidly to disturbed microsites. Alternatively, their ability to orient foliage to side lighting may yield advantages near woodland edges, steam banks, and other side-lit habitats. Indeed, in second-growth forests, I have observed *Polygonatum* and *Smilacina* sporadically, and occasionally in abundance, on level sites. However, such occurrences are rare in virgin forests like those at Porter's Flat, perhaps because umbrella-like and leaning herbs there have had more ecological time in which to interact. *Polygonatum* and *Smilacina* also appear to increase in relative abundance toward woodland edges and roadsides in many forests in southern and central New England. Thus, the distribution of these species with respect to slope inclination shown in Figure 16.15 reflects their competitive ability under only one set of rather constrained conditions – namely, life under relatively undisturbed, closed canopies.

It should be emphasized that although the distribution of leaning shoots corresponds quantitatively to the predictions of a model based on mechanical efficiency as a function of slope inclination, this does not prove that slope inclination per se is the factor directly responsible for the observed pattern of distribution. Other ecological factors that may be correlated with slope inclination (e.g., soil texture and depth, site productivity, or foliage height and density) undoubtedly also play some role in limiting the distribution of these species under closed canopies. For example, *Smilacina* and *Polygonatum* should not occur on slopes with an infertile substrate (where they are indeed rarely found), simply because such sites support sparse herbaceous cover and should favor plants with shorter, less costly shoots (Givnish 1982). Nevertheless, the principles embodied in equation (16.20) demonstrate that on slopes, leaning herbs can gain a mechanical advantage that they lack on horizontal sites. The correspondence between their observed and predicted distributions relative to slope inclination strongly suggests that this mechanical advantage plays an important role in determining their context-specific competitive ability.

Leaf shape in violets

We conclude with a qualitative analysis of the biomechanics and natural history of leaf shape in the genus *Viola*. Violets native to the northeastern United States are herbs of forests, meadows, and fields (Fernald 1950). Many bear leaves with a characteristic cordate, or heart-shaped, base, whereas other species have lanceolate, sagittate, rhombic, orbicular, or deeply divided leaves (Figure 16.16). Violet petioles can emerge either

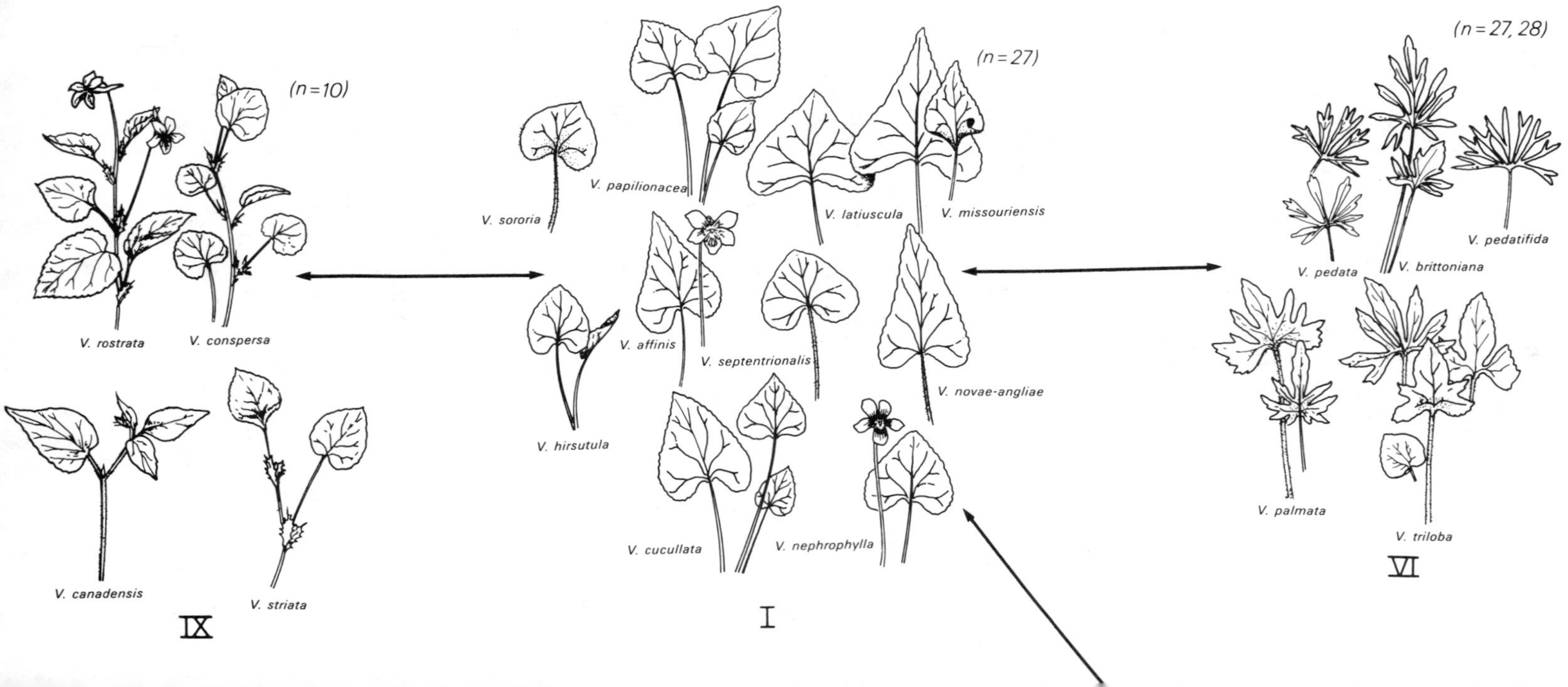

(n=10)
V. rostrata
V. conspersa
V. canadensis
V. striata
IX

(n=27)
V. sororia
V. papilionacea
V. latiuscula
V. missouriensis
V. hirsutula
V. affinis
V. septentrionalis
V. novae-angliae
V. cucullata
V. nephrophylla
I

(n=27, 28)
V. pedata
V. brittoniana
V. pedatifida
V. palmata
V. triloba
VI

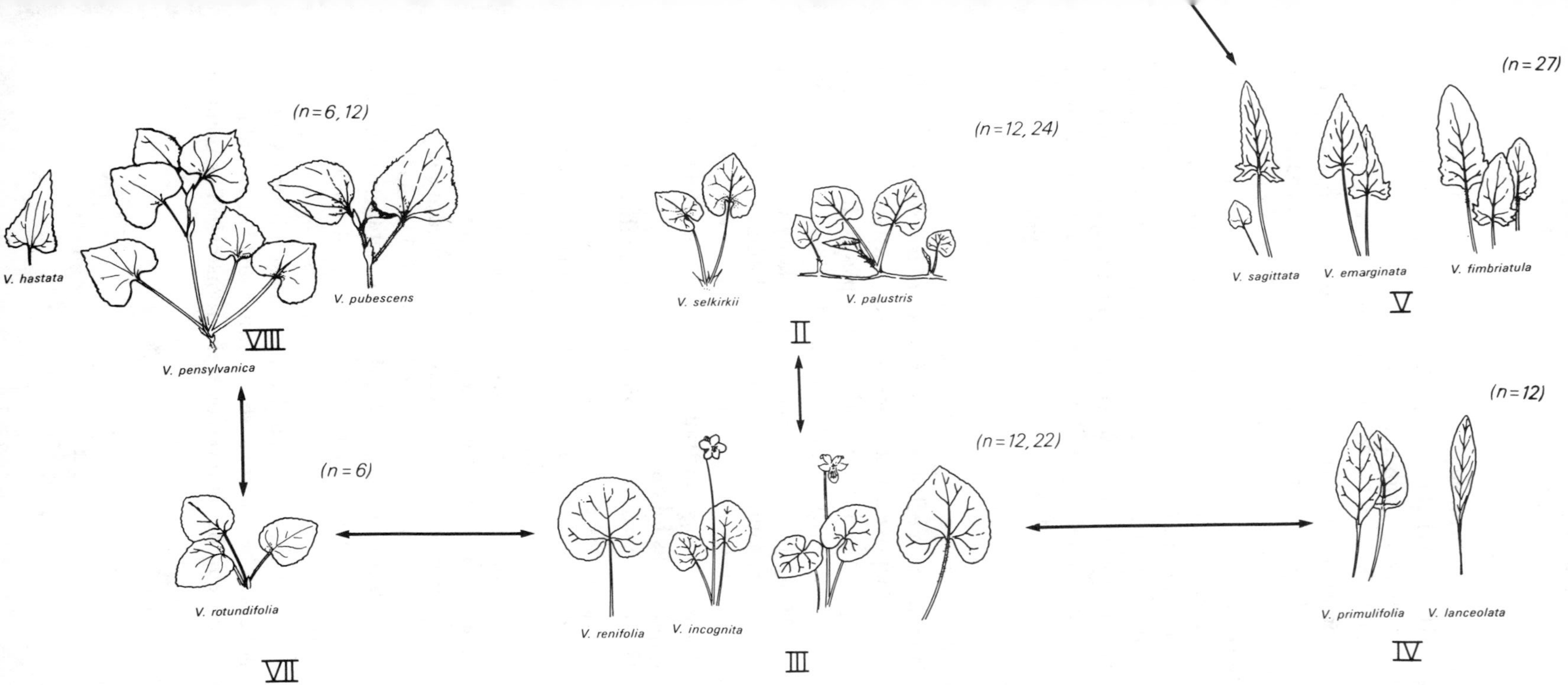

(n=6, 12)

V. hastata

V. pubescens

V. pensylvanica

VIII

(n=12, 24)

V. selkirkii

V. palustris

II

(n=27)

V. sagittata

V. emarginata

V. fimbriatula

V

(n=6)

V. rotundifolia

VII

(n=12, 22)

V. renifolia

V. incognita

III

(n=12)

V. primulifolia

V. lanceolata

IV

Figure 16.16. Ecological groups of violets native to the northeastern United States, based on leaf form, leaf arrangement, and underground morphology. Arrows indicate phylogenetic relationships inferred from chromosome number (in parentheses) and species crossability. Species in each group are as follows: I (*V. affinis, cucullata, hirsutula, langloisii, latiuscula, missouriensis, nephrophylla, novae-angliae, papilionacea, pectinata, septentrionalis, sororia, villosa*); II (*V. selkirkii*); III (*V. blanda, incognita, pallens, palustris, renifolia*); IV (*V. lanceolata, primulifolia*); V (*V. emarginata, fimbriatula, sagittata*); VI (*V. brittoniana, chalcosperma, egglestoni, esculenta, lovelliana, palmata, pedata, pedatifida, septemloba, stoneana, triloba, viarum*); VII (*V. rotundifolia*); VIII (*V. canadensis, hastata, pensylvanica, pubescens, tripartita*); IX (*V. adunca, conspersa, rostrata, rugulosa*).

directly from underground rhizomes or stolons or from an aboveground leafy stem (Figure 16.16). What is the significance of this variation in leaf shape, and how is it related to habitat?

Leaves of different shapes differ in biomechanical properties, but not necessarily in properties related to gas exchange. Leaves of several different shapes can have the same effective sizes, stomatal conductances, and chlorophyll and protein contents. However, such differently shaped leaves must, almost always, differ in terms of the efficiency with which they can support a given laminar mass. For example, the forest violets (*V. papilionacea, V. pensylvanica*) studied by Givnish et al. (1986) have effective leaf sizes, stomatal conductances, chlorophyll contents, and chlorophyll/N ratios in the same range as those for nearly all other early summer species, but their leaves have much shorter internal cantilevers than those of *Podophyllum* or *Polygonatum,* and their canopies are supported on multiple rather than single petioles.

General principles regarding cordate leaf bases

In assessing the biomechanical significance of leaf shape in *Viola,* we must consider the adaptive significance of cordate leaf bases. Givnish and Vermeij (1976) presented the hypothesis that plants that hold their leaf blades roughly perpendicular to their petioles should have deeply cordate or symmetrically peltate leaves, whereas plants that hold leaf blades and petioles in the same plane should have blunt or convex leaf bases. This argument, which is based on the economy of support in individual leaves free of leaf packing constraints (Givnish 1979, 1984), can be summarized and updated as follows.

Consider plants with leaves of fixed laminar mass and effective size and roughly constant length and width. In this case, the salient remaining difference between leaves with cordate versus blunt or convex bases is the position of petiole attachment (Figure 16.17). Viewed in profile, deeply cordate or peltate leaves have a petiole attachment near the centroid, with short, roughly equal lever arms radiating to other points on the leaf. Leaves with blunt or convex bases are, by contrast, attached at the periphery and have much greater lever arms within the lamina. The mass of veins required to support a leaf of fixed mass and size should increase as the point of petiole attachment shifts from the centroid of the leaf to its periphery.

For an herb with a single horizontal leaf, the most efficient support skeleton should generally be a vertical stem, with radial horizontal lever arms of roughly equal lengths centered over the rootstock. The point of

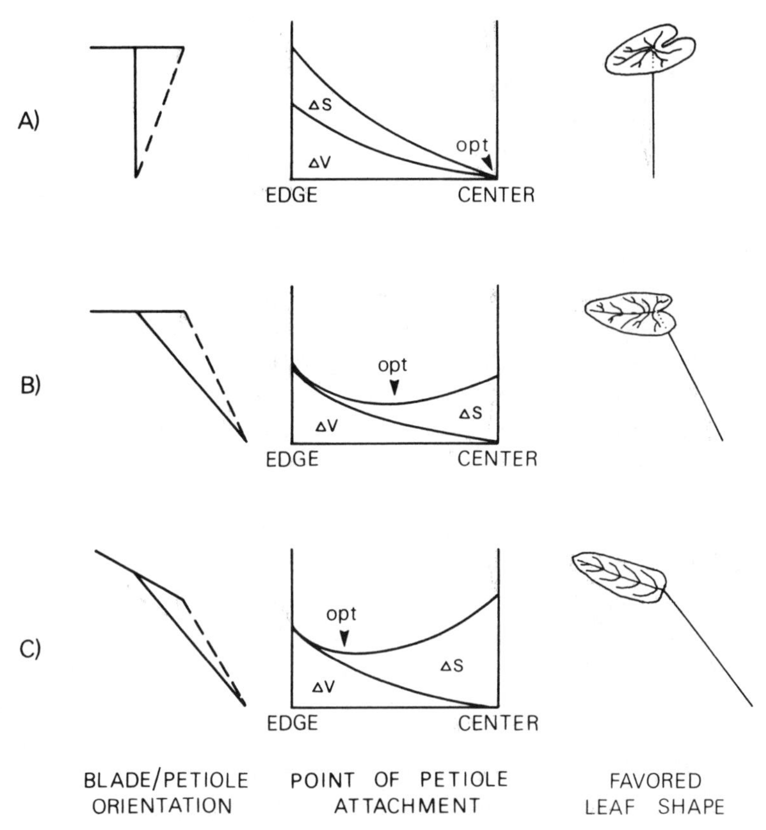

Figure 16.17. Energetic costs and benefits underlying the evolution of cordate leaf bases: (A) In a plant with a single petiole, the marginal costs of both veins (ΔV) and stems/petioles (ΔS) are minimized by a central point of attachment (solid petiole) and maximized by a peripheral attachment point (dashed petiole). Deeply cordate or peltate leaves are thus favored. (B) In rosette herbs with horizontal leaves displaced from the rootstock, marginal vein costs are minimized by a central attachment, whereas marginal stem costs are minimized by a peripheral attachment, favoring shallowly cordate leaves with a petiole attachment midway between leaf center and edge. (C) In rosette herbs with steeply inclined leaves, the steep increase in marginal stem costs as the attachment point recedes from the periphery favors shallowly cordate or auriculate leaves.

petiole attachment favored is the leaf centroid, implying a peltate or deeply cordate leaf. Such an arrangement minimizes the cost of veins within the leaf by reducing the length and inequality of the support arms, as well as minimizing stem length and support cost outside the leaf. The data presented previously for *Podophyllum* support these ideas and for the first time demonstrate the actual magnitude of the costs of lanceolate versus peltate or cordate leaves, of unequal versus equal support arms, and of vertical versus partly horizontal leaf stems/petioles.

For herbs with several leaves attached independently to a single rootstock, the most efficient leaf packing arrangement is a rosette, with all leaves at roughly equal distances from the central rootstock. As a result, the petiole joining each leaf to the rootstock must be inclined rather than vertical. If each leaf is held horizontally, this favors a petiole attachment somewhere between the leaf centroid and periphery (Figure 16.17). Vein mass, as before, is minimized by a central point of attachment and increases as that point moves toward the leaf's periphery. However, for a leaf held at a given height and distance from the rootstock, petiole length and mass should be least for an attachment at the point on the leaf periphery nearest the roots and should increase as the attachment point moves toward the leaf centroid. Hence, the combined mass of petiole and veins should be minimized for a petiole attachment somewhere between the leaf centroid and periphery.

Technically, in order to ensure that such an internal minimum exists, two sufficient conditions must be met: (1) both vein and petiole masses must increase at a greater than linear rate with distance from the attachment points at which they achieve their respective minima, and (2) both rates of mass increase must be of roughly equal magnitudes and opposite signs (Figure 16.17). Both requirements are likely to be met. Veins or petioles must increase disproportionately in mass as their length increases [equation (16.9)], and increases in petiole length and mass accompanying a shift in attachment point are likely to be small, based on the length and nearly vertical orientation of actual violet petioles. Support for these conclusions comes from our demonstration that total petiole and vein masses in *Podophyllum* are minimized at an intermediate branching angle and point of petiole attachment (Figure 16.10), because petiole mass and vein mass increase at greater than linear rates with increasing distance of the attachment from the leaf periphery and centroid, respectively. The principles involved in the models for branching angle in *Podophyllum* and leaf shape in *Viola* are analgous, except that the maximum petiole inclination in *Viola* may be less than vertical because there are several leaves in a *Viola* rosette.

If each leaf is inclined somewhat, so that its plane lies closer to the petiole, the rate at which petiole length and mass increase as the attachment point shifts from periphery to centroid should be greater than the corresponding rate for horizontal leaves. As a result, the increase in petiole mass should balance the decrease in vein mass for an attachment point closer to the leaf periphery, favoring shallowly cordate or sagittate leaves. For leaves in or near the same planes as their petioles, peripheral attachment is favored, because the rate of increase of petiole mass is maximal, and because a cordate or peltate leaf will require longer internal cantilevers to support lamina adjacent to the coplanar petiole than if that lamina were connected directly to the petiole (Givnish and Vermeij 1976). Thus, species that hold their leaves horizontally and roughly perpendicular to their petioles should have deeply cordate or peltate leaves, and the degree to which the leaf base is indented should decrease as the angle between blade and petiole decreases (Givnish and Vermeij 1976).

The case of Viola

These predictions are borne out by the associations among habitat, leaf orientation, and leaf shape seen in violets of the northeastern United States (Figure 16.16). The 47 native species listed by Fernald (1950) can be subdivided into nine ecological groups, based on leaf shape and arrangement and underground morphology. Each ecological group, in turn, shows a characteristic range of habitats and represents a phylogenetic group. Taxonomic affinities of the ecological groups, based on the species arrangement by Fernald (1950) and cytological studies of species and hybrids by Brainerd (1924), Gershoy (1934), and Clausen (1967), are shown in Figure 16.16.

The largest group of species is characterized by stemless plants with deeply cordate leaves and rhizomatous rootstocks. The 13 blue-flowered species of group I are all closely related and mainly inhabit rich deciduous forests and thickets; a few also occur in meadows, bogs, or gravelly shores (Fernald 1950). Typical species include *V. cucullata, V. novae-angliae,* and *V. papilionacea.* The shady, productive habitat of these species favors roughly horizontal leaves atop tall, roughly vertical petioles and thus favors the deeply cordate leaf bases seen in this group. Two other groups bear similar, deeply cordate leaves and occur mainly in woodland habitats, but differ in having stoloniferous rootstocks and either white (group II) or blue (group III) flowers.

At the opposite extreme in leaf shape, the two stoloniferous species of group IV (*V. lanceolata, V. primulifolia*) have long, lanceolate leaves with blunt or convex leaf bases. Both occur in brightly lit, rather open habitats (such as damp meadows, bogs, open shores, and thin woods), with their leaves held vertically and in the same plane as their petioles. Comparative photosynthetic studies by Curtis (1984) have shown that *Viola* species characteristic of open and wooded habitats have roughly the same, shade-adapted photosynthetic responses to light. Species growing in open habitats thus suffer no disadvantage by holding their leaves vertically and may thereby reduce transpirational costs and the potential for photobleaching and photoinhibition. Thus, *Viola* species of open habitats are expected to have vertical leaves, lack cordate leaf bases, and have a peripheral point of attachment, as observed.

Group V consists of three stemless species (*V. emarginata, V. fimbriatula,* and *V. sagittata*) with sagittate leaves that are essentially intermediate in form between the deeply cordate leaves of groups I–III and the lanceolate leaves of group IV. Not surprisingly, these occur mainly in habitats of intermediate openness, such as meadows, thickets, wood edges, and open woods, that may favor somewhat inclined leaves and a petiole attachment near the leaf base. Group VI is another blue-flowered group of stemless rhizomatous violets, apparently closely related to the sagittate- and cordate-leaved groups I and V, and is characterized by deeply divided palmatifid or pedatifid leaves. Characteristic species include *V. brittoniana, V. pedata,* and *V. septemloba.* This group typically occupies nutrient-poor sites that favor effectively narrow leaves (Givnish 1979, 1984), such as peaty or sandy soils of the coastal plain, pine woods, prairies, and calcareous soils (Fernald 1950). The bizarre leaf form of species in this group may partly reflect developmental constraints (the lobes form around a venation system typical of cordate-leaved *Viola* species) and partly reflect adaptation. Narrowly lobed, horizontal leaves should be favored over narrow, erect, undivided leaves (e.g., group IV) in wooded habitats that become partly shaded in summer. In such woody, sterile habitats, pedatifid leaves may also be favored over undivided, horizontal leaves of the same effective size on mechanical grounds. Lobed leaves should require less vein mass for support than should a single, long, undivided leaf of comparable laminar mass and should require less petiole mass for support than should a collection of shorter, undivided leaves of the same total laminar mass.

Group VII consists of a single species (*V. rotundifolia*) with orbicular, round leaves (Figure 16.16). This yellow-flowered species occurs in rich deciduous woods, often in microsites with sparse coverage by other herbs, as on shallow soil or moss-covered rocks (personal observation). As expected, it holds its horizontal leaves roughly at ground level. As a result, its horizontal petioles and leaves are coplanar, and its leaves are only shallowly notched at the base.

Finally, groups VIII and IX are composed of stemmed violets that bear cauline leaves on an aboveground stem; some species also bear basal leaves, like those of the stemless violets, that spring directly from the underground rootstock. Yellow- or white-flowered species in group VIII include *V. canadensis, V. hastata, V. pensylvanica,* and *V. pubescens; V. tripartita* sometimes bears divided leaves. Typical blue-flowered species in group IX include *V. rostrata* and *V. striata.* All species in these groups are native to rich deciduous forests. As expected from the low light availability in such forests, both basal and cauline summer-active leaves in groups XIII and IX tend be held roughly horizontally. Basal leaves are invariably cordate (e.g., as in *V. pensylvanica* and *V. pubescens*), as expected from the horizontal posture of the leaves and the erect posture of their petioles. Cauline leaves, however, are borne on horizontal rather than vertical petioles in several species and should have blunter leaf bases for mechanical reasons, as well as to avoid self-shading among leaves packed in a spiral about a vertical axis (Givnish 1979, 1984). Indeed, several stemmed species (e.g., *V. adunca, V. conspersa, V. hastata, V. pensylvanica, V. pubescens, V. rostrata, V. striata*) bear rhombic or hastate leaves with blunt or only shallowly notched leaf bases (Fernald 1950; Ricketts 1966).

The model for leaf shape in *Viola* applies to other forest herbs as well. Genera with horizontal foliage on an erect petiole arrange it in peltate (*Diphylleia, Podophyllum*) or deeply cordate leaves (*Asarum, Sanguinaria*), or in functionally equivalent leaf whorls (*Anemonella, Isotria, Medeola, Trillium*) or ternately compound leaves (*Actaea, Botrychium, Caulophyllum, Cimicifuga, Dicentra*). Rosette herbs with leaves on multiple petioles have more shallowly cordate leaves (*Hepatica, Heuchera, Tiarella*); species with recumbent foliage and petioles have ovate leaves (*Pyrola*).

Discussion

The studies summarized here address the adaptive significance of four fundamental aspects of canopy geometry in forest herbs: canopy height, branching pattern, canopy shape and mode of leaf arrangement, and leaf shape. Each of these traits has important implications for the mechanical costs associated with supporting a canopy of given leaf mass and photosynthetic characteristics. The analyses in this chapter suggest how selection on these traits may favor biomechanical adaptations in canopy geometry that minimize support costs in various environments. These studies complement those of Givnish et al. (1986) on traits that directly affect gas exchange (such as effective leaf size, leaf thickness, stomatal conductance, leaf N content, and chlorophyll/N ratio) and begin to relate the remarkable variations in leaf and canopy forms in forest herbs to underlying species differences in habitat and seasonal period of leaf activity.

The techniques used to study optimal canopy geometry in *Podophyllum* may have general applicability. The approach focuses on the effects of branching pattern per se by considering different means of supporting a given leaf mass with fixed photosynthetic characteristics and height above the ground. It thus compares canopy geometries that differ only in terms of allocation to support tissues of relatively uniform compositions, and it asks which geometry minimizes support biomass, thus circumventing many of the problems regarding the choice of energetic currency raised by Harper (1977) and Lovett-Doust and Harper (1980). The resulting quantitative approach is based on biomechanical theory, the observed allometry of support organs, and comparisons within a space of photosynthetically equivalent but mechanically disparate canopies. The analysis of optimal branching angle and height given for *Podophyllum* precisely predicts the trends observed; it is the only such model based on a rigorous minimization of support costs, and it could serve as a guide for studies of other aspects of optimal canopy geometry. Other models for optimal branching angle [Honda and Fisher (1978a, 1978b) and Honda et al. (1981, 1982) for *Terminalia;* Borchert and Tomlinson (1984) for *Tabebuia*] are difficult to assess because they are based on minimizing (1) a parameter that does not incorporate support costs and may not specify a single branching pattern (e.g., leaf overlap in *Terminalia*) or (2) a biomechanically inappropriate index of support mass [e.g., branch mass in *Tabebuia* estimated from the Shinozaki et al. (1964) pipe-model theory]. Future analyses of optimal branching angles in tree crowns based on more appropriate measures of support costs, as proposed by Borchert and Tomlinson (1984), must carefully address the assignment and amortization of such costs in plants of indeterminate growth. The superb study of King (1981) on optimal allocation between crown and trunk tissue to maximize height growth in *Populus tremuloides,* based on allometric measurements and the calculus of variations, illustrates how the complications introduced by indeterminate growth might be addressed.

The general approach taken in this chapter is unique in that it asks how optimal biomechanical design should vary with a plant's competitive context. Almost all studies of biomechanical adaptation have asked simply how a plant's form and/or structural materials must be designed so that it can support itself, or how its form and materials must vary to withstand different mechanical stresses (e.g., Howland 1962; McMahon 1973; McMahon and Kronauer 1976; Wainwright et al. 1976; Wilson and Archer 1977; Koehl 1982). Some studies have also addressed the link between natural selection and biomechanics and have asked how plant form should vary in order to minimize support costs and/or maximize growth rate (e.g., King 1981; Niklas and O'Rourke 1982; Borchert and Tomlinson 1984). However, in such studies, optimality in design is generally taken to be absolute and is not related to differences in selective pressures and competitive milieu among ecological gradients. This chapter demonstrates that optimal biomechanical design depends on competitive context: Optimal leaf height increases with total herb cover; optimal canopy shape shifts from umbrella-like to feather-like with increasing slope inclination; optimal leaf shape in rosette herbs shifts from deeply cordate (or peltate) to lanceolate with increasing light availability. Furthermore, optimal biomechanical design can change qualitatively with plant size: Two-leaved, branched canopies are more efficient at high total laminar masses in *Podophyllum,* and one-leaved, unbranched canopies are more efficient at lower masses. This dependence of biomechanical adaptations on competitive context and plant size is expected, in that both can influence the energetic tradeoff between a trait's photosynthetic benefits and mechanical costs and thus influence the trait's impact on plant competitive ability.

The opportunities for research on context-specific advantages of biomechanical adaptations are today as exciting as those facing physiological ecology in the 1960s and 1970s. The great strength of physiological ecology is that it provides an approach to the proximal mechanisms responsible for species differences in habitat-specific competitive ability. An understanding of the basis for some of these differences is potentially quite powerful, in that it can help relate variation in plant morphology and physiology to ecological patterns in species distribution, community composition, and community structure (e.g., Björkman et al. 1972a, 1972b, 1975; Mooney et al. 1974; Ehleringer and Mooney 1978; Miller and Stoner 1979; Teeri 1979; Robichaux and Pearcy 1980; Björkman 1981; Pearcy and Ehleringer 1984; Givnish et al. 1986). Studies of the biomechanical economy of plant form may yield similar kinds of insights into the ties among plant form, function, and ecology. An emerging field of biomechanical ecology might address such issues as the mechanical constraints on (1) plant population dynamics in self-thinning stands (Givnish 1986), (2) the depth zonation of floating and emergent aquatic plants, and (3) the distribution and ecology of woody plants with different branching patterns (Horn 1971; Givnish 1978, 1979, 1984; Hallé et al. 1978; Pickett and Kempf 1980; Fisher 1984).

Summary

Forest herbs vary in several traits that affect the amount of unproductive tissue needed to support a canopy with given photosynthetic characteristics. Selection on such traits involves an energetic tradeoff between the photosynthetic benefits and mechanical costs associated with different canopy geometries and should favor plants whose canopy geometry maximizes the difference between these benefits and costs in a given environment.

Here I present four case studies of biomechanical adaptation in forest herbs. These focus on the adaptive significance of (1) leaf height; (2) branched versus unbranched canopies, branching angle, and branching height in *Podophyllum peltatum,* (3) the arching growth habit in *Polygonatum biflorum* and *Smilacina racemosa;* and (4) leaf shape in the genus *Viola.* In each case, I examine how the amount of mechanical tissue required to support a given amount of leaf tissue varies with plant size and geometry. In the first study, these mechanical costs are balanced against the photosynthetic benefits associated with differences in leaf height to predict optimal leaf height. The game-theoretic model of Givnish (1982) is reviewed, and predicts that ESS leaf height should increase with the average herb coverage to which a species is exposed. Data are presented to show that species leaf height increases with ambient herbaceous cover and that the observed relationship of leaf height to cover roughly corresponds to that predicted.

In the remaining three studies, data on the allometry of support tissue is used to analyze which canopy geometry would minimize the cost of supporting a given amount of photosynthetic tissue at a given height. For *Podophyllum,* I propose that sexual shoots branch and bear two leaves rather than a single leaf of equal mass in order to minimize total support costs of vein and stem tissue above a threshold laminar mass. To test this hypothesis quantitatively, data on the allometry of vein and stem tissue are analyzed as functions of laminar mass in sexual and asexual shoots. The marginal vein costs of unbranched versus branched canopies are quantified and expressed in terms of the costs associated with one- versus two-leaved canopies, with unequal versus equal leaves, and with leaf asymmetry versus symmetry. Vein mass shows the expected allometric relationships to lobe length, leaf diameter, and lobe and leaf laminar masses. The marginal stem benefits of unbranched versus branched canopies are calculated using the observed height of foliage above the branchpoint in sexual shoots, the length and mass of their branches, and the allometry of stem mass per unit length as a function of laminar mass in asexual shoots. The marginal stem benefits of unbranched canopies increase less rapidly with total laminar mass than do the corresponding marginal vein costs, leading to a predicted shift from unbranched to branched canopies at a laminar mass of 4.6 g. Though the model agrees qualitatively with the hypothesis in predicting a shift to branched canopies at high laminar masses, it fails to predict the observed transition mass of roughly 1.2 g. Incorporation of two overlooked factors (the need to counterbalance the fruit in a single-leaved sexual shoot, and the greater variance in the height – laminar-mass relationship in sexual shoots) leads to an expected transition mass of 1.6 g, considerably closer to that observed.

Two additional models predict rather precisely the mean observed branching angle and branching height in sexual *Podophyllum* shoots. The first fixes branching height at the mean observed, and then minimizes total support costs above the branchpoint as branching angle is varied in a sexual shoot of mean laminar mass. As branching angle increases, vein costs decrease because the two leaves can become more symmetrical without overlapping, whereas stem costs increase because the length of each stem branch increases. A quantitative model shows that total support costs are minimized at a branch angle of 24.0° from the vertical, compared with observed mean angles of 23.8° and 25.2°, based on two different ways of measuring the branching angle. A second model fixes the distance between the centers of two leaves at that observed, and then minimizes total support costs as branching height is varied. The model is based on pipeline theory in economics and predicts optimal branching angle and height as functions of linear mass density of branched and unbranched segments of the stem. The predicted branching angle and height do not differ significantly from the means observed.

Polygonatum, Smilacina, and related genera arrange their foliage along an arching stem, rather than in the umbrella-like canopy of many other forest herbs. But an arching canopy geometry requires more mechanical tissue to support a given laminar mass at a given height above a horizontal surface, and so should be outcompeted by umbrella-like forms. Ecological data from virgin forests in the Great Smoky Mountains show that arching herbs orient downslope. As a result, they can obtain a mechanical advantage on slopes steeper than a critical angle, because umbrella-like shoots must have increasingly tall stems on steeper slopes in order to hold their leaves in the same position as arching herbs rooted uphill of them. I present a model for the critical slope inclination above which *Polygonatum biflorum* and *Smilacina racemosa* should be favored, expressed in terms of the allometry of support tissue in umbrella-like and arching herbs and in terms of the ratio of vertical to horizontal distances of the foliar centroid from the base of a shoot in arching herbs. The model predicts that *P. biflorum* should increase sharply in relative abundance at slope inclinations of 13.6° to 25.5° and that *S. racemosa* should increase on slopes steeper than 17.7° to 28.6°. Distributional data from a gradient of slope inclinations at Porter's Flat show that dominance by both species increases sharply at slope inclinations exceeding 20°, with each species becoming increasingly common on steeper slopes, as expected.

Finally, a qualitative model relating leaf shape to environment in rosette herbs is presented and applied to the genus *Viola*, based on a refinement of an earlier model for the evolution of cordate leaves and on substantiation of that model's assumptions by the *Podophyllum* data. The model successfully accounts for the adaptive radiation in leaf form of violets inhabiting sites ranging from meadows to thickets, rich deciduous forests, and densely shaded, mossy forests.

Acknowledgments

I wish to thank Stephen Bartz, Ann Bublitz, Elizabeth Burkhardt, and Victoria Elliott for invaluable help in gathering the *Podophyllum* data. Peter Ashton permitted access to the study site at the Case Estates of the Arnold Arboretum in Weston, Massachusetts. David Policansky provided useful discussion, encouragement, and access to an unpublished manuscript on *Podophyllum* reproduction.
Ann Bublitz, Nigel Franks, and Michael Tsai provided important assistance in gathering the *Polygonatum* and *Smilacina* data in Great Smoky Mountains National Park. Donald DeFoe and Peter White of the National Park Service helped in obtaining permits and suggesting potential field sites. Victoria Elliott, Stephen Bartz, and Mark Patterson read the mansucript and helped improve it substantially. I wish to thank the University of Chicago Press for permission to reprint portions of my 1982 article on the evolution of leaf height from the *American Naturalist.*

Appendix I: Vein cost of one versus two leaves in *Podophyllum*

1. Vein mass in a single asexual leaf of mass L.

$$V_{\mathrm{I}} = 0.1622 \cdot L^{1.464}$$

2. Vein mass in two asexual leaves each having mass $L_i = L/2$:

$$\begin{aligned} V_{\mathrm{II}} &= 0.1622 \cdot \Sigma L_i^{1.464} \\ &= 0.1622 \cdot 2(L/2)^{1.464} \\ &= 0.1179 \cdot L^{1.464} \end{aligned}$$

3. Proportional cost of arranging lamina in one versus two leaves:

$$\frac{V_{\mathrm{I}} - V_{\mathrm{II}}}{V_{\mathrm{II}}} = 0.378$$

Appendix II: Vein cost of leaf inequality in *Podophyllum*

1. In actual sexual shoots, the laminar mass L_1 of the larger leaf is, on average, 1.634 times the laminar mass L_2 of the smaller leaf. Thus, the total laminar mass L of a sexual shoot is

$$L = \Sigma L_i = 2.634 \cdot L_2$$

2. For individual sexual leaves of laminar biomass L_i, their vein mass is

$$V_i = 0.1882 \cdot L_i^{1.355}$$

3. Thus, the total biomass V_u of veins in the unequal leaves of a sexual shoot is

$$\begin{aligned} V_u = \Sigma V_i &= 0.1882 \cdot [L_2^{1.355} + (1.634 \cdot L_2)^{1.355}] \\ &= 0.1882 \cdot L^{1.355}(1 + 1.634^{1.355})/2.634^{1.355} \\ &= 0.1492 \cdot L^{1.355} \end{aligned}$$

4. For leaves of equal size, $L_1 = L_2 = L/2$, the total vein mass V_e equals

$$\begin{aligned} V_e = \Sigma V_i &= 0.1882 \cdot L^{1.355} \cdot 2/2^{1.355} \\ &= 0.1471 \cdot L^{1.355} \end{aligned}$$

5. Proportional cost of arranging lamina in unequal versus equal leaves:

$$\frac{V_u - V_e}{V_e} = 0.014$$

Appendix III: Vein cost of leaf asymmetry in *Podophyllum*

1. For an asymmetric leaf with lobe lengths $x_i = \bar{x} + \epsilon_i$ and n lobes, vein and laminar biomasses are given by

$$V_A = b_1 \Sigma x_i^{a_1} \simeq b_1\left(n\bar{x} + \frac{a_1(a_1 - 1)}{2} \cdot \bar{x}^{a_1 - 2} \cdot \Sigma \epsilon_i^2\right)$$

$$L_A = b_2 \Sigma x_i^{a_2} \simeq b_2\left(n\bar{x} + \frac{a_2(a_2 - 1)}{2} \cdot \bar{x}^{a_2 - 2} \cdot \Sigma \epsilon_i^2\right)$$

The approximations are based on Taylor-series expansions of the allometric power functions near $\bar{x}$ and on the fact that $\Sigma \epsilon_i = 0$.

2. A symmetric leaf having the same laminar mass L_A and lobe length y satisfies

$$y^{a_2} \simeq \bar{x}^{a_2} + \frac{a_2(a_2 - 1)}{2} \cdot \bar{x}^{a_2 - 2} \cdot \frac{\Sigma \epsilon_i^2}{n}$$

3. Based on this expression for lobe length y and the Taylor-series expansion of the allometric function relating vein mass to lobe length, the total vein

mass V_S for a symmetric leaf equal in laminar mass to the asymmetric leaf is

$$V_S \simeq b_1 n\left(\bar{x}^{a_1} + \frac{(a_1 - a_2)(a_2 - 1)}{2} \cdot \bar{x}^{a_1 - 2} \cdot \frac{\Sigma\epsilon_i^2}{n}\right)$$

4. Proportional cost of arranging lamina in asymmetric versus symmetric leaves:

$$\frac{V_A - V_S}{V_S} \simeq \frac{[a_1(a_1 - 1) - (a_1 - a_2)(a_2 - 1)] \cdot \frac{1}{2} \cdot \mathrm{Var}(\epsilon_i/\bar{x})}{1 + (a_1 - a_2)(a_2 - 1) \cdot \frac{1}{2} \cdot \mathrm{Var}(\epsilon_i/\bar{x})}$$

$$= \frac{2.698 \cdot \mathrm{Var}(\epsilon_i/\bar{x})}{1 + 0.550 \cdot \mathrm{Var}(\epsilon_i/\bar{x})} = 0.032$$

given that $a_1 = 3.097$, $a_2 = 2.035$, and $\mathrm{Var}(\epsilon_i/\bar{x}) = 0.0121$.

References

Björkman, O. 1981. Photosynthetic responses to different quantum flux densities. Pp. 57–107 *in* O. L. Lange, P. S. Novel, C. B. Osmond, and H. Ziegler (eds.), Physiological plant ecology. I. Encyclopedia of plant physiology, new series, vol. 12A. Springer-Verlag, Berlin.

Björkman, O., N. K. Boardman, J. N. Anderson, S. W. Thorne, D. J. Goodchild, and N. A. Pyliotis. 1972a. Effect of light intensity during growth of *Atriplex patula* on the capacity of photosynthetic reactions, chloroplast components and structure. Carnegie Inst. Washington Yearbook 1971:115–135.

Björkman, O., M. M. Ludlow, and P. A. Morrow. 1972b. Photosynthetic performance of two rainforest species in their native habitat and analysis of their gas exchange. Carnegie Inst. Washington Yearbook 1971:94–102.

Björkman, O., H. A. Mooney, and J. Ehleringer. 1975. Photosynthetic response of plants from habitats with contrasting thermal regimes: comparison of photosynthetic characteristics of intact plants. Carnegie Inst. Washington Yearbook 1974:743–748.

Borchert, R., and P. B. Tomlinson. 1984. Architecture and crown geometry in *Tabebuia rosea* (Bignoniaceae). Am. J. Bot. 71:958–969.

Brainerd, E. 1924. Some natural violet hybrids of North America. Vermont Agric. Expt. Stat. Bull. 239:1–205.

Bratton, S. P. 1976. Resource division in an understory herb community: responses to temporal and microtopographic gradients. Am. Nat. 110:679–693.

Chabot, B. F., and D. F. Hicks. 1982. The ecology of leaf life spans. Ann. Rev. Ecol. Syst. 13:229–259.

Clausen, J. 1967. Stages in the evolution of plant species. Hafner, New York.

Curtis, W. 1984. Photosynthetic light response in the genus *Viola*. Can. J. Bot. 62:1273–1278.

Ehleringer, J. R., and H. A. Mooney. 1978. Leaf hairs: effects on physiological activity and adaptive value to a desert shrub. Oecologia 37:183–200.

Ernst, W. R. 1964. The genera of Berberidaceae, Lardizabalaceae, and Menispermaceae in the southeastern United States. J. Arnold Arbor. 45:1–35.
Fernald, M. L. 1950. Gray's manual of botany. Van Nostrand, New York.
Fisher, J. B. 1984. Tree architecture: relationships between structure and function. Pp. 541–589 *in* R. A. White and W. C. Dickison (eds.), Contemporary problems in plant anatomy. Academic Press, Orlando.
Foerste, A. F. 1884. The may apple. Bull. Torrey Bot. Club 11:62–64.
Gershoy, A. 1934. Studies in North American violets. III. Chromosome numbers and species characters. Vermont Agric. Expt. Stat. Bull. 367:1–92.
Givnish, T. J. 1976. Leaf form in relation to environment. PhD dissertation, Princeton University.
– 1978. On the adaptive significance of compound leaves, with particular reference to tropical trees. Pp. 351–380 *in* P. B. Tomlinson and M. H. Zimmerman (eds.), Tropical trees as living systems. Cambridge University Press, Cambridge.
– 1979. On the adaptive significance of leaf form. Pp. 375–407 *in* O. T. Solbrig, P. H. Raven, S. Jain, and G. B. Johnson (eds.), Topics in plant population biology. Columbia University Press, New York.
– 1982. On the adaptive significance of leaf height in forest herbs. Am. Nat. 120:353–381.
– 1984. Leaf and canopy adaptations in tropical forests. Pp. 51–84 *in* E. Medina, H. A. Mooney, and C. Vásquez-Yánez (eds.), Physiological ecology of plants of the wet tropics. Dr. Junk, The Hague.
– 1986. Biomechanical constraints on self-thinning in plant populations. J. Theor. Biol. (in press).
Givnish, T. J., J. W. Terborgh, and D. M. Waller. 1986. Plant form, temporal community structure, and species richness in forest herbs of the Virginia Piedmont. Ecol. Monogr. (in review).
Givnish, T. J., and G. J. Vermeij. 1976. Sizes and shapes of liane leaves. Am. Nat. 110:743–778.
Hallé, F., R. A. A. Oldemann, and P. B. Tomlinson. 1978. Tropical trees and forests. Cambridge University Press, Cambridge.
Harper, J. L. 1977. Population biology of plants. Academic Press, London.
Harris, J. A. 1909. The leaves of *Podophyllum*. Bot. Gaz. 47:438–444.
Hicks, D. F. 1980. Intrastand distribution patterns of southern Appalachian cove forest herbaceous species. Am. Midl. Nat. 104:209–223.
Holms, T. 1889. *Podophyllum peltatum*. Bot. Gaz. 27:419–433.
Honda, H., and J. B. Fisher. 1978a. Tree branch angle: maximizing effective leaf area. Science 199:888–890.
– 1978b. Ratio of tree branch lengths: the equitable distribution of leaf clusters on branches. Proc. Natl. Acad. Sci. USA 76:3875–3879.
Honda, H., P. B. Tomlinson, and J. B. Fisher. 1981. Computer simulation of branch interaction and regulation by unequal flow rates in botanical trees. Am. J. Bot. 68:569–585.
– 1982. Two geometrical models of branching in botanical trees. Ann. Bot. 49:1–11.
Horn, H. S. 1971. The adaptive geometry of trees. Princeton University Press.

Howland, H. 1962. Structural, hydraulic, and "economic" aspects of leaf venation and shape. Pp. 183–191 *in* E. E. Bernard and M. R. Kare (eds.), Biological prototypes and synthetic systems. Plenum, New York.

Kana, T. M. 1982. The influence of spatial heterogeneity on the growth and demography of *Maianthemum canadense.* PhD dissertation, Harvard University.

Kawano, S., J. Masuda, H. Takasu, and F. Yoshie. 1983. The productive and reproductive biology of flowering plants. XI. Assimilation behavior of several evergreen temperate woodland plants and its evolutionary-ecological significance. J. Coll. Lib. Arts, Toyama Univ. (Nat. Sci.) 16:31–65.

King, D. 1981. Tree dimensions: maximizing the rate of height growth in dense stands. Oecologia 51:351–356.

Koehl, M. A. R. 1982. The interaction of moving water and sessile organisms. Sci. Am. 247:124–134.

Lovett-Doust, J., and J. L. Harper. 1980. The resource costs of gender and maternal support in an andromonoecious umbellifer, *Smyrnium olusastrum.* New Phytol. 85:251–264.

McMahon, T. A. 1973. Size and shape in biology. Science 179:1201–1204.

McMahon, T. A., and R. E. Kronauer. 1976. Tree structures: deducing the principle of mechanical design. J. Theor. Biol. 59:443–466.

Miller, P. C., and W. H. Stoner. 1979. Canopy structure and environmental interactions. Pp. 428–458 *in* O. T. Solbrig, P. H. Raven, S. Jain, and G. B. Johnson (eds.), Topics in plant population biology. Columbia University Press, New York.

Monk, C. D. 1966. An ecological significance of evergreenness. Ecol. 47:504–509.

Mooney, H. A., and S. L. Gulmon. 1979. Environmental and evolutionary constraints on the photosynthetic characteristics of higher plants. Pp. 316–337 *in* O. T. Solbrig, P. H. Raven, S. Jain, and G. B. Johnson (eds.), Topics in plant population biology. Columbia University Press, New York.

Mooney, H., J. H. Troughton, and J. A. Berry. 1974. Arid climates and photosynthetic systems. Carnegie Inst. Washington Yearbook 1973:793–805.

Muller, R. N. 1978. The phenology, growth, and ecosystem dynamics of *Erythronium americanum* in the northern hardwood forest. Ecol. Monogr. 48:1–20.

Niklas, K. J., and T. D. O'Rourke. 1982. Growth patterns of plants that maximize vertical growth and minimize internal stress. Am. J. Bot. 69:1367–1374.

Orians, G. H., and O. T. Solbrig. 1977. A cost-income model of leaves and roots with special reference to arid and semi-arid areas. Am. Nat. 111:677–690.

Pearcy, R. W., and J. Ehleringer. 1984. Comparative ecophysiology of C_3 and C_4 plants. Plant. Cell Envir. 7:1–13.

Pickett, S. T. A., and J. S. Kempf. 1980. Branching patterns in forest shrubs and understory trees in relation to habitat. New Phytol. 86:219–228.

Ricketts, H. W. 1966. Wild flowers of the United States. Vol I: The northeastern States. McGraw-Hill, New York.

Robichaux, R. H., and R. W. Pearcy. 1980. Environmental characteristics, field water relations, and photosynthetic responses of C_4 Hawaiian *Euphorbia* species from contrasting habitats. Oecologia 147:99–105.

Shinozaki, K., K. Yoda, K. Hozumi, and T. Kira. 1964. A quantitative analysis of plant form – the pipe model theory. I. Basic analysis. Jpn. J. Ecol. 14:97–105.
Sohn, J. J., and D. Policansky. 1977. The costs of reproduction in mayapple *Podophyllum peltatum* (Berberidaceae). Ecology 58:1366–1374.
Sokal, R. R., and F. J. Rohlf. 1981. Biometry, 2nd ed. Freeman, San Francisco.
Taylor, S. E. 1975. Optimal leaf form. Pp. 73–86 *in* D. M. Gates and R. B. Schmerl (eds.), Perspectives in biophysical ecology. Springer-Verlag, New York.
Teeri, J. A. 1979. The climatology of the C_4 photosynthetic pathway. Pp. 356–374 *in* O. T. Solbrig, P. H. Raven, S. Jain, and G. B. Johnson (eds.), Topics in plant population biology. Columbia University Press, New York.
Wainwright, S. A., W. D. Biggs, J. D. Currey, and J. M. Gosline. 1976. Mechanical design in organisms. Princeton University Press.
White, P. S. 1983. Corner's rules in eastern deciduous trees: allometry and its implications for the adaptive architecture of trees. Bull. Torrey Bot. Club 110:203–212.
– 1984. The architecture of devil's walking stick, *Aralia spinosa* (Araliaceae). J. Arnold Arbor. 65:403–418.
Whittaker, R. H. 1956. Vegetation of the Great Smoky Mountains. Ecol. Monogr. 26:1–80.
Wilson, B. F., and R. R. Archer. 1977. Reaction wood: induction and mechanical action. Ann. Rev. Plant Physiol. 28:23–43.

17 The roles of carbon balance and branching pattern in the growth of woody species

E.-DETLEF SCHULZE, MANFRED KÜPPERS, AND RAINER MATYSSEK

Introduction

The terrestrial vegetation of the earth is characterized by plant formations that differ in terms of biomass accumulation and partitioning. Forest vegetation maintains the largest biomass of all vegetation types, but it maintains the smallest fraction of biomass as living leaves (Table 17.1). These data show that despite a 10-fold-larger photosynthetic capacity in herbaceous species than in woody species, the values for net primary production per month during the growing season are very similar for vegetations dominated by either growth form as long as drought or cold does not govern plant performance. The most extreme contrast in the relationship between biomass and productivity may be between evergreen boreal forest and tropical grasslands. The former accumulates almost 20 times the biomass of tropical grasslands at a photosynthetic rate about one-tenth of that in tropical C_4 grasses, but net primary production is the same in the two vegetation types. Obviously, the relationships among CO_2 assimilation, net primary production, and biomass need further explanation.

A great variety of plant forms exist that in response to their growth environments show different potential solutions to the problem of carbon accumulation and growth under conditions of limited access to water and nutrients, as reviewed by Schulze (1982). Also, within a single life form (e.g., woody plants), large ranges in photosynthetic rates exist, despite similarities in biomass and net primary production.

The following study explores the relationships among growth, primary production, and leaf photosynthesis using evergreen and deciduous trees and shrubs as examples. Carbon relations constitute an important factor, but not the only one governing plant life. Therefore, we shall examine the extent to which leaf longevity, plant architecture (i.e., branching patterns)

Table 17.1. *Biomass per stand area, leaf biomass as proportion of total biomass, photosynthetic capacity and net primary production per month of growing season*[a]

Plant formation	Total biomass (kg m^{-2})	Leaf biomass as proportion of total biomass (%)	Photosynthetic capacity (mg CO_2 dm^{-2} hr^{-1})	Net primary production per month of growing season (kg m^{-2} $month^{-1}$)
Evergreen tropical and subtropical forests	41–65	1–2	18–22	0.15–0.25
Tropical seasonal forest	9–38	1–2	15–25	0.08–0.20
Temperate deciduous forest	18–60	1–2	10–18	0.10–0.30
Evergreen boreal forest	20–90	4–5	5–18	0.11–0.30
Mediterranean scrub	1.2–7.0	10–28	5–15	0.05–0.10
Temperate grassland	0.3–3.0	40–60	20–40	0.08–0.28
Tropical grassland	0.2–5.0	40–60	30–80	0.08–0.35
Deserts	0.4–6.0	1–6	4–20	0.01–0.10
Tundras	1.1–5.8	10–15	3–10	0.03–0.16

[a] Adapted from Larcher (1980) and Schulze (1982).

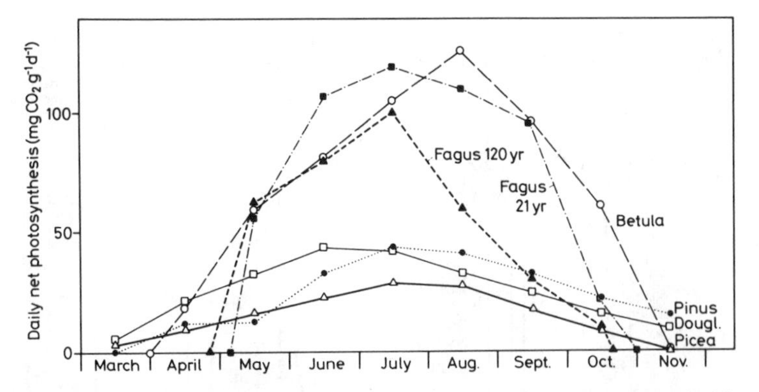

Figure 17.1. Seasonal changes in daily net photosynthesis for winter-deciduous *Fagus sylvatica* and *Betula verrucosa* and evergreen *Pinus silvestris, Pseudotsuga menziesii* (Dougl.), and *Picea abies* (after Schulze 1982).

and water and nutrient relations help to explain the observed patterns of biomass accumulation and growth of woody species.

Results and discussion

Relationships between CO_2 assimilation and primary production in deciduous Fagus sylvatica *and evergreen* Picea abies

Mid-European broad-leaved forest species generally have lower rates of wood production on a stand-area basis (0.6 – 0.9 kg m^{-2} $year^{-1}$) than evergreen conifers (0.8 – 1.6 kg m^{-2} $year^{-1}$) (Bonnemann and Röhrig 1972) and higher maximal rates of photosynthesis (10 – 14 mg CO_2 g^{-1} hr^{-1} in broad-leaved trees versus 4 – 6 mg CO_2 g^{-1} hr^{-1} in conifers) (Larcher 1980). On a daily basis, the net photosynthetic gain in deciduous species is twofold to fourfold higher than in evergreen conifers (Figure 17.1). The low photosynthetic capacity of evergreen needles is not counterbalanced by a prolonged growing season in spring and autumn, because a short photoperiod, low light, and low temperatures restrict photosynthetic carbon uptake in the European spring and fall climates (Fuchs et al. 1977; Schulze 1982). Conditions may be different in summer-dry and winter-mild climates, such as the Pacific Northwest of North America.

The differences in biomass production between deciduous and evergreen forms are consequences of leaf longevity (Figure 17.2). Conifers have a lower carbon investment per season in the production of new leaves per unit area than deciduous trees, but the total leaf biomass of conifers is higher. Needles are maintained for up to 12 years in *Picea abies* (Schulze et

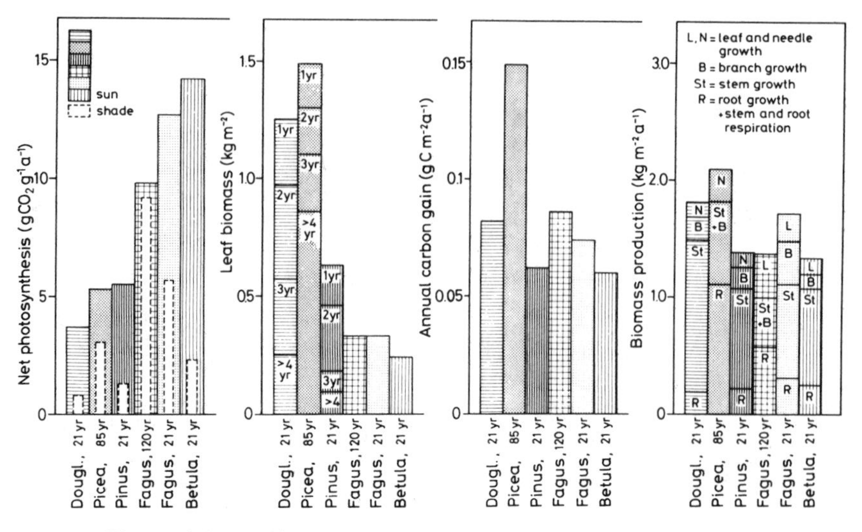

Figure 17.2. Values of annual gain in net photosynthesis per gram leaf dry weight, leaf biomass of the canopy, annual carbon gain, and annual biomass production per square meter ground cover for evergreen *Pseudotsuga menziesii* (Dougl.), *Picea abies, Pinus silvestris,* and winter-deciduous *Fagus sylvatica* and *Betula verrucosa* (after Schulze 1982).

al. 1977b). Even though there is a lower carbon gain per unit leaf weight in conifers, the annual carbon gain for conifers is equal to or greater than that for deciduous trees because of a greater total leaf biomass. Needle longevity increases with elevation and aridity and reaches a maximum of 45 years in *Pinus longaeva* (Ewers and Schmid 1981). A large needle biomass (and therefore larger leaf area) can be sustained in conifers because of the very special clustered foliage arrangement that encloses the twigs (Norman and Jarvis 1974).

Defoliation experiments have shown the significance of the old foliage for growth in conifers. Removal of one- and two-year-old needles causes a 51% decrease in growth of *Pinus radiata* (Rook and Whyte 1976). Schulze et al. (1977b) showed that the annual carbon gain by the evergreen spruce *Picea abies* would decrease 9% if its annual photosynthetic period were the same as that of the deciduous beech *Fagus sylvatica.* If spruce had to exist on carbon fixed by the current-year foliage, its carbon gain would decrease 85% to a level insufficient to cover the respiratory demand of non-green living tissues. The current-year needles in spruce assimilate only 15% of the annual carbon gain of the tree crown (Schulze et al. 1977a).

The implications of these findings appear to be significant for an understanding of successional stages and geographical distributions of tree forms. Broad-leaved trees (e.g., successional species *Populus, Betula, Alnus,* and others) grow more rapidly than evergreens during the sapling stage because conifers have to accumulate leaf biomass over many years before CO_2 uptake per tree reaches its highest rates. With decreasing length of the growing season, and with unfavorable habitats or climates, the annual costs of construction for deciduous leaves increase relative to their total carbon gain. Hence, evergreen trees are dominant in boreal climates. Deciduous trees occur at these latitudes, but only on very nutritious soils or after disturbance (Schulze 1982).

Growth and competition as related to plant architecture

The partitioning of biomass into organs varies among different tree species (Figure 17.2). Conifers reach greater tree dimensions, although rates of stem growth in terms of carbon partitioning may not be very different. All biomass studies have expressed their results as "carbon gain per ground area," which is a convenient basis for comparisons, but it is not sufficient to fully understand growth and competition of woody species. Plants do not accumulate carbon as "mass per ground area"; they have complex means of biomass exposure that are related to plant architecture and to branching patterns. The interactions among carbon relationships, growth, and plant architecture were studied in hedgerows, a vegetation with strong competitive interactions and a successional replacement of woody species (Küppers 1982; Schulze et al. 1982). Hedgerows in south Germany are linear arrangements of woody species (2 – 5 m broad) that follow property boundaries between fields. They stabilize slopes and define the borders between forests and managed fields. They are generally not planted, but occupy space that is not cultivated, and they are managed by cutting every 30 to 40 years. Hedgerows are very rich in woody species (about 100 species). They undergo a regular succession, starting with the thorny shrub *Prunus spinosa,* which produces underground suckers for radial occupation of new land (Reif 1982). Under this thorny protection from animal herbivory, *Crataegus macrocarpa* and several other shrubs germinate, and they are soon taller than *Prunus spinosa. Prunus* is soon thereafter outshaded and disappears from later successional stages. Shrubs are eventually suppressed by the establishment of late successional trees, especially *Acer campestre* (Figure 17.3). If the hedge is not cut at this stage, "climax" species (*Quercus robur* or *Fagus sylvatica*) become established.

The maximal rates of CO_2 assimilation of sun leaves are quite similar in all hedgerow species and therefore cannot explain the success of various species during different successional stages (Table 17.2). During the day

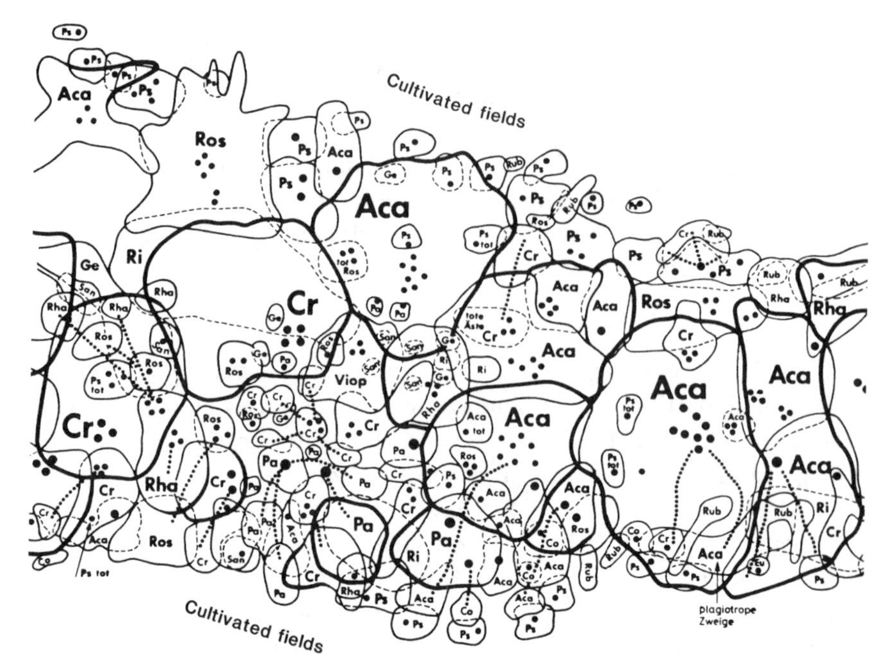

Figure 17.3. Map of projected crown areas for woody and herbaceous species in a portion of a hedgerow. Thick solid lines, dominant canopy; thin solid lines, subcanopy; broken lines, understory; Aca, *Acer campestre;* Co, *Cornus sanguinea;* Cr, *Crataegus macrocarpa;* Eu, *Euonymus europaeus;* Ge, *Geranium robertianum;* Pa, *Prunus avium;* Ps, *Prunus spinosa;* Rha, *Rhamnus cathartica;* Ri, *Ribes uva-crispa;* Ros, *Rosa canina, R. subcanina, R. vosagiaca;* Rub, *Rubus corylifolius;* Sam, *Sambucus nigra;* tot, dead.

and throughout the season, plants operate at maximal rates for only very short periods (Schulze and Hall 1982). Generally, CO_2 uptake is suppressed by low light at dawn and dusk and by clouds during midday. Additionally, photosynthetic rates are lower during leaf expansion and with aging. The climax species *Fagus sylvatica* has a lower annual carbon gain per leaf area than either the late successional *Acer campestre* or the early successional *Prunus spinosa.* Therefore, progressive increases in annual carbon gain do not correlate with the pattern of succession. This is also the case for water-use efficiency (not corrected for differences in vapor-pressure deficit), which is greater in *Acer* than in *Prunus,* but lower in *Crataegus* and *Fagus.* Even at the primary production level, the early successional species are superior to the late successional ones.

Although higher rates of productivity would tend to enhance the competitive ability of species by providing more carbon in order to produce leaves and branches, this fails to take into account competitive effects due

Table 17.2. *Carbon relations of woody species (early successional species,* Prunus spinosa; *late successional species,* Crataegus macrocarpa, Acer campestre; *climax species,* Fagus sylvatica)[a]

Parameter	*P. spinosa*	*C. macrocarpa*	*A. campestre*	*F. sylvatica*
Maximal rate of CO_2 assimilation (μmol m^{-2} s^{-1})	9.0–12.0	9.0–12.0	9.0–12.0	3–4
Annual carbon gain (mol m^{-2} $year^{-1}$)	31	43	27	9
Annual transpiration (kmol m^{-2} $year^{-1}$)	5.9	13.1	3.8	2.8
Water-use efficiency [mol CO_2 (kmol $H_2O)^{-1}$]	5.3	3.3	7.1	3.4
Aboveground primary production per leaf (g g^{-1} $year^{-1}$)	12.6	12.1	9.7	2.4

[a] Data from Schulze (1970) and Küppers (1982).

to interspecific differences in three-dimensional crown structure (i.e., architecture and branching pattern). Branching and leaf exposure are principal mechanisms in competition. Short internodes and thorns may be important for early successional species as protection against herbivory, whereas internodes and leaves of late successional species do not simply serve the carbon assimilation process but may act also as light interceptors, shading the competing neighbor below. Strong shading of a neighbor with a minimum of self-shading should yield a competitive advantage. Therefore, the capacity for occupying new, higher aerial space by growth of branches, coupled with maintenance of a closed leaf cover above the presently occupied space, could be of successional significance.

The growth of trees and shrubs (Troll 1937) can be characterized by (1) longitudinal symmetry (along the vertical axis) and (2) lateral symmetry (along the horizontal axis) (Figure 17.4). Longitudinal symmetry is characterized by the development of main lateral branches at the top (acrotony) or at the base (basitony) of the main axis, and lateral symmetry is characterized by the polarity of growth of second-order branches at the upper (epitony) or lower (hypotony) side of the lateral branch.

The longitudinal and lateral symmetries of branching determine the plant growth form. Each species has a different genetically fixed growth form depending on its longitudinal and lateral symmetries (Figure 17.5). Shrubs are characterized by basitonic branching. The main axis terminates growth, and the buds at the base develop lateral branches of greatest

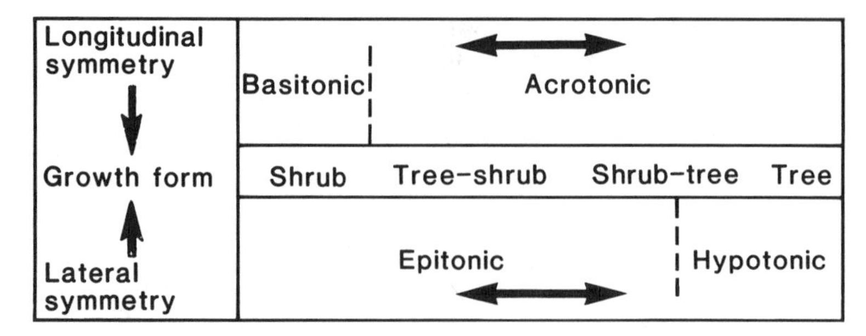

Figure 17.4. Principles of bud development and growth-form symmetry in woody plants (after Troll 1937; Küppers 1982).

length. The lateral branches at the tip of the main axis remain short (e.g., *Rosa*). This contrasts with the situation for shrub-trees, in which the greatest lateral branch lengths occur in the middle or apical part of the main axis (*Crataegus*), and trees in which lateral branches grow largest at the top of the main axis (*Acer*). Plant form is further determined by the dominance of branch development. Branches may reach their greatest lengths if they develop from buds on the upper side of a stem or main lateral branch. In this case, the stem or first-order branch may terminate growth, and long vertical (epitonic) shoots may continue foliage development (*Rubus, Rosa,* and shade-grown *Crataegus*). This appears to be important in shrub competition. Epitonic shoots may grow through the competing canopy of a neighboring plant and produce a new foliage layer above an existing canopy. This contrasts with hypotonic branching, in which buds of the lateral first-order branch develop only at the lower side, and these branches are capable of extending only the lateral range. Generally, they do not produce a vertical stem in order to overtop an existing canopy.

Prunus spinosa, Crataegus macrocarpa, and *Acer campestre* show characteristic differences in terms of competitive ability because of different branching patterns (Figure 17.6). In *Prunus spinosa,* the main axis terminates growth, and lateral branches from a mesotonic position to an acrotonic position continue growth (Figure 17.6). Second-order lateral shoots develop at an epitonic position, but these terminate growth after a brief branch development. Again, third-order mesotonic and fourth-order epitonic branches occur with decreasing lengths. This results in the appearance of a short, stunted shrub-tree that is not capable of producing long vertical sprouts in order to compete with higher-growing species. Shading may increase elongation growth of individual branches, but it does not change the basic branching and developmental pattern.

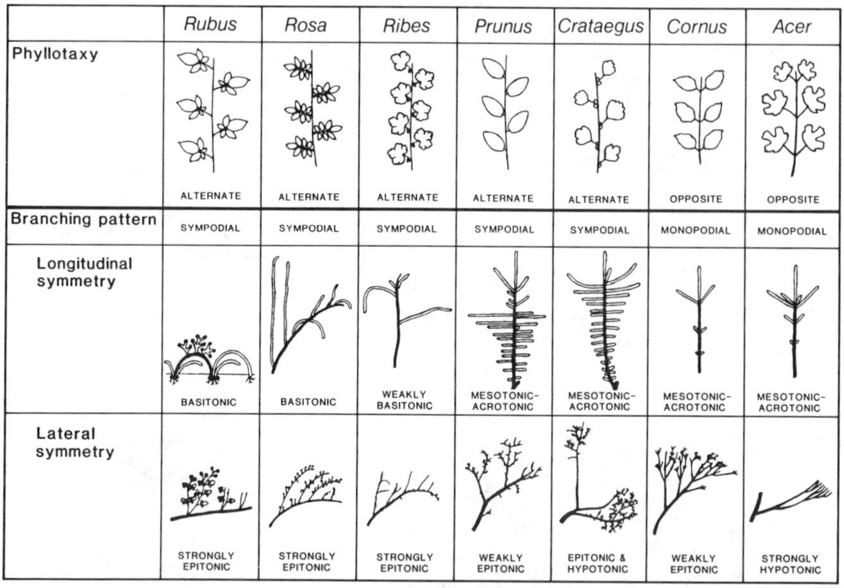

	Rubus	*Rosa*	*Ribes*	*Prunus*	*Crataegus*	*Cornus*	*Acer*
Phyllotaxy	ALTERNATE	ALTERNATE	ALTERNATE	ALTERNATE	ALTERNATE	OPPOSITE	OPPOSITE
Branching pattern	SYMPODIAL	SYMPODIAL	SYMPODIAL	SYMPODIAL	SYMPODIAL	MONOPODIAL	MONOPODIAL
Longitudinal symmetry	BASITONIC	BASITONIC	WEAKLY BASITONIC	MESOTONIC-ACROTONIC	MESOTONIC-ACROTONIC	MESOTONIC-ACROTONIC	MESOTONIC-ACROTONIC
Lateral symmetry	STRONGLY EPITONIC	STRONGLY EPITONIC	STRONGLY EPITONIC	WEAKLY EPITONIC	EPITONIC & HYPOTONIC	WEAKLY EPITONIC	STRONGLY HYPOTONIC

Figure 17.5. Phyllotaxy, branching patterns, and branching symmetry for different woody hedgerow species. Longitudinal symmetry is determined by the branching pattern of a shoot in its second year. Filled branches indicate one-year branches; open branches are current-year branches (after Troll 1937; Küppers 1982).

The development of *Crataegus* is very similar (Figure 17.6), but the epitonic second-order branches may reach apical dominance and therefore considerable length (3 – 4 m), which enables this species to compete with other species during crown development. A different pattern is exhibited by *Acer campestre* (Figure 17.6), which maintains growth of the main axis and produces a regular tree structure by acrotonic branch development. No vertical second-order branches compete with the main trunk because bud development is hypotonic (i.e., on the lower side of lateral branches). This causes the crown to become rather broad and, under competitive pressure, to maintain acrotonic development (i.e., the capability to overtop rather than to evade neighboring shrubs). It appears that epitonic bud development is capable of responding quickly to changing growth conditions, but that only acrotony paired with hypotonic branching is capable of developing a permanent and dominating canopy. Several other branching patterns exist (Figure 17.5) (e.g., *Rosa, Cornus*), demonstrating a continuum of plant growth forms that occupy certain stages in hedge development.

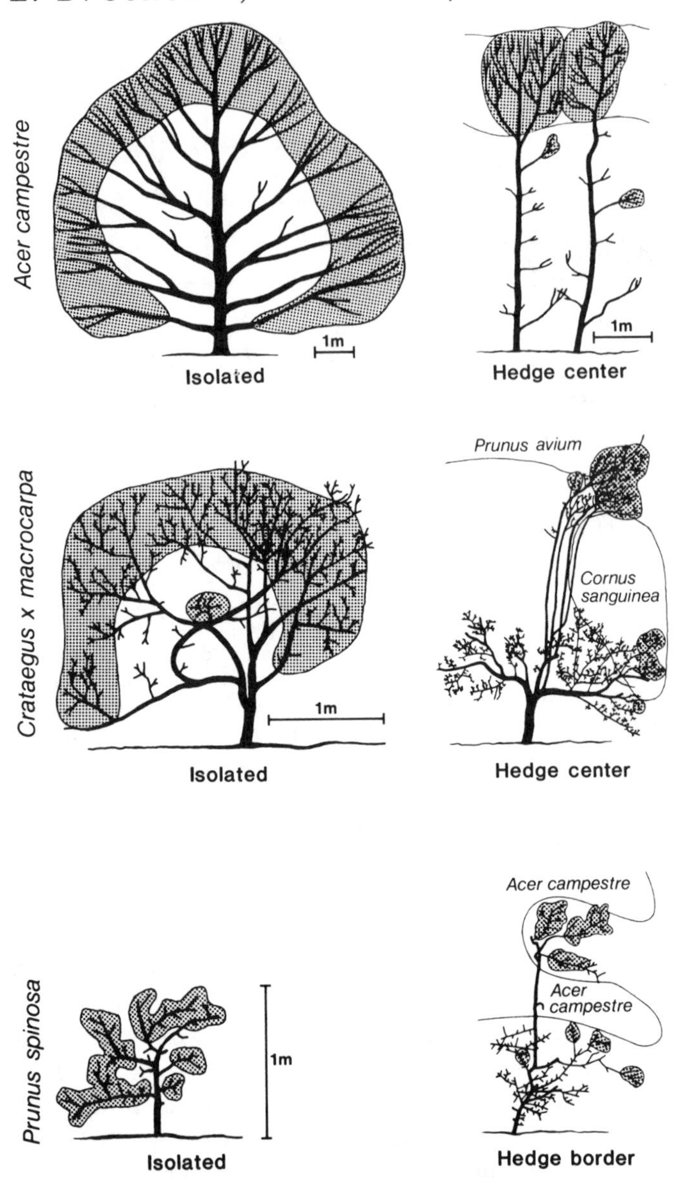

Figure 17.6. Growths of *Acer campestre, Crataegus macrocarpa,* and *Prunus spinosa* as isolated plants and under competitive pressure of neighboring plants. Shaded area: region of leaf development.

Table 17.3. *Occupation of space by woody species*[a]

Parameter	*Prunus spinosa*	*Crataegus macrocarpa*	*Acer campestre*	*Fagus sylvatica*
Volume gain per dry-matter investment of single twigs ($m^3\ kg^{-1}$)	0.038	0.078	0.134	0.549
Volume gain per photosynthetic carbon gain ($m^3\ 10^{-3}\ g^{-1}$)	1.1	1.1	2.3	—[b]

[a] Data from Küppers (1982) and Schulze (unpublished).
[b] Data not available.

Neither carbon balance nor branching pattern alone can explain the success of a species during competition. The carbon balance determines how much carbon can be partitioned into growth of trunks and branches. Therefore, carbon relationships must be integrated with the growth pattern in order to understand the efficiency with which a certain species exploits its living space. In order to do this, we calculated the volume gain in space for each species in two different ways. First, we determined the volume of aerial space that was occupied by the growth of a single branch system of a specific length (70 cm) and its leaves at the outer edge of the canopy. We assumed that leaves occupied a radial cylinder, because leaves were either alternate or decussate. Second, we determined the annual radial volume increase of the crown per existing crown surface and the corresponding dry weight occupying this volume. In both cases, the later successional *Acer campestre* achieved a greater volume gain per dry matter investment than any other species (Table 17.3). Furthermore, it also reached the greatest volume gain per unit of photosynthetic carbon fixation. It is conceivable that a late successional species could have a lower assimilation rate and a lower nitrogen requirement than an early successional species and still succeed during competition.

Competition is governed not only by mechanisms of volume gain but also by additional factors that have been neglected in the foregoing analysis. *Acer* cannot become established without the thorny protection from herbivory, but it has to start from seeds, whereas *Prunus* is capable of vegetative regeneration, being protected by thorns. Once a stable hedge canopy is developed, new microhabitats are created that may be filled by different species. For instance, *Ribes* cannot stabilize its water budget when exposed to dry air and full sunlight, but also it is not capable of growing in full shade. This shrub produces runners near the ground that grow most successfully in the partial shade at the edge of a hedge. In contrast to the *Ribes* pattern is that of *Rubus,* a shrub that produces 3- to 4-m-long aerial shoots that overgrow other species, in a manner similar to that of vines. *Rubus* avoids competition by always growing out of the shaded area, because shoots build new roots at their tips and vegetatively regenerate from these points. The old aboveground sprouts die back two years after flowering, resulting in a constant migration into full sunlight.

Growth as related to plant water relations

A positive carbon balance is a prerequisite for growth, but it can be discussed only in relation to various "costs" (such as the costs of water and nutrients) that the plant must incur during the process of growth and leaf development. Increases in plant biomass are mainly balanced between growth of new leaves and growth of new roots. The new leaves have a positive feedback on the production process, and the new roots have a positive feedback on the plant water status (Schulze et al. 1983). For the annual herbaceous plant *Vigna unguiculata*, the gain in net primary production may be quite small with an increase in foliage development, but the same increase in foliage development may have a very pronounced negative impact on the water balance at a given amount of root area. Therefore, growth should be expressed not only in terms of carbon balance but also in terms of water balance.

The effects of water relationships on the growth of woody species can be discussed in regard to a growth study of similarly aged trees, including the evergreen-leaved conifer *Picea abies* and three deciduous-leaved conifers *Larix leptolepis* from Japan, *Larix decidua* from Europe, and the F_1 hybrid *Larix decidua* × *leptolepis*. In this case, the deciduous/evergreen comparison can be made within conifer species of similar wood structures (Table 17.4). In similarity to the *Fagus*/*Picea* example, the deciduous-leaved conifers have higher maximal photosynthetic rates, but less total leaf biomass, than their evergreen-leaved counterpart. The evergreen-leaved spruce has an aboveground primary production similar to that of the deciduous-leaved *Larix decidua* and *L. leptolepis*. However, the *Larix* hybrid has a much larger net primary production and produces not only the tallest but also the thickest tree trunk. The *Larix* hybrid produces a smaller number of main branches than spruce or European larch, but it has higher photosynthetic rates and a higher leaf biomass than either of the parental species. The high needle biomass results mainly from a greater needle density per twig length in the *Larix* hybrid. High assimilation rates and high needle biomass result in a higher carbon gain than in the parental plants. The higher carbon gain will lead to a higher biomass, but it does not explain the larger trunk dimensions by itself.

Table 17.4. *Growth parameters for evergreen* Picea abies *and summer-deciduous* Larix leptolepis, Larix decidua, *and hybrid* Larix decidua × leptolepis

Parameter	*P. abies*	*L. leptolepis*	*L. decidua* × *leptolepis*	*L. decidua*
Age (years)	34	32	32	32
Height (m)	16.6	16.2	19.8	15.3
Diameter (cm in 1.3 m height)	19.3	17.1	26.1	16.5
Maximal rates of CO_2 assimilation (μmol m^{-2} s^{-1})	2–3	3.5–4.5	4–6	3.5–4.5
Aboveground net primary production per tree (kg $year^{-1}$)	10.3	11.1	23.4	11.4
Needle biomass per tree (g)	16.5	4.8	9.8	5.1
Net primary production per needle biomass (g g^{-1})	0.6	2.3	2.4	2.2
Needle biomass of 1-year-old twigs (g m^{-1})	9.5	6.5	11	7.5
No. of living branches inserted to the trunk	17.5	12.1	13.6	22.9
Minimum daily water potential in June (bar)	13.0	15.5	17.5	15.3
Sapwood area per leaf dry weight (cm^2 kg^{-1})	—[a]	15.5	20.8	15.5

[a] Data not available.

The hybrid requires about 30% more sapwood area than the parent species to supply the same amount of needle biomass with water (Table 17.4). This is true for all crown heights: When needle biomass is related to sapwood area for 1-m increments in crown height, the hybrid consistently requires more sapwood per needle biomass than the parent species. Stomatal responses to humidity and stomatal conductances at maximal rates of CO_2 uptake are similar in all *Larix* species. Despite a larger sapwood area per needle biomass, on a clear summer day the *Larix* hybrid reaches lower leaf water potentials than its parents at similar leaf conductances and crown heights. This suggests that the hybrid has a lower hydraulic conductance per xylem area than the parental species. This could lead to a greater demand for stem growth and, compounded over years, could lead to a larger trunk size in the hybrid than in the parent plant.

A similar comparison can be made with the hedgerow species (Küppers 1982, 1984). *Acer, Crataegus,* and *Prunus* have similar hydraulic conductances through the soil–plant continuum (0.25–0.28 mmol H_2O m^{-2} s^{-1} bar^{-1}), but very different trunk dimensions. The late successional tree species *Acer* requires 3.14 cm^2 sapwood area per 1 m^2 leaf area, as compared with 0.90, 1.53, and 1.05 $cm^2 m^{-2}$ for *Rubus, Prunus,* and *Crataegus,* respectively. Hydraulic conductance per sapwood area is higher in the shrub species than in tree species (*Prunus,* 0.16 mmol H_2O m^{-2} leaf area s^{-1} bar^{-1} cm^{-2} sapwood area; *Crataegus,* 0.27; *Acer,* 0.08). The tree *(Acer)* must allocate more carbon to the wood of the main trunk in order to sustain transpiration and leaf area without perturbing the plant water relationships.

Growth as related to water and nutrient relations

The water and nutrient relationships of plants and their effects on growth are generally studied independently. This occurs because these parameters show different uptake and storage characteristics and because the processes of uptake, growth, and turnover in the roots are difficult to observe. Therefore, the partitioning of carbohydrates has rarely been analyzed with respect to water and nutrient flows (Schulze 1982). The effects of water and nutrient flows on growth processes can be studied in a simpler system, namely, xylem parasites (Schulze et al. 1984), although the observations made on such a system may not be generalizable to soil-rooted species. Xylem parasites have very clearly defined inputs from the host: the water and nutrients from the host's xylem. These parasites receive no input from the phloem of the host (Glatzel 1983).

During the course of a day, *Loranthus europaeus* has a CO_2 assimilation rate that is very similar to that of its host, *Quercus robur* (Figure 17.7). Maximum rates of CO_2 uptake in the morning are slightly higher in *Loranthus* than in *Quercus.* Both species show reduced photosynthetic rates in the afternoon, although the mistletoe maintains CO_2 uptake slightly longer than its host. Because of a higher leaf conductance to water vapor, but a similar rate of CO_2 uptake, *Loranthus* operates at a higher mesophyll internal CO_2 concentration than its host. A pronounced difference exists in transpiration rates. The mistletoe has about twice the transpiration rate of its host because of higher leaf conductance. Although water loss is high, *Loranthus* shows a definite stomatal regulation of water loss in the afternoon, probably in response to dry air. Obviously the parasite regulates water loss, but the oak leaves close stomata earlier in the day than the mistletoe. Schulze et al. (1984) integrated the water, nutrient, and carbon balances in *Loranthus* and suggested that the mistletoe maintains a higher rate of water loss in order to channel sufficient nitrogen from the host xylem into the parasite for growth. Table 17.5 shows that the daily transpiration by the mistletoe is about two times higher than that in the host, but the leaf nitrogen contents are very similar. The nitrogen contents of the xylem sap were similar during the early and late stages of the oak leaf

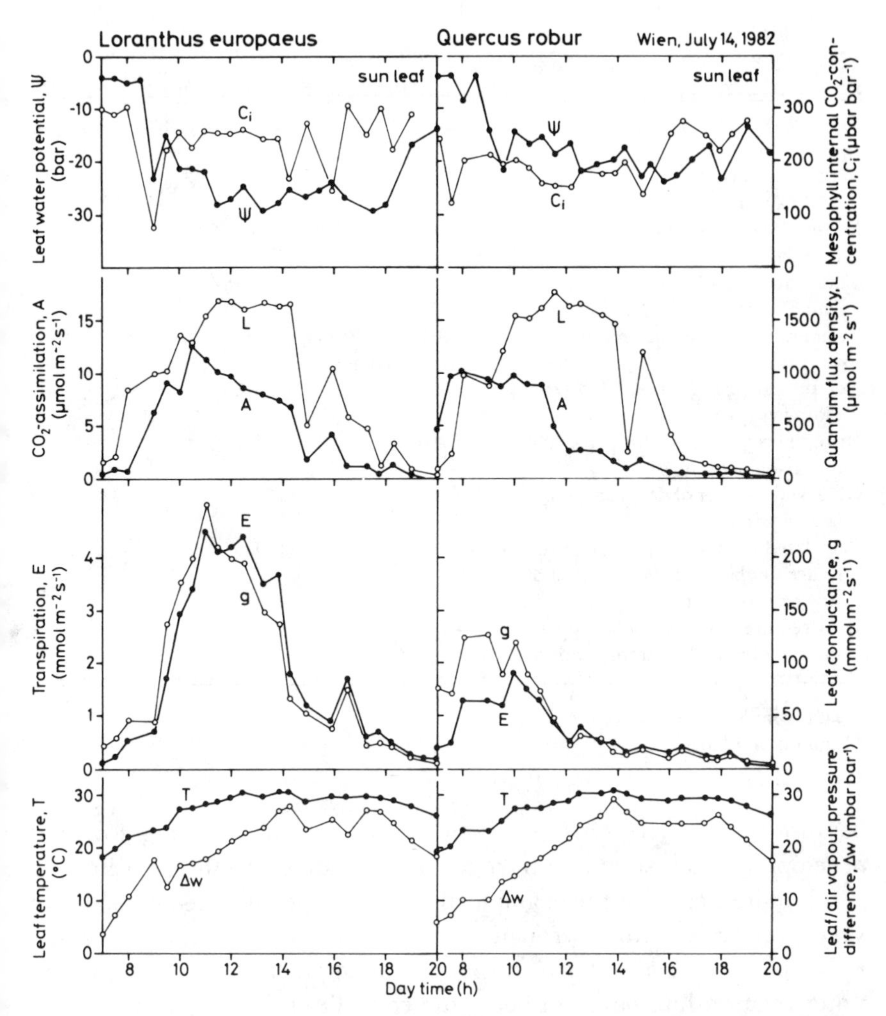

Figure 17.7. Daily course of microclimate, water-potential, and gas-exchange parameters for *Loranthus europaeus* and *Quercus robur* sun leaves (after Schulze et al. 1983b).

development (May-June). The biomass produced by *Loranthus* requires a certain amount of nitrogen. At the observed rate of transpiration, it is calculated that a three-year-old *Loranthus* twig needs four to five months (close to the full growing season) in order to accumulate the required nitrogen. The hypothesis can be checked by an independent assessment of potassium, which is passively accumulated from the xylem sap. The measured accumulated potassium values in *Loranthus* represent about seven months of transpiration. The oak leaves do not need to transpire as much

Table 17.5. *Carbon, water, and nutrient relationships for the xylem parasite* Loranthus europaeus *and its host* Quercus robur[a]

Parameter	Host: *Q. robur*	Xylem parasite: *L. europaeus*
Midsummer daily transpiration (mol m^{-2} day^{-1})	37.6	82.4
Midsummer daily CO_2 uptake (mol m^{-2} day^{-1})	0.199	0.196
Total nitrogen content of leaves (mmol g^{-1})	1.32	1.12
Xylem nitrogen content (mmol)	1.01 (July 82) 1.23 (May 83)	— —
Net primary production of 3-year-old twigs (g)	5.5	7.30
Nitrogen requirement for growth of 3-year-old twigs (g)	0.102	0.115
Observed number of days to accumulate required nitrogen	150	150
Calculated number of days required to accumulate nitrogen at 60% of July transpiration	—[b]	141
Days required to accumulate potassium at 60% of July transpiration	—	209

[a] Data from Schulze et al. (1984).
[b] Data not available.

because they receive nitrogen from the phloem in early stages of bud development; this source of nitrogen is not available to the parasite.

It is quite clear that the example of the mistletoe–host relationship is a very special case of interaction of nutrient/water relationships. But, at the whole-plant level of soil-grown species, the interactions of nitrogen and water relationships have not been studied sufficiently.

Conclusions

Plants have a great variety of mechanisms to tolerate low photosynthetic rates and still be successful with respect to net primary production and competition. One important mechanism appears to be the deciduous versus evergreen habit of leaves, which cannot be fully understood until we know more about the costs of evergreen foliage development. Net primary production by itself is not sufficient to explain competitive relationships between woody species. It is necessary to understand the functional basis for species-specific growth patterns. However, plant architecture, in combination with carbon input, can lead to a closer understanding of plant success. It appears that foliage not only serves carbon assimilation but also has a significant role in competition and species replacement. However, it is quite clear that the succession of vegetation is controlled by additional factors, such as herbivory, that were not taken into account in this analysis.

Plant carbon relationships cannot be discussed without giving consideration to the demands for water and nutrients. The growth of new leaves has positive feedback on the carbon balance, but at the same time it has negative feedback on water and nutrient relationships. In woody species, a permanent transfer of living tissue occurs during formation of wood that requires little maintenance respiration but serves permanently as supporting structure and as water-conducting system. If the hydraulic properties of the wood are low, the plant is forced to put more carbon into wood per unit leaf biomass, which at the same time will contribute to a larger supporting structure.

Very little is known about the interactions of water uptake and nutrient uptake. The mistletoe example shows that special cases exist in which high transpiration rates are maintained in order to accumulate sufficient nitrogen for growth. Calculating the water-use efficiency of mistletoes and neglecting the nutrient part cannot explain the ecological significance of the high transpiration rates. Although the nutrient/water interactions are more complicated in roots, these interactions should be studied more thoroughly in the future.

References

Bonnemann, A., and E. Röhrig. 1972. Baumartenwahl, Bestandesgründung und Bestandespflege, Waldbau auf ökologischer Grundlage, vol. 2. Parey, Berlin.

Ewers, F. W., and R. Schmid. 1981. Longevity of needle fascicles of *Pinus longaeva* (brittlecone pine) and other North American pines. Oecologia 51:107–115.

Fuchs, M., E.-D. Schulze, and M. I. Fuchs. 1977. Spacial distribution of photosynthetic capacity and performance in a mountain spruce forest of northern Germany. II. Climatic control of carbon dioxide uptake. Oecologia 29:329–340.

Glatzel, G. 1983. Mineral nutrition and water relations of hemiparasitic mistletoes: a question of partitioning. Experiments with *Loranthus europaeus* on *Quercus petraea* and *Quercus robur*. Oecologia 56:193–201.

Küppers, M. 1982. Koheenstoffhaushalt, Wasserhaushalt, Wachstum und Wuchsform von Holzgewächsen im Konkurrenzgefüge eines Heckenstandortes. Doctoral thesis, Bayreuth.

– 1984. Carbon relations and competition between woody species in a hedge row. II. Stomatal response, water use and conductivity to liquid water in the soil/plant pathway. Oecologia 64:344–354.

Larcher, W. 1980. Ökologie der Pflanzen, UTB 232, 3rd ed. Ulmer, Stuttgart.

Norman, J. M., and P. G. Jarvis. 1974. Photosynthesis in Sitka spruce [*Picea sitchensis* (Bong.) Carr.]. III. Measurements of canopy structure and interception of radiation. J. Appl. Ecol. II:375–398.

Reif, A. 1982. Vegetationskundliche Gliederung und standortliche Kennzeichnung nordbayerischer Heckengesellschaften. Doctoral thesis, Bayreuth.

Rook, D. A., and A. G. D. Whyte. 1976. Partial defoliation and growth of 5-year-old *Radiata* pine. NZJ For. Sci. 6:40–56.

Schulze, E.-D. 1970. Der CO_2-Gaswechsel der Buche (*Fagus sylvatica* L.) in Abhängigkeit von den Klimafaktoren im Freiland. Flora 159:177–232.

– 1982. Plant life forms and their carbon, water and nutrient relations. Pp. 616–676 *in* O. L. Lange, P. S. Nobel, C. B. Osmond, and H. Ziegler (eds.), Encyclopedia of plant physiology, new series, vol. 12B. Springer-Verlag, Berlin.

Schulze, E.-D., M. I. Fuchs, and M. Fuchs. 1977a. Spacial distribution of photosynthetic capacity and performance in a montane spruce forest of northern Germany. I. Biomass distribution and daily CO_2 uptake in different crown layers. Oecologia 29:43–61.

– 1977b. Spacial distribution of photosynthetic capacity and performance in a montane spruce forest of northern Germany. III. The significance of the evergreen habit. Oecologia 30:239–248.

Schulze, E.-D., and A. E. Hall. 1982. Stomatal responses, water loss and CO_2-assimilation rates of plants in contrasting environments. Pp. 181–230 *in* O. L. Lange, P. S. Nobel, C. B. Osmond, and H. Ziegler (eds.), Encyclopedia of plant physiology, new series, vol. 12B. Springer-Verlag, Berlin.

Schulze, E.-D., A. Reif, and M. Küppers. 1982. Ökologishe Funktionsanalyse von Hecken and Flurgehölzen – Ökologisch Untersuchungen über Strukturen und Funktionen der Pflanzen in Feldhecken und deren Beziehung zu angrenzenden Biotopen. Schlussbericht, Bayerishes Landesamt für Umweltschutz, München.

Schulze, E.-D., K. Schilling, and S. Nagarajah. 1983. Carbohydrate partitioning in relation to whole plant production and water use of *Vigna unguiculata* (L.) Walp. Oecologia 58:169–177.

Schulze, E.-D., N. C. Turner, and G. Glatzel. 1984. Carbon, water and nutrient relations of two mistletoes and their hosts: a hypothesis. Plant Cell Environ. 7:293–299.

Troll, W. 1937. Vergleichende Morphologie der höheren Pflanzen. Bd 1. Teil 1. Vegetationsorgane. Berlin; reprint Koeltz, Koenigstein.

18 Seaweeds in moving water: form and mechanical function

MIMI A. R. KOEHL

Most macroalgae spend part of their lives attached to the substratum. The water moving around these sessile organisms carries dissolved gases and nutrients to them. Moving water can also remove waste products or sediments from the vicinity of a plant, and can disperse gametes or spores. Although seaweeds benefit from such external transport by ambient currents, they also risk being torn or dislodged by moving water. Hence, there are tradeoffs between living at sites exposed to rapid water flow and living at sites characterized by calm water.

Do morphological features that enhance an alga's ability to withstand moving water carry with them "costs" of reduced performance of other tasks, such as photosynthesis? The effects of specific structural features on defined aspects of an alga's performance must be quantitatively described before the costs and benefits of various morphologies can be evaluated for different environments. This chapter discusses ways in which certain mechanical aspects of an alga's performance in moving water depend on its structure. Specifically, I analyze how the morphology of a seaweed affects (1) the water flow it encounters, (2) the hydrodynamic forces it experiences, and (3) its deflection and breakage in flowing water. After the biomechanics of macroalgae are discussed, some examples of tradeoffs between decreasing the susceptibility of algae to mechanical damage and increasing their photosynthetic performance will be mentioned.

Water flow encountered by benthic algae

Flow habitats

There are several types of water-flow regimes that benthic algae encounter in different marine habitats (Bascom 1964; Carstens 1968; Riedl 1971; Neushul 1972; Wainwright and Koehl 1976). Perhaps the most obvious feature of water movement at many coastal sites is wave action (Figure 18.1). As a wave shape moves along the surface of the ocean,

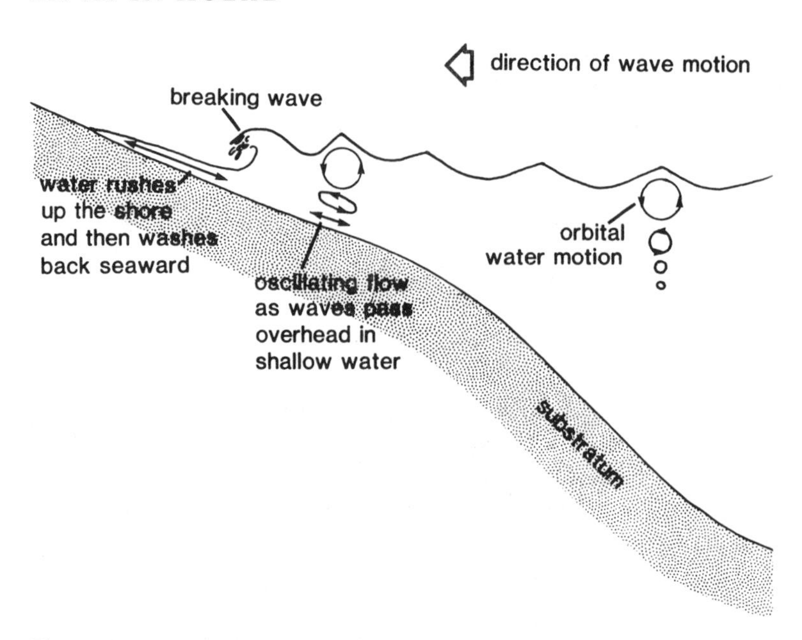

Figure 18.1. Diagram of the gross directions of water movement at a wave-swept shore, based on information presented by Bascom (1964), Carstens (1968), and Riedl (1971).

the water moves around locally in an orbital path. A giant kelp growing offshore with buoyant fronds near the water's surface encounters such orbital water movement. When waves move near a shore where the water depth is less than half the crest-to-crest distance of the waves, the orbital motion of the water is restricted near the substratum such that it oscillates back and forth as waves pass overhead. Shallow benthic algae encounter such bidirectional flows. Intertidal seaweeds are exposed to breaking waves and the consequent rapid surge of water up the shore, followed by seaward backwash. The water velocities and accelerations measured over organisms at low tide in breaking waves are greater than those they encounter when submerged at high tide with nonbreaking waves passing overhead (Koehl 1977a). If algae are growing in deep water or at sites protected from wave action, they may be in habitats subjected either to periodic tidal currents or to steady unidirectional currents.

Flow microhabitats

Within the various gross water-movement habitats described, the flow microhabitat of a particular alga depends on the form of the alga (Neushul 1972; Koehl 1984).

When water moves across the substratum, a velocity gradient exists between the bottom and the free-stream flow. Very small algae, such as newly settled spores or benthic diatoms, and very flat plants hugging the substratum, such as crustose coralline algae, may well be living in this boundary layer of slowly moving water (Neushul 1972; Wheeler and Neushul 1981; Norton et al. 1982). Immediately adjacent to the substratum, a viscous sublayer develops in which velocity is proportional to distance above the bottom. In most marine habitats, there is, above this sublayer, a turbulent boundary layer in which the average velocity increases with height above the substratum and in which flow is dominated by turbulent eddies. [For more detailed, quantitative descriptions of boundary-layer flow profiles, see Batchelor (1967) Schlichting (1968) Jumars and Nowell (1984), and Nowell and Jumars (1984).] The flow microhabitat encountered by an alga of a given height depends on a number of features of the site. For example, the faster or more unsteady the ambient flow and the rougher the substratum, the thinner the viscous sublayer. Furthermore, because a boundary layer builds up as water moves across a surface, the greater the distance between an alga and the leading edge of a structure (such as a rock or another alga) on which it sits, the thicker the boundary layer it encounters.

Even large algae can be in microhabitats that are protected from rapid ambient flow if they "hide" behind rocks or in crevices, or if they are surrounded by other organisms of similar size (Koehl 1976, 1977a, 1982). For example, water velocities within kelp beds (Wheeler 1980a, 1980b; Jackson and Winant 1983; M. Koehl, unpublished data) and seagrass beds (e.g., Peterson 1984) can be significantly lower than those outside the beds. Studies of air movement in plant canopies (e.g., Raupach and Thom 1981) and of water movement through arrays of coral branches (Chamberlain and Graus 1975), worm tubes (Eckman 1982; Jumars and Nowell 1984), and marsh grass (Eckman 1982, 1983; Fonseca et al. 1982) indicate that the size and spacing of neighbors can have profound effects on flow within a group of such structures. Furthermore, if organisms in an aggregation are flexible, they are bent over (and hence packed together more closely) in flowing water; such flexibility can reduce in-canopy flow velocities in seagrass meadows as water currents are redirected over, rather than through, the array of plants (Fonseca et al. 1982). Similar studies for arrays of structures of the sizes, shapes, flexibilities, and spacings of macroalgae remain to be done, although Anderson and Charters (1982) have described the effects on water flow of a single thallus of the bushy alga *Gelidium nudifrons:* Ambient turbulence is suppressed, but microturbulence is generated in the alga's wake if the current is above a critical speed (which increases as the ratio of branch spacing to diameter increases).

Note that if algae are very short or are hidden among other structures, they not only avoid rapid water flow but also may be shaded from light by other organisms or topographic features. It should also be kept in mind that as an alga grows, it can encounter different water-flow microhabitats even though its position on the substratum is fixed (Neushul 1972; Wainwright and Koehl 1976; Norton et al. 1982).

Hydrodynamic forces on algae

Drag on algae in water currents

The shape and size of an alga affect not only the water flow it encounters but also the magnitudes of the forces it bears in that flow. The pattern of water flow around a body of a given shape, and hence the hydrodynamic force the body experiences, depends on the relative importance of inertia and viscosity, as given by the Reynolds number (Re) for that flow situation [$\text{Re} = (\rho UL)/\mu$, where ρ is the density and μ the viscosity of the water, U is the flow velocity, and L is a characteristic linear dimension of the body]. Because water resists being deformed in shear as it moves across the surface of an organism, the organism is subjected to a force, skin-friction drag, tending to move it downstream. The drag on microscopic (i.e., low Re) and encrusting algae in moving water is due to skin friction. A low-pressure wake forms on the downstream side of macroscopic (i.e., high Re) upright algae in currents; hence, these plants are subjected to pressure drag (form drag) in addition to skin-friction drag (Hoerner 1965; Batchelor 1967; Vogel 1981). Any morphological feature of a macroscopic marine organism that reduces the size of the wake will reduce drag.

The magnitude of the total drag force (F_D) on a macroscopic organism is given by

$$F_D \propto C_D \rho U^2 S \qquad (18.1)$$

where C_D is the drag coefficient of the organism (C_D depends in part on the shape of the plant), ρ is the density of the fluid, U is the velocity of water flow, and S is some characteristic area of the organism. Because drag is proportional to velocity squared, any of the features mentioned earlier that put an alga in a protected microhabitat can significantly reduce drag. Furthermore, an increase in size (S) is accompanied by an increase in drag for an alga in a unidirectional current. Algae of the same size, however, can experience different drags in the same velocity of flow if they are of different shapes (C_D).

Most macroscopic algae are flexible and are deformed in moving water; such flexibility can reduce drag in several ways. For example, algal blades are bent over parallel to the direction of water movement and experience much less drag than would blades of the same area rigidly held normal to the flow. The C_D for a long, flat plate in the same range of Re values as macroalgae in tidal currents is more than 200 times greater when the plate is normal to the flow than when it is parallel (Vogel 1981). Furthermore, the blades on many algae can collapse on top of each other into a more compact shape as ambient velocities increase. Such passive streamlining has been shown to reduce drag in such diverse organisms as sea anemones (Koehl 1976, 1977a) and trees (Fraser 1962; Vogel 1984), as well as in algae (Charters et al. 1969; M. Koehl and R. Alberte, unpublished data). Even algae that do not fold up on top of themselves in a current can experience a reduction in drag if they are blown over close to the substratum, as illustrated by the red alga *Gigartina exasperata*. T. Mumford and I (unpublished data) found that the drag on *G. exasperata* near the substratum could be as low as half of that on the same plants encountering the same velocities, but not near the substratum (Figure 18.2). In both cases the algae flopped over parallel to the flow: Pressure drag was much reduced, but the plants still experienced skin friction. However, only the upper surfaces of the plants that were near the substratum were subjected to rapid flow and high skin friction, whereas both the upper and lower surfaces of plants held away from the substratum encountered rapid flow. In addition to drag reduction, another advantage of flexibility for algae in water currents is that thalli blown over prostrate on the substratum are less likely than are more upright, stiff thalli to be damaged by drifting objects, such as holdfasts of ripped-up algae (Dayton and Tegner 1984).

Mucilage secreted by seaweeds can reduce hydrodynamic forces in a number of ways. (1) Polymer coatings can reduce skin-friction drag (e.g., Wells 1969; Cox et al. 1974), as mucus does on fish (e.g., Daniel 1981). Although Norton et al. (1981, 1982) have suggested that mucilage on algal surfaces may reduce skin friction, my comparisons of drag measured on mucilage-covered kelp (Koehl, unpublished data) and on mucus-covered sea anemones (Koehl 1977a) with drag on nonslimy models of the organisms indicate that such coatings have a negligible effect on the total drag. However, other species must be studied at a range of Reynolds numbers before the importance of mucilage coatings in reducing drag on macroalgae can be evaluated. (2) The hydrodynamic forces on seaweeds can be increased when they bear epiphytes (e.g., Sousa 1979). Therefore, if a

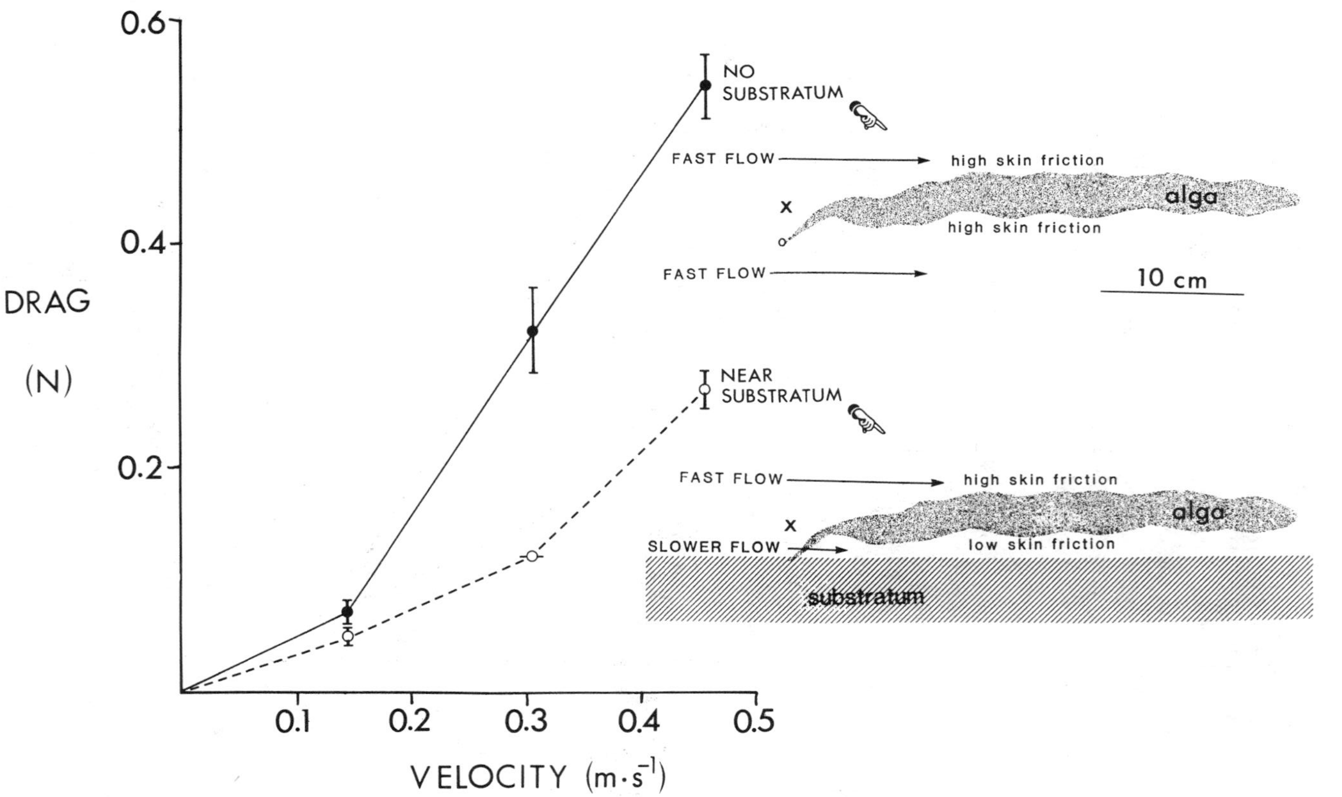

Figure 18.2. Drag forces measured on a *Gigartina exasperata* when next to and not next to a substratum. Water velocities were measured just upstream from the alga at the height of the blade when it was blown over (position indicated by ×). Error bars indicate one standard deviation (graph redrawn from Koehl 1984).

mucilaginous secretion reduces the attachment of epiphytes on an alga's surface, the secretion can indirectly reduce the drag the host plant has to bear. (3) Films of organic surfactants on the water surface can suppress wave motions (e.g., Van Dorn 1953; Vines 1960). Deacon (1979) has suggested that mucus secreted by corals can form a surface film that will reduce wave damage to reefs by this mechanism. Perhaps the surface slicks that sometimes form over kelp beds (Sturdy and Fischer 1966) can reduce the wave-associated water flow encountered by the kelp.

Forces on algae in waves

Intertidal and high subtidal seaweeds exposed to wave action experience accelerating water. Wave-swept organisms are subjected to acceleration reaction forces as well as to drag (Keulegan and Carpenter 1958; Wiegel 1964; Batchelor 1967; Carstens 1968). The pressure differential on a body in a fluid subjected to gravity gives rise to a buoyant force on that body that is proportional to the mass of the fluid displaced by the body. Similarly, the pressure gradient on a sessile body in accelerating water gives rise to "virtual buoyancy," a force proportional to the mass of water the body displaces. Furthermore, when an animal accelerates through a fluid, its motion affects some volume of fluid such that the animal exerts force to accelerate not only its own mass but also an "added mass" of fluid. Similarly, when water accelerates with respect to a stationary plant, this added mass contributes to the force the plant experiences.

The acceleration reaction force (F_I), which is a combination of the virtual buoyancy and the force due to added mass, is given by

$$F_I \propto C_M \rho V \, dU/dt \tag{18.2}$$

where C_M is the added-mass coefficient of the body (dependent in part on shape), ρ is the density of the water, V is the volume of the organism, and dU/dt is the acceleration of the water. The acceleration reaction acts in the direction the water is accelerating, that is, in the direction of water movement when it is speeding up, but in the opposite direction when it is slowing down. The force on a wave-swept organism at any instant in time is the sum of the drag force at that instant and the acceleration reaction at that instant.

Because the acceleration reaction is proportional to the volume of an organism, we might expect it to become increasingly important with respect to drag (which is proportional to area) as organisms get larger, and to impose a physical upper limit to the sizes wave-swept organisms can attain (Denny et al. 1985). Macroalgae, which are often the largest organisms on

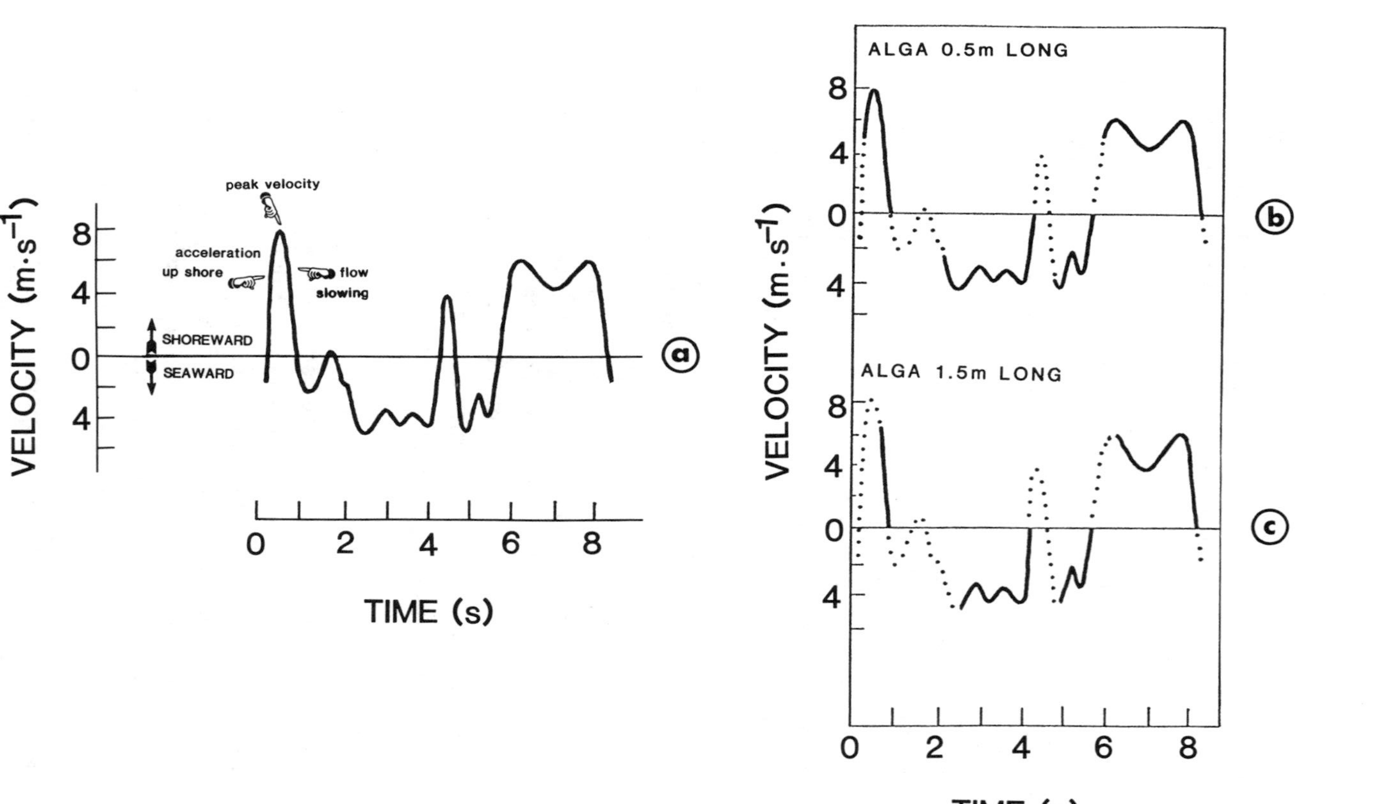

Figure 18.3. Water velocity versus time measured 15 cm above the substratum at a rocky shore on Tatoosh Island, Washington, where *Postelsia palmaeformis* and *Lessoniopsis littoralis* were abundant. (a) Water movement relative to a rigid structure. (b) Dotted lines indicate water movement at times when a perfectly flexible alga 0.5 m long would be moving with the water. Solid lines indicate water movement relative to the thallus at times after the plant would be fully laid out in the direction of flow. The thallus must resist forces produced by the water movement indicated by solid lines. (c) Notation as in (b), but for a plant 1.5 m long.

temperate wave-beaten rocky shores, have several morphological features that tend to reduce the acceleration reaction. For example, the blades of most macroalgae are thin; hence, their volumes are low even though they can be quite large in area. Furthermore, the C_M for flattened bodies oriented parallel to the acceleration is lower than the C_M for spherical bodies or for flattened objects normal to the acceleration (Daniel 1983, 1984). Most macroalgae not only have relatively flat blades but also are flexible and become passively oriented parallel to flow, as described earlier.

The flexibility of large algae at wave-swept sites also provides a mechanism by which the thalli can avoid bearing flow-induced forces at times during a wave when they are likely to be highest. An example of water velocities recorded among seaweed blades at a wave-swept rocky intertidal site is shown in Figure 18.3a; the maximum drag should occur when such a plot of velocity versus time peaks ($F_D \propto U^2$), whereas the maximum acceleration reaction should occur when the slope is steepest ($F_I \propto dU/dt$). However, as the water in a wave accelerates, the forces it imposes on a floppy alga move the plant along with the water; not until the attached plant is strung out in the direction of flow and is no longer free to move with the water does its thallus have to resist substantial force. If a flexible alga is long enough, the water may slow down or move back in the opposite direction before the slack in the plant has been taken up. Figure 18.3b,c illustrates that an increase in the length of a flexible seaweed can lead to a decrease in the forces its thallus must resist in oscillating flow. [In contrast, an increase in the length of an alga in unidirectional flow leads to an increase in the load (drag force) the thallus must bear.] Furthermore, even if a flexible plant is not long enough to totally avoid the peak accelerations and velocities of water in waves, the duration of flow relative to the plant (and hence the duration of high loads on the thallus) is less than if the alga were rigid (the consequences of this for plant breakage will be described later). Of course, the stiffer a seaweed of a given length, the sooner there will be water movement relative to its surfaces as a wave rushes in to shore.

Deflection and breakage

Just as water-flow patterns around and flow-induced forces on macroalgae depend on their morphology, so do their deformation and possible breakage. Engineering analyses can be used to elucidate aspects of algal structure that affect the deflection or breakability of the plants. Nonetheless, how these two aspects of mechanical performance relate to the success of various seaweeds in different habitats can be assessed only if the physiology, reproductive biology, and ecological interactions of the algae are also known. Is a stiff plant able to hold its blades above those of its more flexible neighbors, thereby out-competing them for light? Is a flexible plant flailing back and forth in waves able to lash off epiphytes (Fletcher and Day 1983) or fling off herbivores, such as snails or sea urchins, more readily than stiffer algae? Is a flexible seaweed more effective than a stiff one at sweeping the surrounding substratum clear of newly settled potential competitors (e.g., Dayton 1975; Velimirov and Griffiths 1979)? Does breakage mean death for a plant, or can the stump regrow? Can the broken-off fragments, or the spores they carry, attach to the substratum and grow [i.e., is fracture a mechanism of asexual reproduction and dispersal for seaweeds (e.g., Norton et al. 1982) as it is for many corals (e.g., Highsmith 1982)]? Has an alga released its spores or gametes before seasonally predictable periods of high mechanical stress occur?

In analyzing both deflection and breakage of seaweeds, one needs to consider how the shape of an alga affects the magnitude and distribution of mechanical stresses in its thallus and how the material properties (e.g., stiffness, strength, toughness, resilience) of its tissues determine an alga's response to those stresses. Mechanical stress is defined as the force per cross-sectional area of tissue bearing that force.

Stress in an algal thallus

Sessile organisms in moving water can be deformed in a number of ways, as illustrated in Figure 18.4. The stresses in the organism for a given load, such as drag force, depend on the shape and size of the organism (Koehl 1977b, 1982, 1984). Some basic relationships between stress and certain morphological parameters will be presented later. For simplicity of discussion, we shall assume that the force is applied to an alga at a point and that the alga undergoes small deformation. [Of course, real algae can bear distributed loads and can deform a great deal. Specific equations describing stress distributions in structures of various shapes under different loading regimes can be found in engineering books (e.g., Popov 1968; Timoshenko and Gere 1972; Roark and Young 1975; Faupel and Fisher 1981), and an example of stress analysis of an algal stipe undergoing large deformations can be found in the work of Charters et al. (1969).]

If an alga is simply pulled by water currents (Figure 18.4), the magnitude of the tensile stress (σ) in its tissues is

$$\sigma \propto 1/A \qquad (18.3)$$

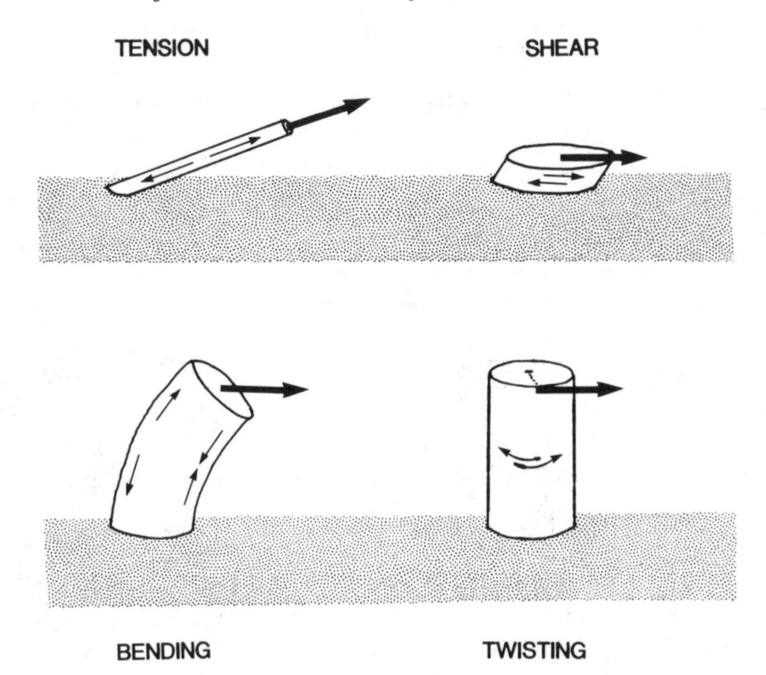

Figure 18.4. Diagrams of some of the ways a sessile organism can deform when bearing a load such as a hydrodynamic force. Heavy arrows indicate the direction of force application, and fine arrows indicate the deformation of the organism's tissues.

where A is the cross-sectional area of the plant. Similarly, if an alga is deformed in shear (Figure 18.4), the magnitude of the shear stress (τ) in its tissues is

$$\tau \propto 1/A \qquad (18.4)$$

In both cases, the greater the cross-sectional area of the plant, the lower the stress, but cross-sectional shape is relatively unimportant. Tissues in narrow regions of a plant will be subjected to greater stresses than tissues in wide regions of the thallus.

If an alga is bent in a current (Figure 18.4), the tissue on its upstream side is stretched, and that on its downstream side is compressed. The tensile or compressive stress (σ) in the plant is

$$\sigma \propto hy/I \qquad (18.5)$$

where h is the distance of the bit of tissue being considered from the free end of the sessile organism, y is the distance from the axis of bending, and I is the second moment of area of the cross-section of the organism (Alexander 1968; Wainwright et al. 1976). Because stress is proportional to h, the stresses in tissues near the attached end of an alga bent by flowing water are greater than those near its free end (if it has the same cross-sectional shape and area, base to tip). Similarly, if two stipes of the same cross-sectional shape and area are bent by loads of the same magnitude, but one stipe is taller than the other, the maximum stress in the tall stipe will be greater than that in the short one.

Tissues on the upstream and downstream surfaces of a bent alga are stretched or compressed more than tissues closer to the axis of bending of the plant ($\sigma \propto y$). Because such peripheral tissue contributes more to bearing the load than does tissue near the axis of bending, the more tissue a plant has far from the axis of bending, the lower the stress that any little bit of tissue within the plant has to bear. The second moment of area (I) is a measure of the distribution of tissue around the axis of bending of an organism and is roughly proportional to the radius raised to the fourth power. Therefore, the cross-sectional shape of a bent alga can have a big effect on the stresses in its tissues. For example, if an alga invests a given area of tissue in a cross section of its stipe, the stress for a given load will be lower if the stipe is elliptical in cross-sectional shape with its major axis parallel to the direction of ambient currents than if the stipe is circular in cross section. Similarly, for an alga in a habitat in which flow direction is not predictable, for a given cross-sectional area the stress produced by a particular load will be lower if the stipe is hollow and wide than if it is solid and narrow (Currey 1970). Furthermore, because stress in a bit of tissue is roughly inversely proportional to the radius raised to the third power ($\sigma \propto y/I$), a small increase in the width of an organism can lead to a large decrease in the stress in its tissues for a given load. Similarly, a narrow "waistline" in an algal thallus can be a region of much higher stress than wider parts of the thallus.

If an alga is twisted by moving water (Figure 18.4), the shear stress (τ) in a bit of its tissue is

$$\tau \propto My/J \tag{18.6}$$

where M is the distance (moment arm) of force application from the axis of torsion, and J is the polar second moment of area of the cross section of the plant. J, like I, is roughly proportional to the radius raised to the fourth power (Wainwright et al. 1976). As in the case of bending, tissues around the periphery of a twisted alga bear the largest stresses ($\tau \propto y$), and a small decrease in thallus width can lead to a large increase in stress. Furthermore, asymmetrical plants and plants with wide, stiff blades on narrow stipes are likely to experience higher torsional shear stresses than are plants with lower M values.

Although stresses in hollow, wide organisms are lower than those in solid, narrow ones under a given load in bending or twisting, hollow, thin-walled structures tend to undergo local buckling (i.e., kink like a beer can) (Currey 1970; Wainwright et al. 1976; Koehl 1977b). The critical stress (σ_{crit}) to produce local buckling in a bending hollow cylinder is

$$\sigma_{crit} \propto t/d \qquad (18.7)$$

where d is the diameter and t is the wall thickness (i.e., wide, thin-walled cylinders kink at lower stresses than do narrow, thick-walled ones). Hence, the "benefit" of stress reduction for a given investment in material obtained by an increase in d (and hence a reduction in t) is offset at some point by the risk of local buckling; see Currey (1970) for a discussion.

Keeping in mind these simple relationships between the magnitude of stress in an alga's tissues and its size and shape, consider some examples:

Algae often break at regions where grazers have taken bites out of them (e.g., Black 1976; Koehl and Wainwright 1977; Koehl 1979; Santelices et al. 1980). Because removal of a small amount of peripheral tissue can lead to large decreases in I and J, it should not be surprising that stresses in regions of a thallus where bites have been taken can reach the breaking point under bending or twisting loads that would not damage the intact plant. Because of this simple mechanical fact of life, the loss of algal biomass from a stretch of shore because of the activities of herbivores can be much greater than the plant tissue they consume. Nonetheless, "pruning" by grazers can in certain cases decrease the likelihood that an entire plant will be ripped off the substratum during a storm, because as pieces of an alga are broken off, the flow force on the remaining smaller thallus can be reduced (Black 1976; Santelices et al. 1980).

Stresses associated with bending depend on length and cross-sectional shape, whereas those associated with pulling do not. Therefore, if the long, slim stipe of an alga bears a load in tension, the stresses in its tissues can be lower than if it bears that same load in bending. An example of this point is provided by two species of macroalgae abundant on the wave-swept rocky shores of central and southern Chile, *Lessonia nigrescens* and *Durvillaea antarctica*. These seaweeds experience similar ranges of flow forces in situ (Koehl 1979, 1982, unpublished data). Santelices et al. (1980) have pointed out that a *Lessonia* bears its blades (and hence the hydrodynamic force on the blades) on several stipes, whereas a *Durvillaea* bears the whole force on only one stipe; they suggest, therefore, that *Lessonia* has a morphology better suited to withstanding wave action. However, *Lessonia* stipes, which are widest near their bases and are relatively stiff, are bent by flowing water, whereas *Durvillaea* stipes flop over at a flexible joint at the holdfast and are pulled by ambient currents. Calculation of the maximum stresses in stipes of a *Lessonia* and a *Durvillaea* of the same length and mean diameter indicates that stresses in the bent stipe are over 800 times greater than those in the pulled one when they bear the same load (Koehl 1982). It appears, therefore, that both species have stress-reducing thallus shapes.

Deformation of an alga in moving water

The amount of deformation an alga undergoes when bearing a flow-induced force depends not only on the shape of the thallus but also on the stiffness of its tissues. For example, the lateral deflection (δ) of the free end of a sessile organism bent by a force is

$$\delta \propto h^3/EI \tag{18.8}$$

where h is the height of the organism, and E is the tensile elastic modulus of the tissues, a measure of their stiffness. Similarly, the angular deflection (θ) of the free end of an alga twisted by a load (F) is

$$\theta \propto Mh/GJ \tag{18.9}$$

where G is the shear elastic modulus. Algae with tall, slim thalli (high h, low I and J) have deflection-increasing morphologies, whereas algae with short, wide thalli (low h, high I and J) have deflection-decreasing body shapes.

An alga can decrease the amount it deforms for a given load by changing its shape (i.e., decreasing h/I and h/J) and/or by increasing the stiffness of its tissues (E and G). Several species of macroalgae have been observed to produce shorter, wider stipes at sites exposed to rapid water flow than they produce at more protected sites (e.g., Mann 1971; Chapman 1973; Gerard and Mann 1979). Similarly, the tissues of exposed algae of some species have been found to have higher elastic moduli than the tissues of protected members of the same species (e.g., Charters et al. 1969; Armstrong 1982).

If localized portions of a thallus are made of easily deformable tissue (i.e., of low E or G) and/or if certain sections of an alga are narrower (i.e., have lower I and J) than the rest of the thallus, these regions can behave as flexible joints (Koehl 1977b). Examples of the former are the deformable genicula between the heavily calcified rigid sections of upright coralline algae such as *Calliarthron tuberculosum* (R. B. Emlet, unpublished data). Examples of the latter are the joints between the stipes and holdfasts of *Durvillaea antarctica* (Koehl 1982; Norton et al. 1982). Therefore, the

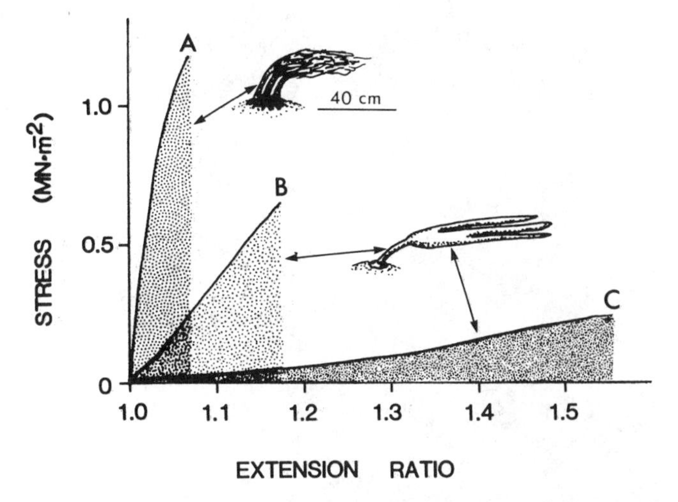

Figure 18.5. Stress–extension-ratio curves for tissues from *Lessonia nigrescens* stipe (A) and *Durvillaea antarctica* stipe (B) and blade (C) (redrawn from Koehl 1982). Both seaweeds are drawn to the same scale. Extension ratio is the ratio of the extended length of a specimen to its unstretched length.

structure of an alga can determine the location on a thallus at which deformation will occur, as well as the amount of deflection produced by a given load.

The stiffness of algal tissue can be measured by pulling on a sample of the tissue at a defined rate and simultaneously measuring the force with which the tissue resists the pull (Koehl and Wainwright in press). Results of such measurements for tissues from the Chilean seaweeds *Lessonia nigrescens* and *Durvillaea antarctica* are presented in Figure 18.5. The more resistant a tissue is to deformation, the steeper will be the slope of such a stress–extension-ratio curve; E, which is the slope of a stress–extension-ratio curve, is a measure of tissue stiffness. (Similar tests can be done in shear to measure G.) The stipe tissue of *Lessonia* is stiffer than that of *Durvillaea,* which in turn is stiffer than its blade tissue.

The stiffness of a tissue depends on its microscopic structure and molecular structure. Much remains to be learned about the ways in which algae build stiff or stretchy tissues out of fiber-reinforced cell walls and intercellular matrix. The ultrastructure, chemical composition, and mechanical properties of algal cell walls are beyond the scope of this chapter, but reviews of these topics can be found in the work of Preston (1974), Wainwright et al. (1976), Sellen (1980), McCandless (1981), and Vincent (1982). I shall briefly point out, however, some patterns I have noticed in the microscopic structures of stiff versus stretchy seaweed tissues I have

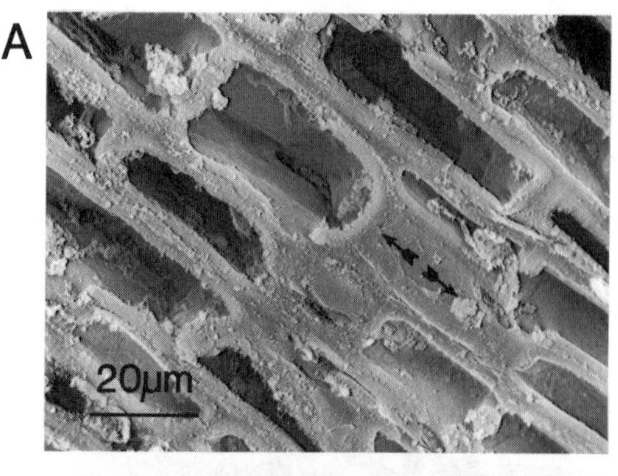

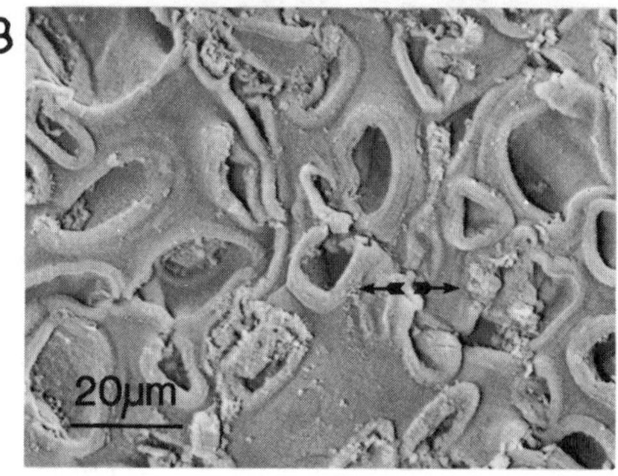

Figure 18.6. Scanning electron micrographs of transverse sections of stipe tissue from (A) *Lessonia nigrescens* and (B) *Durvillaea antarctica*. Arrows indicate radial direction.

studied. Stiff tissues, such as from the stipes of *Lessonia,* have closely packed cells with thick cell walls that appear to be continuous from cell to cell parallel to the major stress axes in the tissue (Figure 18.6A). In contrast, stretchy tissues, such as from the stipes of *Durvillaea,* tend to have thin-walled cells with large areas of intercellular matrix between them (Figure 18.6B). The tremendous extensions that such stretchy algal tissues can undergo may occur by movement of cells relative to each other as the intercellular matrix is sheared. The incredibly stretchable blades of *Durvillaea* are composed of a honeycomb of air sacs that are easily deformed from roughly spherical to roughly ellipsoidal in shape as the tissue is stretched.

The chemistry of the intercellular matrix can affect its resistance to deformation. For example, the polysaccharide alginate is the major intercellular substance in most brown algae. L-guluronic acid residues in algin-

Table 18.1. *Strength and toughness of seaweed tissues and other biomaterials*[a]

Material	Strength (MN/m)	Toughness (kJ/m)
Seaweed tissues		
Durvillaea antarctica	0.7_1	4_1
Laminaria digitata	0.9_2	—[b]
Lessonia nigrescens	1.2_1	4_1
Ascophyllum nodosum	1.5_2	—
Nereocystis luetkeana	3.6_3	—
Fucus serratus	4.2_2	—
Other biomaterials		
Mammalian skin	12 (cat)$_4$	20 (rabbit)$_{5,t}$
Whale tympanic bulla	$33_{5,b}$	$0.2_{5,b}$
Insect cuticle	95 (*Schistocerca* tibia)$_4$	1.4 (*Rhodnius* tergum)$_{5,t}$
Deer antler	$179_{5,b}$	$6.2_{5,b}$
Softwood	180_5	12_5
Tooth enamel	$200_{5,c}$	$0.2_{5,b}$
Cow leg bone	$247_{5,b}$, 184_4	$1.7_{5,b}$
Hardwood	240_5	11_5
Tooth dentine	$300_{5,c}$	$0.6_{5,b}$

[a] All materials were tested in tension except where indicated by one of the following subscripts: b = bending, c = compression, and t = tearing. Results from specimens tested in different ways are not strictly comparable. Numerical subscripts indicate the source of the values reported: 1 = Koehl (1979); 2 = Wheeler and Neushul (1981); 3 = Koehl and Wainwright (1977); 4 = Wainwright et al. (1976); 5 = Vincent (1982).
[b] Data not available.

ate are responsible for the formation of junctions in alginate gels produced in the lab; within- and between-species comparisons indicate that brown algal tissues are stiffer when their alginates are rich in L-guluronic acid residues than when their alginates contain small proportions of these residues (Haug, Larsen, and Baardseth 1969; Andreson et al. 1977).

The stiffnesses of many biological materials and the permanence of deformations imposed on them can depend on the rate at which they are deformed as well as on their history of load bearing. Little information is available on the possible viscoelastic, stress-softening, or plastic behaviors of seaweed tissues of different microarchitectures.

Breakage of an alga in moving water

The strength of a tissue is defined as the stress required to break it. Compared with other biomaterials, the tensile strengths of seaweeds are low (Table 18.1) (Delf 1932; Koehl and Wainwright 1977; Norton et al. 1981; Wheeler and Neushul 1981; Koehl 1984). This observation seems puzzling in light of the water currents and waves that seaweeds can withstand, until the toughness of algal tissues is considered. The amount of work that must be performed by moving water to break an alga depends on the toughness of the plant.

The area under the force–extension curve of a tissue pulled until it breaks represents the work required to break the specimen; a measure of a tissue's toughness is its work of fracture, the quantity of energy required to break a given cross section of the material. The algal tissues described in Figure 18.5 were all pulled until they broke. *Lessonia* stipe tissue was stronger than that of *Durvillaea* stipe or blade. However, *Durvillaea* blade tissue stretched farther before breaking than did tissue from *Durvillaea* or *Lessonia* stipes. The values for the work of fracture for these three types of tissues are not significantly different (Koehl 1979), and compared with other biomaterials, these particular algal tissues are relatively tough (Table 18.1). These tissues illustrate that there are several ways to be tough: An alga can be tough by having stiff, strong tissues like *Lessonia* stipe, but an alga can also be tough by having very stretchy tissues, even if those tissues are not strong, like *Durvillaea* blade (Koehl 1982, 1984). I do not mean to imply, however, that all such materials are tough. If a material is extremely stiff, the area under its force–extension curve can be small; an example of such a brittle material is glass. Similarly, if a tissue is stretchy but very weak, the area under its stress–extension curve can also be small.

One can speculate about the constraints on algae made of extensible, weak material rather than stiff, strong tissue. If a load is applied for a period of time long enough to stretch an alga's tissue to its maximum extension, the tissue will then break if the stress in it is greater than its strength. Therefore, one might predict that algae with extensible, weak tissues could survive at sites where flow forces on them are high only if the duration of high loads is brief (i.e., shorter than the time required to stretch them to their breaking lengths) and if the plants' tissues are resilient enough to bounce back to their undeformed shapes between pulses of high load. *Nereocystis* stipes are weak, but are extremely extensible (Koehl and Wainwright 1977). Recordings of water movement encountered by *Nereocystis* blades reveal that peak velocities last only a fraction of a second (Koehl 1984). Furthermore, *Nereocystis* stipes are extremely resilient (Koehl 1982), as predicted. One might also predict that extensible, but weak, algae would be less likely to survive storms (when the magnitude and duration of flow forces are higher than under non-storm conditions) than would stiff, strong seaweeds of similar toughness. This prediction seems to be borne out in the case of *Durvillaea* and *Lessonia:* Greater proportions of *Durvillaea* are ripped off the shore during winter storms than are *Lessonia*

(Santelices et al. 1980). One could further speculate that extensible, weak algae might grow rapidly and produce their propagules before seasonally predictable storms rip them off the substratum. Both *Nereocystis* (Foreman 1970; Nicholson 1970; Vadas 1972; Duncan and Foreman 1980) and *Durvillaea* (Santelices et al. 1980) show these characteristics, in contrast to stiff, strong plants like *Lessonia* that grow more slowly, take longer to reach the spore-producing stage, and live for several years (Santelices et al. 1980). Of course, both the biomechanics and the population dynamics of many more species of algae must be studied before we can assess the generality of these predictions about the habitats or life histories of algae whose tissues possess particular mechanical properties.

No matter how strong or tough an alga's tissues are, the plant will be ripped off the substratum by moving water if the adhesion of its holdfast to the substratum fails, or if the rock [or mussel (Dayton 1973; Paine 1979), dead barnacle, etc.] to which it has attached breaks. Surveys of seaweeds washed up on beaches reveal that such holdfast or substratum failure can be quite common (e.g., Koehl and Wainwright 1977; Witman and Suchanek 1984; M. Koehl unpublished data). A number of people have suggested that seaweeds that are flexible, that have coiled stipes, or that have highly extensible tissues deform before they transfer to their holdfasts the full load hitting their thalli (e.g., Koehl and Wainwright 1977; Kain 1979; Norton et al. 1981, 1982; Koehl 1984; M. Denny, unpublished data). Such morphological features that render plants good shock-absorbers can protect the holdfast-substratum attachment from high loads at sites subjected to short pulses of rapid water flow, such as wave-beaten shores.

One important aspect of the breakability of macroalgae concerning which little is known is fatigue fracture. Many materials, when subjected to repeated stresses that are too low to break them initially, will eventually fail (e.g., Andrews 1968), as a coat hanger does when bent back and forth repeatedly. Fatigue failure appears to be involved in the breakage of *Durvillaea* stipes, which accumulate cracks that enlarge as the plants are repeatedly pulled by waves (Santelices et al. 1980). What roles do tissue microarchitecture, growth, and repair play in the susceptibility of different algae to fatigue fracture?

Epiphytes have been reported to make the blades of giant kelp, *Macrocystis* species, more brittle than unfouled blades (Lobban 1978; Dixon, Schroeter, and Kastendiek 1981). However, the reductions in strength and toughness of fouled blades have not, to my knowledge, been measured, and the mechanisms responsible for this effect have not been elucidated.

The mechanical properties of algal tissues may affect their susceptibility to herbivores as well as to damage by moving water. For example, Padilla (1982) found that for certain limpet radulae, more force was required to rasp tissue from the deformable blades of *Iridaea cordata* than from the more "leathery" blades of *Hedophyllum sessile* or from the stiff calcified crusts of *Pseudolithophyllum* species. These surprising results warn against making assumptions about mechanical susceptibility to predation based on measures of the strengths of tissues that have been broken in a manner that is very different from the way in which they are attacked by the harvesting tools of specific herbivores. The mechanical properties that render algal tissues most effective at resisting the teeth of other herbivores (such as fish or sea urchins) remain to be studied.

Tradeoffs

Thus far, we have discussed ways in which the mechanical behaviors of macroalgae depend on their structures. However, a morphological feature that improves some aspect of the mechanical performance of a seaweed may worsen the alga's performance of some other function. We shall mention a few examples of such tradeoffs.

Tradeoff between drag reduction and photosynthesis

Many sessile animals and plants that make their living by extracting food or dissolved substances from the water around them have morphological features that increase the degree to which they interfere with the water flowing by them (Koehl 1982, 1984). We might expect that such organisms would bear larger flow-induced forces than would other organisms that interfere less with ambient currents.

The giant bull kelp of the northwest coast of North America, *Nereocystis luetkeana,* provides an example of a sessile organism that "compromises" between increasing and decreasing the effects that ambient currents exert on it. A *Nereocystis,* which generally lives in water 10 to 17 m deep, is attached to the substratum by a holdfast and has a long, slim stipe terminating in a gas-filled float that holds the blades (which can be up to 4 m long) near the water's surface (Abbott and Hollenberg 1976). *Nereocystis* can be found at sites exposed to nonbreaking waves or to rapid tidal currents, as well as at sites protected from both wave action and fast currents (Koehl 1984, unpublished data). The blades of plants from exposed sites are narrow and flat, whereas those of plants from protected sites are wider and have undulate (ruffled) edges (Figure 18.7) (M. Koehl and R. Alberte,

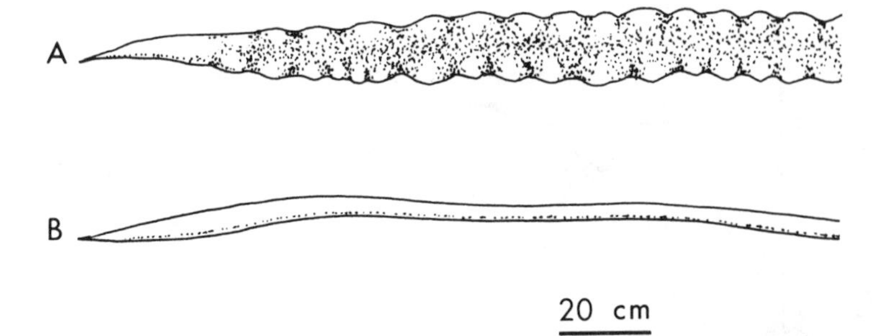

Figure 18.7. Proximal end of a blade of *Nereocystis luetkeana* from a site protected from rapid currents (A) and of *N. luetkeana* from a site exposed to rapid water flow (B) (drawings traced from photographs).

unpublished data). *Nereocystis* are flexible and stream out parallel to the direction of water flow, thereby reducing form drag, as described earlier. However, thin flexible structures such as algal blades flap in water currents as flags do in the wind. Such flapping increases the drag force on the blade (Hoerner 1965; Raupach and Thom 1981; Vogel 1981; Witman and Suchanek 1984).

The drag on *Nereocystis* with wide, undulate blades from protected habitats can be compared with the drag on *Nereocystis* with narrow, flat blades from exposed sites; the drag per blade area is about twice as large for the protected blades (M. Koehl and R. Alberte, unpublished data). Some of the increase in drag on protected plants is due to the undulate edges of their blades; the drag on a protected plant is greater than that on the same plant that has had its ruffled blades replaced by flat plastic models of its blades (Figure 18.8). However, the greater width of protected blades is also responsible for some of the increase in drag; the drag per blade area is greater for wide, flat models of protected blades than for real exposed blades, which are narrow and flat (M. Koehl and R. Alberte, unpublished data). Drag is higher on wide, undulate blades than on narrow, flat ones because the ruffled blades flap with greater amplitude than do the flat ones, and because the flat blades stack up on top of each other in skinnier streamlined bundles than do the undulate ones. If exposed *Nereocystis* had wide ruffled blades rather than narrow flat ones, and if the area of blades per plant was the same as that measured for exposed *Nereocystis,* the drag on the plants would be sufficient to break their stipes, even under non-storm conditions (M. Koehl and R. Alberte, unpublished data). Therefore, flatness and narrowness of blades could be considered morphological features that permit *Nereocystis* to attain large surface areas for photosynthesis at current-swept sites.

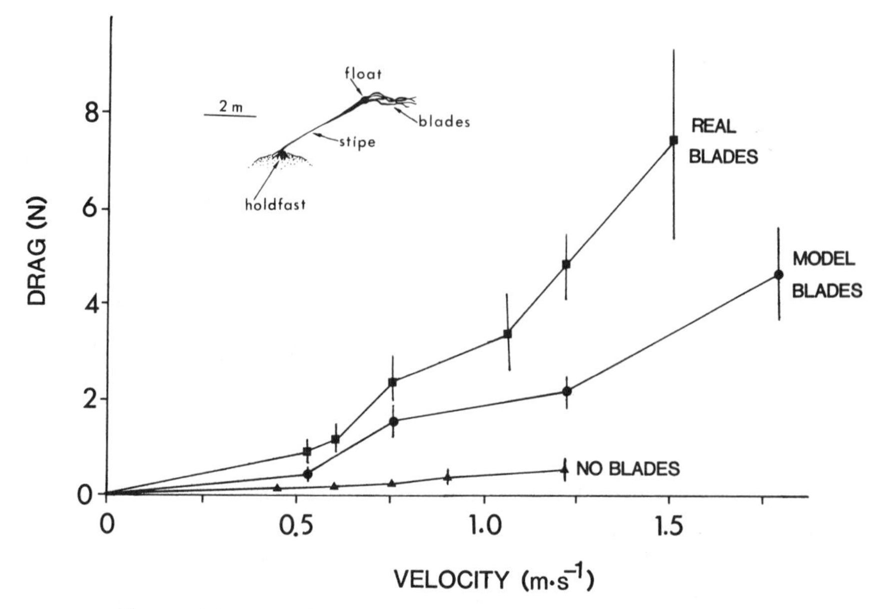

Figure 18.8. Drag force measured on a *Nereocystis luetkeana* (squares), on the stipe and pneumatocyst (float) after the fronds had been removed (triangles), and on the stipe and pneumatocyst plus plastic fronds (circles) (which were exact tracings of the removed fronds, but were flat rather than undulate). The alga was towed outside the wake of a boat, velocities were measured using an electromagnetic flow meter (EPCO water current meter, model 6130), and forces were measured using a spring dynamometer. Error bars represent one standard deviation. That this reduction in drag was due to lack of ruffles rather than to some other difference between real and plastic blades is supported by the observation that drag per blade area was greater for wide plastic blades than for narrow real blades (M. Koehl and R. Alberte, unpublished data).

Are there any advantages to plants bearing drag-increasing ruffled blades? R. Alberte and I (unpublished data) found that the photosynthetic rates of blades with undulate edges can be enhanced:

1. Algae take up bicarbonate to support photosynthesis and release dissolved organics and oxygen by diffusion. The boundary layer of slowly moving water along the surface of a blade can be thought of as a barrier across which diffusion, which can be the rate-limiting step in algal photosynthesis, takes place. Boundary layers are thinner and hence photosynthetic rates are higher the faster or more turbulent the flow across the surface of an alga; however, above some water-velocity saturation point, diffusion is no longer limiting to photosynthesis (Munk and Riley 1952; Whitford 1960; Whitford and Schumacher 1961, 1964; Schumacher and Whitford 1965; Westlake 1967; Conover 1968; Sperling and Grunewald

1969; Neushul 1972; Pasciak and Gavis 1974; Canelli and Fuhs 1976; Gavis 1976; Lock and John 1979; Smith and Walker 1980; Wheeler 1980a, 1980b; Wheeler and Neushul 1981; Anderson and Charters 1982; M. Koehl and R. Alberte, unpublished data). Alberte and I found that flapping, which is more pronounced for ruffled *Nereocystis* blades than for flat ones, stirs the water near blade surfaces and can nearly double photosynthetic rates in slow ambient currents such as protected *Nereocystis* often encounter. Other morphological features of algae that have been suggested as introducing turbulence near blade surfaces, and thus enhancing photosynthesis, include rugosities, perforations, marginal spines, and floats (which cause fronds to bounce up and down in waves) (Neushul 1972; Wheeler 1980a, 1980b; Norton et al. 1981; Gerard 1982).

2. Alberte and I found that ruffles can also enhance the photosynthesis of *Nereocystis* blades in water currents above saturation velocities. Undulate edges help prevent self-shading when blades are in moving water, not only by keeping the blades spread apart but also by increasing the amplitude and variability of blade motion such that the amount of time that one blade finds itself under another is reduced.

The advantages of wide ruffled blades carry with them the cost of increased drag forces to be sustained by stipes and holdfasts. It should not be surprising, therefore, that some macroalgae (such as *Hedophyllum sessile* and several species of *Laminaria*) have flat, narrow blades at current-swept sites, but undulate, wide blades at protected sites (e.g., Sundene 1964; Russell 1978; Gerard and Mann 1979; Kain 1979; Armstrong 1982). Whether or not the other structures (e.g., spines, rugosities) thought to enhance the movement of substances to algal surfaces also carry with them the cost of increased drag remains to be studied.

Tradeoff between unbreakability and photosynthesis

A number of people have suggested that a tradeoff exists for seaweeds between mechanical sturdiness and photosynthetic performance or productivity (e.g., Gerard and Mann 1979; Littler 1979; Littler and Littler 1980; Norton et al. 1981, 1982; Mann 1982; Littler et al. 1983). The investment of materials and energy in structural components means that less is available for producing photosynthetic machinery (or propagules). Furthermore, a large ratio of structural material to photosynthetic material in a thallus can cause internal self-shading. Similarly, the low surface-to-volume ratios of robust plants can limit the rates at which they take up nutrients from the surrounding water. However, plants with such investments in mechanical support can survive at wave-swept sites where the supply of dissolved substances by moving water is enhanced, where the activity of herbivores may be restricted, and where other organisms that could otherwise out-compete these plants for space may be ripped away. A striking example of tradeoffs between the mechanical and photosynthetic performances of seaweeds of different shapes is provided by *Laminaria longicruris,* which is plastic in morphology (Gerard and Mann 1979). As described earlier, photosynthetic rates of algae can be increased by ambient water flow, as can growth rates (e.g., Matsumoto 1959; Boalch 1961; Whitford and Kim 1966; Norton et al. 1982). Nonetheless, *L. longicruris* shows higher growth rates at protected sites (where plants produce thin, wide, undulate blades) than at exposed sites (where plants have thick, narrow, flat blades, and where transplants from protected sites are torn apart by wave action) (Gerard and Mann 1979).

Conclusions

The distributions of various species and morphological types of seaweeds in different water-flow habitats have been extensively studied; see the reviews by Schwenke (1971), Neushul (1972), Mann (1973), Chapman and Chapman (1976), Russell (1978), and Norton et al. (1981, 1982). Furthermore, a number of cases of environmentally correlated within-species variations in seaweed morphology have been documented; see the reviews by Russell (1978), Kain (1979), Mathieson et al. (1981), Norton et al. (1981, 1982), and Mann (1982). One purpose of this chapter has been to illustrate that an engineering approach can be used to study, in a quantitative experimental manner, the ways in which such observed differences in morphology of macroalgae affect defined aspects of their mechanical performance in various water-flow habitats.

A second purpose of this chapter has been to advertise the field of algal functional morphology as one that offers a rich array of problems for which the sorts of quantitative optimization studies described elsewhere in this volume could be done. We have focused in this chapter on biomechanical analyses of ways in which the shapes of macroalgae and the mechanical properties of their tissues affect the flow forces they experience, their deformation and breakage in moving water, and their uptake of dissolved substances. Of course, the morphologies of seaweeds also affect their performance of other functions such as capturing light (e.g., Ramus 1978; Littler 1979; Littler and Littler 1980; Hay 1981a; Jeffrey 1981; Norton et al. 1981, 1982; Wheeler and Neushul 1981; Littler et al. 1983), resisting desiccation (e.g., Kristensen 1968; Hay 1981a; Norton et al.

1982), dispersing spores or sperm (e.g., Fetter and Neushul 1981; Norton et al. 1982), and resisting herbivores (e.g., Lubchenco 1978; Lubchenco and Cubit 1980; Slocum 1980; Dethier 1981; Hay 1981a, 1981b; Norton et al. 1982; Padilla 1982; Littler et al. 1983). Thallus designs that improve the performance of some of these functions appear to worsen the performance of others (e.g., Littler and Littler 1980).

A third purpose of this chapter has been to show that there are different morphological ways of performing a particular mechanical task, such as withstanding breaking waves. Furthermore, even plants with "bad" mechanical designs (such as drag-increasing shapes, or easily breakable tissues) can survive on wave-swept shores if they use the appropriate life-history "strategy." Of course, many biological interactions and physical factors affect the abundance and distribution of macroalgae of various morphologies. Nonetheless, I hope that the observations concerning seaweeds and the mechanical rules by which they operate that I have listed in this chapter will provide food for thought for those analyzing constraints on plant form and function.

Acknowledgments

I am grateful to the following people for the many discussions that have influenced the way I think about macroalgae: R. S. Alberte, T. L. Daniel, M. W. Denny, M. LaBarbera, R. T. Paine, W. P. Sousa, S. Vogel, and S. A. Wainwright. I thank T. Givnish, T. Hunter, A. Johnson, J. Kingsolver, B. Okamura, M. Patterson, and R. Robichaux for their helpful comments on this chapter. Some of my work reported here was supported by grants from the American Philosophical Society, the Cocos Foundation, and the National Science Foundation (#DES75-14378 to R. T. Paine, and Presidential Young Investigator Award to M. Koehl).

Appendix I: Notation

A = cross-sectional area (m^2)
C_D = coefficient of drag
C_M = added-mass coefficient
d = diameter of a cylinder (m)
E = tensile elastic modulus of a tissue ($N\ m^{-2}$)
F_D = drag force (N)
F_I = acceleration reaction force (N)
G = shear elastic modulus of a tissue ($N\ m^{-2}$)
h = height of an organism
I = second moment of area of a cross section of a bent organism (m^4)
J = polar second moment of area of a cross section of a twisted organism (m^4)

l = distance of a bit of tissue from the free end of a sessile organism (m)
M = perpendicular distance between the point of load application and the axis of rotation of a twisted organism (m)
S = characteristic area of a body (usually plan area, surface area, or projected area normal to the direction of water flow) (m^2)
t = thickness of the wall of a hollow cylinder (m)
U = velocity of water flow ($m\ s^{-1}$)
V = volume of an organism (m^3)
y = distance of a bit of tissue from the axis of bending of a sessile organism (m)
δ = linear deflection of the free end of a bent sessile organism (m)
λ = extension ratio of a stretched piece of tissue (i.e., ratio of its extended length to its unstretched length)
μ = viscosity of seawater ($1.39 \times 10^{-3}\ N\ s\ m^{-2}$ at 10°C)
ρ = density of seawater ($1.026 \times 10^3\ kg\ m^{-3}$ at 10°C)
σ = tensile or compressive stress in an organism's tissues ($N\ m^{-2}$)
θ = angular rotation of the free end of a twisted sessile organism (rad)
τ = shear stress in an organism's tissues ($N\ m^{-2}$)

References

Abbott, I. A., and G. J. Hollenberg. 1976. Marine algae of California. Stanford Univesity Press, Stanford, Calif.

Alexander, R. M. 1968. Animal mechanics. University of Washington Press, Seattle.

Anderson, S. M., and A. C. Charters. 1982. A fluid dynamic study of seawater flow through *Gelidium nudifrons.* Limnol. Oceanogr. 27:399–412.

Andreson, I.-L., O. Skipnes, O. Simsrod, K. Ostgaard, and P. C. Hemmer 1977. Some biological functions of matrix components in benthic algae in relation to their chemistry and the composition of seawater. Pp. 361–381 *in* J. C. Arthur, Jr. (ed.), Cellulose chemistry and technology. American Chemical Society Symposium Series no. 48, Washington, D.C.

Andrews, E. H. 1968. Fracture in polymers. Oliver and Boyd, Edinburgh.

Armstrong, S. L. 1982. Mechanical behavior of two morphs of *Hedophyllum sessile* (Phaeophyta, Laminariales) from exposed and protected habitats. Am. Zool. 22:907.

Bascom, W. 1964. Waves and beaches: the dynamics of the ocean surface. Doubleday, Garden City, N.Y.

Batchelor, G. K. 1967. An introduction to fluid mechanics. Cambridge University Press, London.

Black, R. 1976. The effects of grazing by the limpet. *Acmaea insessa,* on the kelp *Egrigia laevigata,* in the intertidal zone. Ecology 57:265–277.

Boalch, G. T. 1961. Studies of *Ectocarpus* in culture. II. Growth and nutrition of a bacteria-free culture. J. Mar. Biol. Ass. U.K. 41:287–304.

Canelli, E., and G. W. Fuhs. 1976. The effect of the sinking rate of two diatoms (*Thalassiosira* spp.) on uptake from low concentrations of phosphate. J. Phycol. 12:93–99.

Carstens, T. 1968. Wave forces on boundaries and submerged bodies. Sarsia 34:37–60.
Chamberlain, J. A., and R. R. Graus. 1975. Water flow and hydromechanical adaptations of branched reef corals. Bull. Mar. Sci. 25:112–125.
Chapman, A. R. O. 1973. Phenetic variability of stipe morphology in relation to season, exposure, and depth in the non-digitate complex of *Laminaria* Lamour (Phaeophyta, Laminariales) in Nova Scotia. Phycologia 12:53–57.
Chapman, V. J., and D. J. Chapman. 1976. Life forms in the algae. Bot. Mar. 19:65–74.
Charters, A. C., M. Neushul, and C. Barilotti. 1969. The functional morphology of *Eisenia arborea.* Proc. Int. Seaweed Symp. 6:89–105.
Conover, J. T. 1968. The importance of natural diffusion gradients and transport of substances related to benthic marine plant metabolism. Botanica Mar. 11:1–9.
Cox, L. R., E. H. Dunlop, and A. M. North. 1974. Role of molecular aggregates in liquid drag reduction by polymers. Nature 249:243–245.
Currey, J. D. 1970. Animal skeletons. Edward Arnold, London.
Daniel, T. L. 1981. Fish mucus: *in situ* measurements of polymer drag reduction. Biol. Bull. 60:376–382.
– 1983. Mechanics and energetics of medusan jet propulsion. Can. J. Zool. 61:1406–1420.
– 1984. Unsteady aspects of aquatic locomotion. Am. Zool. 24:121–134.
Dayton, P. K. 1973. Dispersion, dispersal, and persistence of the annual intertidal alga, *Postelsia palmaeformis* Ruprecht. Ecology 54:433–438.
– 1975. Experimental evaluation of ecological dominance in a rocky intertidal alga, *Postelsia palmaeformis* Ruprecht. Ecology 54:433–438.
Dayton, P. K., and M. J. Tegner. 1984. Catastrophic storms, El Niño, and patch stability in a southern California kelp community. Science 224:283–285.
Deacon, E. L. 1979. The role of coral mucus in reducing the wind drag over coral reefs. Boundary-Layer Meterol. 17:517–521.
Delf, E. M. 1932. Experiments with the stipes of *Fucus* and *Laminaria.* J. Exp. Biol. 9:300–313.
Denny, M. W., T. L. Daniel, and M. A. R. Koehl. 1985. Mechanical limits to size in waveswept organisms. Ecol. Monogr. 55:69–102.
Dethier, M. N. 1981. Heteromorphic algal life histories: the seasonal pattern and response to herbivory of the brown crust *Ralfsia californica.* Oecologia 49:333–339.
Dixon, J., S. C. Schroeter, and J. Kastendiek. 1981. Effects of the encrusting bryozoan, *Membranipora membranacea,* on the loss of blades and fronds by the giant kelp, *Macrocystis pyrifera* (Laminariaes). J. Phycol. 17:341–345.
Duncan, M. J., and R. E. Foreman. 1980. Phytochrome-mediated stipe elongation in the kelp *Nereocystis luetkeana* (Phaeophyceae). J. Phycol. 16:138–142.
Eckman, J. E. 1982. Hydrodynamic effects exerted by animal tubes and marsh grasses and their importance to the ecology of soft-bottom, marine benthos. PhD. dissertation, University of Washington.
– 1983. Hydrodynamic processes affecting benthic recruitment. Limnol. Oceanogr. 28:241–247.

Faupel, J. H., and F. E. Fisher. 1981. Engineering design, 2nd ed. Wiley, New York.

Fetter, R., and M. Neushul. 1981. Studies on developing and released spermatia in the red alga, *Tiffaniella snyderae* (Rhodophyta). J. Phycol. 17:141–159.

Fletcher, W. J., and R. W. Day. 1983. The distribution of epifauna on *Ecklonia radiata* (C. Agardh). J. Agardh and the effect of disturbance. J. Exp. Mar. Biol. Ecol. 71:205–220.

Fonseca, M. S., J. S. Fisher, J. C. Zieman, and G. W. Thayer. 1982. Influence of the seagrass *Zostera marina* L., on current flow. Est. Coast. Shelf Sci. 15:351–354.

Foreman, R. E. 1970. Physiology, ecology and development of the brown alga *Nereocystis luetkeana* (Mertens) P & R. PhD dissertation, University of California, Berkeley.

Fraser, A. I. 1962. Wind tunnel studies of the forces acting on the crowns of small trees. Rep. For. Res. H.M.S.O. Lond. pp. 178–183.

Gavis, J. 1976. Munk and Riley revisited: Nutrient diffusion transport and rates of phytoplankton growth. J. Mar. Res. 34:161–179.

Gerard, V. A. 1982. *In situ* water motion and nutrient uptake by the giant kelp *Macrocystis pyrifera*. Mar. Biol. 69:51–54.

Gerard, V. A., and K. H. Mann. 1979. Growth and production of *Laminaria longicruris* (Phaeophyta) populations exposed to different intensities of water movement. J. Phycol. 15:33–41.

Haug, A., B. Larsen, and E. Baardseth. 1969. Comparison of the constitution of alginates from different sources. Proc. Int. Seaweed Symp. 6:443–451.

Hay, M. E. 1981a. The functional morphology of turf-forming seaweeds: persistence in stressful marine habitats. Ecology 62:739–750.

– 1981b. Herbivory, algal distribution, and the maintenance of between-habitat diversity on a tropical fringing reef. Am. Nat. 118:520–540.

Highsmith, R. C. 1982. Reproduction by fragmentation in corals. Mar. Ecol. Prog. Ser. 7:207–226.

Hoerner, S. F. 1965. Fluid-dynamic drag. S. F. Hoerner. Brick Town, N.J.

Jackson, G. A., and C. D. Winant. 1983. Effect of a kelp forest on coastal currents. Continental Shelf Research 2:75–80.

Jeffrey, S. W. 1981. Responses to light in aquatic plants. Pp. 249–277 *in* O. L. Lange, P. S. Nobel, C. B. Osmond, and H. Ziegler (eds.), Physiological plant ecology, I, Responses to the physical environment, vol. 12A, Encyclopedia of plant physiology, new series. Springer-Verlag, New York.

Jumars, P. A., and A. R. M. Nowell. 1984. Fluid and sediment dynamic effects on marine benthic community structure. Am. Zool. 24:45–55.

Kain, J. M. 1979. A view of the genus *Laminaria*. Oceanogr. Mar. Biol. Ann. Rev. 17:101–161.

Keulegan, G. H., and L. H. Carpenter. 1958. Forces on cylinders and plates in an oscillating fluid. J. Res. Natn. Bur. Stan. 60:423–440.

Koehl, M. A. R. 1976. Mechanical design of sea anemones. Pp. 23–31 *in* G. O. Mackie (ed.), Coelenterate ecology and behavior. Plenum, New York.

– 1977a. Effects of sea anemones on the flow forces they encounter. J. Exp. Biol. 69:87–105.

– 1977b. Mechanical organization of cantilever-like sessile organisms: Sea anemones. J. Exp. Biol. 69:127–142.

– 1979. Stiffness or extensibility of intertidal algae: a comparative study of modes of withstanding wave action. J. Biomech. 12:634.

– 1982. The interaction of moving water and sessile organisms. Sci. Am. 247:124–134.

– 1984. How do benthic organisms withstand moving water? Am. Zool. 24:57–70.

Koehl, M. A. R., and S. A. Wainwright. 1977. Mechanical adaptations of a giant kelp. Limnol. Oceanogr. 22:1067–1071.

– In press. Biomechanics. *In* M. M. Littler (ed.), Ecological field methods: macroalgae, phycological handbook, vol. 4. Cambridge University Press, London.

Kristensen, I. 1968. Surf influence on the thallus of fucoids and the rate of dessication. Sarsia 34:69–82.

Littler, M. M. 1979. Morphological form and photosynthetic performance of marine macroalgae: tests of a functional/form hypothesis. Botanica Mar. 22:161–165.

Littler, M. M., and D. S. Littler. 1980. The evolution of thallus form and survival strategies in benthic marine macroalgae: field and laboratory tests of a functional form model. Am. Nat. 116:25–44.

Littler, M. M., D. S. Littler, and P. R. Taylor. 1983. Evolutionary strategies in a tropical barrier reef system: functional form groups of marine macroalgae. J. Phycol. 19:229–237.

Lobban, C. S. 1978. Growth of *Macrocystis integrifolia* in Barkley Sound, Vancouver Island, B. C. Can. J. Bot. 56:2701–2711.

Lock, M. A., and P. H. John. 1979. The effect of flow patterns on uptake of phosphates by river periphyton. Limnol. Oceanogr. 24:376–383.

Lubchenco, J. 1978. Plant species diversity in a marine intertidal community: Importance of herbivore food preference and algal competitive abilities. Am. Nat. 112:23–39.

Lubchenco, J., and J. Cubit. 1980. Heteromorphic life histories of certain marine algae as adaptations to variations in herbivory. Ecology 61:676–687.

McCandless, E. L. 1981. Polysaccharides of the seaweeds. *In* C. S. Lobban and M. J. Wynne (eds.), The biology of the seaweeds. Botanical monographs, vol. 17. University of California Press, Berkeley.

Mann, K. H. 1971. Relation between stipe length, environment, and the taxonomic characters of *Laminaria.* J. Fish. Res. Bd. Can. 28:778–780.

– 1973. Seaweeds: their productivity and strategy for growth. Science 182:975–981.

– 1982. Ecology of coastal waters, a systems approach. Studies in ecology, vol. 8. University of California Press, Berkeley.

Mathieson, A. C., T. A. Norton, and M. Neushul. 1981. The taxonomic implications of genetic and environmentally induced variations in seaweed morphology. Bot. Rev. 47:313–347.

Matsumoto, F. 1959. Studies on the effect of environmental factors on the growth of nori (*Porphyra tenera* Kjellmi) with special reference to water current. Hiroshima Univ. Fac. Fish. Anim. Husb. 2:329–332.

Munk, W. H., and G. A. Riley. 1952. Absorption of nutrients by aquatic plants. J. Mar. Res. II:215–240.

Neushul, M. 1972. Functional interpretation of benthic marine algal morphology. Pp. 47–74 *in* I. A. Abbott and M. Kurogi (eds.), Contributions to the systematics of benthic marine algae of the North Pacific. Japanese Society of Phycology, Kobe.

Nicholson, N. L. 1970. Field studies on the giant kelp *Nereocystis*. J. Phycol. 6:177–182.

Norton, T. A., A. C. Mathieson, and M. Neushul. 1981. Morphology and environment. Pp. 421–451 *in* C. S. Lobban and M. J. Wynne (eds.), The biology of seaweeds. Botanical monographs, vol. 17. University of California Press, Berkeley.

– 1982. A review of some aspects of form and function in seaweeds. Botanica Mar. 25:501–510.

Nowell, A. R. M., and P. A. Jumars. 1984. Flow environments of aquatic benthos. Ann. Rev. Ecol. Syst. 15:303–328.

Padilla, D. K. 1982. Limpet radulae as tools for removing tissue from algae with different morphologies. Am. Zool. 22:968.

Paine, R. T. 1979. Disaster, catastrophe, and the local persistence of the sea palm, *Postelsia palmaeformis*. Science 205:685–687.

Pasciak, W. J., and J. Gavis. 1974. Transport limitation of nutrient uptake in phytoplankton. Limnol. Oceanogr. 19:881–888.

Peterson, C. H., H. C. Summerson, and P. B. Duncan. 1984. The influence of seagrass cover on population structure and individual growth rate of a suspension-feeding bivalve, *Mercenaria mercenaria*. J. Mar. Res. 42:123–138.

Popov, E. P. 1968. Introduction to the mechanics of solids. Prentice-Hall, Englewood Cliffs, N.J.

Preston, R. D. 1974. The physical biology of plant cell walls. Chapman & Hall, London.

Ramus, J. 1978. Seaweed anatomy and photosynthetic performance: the ecological significance of light guides, heterogeneous absorption and multiple scatter. J. Phycol. 14:352–362.

Raupach, M. R., and A. S. Thom. 1981. Turbulence in and above plant canopies. Ann. Rev. Fluid Mech. 13:97–129.

Riedl, R. 1971. Water movement: animals. Pp. 1123–1156 *in* O. Kinne (ed.), Marine ecology, vol. 1, part 2. Wiley, New York.

Roark, R. J., and W. C. Young. 1975. Formulas for stress and strain, 5th ed. McGraw-Hill, New York.

Russell, G. 1978. Environment and form in the discrimination of taxa in brown algae. Pp. 339–369 *in* D. E. G. Irvine and J. H. Price (eds.), Modern approaches to the taxonomy of red and brown algae. Systematics Association special volume no. 10. Academic Press, New York.

Santelices, B., J. C. Castilla, J. Cancino, and P. Schmiede. 1980. Comparative ecology of *Lessonia nigrescens* and *Durvillea antarctica* (Phaeophyta) in central Chile. Mar. Biol. 59:119–132.

Schlichting, J. 1968. Boundary layer theory. McGraw-Hill, New York.

Schumacher, G. J., and L. A. Whitford. 1965. Respiration and ^{32}P uptake in various species of freshwater algae as affected by current. J. Phycol. 1:78–80.

Schwenke, H. 1971. Water movement: plants. Pp. 1091–1121 *in* O. Kinne (ed.), Marine ecology, vol. 1, part 2. Wiley, New York.

Sellen, D. B. 1980. The mechanical properties of plant cell walls. Pp. 315–329 *in* J. F. V. Vincent and J. D. Currey (eds.), The mechanical properties of biological materials, 34th symposium of the Society for Experimental Biology. Cambridge University Press, London.

Slocum, C. J. 1980. Differential susceptibility to grazers in two phases of an intertidal alga: advantages of heteromorphic generations. J. Exp. Mar. Biol. Ecol. 46:99–110.

Smith, F. A., and N. A. Walker. 1980. Photosynthesis by aquatic plants: the effects of unstirred layers in relation to assimilation of CO_2 and HCO_3 and to carbon isotope discrimination. New Phytol. 86:245–259.

Sousa, W. P. 1979. Experimental investigations of disturbance and ecological succession in a rocky intertidal algal community. Ecol. Monogr. 49:227–254.

Sperling, J. A., and R. Grunewald. 1969. Batch culturing of thermophilic benthic algae and phosphorous uptake in a laboratory stream model. Limnol. Oceanogr. 14:944–949.

Sturdy, G., and W. H. Fischer. 1966. Surface tension of slick patches near kelp beds. Nature 21:951–952.

Sundene, O. 1964. The ecology of *Laminaria digitata* in Norway in view of transplant experiments. Nytt. Mag. Bot. 11:83–107.

Timoshenko, S. P., and J. M. Gere. 1972. Mechanics of materials. D. Van Nostrand, New York.

Vadas, R. L. 1972. Ecological implications of culture studies on *Nereocystis luetkeana*. J. Phycol. 8:196–203.

Van Dorn, W. G. 1953. Wind stress on an artificial pond. J. Mar. Res. 12:249–276.

Velimirov, B., and C. L. Griffiths. 1979. Wave-induced kelp movement and its importance for community structure. Bot. Mar. 22:169–172.

Vincent, J. F. V. 1982. Structural biomaterials. Wiley, New York.

Vines, R. G. 1960. The damping of water waves by surface films. Austral. J. Phys. 13:43–51.

Vogel, S. 1981. Life in moving fluids: the physical biology of flow. Willard Grant Press, Boston.

– 1984. Drag and flexibility in sessile organisms. Am. Zool. 24:37–44.

Wainwright, S. A., W. D. Biggs, J. M. Currey, and J. D. Gosline. 1976. Mechanical design in organisms. Wiley, New York.

Wainwright, S. A., and M. A. R. Koehl. 1976. The nature of flow and the reaction of benthic cnidaria to it. Pp. 5–21 *in* G. O. Mackie (ed.), Coelenterate ecology and behavior. Plenum, New York.

Wells, C. S. (ed.). 1969. Viscous drag reduction. Plenum Press, New York.

Westlake, D. F. 1967. Some effects of low-velocity currents on the metabolism of aquatic macrophytes. J. Exp. Bot. 18:187–205.

Wheeler, W. N. 1980a. Effect of boundary layer transport on the fixation of carbon by the giant kelp *Macrocystis pyrifera*. Mar. Biol. 56:103–110.

– 1980b. Laboratory and field studies of photosynthesis in the marine crop plant *Macrocystis*. Proc. Int. Seaweed Symp. 8:264–272.

Wheeler, W. N., and M. Neushul. 1981. The aquatic environment. Pp. 229–247 *in* O. L. Lange, P. S. Nobel, C. B. Osmond, and H. Ziegler (eds.),

Physiological plant ecology, I, vol. 12A, Encyclopedia of plant physiology, new series. Springer-Verlag, New York.
Whitford, L. A. 1960. The current effect and growth of fresh-water algae. Trans. Am. Micros. Soc. 79:302–309.
Whitford, L. A., and C. S. Kim. 1966. The effect of light and water movement on some species of marine algae. Rev. Algal 8:251–254.
Whitford, L. A., and G. J. Schumacher. 1961. Effect of current on mineral uptake and respiration by a freshwater alga. Limnol. Oceanogr. 6:423–425.
– 1964. Effect of current on respiration and mineral uptake in *Spirogyra* and *Oedogonium*. Ecology 45:168–170.
Wiegel, R. L. 1964. Oceanographical engineering. Prentice-Hall, Englewood Cliffs, N.J.
Witman, J. D., and T. H. Suchanek. 1984. Mussels in flow: drag and dislodgement by epizoans. Mar. Ecol. Prog. Ser. 16:259–268.

19 Functional geometry of seaweeds: ecological consequences of thallus layering and shape in contrasting light environments

MARK E. HAY

Introduction

In this study I apply Horn's model (1971) for adaptive patterns of light capture in forest trees to predict how thallus shape and layering affect growth and competitive ability in seaweeds as a function of light intensity. Seaweeds present a number of interesting evolutionary problems regarding energy capture. They display considerable intraspecific, interspecific, and interclass variations in photochemistry, stature, thallus shape, and branching pattern (Lobban and Wynne 1981; Ramus and Van der Meer 1983) (Table 19.1 and Figure 19.1) and occur throughout a wide range of habitats that differ greatly in physical characteristics.

In recent years, ecologists have begun to investigate the functional significance of seaweed growth form in regard to blade-level photosynthesis (Littler and Littler 1980; Lubchenco and Cubit 1980; Hay 1981a; Norton et al. 1981; Littler and Arnold 1982; Taylor and Hay 1984) (see Chapters 14 and 18), resistance to mechanical stress in various flow regimes (Neushul 1972; Koehl and Wainwright 1977; Koehl 1979; Norton et al. 1981), and herbivore defense and/or mutualism (Littler and Littler 1980; Lubchenco and Cubit 1980; Slocum 1980; Hay 1981a; Steneck 1982, 1983; Steneck and Watling 1982; Littler, Taylor, and Littler 1983). The morphological plasticity of seaweeds, which is often viewed merely as a taxonomic nemesis, was beneficial in many of these studies, allowing experimental investigation of the selective factors influencing seaweed growth form. Several of these studies pointed out that evolution of the forms that decrease losses to herbivores or physical stresses often mandates large tradeoffs in photosynthetic rate. For example, seaweeds with upright blades that have large surface-to-volume ratios are highly productive, but also very susceptible to removal by herbivores. Increasing the amount of structural material relative to photosynthetic material (Littler and Littler 1980, 1983), growing as a crustose form that is tightly adherent to the

Table 19.1. *Examples of morphological types discussed in this study*[a]

Growth form	Species	Class	Family	Reference[b] showing an illustration
I. Monolayers with flat, opaque thalli	*Weberella peltata*[c]	Rhodophyta	Rhodymeniaceae	(1) p. 771
	Maripelta rotata	Rhodophyta	Rhodymeniaceae	(2) p. 548
	Sciadophycus stellatus	Rhodophyta	Rhodymeniaceae	(2) p. 548
	**Gracilaria* species 1	Rhodophyta	Gracilariaceae	(5) p. 65; both are similar to Figure 19.1B
	**Gracilaria* species 2	Rhodophyta	Gracilariaceae	
	Johnson-sea-linkia profunda	Chlorophyta	Codiaceae?	(3) pp. 2–3
	Udotea cyathiformis	Chlorophyta	Udoteaceae	(1) p. 707
II. Multilayers with translucent thalli	*Halymenia* species	Rhodophyta	Cryptonemiaceae	(2) pp. 426–428
	Porphyra species	Rhodophyta	Bangiaceae	(2) pp. 295–305
	**Ulva* species	Chlorophyta	Ulvaceae	(2) p. 80
	**Dictyota* species	Phaeophyta	Dictyotaceae	(1) pp. 723–727
	**Padina* species	Phaeophyta	Dictyotaceae	(1) p. 731
III. Multilayers with flat but narrowly dissected thalli	*Gelidiella acerosa*	Rhodophyta	Gelidiaceae	(1) p. 755
	**Gracilaria domingensis*	Rhodophyta	Gracilariaceae	(1) p. 776
	Caulerpa mexicana	Chlorophyta	Caulerpaceae	(1) p. 687
	**Sargassum* species	Phaeophyta	Sargassaceae	(1) pp. 737–743
IV. Multilayers with midribs supporting thin blades	*Hypoglossum* species	Rhodophyta	Delesseriaceae	(1) p. 799
	Cryptonemia luxurians	Rhodophyta	Grateloupiaceae	(1) p. 779
	Dictyopteris species	Phaeophyta	Dictyotaceae	(1) pp. 727–729
	**Sargassum* species	Phaeophyta	Sargassaceae	(1) pp. 737–743
V. Multilayers with thin, terete branches	**Solieria tenera*	Rhodophyta	Solieriaceae	(4) p. 33
	**Spyridia aculeata*	Rhodophyta	Ceramiaceae	(1) p. 805
	Hypnea species	Rhodophyta	Hypneaceae	(1) p. 809
	Cladophora species	Rhodophyta	Cladophoraceae	(1) p. 669

[a] Asterisk indicates species investigated in this study.
[b] Code for references: (1) Taylor (1960); (2) Abbott and Hollenberg (1976); (3) Eiseman and Earle (1983); (4) Gabrielson and Hommersand (1982); (5) Hay and Norris (1984).
[c] Taylor (1960) lists *Weberella peltata* as *Fauchea peltata*.

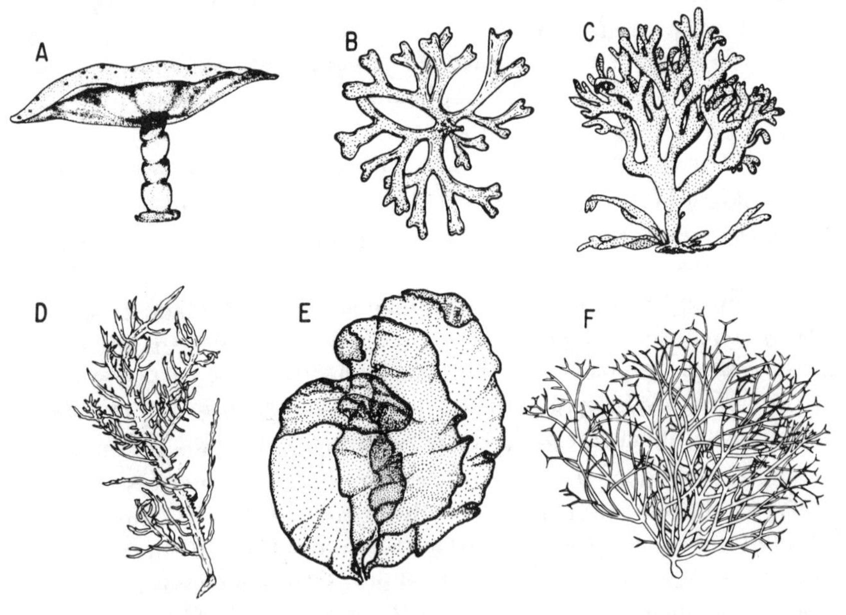

Figure 19.1. A variety of seaweed morphologies. (A) and (B) are flat monolayered species with opaque thalli. If species with flat thalli are multilayered, they should have translucent thalli (E), thalli that are oriented vertically (C), or thalli that are highly dissected (D). In situation C, thalli in the upper most portions of the plant should be vertical; those in lower (and thus more shaded) portions should be arranged more horizontally. Species with narrow, terete thalli should be multilayered (F). On the sand plain at Galeta, the most common species have morphologies similar to either B or F. *Gracilaria* species 1 and 2 are shaped like example B. *Solieria tenera* and *Spyridia aculeata* are shaped like F.

substrate (Steneck and Adey 1976; Steneck 1982, 1983), or growing in tightly compressed mats, or turfs (Hay 1981a), all have been shown to reduce losses to herbivores; however, all of these adaptations also involve reductions in photosynthetic rate. Several seaweeds that are heteromorphic appear to have minimized the morphological costs of herbivore deterrence. These species grow as highly productive, upright blades during seasons when herbivores are inactive, and they grow as herbivore-resistant, but slow-growing, crustose forms during seasons when herbivores are actively feeding (Lubchenco and Cubit 1980; Slocum 1980; Littler and Littler 1983). Turf-forming seaweeds show a similar plasticity. When protected from desiccation, they grow as separated individuals or as loosely arranged aggregates. When exposed to desiccation, these species grow as tightly compacted aggregates, or turfs. When compared with the loosely arranged forms, the turf morphology reduces mortality and tissue loss associated with desiccating low tides (Hay 1981a; Taylor and Hay 1984) but also leads to a reduced rate of production that appears to result from increased self-shading and diffusion gradients. Although algal morphology clearly affects susceptibility to herbivores and to certain physical stresses, these interactions will not be considered in this investigation. Here we shall examine, both theoretically and experimentally, the implications of seaweed growth form for light capture, competitive ability, and algal distribution. Other implications of seaweed form are discussed in Chapters 14 and 18.

The central hypothesis of this chapter is that in well-lighted habitats, seaweeds that produce photosynthetic surface areas that greatly exceed the ground areas they shade (i.e., a high thallus/area ratio) will have a growth-rate advantage over seaweeds that have thallus/area ratios near unity. I assume that this growth-rate advantage will confer a competitive advantage and will be selected for in environments in which resources are limiting. Efficient production of high thallus/area ratios demands that seaweeds display multiple layers of thalli while minimizing the shade that upper layers cast on lower layers. I propose that many multilayered seaweeds have solved this conflict in one of two basic ways. First, they can be composed of multiple layers of branches with narrow diameters. This will minimize the distance that penumbras of upper branches are cast toward lower branches. Familiar examples are species of *Hypnea, Chondria, Spyridia, Solieria,* and *Dasya* (Figure 19.1F). Second, multilayered seaweeds can be composed of broad, flat thalli that are somewhat translucent; this allows light to penetrate to lower layers by passing directly through the thalli of upper layers. Examples of multilayered plants with this type of construction are *Ulva, Porphyra, Padina,* and *Dictyota* (Figure 19.1E). Either of these types of construction would allow upper layers of thalli to act as density filters, not as shades.

In habitats with very low light levels, any filtering of light by upper layers of thalli may cause lower thalli to be below the compensation point (where production just balances respiration). In these low-light environments, I hypothesize that seaweeds should be monolayered and optically dense. Optically dense thalli should allow plants to capture all of the light that strikes them. To avoid the self-shading that would be caused by overlapping thalli, monolayered plants should be composed of a very few blades that are broad and flat and are held perpendicular to the incoming light. Such plants would be shaped somewhat like umbrellas or like blades formed along the frame of an umbrella's ribs. Although the deepest seaweeds have rarely been studied, recent investigations using submersibles have shown that several of the most common upright seaweeds from deep habitats are indeed umbrella-shaped (Eiseman 1978; Eiseman and Earle 1983). *Johnson-sea-linkia profunda, Weberella* (= *Fauchea*) *peltata, Sciadophycus stellatus,* and *Maripelta rotata* are examples of this form (Figure 19.1A).

This study is restricted to the interrelationships among seaweed shape, growth rate, and competitive ability in environments of high or low light. Several other important aspects of plant morphology or physiology will not be considered here, or will be mentioned only briefly. For example, in seaweeds, external morphology works in concert with other characteristics such as chromatophore displacement (Nultsch and Pfau 1979), internal morphology and plastid layering (Colombo and Orsenigo 1977), pigment concentration and type (Peterson 1972; Ramus et al. 1976a, 1976b, 1977), and ratio of photosynthetic material to structural material (Littler and Littler 1980) to affect energy acquisition and thus fitness. The potential importance of these factors should not be forgotten.

Experimental investigation of the costs and benefits associated with various seaweed shapes is a relatively new field (Littler and Littler 1980). I shall therefore depend heavily on the more extensive terrestrial literature to provide a background for advancing hypotheses concerning the functional morphology of seaweeds.

Morphological patterns of terrestrial vegetation

Within a forest, tree species may have differently shaped leaves, some being entire, others lobed, and still others pinnate. Even on an individual tree, one finds that leaves in the direct sunlight are more divided or smaller than those from shaded portions of the tree (Vogel 1968; Horn 1971). Theories attempting to explain the adaptiveness of this variability have concentrated primarily on either (1) leaf temperature, CO_2 acquisition, water loss, and the effect on productivity (Vogel 1968, 1970; Taylor and Sexton 1970; Parkhurst and Loucks 1972; Taylor 1975; Givnish and Vermeij 1976; Orians and Solbrig 1977; Givnish 1978, 1984), including the cost of roots (Givnish 1979), or (2) leaf arrangement, self-shading, and their effects on productivity (Kramer and Clark 1947; Horn 1971; Givnish 1979, 1984).

Because the shapes of subtidal seaweeds will not be constrained by water loss, temperature of the thallus, or root production, these aspects will not be considered. Additionally, acquisition of carbon should not affect the shape of most subtidal seaweeds, because inorganic carbon is abundant relative to essential nutrients such as nitrogen, which often limits the rate of plant growth in marine habitats.

Kremer (1981) has recently reviewed carbon metabolism by seaweeds. Inorganic carbon is available in seawater as CO_2, H_2CO_3, HCO_3^-, and CO_3^{2-}. Many seaweeds utilize HCO_3^- as a carbon source, although some species have been reported to utilize only CO_2. At normal pH values of 7.8–8.2, HCO_3^- makes up more than 90% of the total inorganic carbon and occurs at a concentration of more than 2 mmol liter^{-1}. The concentration of unhydrated CO_2 is only about 10 μmol liter^{-1}. This can be compared with the situation for terrestrial plants that live in an atmosphere of 0.03% CO_2, or about 13 μmol liter^{-1} (assuming 20°C and a pressure of 1 bar). Although growth of seaweeds that utilize HCO_3^- should not be limited by carbon concentration under light-saturating conditions, seaweeds that must utilize CO_2 might experience carbon limitation, because the diffusion constant for CO_2 in still water is about 10^4 less than that in air. From the limited data set available (Kremer 1981), it appears that seaweeds that must use CO_2 occur primarily in the intertidal area, where they are often exposed to air during low tides. Most subtidal seaweeds, like those being examined here, utilize HCO_3^- as a carbon source. This suggests that uptake of carbon should rarely limit photosynthesis in subtidal seaweeds and should therefore have little effect on the evolution of thallus shape. However, low nutrient concentrations can limit algal growth, and this may affect the evolution of seaweed shape, because nutrients are taken up by the thallus, and the efficiency of uptake increases with increasing surface-to-volume ratio. Thus, the factors listed under item (1) should have less effect on seaweed shape than the considerations listed under item (2).

Horn (1971) analyzed the effects of canopy geometry and leaf shape and size on net income (and hence competitive ability) of plants as a function of light intensity. He contrasted plants with two different types of leaf arrangement: monolayers and multilayers. Monolayers have their leaves packed in a single shell around the perimeter of the crown; multilayers have their leaves scattered diffusely within the several layers of leaves held by the plant. Horn calculated the total rate of photosynthetic return accruing from idealized plants with each of these canopy arrangements, based on (1) the pattern of light interception, (2) the nonlinear photosynthetic response of individual leaves to light intensity, plateauing at moderate to high intensities, (3) the rate of dark respiration for completely shaded leaves, and (4) the light intensity at the top of the plant crown. In a monolayer, all leaves are exposed to the full intensity of light striking the top of the plant. If one considers a monolayer exposed to full sunlight, with unit crown area and saturation photosynthetic rate *P*, its total gain per unit area is just *P*. In a multilayer under the same conditions, only the top leaves are exposed to full sunlight, whereas the lower leaves are exposed to lesser light intensities and thus may photosynthesize at a lower rate. There are two central questions: (1) How many layers of leaves can a multilayer retain and still obtain a net profit on each? (2) Is the most profitable multilayer under a given set of conditions more productive than a monolayer occupying a comparable area of ground?

Critical to the first calculation are the facts that photosynthesis saturates at a small fraction of full sunlight [ca. 20% in many forest trees and 5–30% in seaweeds (Luning 1981)] and that gross photosynthesis just balances leaf respiration at some minimal light level, known as the compensation point (ca. 5% in many forest trees, and 0.3–1.5% of surface irradiance for many seaweeds). A final critical assumption is that each leaf layer acts as a uniform-density filter, passing a light intensity whose magnitude has been diminished by the fraction of area covered by leaves in that layer; this assumption is valid if the leaves are small enough (width of leaves, lobes, or leaflets less that 1% of the distance between "layers") so that their shadow penumbras largely diffuse together before the next lower layer is reached. The distance between layers may be less critical for marine than for terrestrial plants, because light scatter may be greater in subtidal environments.

As an example of the effect of layering, consider a tree with leaves that are photosynthetically saturated at 20% of full sunlight, reach compensation at 5% of full sunlight, and cover 50% of the canopy area within each layer. The uppermost layer of leaves will be photosynthetically saturated and will present a leaf area that is half that presented by a monolayer; light levels below the top layer will be 50% of full sunlight. The second layer of leaves will also be light-saturated and will decrease light levels to 25% of full sunlight. The third layer of leaves will also be saturated. The fourth and fifth layers will be below saturation intensity but above compensation. In this example, when monolayered and multilayered plants are compared, the multilayer has 50% more leaf area that is above light saturation, and 150% more leaf area that is above compensation.

Therefore, in an open, well-lighted habitat, a multilayered tree can contain more leaf area than a monolayered tree, thereby significantly increasing the rate of whole-plant energy acquisition. Under conditions of lower light, layering will not be energetically advantageous, and trees should grow as monlayers. Competitive interactions between monolayers and multilayers should be largely dependent on light levels. At high light levels, multilayers can expose more leaf area and grow faster than monolayers, but at low light levels the extra, shaded leaves will be an energetic

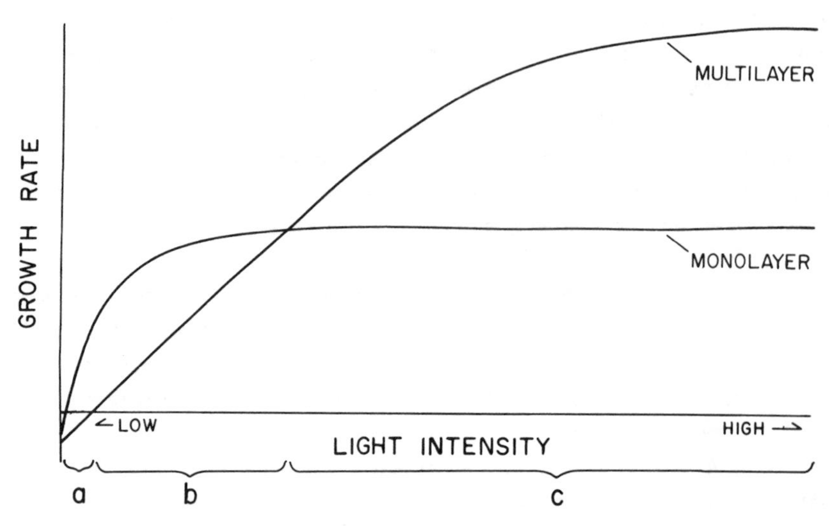

Figure 19.2. Hypothesized effects of light intensity on net photosynthesis for whole plants for monolayered and multilayered seaweeds (modified from Horn 1971). At light intensities in range (a), monolayers will decrease less rapidly than multilayers, or show positive growth while multilayers decline. In range (b), monolayers will be favored over multilayers. In range (c), multilayers will be favored over monolayers.

liability, and monolayered trees will be most efficient. This reasoning is depicted graphically in Figure 19.2. Data presented by Kramer and Clark (1947) and Horn (1971) are consistent with the theory but are largely correlative.

Functional morphology of seaweeds

Previous studies have shown that algal morphology can affect productivity, herbivore resistance, and resistance to various physical stresses (Odum et al. 1958; Steneck and Adey 1976; Koehl 1979; Littler 1980; Littler and Littler 1980; Lubchenco and Cubit 1980; Slocum 1980; Hay 1981a; Steneck 1982; Steneck and Watling 1982). These studies have not assessed how the arrangement and shape of algal thalli will affect productivity and competitive interactions under varying light regimes.

Thallus arrangement, shape, and transparency

For algae, productivity per unit biomass increases both with increasing surface-to-volume ratio and with increasing ratio of photosynthetic material to structural material (Odum et al. 1958; Littler 1980; Littler and Littler 1980; Littler and Arnold 1982). Thus, in the absence of counterselective forces, those algae that have the greatest proportion of their cells in direct contact with the physical environment should reap the greatest energetic rewards. Growing as uniserate filaments or as sheets that are one to two cell layers thick is one of the most obvious ways to do this. Although certain physical and biotic forces (waves, sand scouring, and grazers) may select for greater strength, and thus a thickening of the thallus, flattened thalli will generally have a greater surface-to-volume ratio than cylindrical thalli that are more than twice as thick. This will result in a growth-rate advantage for species with thin, flat thalli. Flat thalli have a second advantage in low-light environments, because a greater proportion of the cells can be oriented perpendicular to the light source.

Self-shading in multilayers will select against broad, opaque thalli and for thalli that are narrow (Figure 19.1F), highly dissected (Figure 19.1D), or broad but translucent (Figure 19.1E). If multilayered plants produced branches that were both narrow (consistent with minimizing self-shading) and thin (consistent with high productivity), these would be flaccid and would droop onto one another, resulting in increased shading. Branches that are both narrow and turgid can be held perpendicular to the main axis of the plant and arranged along the axis so as to minimize the shading of lower branches. This can be accomplished if the branches are cylindrical in cross section (= terete) (e.g., *Solieria* and *Hypnea*) or, if flattened, have a strengthening midrib (e.g., *Sargassum* and *Hypoglossum*). Under certain selective regimes, multilayered algae may evolve broad, thin blades (increasing surface area and the ability for nutrient uptake) that overlap (increasing biomass per unit area), so long as these blades are somewhat translucent and allow sufficient light to pass through to underlying thalli (e.g., *Ulva, Padina, Porphyra, Halymenia*). Table 19.1 lists the species studied here, as well as other common seaweeds, that are typical of these morphological types. Thus, monolayered seaweeds should have flat opaque thalli; multilayered seaweeds should have thalli that are terete, or, if flattened, the thallus should have a midrib or be translucent relative to monolayered species.

These contentions are difficult to assess adequately because few measurements of thallus area per unit of shaded surface have been made for seaweeds. However, I have subjectively classified the nonfilamentous, upright seaweeds from the Caribbean as monolayers (having a thallus-area index of about 1) or multilayers (having a thallus-area index of considerably more than 1) in order to allow a preliminary assessment of interactions between layering, thallus shape, and distribution of various morphologies

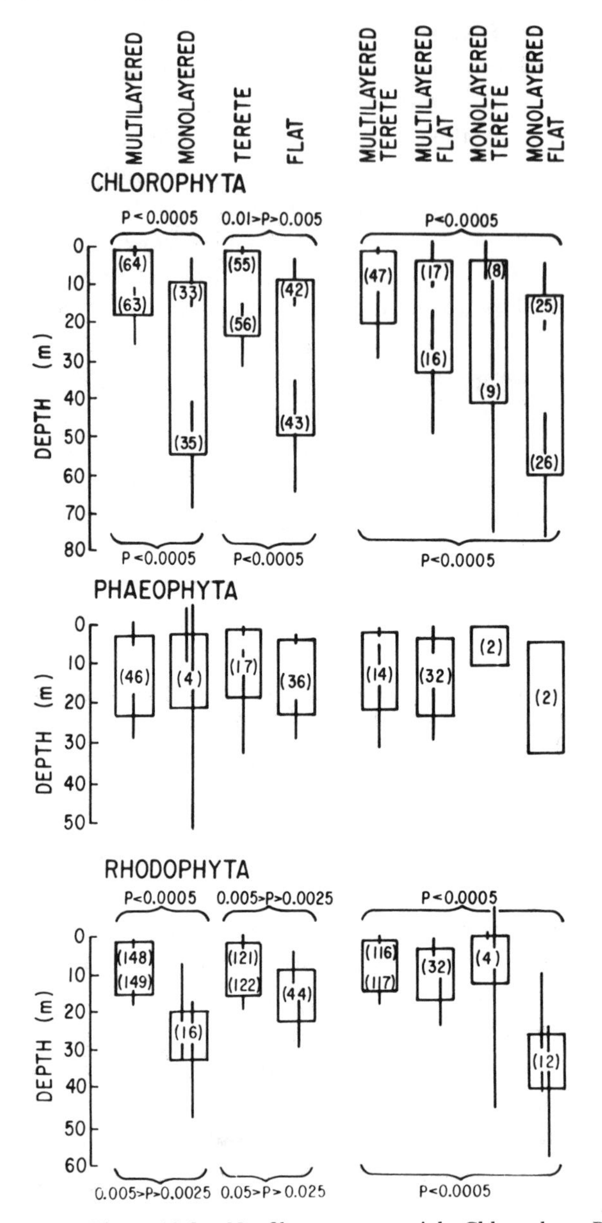

Figure 19.3. Nonfilamentous, upright Chlorophyta, Phaeophyta, and Rhodophyta for which Taylor (1960) gave depth data. Mean minimum and mean maximum depths for each morphological group are shown with 95% confidence interval (*P* values are by the Mann-Whitney U-test). Numbers in parentheses give sample sizes for each category.

along a depth gradient (Figure 19.3). I have restricted my data set to species listed in Taylor's *Marine Algae of the Eastern Tropical and Subtropical Coasts of the Americas* (1960). Filamentous species, the Bangiophyceae, simple encrusting forms, and species for which depth data were not given were omitted from consideration. No effort was made to include new species described from the Caribbean since 1960 or to update the depth ranges given by Taylor. Although new information is available for certain sites within the Caribbean, the inclusion of these sites could bias the data toward patterns that prevail only within a few unique areas, and I believe it is important to assess, as fairly as possible, the predominant patterns of a large area. The Caribbean flora was chosen because I am most familiar with the algae of that region, and it is often difficult to assign species to multilayered or monolayered categories unless one has seen them under various conditions in the field. My observations come from more than 1,000 SCUBA dives on reefs in Panama, Nicaragua, Honduras, Belize, Mexico, Haiti, the Bahamas, the United States and British Virgin Islands, and the Florida Keys.

Many algae (*Anadyomene stellata*, most of the Dictyotales, and several *Gracilaria* species are a few examples) occur as overlapping layers in shallow water and single layers in deeper water, as would be predicted by the previously outlined hypotheses. In order to be as conservative as possible in testing the hypotheses, these species are always listed as multilayers. In categorizing species with which I was unfamiliar, I consulted several descriptions whenever possible and did not depend completely on Taylor (1960). Plants described as "bushy" or "highly gregarious" were always listed as multilayers. In cases in which I could not be reasonably certain of the degree of layering, I omitted the species from all data sets used to assess the effects of layering (Hay 1980).

For both the Chlorophyta and Rhodophyta (Table 19.2) there is a significant association between thallus form and degree of layering ($P < 0.005$, chi-square test). Seventy-four percent of the monolayered Chlorophyta and 75% of the monolayered Rhodophyta have flattened thalli; 74% of the multilayered Chlorophyta and 79% of the multilayered Rhodophyta have terete thalli (Table 19.2).

The pattern does not hold for the Phaeophyta. One reason for this might be that many species in the Phaeophyta have flattened thalli that are relatively translucent (e.g., *Dictyota, Padina, Dictyopteris*); this may account for the high proportion of this group (70%) that are both flat and multilayered. Measuring the percentage of photosynthetically active light that passes through the thalli of flat species found near Carrie Bow Cay, Belize (Table 19.3), shows that flat monolayered species are relatively opaque

Table 19.2. *Numbers and proportions of multilayered and monolayered species that have flattened thalli and terete thalli*[a]

Thalli	Monolayered number of species (%)	Multilayered number of species (%)	Chi-Square Analysis
Chlorophyta			
Flattened	26 (74)	17 (26)	$P < 0.001$
Terete	9 (26)	47 (74)	
Phaeophyta			
Flattened	2 (50)	32 (70)	$0.50 > P > 0.25$
Terete	2 (50)	14 (30)	
Rhodophyta			
Flattened	12 (75)	32 (21)	$P < 0.001$
Terete	4 (25)	117 (79)	

[a] All data are derived from Taylor (1960). Multilayered Chlorophyta and Rhodophyta were significantly ($P < 0.005$, chi-square test) skewed toward rounded thalli. Other ratios were not significantly different from a 50 : 50 proportion.

(transmitting 2–12% of the ambient light) and that flat multilayered species are relatively translucent (transmitting 19–70%). Differences in transmission by monolayered and multilayered species were significant ($P < 0.05$, Mann-Whitney U-test).

In addition, assessing the proportions of flat monolayered and flat multilayered Rhodophyta in the Caribbean that have midribs shows that 20% (8 of 39) of the multilayered species have midribs, whereas none (0 of 21) of the monolayered species have midribs. Thus, midribs are significantly more common in multilayers ($P = 0.024$, Fisher's exact test), suggesting that midribs function to separate and orient thin blades in order to decrease self-shading. The Chlorophyta do not produce midribs, and the Phaeophyta had only four monolayered species; this precludes analysis for interrelationships between midribs and layering in these classes.

Among Caribbean seaweeds, only *Sargassum* has leaf-like blades attached to an upright stem. In gross morphology, *Sargassum* species are similar to many terrestrial plants and thus might be expected to have small blades (sun leaves) in shallow, well-lit waters and large blades (shade leaves) in deep waters. Indeed, the maximum width of *Sargassum* blades shows a significant positive correlation (Figure 19.4) to the maximum depth of occurrence for each species listed by Taylor (1960). Species that occur in deeper, and presumably darker, habitats have larger blades than species that grow in shallow, and presumably well-lit, areas. Within a plant, upper leaves are also smaller than lower leaves (Taylor 1960). Both of these

Table 19.3. *Proportions of ambient photosynthetically active sunlight* (μE m^{-2} *sec*$^{-1}$) *that pass through different algal species with flattened thalli*[a]

Species	N	Monolayered	Multilayered	Mean percentage transmitted	Standard deviation
Udotea flabellum	8	×		2.0	0.5
Avrainvillea levis	14	×		2.5	1.1
Lobophora variegata	10	×		12.1	3.4
Padina sanctae-crucis	8		×	19.0	5.0
Stypopodium zonale	10		×	27.3	4.4
Sargassum species	10		×	30.4	7.6
Anadyomene stellata	10		×	43.9	7.9
Dictyota bartayresii	10		×	61.0	9.1
Ulva species	10		×	70.4	11.0

[a] Transmission was determined in the field by laying algal fronds over the quantum sensor of a Licor 185 quantum meter. Significantly less light passes through the monolayered species ($P < 0.05$, Mann-Whitney U-test).

patterns are similar to those seen in terrestrial plants (Vogel 1968; Horn 1971; Givnish and Vermeij 1976). In terrestrial habitats, small sun leaves and larger shade leaves are hypothesized to result from selection for maintenance of lower leaf temperature or increased water-use efficiency. In subtidal marine communities, neither of these factors can be operating.

Field tests of the hypotheses

Leaf size, thallus transparency, and presence or absence of midribs may affect the potential productivity of many seaweeds. However, most of the data presented so far are strictly correlative, and several of the hypotheses presented may not be amenable to direct experimentation.

For the remainder of this chapter, I shall concentrate on the interrelationships of flat monolayered and terete multilayered morphologies with (1) growth rate, (2) photosynthetic physiology, (3) responses to seasonally impressed gradients in light intensity, (4) colonization ability, and (5) competitive ability. To evaluate these interrelationships, I draw on an intensive study of two monolayered and two multilayered seaweed species that grow on subtidal sand plains in the tropical Caribbean.

Study site and organisms

The study was conducted between December 1977 and September 1979 at the Smithsonian Tropical Research Institute's Galeta

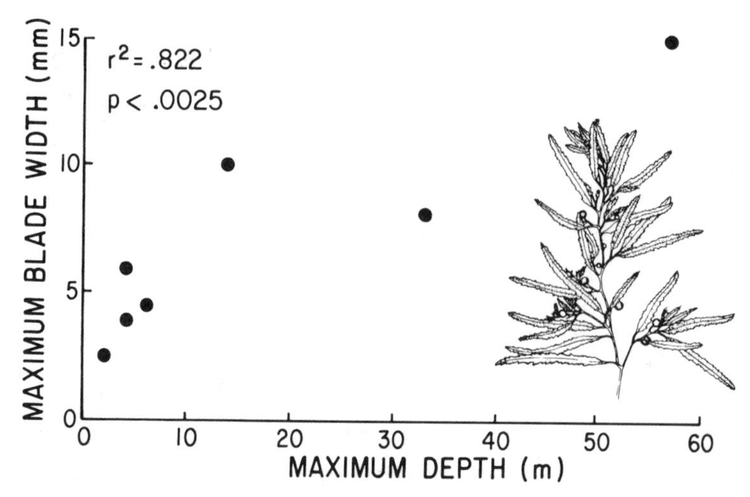

Figure 19.4. Maximum blade width versus maximum depth of occurrence for the seven species of *Sargassum* for which Taylor (1960) gave depth data (*P* and *r* values are for the correlation analysis).

Point Laboratory located on the Caribbean coast of Panama at 9°24′ N, 79°52′ W. The reef at Galeta is representative of fringing reefs found along the northern coast of Panama (Glynn 1972; MacIntyre and Glynn 1976). At a depth of 10–14 m, the reef abuts an extensive sand plain that includes sparse fragments of hard substrate to which upright algae are attached. These hard substrates cover only 4% of the area of the sand plain (Hay 1981b). Red seaweeds, such as *Gracilaria, Solieria, Spyridia,* and *Halymenia,* compose most of the cover and are characteristically attached to small coral or shell fragments partially buried in the sand. Siphonaceous green seaweeds such as *Halimeda* and *Udotea,* which can anchor directly in the sand, are present, but relatively uncommon. Algal density and percentage cover are limited both by the scarcity of hard substrates and by low light levels (Hay 1981b).

During the dry season, usually December–April, winds blow consistently from the north at 24–27 km hr^{-1} (Hendler 1976), which is about three times the mean wet-season velocity. Waves generated by these strong, unidirectional winds cause extreme turbidity that reduces light penetration markedly, leaving the sand plain in near darkness for a majority of this time. During the turbid dry season, mean midday levels of photosynthetically active light ($\mu E\ m^{-2}\ sec^{-1}$) measured at 20 permanently marked locations scattered throughout the sand-plain study area are about 60% below ($\bar{x} = 33$, SE = 12, $N = 29$ separate days) levels recorded during the wet season ($\bar{x} = 81$, SE = 26, $N = 12$). Light measure-

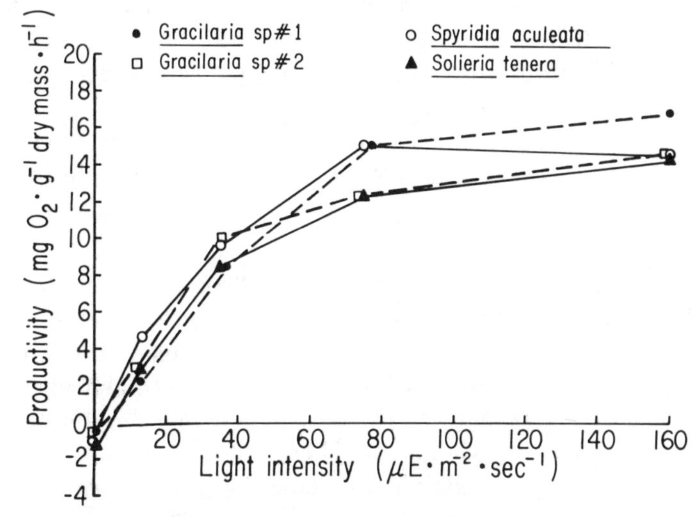

Figure 19.5. Apparent photosynthesis as a function of light quantity for two common flat monolayered (*Gracilaria* species 1 and 2) and two common terete multilayered (*Solieria tenera* and *Spyridia aculeata*) species from the Galeta sand plain. All thalli were incubated in a nonoverlapping arrangement so that differences in layering would not confound determination of saturation intensities. $N = 6$ for each point. At 13 μE m^{-2} sec^{-1}, the productivity of *Spyridia* was significantly higher than that for any of the other species. Additionally, the dark respiration rates for both *Spyridia* and *Solieria* were significantly higher than that for either of the *Gracilaria* species ($P < 0.05$, ANOVA and Student-Newman-Keuls tests). There were no other significant differences.

ments for the dry season are biased toward high readings, because all measurements were taken while SCUBA diving, and dives were generally made only on days when conditions were relatively calm. During the few dives conducted under the normally rough conditions of dry season, light levels were so low that I could not see my light meter even when holding it against my mask. Because I can easily see and make readings at light levels of 1 μE m^{-2} sec^{-1}, light levels must be at, or near, zero for much of this time. For the six sand-plain species that have been studied (Hay 1981b), photosynthetic saturation occurs at 80 – 100 μE m^{-2} sec^{-1}, and compensation occurs at 2 – 6 μE m^{-2} sec^{-1} (Figure 19.5). Thus, during the relatively calm wet season, midday light levels on the sand plain are usually at, or above, photosynthetic saturation. During the turbid dry season, midday light levels may average as much as one-third of saturation but will often be below compensation. Percentage cover of seaweeds on the sand plain may decrease by as much as 80% during this turbid season (Hay 1981b). In some years, a short period of windy weather, very like that of the dry season, occurs in July or August and markedly increases turbidity; I shall refer to this period as the short dry season. Other physical factors, such as water temperature and salinity, change very little between seasons (Hendler 1976, 1977). Nutrient limitation in this community seems relatively unimportant compared with light limitation, because growing plants at 3 m versus 9 m deep can result in an increase of as much as 400-fold in algal abundance (Hay 1981b), even though the plants are only 6 m apart and the water column is well mixed, preventing an accumulation of nutrients at one depth.

Herbivorous fish (primarily Scaridae and Acanthuridae) and sea urchins (primarily *Diadema, Echinometra,* and *Eucidaris*), which occur in abundance on the reef, are rare in the sand-plain habitat. During a typical dive on the reef slope, one may see hundreds of herbivorous fish and many sea urchins. If the same time is spent on the sand plain, a diver will usually encounter no more than three immature herbivorous fish and no sea urchins. When palatable seaweeds are simultaneously transplanted onto the sand plain and nearby reef slope, those individuals placed on the reef slope are consumed at one to several orders of magnitude faster than those placed on the sand plain (Hay 1981b; Hay et al. 1983). The sand plain thus provides seaweeds a spatial escape from the high herbivore activity that is typical of most tropical marine habitats.

Of the five most abundant seaweeds on the sand plain, two (*Gracilaria* species 1 and 2) are flat monolayers, and two (*Solieria tenera* and *Spyridia aculeata*) are terete multilayers (Table 19.1 and Figure 19.1). All four species are in the class Rhodophyta and thus have similar photochemistries. I shall use these species to compare changes in abundance and potential interaction between these forms as light availability changes seasonally.

Growth rate and seasonality

If flat monolayered species and terete multilayered species saturate at similar light levels and exhibit similar photosynthetic rates at saturation, then experiments can be designed to assess how shape, as opposed to physiology, affects growth under different light levels. Photosynthetic rates at different light levels were determined in the laboratory using freshly collected plants. All plants were collected at dawn and incubated in a series of increasing light intensities ranging from 0 to 160 μE m^{-2} sec^{-1}. All thalli were incubated in a nonoverlapping arrangement so that differences in layering would not confound determination of blade-level saturation or compensation intensity. Each sample was constantly stirred by magnetic stirring bars, and oxygen determinations were made using a YSI model 58 dissolved-oxygen meter and polarographic probe.

Relationships between photosynthesis and light level show few significant differences for the two most common terete multilayers (*Solieria tenera* and *Spyridia aculeata*) and the two most common flat monolayers (*Gracilaria* species 1 and 2) (Figure 19.5). The multilayered species, *Spyridia aculeata,* was slightly, but significantly, more productive than other species at 13 $\mu E\ m^{-2}\ sec^{-1}$, and both *Spyridia* and *Solieria* exhibited respiration rates that were slightly, but significantly, higher than those of the monolayered *Gracilaria* species ($P < 0.05$, ANOVA and Student-Newman-Keuls test). Therefore, small differences in growth rates might result from physiological differences between species. If large differences occur, these should depend in large part on the relationship between morphology and ambient light levels. At high light levels, terete multilayers should be able to expose more thallus area and grow faster than flat monolayers.

As light decreases toward photosynthetic compensation (about 2–6 $\mu E\ m^{-2}\ sec^{-1}$ for these species), the cost of self-shading in terete multilayers should outweigh the benefit of increased thallus area. If the waves and sand movement that accompany the dry season cause frond loss due to breakage or burial, then the ecological compensation point (*sensu* Givnish 1984) may be considerably higher than the photosynthetic compensation point documented here. Later in this chapter, I shall show that multilayers may be more susceptible to some of these physical stresses than are monolayers; this should result in an increased ecological compensation point for the multilayers. Therefore, at low light levels, terete multilayers should decrease rapidly, while flat monolayers maintain their populations or even grow (part a of Figure 19.2).

Approximately monthly assessments of percentage cover and number of individuals per square meter for 100 1.0-m^2 quadrants located on the sand plain (Hay 1981b) showed that the terete multilayered species grow much faster than flat monolayered species during periods of increased water clarity (Figure 19.6). Between April and June, the coverages of *Solieria* and *Spyridia* increased 18- and 11-fold, respectively; during this same time period, coverage of *Gracilaria* species 1 doubled, and *Gracilaria* species 2 increased by only 50%. Although monolayered species grew slower than multilayered forms when light was abundant, they also showed very little change in cover (+11% to −34%) during the turbid dry season (November–April) when the two multilayered species decreased by 96% and 97%. During the small dry season (June–August), this pattern was repeated; cover of the monolayers decreased by only 12% and 18%, and cover of the multilayers decreased by 80% and 93% (Figure 19.6). Thus,

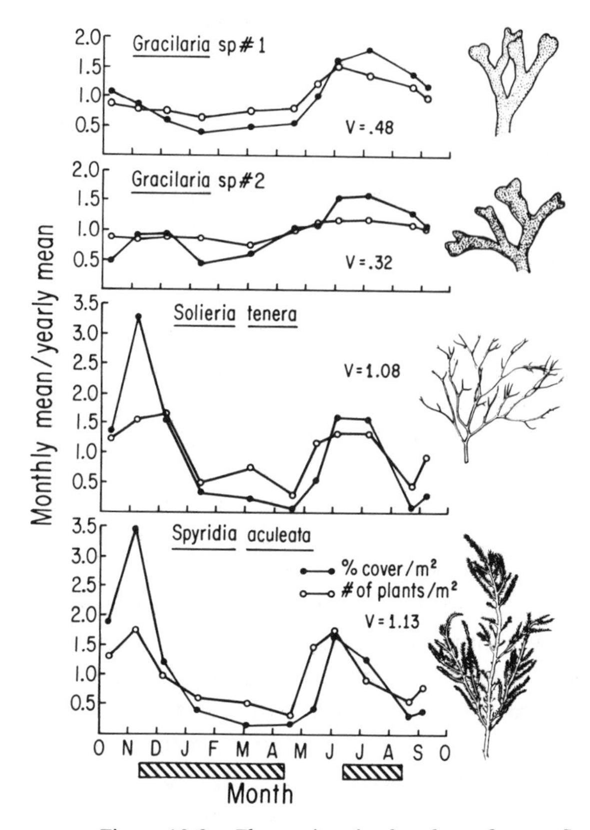

Figure 19.6. Fluctuations in abundance for two flat monolayered (*Gracilaria* species 1 and 2) and two terete multilayered (*Solieria tenera* and *Spyridia aculeata*) species on the Galeta sand plain during 1978–79. The coefficient of variation (standard deviation/mean = V) was calculated from percentage coverage data. Timings of the turbid dry season and short dry season are indicated by the striped bars below the x axis. Illustrations to the right of each graph show individual branches of each species.

flat monolayered species are relatively resistant to the seasonal stress of low light levels; however, they are also unable to make efficient use of higher light levels when these occur. Terete multilayered species are not resistant to low light levels, but they are resilient and thus show rapid recovery following the return of adequate light. Differences in the degrees to which these forms are affected by seasonally changing light levels are summarized by the coefficient of variation (V) calculated using monthly measurements of percentage cover. For the flat monolayers, $V = 0.32$ and 0.48; for the terete multilayers, $V = 1.08$ and 1.13.

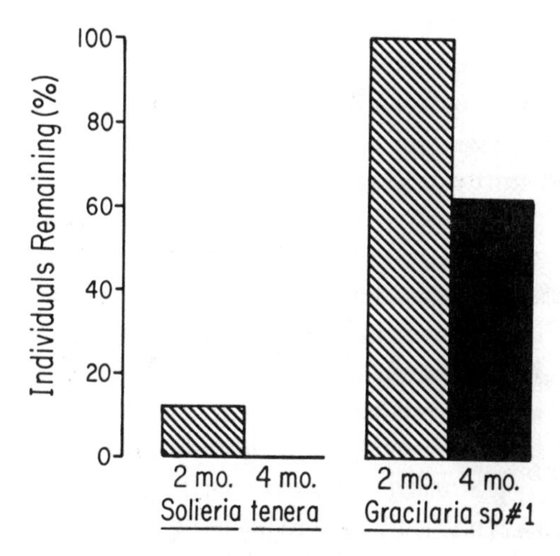

Figure 19.7. Percentages of individuals still alive after two months and four months of burial 15 cm below the surface of the sand plain at Galeta. The flat monolayered species (*Gracilaria*) was very resistant to damage. The terete multilayered species (*Solieria*) exhibited little resistance. $N = 10$ for each species during each period. During both periods, *Solieria*'s tolerance to burial was significantly less than that of *Gracilaria* ($P < 0.05$, Fisher's exact test).

Physiology

The two major disturbance agents affecting seaweeds on the sand plain appear to be long periods of darkness caused by wave-generated turbidity and periodic burial caused by wave-generated sand movement. If monolayered species are to persist during turbid periods, they must not only be more efficient than multilayered species when light levels are low but also be more tolerant of burial and long periods of darkness.

Effects of burial beneath the sand plain were assessed by placing 10 coral fragments with *Solieria tenera* and 10 with *Gracilaria* species 1 in lengths of three-stand rope and burying these 15 cm below the sand surface. After two months and four months, the ropes were unearthed and placed on the surface of the sand plain, and notes were taken on the condition of the plants. The ropes were monitored again after two weeks to check for plant growth or regrowth from basal sections that were not apparent when the ropes were first uncovered. A sterilized coral fragment had been placed next to each occupied fragment before burial so that regrowth from cryptic basal sections could be differentiated from settlement and growth following the unearthing. The monolayer, *Gracilaria* species 1, was resistant to damage, but the multilayer, *Solieria tenera,* showed large losses (Figure 19.7). After both two months and four months of burial, significantly

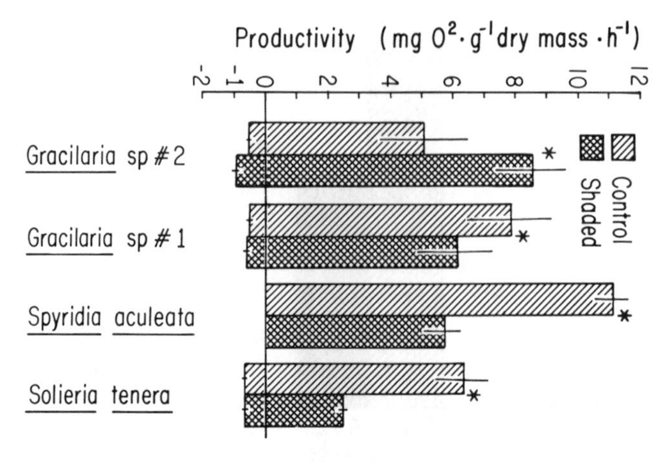

Figure 19.8. Rates of apparent photosynthesis (at 75 μE m^{-2} sec^{-1}) and respiration of plants that had been beneath black plastic bags (shaded) and clear plastic bags (control) for two weeks. Black bags reduced light by 98%, clear bags by 2%. Asterisk indicates significant at $P < 0.05$, Mann-Whitney U-test. $N = 6$ for each determination. Vertical bars represent 95% confidence intervals.

more individuals of *Gracilaria* species 1 were still alive ($P < 0.05$, Fisher's exact test).

A repetition of this experiment using a highly multilayered species with broad, translucent blades, *Halymenia floresia,* and a slightly multilayered species with opaque but somewhat dissected blades, *Gracilaria domingensis,* showed similar results. No individuals of *H. floresia* survived either two or four months of burial; seven of eight individuals of *G. domingensis* survived burial for two months, and three of eight survived burial for four months. These data are consistent with the hypothesis that less-layered species will be more resistant to damage from burial than more-layered species, but more species need to be tested before this can be adequately evaluated.

To simulate dark periods, plants from the sand plain were placed in Vexar cages that were covered with either black plastic bags (decreasing photosynthetically active light levels by 98–99%) or clear plastic bags (decreasing light levels by about 2%) for two weeks. Water could circulate to the plants because the bottom of the bag remained open and the cage and bag were suspended 15 cm above the sand bottom on metal poles. The photosynthetic rate, at about saturation (75 μE m^{-2} sec^{-1}), for six plants from the darkened bag was then compared with that for six plants from the control (clear bag).

Gracilaria species 2 showed a significant increase, and *Gracilaria* species 1 showed a significant, but small (21%), decrease in apparent photosynthe-

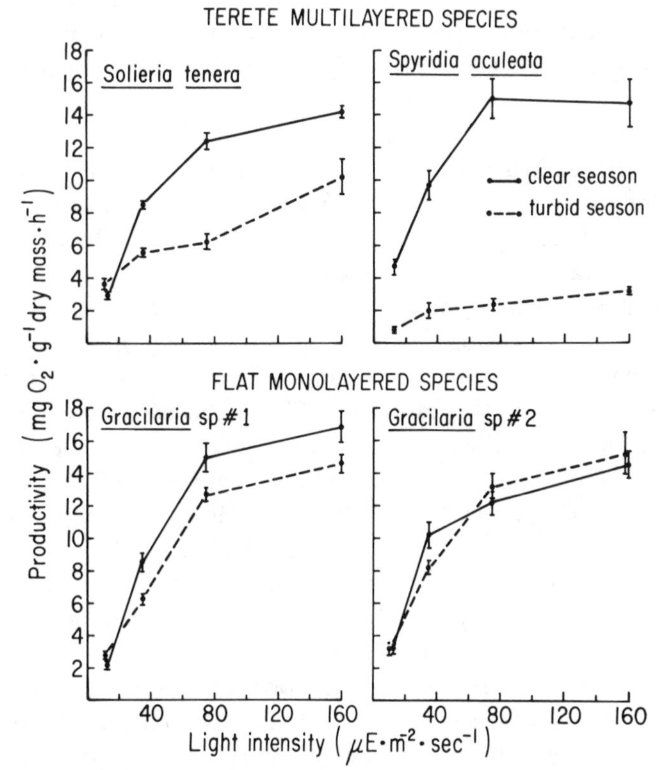

Figure 19.9. Laboratory determinations of apparent photosynthesis versus light quantity for two common multilayered and two common monolayered species taken from the Galeta sand plain during the clear wet season and the turbid dry season. $N = 6$ for each point, and vertical lines show 95% confidence intervals.

sis when compared with control plants (Figure 19.8). When subjected to the same conditions, *Solieria* and *Spyridia* showed significant, and larger, decreases of 61% and 49%, respectively. Thus, physiological resistance to major physical disturbances appears to be a correlate of the flat monolayered morphology. This pattern is also apparent in clear-season-versus-turbid-season comparisons of photosynthetic output across a light gradient. The terete multilayered species show large decreases in photosynthetic ability during the turbid, dry season, and flat monolayered species show little, if any, change between seasons (Figure 19.9).

Colonization

Between periods of prolonged low light, multilayered species must spread throughout the habitat, find suitable attachment sites, and grow to maturity. Therefore, they should have greater colonizing ability than the more stable monolayered forms. Evidence of rapid colonization by terete multilayers can be seen in Figure 19.6. After the end of the turbid season in April, *Solieria* and *Spyridia* increased their numbers by 316% and 452%, respectively, within six weeks. During this same time period, *Gracilaria* species 1 increased in density by 89%, and *Gracilaria* species 2 by only 16%. This difference could result from higher fecundity via spores or from vegetative fragmentation, or both (Dixon 1965, 1973). Spore production has three disadvantages. First, most spores will fall into inappropriate microhabitats, because 96% of the sand plain is sand, and these species must attach to solid substrate (Hay 1981b). Second, many spores that do land on hard substrate may become rapidly buried, because sedimentation rates are high on the sand plain. Third, because terete multilayers are rare at the end of the turbid dry season, males and females may be widely separated, making fertilization of females inefficient (Levin and Kerster 1974; Bobisud and Neuhaus 1975; Schaal 1980). Although completely unstudied, the effects of plant density on efficiency of fertilization may be especially important for red seaweeds, because male gametes are nonmotile, and seaweeds are not known to have animal symbionts that could function as the equivalent of insect pollinators.

An initial assessment of reproductive condition throughout the year has been completed for several species (Hay and Norris 1984) (Figure 19.10). On an average yearly basis, the percentage of terete multilayers producing spores is significantly less, not more, than the percentage of flat monolayers producing spores ($P < 0.05$, Kruskal-Wallis and Student-Newman-Keuls tests). The yearly means are as follows (mean $\pm$ standard error for the 11 sample periods): *Gracilaria* species 1, 40 $\pm$ 6%; *Gracilaria* species 2, 43 $\pm$ 7%; *Solieria*, 16 $\pm$ 6%; *Spyridia*, 18 $\pm$ 3%. By multiplying mean plant density for each month by the percentages that are reproductive during each month, we can also estimate the monthly density of reproductive plants. This is plotted along with percentage reproduction in Figure 19.10. The yearly averages for reproductive plants per square meter are as follows ($\bar{x} \pm$ SE): *Gracilaria* species 1, 2.25 $\pm$ 0.41; *Gracilaria* species 2, 0.58 $\pm$ 0.10; *Solieria*, 0.21 $\pm$ 0.08; *Spyridia*, 0.15 $\pm$ 0.03. *Gracilaria* species 1 shows a significantly higher density of reproductive plants than does any of the other species ($P < 0.05$, ANOVA and Student-Newman-Keuls tests), but this is in large part due to its high population density. It is, on yearly average, the most common sand-plain species.

The actual number of spores occurring on each plant could not be determined because of their small size and scattered distribution throughout the plant thallus. However, qualitative observations indicated that

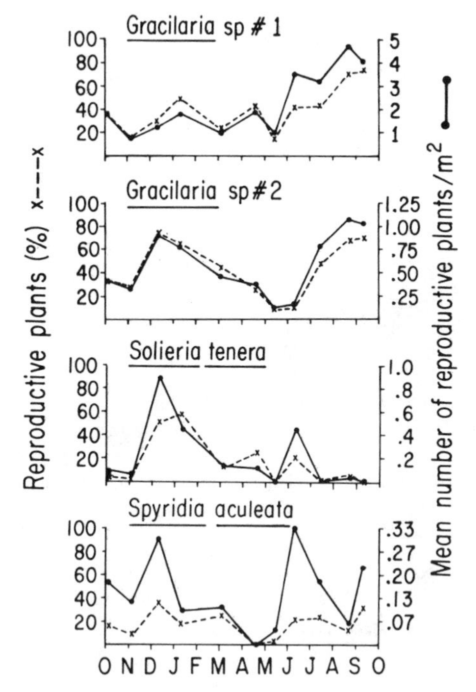

Figure 19.10. Seasonal patterns of reproduction in two flat monolayered species (*Gracilaria* species 1 and 2) and two terete multilayered species (*Solieria* and *Spyridia*) on the sand plain at Galeta, Panama. Dash line indicates percentage of the population producing gametes or spores; solid line indicates average number of reproductive plants per square meter (i.e., % reproductive multiplied by mean plant density). Plant portions were collected and preserved at approximately monthly intervals. These were later sectioned and examined microscopically to determine reproductive condition (Hay and Norris, 1984). During the year-long study, more than 2,200 plants of these four species were examined.

there were no striking differences in the densities of spores held by the different species. These observations were not quantified but are based on more than 2,200 individual plants of these four species that were sectioned and examined microscopically during the course of the study. Overall, differences in spore production between monolayers and multilayers do not appear to explain the large differences in colonization rates that can be seen in Figure 19.6.

Terete multilayered species can also colonize vegetatively via fragmentation and reattachment. This has an advantage in that the persistent monolayered species may act as structures marking the locations of suitable pieces of hard substrate. If fragments of the multilayered species can move along the sand surface via currents or wave surge until striking

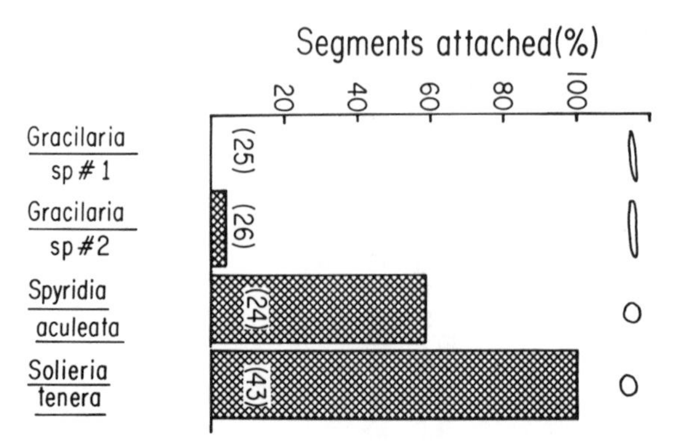

Figure 19.11. Percentages of seaweed portions that atached after being held against coral fragments for 10 days by cable ties. Numbers in parentheses show *N* for each species. The cross-sectional shape of each species is shown at the top of the figure. Terete multilayered species showed significantly higher rates of attachment than did flat monolayered species ($P < 0.05$, contingency-table analysis).

another plant or a suitable piece of hard substrate, and can then reattach, this could be a profitable strategy. During the clear season, I often observed this happening in the field; flat monolayers are commonly overgrown by terete multilayers that colonize vegetatively and function as facultative epiphytes. Monolayered species do not appear to have this ability. As an example, in June 1979, other plants served as the attachment sites for 26% of the *Solieria tenera* individuals ($N = 134$) and 0% of the *Gracilaria* species 1 individuals ($N = 178$) that occurred in 30 1.0-m^2 quadrants positioned randomly on the sand plain. This difference was highly significant ($P \ll 0.001$, Fisher's exact test). In contrast to the interactions between sessile marine invertebrates (Jackson 1979; Buss and Jackson 1979), overgrowth of flat monolayers by terete multilayers does not result in rapid death of the overgrown individual.

Experiments evaluating the potential for vegetative colonization, via fragmentation, of terete multilayers versus flat monolayers were conducted by affixing 24 to 43 fragments of each species to bare coral fragments using cable ties. These coral fragments were anchored on the sand plain by placing them between the strands of a weighted section of three-strand rope. After 10 days, the cable ties were cut and the coral fragments shaken to determine whether or not the branches had attached. *Solieria* and *Spyridia* attached well, 100% and 58%, respectively; *Gracilaria* species 1 never attached, and only 1 of 26 individuals of *Gracilaria* species 2 attached (Figure 19.11). Attachment rates for the monolayered species were significantly lower than for either of the terete multilayers ($P < 0.05$, contingency-table analysis). Expansion of this experiment to include seven flat-bladed and five terete species has shown that only the terete forms reattach at any appreciable rate (Hay, unpublished data). Selection for this ability may be a direct result of morphology, because terete forms should fragment more easily. When small herbivorous fish feed on the terete forms, they often produce fragments, because the diameters of branches are smaller than the length of the fish's bite. This is not the case for flat forms, because they are much wider than the bite sizes produced by the small herbivorous fish that occasionally occur in this environment. In addition, branches of several of the terete species (*Chondria tenuissima, C. littoralis,* and *Gracilaria blodgetii* are good examples) are constricted at the base, producing a weak point that should facilitate fragmentation. These species appear to function like cacti, such as cholla (*Opuntia*), that fragment easily and spread rapidly via vegetative propagation (Benson 1982).

Competitive ability

Terete multilayered species grow much faster than flat monolayered species during periods of adequate light (Figure 19.6). In addition, they tend to be much taller than monolayered species (*Gracilaria* species 1 and 2 are no more than 5–10 cm tall; *Spyridia* and *Solieria* are often 20–30 cm tall) and can invade either occupied or unoccupied pieces of substrate, often as vegetative fragments (Figure 19.11). When light levels are high, terete multilayered species should be competitively superior to flat monolayered species. As light decreases toward the compensation point, the cost of self-shading in terete multilayers will outweigh the advantages of increased thallus area, and flat monolayered species will be favored (Figure 19.2). Because of their lesser height, it is unlikely that flat monolayered adults will shade out adult multilayers during periods of low light. However, monolayers should be able to survive, or even grow, under conditions of very low light that would physically exclude multilayered species. The seasonal patterns shown in Figure 19.6 are consistent with this contention.

Additional support for the importance of morphology in determining competitive outcomes under differing light levels comes from correlations between depth distribution and morphology for Caribbean seaweeds. Mean minimum and maximum depths of various forms of nonfilamentous, upright seaweeds for which Taylor (1960) gives depth ranges are shown in Figure 19.3; see Hay (1980) for methods and limitations of the data set. In the Chlorophyta and Rhodophyta, multilayered algae, terete algae, and algae that are both multilayered and terete occur in shallow water; monolayered algae, flat algae, and algae that are both monolayered and flat occur in deeper water. This is consistent with the hypothesis that multilayers should be competitively superior in shallow, well-lit areas. The Phaeophyta showed no significant interrelationships between morphology and depth.

On the sand plain at Galeta, removal experiments designed to test the effects of multilayers on monolayers were inconclusive, because a short dry season occurred between June and August and removed most multilayers from the control plots (Figure 19.6). However, given the interrelationships among morphology, growth rate, and colonization ability, it is probable that terete multilayered species would competitively exclude flat monolayered species from the sand plain if periods of low light did not continually interrupt this process. Considering the high tolerances to darkness (Figures 19.6, 19.8, and 19.9) and burial (Figure 19.7) that are characteristic of the flat monolayered species, several years of continuously high light might be required before monolayers could be excluded by terete multilayers. Because dry-season winds are very predictable in Panama (Meyer and Birkeland 1974; Hendler 1976, 1977), the probability of this occurring at Galeta seems remote. However, sand plains in the U.S. Virgin Islands do not undergo these seasonally predictable periods of low light, and multilayered species completely dominate hard substrates in these habitats [see Earle (1972) for one site on St. John; Hay (personal observation) for four sites on St. Thomas].

The maintenance of species richness on the sand plain at Galeta appears to result because prolonged periods of low light cause selective mortality of the species that are capable of growing most rapidly when light is high. These fast-growing species appear to be competitively superior to the slower-growing, monolayered forms. The relationship between increasing abundance during the clear season and decreasing abundance during the turbid season is shown in Figure 19.12. Outcomes of competitive interactions and differential patterns of mortality appear to be predictable correlates of algal morphology. This is very similar to patterns previously documented for reef corals. In shallow habitats, branching multilayered corals are competitively superior to massive monolayered forms during benign periods, but they experience much higher mortality during infrequent hurricanes (Connell 1978; Porter et al. 1981; Woodley et al. 1981).

In terrestrial environments, monolayered species come to dominate late successional forests (Horn 1971) by growing up through multilayered species, which do not cast dense shade, overtopping them and producing a shade so dense that multilayered seedlings cannot grow. This appears to be uncommon in marine habitats, possibly because wave action moves plant canopies so much that monolayers cannot cast complete shade and inhibit

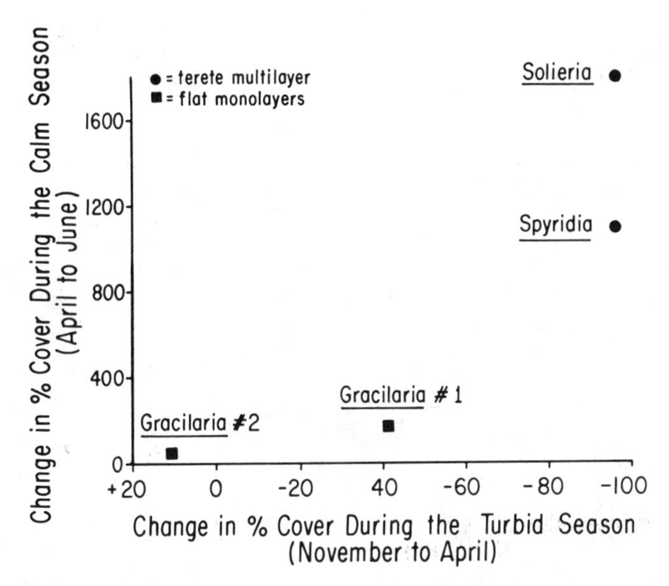

Figure 19.12. Relationships between percentage change in cover during the most physically benign portion of the year (April–June) and percentage change in cover during the most physically stressful portion of the year (November–April) for two common flat monolayers and two common terete multilayers on the Galeta sand plain. Correlation analysis shows a significant interaction between the two variables ($r = 0.907, P < 0.0025$).

the establishment of multilayered species. The potential for canopy domination by monolayered species occurs in the kelp beds of higher latitudes, but both biological (Leighton, Jones, and North 1966; Black 1976; Breen and Mann 1976) and physical (Dayton 1973; Rosenthal, Clarke, and Dayton 1974; Koehl and Wainwright 1977) disturbance rates are high and make many openings in the canopy or remove it completely (Chapman 1981; Dayton and Tegner 1984). In marine habitats, the competitively dominant forms appear to be multilayered, with monolayered forms being more prevalent in deeper areas that are physically marginal for multilayered species.

Summary

A simple model is proposed that predicts the effects of seaweed morphology on growth rate and competitive ability in environments of high or low light. The model is in large part an elaboration of one proposed by Horn (1971) to explain the geometry of forest trees. In general, the model predicts that monolayered plants should occur primarily in habitats where light is very low. They should also have flattened thalli that are optically dense and grow slower than those of multilayered plants when light levels are high. Multilayered plants should occur primarily in habitats with high light and, under such conditions, should be competitively superior to monolayers. Multilayers should have terete thalli, or flattened thalli that are translucent, highly dissected, or supported by midribs. They should grow more rapidly than monolayers when light levels are high and should show large and rapid changes in abundance when growing in habitats where light levels are highly variable throughout the season.

Interrelationships among morphologies, distributions, growth rates, seasonal patterns of abundance, and interrelated physiological attributes of Caribbean seaweeds are consistent with these hypotheses. Those patterns support the contention that selection for efficient light capture has been an important factor affecting the external morphology of seaweeds.

Acknowledgments

This study was funded by a grant from the Environmental Sciences Program of the Smithsonian Institution to Jim Norris and John Cubit, by a University of California dissertation fellowship, and by the University of North Carolina's Institute of Marine Sciences. Thomas Givnish, Egbert Leigh, and Randy Olson provided comments that improved the manuscript. Joe Ramus provided insight to several aspects of seaweed physiology.

References

Abbott, I. A., and G. J. Hollenberg. 1976. Marine algae of California. Stanford University Press. Stanford, Calif.

Benson, L. 1982. The Cacti of the United States and Canada. Stanford University Press, Stanford, Calif.

Black, R. 1976. The effect of grazing by the limpet, *Acmaea insessa,* on the kelp, *Egregia laevigata,* in the intertidal zone. Ecology 57:265–277.

Bobisud, L. E., and R. J. Neuhaus. 1975. Pollinator constancy and survival of rare species. Oecologia 21:263–272.

Breen, P. A., and K. H. Mann. 1976. Destructive grazing of kelp by sea urchins in Eastern Canada. J. Fish. Res. Bd. Can. 33:1278–1283.

Buss, L. W., and J. B. C. Jackson. 1979. Competitive networks: nontransitive competitive relationships in cryptic coral reef environments. Am. Nat. 113:223–234.

Chapman, A. R. O. 1981. Stability of sea urchin dominated barren grounds following destructive grazing of kelp in St. Margaret's Bay, Eastern Canada. Mar. Biol. 62:307–311.

Colombo, P. M., and M. Orsenigo. 1977. Sea depth effects on the algal photosynthetic apparatus. II. An electron microscopic study of the photosynthetic apparatus of *Halimeda tuna* (Chlorophyta, Siphonales) at −0.5 m and −6.0 m sea depths. Phycologia 16:9–17.

Connell, J. H. 1978. Diversity in tropical rainforests and coral reefs. Science 199:1302–1310.

Dayton, P. K. 1973. Dispersion, dispersal, and persistence of the annual intertidal alga, *Postelsia palmaeformis* Ruprecht. Ecology 54:433–438.

Dayton, P. K., and M. J. Tegner. 1984. Catastrophic storms, El Niño, and patch stability in a southern California kelp community. Science 224:283–285.

Dixon, P. S. 1965. Perennation, vegetative propagation and algal life histories, with special reference to *Asperagopsis* and other Rhodophyta. Bot. Gothoburg. 3:67–74.

– 1973. The biology of the Rhodophyta. Oliver and Boyd, Edinburgh.

Earle, S. A. 1972. The influence of herbivores on the marine plants of Great Lameshur Bay, with an annotated list of plants. Sci. Bull. Los Ang. Cty. Nat. Hist. Mus. 14:17–44.

Eiseman, N. J. 1978. Observations of the marine algae occurring from 30 to 100 m depths off the east coast of Florida. J. Phycol. 14:25.

Eiseman, N. J., and S. A. Earle. 1983. *Johnson-sea-linkia profunda,* a new genus and species of deep water Chlorophyta from the Bahama Islands. Phycologia 22:1–6.

Gabrielson, P. W., and M. H. Hommersand. 1982. The Atlantic species of *Solieria* (Gigartinales, Rhodophyta): their morphology, distribution and affinities. J. Phycol. 18:31–45.

Givnish, T. J. 1978. On the adaptive significance of compound leaves, with particular reference to tropical trees. Pp. 351–380 *in* P. B. Tomlinson and M. H. Zimmerman (eds.), Tropical trees as living systems. Cambridge University Press, London.

– 1979. On the adaptive significance of leaf form. Pp. 375–407 *in* O. T. Solbrig, P. H. Raven, S. Jain, and G. B. Johnson (eds.), Topics in plant population biology. Columbia University Press, New York.

– 1984. Leaf and canopy adaptations in tropical forests. Pp. 51–84 *in* E. Medina, H. A. Mooney, and C. Vásques-Yánez (eds.), Physiological ecology of plants of the wet tropics. Dr. Junk, The Hague.

Givnish, T. J., and G. J. Vermeij. 1976. Sizes and shapes of liane leaves. Am. Nat. 110:743–778.

Glynn, P. W. 1972. Observations on the ecology of the Caribbean and Pacific coasts of Panama. Biol. Soc. Wash. Bull. 2:13–20.

Hay, M. E. 1980. Algal ecology on a Caribbean fringing reef. PhD dissertation, University of California, Irvine.

– 1981a. The functional morphology of turf forming seaweeds: persistence in stressful marine habitats. Ecology 62:739–750.

– 1981b. Herbivory, algal distribution and the maintenance of between habitat diversity on a tropical fringing reef. Am. Nat. 118:520–540.

Hay, M. E., T. Colburn, and D. Downing. 1983. Spatial and temporal patterns in herbivory on a Caribbean fringing reef: the effects on plant distribution. Oecologia 58:299–308.

Hay, M. E., and J. N. Norris. 1984. Seasonal reproduction and abundance of six sympatric species of *Gracilaria* Grev. (Gracilariaceae; Rhodophyta) on a Caribbean subtidal sand plain. Hydrobiologia 116–117:63–72.

Hendler, G. L. 1976. Marine studies – Galeta Point. Pp. 131–249 *in* D. M. Windsor (ed.), 1975 environmental monitoring and baseline data compiled under the Smithsonian Institution environmental sciences program: tropical studies. Smithsonian Institution, Washington, D.C.

– 1977. Marine studies – Galeta Point. Pp. 138–267 *in* D. M. Windsor (ed.), Environmental monitoring and baseline data from the Isthmus of Panama – 1976, vol. IV. Smithsonian Institution, Washington, D.C.

Horn, H. S. 1971. The adaptive geometry of trees. Princeton University Press, Princeton, N.J.

Jackson, J. B. C. 1979. Overgrowth competition between encrusting Cheilostomes in a Jamaican cryptic reef environment. J. Anim. Ecol. 48:805–823.

Koehl, M. A. R. 1979. Stiffness or extensibility of intertidal algae: a comparative study of modes of withstanding wave action. J. Biomech. 12:634 (abstract).

Koehl, M. A. R., and S. A. Wainwright. 1977. Mechanical adaptations of a giant kelp. Limnol. Oceanogr. 22:1067–1071.

Kramer, P. J., and W. S. Clark. 1947. A comparison of photosynthesis in individual pine needles and entire seedlings at various light intensities. Plant Physiol. 22:51–57.

Kremer, B. P. 1981. Aspects of carbon metabolism in marine macroalgae. Oceanogr. Mar. Biol. Ann. Rev. 19:41–94.

Leighton, D. L., L. G. Jones, and W. J. North. 1966. Ecological relationships between giant kelp and sea urchins in Southern California. Pp. 141–153 *in* E. G. Young and J. L. McLachlan (eds.), Proceedings of the 5th International Seaweed Symposium. Oxford, New York.

Levin, D. A., and H. W. Kerster. 1974. Gene flow in seed plants. Pp. 139–220 *in* T. Dobzhansky, M. K. Hecht, and W. C. Steeve (eds.), Evolutionary biology, vol. 7. Plenum Press, New York.

Littler, M. M. 1980. Morphological form and photosynthetic performances of marine macroalgae: tests of a functional/form hypothesis. Bot. Mar. 22:586–590.

Littler, M. M., and K. E. Arnold. 1982. Primary productivity of marine macroalgal functional-form groups from southwestern North America. J. Phycol. 18:307–311.

Littler, M. M., and D. S. Littler. 1980. The evolution of thallus form and survival strategies in benthic marine macroalgae: field and laboratory tests of a functional-form model. Am. Nat. 116:25–44.

– 1983. Heteromorphic life history strategies in the brown alga *Scytosiphon lomentaria* (Lyngb.) Link. J. Phycol. 19:425–431.

Littler, M. M., P. R. Taylor, and D. S. Littler. 1983. Algal resistance to herbivory on a Caribbean barrier reef. Coral Reefs 2:111–118.

Lobban, C. S., and M. J. Wynne. 1981. The biology of seaweeds. Blackwell, Oxford.

Lubchenco, J., and J. Cubit. 1980. Heteromorphic life histories of certain marine algae as adaptations to variations in herbivory. Ecology 61:676–687.

Luning, K. 1981. Light. Pp. 326–355 *in* C. S. Lobban and M. J. Wynne (eds.), The biology of seaweeds. Blackwell, Oxford.
MacIntyre, I. G., and P. W. Glynn. 1976. Evolution of a modern Caribbean fringing reef, Galeta Point, Panama. Am. Assoc. Pet. Geol. Bull. 60:1054–1072.
Meyer, D. L., and C. Birkeland. 1974. Marine studies – Galeta Point. Pp. 129–253 *in* R. W. Rubinoff (ed.), 1973 environmental monitoring and baseline data compiled under the Smithsonian Institution environmental sciences program: tropical studies. Smithsonian Institution, Washington, D.C.
Neushul, M. 1972. Functional interpretation of benthic marine algal morphology. Pp. 47–74 *in* I. A. Abbott (ed.), Contributions to the benthic marine algae of the north Pacific. Japanese Society of Phycology, Kobe.
Norton, T. A., A. C. Mathieson, and M. Neushul. 1981. Morphology and environment. Pp. 421–451 *in* C. S. Lobban and M. J. Wynne (eds.), The biology of seaweeds. Blackwell, Oxford.
Nultsch, W., and J. Pfau. 1979. Occurrence and biological role of light-induced chromatophore displacements in seaweeds. Mar. Biol. 51:77–82.
Odum, E. P., E. J. Kuenzler, and M. X. Blunt. 1958. Uptake of P^{32} and primary productivity in marine benthic algae. Limnol. Oceanogr. 3:340–345.
Orians, G. H., and O. T. Solbrig. 1977. A cost-income model of leaves and roots with special reference to arid and semiarid areas. Am. Nat. 111:677–690.
Parkhurst, D. F., and O. L. Loucks. 1972. Optimal leaf size in relation to environment. J. Ecol. 60:505–537.
Peterson, R. D. 1972. Effects of light intensity on the morphology and productivity of *Caulerpa racemosa* (Forsskal). J. Agardh. Micronesia 8:63–86.
Porter, J. W., J. D. Woodley, G. J. Smith, J. E. Neigel, J. F. Battey, and D. G. Dallmeyer. 1981. Population trends among Jamaican reef corals. Nature 294:249–250.
Ramus, J., S. I. Beale, and D. Mauzerall. 1976a. Correlations of changes in pigment content with photosynthetic capacity of seaweeds as a function of water depth. Mar. Biol. 37:231–238.
Ramus, J., S. I. Beale, D. Mauzerall, and K. L. Howard. 1976b. Changes in photosynthetic pigment concentrations in seaweeds as a function of water depth. Mar. Biol. 37:223–229.
Ramus, J., F. Lemons, and C. Zimmerman. 1977. Adaptation of light harvesting pigments to downwelling light and the consequent photosynthetic performance of the eulittoral rockweeds *Ascophyllum nodosum* and *Fucus vesiculosus*. Mar. Biol. 42:293–303.
Ramus, J., and J. P. Van der Meer. 1983. A physiological test of the theory of complementary chromatic adaptation. I. Color mutants of a red seaweed. J. Phycol. 19:86–91.
Rosenthal, R. J., W. D. Clarke, and P. K. Dayton. 1974. Ecology and natural history of a stand of giant kelp, *Macrocystis pyrifera*, off Del Mar, California. Fish. Bull. 72:670–684.
Schaal, B. A. 1980. Measurement of gene flow in *Lupinus texensis*. Nature 284:450–451.

Slocum, C. J. 1980. Differential susceptibility to grazers in two phases of an intertidal alga: advantages of heteromorphic generations. J. Exp. Mar. Biol. Ecol. 46:99–110.

Steneck, R. S. 1982. A limpet-coralline alga association: adaptations and defenses between a selective herbivore and its prey. Ecology 63:507–522.

– 1983. Escalating herbivory and resulting adaptive trends in calcareous algae. Paleobiology 9:45–63.

Steneck, R. S., and W. H. Adey. 1976. The role of environment in control of morphology in *Lithophyllum congestum,* a Caribbean algal ridge builder. Bot. Mar. 19:197–215.

Steneck, R. S., and L. Watling. 1982. Feeding capabilities and limitation of herbivorous molluscs: a functional group approach. Mar. Biol. 68:299–319.

Taylor, P. R., and M. E. Hay. 1984. The functional morphology of intertidal seaweeds: the adaptive significance of aggregate vs. solitary forms. Mar. Ecol. Prog. Ser. 18:295–302.

Taylor, S. E. 1975. Optimal leaf form. Pp. 73–86 *in* D. M. Gates and R. B. Schmerl (eds.), Perspectives of biophysical ecology. Springer-Verlag, New York.

Taylor, S. E., and O. J. Sexton. 1970. Some implications of leaf tearing in Musaceae. Ecology 53:143–149.

Taylor, W. R. 1960. Marine algae of the eastern tropical and subtropical coasts of the Americas. University of Michigan Press, Ann Arbor.

Vogel, S. 1968. "Sun leaves" and "shade leaves": differences in connective heat dissipation. Ecology 49:1203–1204.

– 1970. Convective cooling at low airspeeds and the shapes of broad leaves. J. Exp. Bot. 21:91–101.

Woodley, J. D., E. A. Chornesky, P. A. Clifford, J. B. C. Jackson, L. S. Kaufman, N. Knowlton, J. C. Lang, M. P. Pearson, J. W. Porter, N. C. Rooney, K. W. Rylaorsdam, V. J. Tunniclitte, C. M. Wahle, J. L. Wulff, A. S. G. Curtis, M. D. Dahlmeyer, B. P. Jupp, M. A. R. Koehl, J. Neigel, and E. M. Sides. 1981. Hurricane Allen's impact on Jamaican coral reefs. Science 214:749–755.

Part III

Economics of biotic interactions

Many traits that enhance a plant's potential photosynthetic rate or competitive dominance – such as high leaf nitrogen content, heavy allocation to foliage, or an erect growth form – also may make it more attractive to herbivores. Chemical, physical, or biological defenses, lower leaf N levels, or a more compact growth form may reduce a plant's losses to herbivores, but do so at the cost of reduced allocation to photosynthetic tissue, reduced rate of energy capture, and/or reduced competitive dominance (Janzen 1979; Rhoades 1979; Lubchenco and Cubit 1980; Mattson 1980; Slocum 1980; Baines et al. 1982; Dirzo and Harper 1982a, 1982b; Lincoln et al. 1982; Mooney and Gulmon 1982; Dirzo 1984; Snaydon 1984). These tradeoffs create an *economics of biotic interactions,* linking the photosynthetic benefits of traits to their associated costs of herbivory, and the benefits of reduced herbivory to associated photosynthetic costs.

The net effect of these tradeoffs on competitive ability depends not only on the direct impact of plant–animal interactions on a plant's own rate of energy capture but also on the indirect effects of such interactions on competitors, through differences in animal feeding preference, plant responses to herbivory, local abundance of plants with various defenses, plant growth and competitive ability in the absence of herbivores, indirect competitive effects among plants sharing herbivores in common, visual mimicry, and mutualistic interactions (Paine 1966, 1969; Harper 1969, 1977; Tahvanainen and Root 1972; Atsatt and O'Dowd 1976; Barlow and Wiens 1977; Greenwood and Atkinson 1977; Lubchenco 1978; Rausher 1978; Gilbert 1980; Lubchenco and Cubit, 1980; Slocum 1980; Lubchenco and Gaines 1981; Dirzo and Harper 1982a, 1982b; Janzen and Martin 1982; Steneck 1982; Krischik and Denno 1983; McNaughton 1983; Whitham 1983; Dirzo 1984; Holt 1984). The economics of biotic interactions thus depend intimately on the characteristics of a plant's competitors, predators, and mutualists. Hence, plant defensive behavior should depend not only on environmental conditions but also on the specific web of interactions between plants and animals in a particular area and their detailed ecological characteristics.

Quantitative studies on the economics of biotic interactions are in their infancy. However, four questions seem likely to be crucial among those involving traits that entail an interplay between plant energy capture and plant–animal interactions. These are reviewed here to provide a context for the single chapter in this section (Gulmon and Mooney, Chapter 20), which makes a fundamental theoretical contribution to resolving the first of these issues:

1. *How can the costs and benefits associated with various defensive strategies be assessed?* To evaluate the net return on any defensive adaptation, the energy costs and benefits associated with that trait must be analyzed. Costs are perhaps easier to evaluate and can be divided roughly into two classes: direct and indirect.

Direct costs include the metabolic energy needed to construct mechanical defenses such as spines or leaf trichomes, to feed and/or house ants or other bodyguards, and to synthesize defensive compounds. Although assessing such costs would seem straightforward, in practice several complications arise. For example, if the biochemical pathway leading to a defensive compound and its standing pool size are known, the ATP cost of synthesizing the compound could be calculated. However, this cost would not include the costs (or benefits) of (1) maintaining the biochemical pathway itself, (2) obtaining minerals needed to synthesize the compound, (3) avoiding autotoxicity, (4) sharing the compound's biochemical pathway with other synthetic pathways, (5) multiple functions of the compound, (6) turnover in the defensive compound pool, or (7) later reincorporation of valuable substances from that pool into other compounds (Janzen 1979; Chew and Rodman 1979; Mooney and Gulmon 1982; Mooney et al. 1983).

Indirect costs include the photosynthetic benefits foregone as a result of adopting a given defensive strategy. Such opportunity costs could arise from diverting energy from productive tissue into physical, chemical, or biological defenses, producing less attractive but less productive foliage with low N levels, or diverting photosynthate into storage to allow later, rapid expansion of foliage with a short period of vulnerability. These indirect costs are intrinsically difficult to measure, requiring either (intractable) experimental manipulations of plant energy allocation or theoretical calculations based on growth models (see Chapter 20). Plants with facultative defenses that are induced by herbivore activity (Ryan 1978; Janzen 1979; Rhoades 1979, 1983; Schultz and Baldwin 1982; Baldwin and Schultz 1983; Schultz 1983) may escape direct and indirect defense costs much of the time, but even they must maintain the capacity to start up biochemical production lines and detect herbivore damage.

The *benefits* of a plant defense ordinarily result from repelling, killing, or sterilizing its herbivore enemies, or reducing their feeding rate or feeding efficiency. Numerous studies have shown that various plant secondary compounds act as toxins, feeding deterrents, or digestibility reducers (e.g., Feeny 1968, 1976; Whittaker and Feeny 1971; Erickson and Feeny 1974; Gilbert 1975; Rhoades and Cates 1976; Rosenthal and Janzen 1979; Scriber and Feeny 1979; Bernays 1981; Berenbaum 1983). Reductions in leaf nitrogen level and water content, although likely to reduce photosynthetic capacity, also reduce the feeding rate and/or growth of folivores (McNeill and Southwood 1978; Scriber and Feeny 1979; Mattson 1980; McClure 1980; Lincoln et al. 1982). Extrafloral nectaries and hollow leaf bases, petioles, or stems attract ants that protect their plant hosts from vertebrate and invertebrate attack, and sometimes from overgrowth by competing plants (Wheeler 1942; Janzen 1966, 1969, 1983; Gilbert 1975; Bentley 1977; Tilman 1978; Messina 1981; Skinner and Whittaker 1981; Koptur et al. 1982).

The advantages of such defenses are strongly context-dependent. For example, the rate at which specialist herbivores can locate their host is greater when it grows among visually dissimilar plants (Rausher 1981) and when it grows in monoculture rather than among a mixture of species with different defenses (Tahvanainen and Root 1972; Root 1973; Atsatt and O'Dowd 1976). Some toxins may be effective only in certain environments. Many plants of the Umbelliferae possess linear furanocoumarins that cross-link herbivore DNA after exposure to ultraviolet light (Berenbaum 1981). As Berenbaum (1981, 1983) noted, such furanocoumarins are thus useful only in sunlit habitats, and shade-adapted species generally do not possess them. Oecophorid caterpillars apparently can avoid phototoxic effects by shading themselves inside a rolled-up leaf while feeding (Berenbaum 1981). However, many umbellifers may avoid this tactic by possessing unusually finely divided leaves. In a similar vein, the beneficial effect of defensive compounds that reduce herbivore feeding efficiency may depend on the presence of a herbivore's natural enemies. For example, a "resistant" strain of *Glycine max* actually sustains heavier damage from the Mexican bean beetle *Epilachna* than does a susceptible variety in the absence of *Epilachna*'s predators, apparently because the beetle has to eat more of the resistant strain to complete its development (Price et al. 1980; Thompson 1982; Strong et al. 1984). However, because beetle larvae grow more slowly on the resistant strain, in the presence of a predaceous bug they suffer much heavier mortality. As a result, in the presence of the bug the resistant strain suffers 70% less damage than does the susceptible strain (Price et al. 1980).

The "benefits" associated with plant defenses may also have contrasting effects on different components of plant fitness. Louda (1982) showed that spiders that inhabit the inflorescences of *Haplopappus venetus,* apparently drawn in part by inflorescence shape, affect seed set in two ways: by reducing pollination by killing floral visitors, but also decreasing seed predation by invertebrates. Although not directly related to plant energy capture, this study showed that the net effect of spiders (and, presumably, inflorescence shape) on plant fitness is positive. One fundamental problem with most studies of the benefits of defensive traits is that they are rarely expressed in terms that make them comparable to the costs – in terms of energy or fitness – of those traits. Such a common currency is needed for any cost–benefit analysis of optimal defensive behavior.

2. *What are the ecological determinants of the nature of, and amount of energy allocated to, antiherbivore defenses?* Two principal hypotheses have emerged in recent years concerning selection for different kinds and amounts of plant defenses in different contexts. The first is based on cost–benefit analysis. Investment in defensive compounds or other traits should reflect their cost, the likelihood of herbivory, and the cost of losing and replacing leaf tissue; selection should favor heavier investment until the marginal profit associated with further defensive increments is zero. Plants should thus be defended more heavily in unproductive habitats or in slow-growing forms, in which a leaf is more costly to replace, in terms of nutrients or the photosynthetic period needed to repay its construction cost (Janzen 1974a, 1979; Gilbert 1975; McKey et al. 1978; Mooney and Gulmon 1982; Coley 1983). Furthermore, nitrogenous defensive compounds, such as alkaloids, should be selected against in nitrogen-poor habitats.

A second approach is based on the concepts of *plant apparency* and *qualitative versus quantitative defenses* (Feeny 1976; Rhoades and Cates 1976; Cates and Rhoades 1977; Rhoades 1979). Plants or plant parts are said to be apparent if they are predictable in space and time and thus readily found by herbivores. Thus, long-lived perennials, late successional plants, and woody tissue would be considered apparent, whereas annuals, early successional plants, and ephemeral buds, flowers, and young leaves would, at least for some herbivores, be unapparent. *Qualitative defenses* are toxins, like alkaloids or cardiac glycosides, that interfere with specific steps of intermediary metabolism and are effective in small doses. *Quantitative defenses* are digestibility reducers, like tannins or resins, that reduce the nutritional value of plant parts ingested, often by complexing with proteins or digestive enzymes, and are most effective at high concentrations.

The plant-apparency hypothesis states that unapparent plants should use qualitative defenses and that apparent plants should use quantitative defenses. Qualitative defenses cost little and can protect an unapparent plant from most of the few herbivores that locate it, but they will be ineffective against specialists that can detoxify or bypass the poison. Quantitative defenses, although costly, are essentially economic and thus effective against all herbivores [but see Fox and Macauley (1977), Berenbaum (1980), and Bernays (1981) on insects resistant to tannins]. They can protect an apparent plant from enemies that will surely locate it, and they may make such herbivores apparent to their own natural enemies.

The predictions of the plant-apparency hypothesis are in general accord with observed defensive patterns as a function of growth form and successional status (Feeny 1976; Rhoades and Cates 1976; Rhoades 1979; Langenheim 1984). Furthermore, as expected from qualitative cost–benefit analysis, plants from nutrient-poor sites are heavily defended by high concentrations of leaf tannins and phenols (Janzen 1974a; McKey et al. 1978), and fast-growing early successional trees are only lightly defended and suffer relatively high rates of herbivore damage by comparison with late successional trees in the same habitat (Coley 1983). Plants at lower latitudes and altitudes, presumably exposed to higher rates of herbivory because of the relative constancy of the physical environment, appear to be more heavily defended (Hartley et al. 1973; Bentley 1977; Levin and York 1978).

To test and distinguish between the predictions of these hypotheses, we urgently need quantitative studies of the energy costs and benefits of different defensive traits. Such studies must trace the effects of a defense on (1) a herbivore's ability to locate, feed, and reproduce on a given host, (2) its susceptibility to natural enemies, (3) plant growth and competitive ability, independent of herbivory, (4) a plant's own growth as influenced by herbivory, and (5) the growth of other plants as influenced by herbivory. Lincoln et al. (1982) have taken a first step toward this goal by comparing the effects of different levels of leaf nitrogen versus defensive resin in the chaparral shrub *Diplacus aurantiacus* on feeding behavior and larval growth by its specialist herbivore *Euphydryas chalcedona.* Lower leaf nitrogen levels decrease larval growth by *Euphydryas,* but also strongly reduce photosynthesis (see Chapter 1) (Gulmon and Chu 1981) in comparison with the costs of resin having similar antiherbivore effects. However, this analysis remains incomplete, because it does not express the benefits of different levels of either defense in terms of their effects on plant energy capture or measure the costs of obtaining nitrogen.

3. Under what circumstances can herbivore attack be beneficial to a plant? Herbivores can favor plants that are avoided by, unpalatable to, or resistant to grazers (Darwin 1859; Tansley and Adamson 1925; Harper 1969; Lubchenco 1978). But when can being eaten be advantageous? If we leave aside plant mutualisms with pollinators or seed dispersers, which affect energy capture only indirectly, two advantages of being eaten have been suggested. These involve (1) indirect effects of competing plants on shared herbivores or (2) postulated increases in plant productivity caused by grazing.

Indirect effects on competitors through shared herbivores: Holt (1984) proposed that an energetic subsidy of herbivores by "tolerant" plants, which can withstand some grazing damage, can lead to the elimination of nontolerant plants that might not be attacked in the absence of the former. To show that is plausible, he cited studies on the biological control of *Opuntia* cactus by the moth *Cactoblastus.* Attempts to eradicate introduced *Opuntia* in Australia revealed that stands of chlorotic, "yellow-pear" strains of *O. inermis* and *O. stricta* were not affected by *Cactoblastus* attacks, apparently as a result their low nutrient content. However, the presence of nonchlorotic, "green-pear" strains at low frequencies, themselves attacked by *Cactoblastus,* led to *Cactoblastus* infestation and death of the yellow-pear population. The green-pears thus obtained a competitive advantage over the yellow-pears (albeit short-lived) through interactions with a shared predator, independent of any direct interaction. Such "apparent competition" (Holt 1977) can result from a predator's differing functional responses to competing prey species and the competitors' differential responses to consumption.

Apparent competition is most likely to favor the evolution of palatability in plants if competing species (1) are palatable to, but not the preferred prey of, a herbivore (or class of herbivores), (2) would not, alone, support the herbivore population, but (3) would suffer debilitating damage if attacked by the herbivore. In this case, a plant that is palatable, *preferred* by the herbivore, and can support its population growth could gain a competitive advantage if (4) it itself does not suffer grave damage from herbivory. Evolution of palatability as a means of overcoming competitors through effects on shared herbivores has been proposed for systems involving coralline algae (Steneck 1982, 1983), grasses and graminoids (Owen and Wiegert 1981; cf. Silvertown 1982), and fire-adapted woody plants (Mutch 1965; Leigh 1971).

The most convincing of these arguments is that for coralline algae (Steneck 1982, 1983). Several encrusting forms depend on constant grazing by limpets, urchins, or fish to exclude settlement by taller competitors like barnacles or foliose algae (Paine 1980; Steneck 1982). Some species preferred by limpets have an exceptionally smooth surface and an unusually thick, protective epithallus overlying meristematic and photosynthetic tissues; the epithallus cells are sometimes only lightly calcified and filled with starch (Steneck 1982). These traits appear to facilitate limpet grazing by helping prevent limpet removal by waves or starfish; they also provide a readily grazed, inexpensive "turf" with well-protected meristems. These traits thus appear to be adaptations by which algae employ herbivores to exclude taller or more rapidly growing competitors. This conclusion is strengthened by the explosive radiation of coralline algae at the end of the Mesozoic, contemporaneous with a rise in organisms with mouthparts capable of excavating calcareous substrates, and an increase in damage by such herbivores (Steneck 1983).

Fire is also an herbivore, albeit more omnivorous than most. Many terrestrial plants have traits (e.g., thick bark, epicormic buds, belowground storage organs, serotinous cones) that enhance an individual's chance of surviving a fire or reproducing in sites cleared by fire elsewhere. Mutch (1965), Leigh (1971), and others have suggested that fire-adapted species would gain a further advantage by having traits that encourage the start or spread of fire that would consume unadapted competitors. Many fire-adapted species do have highly flammable, resinous leaves and thin stems, although these may simply be side effects of adaptations to sterile soils (Janzen 1974a) or direct consequences of frequent disturbance. However, an alternative explanation is difficult to find for the "pyres" created by the massive amounts of bark shed by certain fire-adapted *Eucalyptus* species.

Another situation in which plants may benefit from herbivore attack occurs in hollow, ant-defended trees of the genera *Barteria* and *Cecropia* (Janzen 1972, 1973), inside of which ants maintain colonies of homopterans that feed on the plant and secrete honeydew. Such homopterans are analogous to the extrafloral nectaries that other plants use to attract ants that assail folivores; we would not expect plants to evolve strong defenses against them (Janzen 1979). Their small negative effect on one component of plant fitness is likely outweighed by the large positive effects of ants on others.

Effects on plant productivity: Several studies by McNaughton and his colleagues (McNaughton 1979, 1983, 1984; McNaughton et al. 1983) suggest that grazing by ungulates can increase the productivity of tropical savanna grasses. Yet these studies may be misleading, because they base productivity on measures of aboveground biomass only. If productivity is expressed per unit aboveground biomass, a spurious increase in response to grazing is expected if regrowth occurs through a mobilization of underground reserves. Grazing potentially can, of course, stimulate productivity on a whole-plant basis by mobilizing reserves into productive tissue (Ericsson et al. 1980) or by reducing self-shading and proportional allocation to support tissue. However, the possession of energetic reserves itself entails a large opportunity cost (see Chapter 20).

Even if grazing increases a plant's own productivity (not counting energy harvested by herbivores), such an increase will almost certainly be context-specific. It is known that small amounts of insect damage can increase cotton and bean yields, apparently by removing apical dominance and helping create bushier plants (Janzen 1979). However, this benefit is likely only in open situations with considerable side lighting, as found in row crops; loss of apical dominance and height growth would be catastrophic in crowded stands (Janzen 1979).

4. What are the costs and benefits of other plant–animal interactions that affect plant energy capture? A small minority of plants obtain mineral nutrients crucial to growth and/or reproduction through carnivory or ant-fed myrmecophily (Janzen 1974b; Huxley 1978; Rickson 1979; Thompson 1981; Dixon et al. 1982; Lüttge 1983; Givnish et al. 1984). Prey or symbionts feed plants with minerals, but require energy input themselves, in the form of lures, traps, and/or digestive enzymes in carnivores, or shelters in ant-fed myrmecophytes. Cost–benefit analysis would thus suggest that such plant–animal interactions are most likely to evolve in habitats where the nutrients thus obtained – and no other factor – limit photosynthesis (Givnish et al. 1984). This may partly explain the general restriction of carnivorous plants to habitats that are not only nutrient-poor but also sunny and at least seasonally moist. To test such predictions rigorously, we need quantitative studies of the photosynthetic costs and benefits associated with carnivory, as well as those associated with similar systems involving ant-fed myrmecophily, nitrogen fixation, and mycorrhizae.

Gulmon and Mooney (Chapter 20) provide a valuable framework for assessing the direct and indirect energetic costs of plant defenses. They focus particularly on indirect costs, using a growth model to analyze the amount of photosynthesis foregone as a result of investment in defensive compounds. This model is used to determine the indirect costs of different time courses of defensive allocation, as a first step toward predicting the optimal time course of such allocation to organs during development. These costs increase with leaf photosynthetic capacity, suggesting an alternative to the plant-apparency hypothesis for the light investment in defensive compounds seen in rapidly growing, early successional plants. The model is also used to assess the indirect costs of energy storage, which might be employed as a defensive tactic to sequester energy from herbivores, or to allow a later, rapid expansion of foliage with a narrow window of vulnerability. These results are important initial contributions to the quantitative cost–benefit analysis of defensive traits.

References

Atsatt, P. R., and D. J. O'Dowd. 1976. Plant defense guilds. Science 193:24–29.

Baines, R. N., J. H. Grieshaber-Otto, and R. W. Snaydon. 1982. Factors affecting the performance of white clover in swards. Pp. 217–221 *in* A. J. Corrall (ed.), Efficient grassland farming. British Grassland Society, Hurley.

Baldwin, I. T., and J. C. Schultz. 1983. Rapid changes in tree chemistry induced by damage: evidence for communication between plants. Science 221:277–278.

Barlow, B. A., and D. Wiens. 1977. Host-parasite resemblance in Australian mistletoes: the case for cryptic mimicry. Evolution 31:69–84.

Bentley, B. L. 1977. Extrafloral nectaries and protection by pugnacious bodyguards. Ann. Rev. Ecol. Syst. 8:407–427.

Berenbaum, M. 1980. Adaptive significance of midgut pH in larval Lepidoptera. Amer. Nat. 115:138–146.

– 1981. Patterns of furanocoumarin distribution and insect herbivory in the Umbelliferae: plant chemistry and community structure. Ecology 62:1254–1266.

– 1983. Coumarins and caterpillars: a case for coevolution. Evolution 37:163–179.

Bernays, E. A. 1981. Plant tannins and insect herbivores: an appraisal. Ecol. Entom. 6:353–360.

Cates, R. G., and D. F. Rhoades. 1977. Patterns in the production of antiherbivore chemical defenses in plant communities. Biochem. Syst. Ecol. 5:185–193.

Chew, F. S., and J. E. Rodman. 1979. Plant resources for chemical defense. Pp. 271–307 *in* G. A. Rosenthal and D. H. Janzen (eds.), Herbivores: their interaction with secondary plant metabolites. Academic Press, New York.

Coley, P. D. 1983. Herbivory and defensive characteristics of tree species in a lowland tropical forest. Ecol. Monogr. 53:209–233.

Darwin, C. 1859. On the origin of species by means of natural selection. John Murray, London.

Dirzo, R. 1984. Herbivory: a phytocentric overview. Pp. 141–165 *in* R. Dirzo and J. Sarukhán (eds.), Perspectives on plant population ecology. Sinauer, Sunderland Mass.

Dirzo, R., and J. L. Harper. 1982a. Experimental studies on slug–plant interactions. III. Differences in the acceptability of individual plants of *Trifolium repens* to slugs and snails. J. Ecol. 70:101–118.

– 1982b. Experimental studies on slug–plant interactions. IV. The perform-

ance of cyanogenic and acyanogenic morphs of *Trifolium repens* in the field. J. Ecol. 70:119–138.
Dixon, K. W., P. S. Pate, and W. J. Bailey. 1982. Nitrogen nutrition of the tuberous sundew *Drosera erythrorhiza* Lindl. with particular reference to catch of arthropod fauna by its glandular leaves. Austral. J. Bot. 28:283–297.
Erickson, J. M., and P. Feeny. 1974. Sinigrin: a chemical barrier to the black swallowtail butterfly, *Papilio polyxenes.* Ecology 55:103–111.
Ericsson, A., J. Hellkvist, K. Hillerdal-Hagstromer, S. Larsson, E. Mattson-Djos, and O. Tenow. 1980. Consumption and pine growth – hypothesis of effects on growth processes by needle-eating insects. Pp. 1–9 *in* T. Persson (ed.), Structure and function of northern coniferous forest: an ecosystem study. Swedish Natural Science Research Council, Stockholm.
Feeny, P. 1968. Effect of oak leaf tannins on larval growth of the winter moth *Operophtera brumata.* J. Insect Physiology 14:805–817.
– 1976. Plant apparency and chemical defense. Pp. 1–40 *in* J. W. Wallace and R. L. Mansell (eds.), Interactions between plants and insects. Vol. 10. Recent advances in phytochemistry. Plenum Press. New York.
Fox, L. R., and B. J. Macauley. 1977. Insect grazing on *Eucalyptus* in response to variation in leaf tannins and nitrogen. Oecologia 29:145–162.
Gilbert, L. E. 1975. Ecological consequences of a coevolved mutualism between butterflies and plants. Pp. 210–240 *in* L. E. Gilbert and P. H. Raven (eds.), Coevolution between animals and plants. University of Texas Press, Austin.
– 1980. Food web organization and the conservation of neotropical diversity. Pp. 11–33 *in* M. E. Soulé and B. A. Wilcox (eds.), Conservation biology. Sinauer, Sunderland, Mass.
Givnish, T. J., E. L. Burkhardt, R. E. Happel, and J. D. Weintraub. 1984. Carnivory in the bromeliad *Brocchinia reducta,* with a cost/benefit model for the general restriction of carnivorous plants to sunny, moist, nutrient-poor habitats. Amer. Nat. 124:479–497.
Greenwood, R. M., and I. A. E. Atkinson. 1977. Evolution of divaricating plants in New Zealand in relation to moa browsing. Proc. N. Zeal. Ecol. Soc. 24:21–33.
Gulmon, S. L., and C. C. Chu. 1981. The effects of light and nitrogen on photosynthesis. leaf characteristics, and dry matter allocation in the chaparral shrub *Diplacus aurantiacus.* Oecologia 49:207–212.
Harper, J. L. 1969. The role of predation in vegetational diversity. Brookhaven Symp. Biol. 22:48–61.
– 1977. Population biology of plants. Academic Press, London.
Hartley, T. G., E. A. Dunstona, J. S. Fitzgerald, S. R. Johnson, and J. A. Lamberton. 1973. A survey of New Guinea plants for alkaloids. Lloydia 36:217–319.
Holt, R. E. 1977. Predation, apparent competition, and the structure of prey communities. Theor. Pop. Biol. 12:197–229.
– 1984. Spatial heterogeneity, indirect interactions, and the coexistence of prey species. Amer. Nat. 124:377–406.
Huxley, C. R. 1978. The ant plants *Myrmecodia* and *Hydnophytum* (Rubiaceae) and the relationships between their morphology, ant occupants, physiology, and ecology. New Phytol. 80:231–268.

Janzen, D. H. 1966. Coevolution of mutualism between ants and acacias in Central America. Evolution 20:249–275.
– 1969. Allelopathy by myrmecophytes: the ant *Azteca* as an allelopathic agent of *Cecropia*. Ecology 50:147–153.
– 1972. Protection of *Barteria* (Passifloraceae) by *Pachysima* ants (Pseudomyrmecinae) in a Nigerian rain forest. Ecology 53:885–892.
– 1973. Dissolution of mutualism between *Cecropia* and its *Azteca* ants. Biotropica 5:15–28.
– 1974a. Tropical blackwater rivers, animals, and mast fruiting in the Dipterocarpaceae. Biotropica 6:69–103.
– 1974b. Epiphytic myrmecophytes in Sarawak: mutualism through the feeding of plants by ants. Biotropica 6:237–259.
– 1979. New horizons in the biology of plant defenses. Pp. 331–350 *in* G. A. Rosenthal and D. H. Janzen (eds.), Herbivores: their interaction with secondary plant metabolites. Academic Press, New York.
– 1983. Food webs: who eats what, why, how, and with what effects in a tropical forest? Pp. 167–182 *in* F. B. Golley (ed.), Tropical rain forest ecosystems: structure and function. Springer-Verlag, New York.
Janzen, D. H., and P. S. Martin. 1982. Neotropical anachronisms: the fruits the gomphotheres ate. Science 214:19–27.
Koptur, S., A. R. Smith, and I. Baker. 1982. Nectaries in some neotropical species of *Polypodium* (Polypodiaceae): preliminary observations and analysis. Biotropica 14:108–113.
Krischik, V. A., and R. F. Denno. 1983. Individual, population, and geographic patterns in plant defense. Pp. 463–512 *in* R. F. Denno and M. S. McClure (eds.), Variable plants and herbivores in natural and managed systems. Academic Press, New York.
Langenheim, J. H. 1984. The roles of plant secondary chemicals in wet tropical ecosystems. Pp. 189–208 *in* E. Medina, H. A. Mooney, and C. Vásquez-Yánez (eds.), Physiological ecology of plants of the wet tropics. Dr. Junk, The Hague.
Leigh, E. G. 1971. Adaptation and diversity. Freeman, Cooper & Co., San Francisco.
Levin, D. A., and B. M. York. 1978. The toxicity of plant alkaloids: an ecogeographic perspective. Biochem. Syst. Ecol. 6:61–76.
Lincoln, D. E., T. S. Newton, P. R. Ehrlich, and K. S. Williams. 1982. Coevolution of the checkerspot butterfly *Euphydryas chalcedona* and its larval food plant *Diplacus aurantiacus:* larval response to protein and leaf resin. Oecologia 52:216–223.
Louda, S. M. 1982. Inflorescence spiders: a cost/benefit analysis for the host plant, *Haplopappus venetus* Blake (Asteraceae). Oecologia 55:185–191.
Lubchenco, J. 1978. Plant species diversity in a marine intertidal community: importance of herbivore food preference and algal competitive abilities. Amer. Nat. 112:23–39.
Lubchenco, J., and J. Cubit. 1980. Heteromorphic life histories of certain marine algae as adaptations to variations in herbivory. Ecology 61:676–687.
Lubchenco, J., and S. D. Gaines. 1981. A unified approach to marine plant–

herbivore interactions. I. Populations and communities. Ann. Rev. Ecol. Syst. 12:405–437.
Lüttge, U. 1983. Ecophysiology of carnivorous plants. Pp. 489–517 *in* O. L. Lange, P. S. Nobel, C. B. Osmond, and H. Ziegler (eds.), Physiological ecology IV. Vol. 12D. Encyclopedia of plant physiology, new series. Springer-Verlag, New York.
McClure, M. S. 1980. Foliar nitrogen: a basis for host suitability for elongate hemlock scales. *Fiorinia externa* (Homoptera: Diaspididae). Ecology 61:72–79.
McKey, D. B., P. G. Waterman, C. H. Msi, J. S. Gartlan, and T. T. Struhsaker. 1978. Phenolic content of vegetation in two African rain forests: ecological implications. Science 202:61–63.
McNaughton, S. J. 1979. Grazing as an optimization process: grass–ungulate relationships in the Serengeti. Amer. Nat. 113:691–703.
– 1983. Compensatory plant growth as a response to herbivory. Oikos 40:329–336.
– 1984. Grazing lawns: animals in herds, plant form, and coevolution. Amer. Nat. 124:863–886.
McNaughton, S. J., L. L. Wallace, and M. B. Coughenour. 1983. Plant adaptation in an ecosystem context: effects of defoliation, nitrogen and water on growth on an African C_4 sedge. Ecology 64:307–318.
McNeill, S., and T. R. E. Southwood. 1978. The role of nitrogen in the development of insect/plant relationships. Pp. 77–98 *in* J. B. Harborne (ed.), Biochemical aspects of plant and animal coevolution. Academic Press, London.
Mattson, W. J. 1980. Herbivory in relation to plant nitrogen content. Ann. Rev. Ecol. Syst. 11:119–161.
Messina, F. J. 1981. Plant protection as a consequence of an ant–membracid mutualism: interactions on goldenrod (*Solidago* sp.). Ecology 62:1433–1440.
Mooney, H. A., and S. L. Gulmon. 1982. Constraints on leaf structure and function in reference to herbivory. BioScience 32:198–206.
Mooney, H. A., S. L. Gulmon, and N. D. Johnson. 1983. Physiological constraints on plant chemical defenses. Pp. 21–36 *in* P. A. Hedin (ed.), Plant resistance to insects. American Chemical Society Symposium Series 208, Washington, D.C.
Mutch, R. W. 1965. Wildland fires and ecosystems – a hypothesis. Ecology 51:1046–1051.
Owen, D. F., and R. G. Wiegert. 1981. Mutualism between grasses and grazers: an evolutionary hypothesis. Oikos 36:376–378.
Paine, R. T. 1966. Food web complexity and species diversity. Amer. Nat. 100:65–75.
– 1969. The *Pisaster–Tegula* interaction: prey patches, predator food preference and intertidal community structure. Ecology 50:950–961.
– 1980. Food webs: linkage, interaction strength and community infrastructure. J. Anim. Ecol. 49:667–685.
Price, P. W., C. E. Bouton, P. Gross, B. A. McPheron, J. N. Thompson, and A. E. Weis. 1980. Interactions among three trophic levels: influence of plants on interactions between insect herbivores and natural enemies. Ann. Rev. Ecol. Syst. 11:41–66.

Rausher, M. D. 1978. Search image for leaf shape in a butterfly. Science 200:1071–1073.
– 1981. The effect of native vegetation on the susceptibility of *Aristolochia reticulata* (Aristolochiaceae) to herbivore attack. Ecology 62:1187–1195.
Rhoades, D. F. 1979. Evolution of plant chemical defense against herbivores. Pp. 4–54 *in* G. A. Rosenthal and D. H. Janzen (eds.), Herbivores. Academic Press, New York.
– 1983. Responses of alder and willow to attack by tent caterpillars and webworms: evidence for pheromonal sensitivity of willows. Pp. 55–68 *in* P. A. Hedin (ed.), Plant resistance to insects. American Chemical Society Symposium Series 208, Washington, D.C.
Rhoades, D. F., and R. G. Cates. 1976. A general theory of plant anti-herbivore chemistry. Pp. 168–213 *in* J. W. Wallace and R. L. Mansell (eds.), Interactions between plants and insects. Vol. 10. Recent advances in phytochemistry. Plenum Press, New York.
Rickson, F. R. 1979. Absorption of animal tissue breakdown products into a plant stem – the feeding of a plant by ants. Amer. J. Bot. 66:87–90.
Root, R. B. 1973. Organization of a plant-anthropod association in simple and diverse habitats: the fauna of collards *(Brassica oleracea)*. Ecol. Monogr. 43:95–124.
Rosenthal, G. A., and D. H. Janzen (eds.). 1979. Herbivores: their interaction with secondary plant metabolites. Academic Press, New York.
Ryan, C. A. 1978. TIBS Trends in Biochemical Research July: 148–150.
Schultz, J. C. 1983. Impact of variable plant defensive chemistry on susceptibility of insects to natural enemies. Pp. 39–54 *in* P. A. Hedin (ed.), Plant resistance to insects. American Chemical Society Symposium Series 208, Washington, D.C.
Schultz, J. C., and I. T. Baldwin. 1982. Oak leaf quality declines in response to defoliation by gypsy moth larvae. Science 217:149–151.
Scriber, J. M., and P. Feeny. 1979. Growth of herbivorous caterpillars in relation to feeding specialization and to growth form of their food plants. Ecology 60:829–850.
Silvertown, J. W. 1982. No evolved mutualism between grasses and grazers. Oikos 38:353–354.
Skinner, G. J., and J. B. Whittaker. 1981. An experimental investigation of inter-relationships between the wood-ant *(Formica rufa)* and some tree-canopy herbivores. J. Anim. Ecol. 50:313–326.
Slocum, J. C. 1980. Differential susceptibility to grazers in two phases of an intertidal alga: advantages of heteromorphic generations. J. Exp. Mar. Biol. Ecol. 46:99–110.
Snaydon, R. W. 1984. Plant demography in an agricultural context. Pp. 389–407 *in* R. Dirzo and J. Sarukhán (eds.), Perspectives in plant population ecology. Sinauer, Sunderland, Mass.
Steneck, R. S. 1982. A limpet-coralline alga association: adaptations and defenses between a selective herbivore and its prey. Ecology 63:507–522.
– 1983. Escalating herbivory and resulting adaptive trends in calcareous algae. Paleobiology 9:45–63.
Strong, D. R., J. H. Lawton, and R. Southwood. 1984. Insects on plants:

community patterns and mechanisms. Harvard University Press, Cambridge, Mass.

Tahvanainen, J. O., and R. B. Root. 1972. The influence of vegetation diversity on the population ecology of a specialized herbivore. *Phyllotreta cruciferae* (Coleoptera: Chrysomelidae). Oecology 10:321–346.

Tansley, A. G., and R. S. Adamson. 1925. Studies on the vegetation of the English chalk. III. The chalk grasslands of the Hampshire-Sussex border. J. Ecol. 13:177–223.

Thompson, J. N. 1981. Reversed animal–plant interactions: the evolution of insectivorous and ant-fed plants. Biol. J. Linn. Soc. 16:147–155.

– 1982. Interaction and coevolution. Wiley, New York.

Tilman, D. 1978. Cherries, ants and tent caterpillars: timing of nectar production in relation to susceptibility of caterpillars to ant predation. Ecology 59:686–692.

Wheeler, W. M. 1942. Studies of neotropical ant-plants and their ants. Bull. Mus. Comp. Zool. 90(1):1–262.

Whitham, T. G. 1983. Host manipulation of parasites: within-plant variation as a defense against rapidly evolving pests. Pp. 15–42 *in* R. F. Denno and M. S. McClure (eds.), Variable plants and herbivores in natural and managed systems. Academic Press, New York.

Whittaker, R. H. 1970. The biochemical ecology of higher plants. Pp. 43–70 *in* E. Sondheimer and H. Simeone (eds.), Chemical ecology. Academic Press, New York.

Whittaker, R. H., and P. Feeny. 1971. Allelochemics: chemical interactions between species. Science 71:757–770.

20 Costs of defense and their effects on plant productivity

SHERRY L. GULMON AND
HAROLD A. MOONEY

Introduction

The net carbon gained by a leaf is equal to its photosynthetic capacity times its lifetime minus the costs of construction and maintenance. This carbon gain is proportionately reduced if the leaf lifetime is truncated by herbivory. Earlier (Mooney and Gulmon 1982), we discussed the interactions among these components and how they are affected by habitat characteristics. For example, photosynthetic capacity is generally inversely related to leaf longevity and positively related to the abundance of habitat resources. Maintenance costs are proportional to productive capacity, and construction costs are positively correlated with longevity. The potential intensity of herbivory is related to productive capacity, because both are correlated with high nitrogen concentration.

To reduce herbivory losses, most plants allocate a certain amount of fixed carbon to defensive chemicals that deter potential herbivores or reduce the fitness of adaptive herbivore populations. Here we consider the factors that affect the cost of chemical defense on a whole-plant basis and how they lead to predictions of the patterns of chemical defense in different plant types.

There are two components to defense cost, the first being the direct carbon cost of construction of the molecules and the cost of maintenance of the cellular machinery needed to construct them. The second is the indirect cost, which involves the reductions in plant growth and reproduction at some future time because of the allocation of carbon to defense during the present.

Direct costs of defense

Considering first the direct cost, a survey of different types of defensive chemicals (Table 20.1) indicates that the specific cost will usually

Table 20.1. *Costs of construction of leaves and various leaf constituents*

Type	Species example	Compound	Formula	Cost ($g\ CO_2\ g^{-1}$)	Content (% leaf weight)
Phenolic resin	*Diplacus aurantiacus*[a]	Diplacol	$C_{22}H_5O_7$	2.58	29
Cyanogenic glucoside	*Heteromeles arbutifolia*[b]	Prunasin	$C_{14}H_{17}NO_6$	2.79	6
Alkaloid	*Nicotiana tabacum*[c]	Nicotine	$C_{10}H_{14}N_2$	5.00	0.2–0.5
Long-chain hydrocarbon	*Lycopersicum hirsutum*[d]	2-tridecanone	$CH_3(CH_2)_{10}COCH_3$	4.78	0.9–1.7
Terpene array	*Salvia mellifera*[e]	Camphor (50%) + $C_{10}H_{16}O$ + others		4.65	1.3
Leaves	Shrub species[f]			1.93–2.69	

[a] Lincoln (1980).
[b] Dement and Mooney (1974).
[c] Tso (1972).
[d] Kennedy et al. (1981).
[e] Tyson, Dement, and Mooney (1974); Winner (1981).
[f] Miller and Stoner (1979); Merino, Field, and Mooney (1982).

range from 2.5 to 5.0 g CO_2 g^{-1} product, utilizing cost calculations as given by McDermitt and Loomis (1981). This range of variation in specific cost among a diversity of compounds is fairly small compared with the variation in their concentrations found in leaves (Table 20.1). Consequently, a rough comparative index among plants of energy spent on defense can, in most cases, be derived by comparing total concentrations of defensive compounds, regardless of their specific structures. Leaf costs (independent of supporting tissues) of shrub species vary from 1.93 to 2.69 g CO_2 g^{-1} leaf. Thus, the concentration by dry weight of defensive chemicals in a leaf represents, within a factor of two, the percentage of total leaf construction cost attributable to defense. In many cases, putative defensive compounds are present at concentrations of less than 1% of leaf weight (e.g., Rhoades and Cates 1976; Van Etten and Tookey 1979). In such cases, defense cost may be unmeasurable in terms of total productivity, reproductive output, or even competitive outcome, given the generally observed variance in such parameters.

Defense compounds are synthesized continuously at various rates as a plant grows; so the direct cost is a function of time. We can potentially evaluate this cost in photosynthesis units of g CO_2 g^{-1} leaf day^{-1} by considering the increase through time of total leaf weight and total weight of chemical defense in the entire plant (Figure 20.1). The defense cost $C(t)$ is calculated as follows, where the specific cost, C_{sp}, is the cost in grams CO_2 fixed per gram defensive chemical:

$$C(t) = \frac{C_{sp}(d/dt)S(t)}{W_l(t)} \qquad (20.1)$$

$S(t)$ and $W_l(t)$ are the total weights, respectively, of defense compound and leaf biomass.

The direct cost, $C(t)$, as calculated here does not include the added maintenance costs of sequestration and turnover. In the case of turnover, alkaloids and terpenes, for example, have been shown to turn over completely in hours or days (Burbott and Loomis 1969; Waller and Nowacki 1978). This will increase the construction cost to the extent that energy is lost in the chemical transformations. However, this cannot be assigned to defense without understanding the role that such turnover plays in total cellular metabolism, because many secondary metabolites are involved in primary metabolic pathways (Seigler 1977). Similarly, the cost of sequestration will be difficult to quantify if the sequestering structures, such as vacuoles, serve multiple functions in the cell. In the discussion that follows, we assume that $C(t)$ can be determined.

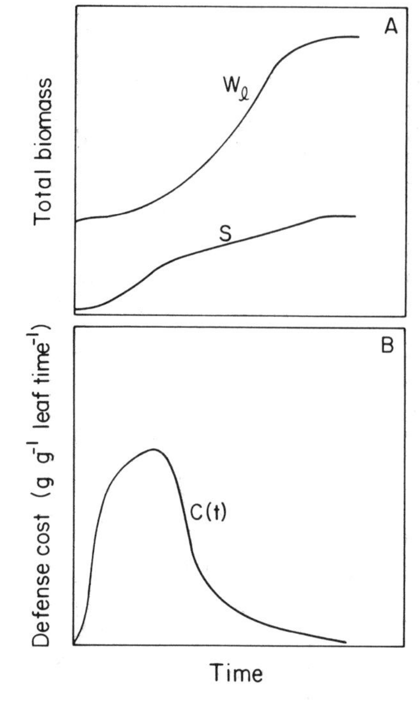

Figure 20.1. Sample determination of direct cost of defense, $C(t)$. (A) Hypothetical measured time courses of total leaf biomass, W_l, and total weight of defense compounds, S, per plant. (B) Time course of direct defense cost, $C(t)$, calculated from curves in (A) according to equation (20.1).

Indirect costs of defense

The indirect, future cost of chemical defense at a specified time, $D(t)$, can be defined as the difference between potential total leaf weight of an undefended plant and potential total leaf weight of a defended plant at that time:

$$D(t) = [W_l(t)]_{\mathrm{undef}} - [W_l(t)]_{\mathrm{def}} \qquad (20.2)$$

Of course, the true cost of defense, in evolutionary terms, is reduced reproductive success. However, leaves are the productive organs, and to the extent that reproductive success is affected by reproductive output, total leaf growth will be its primary determinant.

The formulation in equation (20.2) does not explicitly consider the loss in competitive status that can accrue because of reduced leaf growth and hence reduced carbon gain. This can be modeled if the specific microhabitat of a plant is quantified.

We can examine the future cost of defense as a function of basic physiological parameters by assuming that exponential growth is occurring and that the rate of increase in total leaf weight per plant is proportional to existing leaf weight, average photosynthetic rate, fractional carbon allocation to leaf production, and conversion efficiency. The increase in total leaf weight per plant with time is given by

$$dW_l/dt = W_l kPL \tag{20.3}$$

$$W_l(t) = WS_0^t \frac{dW_l}{dt}\, df \tag{20.4}$$

where P is photosynthetic rate, L is allocation to leaf, and k is conversion efficiency.

To compute the cost of defense, we assume that the growth pattern (i.e., root : shoot : stem : storage ratio) is not altered by diversion of carbon to defense. The basis for this assumption is that these parameters are determined by environmental constraints such as drought, light intensity, temperature, nutrient status, or competitive interactions (e.g., Davidson 1969a, 1969b; Fick et al. 1971; Raper et al. 1978; Gales 1979; Caloin et al. 1980) and will not be altered by the production of chemical defense. Thus,

$$dW_l/dt = W_l(P - C)Lk \tag{20.5}$$

and

$$(d/dt)S = W_l C/C_{sp} \tag{20.6}$$

The cost of defense at time t, $D(t)$, and the quantitative production of defense, $S(t)$, are determind by integrating these equations over time. If P, C, and L are constant,

$$D(t) = W_l \exp[ekPlt - ek(P - C)Lt] \tag{20.7}$$

$$S(t) = CW_{l_0} \exp[\{[ek(P - C)Lt - 1]/Lt\}(P - C)]/C_{sp} \tag{20.8}$$

In all the illustrative examples that follow, k is 0.4 g per gram CO_2, and C_{sp} is 2.5 g CO_2 per gram defense compound.

These calculations are illustrated in Figure 20.2 for a hypothetical plant growing for 40 days. Defensive compounds accumulate rapidly to 29% of the total leaf weight thereafter. Because of exponential leaf growth, the cost of defense, $D(t)$, increases with time. The cost of defending from day 0 is compared with the cost of starting at day 30. By day 40, both plants have equal proportionate levels of chemical defense. The early defended plant has accumulated 21% less leaf weight than the late defended plant. This proportion will remain constant and represents the relative cost of defending the leaves during the first 30 days of growth.

The two primary parameters in the calculation of leaf growth are photo-

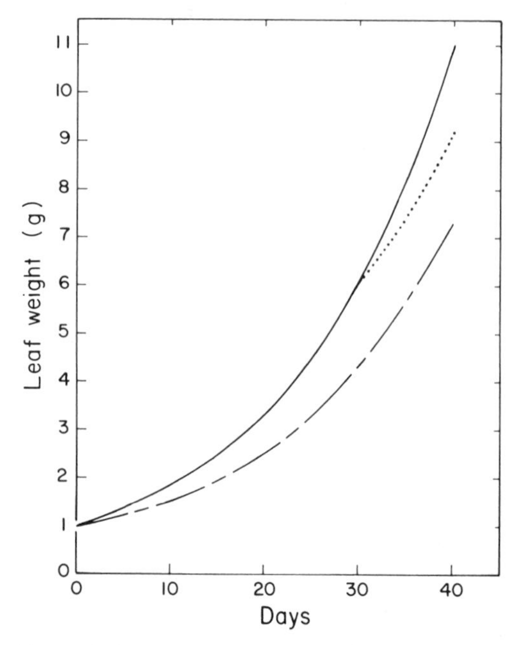

Figure 20.2. Determination of defense cost $D(t)$ for a hypothetical plant [equation (20.7)]. $P = 0.3$ g CO_2 g^{-1} leaf day^{-1}; $L = 0.5$. Solid line is for an undefended plant, $C = 0$. Broken line is for a plant defended from days 0 to 10 at $C = 0.09$ g CO_2 g^{-1} leaf day^{-1} and from days 10 to 40 at $C = 0.038$. Dotted line is for a plant undefended until day 30, then defended at $C = 0.09$. At day 40, both broken-line and dotted-line plants have accumulated defense compounds equal to 29% of total leaf weight [calculated from equation (20.8)]. $D(t)$ is the difference in total leaf weight, at time t, between any two plants.

synthetic rate, P, and leaf allocation fraction, L. Photosynthetic rates vary by more than one order of magnitude (Table 20.2). In general, leaves of trees and shrubs have lower rates than herbaceous plants, and sclerophyllous leaves have lower rates than mesophytic ones (Larcher 1980). Also, early successional species tend to have higher photosynthetic rates than climax species (Bazzaz 1979).

Defense cost, $D(t)$, is shown as a function of photosynthetic rate in Figures 20.3 and 20.4. In Figure 20.3A,B, C/P, the proportion of photosynthate allocated to defense, is constant. All plants attain the same relative level of defense, shown as a percentage of total leaf weight, but plants with lower photosynthetic rates reach this level more slowly (Figure 20.3A). The cost, in terms of loss in potential leaf production, is lower in plants with lower photosynthesis rates (Figure 20.3B). In Figure 20.4, the direct cost, $C(t)$, is varied with time, so that the levels of defense are equal through time for plants with different photosynthetic rates (Figure 20.4A,B). The

Table 20.2. *Daily photosynthesis rates for various plant types*

Plant type	Calculated daily rate[a] (g CO_2 g^{-1} leaf day^{-1})
Death Valley annuals	0.35–0.92
Old-field annuals	0.27–0.55
Deciduous chaparral shrubs	0.11–0.46
Evergreen trees and shrubs	0.03–0.16
South African shrubs	0.02–0.09

[a] Rates from Field and Mooney (Chapter 1). Daily values are approximations taking light-saturated rates for a 10-hr period.

lower-P plant still has a lower total defense cost (Figure 20.4C), despite having to allocate proportionately more photosynthate to defense during the initial growth period.

The leaf allocation fraction (L) also varies among plants (Table 20.3). In general, rapidly growing plants and herbaceous plants will have high L values. Some herbaceous perennials and woody plants that endure a favorable season in the dormant state will have low L values because of high allocation to storage tissue. Climax tree species may have low L values because of high allocation to support tissue. Even in annuals, L can vary considerably in different habitats (Hickman 1975; Clark and Burk 1980; Gulmon et al. 1983).

Figures 20.5 and 20.6 illustrate the effects of varying leaf allocation fraction, L. In Figure 20.5A,B, with defense effort, C, constant, the lower-L plant produces a higher level of defense at lower total cost. When the proportional levels of defense are equalized by varying C with time (Figure 20.6A), the total defense cost of the low-L plant is much smaller than that of the high-L plant (Figure 20.6B).

The foregoing results can be summarized as follows: The cost of defense in terms of loss of potential leaf growth varies directly with photosynthetic rate and fractional allocation to leaf growth. These two parameters determine the rate of compound-interest growth. Because of the compounding effect, a given proportional decrease in growth rate will have a greater relative effect on "capital accumulation" at the higher intrinsic growth rate. This assumes that these parameters (P and L) are constrained by other environmental factors. That is, we do not suggest that low rates of photosynthesis or leaf growth are intrinsically advantageous because of their effects on the cost of chemical defense.

This conclusion can be compared with the apparent–unapparent hypothesis introduced by Feeny (1976); see also Rhoades and Cates (1976).

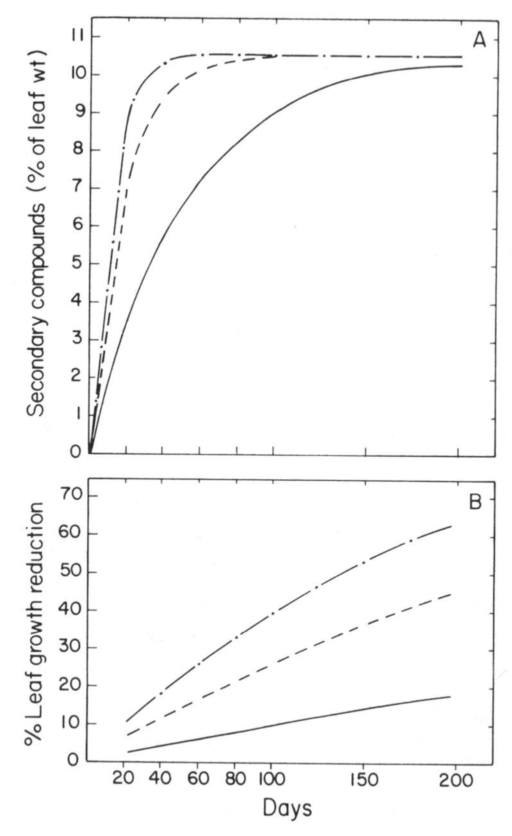

Figure 20.3. (A) Total weight of defense compounds (as percentage of total leaf weight) through time at different photosynthesis rates and constant $C/P = 0.05$. $W_{l_0} = 1$ g; $L = 0.5$. Dot-dash line: $P = 0.5$ g CO_2 g^{-1} leaf day^{-1}; $C = 0.025$ g CO_2 g^{-1} leaf day^{-1}. Dash line: $P = 0.3$; $C = 0.015$. Solid line: $P = 0.1$; $C = 0.005$. (B) $D(t)$ expressed as a percentage of the undefended ($C = 0$) leaf weight under the same conditions as in (A).

Feeny proposed that quantitative chemical defenses are produced by long-lived plants because such plants will always be discovered by specialist herbivores sooner or later. Short-lived plants, on the other hand, can escape specialists in time, and they are able to deter generalist herbivores with small quantities of highly toxic, qualitative defensive compounds.

Our conclusions do not contradict this hypothesis, but they do suggest an alternative evolutionary basis for the generally observed differences in chemical defense patterns between short-lived and long-lived plants. That is, short-lived plants are usually characterized by high initial growth rates, which in turn are correlated with high P and L values. For such plants,

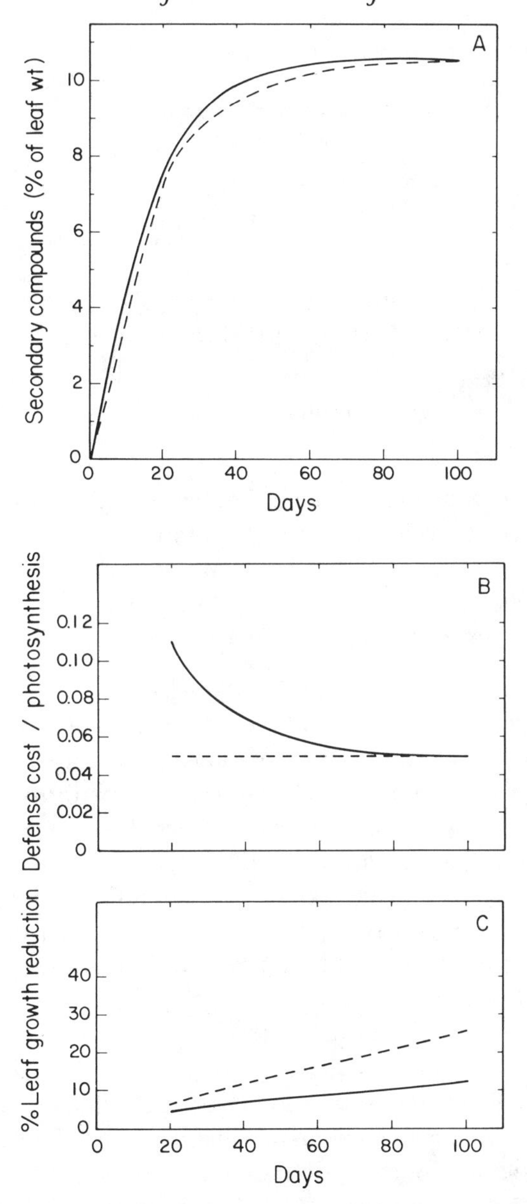

Figure 20.4. (A) Total weight of defense compounds (as percentage of total leaf weight) through time with different photosynthesis rates and similar proportional rates of chemical buildup. $W_{l_0} = 1$ g; $L = 0.5$. Dash line: $P = 0.3$ g CO_2 g^{-1} leaf day^{-1}; $C = 0.015$ g CO_2 g^{-1} leaf day^{-1}. Solid line: $P = 0.1$; C is a function of time. (B) Defense-cost/photosynthesis-rate (C/P) for the two cases in Figure 20.4A. (C) $D(t)$ as percentage of undefended ($C = 0$) leaf weight under the same conditions. Note difference in time scale of the x axis from that of Figure 20.3.

Table 20.3. *Distributions of net production into leaves of various plant types*

Plant type	Allocation (%)
Annual grasses and forbs[a]	55–70
Crop plants[b]	29–54
Chaparral shrubs[c]	34
Temperate trees[d]	18–33

[a] Hickman (1975); Gulmon (1979); Gulmon et al. (1983).
[b] Kimura (1975).
[c] Oechel and Lawrence (1981).
[d] Satoo (1970); Whittaker and Woodwell (1971); Satoo and Madgwick (1982).

allocating a high proportion of carbon to chemical defense would cause a significant loss in productivity and hence competitive status.

Effects of varying carbon acquisition and allocation

In the foregoing simulations we kept photosynthetic rate and leaf allocation pattern constant over the time period of interest. This does not generally occur except over short time periods, but complex patterns of leaf growth can be simulated by varying the parameters P and L. For example, a typical leaf growth pattern for a deciduous tree would be an initial phase of rapid linear growth, when storage is used, followed by a period when both P and L are relatively high. In midseason, L begins to decline, and toward the end of the season both P and L decline to zero (Satoo 1970; Flint 1974; Dougherty et al. 1979).

The time course of defense cost, D, will vary with P and L. In the case of the deciduous tree just described, defending the young leaves early in the season will cost much more, in terms of total seasonal growth, than building up defenses after the initial flush of leaf growth has occurred. This is due in part to the increased cost of starting defensive-compound production sooner, as shown in Figure 20.2. However, D is increased still further by early defense production because it occurs during a period of high photosynthetic rates and leaf allocation fraction. Thus, we would expect strong selective pressure for deciduous trees to delay production of significant quantities of chemical defenses until after leaf expansion.

The calculations of D in terms of P and L imply that maximal production of defense products will, in general, occur out of phase with periods of

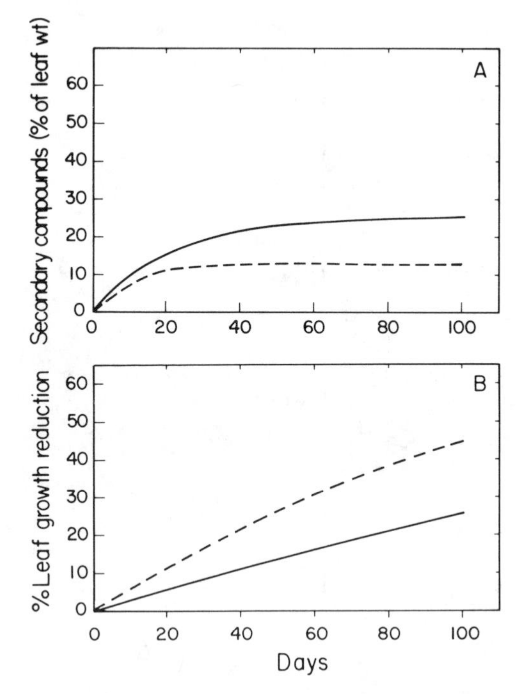

Figure 20.5. (A) Total weight of defense compounds (as percentage of total leaf weight). Dash line: $L = 0.5$. Solid line: $L = 0.25$. $W_{l0} = 1$ g; $P = 0.5$ g CO_2 g^{-1} leaf day^{-1}; $C = 0.03$ g CO_2 g^{-1} leaf day^{-1}. (B) $D(t)$ as percentage of undefended leaf weight at two different leaf allocation fractions under conditions in (A).

peak photosynthesis rates and rapid growth. However, the cost of defense must ultimately be related to the benefit, or the cost of tissue eaten. A given amount of leaf tissue eaten early in the season will have a higher cost than the same quantity eaten later, because of the compounding effect of exponential growth. To quantify benefit, equation (20.5) must be rewritten as

$$dW_l/dt = W_l(P - C)Lk - H \tag{20.9}$$

where H is the rate of leaf herbivory on the plant per unit time. Herbivory may be a function of protein content (and hence photosynthetic rate), time, defense production, or total leaf weight, or it may be independent of these. Where H is unpredictable, plants may be able to respond by increasing defense in response to current values of H. For example, some deciduous tree species have been shown to increase defensive-compound concentrations in new leaves after an initial defoliation has occurred (e.g., Schultz and Baldwin 1982).

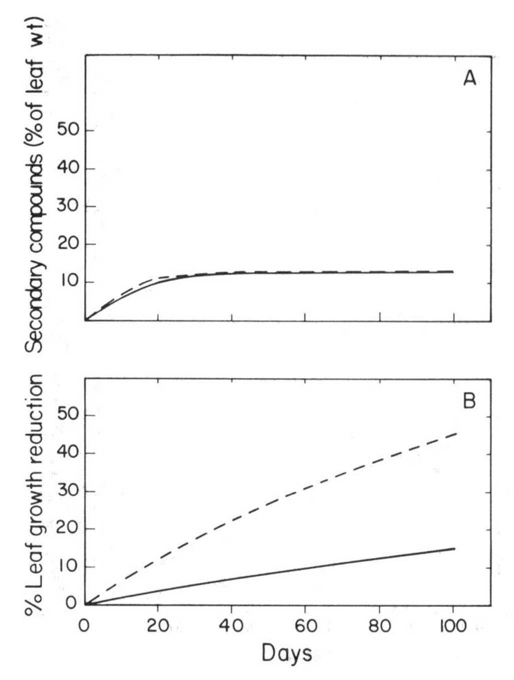

Figure 20.6. (A) total weight of defense compounds (as percentage of total leaf weight) with different leaf allocation fractions and similar proportional rates of chemical accumulation. $W_{l0} = 1$ g; $P = 0.5$ g CO_2 g^{-1} leaf day^{-1}. Dash line: $L = 0.5$; $C = 0.03$ g CO_2 g^{-1} leaf day^{-1}. Solid line: $L = 0.25$; C is a function of time, varying from 0.025 to 0.0155. (B) $D(t)$ as percentage of undefended leaf weight under the same conditions as in (A).

We illustrate this last point with a simplified example of an annual plant (*Hemizonia luzulaefolia*) that grows as a vegetative rosette during the winter and spring wet period and produces a flowering stalk with small caudal leaves during the dry summer. The reproductive plant is covered with sticky, aromatic resin that is completely absent during the vegetative phase.

The parameters we use to simulate the impact of changing photosynthetic rates and allocation patterns in *Hemizonia* are shown in Table 20.4. The vegetative phase has a high photosynthetic rate and equal allocations to leaf and root. In the reproductive phase, the photosynthetic rate is lower because of drought stress, and half of the carbon gained is allocated to reproduction, which includes stems, flowers, and fruits. Also, half of the vegetative leaves senesce at the end of the wet period.

The cost of the resin, a putative chemical defense (Figure 20.7), is 37% of potential leaf weight at the end of the vegetative period. If the defense

Table 20.4. *Inputs utilized in simulation of growth and allocation of defense in* Hemizonia *(Figure 20.7)*[a]

	Vegetative phase	Reproductive phase
Duration (days)	150	120
Photosynthetic rate ($g\ g^{-1}\ day^{-1}$)	0.3	0.15
Allocation to leaf	0.5	0.2
Allocation to root	0.5	0.3
Allocation to reproduction	0.0	0.5

50% of vegetative leaves lost due to drought stress at the start of reproduction

Consider in simulation:

1. no resin production
2. constant resin production of 0.015 g $CO_2\ g^{-1}$ leaf day^{-1}
3. resin production during reproductive period of 0.015 g $CO_2\ g^{-1}$ leaf day^{-1}

[a] From Gulmon et al. (1983).

were maintained throughout reproduction, the final cost would be 49% of the reproductive weight. If the plant produces defensive compounds only during the reproductive period, the cost will be reduced to 18% of the potential reproductive weight. Thus, the cost of defending only the reproductive phase is relatively modest compared with the cost of defending the vegetative phase or the entire life cycle. Incurring the late-season cost probably confers a significant selective advantage to *Hemizonia,* because it is virtually the only green plant during the dry summer.

Nitrogen and defense costs

We now turn to some general considerations of the cost of defense. For defensive compounds containing nitrogen, such as alkaloids, cyanogenic glucosides, toxic amino acids, and so forth, the additional cost of diverting nitrogen from photosynthesis and growth must be considered. Returning to equation (20.5),

$$(d/dt)W_l = W_l[P(N) - C]L(N)k \tag{20.10}$$

both photosynthetic rate, P, and leaf allocation fraction, L, are functions of the plant nitrogen content, N (Natr 1972; Gulmon and Chu 1981). Diversion of nitrogen to defensive compounds will generally reduce one or both of these parameters, which in turn will reduce productivity.

Measurements on many species have shown that for individual light-saturated leaves, the photosynthetic rate (per unit weight) is a linear function of leaf nitrogen content (Mooney et al. 1978; Gulmon and Chu 1981; Mooney et al. 1981, 1983a). However, when entire plants in a community

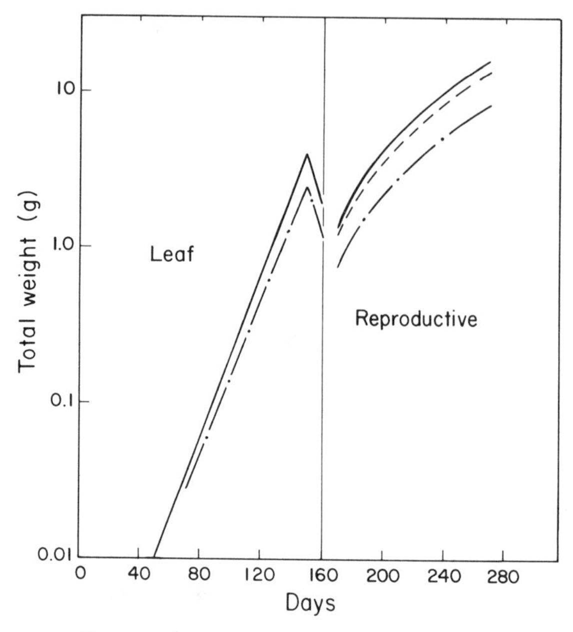

Figure 20.7. Simulation of growth of *Hemizonia* with and without defense production. Left-hand side is total leaf weight during vegetative growth phase; right-hand side is total reproductive weight during reproductive growth phase. Solid line: Undefended plant. Dot-dash line: Plant defended throughout life cycle. Dash line: Plant defended during reproductive phase only. Model parameters given in Table 20.4.

are considered, intraplant and interplant shading will necessarily cause this dependence to level off at higher nitrogen contents (Figure 20.8). Similarly, leaf allocation fraction, L, will, in most cases, increase with nitrogen content, but the effects of self-shading, support requirements, and water use will cause this dependence to level off at higher nitrogen contents (Burton and DeVane 1952; Prine and Burton 1956; Bradshaw et al. 1964).

The cost of nitrogen diversion is a positive function of the slopes dP/dN and dL/dN of the photosynthesis-versus-nitrogen and leaf-allocation-versus-nitrogen curves (Figure 20.8). Thus, nitrogen-containing defensive compounds will be relatively more expensive to produce for plants with low nitrogen contents. Consequently, one would expect chemical-defense compounds containing large amounts of nitrogen to be common only in plants with high photosynthetic rates and high leaf allocation fractions, implying abundant available nitrogen (Bryant et al. 1983). However, as shown earlier, the carbon cost of defense is very high for such plants; so even nitrogenous compounds cannot generally be produced in large quantities.

A significant exception to this conclusion may occur if nitrogen becomes

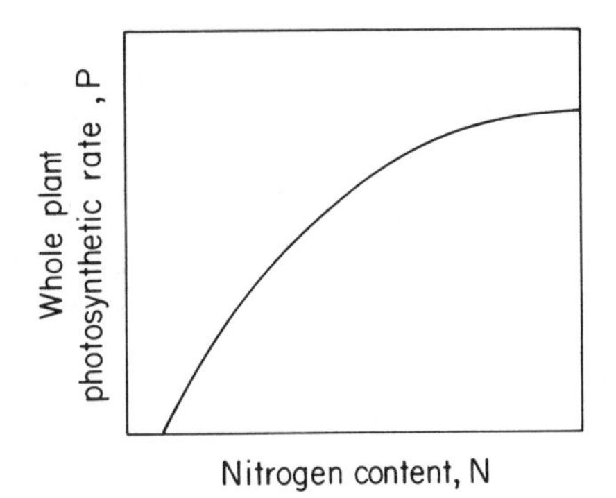

Figure 20.8. Expected dependence of the average photosynthesis rate of the whole plant on total nitrogen content. Under nitrogen-limiting conditions, the dependence will be nearly linear and dP/dN high. At high nitrogen content, light limitation becomes significant, and dP/dN approaches zero.

temporarily abundant when photosynthesis and growth are constrained by other environmental factors such as temperature or light availability. Because nitrogenous secondary compounds appear to be very labile in the plant (Mooney et al. 1983b), high quantities of such compounds could be produced and subsequently reused for photosynthesis and growth.

Storage and defense costs

Another form of defense against herbivory is high allocation to storage tissue that is unavailable to herbivores. If lost leaf tissue can be rapidly replaced by mobilizing stored carbon and nitrogen, then the ultimate loss in productivity will be less than proportional to the initial leaf loss (Detling et al. 1979; Stanton 1983). It is possible that some species, particularly grasses, that have evolved under conditions of continuous grazing pressure allocate so much carbon (and nitrogen) to storage that productivity under ungrazed conditions is substantially reduced. When such plants are harvested, total productivity may increase, although factors other than storage may be involved (McNaughton 1983).

The cost of storage as a defense is exactly analogous to the carbon cost of chemical defenses already described. In general, we would expect a higher allocation to storage under less productive conditions, when both P and L are low.

Summary

The cost, in terms of growth reduction, of the carbon and energy for defensive compounds is proportional to photosynthetic rate and leaf allocation fraction. Therefore, we expect quantitative defenses (i.e., high amounts) only in plants with lower photosynthetic rates and leaf allocation fractions. The cost of nitrogen, when it is diverted from growth and metabolic functions to defense compounds, is highest in plants with low N contents (and low photosynthetic rates and leaf allocation fractions). Thus, we would expect to see nitrogenous compounds produced only in plants with high overall N contents. However, such plants are *most sensitive* to the *carbon* costs of defense; so we would expect these nitrogenous compounds to be produced in small quantities in the plant, unless they have other functions.

References

Bazzaz, F. A. 1979. The physiological ecology of plant succession. Ann. Rev. Ecol. Syst. 10:351–372.

Bradshaw, A. D., M. J. Chadwicks, D. Jonett, and R. W. Snaydon. 1964. Experimental investigations into the mineral nutrition of several grass species IV. Nitrogen level. J. Ecol. 52:665–676.

Bryant, J. P., F. S. Chapin III, and D. R. Klein. 1983. Carbon/nutrient balance of boreal plants in relation to vertebrate herbivory. Oikos 40:357–368.

Burbott, A. J., and W. D. Loomis. 1969. Evidence for metabolic turnover of monoterpenes in peppermint. Plant Phys. 44:173–179.

Burton, G. W., and E. H. DeVane. 1952. Effect of rate and method of applying different sources of nitrogen upon the yield and chemical composition of Bermuda grass, *Cynodon dactylon* (L) Pers. Hay. Agronomy J. 44:128–132.

Caloin, M., A. El Khodre, and M. Atry. 1980. Effect of nitrate concentration on the root : shoot ratio in *Dactylis glomerata* L. and on the kinetics of growth in the vegetative phase. Ann. Bot. 46:165–173.

Clark, D. D., and J. H. Burk. 1980. Resource allocation patterns of two California-Sonoran Desert ephemerals. Oecologia 46:81–91.

Davidson, R. L. 1969a. Effect of root/leaf temperature differentials on root/shoot ratios in some pasture grasses and clover. Ann. Bot. 33:561–569.

– 1969b. Effects of soil nutrients and moisture on root/shoot ratios in *Lolium perenne* L. and *Trifolium refens* L. Ann. Bot. 33:571–577.

Dement, W. A., and H. A. Mooney. 1974. Seasonal variation in the production of tannins and cyanogenic glucosides in the chaparral shrub, *Heteromeles arbutifolia*. Oecologia 15:65–76.

Detling, J. K., M. I. Dyer, and D. T. Winn. 1979. Net photosynthesis, root respiration, and regrowth of *Bouteloua gracilis* following simulated grazing. Oecologia 41:127–134.

Dougherty, P. M., R. O. Teskey, J. E. Phelps, and T. M. Hinckley. 1979. Net photosynthesis and early growth trends of a dominant white oak (*Quercus alba* L.). Plant Phys. 64:930–935.

Feeny, P. 1976. Plant apparency and chemical defense. Rec. Adv. Phyto. 10:1–40.

Fick, G. W., W. A. Williams, and R. S. Loomis. 1971. Recovery from partial defoliation and root pruning in sugar beet. Crop Sci. 11:718–721.

Flint, H. L. 1974. Phenology and genecology of woody plants. Pp. 83–97 *in* H. Lieth (ed.), Phenology and seasonality modeling. Springer-Verlag, New York.

Gales, K. 1979. Effects of water supply on partitioning of dry matter between roots and shoots of *Lolium perenne*. J. Appl. Ecol. 16:863–877.

Gulmon, S. L. 1979. Competition and coexistence: three annual grass species. Amer. Mid. Nat. 101:403–415.

Gulmon, S. L., N. R. Chiarello, H. A. Mooney, and C. C. Chu. 1983. Phenology and resource use in three co-occurring grassland annuals. Oecologia 58:33–42.

Gulmon, S. L., and C. C. Chu. 1981. The effects of light and nitrogen on photosynthesis, leaf characteristics, and dry matter allocation in the chaparral shrub *Diplacus aurantiacus*. Oecologia 49:207–212.

Hickman, J. C. 1975. Environmental unpredictability and plastic energy allocation strategies in the annual *Polygonum cascadense* (Polygonaceae). J. Ecol. 63:689–701.

Kennedy, G. G., R. T. Yamamoto, M. B. Dimock, W. G. Williams, and J. Bordner. 1981. Effect of daylength and light intensity on 2-tridecanone levels and resistance in *Lycopersicon hirsutum* f. *glabratum* to *Marduca sexta*. J. Chem. Ecol. 7:707–716.

Kumura, A. 1975. Dry matter production and climatic factors. Pp. 49–59 *in* Y. Murata (ed.), Crop productivity and solar energy utilization in various climates in Japan. University of Tokyo Press, Tokyo.

Larcher, W. 1980. Physiological plant ecology, pp. 94–95. Springer-Verlag, Berlin.

Lincoln, D. E. 1980. Leaf resin flavonoids of *Diplacus aurantiacus*. Biochem. Syst. Ecol. 8:397–400.

McDermitt, D. K., and R. S. Loomis. 1981. Elemental composition of biomass and its relation to energy content, growth efficiency, and growth yield. Ann. Bot. 48:275–290.

McNaughton, S. J. 1983. Compensatory plant growth as a response to herbivory. Oikos 40:329–336.

Merino, J., C. Field, and H. A. Mooney. 1982. Construction and maintenance costs of Mediterranean-climate evergreen and deciduous leaves. I. Growth and CO_2 exchange analysis. Oecologia 53:208–213.

Miller, P. C., and W. A. Stoner. 1979. Canopy structure and environmental interactions. Pp. 428–458 *in* O. T. Solbrig, S. Jain, G. B. Johnson, and P. H. Raven (eds.), Topics in plant population biology. Columbia University Press, New York.

Mooney, H. A., P. J. Ferrar, and R. O. Slatyer. 1978. Photosynthetic capacity and carbon allocation patterns in diverse growth forms of *Eucalyptus*. Oecologia 36:103–111.

Mooney, H. A., C. Field, S. L. Gulmon, and F. A. Bazzaz. 1981. Photosynthetic capacity in relation to leaf position in desert versus old-field annuals. Oecologia 50:109–112.

Mooney, H. A., C. Field, S. L. Gulmon, P. Rundel, and F. J. Kruger. 1983a. Photosynthetic characteristics of South African sclerophylls. Oecologia 58:398–401.

Mooney, H. A., and S. L. Gulmon. 1982. Constraints on leaf structure and function in reference to herbivory. BioScience 32:189–206.

Mooney, H. A., S. L. Gulmon, and N. D. Johnson. 1983b. Physiological constraints on plant chemical defenses. Pp. 21–36 *in* P. A. Hedin (ed.), Plant resistance to insects. American Chemical Society Symposium Series 208, Washington, D.C.

Natr, L. 1972. Influence of mineral nutrients on photosynthesis of higher plants. Photosynthetica 6:80–99.

Oechel, W. C., and W. Lawrence. 1981. Carbon allocation and utilization. Pp. 185–235 *in* P. C. Miller (ed.), Ecological studies. Vol. 39. Resource use by chaparral and matorral. Springer-Verlag, New York.

Prine, G. M., and G. W. Burton. 1956. The effect of nitrogen rate and clipping frequency upon the yield, protein content and certain morphological characteristics of coastal Bermuda grass (*Cynodon dactylon* (L) Pers.) Agronomy J. 48:296–301.

Raper, C. D., D. L. Osmond, M. Wann, and W. W. Weeks. 1978. Interdependence of root and shoot activities in determining nitrogen uptake rate of roots. Bot. Gaz. 139:289–294.

Rhoades, D. F., and R. G. Cates. 1976. A general theory of plant antiherbivore chemistry. Rec. Adv. Phyto. 10:168–213.

Satoo, T. 1970. A synthesis of studies by the harvest method. *In* D. E. Reichle (ed.), Analysis of temperate forest ecosystems. Springer-Verlag, New York. Pp. 55–72.

Satoo, T., and H. A. Madgwick. 1982. Forest biomass. Dr. W. Junk, The Hague.

Schultz, J. C., and I. T. Baldwin. 1982. Oak leaf quality declines in response to defoliation by gypsy moth larvae. Science 217:149–151.

Seigler, D. S. 1977. Primary roles for secondary compounds. Biochem. Syst. Ecol. 5:195–199.

Stanton, N. L. 1983. The effect of clipping and phytophagous nematodes on net primary production of blue grama, *Bouteloua gracilis.* Oikos 40:249–257.

Tso, T. 1972. Physiology and biochemistry of tobacco plants. Dowden, Hutchinson and Ross, Stroudsburg, Pa.

Tyson, B. J., W. A. Dement, and H. A. Mooney. 1974. Volatilization of terpenes from *Salvia mellifera.* Nature 252:119–120.

Van Etten, C. H., and H. L. Tookey. 1979. Chemistry and biological effects of glucosinolates. Pp. 471–496 *in* G. A. Rosenthal and D. H. Janzen (eds.), Herbivores and their interaction with secondary plant metabolites. Academic Press, New York.

Waller, G. R., and E. K. Nowacki. 1978. Alkaloid biology and metabolism in plants, pp. 190–196. Plenum Press, New York.

Whittaker, R. H., and G. M. Woodwell. 1971. Measurement of net primary production of forests. Pp. 159–179 *in* P. Duvigneaud (ed.), Productivity of forest ecosystems. UNESCO, Paris.

Winner, W. E. 1981. The effect of SO_2 on photosynthesis and stomatal behavior of Mediterranean-climate shrubs and trees. *In* N. S. Margaris and H. A. Mooney (eds.), Components of productivity of Mediterranean-climate regimes. Dr. W. Junk, The Hague.

Index